COMPUTERIZED ENGINE CONTROLS

Sixth Edition

Steve V. Hatch
Denver Automotive and Diesel College

Dick H. King
Glendale Community College

THOMSON
DELMAR LEARNING™

Australia Canada Mexico Singapore Spain United Kingdom United States

Computerized Engine Controls, Sixth Edition
Steve V. Hatch and Dick H. King

Vice President, Technology and Trades SBU:
Alar Elken

Editorial Director:
Sandy Clark

Acquisitions Editor:
David Boelio

Developmental Editor:
Alison Weintraub

Marketing Director:
Cyndi Eichelman

Channel Manager:
Fair Huntoon

Marketing Coordinator:
Sarena Douglass

Production Director:
Mary Ellen Black

Production Editor:
Ruth Fisher

Art/Design Coordinators:
David Arsenault
Rachel Baker

Cover Design:
Julie Moscheo

Printed in Canada
2 3 4 5 XXX 07 06 05 04

For more information contact
Delmar Learning
Executive Woods
5 Maxwell Drive, PO Box 8007,
Clifton Park, NY 12065-8007
Or find us on the World Wide Web at
http://www.delmarlearning.com

Library of Congress Cataloging-in-Publication Data:

ISBN: 0-7668-5020-X

NOTICE TO THE READER

Publisher does not warrant or guarantee any of the products described herein or perform any independent analysis in connection with any of the product information contained herein. Publisher does not assume, and expressly disclaims, any obligation to obtain and include information other than that provided to it by the manufacturer.

The reader is expressly warned to consider and adopt all safety precautions that might be indicated by the activities herein and to avoid all potential hazards. By following the instructions contained herein, the reader willingly assumes all risks in connection with such instructions.

The publisher makes no representation or warranties of any kind, including but not limited to, the warranties of fitness for particular purpose or merchantability, nor are any such representations implied with respect to the material set forth herein, and the publisher takes no responsibility with respect to such material. The publisher shall not be liable for any special, consequential, or exemplary damages resulting, in whole or in part, from the readers' use of, or reliance upon, this material.

Contents

Preface

The application of the microprocessor with its related components and circuits has made automotive technology exciting, fast paced and more complicated. Recent technological developments and those that follow will require entry level automotive service technicians to be well trained in the principles of automotive technology and to be career-long students. Those who respond to this requirement will find the task challenging, but achievable and rewarding.

This text was written in response to a widely recognized need within the industry: to help student-technicians get a commanding grasp of how computerized engine control systems work and how to diagnose problems within them. The student-technician who studies this text will soon come to realize that no single component or circuit within any given computerized engine control system, other than the computer itself, is complicated.

Computerized Engine Controls is written with the assumption that the reader is familiar with the basic principles of traditional engine, electrical system and fuel system operation. Although everything here is within the grasp of a good technician, this is not a beginner's book.

A computerized engine control system does in fact become an integral part of an engine's electrical and fuel system, but it is much too significant and complex to be taught as just a unit in an engine performance textbook or class. For purposes of instruction, we have taken this topic out of context and examined it as a stand-alone control-level system. Once the student-technician fully understands the system's purpose, operation and diagnostic approach, the diagnostic procedures in the service manual will put the system back in its proper perspective as an integral part of the engine's support system.

Computerized Engine Controls presents each popular, multifunction computer control system in a separate chapter. Each system is fully covered, with enough specific information and detail to enable the reader to get a complete and clear picture of how the system works. This text is written with the premise that understanding how the overall system works and what it should be doing not only makes the diagnostic process easier but also makes the diagnostic literature much easier to understand. Correctly interpreting diagnostic procedure directions is often a problem if the technician is not aware of what the procedure is trying to measure, what normal readings or responses should be or what conditions will cause abnormal readings or responses.

NEW TO THIS EDITION

This revision of this textbook contains some newly added chapters that present an overview of concepts that are applicable to understanding and diagnosing systems on all vehicles, regardless of manufacturer. These new chapters are:

- "Diagnostic Concepts" (Chapter 3).
- "Diagnostic Equipment" (Chapter 4).

- "Exhaust Gas Analysis" (Chapter 5).
- "Multiplexing Concepts" (Chapter 7).

In addition, new to this revision is "Appendix A: Approach to Diagnostics," which is designed to summarize the material covered throughout this textbook to help relate the concepts contained herein to a hands-on diagnostic approach. "Appendix B: Terms and Acronyms," has been added as a central location where the reader can look up many of the terms and acronyms used throughout the automotive computer industry. This appendix replaces the Terms and Acronyms charts that had earlier been found at the end of the General Motors chapters.

The chapters of this edition contain:

- *Objectives.* Objectives are provided at the beginning of each chapter to help the reader identify the major concepts to be presented.
- *Key Terms.* Terms that are unique to computerized engine control systems are provided at the beginning of each chapter as a Key Term and then appear in **boldface** type at their first use in the chapter. These Key Terms are also provided in the Glossary, along with their definitions.
- *Diagnostic and Service Tips.* These tips offer helpful advice for the technician on diagnosing and servicing vehicles, as well as addressing customer concerns.
- *Chapter Articles.* Chapter Articles give additional nice-to-know information about technical topics covered in the chapter.
- *Summary.* Each chapter contains a Summary near the end to review the major concepts presented in that chapter.
- *Diagnostic Exercise.* A Diagnostic Exercise is provided near the end of each chapter that applies a real-life scenario to one of the concepts covered in that chapter.
- *Review Questions.* Review Questions are provided at the end of each chapter to help a reader assess his/her recall and comprehension of the material read in each chapter, as well as to reinforce the concepts covered in each chapter. Without exception, all of these Review Questions have been rewritten for this revision in a multiple-choice format, the type of question that would be found on an ASE test.

Personal safety concerns peculiar to specific computerized engine control systems are highlighted where applicable. The book follows the industry standards for how to use the following terms:

- **Warnings** indicate that failure to observe correct diagnostic or repair procedures could result in personal injury or even death.
- **Cautions** indicate that failure to observe correct diagnostic or repair procedures could result in damage to tools, equipment or the vehicle serviced.

Each student should be aware that while working with computerized controls is not inherently dangerous, failure to observe recognized safety practices is. There are unfortunately many more injuries and accidents in the automotive repair business than there should be. Good safety practices, if learned early in a student's career, literally can be life-saving later on.

SUGGESTIONS ON HOW TO USE THIS TEXT

Read the **Introduction** first. This section provides the background needed to help the reader understand the operation of many components common to most systems. This section also provides an in-depth look at how to use voltage drop testing to verify the integrity of the power and ground circuits used to power up the Powertrain Control Module (PCM) and other load components.

Chapters 1 through 7 should be read next as they contain information that pertains to all makes of vehicles. Reading these chapters will also provide some background that will make understanding specific system designs and diagnostic strategies easier when the product-specific chapters of this book are read. The remaining chapters, which are specific to individual systems, can be read in any order, though most students will find it much easier to read a manufacturer's set of chapters in chronological sequence, the way they appear in the book. Generally systems become more complicated over time, and the later versions are easier to understand with a good background knowledge of the earlier, simpler systems. Many students may not wish to study each specific chapter, or their instructors may not choose to assign the study of each chapter. We suggest, however, that at least three of the specific system chapters be selected for study. Remaining chapters can then be skimmed or serve as a reference for future use. Finally, "Appendix A: Approach to Diagnostics," which offers a strategy-based diagnostic approach, should be read to help the reader apply the information contained within the chapters of this book to a live vehicle.

Acknowledgments

I am deeply honored to have had the chance to revise this textbook, following Dick H. King's passing away. Mr. King had assembled within the confines of this textbook an abundance of information, the likes of which must have taken countless hours of his time. Thank you, Mr. King, for the contribution that you have left to our future automotive technicians.

I would also like to thank my wife, Geralyn, for her continual patience with the many hours that this effort has required of me, as well as for the great help that she was to me in performing many of the tasks associated with this type of work.

And finally, many thanks to the following for their critical reviews and/or answers to questions concerning this text:

Wane Boysun
Montana State University-Northern
Havre, MT

David Byrd
Wayne Community College
Goldsboro, NC

Neal Clark
Erie Community College
Orchard Park, NY

Ron Darby
University of Southern Colorado
Pueblo, CO

Carl Hinkley
Central Maine Technical College
Auburn, ME

Kevin Johnson
Montana State University-Northern
Havre, MT

Donald Lumsdon
Ivy Technical State College
Lafayette, IN

Bryan Ricco
Spokane Community College
Spokane, WA

Introduction

A Review of Electricity and Electronics

The earliest automobiles had little in the way of electrical systems, but as the automobile has become more complicated and as more accessories have been added, *electrical* or *electronic* systems have become a major feature on today's vehicles. Additional electronic control systems have made and will continue to make the automobile comply with government standards and consumer demands. Today, most major automotive systems are controlled by computers.

This increased use of electrical and electronic systems means two things for the automotive service technician: first, all service technicians need skills in electrical diagnosis and repair to be effective, almost regardless of the technician's service specialty; second, technicians with such skills will command significantly greater financial rewards and will deserve them.

There are several principles by which electrical systems operate, but they are all fairly simple; learning them is not difficult. As each principle is introduced to you through your reading or in class, ask questions and/or read until you understand it. Review the principles frequently and practice the exercises that your instructor assigns.

ELECTRICAL CIRCUITS VERSUS ELECTRONIC CIRCUITS

The differences between electrical circuits and electronic circuits are not always clear-cut or definite. This has led to some confusion about the use of those terms and how an electronic system differs from an electrical system. Perhaps the comparisons in the table below will help.

Even though the use of solid-state components may often be used as criteria to identify an electronic circuit, solid-state components such as a *power transistor* may also be used in an electrical circuit. A power transistor is a type of transistor designed to carry larger amounts of amperage

Electrical Circuits	Electronic Circuits
Do physical work: movement, heat, light, magnetism.	Communicate information: voltages or on-off signals.
Use electromechanical devices: switches, solenoids, relays.	Use solid-state devices (semiconductors) with no moving parts, transistors or diodes.
Operate at relatively high current or amperage.	Operate at relatively low current or amperage.
Circuits have relatively low resistance (ohms).	Circuits have relatively high resistance (ohms).

than are normally found in an electronic circuit. A power transistor is essentially a highly reliable relay.

ELECTRON THEORY

Molecules and Atoms

A study of electricity begins with the smallest pieces of matter. All substances—air, water, wood, steel, stone and even the various substances that our bodies are made of—are made of the same bits of matter. Every substance is made of units called *molecules*. A molecule is a unit formed by combining two or more atoms and is the smallest unit that a given substance can be broken down to and still exhibit all of the characteristics of that substance. For example, a molecule of water is made up of two atoms of hydrogen and one atom of oxygen (H_2O: H is the chemical symbol for hydrogen and O is the chemical symbol for oxygen). If a molecule of water is broken down into its component atoms, it is no longer water.

As molecules are made up of atoms, atoms are in turn made up of:

- electrons, or negatively charged particles,
- protons, or positively charged particles,
- neutrons, or particles with no charge; at the level of atomic activity concerning us here, these just add mass to the atom.

The smallest and lightest atom is the hydrogen atom. It contains one proton and one electron (it is the only atom that does not have a neutron), Figure I–1. The next smallest and lightest atom is the helium atom. It has two protons, two neutrons and two electrons, Figure I–2. Since the hydrogen atom is the smallest and lightest, and since it has one electron and one proton, it is given an atomic number of 1. Since helium is the next lightest with 2 electrons, protons and neutrons, it has an atomic number of 2. Every atom has been given an atomic number that indicates its relative size

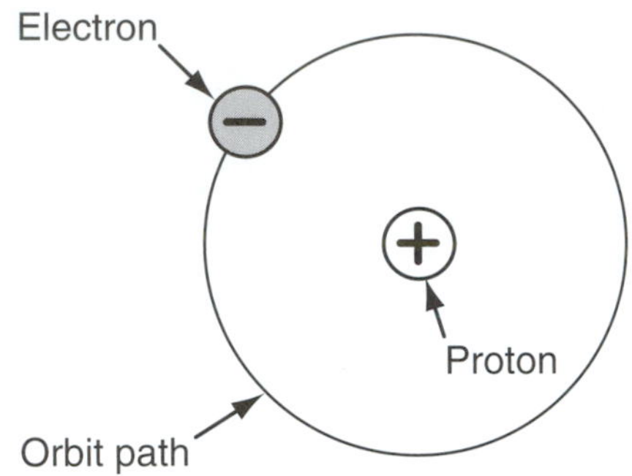

Figure I–1 Hydrogen atom.

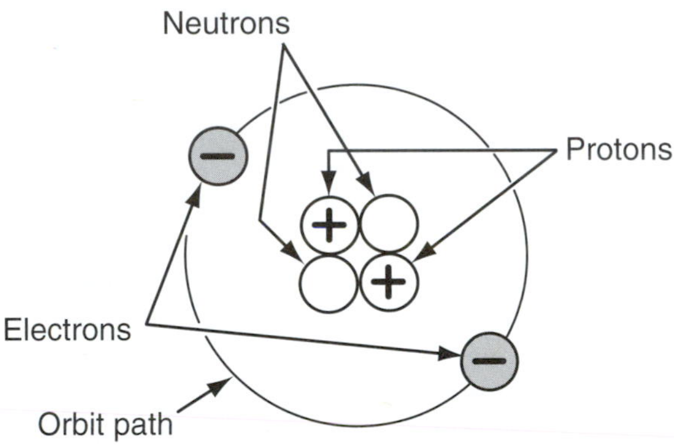

Figure I–2 Helium atom.

and weight (or its mass) and the number of electrons, protons and neutrons it contains. An atom usually has the same number of electrons, protons and neutrons.

Elements

Once the three different bits of matter are united to form an atom, two or more atoms combine to form a molecule. If all of the atoms in the molecule are the same, the molecule is called an *element*. Which element it is depends on how many protons, neutrons and electrons the atoms contain. There are more than 100 different elements. Some examples of elements are hydrogen, oxygen, carbon, iron, lead, gold, silicon and sodium. An element, then, is a pure substance whose molecules contain only one kind of atom. Substances such as water, which contains hydrogen and oxygen atoms, or steel, which contains iron and carbon atoms, are called *compounds*.

Atomic Structure and Electricity

Notice in Figures I–1 and I–2 that the protons and neutrons are grouped together in the center of each atom, called the *nucleus* of the atom. The electrons travel around the nucleus of the atom in an orbit similar to the way that the earth travels around the sun. But because an atom usually has several electrons orbiting around its nucleus, the electrons form in layers, rather than all of them traveling in the same orbit, Figure I–3. Some, however, share the same orbit, as seen in Figure I–3. For the purpose of this text, only the electrons in the last layer are of any real importance. This layer is often called the *outer shell* or *valence ring*. The student should realize that we are speaking very loosely here when we describe electrons in shells having orbits. For our purposes, this simple explanation (a model once called the *Rutherford atom*) satisfactorily conveys the nature of the electron.

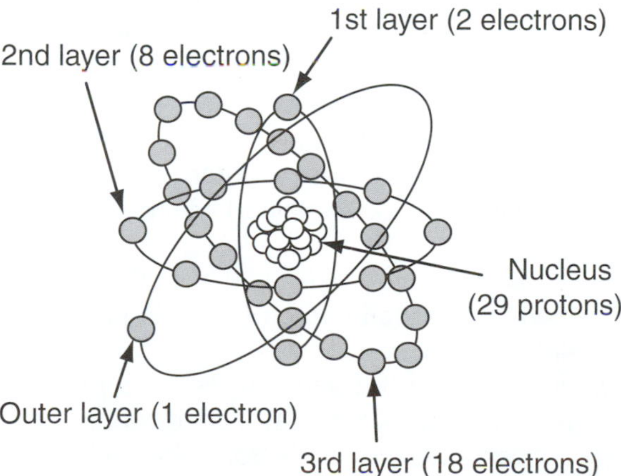

Figure I–3 Layers of electrons around a copper atom nucleus.

As mentioned, electrons are negatively charged and protons are positively charged. You have probably heard or know that like charges repel and unlike charges attract. And electrons are always moving; in fact, they are sometimes said to move at nearly the speed of light. These characteristics work together to explain many of the behaviors of an atom that makes current flow. *Current* is defined as having a mass of free electrons moving in the same direction.

There are two types of current: *direct current (DC)* and *alternating current (AC)*. Direct current always flows in one direction. Current from a battery is the best example. Most of the devices in an automobile use DC. Alternating current circuits continuously switch which end is positive and which end is negative so that current flow (and thus electron movement) reverses direction repeatedly. The power available from commercial utility companies is AC and cycles (changes *polarity*) 60 times per second. One cycle occurs when the current switches from forward to backward to forward again. The car's alternator (an alternating current generator) produces AC current, which is converted to DC before it leaves the alternator.

The fast-moving electron wants to move in a straight line, but its attraction to the proton nucleus makes it act like a ball tied to the end of a string twirled around. The repulsive force between the electrons keeps them spread as far apart as their attraction to the nucleus will allow.

The fewer electrons there are in the outer shell of the atom and the more layers of electrons there are under the outer shell, the weaker is the bond between the outer electrons and the nucleus. If one of these outer electrons can somehow be broken free from its orbit, it will travel to a neighboring atom and fall into the outer shell there, resulting in two unbalanced atoms. The first atom is missing an electron. It is now positively charged and is called a *positive ion*. The second atom has an extra electron. It is negatively charged and is called a *negative ion*. Ions are unstable. They want to either gain an electron or get rid of one so that they are balanced.

Potential

The first atom in the example has positive *potential*. It has more positive charge than negative charge because it has more protons than electrons. Suppose that this atom is at one end of a circuit, Figure I–4. Further suppose that there is an atom at the other end of the circuit that has an extra electron; it has negative potential. Because of the difference in potential at the two ends of the circuit, an electron at the negatively charged end will start moving toward the positively charged end. The greater the difference in potential (the greater the number of opposite charged ions) at each end of the circuit, the greater the number of electrons that will start to flow.

This situation can be created by putting something between the two ends of a circuit that will produce positive and negative ions. This is what a battery or generator does in a circuit, Figure I–4. If you connect the ends of the wire to a battery and put some kind of *resistance*, or opposition to a steady electric current, in the wire (why the resistance is needed is explained later), the example will work perfectly. Remember that it is the difference in potential that makes current flow. Actually, three factors must be present for an electrical circuit to work, that is, for current to flow through it.

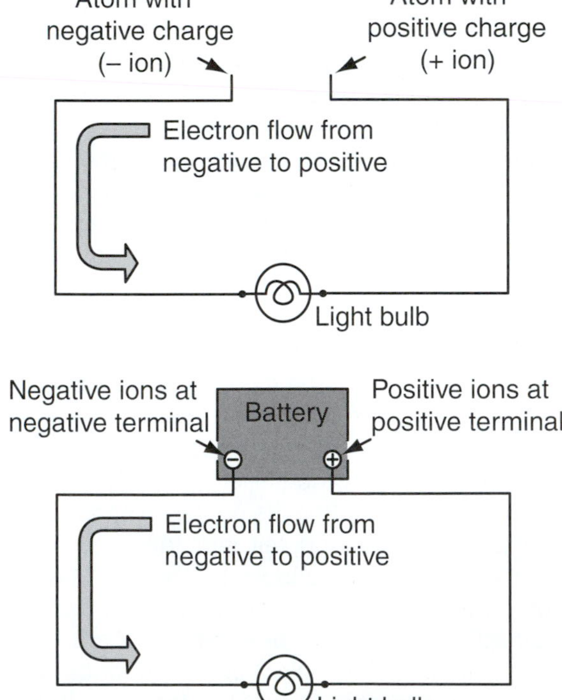

Figure I–4 Negative versus positive potential.

MAGNETISM

Magnetism is closely tied to the generation and use of electricity. In fact, one of the prevailing theories is that magnetism is caused by the movement and group orientation of electrons. Some materials strongly demonstrate the characteristics associated with magnetism and some do not. Those that strongly demonstrate the characteristics of magnetism such as iron are said to have high *permeability*. Those that do not, such as glass, wood and aluminum, are said to have high *reluctance*.

Lines of Force

It is not known whether there actually is such a thing as a *magnetic line of force*. What is known, however, is that magnetism exerts a force that we can understand and manipulate if we assume there are magnetic lines of force. Magnetic force is linear in nature and it can be managed to do many kinds of work. By assigning certain characteristics to these lines of force, the behavior of magnetism can be explained. Magnetic lines of force:

1. have a directional force (north to south outside the magnet).
2. want to take the shortest distance between two poles (just like a stretched rubber band between the two points from which it is held).
3. form a complete loop.
4. are more permeable to iron than air.
5. resist being close together (especially in air).

6. resist being cut.
7. will not cross each other (they will bend first).

Magnetic lines of force extending from a magnet make up what is commonly called a magnetic field and more correctly called magnetic *flux*, Figure I–5. If a magnet is not near an object made of permeable material, the lines of force will extend from the north pole through the air to the south pole (characteristic 1). The lines of force will continue through the body of the magnet to the north pole to form a complete loop (characteristic 3). Every magnet has a north and a south pole. The poles are the two points of a magnet where the magnetic strength is greatest. As the lines of force extend out of the north pole, they begin to spread out. Here you see opposition between characteristics 2 and 5. The lines of force want to take the shortest distance between the poles, but they spread out because of their tendency to repel each other (characteristic 5). The result is a magnetic field that occupies a relatively large area but has greater density near the body of the magnet.

Because the body of the magnet has high permeability, the lines of force are very concentrated in the body of the magnet (characteristic 4). This accounts for the poles of the magnet having the highest magnetic strength.

If there is an object with high permeability near the magnet, the magnetic lines of force will distort from their normal pattern and go out of their way to pass through the object, Figure I–6. The tendency for the lines of force to pass through the permeable object is stronger than their tendency to take the shortest route. The lines of force will, however, try to move the object toward the nearest pole of the magnet.

Electromagnets

Early researchers discovered that when current passes through a conductor, a magnetic field forms around the conductor, Figure I–7. This principle makes possible the use of electromagnets, electric motors, generators and most of the other components used in electrical circuits.

If a wire is coiled with the coils close together, most of the lines of force wrap around the entire coil rather than go between the coils of wire, Figure I–8. This is because if they do try to wrap around each loop in the coil, they must cross each other, which they will not do (characteristic 7), Figure I–9.

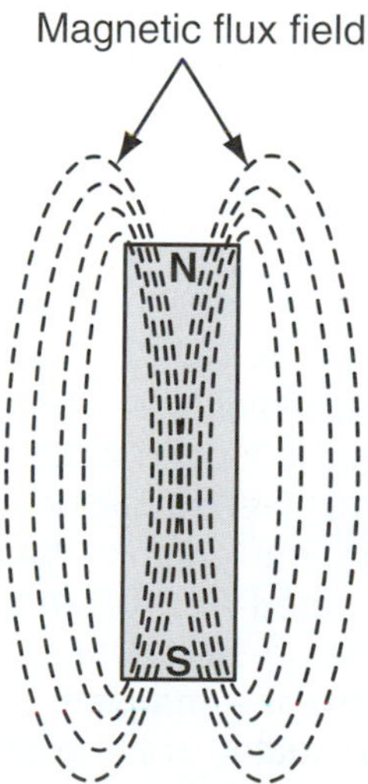

Figure I–5 Magnetic field.

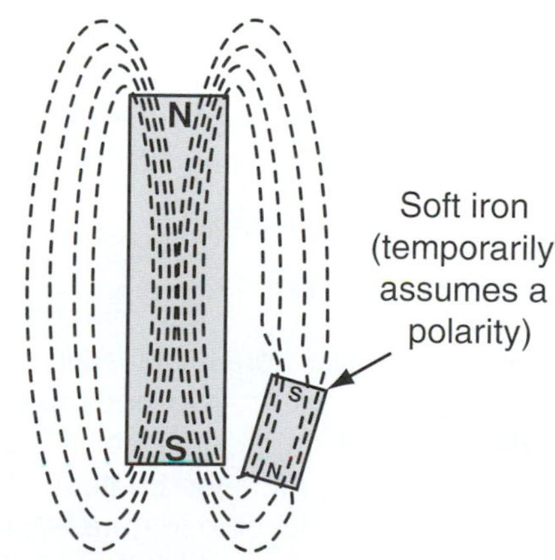

Figure I–6 Magnetic field distortion.

If you place your left hand on the wire with your thumb pointing in the direction of electron flow, your fingers will be pointing in the direction of the directional force of the magnetic lines of force. When thinking of conventional current flow, the same would apply for the right hand.

Figure I–7 Lines of force forming around a conductor.

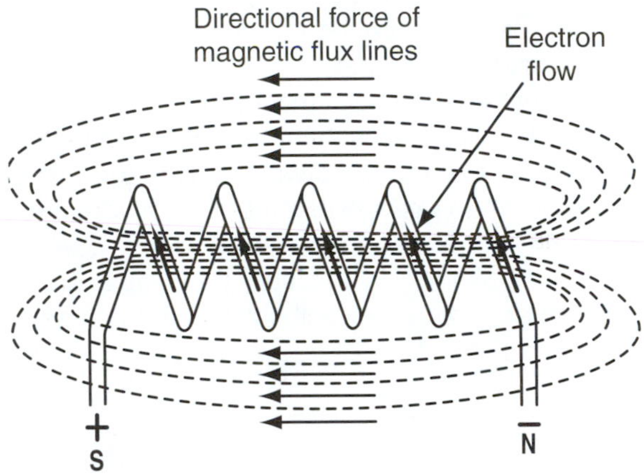

Place your left hand, with thumb extended, around the coil with your fingers pointing in the direction of the electron flow through the coil. Your thumb will point to the north pole of the magnetic field.

Figure I–8 Magnetic field around a coil.

If a high permeability core is placed in the center of the coil, the magnetic field becomes much stronger because the high permeability core replaces low permeability air in the center of the coil and more lines of force develop, Figure I–10. If the core is placed toward one end of the coil, Figure I–11, the lines of force exert a strong force on it to

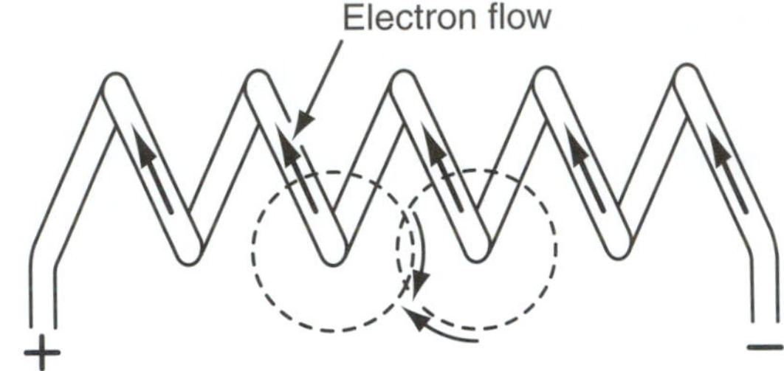

Figure I–9 Lines of force cannot cross.

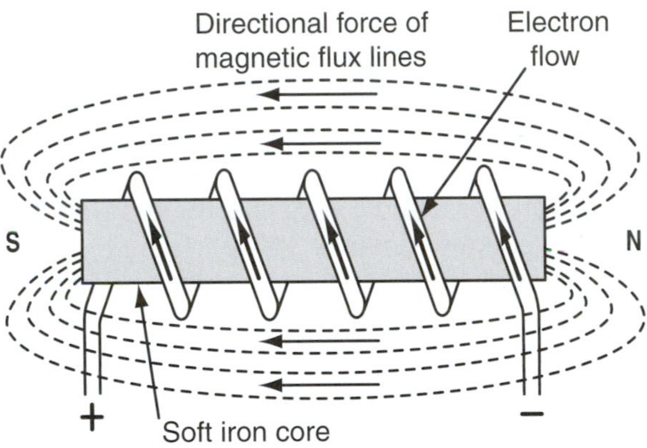

Figure I–10 Electromagnet.

move it toward the center so that they can follow a shorter path. If the core is movable, it will move to the center of the coil. A coil around an off-center, movable, permeable core is a *solenoid.* A spring is usually used to hold the core off center.

Motors

In an electric motor, current is passed through a conductor that is looped around the *armature* core, Figure I–12. The conductor loops are placed in grooves along the length of the core. The core is made of laminated discs of permeable, soft iron that are pressed on the armature shaft. The soft iron core causes the magnetic field that forms around the conductor to be stronger due to its permeability. There are several conductor loops on the armature, but only the loop that is nearest

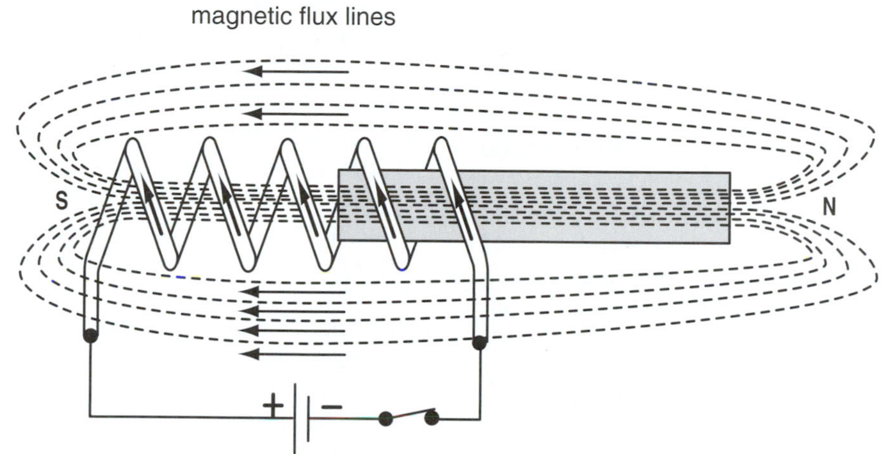

Figure I–11 Solenoid.

the center of the field poles has current passing through it. The loops are positioned so that when one side of a loop is centered on one field pole, its other side is centered on the other field pole.

The field poles are either permanent magnets or pieces of soft iron that serve as the core of an electromagnet. If electromagnets are used, an additional conductor (not shown in Figure I–12) is

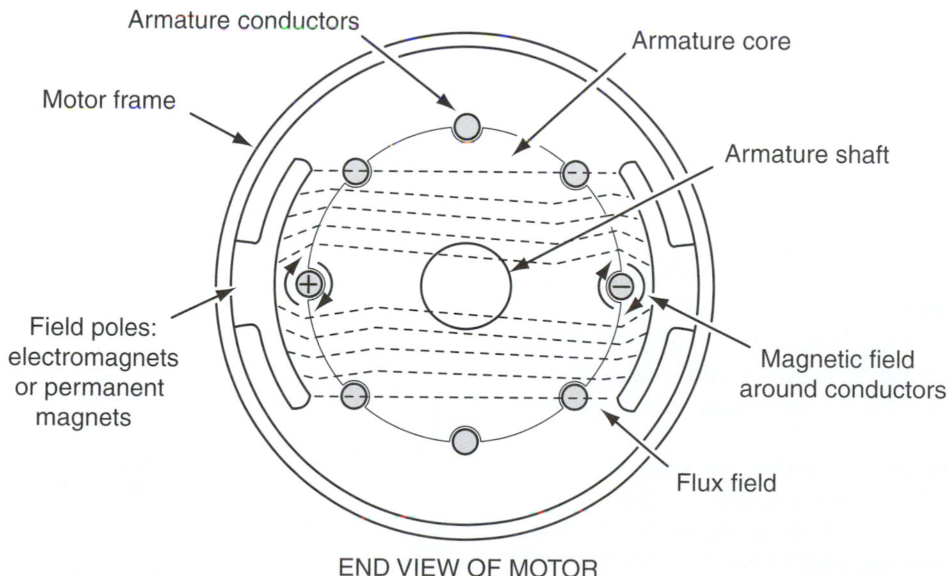

Figure I–12 Electric motor.

wound around each field pole, and current is passed through these field coils to produce a magnetic field between the field poles. The motor frame that the poles are mounted on acts as the magnet body.

Looking at the armature conductor near the north field pole in Figure I–12, you see its magnetic field extending out of the armature core and that it has a clockwise force. The magnetic field between the field poles has a directional force from north pole to south pole. At the top of the armature conductor, the field it has produced has a directional force in the same direction as that of the lines of force between the field poles. The lines of force in this area are compatible, but combining these two fields in the same area produces a high-density field. Remember that magnetic lines of force do not like to be close together.

At the bottom of the armature conductor, the lines of force formed around it have a directional force in the opposite direction from those from the north field pole. The lines of force will not cross each other, so some from the field pole distort and go up and over the conductor into the already dense portion of the field above the conductor, and some just cease to exist. This produces a high-density field above the conductor and a low-density field below it. The difference in density is similar to a difference in pressure. This produces a downward force on the conductor.

The other side of the armature loop, on the other side of the armature, is experiencing the same thing except that the current is now traveling the opposite way. The loop makes a U-shaped bend at the end of the armature. The magnetic field around this part of the conductor has a counterclockwise force. Here, the lines of force around the conductor are compatible with those between the field poles under the conductor but try to cross at the top.

This produces an upward force on this side of the armature loop. The armature rotates counterclockwise. To change the direction in which the armature turns, change either the direction that current flows through the armature conductors, or change the polarity of the field poles.

Magnetic Induction

Passing voltage through a wire causes a magnetic field to form around the wire. However, if lines of force can be formed around a conductor, a voltage is produced in the wire and current starts to flow. This assumes, of course, the wire is part of a complete circuit. Lines of force can be made to wrap around a conductor by passing a conductor through a magnetic field, Figure I–13. This phenomenon occurs because of characteristic 6. As the conductor passes through the magnetic field, it cuts each line of force. Because the lines of force resist being cut, they first wrap around the conductor much like a blade of grass would if struck by a stick, Figure I–14. This principle is used in generators to produce voltage and current flow. The principle will work regardless of whether:

- the conductor is moved through a stationary magnetic field, as in a direct current generator.
- a magnetic field is moved past stationary conductors, as in an alternating current generator.
- the lines of force in an electromagnetic field are moved by having the circuit producing the magnetic field turned on and off, as in an ignition coil.

Note that in each case, movement of either the lines of force or the conductor is needed. A magnetic field around a conductor where both are

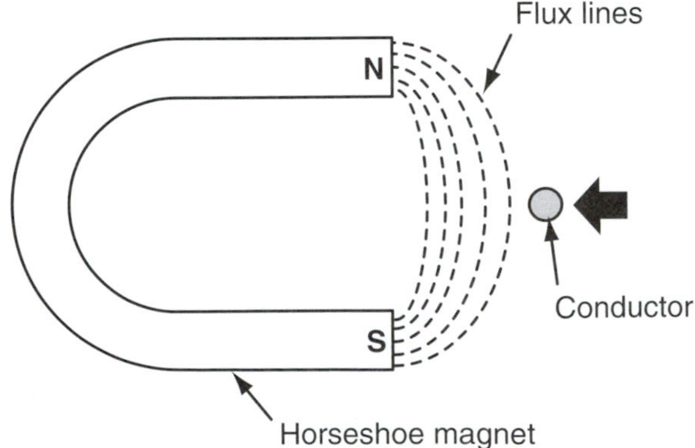

Figure I–13 Magnetic induction.

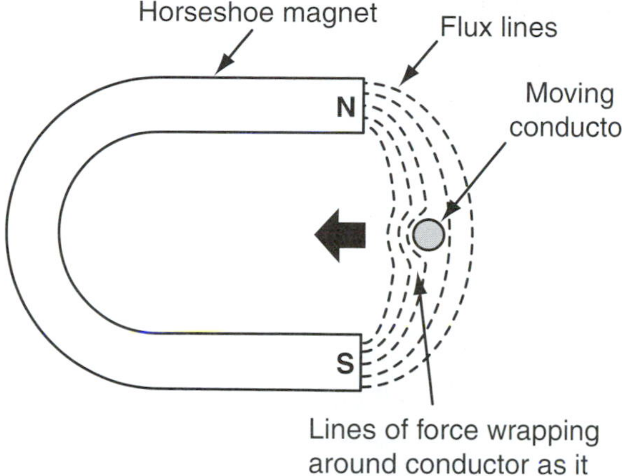

Figure I–14 Cutting lines of force.

steady state will not produce voltage. The amount of voltage and current produced by magnetic induction depends on four factors:

1. The strength of the magnetic field (how many lines of force there are to cut). A tiny amount of voltage is induced in the wire by each line of force that is cut.
2. The number of conductors cutting the line of force. Winding the conductor into a coil and passing one side of the coil through the magnetic field cuts each line of force as many times as there are loops in the coil.
3. How fast and how many times the conductors pass through the magnetic field.
4. The angle between the lines of force and the conductor's approach to them.

Amperage

Amperage is a measure of the amount of current flowing in a circuit. One ampere equals 6,250,000,000,000,000,000 (six and one-quarter billion billion) electrons moving past a given point in a circuit per second. This is often expressed as one *coulomb*.

Voltage

A *volt* is a measure of the force or pressure that caused current to flow; it is often referred to as *voltage*. The difference in potential is voltage. The most common ways of producing voltage are chemically, as in a battery, or by magnetic induction, as in a generator. A more accurate but less used name is *electromotive force*. Note that volts are what drives the electrons through the circuit; voltage is the measurement of that force. Similarly amps are the number of electrons moving; amperage is the measurement of that number.

Resistance

The fact that voltage is required to push current through a circuit suggests that the circuit offers resistance. In other words, you do not have to push something unless it resists moving. Resistance, which is measured in *ohms*, limits the amount of amperage that flows through a circuit, Figure I–15.

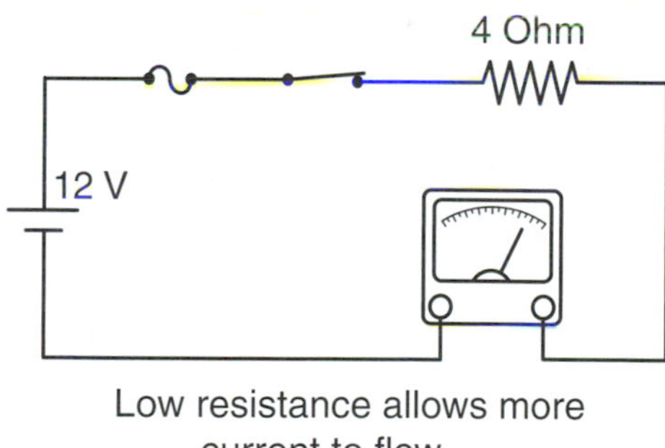

Low resistance allows more current to flow

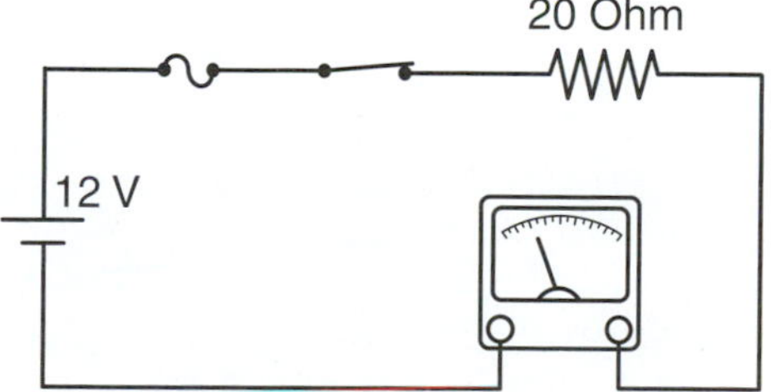

Higher resistance allows less current to flow

Figure I–15 Resistance versus current flow.

If a circuit without enough resistance is connected across a reliable voltage source, wires or some other component in the circuit will be damaged by heat because too much current flows.

As mentioned, a bond exists between an electron and the protons in the nucleus of an atom. That bond must be broken for the electron to be freed so that it can move to another atom. Breaking that bond and moving the electron amount to doing work. Doing that work represents a form of resistance to current flow. This resistance varies from one conductive material to another, depending on the atomic structure of the material. For example, lead has more resistance than iron, and iron has more resistance than copper. It also varies with the temperature of the conductor. Loose or dirty connections in a circuit also offer resistance to current flow. Using current flow to do work (to create heat, light, a magnetic field or to move something) also amounts to resistance to current flow.

Sometimes students confuse voltage and amperage while doing tests on electrical systems. Review these definitions and consider the influence that voltage and amperage have on a circuit. It might also help to remember that voltage can be present in a circuit without current flowing. However, current cannot flow unless voltage is present.

Voltage Loss

When voltage pushes current through a circuit, voltage is lost by being converted to some other energy form (heat, light, magnetism or motion). This is usually referred to as *voltage drop.* Every part of a circuit offers some resistance, even the wires, although the resistance in the wires should be very low, Figure I–16. The voltage drop in each part of the circuit is proportional to the resistance in that part of the circuit. *The total voltage dropped in a circuit must equal the source voltage.* In other words, all of the voltage applied to a circuit must be converted to another energy form within the circuit.

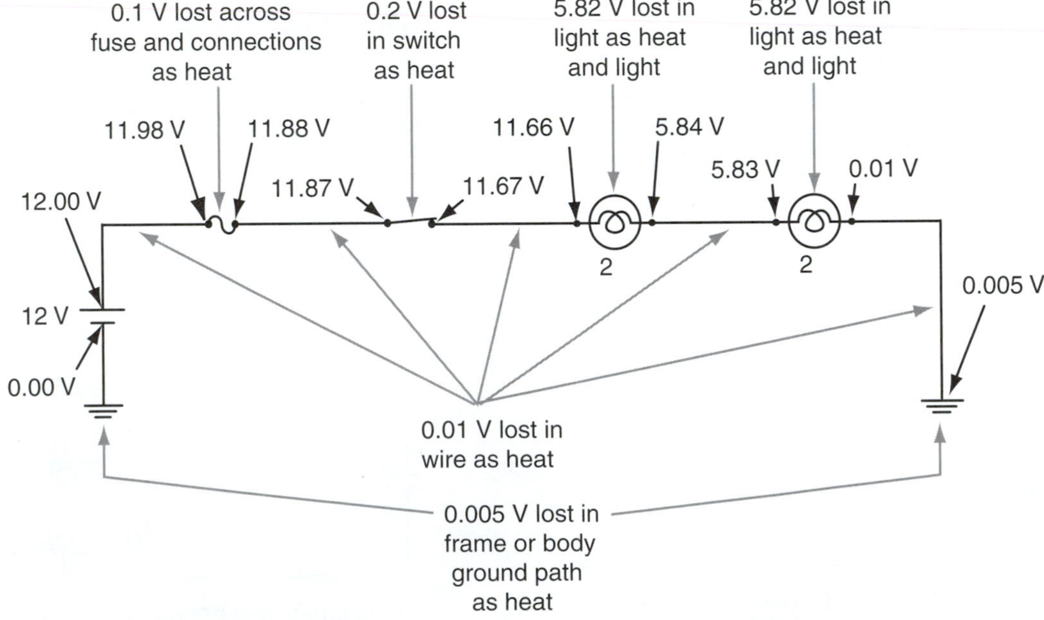

Figure I–16 Voltage drop.

Ohm's Law

Ohm's law defines a relationship between amperage, voltage and resistance. Ohm's law says: It takes one volt to push one ampere through one ohm of resistance. Ohm's law can be expressed in one of three simple mathematics equations:

$$E = I \times R$$
$$I = E/R$$
$$R = E/I$$

where: E = electromotive force or voltage
I = intensity or amperage
R = resistance or ohms

The simplest application of Ohm's law enables you to find the value of any one of the three factors: amperage, voltage or resistance, if the other two are known. For example, in the following diagram, if the voltage is 12 volts and the resistance is 2 ohms, the current flow can be determined as follows:

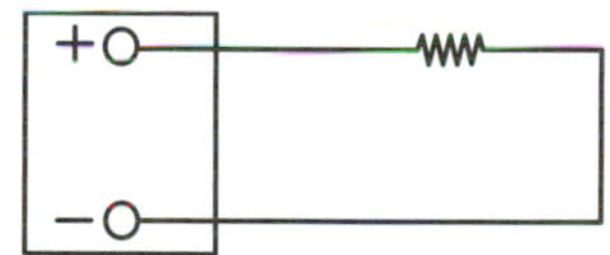

$$I = E/R \text{ or } I = 12\text{ V}/2\ \Omega = 6 \text{ amps}$$

(The Greek letter Ω is often used as a symbol or an abbreviation for ohms, and amps is often used as an abbreviation for amperes.)

If the resistance is 4 ohms and the current is 1.5 amps, the voltage applied can be found by:

$$E = I \times R \text{ or}$$
$$E = 1.5 \text{ amps} \times 4\ \Omega = 6\text{ V}$$

If the voltage is 12 volts and the current is 3 amps, the resistance can be found by:

$$R = E/I \text{ or}$$
$$R = 12\text{ V}/3 \text{ amps} = 4\ \Omega$$

Perhaps the easiest way to remember how to use these equations is to use the following diagram:

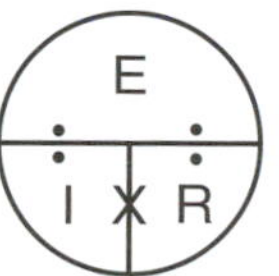

To find the value of the unknown factor, cover the unknown factor with your thumb and multiply or divide the other factors as their positions indicate.

There are many other applications of Ohm's law, some of which are quite complex. (A more complicated application is covered later in this chapter.) An automotive technician is rarely required to directly apply Ohm's law in order to find or repair an electrical problem on a vehicle. But knowing and understanding the relationship of the three factors just covered is a must for the technician who wants to diagnose and repair electrical systems effectively.

CONDUCTORS AND INSULATORS

Earlier, in discussing the electrons in the outer shell of an atom, it was said that the fewer electrons there are in the outer shell the easier it is to break them loose from the atom. If an atom has five or more electrons in its outer shell, the electrons become much harder to break away from the atom, and the substance made up of those atoms is a very poor conductor; so poor, it is classed as an insulator. Rubber, most plastics, glass and ceramics are common examples of insulators. Substances with four electrons in their outer shell are poor conductors but can become good ones under certain conditions. Thus, they are called *semiconductors*. Silicon and germanium are good examples of semiconductors. (Semiconductors are covered in more detail later in this chapter.)

Conductors are substances made up of atoms with three or fewer electrons in their outer shell. These electrons are called free electrons because they can be freed to travel to another atom.

CIRCUITS

Almost all electrical circuits in vehicles will have some type of component to provide each of the following functions:

- *Voltage source,* such as a battery or generator, provides voltage to the power circuit.
- *Circuit protection,* such as a fuse or circuit breaker, serves to open the circuit in the event of excessive current flow.
- *Circuit control,* such as a switch or relay, provides the ability to open and close the circuit.
- *Load,* such as a motor or light, does the function the circuit was designed to do and also provides resistance to limit current.
- *Conduct current,* usually wires or certain parts of the vehicle, such as the frame, body or engine block.

These components are often represented by symbols as shown in Figure I–17.

Circuit Types

There are two distinct types of electrical circuits, plus combinations of the two.

Series Circuits. In a series circuit, there is only one path for current flow, and all of the current flows through every part of the circuit. Parts A and B of Figure I–17 show simple series circuits. Even though there is only one load in each of these circuits, they qualify as series circuits because there is only one path for current flow. Figure I–18 shows a better example of a series circuit. Not only is there just one path for current flow, there are also two loads in series with each other. When there are two or more loads in series, the current must pass through one before it can pass through the next. The characteristics of a series circuit include the following:

- Current flow is the same at all parts of the circuit.
- Resistance units are added together for total resistance.
- Current flow decreases as resistance units are added.
- A portion of source voltage is dropped at each resistance unit in the circuit.
- An open in any part of the circuit disrupts the entire circuit.

The following problems apply Ohm's law to a series circuit. Refer to Figure I–19, which shows a compound series circuit.

Problem 1. Assume that the resistance of R1 is 2 ohms and that R2 is 4 ohms; find the total current flow. In a series circuit, the resistance value of each unit of resistance can be added together because all of the current passes through each resistor.

$$I = E/R = \text{or}$$
$$I = E/(R1 + R2) \text{ or}$$
$$I = 12V/(2\,\Omega + 4\,\Omega) \text{ or}$$
$$12V/6\,\Omega = 2A$$

Problem 2. Assume the resistance values are unknown in Figure I–19, but that the total current flow is 3 amps. To find the total resistance:

$$RT = E/I \text{ or}$$
$$RT = 12\text{ V}/3\text{ amps} = 4\,\Omega$$

Problem 3. Find the voltage drop across R1, applying the same resistance values as in problem 1. Each ohm of resistance value shares equally in the total voltage drop in a series circuit. Therefore, how much voltage each ohm of resistance drops is also equal numerically to how many amps are flowing in the circuit. Thus, because you are concerned with the voltage drop across R1, multiply the resistance of R1 by the circuit's amperage.

$$\text{Voltage drop across } R1 = R1 \times I \text{ or}$$
$$2\,\Omega \times 2A = 4V \text{ (voltage drop across R1)}$$

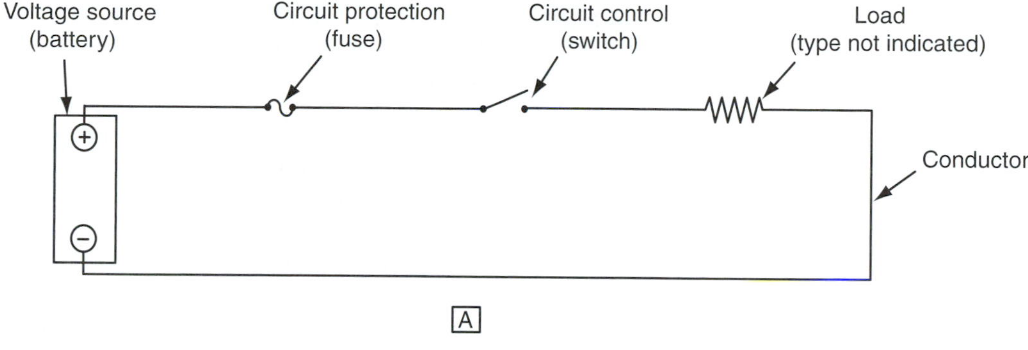

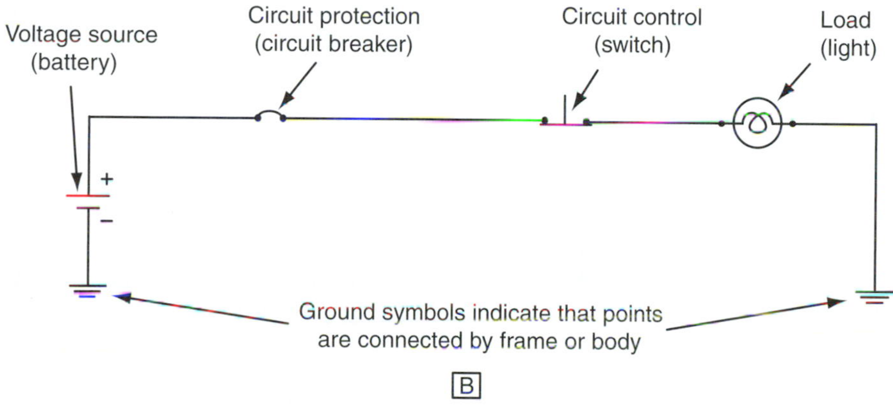

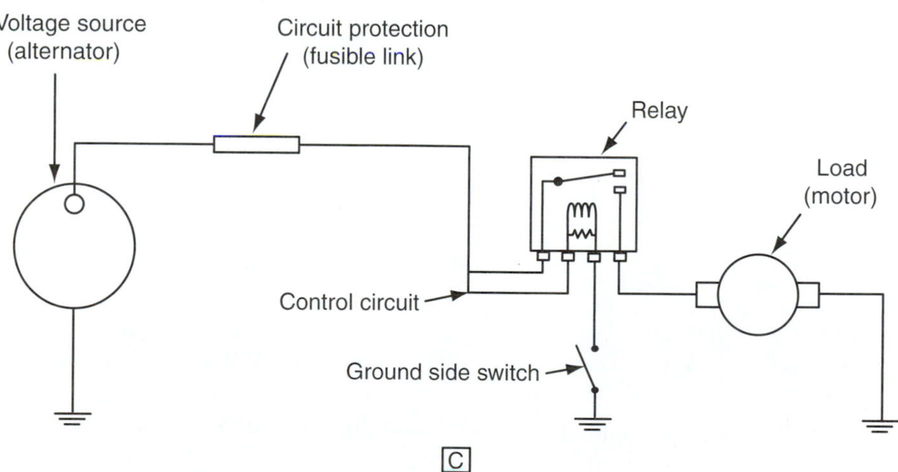

Figure I–17 Circuit components.

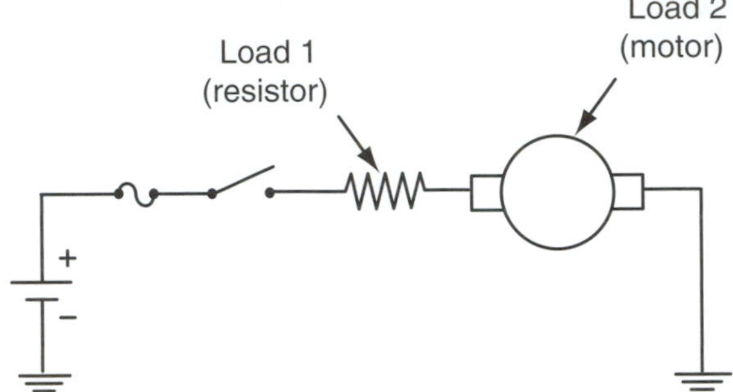

Figure I–18 Series circuit.

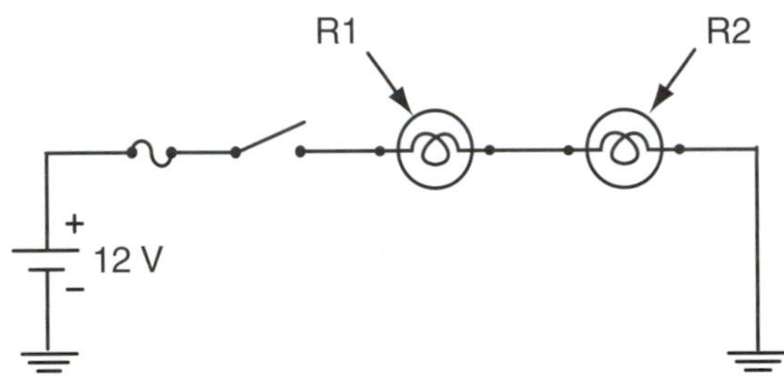

Figure I–19 Series circuit.

Parallel Circuits. In a parallel circuit, the conductors split into branches with a load in each branch, Figure I–17 lower and Figure I–20. Some current will flow through each branch with the most current flowing through the branch with the least resistance. Characteristics of a parallel circuit are:

- current varies in each branch (unless resistance in each branch is equal).
- total circuit resistance goes down as more branches are added and will always be less than the lowest single resistance unit in the circuit.
- total circuit current flow increases as more branches are added.
- source voltage is dropped across each branch.
- an open in one branch does not affect other branches.

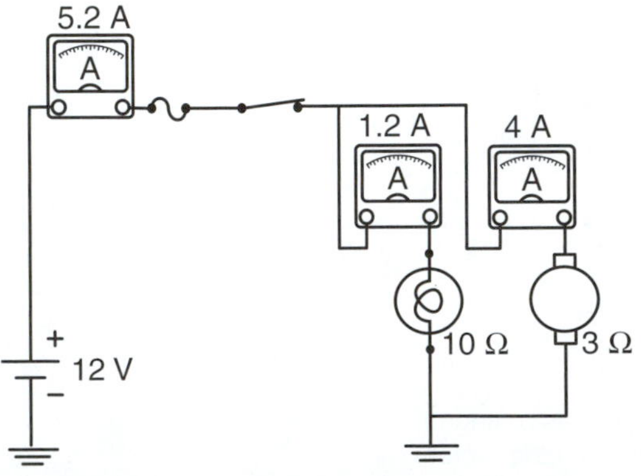

Figure I–20 Parallel circuit.

The following problems apply Ohm's law to a parallel circuit.

Problem 4. In the following diagram, assume that R1 is 3 ohms and R2 is 4 ohms. Find the current flow through the R1 branch. Note that R1 is not affected by R2, so treat R1 like it is a series circuit.

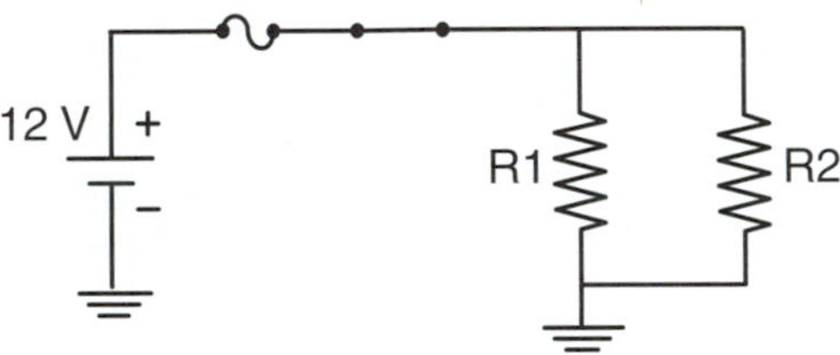

$$I = E/R \text{ or}$$
$$I = 12\ V/3\ \Omega = 4 \text{ amps}$$

Find the current flow through R2.

$$I = E/R \text{ or}$$
$$I = 12\ V/4\ \Omega = 3 \text{ amps}$$

What is the total current flow in the circuit?

$$4 \text{ amps} + 3 \text{ amps} = 7 \text{ amps}$$

Problem 5. Now that you know the current flow, you can easily find the total circuit resistance.

RT = E/I or
RT = 12 V/7 amps = 1.71 Ω

Note that to solve for total resistance you have simply solved each branch for current flow, added the flow of each branch and divided the source voltage by total current. This method will work regardless of how many branches there are in the circuit.

There are other ways to find total resistance in a parallel circuit. If there are only two branches in the circuit, the product over the sum method can be used. In problem 4 there were two resistors in a parallel circuit with values of 3 ohms and 4 ohms. Using those values:

$$\frac{R1 \times R2}{R1 + R2} \text{ or}$$

$$\frac{3\ \Omega \times 4\ \Omega}{3\ \Omega + 4\ \Omega} = \frac{12\ \Omega}{7\ \Omega} = 1.71\ \Omega$$

If there are more than two branches in a circuit, this method can still be used, but it has to be manipulated a bit. Assume there are four branches in a circuit. Select two branches and use the product over the sum method to solve them as just shown. Then take the other two and do the same for them. Now take the two answers and run them through the same formula. The final result is total circuit resistance. There are still other methods, but you have learned all you need to know to solve the problems you are likely to encounter. Many automotive electrical systems texts and electrical texts and reference books present those methods if you want to pursue them.

Series-Parallel Circuits. Some circuits have characteristics of a series as well as those of a parallel circuit. There are two basic types of series-parallel circuit. The most common is a parallel circuit with at least one resistance unit in series with all branches, Figure I–21. All of the current flowing through the circuit in Figure I–21 must pass through the indicator light before it divides to go through the two heat elements, which are in parallel with each other.

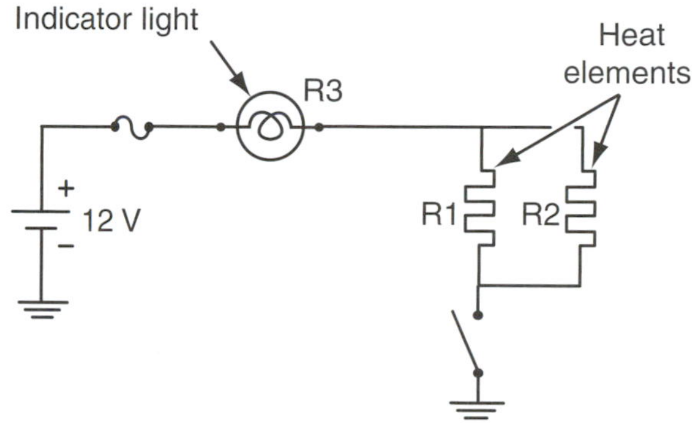

Figure I–21 Series-parallel circuit.

To solve for current, resistance or voltage drop values in a series-parallel circuit, you must identify how the resistance units relate to each other, then use whichever set of formulas (series circuit or parallel circuit) apply. For example:

Problem 6. In Figure I–21, assume a resistance of 4 ohms for the indicator light (R3), 20 ohms for R1 and 30 ohms for R2. Find the total resistance for the circuit.

Branches R1 and R2 are in parallel with each other and in series with R3. The battery sees the resistance of R3 plus the resistance of the two parallel branches. Once you find the total resistance of the branch circuits, you can treat that as a single resistance value that is in series with R3. Using the product over the sum method, first solve for the combined resistance of R1 and R2.

$$R = \frac{R1 \times R2}{R1 + R2} \text{ or}$$

$$\frac{20\ \Omega \times 30\ \Omega}{20\ \Omega + 30\ \Omega} = \frac{600\ \Omega}{50\ \Omega} = 12\ \Omega$$

Now add the total resistance of R1 and R2 (12 ohms) to the resistance of R3 (4 ohms).

RT = R + R3 or
12 Ω + 4 Ω = 16 Ω

Problem 7. Find the total current flow for the circuit.

IT = E/RT or
IT = 12/16 Ω = 0.75 amp

Before you can find the current flow through R1 or R2, you must find the voltage applied to R1 and R2. Remember that R3 is in series with both R1 and R2. Also remember that in a series circuit, a portion of the source voltage is dropped across each resistance unit. Therefore, because there is a volt drop across R3, there is not a full 12 volts applied to R1 and R2.

Problem 8. Find the volt drop across R3.

V3 = IT × R3 or
V3 = 0.75 amp × 4 Ω = 3 V

If 3 volts are dropped across R3 from a source voltage of 12 volts, 9 volts are applied to R1 and R2.

Problem 9. Find the current flow through:

R1: I1 = E/R1 or
I1 = 9 V/20 Ω = 0.45 amp
R2: I2 = E/R2 or
I2 = 9 V/30 Ω = 0.30 amp

The second type of series-parallel circuit is a parallel circuit in which at least one of the branch circuits contains two or more loads in series, Figure I–22. This type of circuit is often referred to as *series string in parallel*. Finding the total current, resistance or voltage drop for this circuit is easy. Identify the branch (or branches) that has loads in series and combine the resistances for each branch. Then treat the circuit like any other parallel circuit. When troubleshooting this kind of circuit in a car, failing to recognize the branch of a circuit you are testing as having loads in series could cause confusion.

CAUTION: If you are not absolutely sure about the circuit you are testing, consult the correct wiring diagram for the vehicle.

POLARITY

Just as a magnet has polarity—a north and a south pole that determine the directional force of the lines of force—an electrical circuit has a polarity. Instead of using north and south to identify the polarity, an electrical circuit's polarity is identified by positive and negative. An electrical circuit's polarity is determined by its power source. The best example is a battery, Figure I–23. The side of the circuit that connects to the positive side of the battery is positive and on most vehicles is the insulated side. The voltage on this side is usually near source voltage. Most of the voltage is dropped in the load where most of the work is

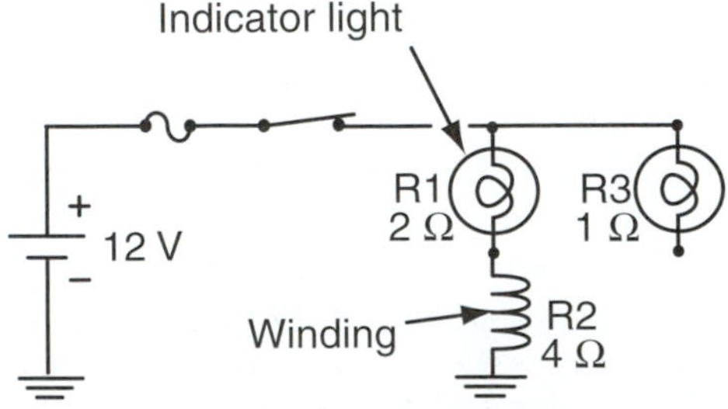

Figure I–22 Series string in parallel.

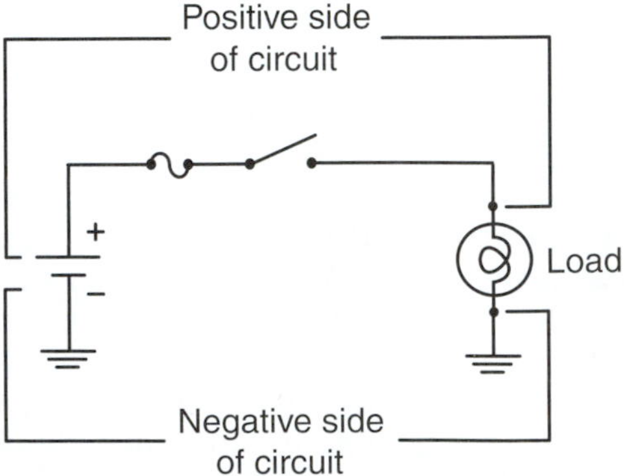

Figure I–23 Circuit polarity.

done. The negative side of the circuit carries the current from the load to the negative side of the battery. There is usually very little resistance in the negative side of the circuit, and the voltage is near zero. Because this side of the circuit has the same potential as the vehicle's frame and sheet metal, this side of the circuit might not be insulated.

CIRCUIT CONTROLS: FAULTS

Three kinds of faults can occur in an electrical or electronic circuit: opens, excessive resistance and shorts.

Opens

By far the most common fault in electrical and electronic circuits is an open circuit, or *open*. An open circuit means that there is a point in the circuit where resistance is so high that current cannot flow. An open can be the result of a broken wire, loose or dirty connection, blown fuse or faulty component in the load device. Experienced technicians know that wires rarely break except in applications in which the wire experiences a lot of flexing. Most opens occur in connections, switches and components. A switch in a circuit provides a way to conveniently open the circuit. Of course, when a circuit is opened deliberately, it is not a fault.

Excessive Resistance

A loose or dirty connection or a partially cut wire can cause excessive resistance in a circuit. Under these conditions, the circuit can still work but not as well as it could because the additional resistance reduces current flow through the circuit, Figure I–24. Excessive resistance can also result from a faulty repair or modification of a circuit in which a wire that is too long or too small in diameter has been installed. The location of excessive resistance in a circuit can be easily found with a series of voltage drop tests, which are discussed under Testing in this chapter.

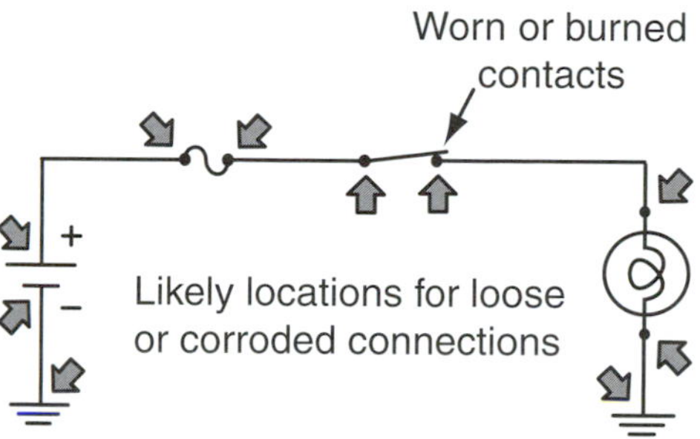

Figure I–24 Possible locations in a circuit for openings or high resistance.

Shorts

A short is a fault in a circuit that causes current to bypass a portion or all of the circuit's resistance. The term *short* as used in electrical terminology means that the current is taking a shortcut rather than following the path it is supposed to take, Figure I–25. Remember Ohm's law: if resistance is reduced, current flow increases.

Short to Positive. In Figure I–25A, electrical contact has developed between two windings of a coil. This causes current to shortcut the loop or loops that exist between the points where the contact occurred, which lowers the coil's resistance. If only a relatively small number of the coil's loops are shorted, the increase in current flow might not be enough to further damage the circuit except possibly to blow a fuse if the circuit is fuse protected. The device might even continue to work but probably at reduced efficiency.

Short to Ground. If a short occurs up circuit from the load (between the load and the positive source), Figure I–25B, it provides a lower resistance path to ground for current flow. Because of the lower resistance, excessive current will flow, and the fuse element will melt, causing the circuit to open. If the circuit does not have a fuse or some other type of circuit protection, the wire between the positive terminal and the short will overheat, possibly melting the wire or starting a fire.

If a short occurs down circuit from the load (between the load and the negative terminal), Figure I–25C, the switch will not be able to provide

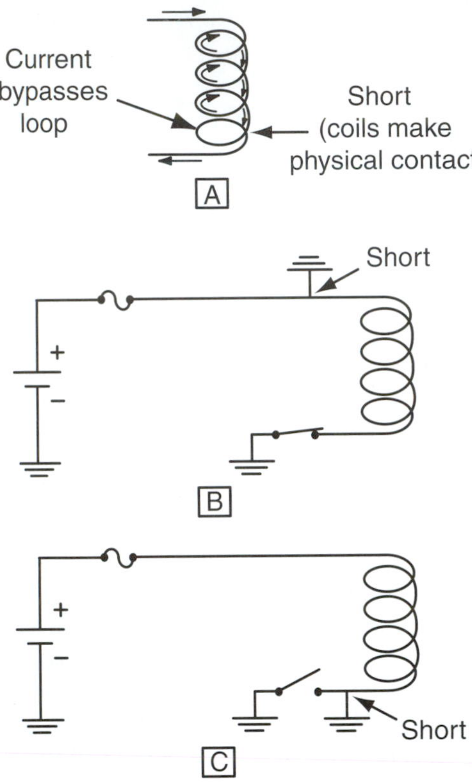

Figure I–25 Shorts.

the needed resistance to open the circuit. This is because the current has found a shortcut to ground that the switch cannot control. If the circuit in question is controlled by an electronic switching circuit, as you will see in following chapters, this condition threatens the solid-state circuit because the additional current flow might damage it.

TESTING

Basic electrical test procedures are easy to learn. They are essential to effective electrical troubleshooting and can save considerable time in identifying a problem. Effective electrical testing requires the following:

- understanding the principles of electricity
- understanding the circuit being tested
- knowing how to use the test equipment
- knowing the three things above well enough to know what the test results mean

Testing Voltage

Although many technicians over the years have used a simple nonself-powered test light to perform this test, a test light, Figure I–26, is considered too inaccurate to really provide reliable results when using the battery's open circuit voltage (OCV) to determine its state of charge. A better method is to test the battery's OCV with a voltmeter, which indicates much more precisely what the voltage is. A digital voltmeter (DVOM) is shown in Figure I–27. See Chapter 4

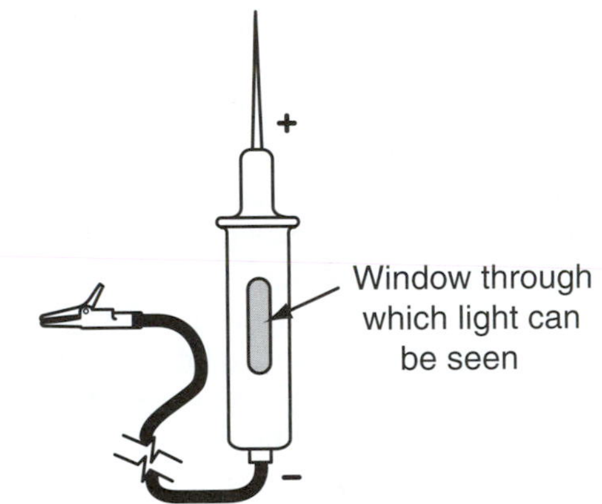

Figure I–26 Nonself-powered test light.

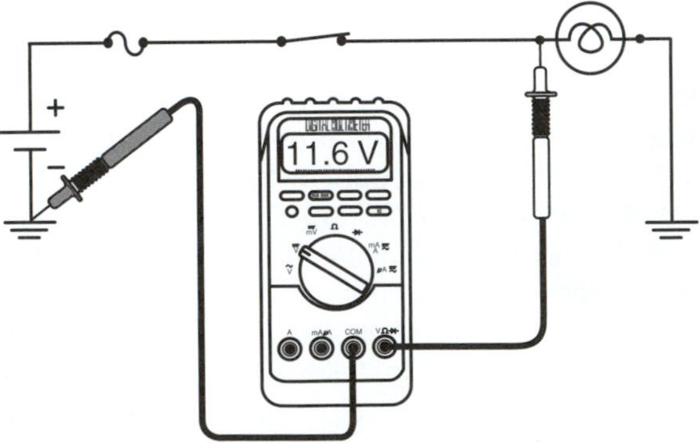

Figure I–27 Available volts test.

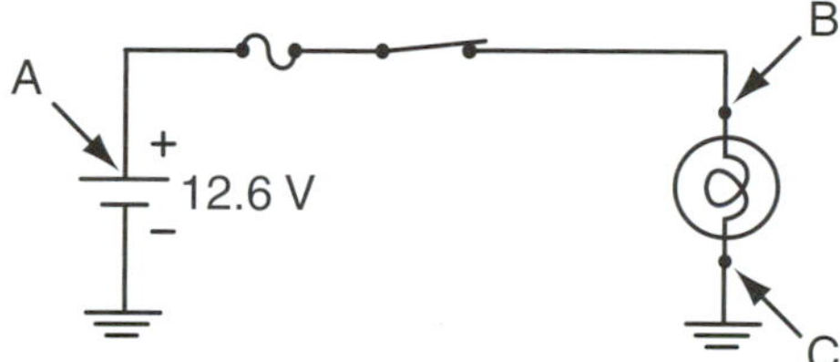

Figure I–28 Series circuit.

for additional information concerning the use of a DVOM.

Figure I–28 shows a circuit with several points indicated. Review the chart below and be sure that you can explain why each set of voltage values given indicates the faults listed.

Voltage Drop

A circuit always consists of three parts: the source, the load component and the connecting circuitry, both on the positive and negative sides of the circuit. When a symptom of poor operation (or even no apparent operation) is encountered, the source should always be checked first. This, of course, is the vehicle's battery if the engine is not running. When the engine is running, then the alternator becomes the voltage source.

The second item that should be tested is the connecting circuitry, which is responsible for supplying the load component with as close to the full amount of the source's voltage potential as possible. There are exceptions to this, however. Some examples are:

- the current limiting resistors used for the low and intermediate speeds of a blower motor control circuit, Figure I–29A.
- the resistor (ballast resistor) used to limit current flow through the ignition coil of many ignition systems a few years ago, Figure I–29B.
- the current-limiting resistor used in a temperature-sensing circuit, Figure I–29C.

When voltage drop testing a circuit, you should remember that:

- In an ideal circuit, all of the voltage drop should take place across the load component.
- In an ideal circuit, none of the voltage drop should take place across the circuitry.
- Voltage drop cannot occur unless the circuit is complete and current is flowing in the circuit.
- Voltage that is dropped (used up) is always the result of encountered resistance when current is flowing in the circuit and therefore proves the existence of resistance in the circuit.
- Voltage measurements can then be used to identify the existence of unwanted, excessive resistance in the circuitry, but only if the circuit is complete and at least some current is flowing in the circuit, even if it is not sufficient current flow to cause the load component to appear to operate.

In order to perform a voltage drop test of a circuit, the DVOM must be connected in *parallel* to the circuitry, either on the positive (insulated) side of the circuit or on the negative (ground) side of the circuit. If performing a voltage drop test on the insulated side of the circuit, the DVOM should be

Points				
A	B	C	Fault	Fault Location
12.6 V	12.6 V	12.6 V	Open	Down circuit from C
12.6 V	12.6 V	0	Open	At load
12.6 V	0	0	Open	Up circuit from B
12.6 V	12.3 V	.3 V	None	

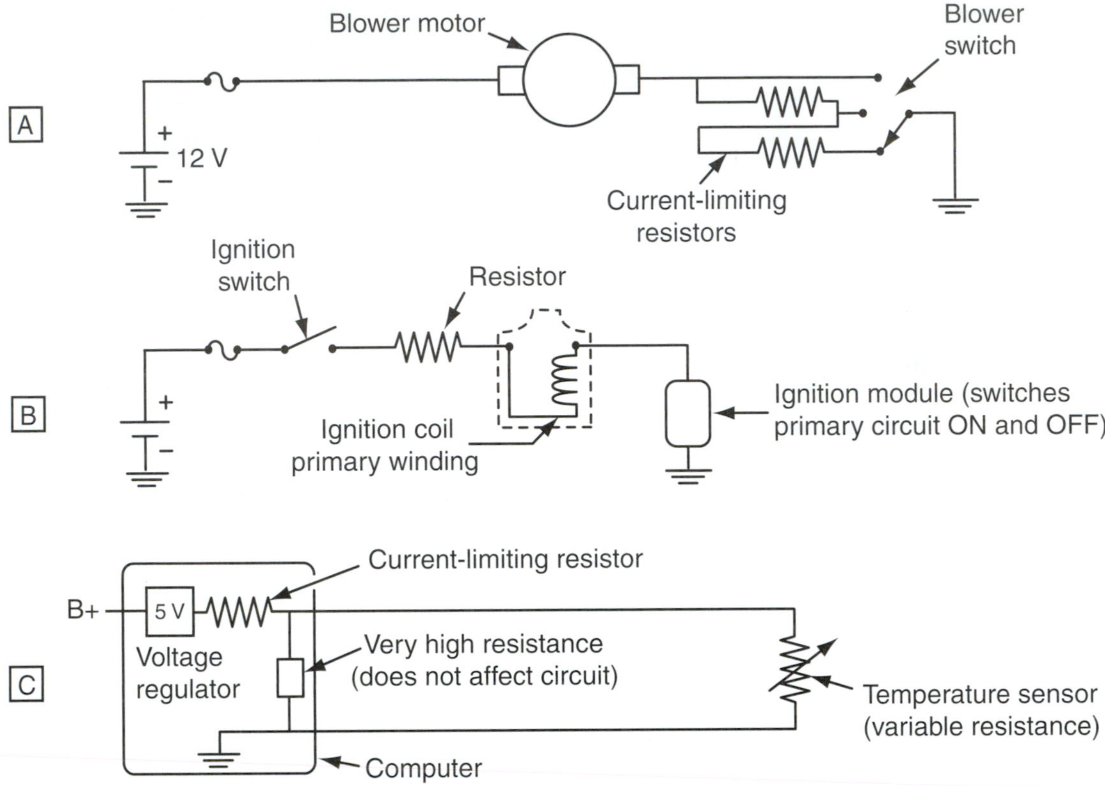

Figure I–29 Current-limiting resistor circuits.

connected *between* the battery positive post and the load component's positive terminal, with the positive DVOM lead connected to the battery. This parallel connection compares the difference in the two voltage values and displays the results, Figure I–30. This difference is the amount of voltage drop that has occurred between the two points where the DVOM's leads are connected. This reading equates to having measured the available voltage at the battery, then having measured the voltage reaching the load and then mathematically figuring the difference, assuming that the battery voltage did not change during the time that the two voltage readings were taken.

If performing a voltage drop test on the ground side of the circuit, the DVOM should be connected *between* the battery negative post and the load component's negative terminal, with the negative DVOM lead connected to the battery. Again, this parallel connection compares the difference in the two voltage values and displays the results, Figure I–31. This difference is, again, the amount of voltage drop that has occurred between the two points where the DVOM's leads are connected and also equates to how much voltage was still left in the circuit after the load due to unwanted resistance in the circuit after the load.

Performing voltage drop tests is the most accurate and efficient method to verify the integrity of the circuitry that connects the source to the load and will alert the technician if excessive unwanted resistance exists in any part of the circuitry, be it in the wiring, any circuit protection devices, any switches (whether relay-operated or not) and any ground connections (either to engine or chassis ground).

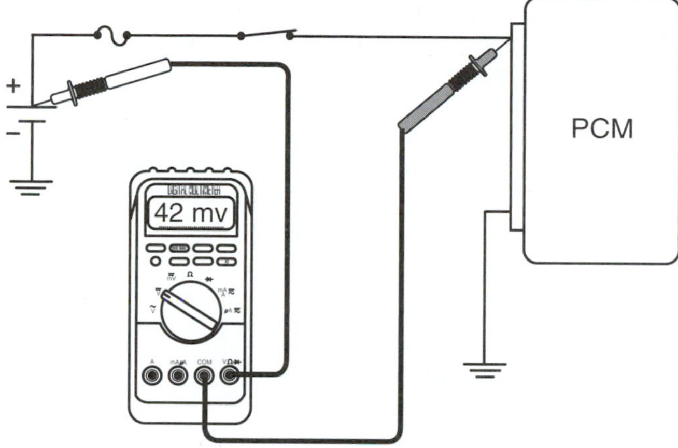

Figure I–30 Positive side voltage drop test. A .042 V reading indicates that insulated circuit resistance is acceptable.

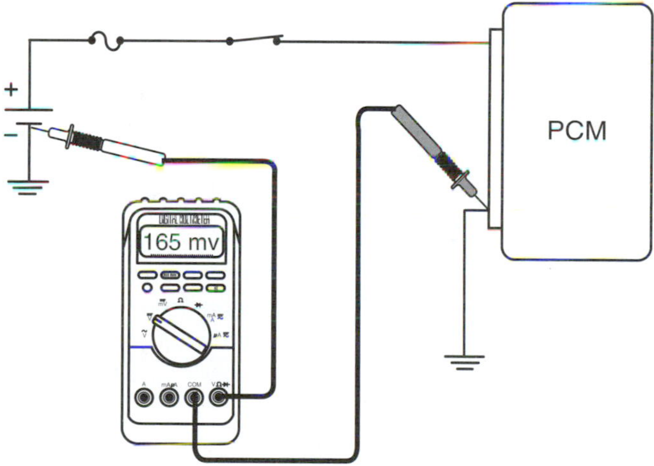

Figure I–31 Negative side voltage drop test. A .165 V reading indicates excessive resistance in the ground-side circuit and is above acceptable limits.

Depending on the type of circuit being tested, the specifications for allowable voltage drop vary, not because we allow more or less resistance for different types of circuits, but primarily because more amperage across a given resistance results in more voltage drop and less amperage across a given resistance results in less voltage drop. This is due to the fact that all of the source voltage will be used up evenly by each ohm of resistance value in a circuit. A high amp circuit has less resistance so each ohm of resistance produces more voltage drop, and a low amp circuit has more resistance so each ohm of resistance produces less voltage drop. As a result, the typical maximum allowable voltage drop specifications are as follows:

- Most circuits = .1 V (100 mV) per connection
- Starter circuits = .2 V (200 mV) per connection
- Computer power and ground circuits = total voltage drop in either the power or ground circuit should never exceed 100 mV, see Figures I–30 and I–31.

Voltage drop readings less than the maximum allowable specification indicate that all components and connections that make up the circuitry have resistance levels low enough to be acceptable. If a high voltage drop reading is encountered, then, with one DVOM lead connected to the battery, the other lead should be drawn back through the circuit toward the battery (as per the applicable electrical schematic) until the voltage drop reading is reduced to acceptable levels. At this point, you know that the high resistance is located between the point in the circuit where the last unacceptable reading was taken and the point in the circuit where the first acceptable reading was taken.

When performing voltage drop tests on circuits that power load components at the rear of the vehicle, a DVOM lead of extra length may be made out of speaker wire. With this long lead connected to the vehicle's battery using an alligator clip, the DVOM may now be carried to the rear of the vehicle to perform the voltage drop tests. The wire gauge is not critical, except for reasons of durability, because the DVOM's high internal resistance does not allow any substantial current flow through the meter when making voltage tests. This extralength lead should be plugged into the DVOM's positive jack and connected to battery positive when performing insulated side voltage drop tests and should be plugged into the DVOM's negative jack and connected to battery negative when performing ground side voltage drop tests.

Diagnostic & Service Tip

For certain kinds of measurements, voltage drop tests are far more reliable than resistance tests. This is particularly true for low-resistance measurements, such as resistance through a starter circuit. It is quite difficult to accurately measure fractions of an ohm, but it is easy to measure fractions of a volt. In consequence, experienced mechanics use a voltage drop test in preference to a direct-resistance measurement in almost every case.

Ammeter Testing

An ammeter is used to measure the current flowing in a circuit. There are two types: the conventional type that must be put in series as part of the circuit, Figure I–32, and the inductive type. The inductive type has an inductive clamp that clamps around a conductor in the circuit to be tested, Figure I–33. The clamp contains a *Hall effect* device that uses the magnetic field around the conductor that it is clamped around to create an electrical signal (Hall effect devices are discussed in Chapter 2). The strength of the signal depends on the strength of the magnetic field around the conductor, which in turn depends on the amount of current flowing through it. The DVOM displays a voltage reading that corresponds to the strength of the signal. The inductive ammeter is more convenient to use because it eliminates having to open the circuit.

When using a conventional ammeter, it is important to select the correct scale. Passing 5 amps through an ammeter with a 2-amp scale selected can damage the meter. Some meters are fuse protected when on an amperage scale. Become familiar with an ammeter before attempting to use it.

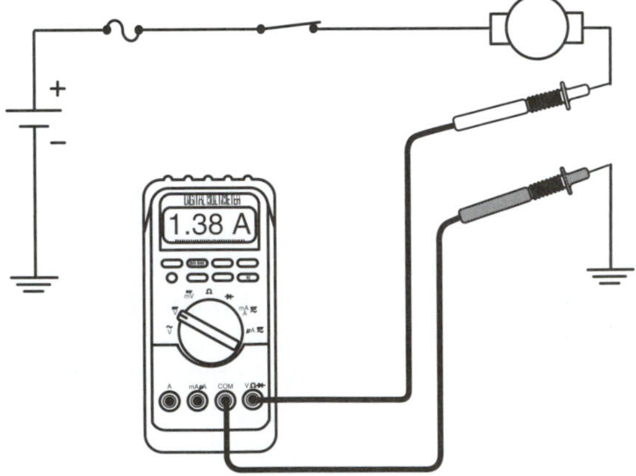

Figure I–32 Ammeter connected in series as part of a circuit.

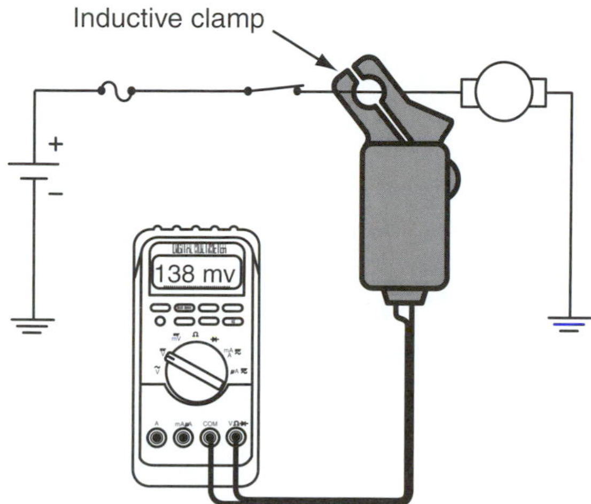

Figure I–33 DVOM used with an amp clamp to perform a nonintrusive current draw test. With the switch on the amp clamp set to 100 mV/Amp, the 138 mV reading equates to 1.38 A.

Resistance Testing

Resistance is measured with an *ohmmeter.* An ohmmeter is self-powered and most have several scales to allow accurate testing of a wide range of resistance values. Most analog ohmmeters (those with a needle) measure from a fraction of 1 ohm to 100,000 ohms. Many digital ohmmeters measure as high as 20 megohms (20,000,000 ohms).

A common mistake when using an ohmmeter is to use a scale that is too low. For example, if you are going to measure the resistance of a circuit that has 1,500 ohms resistance and you use a meter with the scale selector set on R × 1 or R × 10, you will get a reading of infinity; on an analog meter, the needle will not move noticeably from the left side of the scale, and on a digital meter you will get "OL" (open line) or some other indication that the circuit is open. This is because the resistance is higher than the selected scale can measure. To avoid this mistake, start with the highest scale. If you get a low reading, select a lower scale. Continue to select a lower scale until you find the lowest scale that can read the resistance that is present. Some ohmmeters are autoranging. In other words, when the meter is turned on and its leads are contacting a circuit, it automatically selects the correct ohm scale. This type will show an ohm value and display a symbol (K Ω or M Ω) to show what scale is used. Be sure to note the symbol so that you can interpret the resistance value correctly.

When using an ohmmeter it is important to remember that *the circuit or component being tested cannot have voltage in it from any source besides the battery in the ohmmeter.* If there is voltage from another source in a circuit when an ohmmeter is connected to it, the best thing that will happen is an inaccurate reading; the worst is a damaged ohmmeter.

When an analog ohmmeter is turned on, its battery tries to pass current through the meter coil and the two leads, but unless the leads are electrically connected together outside the meter, the circuit is open. If the two leads are touching or if they contact something that has continuity, the circuit is closed. If the leads are connected directly to each other, enough current should flow to cause the meter to swing full scale to the right, which is zero, Figure I–34. If the leads are across a conductor, component or part of a circuit, the resistance of that unit will reduce the current flow

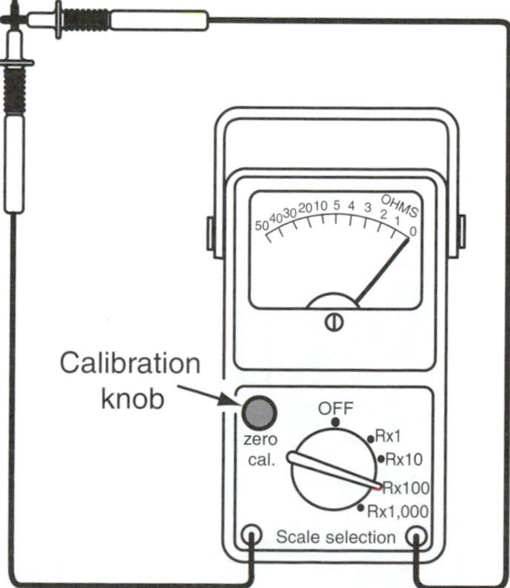

Figure I–34 Calibrating an ohmmeter.

through the meter's circuit. The reduced current flow produces a weaker magnetic field, and the meter's needle will not be moved as far to the right. A digital ohmmeter has an electronic circuit that produces a digital readout in response to the amount of current flowing.

Most needle-type (analog) ohmmeters must be calibrated before each use. Calibration adjusts for voltage variation in the meter's battery due to the discharge that occurs during use. This is done by connecting the leads together and adjusting the zero knob until the needle reads zero, Figure I–34. Doing this establishes that the magnetic field produced by the current flowing through the meter's coil and external leads moves the needle to zero. Any additional resistance placed between the external leads will lower the current flow and will be displayed by the amount the needle lacks in reaching the zero line. Analog ohmmeters also must be recalibrated each time a different scale is selected, because selecting a different scale changes the meter's internal resistance.

Digital ohmmeters do not have a zero adjust knob, but the leads should be shorted together while the DVOM is on the lowest resistance scale, Figure I–35. If any value other than zero is displayed, this value should be subtracted from the displayed reading when making resistance measurements of low-resistance components, Figure I–36. Also, it is important to replace the DVOM's internal battery regularly. Most DVOMs have a "low battery" indicator on the display to indicate when the battery should be replaced.

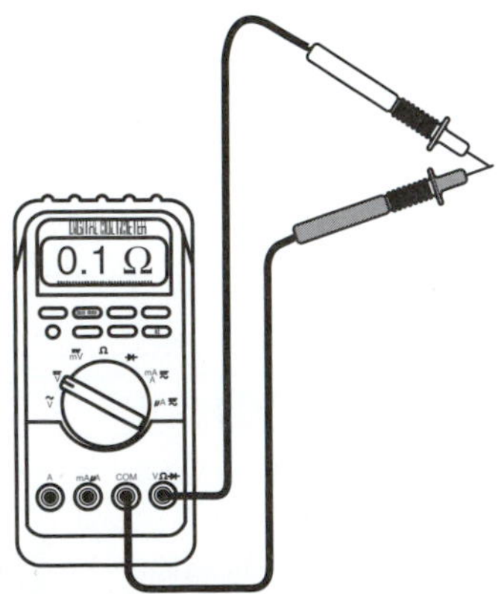

Figure I–35 Testing the calibration of a DVOM's resistance scale.

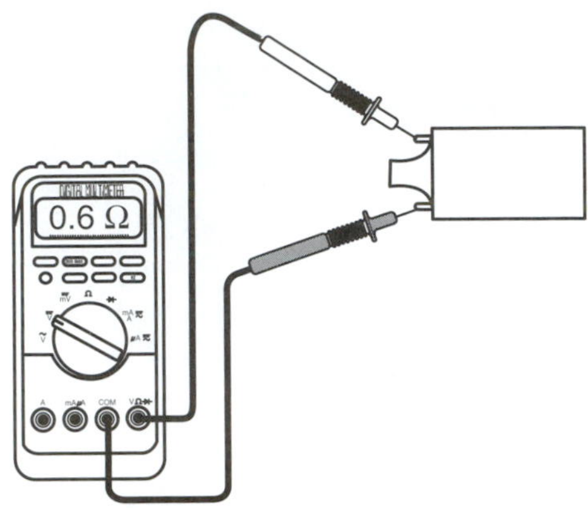

Figure I–36 When using a DVOM to test the resistance of a low-resistance component, the resistance of the leads and meter should be subtracted from the displayed value.

SEMICONDUCTORS

As mentioned earlier in this chapter, a semiconductor is an element with four valence electrons. The two most-used semiconductor materials for solid-state components are silicon and germanium. Of these two, silicon is used much more than germanium. Therefore, most of this discussion will apply to silicon.

As previously stated, an atom with three or fewer electrons in its outer shell easily gives them up. If an atom has more than four but fewer than eight electrons in its outer shell, it exhibits a tendency to acquire more until it has eight. If there are seven electrons in the outer shell, the tendency to acquire another one is stronger than if there are only six. Once there are eight, it becomes very stable; in other words, it is hard to get the atom to gain or lose an electron. In a semiconductor material, for example, a silicon crystal, the atoms share valence electrons in what are called covalent bonds, Figure I–37. Each atom

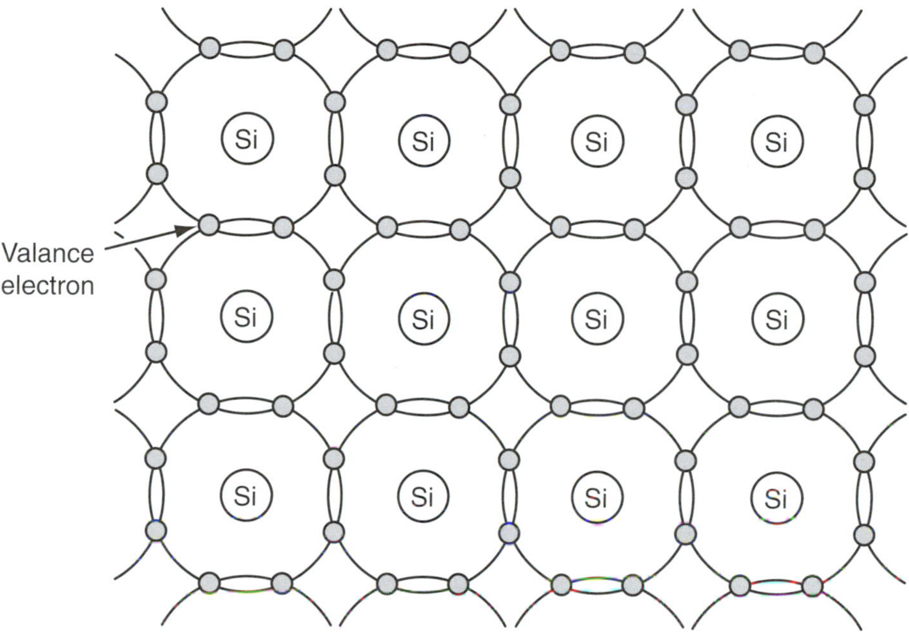

Each silicon atom shares an electron with each neighboring atom.

Figure I–37 Covalent bonds.

positions itself so that it can share the valence electrons of neighboring atoms giving each atom, in effect, eight valence electrons. This lattice structure is characteristic of a crystal solid and provides two useful characteristics:

1. Impurities can be added to the semiconductor material to increase its conductivity; this is called *doping*.
2. It becomes *negative temperature coefficient*, meaning that its resistance goes down as its temperature goes up. (This principle is put to use in temperature sensors, as we will see in Chapter 2.)

Doping

All of the valence electrons in a pure semiconductor material are in valence rings containing eight electrons as shown in Figure I–37. With this atomic structure there are no electrons that can be easily freed, and there are no holes to attract an electron even if some were available. The result is this material has a high resistance to current flow. Adding very small amounts of certain other elements can greatly reduce the semiconductor's resistance. Adding trace amounts (about 1 atom of the doping element for every 100 million semiconductor atoms) of an element with either five or three valence electrons can create a flaw in some of the covalent bonds.

Adding atoms with five valence electrons (referred to as *pentavalent* atoms) such as arsenic, antimony or phosphorus achieves a crystal structure as shown in Figure I–38. In Figure I–38 phosphorus is the doping element and silicon is the base semiconductor element. Four of the phosphorus atom's five valence electrons are shared in the valence rings of neighboring silicon atoms, but the fifth is not included in the covalent bonding with neighboring atoms, and it is held in place only by its attraction to its parent phosphorus

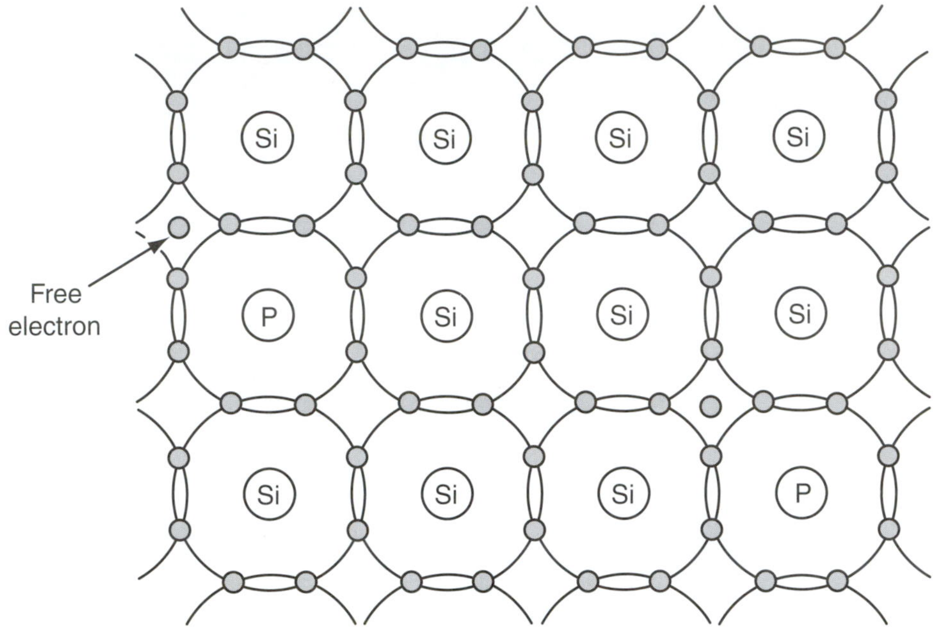

Figure I–38 Silicon doped with phosphorous.

atom—a bond that can be easily broken. This doped semiconductor material is classified as an N-type material. Note that it does not have a negative charge because the material contains the same number of protons as electrons. It does have electrons that can be easily attracted to some other positive potential.

Adding atoms with three valence electrons (referred to as *trivalent* atoms) such as aluminum, gallium, indium or boron achieves an atomic structure as seen in Figure I–39. In Figure I–39, boron is used as the doping material in a silicon crystal. The boron atom's three valence electrons are shared in the valence ring of three of the neighboring silicon atoms, but the valence ring of the fourth neighboring silicon atom is left with a hole instead of a shared electron. Remember that a valence ring of seven electrons aggressively seeks an eighth electron. In fact, the attraction to any nearby free electron is stronger than the free electron's attraction to its companion proton in the nucleus of its parent atom. Thus this material is classified as a P-type.

Doping Semiconductor Crystals

In actuality, a PN junction is not produced by placing a P- and an N-type semiconductor back to back. Rather, a single semiconductor crystal is doped on one side with the pentavalent atom's opposite sides and on the other with trivalent atoms. The center of the crystal then becomes the junction. The doping is done by first bringing the semiconductor crystal to a molten temperature. In a liquid state, the covalent bonds are broken. The desired amount of doping material is then added. As the semiconductor crystal cools, the covalent bonds redevelop with the doping atoms included.

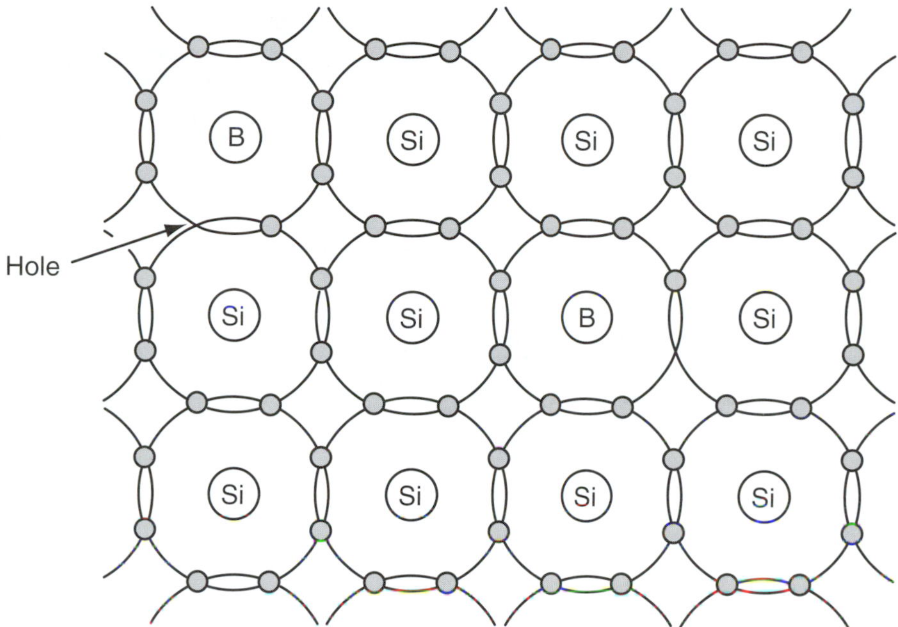

Figure I–39 Silicon doped with boron.

PN Junction

Figure I–40 shows a P-type crystal and an N-type crystal separated from each other. Only the free electrons in the N-type and the holes in the P-type that result from how the doping atoms bond with the base semiconductor material is shown. If the P-type and the N-type are put in physical contact with each other, the free electrons near the *junction* in the N-type cross the junction and fill the first holes they come to in the seven-electron valence rings near the junction of the P-type, Figure I–41. The junction is the area that joins the P-type and N-type. This action quickly creates a zone around the junction in which:

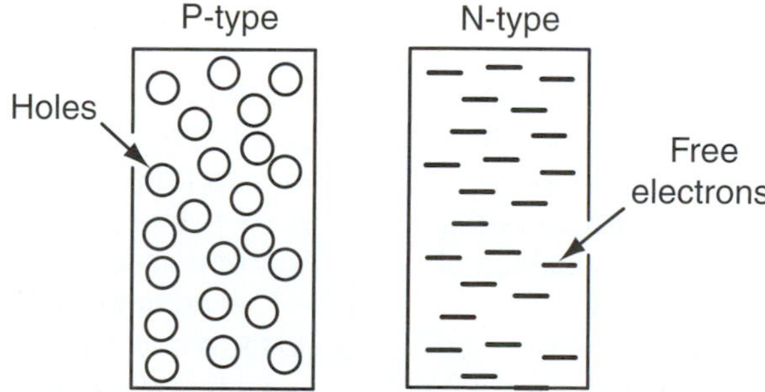

Note: For simplicity, only the holes and free electrons that result from the pentavalent and trivalent atoms are shown; the stable atoms are left out.

Figure I–40 P-type and N-type crystals.

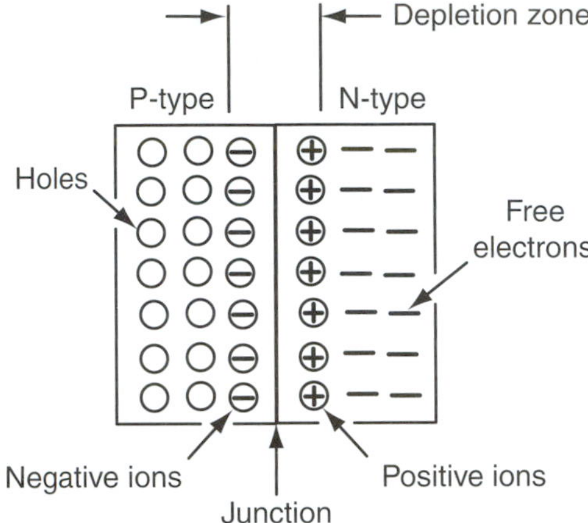

Note: For simplicity, only the holes, free electrons, and ions are shown; the stable atoms are left out.

Figure I–41 PN junction.

- there are no more free electrons in that portion of the N-type.
- there are no more free holes in that portion of the P-type.
- the valence rings near the junction in both the N- and P-type have eight shared electrons so they are reluctant to gain or lose any more.

Another way to state the information in the preceding list is to say that there are no longer any current carriers in the zone around the junction. This zone is often referred to as the *depletion zone*. In addition:

- The phosphorus atoms near the junction in the N-type have each lost an electron, which makes them positive ions (the free electron that crossed the junction to fill a hole in the P-type abandoned a proton in the nucleus of the phosphorus atom).
- The boron atoms near the junction in the P-type have each gained an electron, which makes them negative ions (the electron that dropped into the valence ring around the boron atom is not matched by a proton in the nucleus of the boron atom).

The positive ions near the junction in the N-type are attracted to the free electrons that are farther from the junction, but they are more strongly repulsed by the negative ions just across the junction. Likewise, the negative ions in the P-type are attracted to the holes that are farther from the junction, but those holes are more strongly repulsed by the positive ions on the N-side of the junction. The extra electrons in the negative ion atoms are somewhat attracted to the nearby holes, but they are more strongly bound by the covalent bonding into which they have just dropped.

Keep in mind that a layer of negative ions along the junction in the P-type and a layer of positive ions along the junction of the N-type have been created. The opposing ionic charges on the two sides of the depletion zone create an electrical potential of about 0.6 volt (0.3 volt for germanium). This potential, often referred to as the *barrier potential*, cannot be measured directly, but its polarity prevents current from flowing across the junction unless it is overcome by a greater potential.

Conventional Current Flow versus Electron Current Flow

Before electrons were known, Benjamin Franklin surmised that current was a flow of positive charges moving from positive to negative in a circuit. Franklin's belief became so accepted that even after electrons were discovered and scientists learned that current flow consists of electrons moving from negative to positive, the old idea was hard to give up. As a result, the "idea" of positive charges moving from positive to negative is still often used. It is referred to as *conventional current flow* or less formally as *hole flow*.

The conventional current flow theory has gotten a boost in recent years because it helps explain how semiconductors work: positively charged holes move from positive charges to negative charges as electrons move from negative charges to positive charges.

Diodes

What was just described in the preceding section is in fact the major part of a diode. A *diode* is basically a one-way electrical check valve; it will allow current to pass in only one direction. The best-known automotive application is probably the diodes used in the alternator to rectify the AC current produced to DC current. There are many other applications of diodes in today's vehicles such as in solid-state voltage regulators, electronic modules and circuits where one-way control might be needed. Figure I–42 shows a rectifying diode in a circuit. If switch A is on and C is off, light B will get current, but light D will not because the diode will block it. If switch C is on and A is off, both lights will get current.

Here is how the diode works. By applying voltage with correct polarity from an outside source such as a battery, current will flow across the PN junction, Figure I–43. Applying an external voltage of 0.6 volt or more with the positive to the P-side and the negative to the N-side is called *forward bias voltage*. The higher negative potential introduced by the forward bias voltage at the right side of the crystal repels the free electrons in the N-type. They move toward the junction, canceling the charge of the positive ions. At the same time, the higher positive potential introduced by the forward bias voltage on the left side of the crystal repels the free holes in the P-type. They move toward the junction, canceling the charge of the negative ions. With the barrier potential overcome, current easily flows across the junction, with electrons moving toward the external positive potential, and holes moving toward the external negative potential. When the forward bias voltage is removed, barrier potential redevelops and the diode again presents high resistance to current flow.

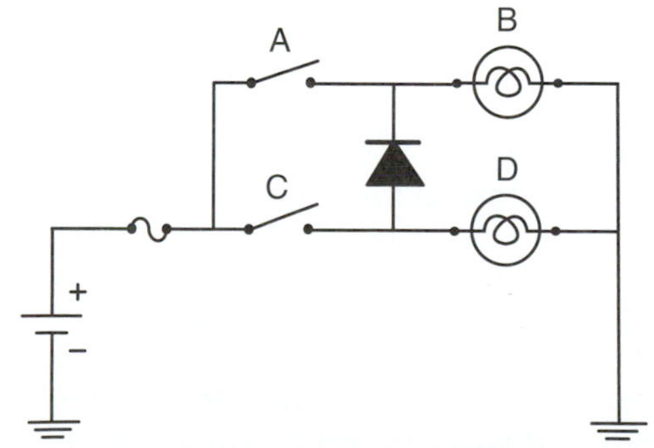

Figure I–42 A rectifying diode application.

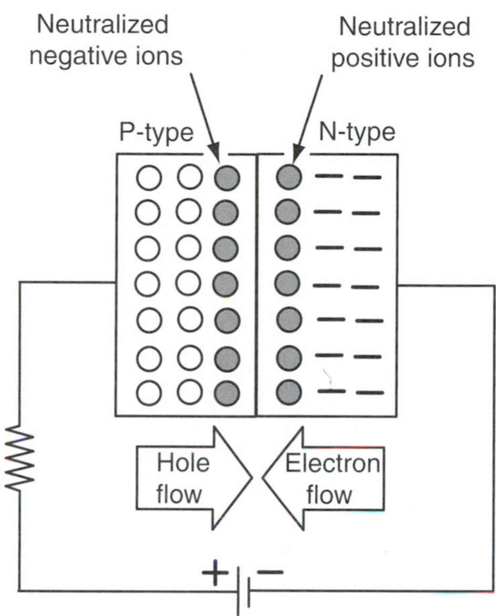

Figure I–43 Forward bias voltage applied to a diode.

The amount of current that a diode or any other type of PN junction semiconductor can safely handle is determined by such things as its size, type of semiconductor and doping material used, heat dissipating ability and surrounding temperature. If the circuit through which forward bias voltage is applied does not have enough resistance to limit current flow to what the semiconductor can tolerate, it will overheat, and the junction will be permanently damaged (open or shorted).

If a reverse bias voltage is applied, with negative external potential to the P-side and positive external potential to the N-side, the positive potential attracts free electrons away from the junction, and the negative potential attracts holes away from the junction, Figure I–44. This causes the depletion zone to be even wider and the resistance across the junction to be even higher.

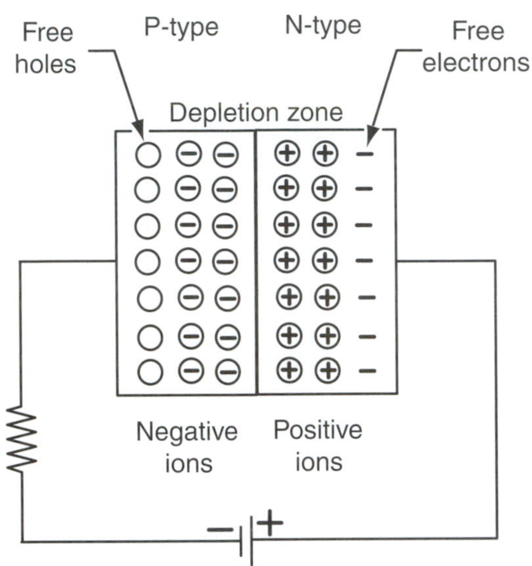

Figure I–44 Reversed bias voltage applied to a diode.

If the reverse bias voltage goes high enough, that is, above 50 volts for most rectifier diodes (those designed to conduct enough current to do work, and the most common type), current will flow. It will rise quickly, and in most cases the diode will be damaged. This is called the *breakdown voltage.*

Zener Diode

A zener diode is one in which the crystals are more heavily doped. Because of this the depletion zone is much narrower, and its barrier voltage becomes very intense when a reverse bias voltage is applied. At a given level of intensity the barrier voltage pulls electrons out of normally stable valence rings, creating free electrons. When this occurs, current flows across the diode in a reverse direction without damaging the diode. The breakdown voltage at which this occurs can be controlled by the amount of doping material added in the manufacturing process. Zener diodes are often used in voltage-regulating circuits.

Diode Symbols

The most commonly used symbol to represent a diode is an arrow with a bar at the point, Figure I–45A. The point always indicates the direction of current flow using conventional current. Electron flow would be opposite to the way the arrow is pointing. The arrow side of the symbol also indicates the P-side of the diode, often referred to as the *anode.* The bar at the end of the arrow point represents the N-side and is often called the *cathode.* Figure I–45B shows a modified diode symbol that represents a zener diode.

On actual diodes, the diode symbol can be printed on the side to indicate the anode and cathode ends, or a colored band can be used instead of the symbol. In this case the colored band will be nearer to the cathode end, Figure I–45C and D. Figure I–45E shows a power diode. A power diode is one large enough to conduct larger amounts of current to power a working device. It will be housed in a metal case, which can serve as either the anode or the cathode connector and will also dissipate heat away from the semiconductor crystals inside. The polarity of a power diode can be indicated by markings or, in the case of specific part number applications such as in an alternator, may be sized or shaped so that it can be installed only one way.

Transistors

The transistor, probably more than any other single component, has made possible the world of modern electronics. Transistors most commonly used in automotive applications are called *bipolar transistors* because they use two polarities—electrons and holes. Bipolar transistors contain three doped semiconductor crystal layers called the *emitter*, the *base* and the *collector*, Figure I–46. The base is always sandwiched between the emitter and the collector. The major difference between a diode and transistor is that a transistor has two PN junctions instead of one. They can be arranged

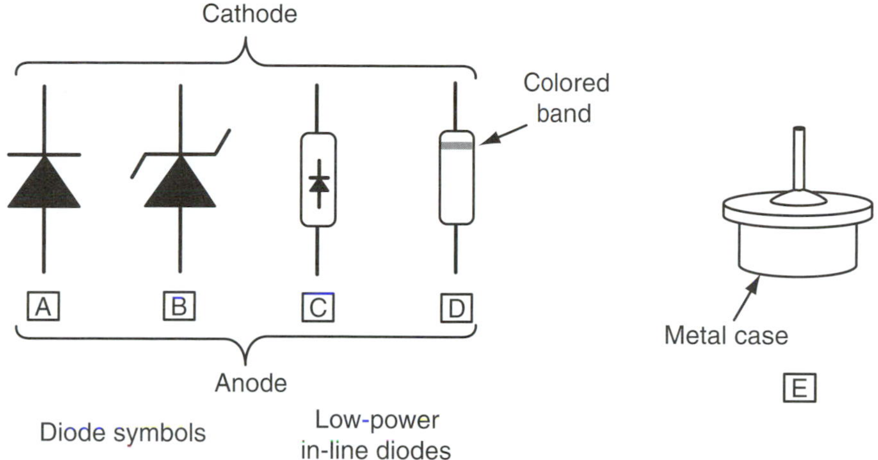

Figure I–45 Diodes.

to have a P-type emitter and collector with an N-type base (a PNP transistor), or an N-type emitter and collector with a P-type base (an NPN transistor). Since most automotive applications are NPN transistors, they are discussed here.

The emitter is heavily doped with pentavalent atoms, and its function is to emit free electrons into the base. The base is lightly doped with trivalent atoms and is physically much thinner than the other sections. The collector is slightly less doped than the emitter but more than the base.

Recalling the discussion of the PN junction (review it if it is not clear), you know that a barrier potential forms at each junction, Figure I–47. If you forward bias the transistor by applying an external voltage of at least 0.6 volt from the emitter to the base with negative potential to the emitter and positive potential to the base, then apply an external voltage of at least 0.6 volt from the collector to the emitter with positive potential to the collector and negative potential to the emitter, you can get some unique results, Figure I–48. A slightly higher voltage is usually used between the collector and emitter, thus two separate voltage sources are shown.

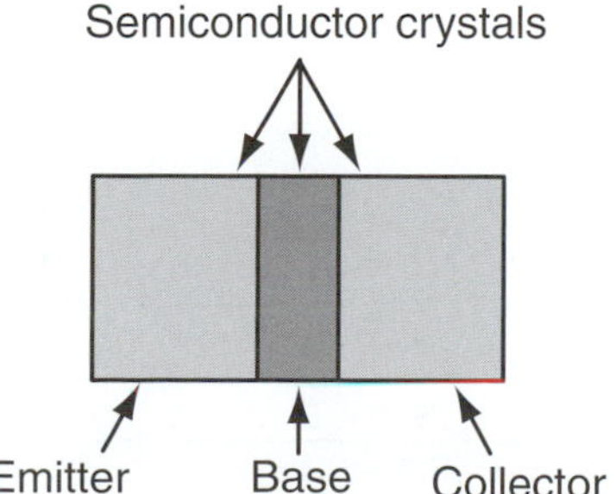

Figure I–46 Components of a bipolar transistor.

Because it is lightly doped, the base does not have many holes for the free electrons to drop into. The holes that do exist are quickly filled. Because the base crystal is thin and has so few holes for free electrons to drop into, the majority of the free electrons coming from the emitter cross the base into the collector. The collector readily accepts them because the free electrons in the collector are attracted to the positive potential at the collector electrode and leaves behind a lot of positive ions.

The positive potential applied to the base electrode is strong enough to attract some of the valence electrons there. As valence electrons near the base electrode flow from the base toward the emitter-base voltage source, nearby valence electrons move over to take their place.

Because primarily only valence electrons move from the base into the base circuit, and because

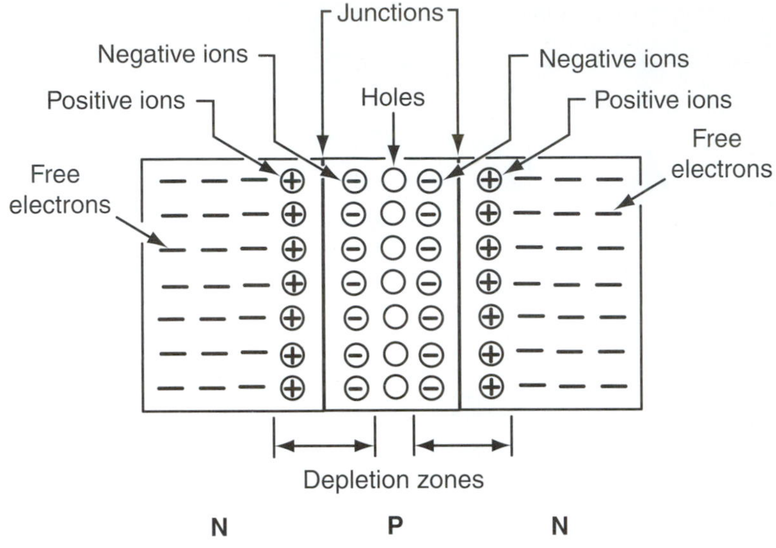

Figure I–47 Unbiased transistor.

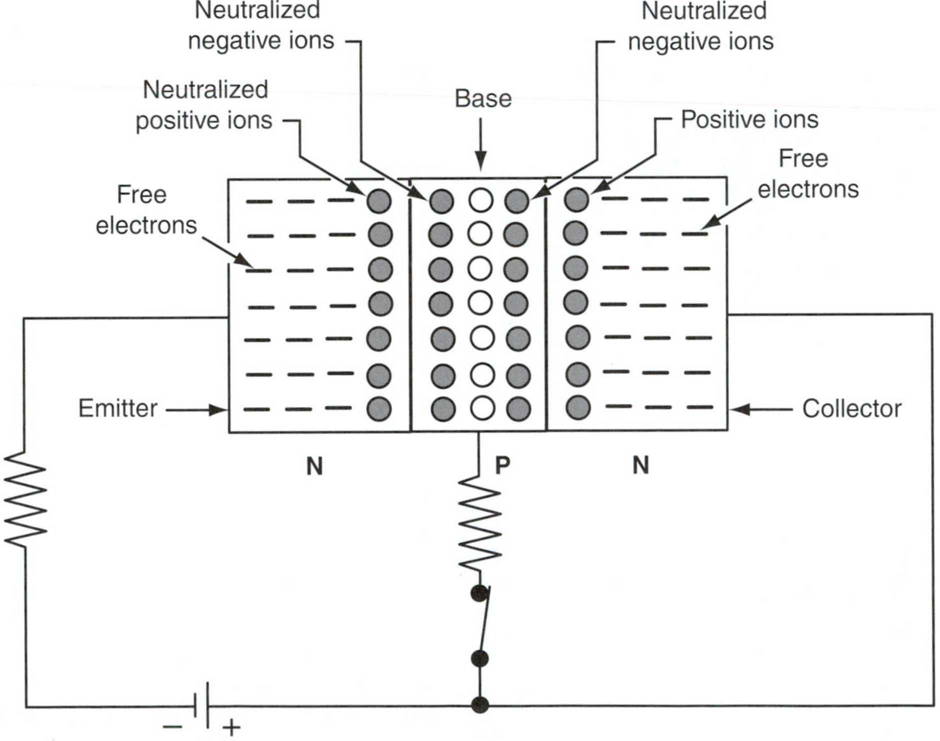

Figure I–48 Forward biased NPN transistor.

there are so few holes that allow free electrons to drop into valence rings in the base (the base is very lightly doped), the current in the base-emitter circuit is quite small compared to the current in the collector-emitter circuit.

If the biasing voltage is removed from the base, the barrier potential is restored at the junctions and both base and collector currents stop flowing. The base current flow caused by the forwarding biasing voltage between the base and the emitter allows current to flow from the emitter to the collector.

Transistors are like switches that can be turned on by applying power to the base circuit. There are many types of transistors, but the most common fall into two categories: power and switching.

Power transistors are larger because the junction areas must be larger to pass more current across them. Passing current across the junctions produces heat; therefore, a power transistor must be mounted on something, ordinarily called a *heat sink*, that can draw heat from the transistor and dissipate it into the air. If not it will likely overheat and fail.

A transistor is also much like a relay in that a relatively small current through the base circuit controls a larger current through the collector. The amount of current through the base is determined by the amount of voltage and resistance in the base circuit. In fact, carefully controlling the voltage applied to the base can control how much current flows through the collector. If enough voltage is applied to the base to just start reducing barrier potential, a relatively small amount of current will begin to flow from the emitter to the collector. This is called *partial saturation*. As base voltage is increased, collector current increases. When enough voltage is applied to the base, full saturation is achieved. There is usually a small voltage spread between minimum and full saturation, and once full saturation is achieved, increasing voltage at the base will not increase collector current. If too much voltage is applied, the transistor will break down. This principle is often used in power transistors. Most power transistors are NPN.

Effect of Temperature on a PN Junction

As the temperature of a semiconductor device goes up, the electrons in the valence ring move at a greater speed. This causes some to break out of the valence ring, creating more free electrons and holes. The increase in the number of free electrons and holes causes the depletion layer to become thinner, reducing the barrier potential. Barrier potential voltages of 0.7 for silicon and 0.3 for germanium semiconductors are true at room temperature only. At elevated temperatures, the lower barrier potential lowers the external bias voltage required to cause current to flow across the junction and raises the current flow.

This means that the expected operating temperature range of a semiconductor device must be considered when the circuit is being designed, and that when in operation, the operating temperature must be kept within that temperature limit. If the operating temperature goes higher, unless some kind of compensating resistance is used, current values will go up, and the semiconductor might be damaged.

Switching transistors are much smaller. They are often used in information processing or control circuits, and conduct currents ranging from a few milliamps (thousandths of an ampere) down to a few microamps (millionths of an ampere). They are more often designed for extremely fast on and off cycling rates, and the base circuits that control them are designed to turn them on at full saturation, or to turn them off.

Transistor Symbols

In electrical schematic illustrations, transistors are usually represented by a symbol. Figure I–49 shows the symbols used most often. Of the two types of symbols, the circle containing the lines for emitter, base and collector are most often used. The arrow is always used to indicate:

- the emitter side of the transistor.
- the direction of conventional current flow.
- whether the transistor is an NPN or PNP. If the arrow points outward, toward the circumference of the circle, it indicates an NPN transistor. If it points inward toward the base, it indicates a PNP transistor.

INTEGRATED CIRCUITS (ICs)

Thanks to scientific research, manufacturers have the capability to produce microscopic transistors, diodes and resistors. As a result, complete circuits are produced containing thousands of semiconductor devices and connecting conductor paths on a chip as small as two or three millimeters across. These integrated circuits operating with current values as low as a few milliamps or less can process information, make logic decisions and issue commands to larger transistors. The larger transistors control circuits that operate on larger current values. Personal computers and the computers in today's vehicles became possible because of the development of ICs. Because the components in ICs are so small, they cannot tolerate high voltages. Care must be taken to avoid creating high-voltage spikes such as those produced by disconnecting the battery while the ignition is on. Care must also be taken when handling a component with ICs in it to avoid exposing it to static discharges such as those you sometimes experience when touching something with ground potential after walking across a carpet.

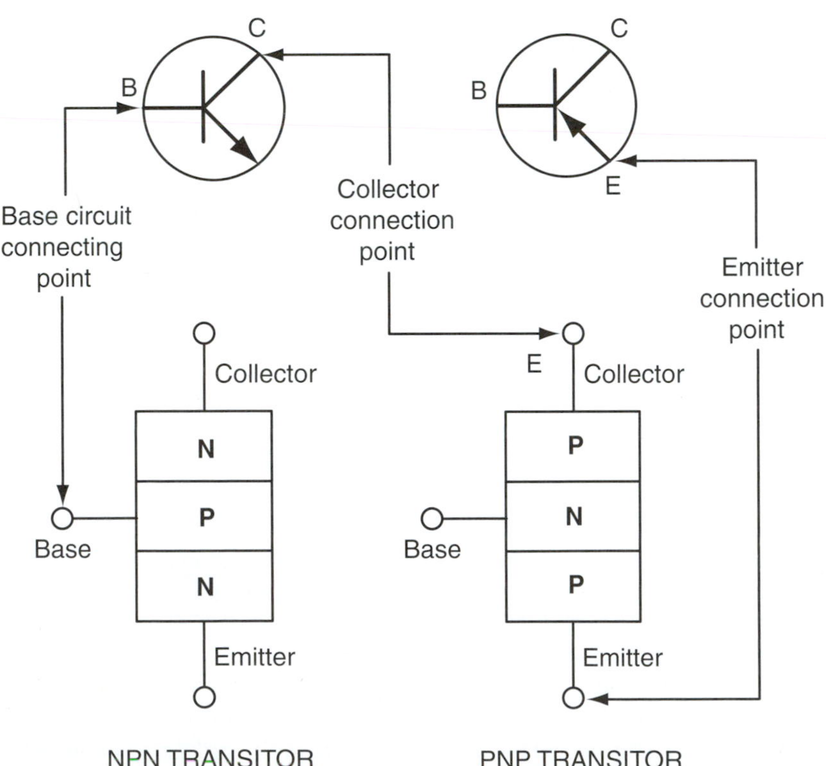

Figure I–49 Transistor symbols.

Chapter 1

Computers in Cars

OBJECTIVES

Upon completion and review of this chapter, you should be able to:

- ❑ Recognize why almost all cars currently being built for use in this country are equipped with a computerized engine control system.
- ❑ Describe the major components of a microcomputer.
- ❑ Understand the factors that influence the production of exhaust gasses.
- ❑ Describe what is meant by a stoichiometric air/fuel ratio.
- ❑ Understand why a stoichiometric air/fuel mixture is needed for best three-way catalytic converter efficiency.
- ❑ Describe the difference between open loop and closed loop operation.
- ❑ Understand the importance of performing a thorough prediagnostic inspection.

KEY TERMS

Analog
Baud Rate
Bit
Byte
Catalyst
Closed Loop
Digital
Driveability
Engine Calibration
Feedback
Interface
KAM
Memory
Microprocessor
Modulate
Open Loop
Palladium
Platinum
PROM
RAM
Rhodium
ROM
Signal
Stoichiometric

WHY COMPUTERS?

In 1963, positive crankcase ventilation systems were universally installed on domestic cars as original equipment. People in the service industry felt that dumping all of those crankcase gasses in the induction system would plug up the carburetor and be harmful to the engine. Instead, engine life doubled.

In 1968 exhaust emission devices were universally applied to domestic cars. Compression ratios began to go down; spark control devices denied vacuum advance during certain driving conditions; thermostat temperatures went up; air pumps were installed; heated air intake systems were used during warmup; and air/fuel ratios began to become leaner. Through the 1970s, evaporation control systems, exhaust gas recirculation systems and catalytic converters were added. Although some of the emissions systems actually tended to improve driveability and even fuel mileage, for the most part, in the early days of

EMISSION REQUIREMENTS FOR PASSENGER CARS (measured in grams per mile)						
	Hydrocarbon (HC)		Carbon monoxide (CO)		Oxides of nitrogen (NOx)	
Model year	**Federal**	**California**	**Federal**	**California**	**Federal**	**California**
1978	1.5	0.41	15.0	9.0	2.0	1.5
1979	0.41	0.41	15.0	9.0	2.0	1.5
1980	0.41	0.39	7.0	7.0	2.0	1.0
1981	0.41	0.39	3.4	7.0	1.0	0.7
1982	0.41	0.39	3.4	7.0	1.0	0.4
1983	0.41	0.39	3.4	7.0	1.0	0.4
1984	0.41	0.39	3.4	7.0	1.0	0.4
1985	0.41	0.39	3.4	7.0	1.0	0.7
1986	0.41	0.39	3.4	7.0	1.0	0.7
1987	0.41	0.39	3.4	7.0	1.0	0.7
1988	0.41	0.39	3.4	7.0	1.0	0.7
1989	0.41	0.39	3.4	7.0	1.0	0.4
1990–2000	0.41	0.39	3.4	7.0	1.0	0.4
1960	10.6*		8.4*		4.1*	

*Typical values before emission controls

Figure 1–1 Federal and California emissions standards compared to the uncontrolled vehicles of 1960.

emission controls, driveability and fuel mileage suffered in order to achieve a dramatic reduction in emissions.

In 1973 and 1974, when domestic car fuel economy was at its worst, we also experienced an oil embargo and an energy shortage. The federal government responded by establishing fuel mileage standards in addition to the already established emissions standards. By the late 1970s, the car manufacturers were hard-pressed to meet the ever more stringent emissions and mileage standards; and the standards set for the 1980s looked impossible, Figures 1–1 and 1–2. To make matters worse, the consumer was getting into the picture, too. Not only did the consumer's car get poor fuel mileage, it had poor **driveability** (idled rough, often hesitated or stumbled during acceleration if the engine was not fully warmed up and had little power). What made the situation so difficult was that the three demands—lower emissions, better mileage and better driveability—were largely in opposition to each other using the technology available at that time, Figure 1–3.

What was needed was a much more precise way to control engine functions, or **engine calibration**. The automotive industry had already looked to **microprocessors**—processors contained on integrated circuits. In 1968, Volkswagen introduced the first large-scale production, computer-controlled electronic fuel injection system, an early version of the Bosch D-Jetronic system. In 1975, Cadillac had introduced a computer-controlled electronic fuel injection (EFI) system. In 1976, Chrysler introduced a computer-controlled electronic spark control system, the Lean-Burn system. And in 1977, Oldsmobile in-

FUEL ECONOMY STANDARDS		
Model year	**MPG**	**Total improvement over 1974 model**
1978	18.0	50%
1979	19.0	58%
1980	20.0	67%
1981	22.0	83%
1982	24.0	100%
1983	26.0	116%
1984	27.0	125%
1985	27.5	129%
1986	26.0*	116%
1987	26.0	116%
1988	26.0	116%
1989	26.5	120%
1990–1999	27.5	129%

*Reduced from 27.5 by the federal government in 1986

Figure 1–2 Federally imposed fuel economy standards.

troduced a computer-controlled electronic spark control system that they called MISAR (microprocessed-sensing automatic regulation). All of these systems had three things in common: they all controlled only one engine function, they all used an **analog** computer (a mechanism that continuously varies within a given range with time used for the change to occur) and none of them started any landslide movement toward computer controls.

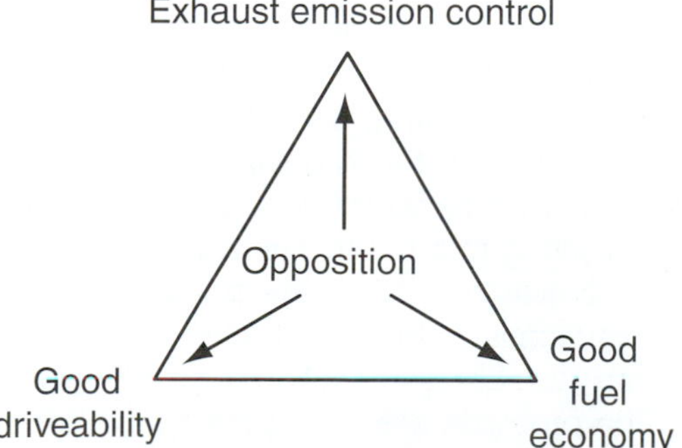

Figure 1–3 Opposition of exhaust emissions, driveability and fuel economy to each other.

By the late seventies, the electronics industry had made great strides with **digital** microprocessors. These small computers with their comparatively low cost, compact size and weight, great speed and application flexibility proved to be ideal for the industry. Probably the most amazing feature of the digital computer is its speed. To put it into perspective, one of these computers controlling functions on an eight-cylinder engine running at 3,000 rpm can send the spark timing command to fire a cylinder; reevaluate input information about engine speed, coolant temperature, engine load, barometric pressure, throttle position, air/fuel mixture, spark knock and vehicle speed; recalculate air/fuel mixture, spark timing, whether to turn on the exhaust gas recirculation (EGR) valve, canister purge valve and the torque converter clutch; and then take a short nap before sending the commands, all before the next cylinder fires. The computer is so much faster than even the fastest engine, it spends most of its time doing nothing but counting time on its internal clock.

At last the automotive manufacturers had the technology to precisely monitor and control, to instantly and automatically adjust, while driving, enough of the engine's calibrations to make the vehicle comply with the government's demands for emissions and fuel economy, and still satisfy the consumer's demands for better driveability. This brought about what is probably the first real revolution in the automotive industry's recent history; other changes have been evolutionary by comparison.

After some experimental applications in 1978 and 1979 and some limited production in 1980, the whole domestic industry was using comprehensive computerized engine controls in 1981. At the same time, General Motors alone produced more computers than anyone else in the world and used half of the world's supply of computer parts.

How Computers Work

Contrary to some "informed" opinions and numerous suggestions from popular movies, computers cannot think for themselves. When properly programmed, however, they can carry out explicit instructions with blinding speed and almost flawless consistency.

Communication Signals. A computer uses voltage values as communication signals, thus voltage is often referred to as a **signal** or a voltage signal. There are two types of voltage signals: analog and digital, Figure 1–4. An analog signal's voltage is continuously variable within a given range and time is used for the voltage to change. An analog signal is generally used to convey information about a condition that changes gradually and continuously within an established range.

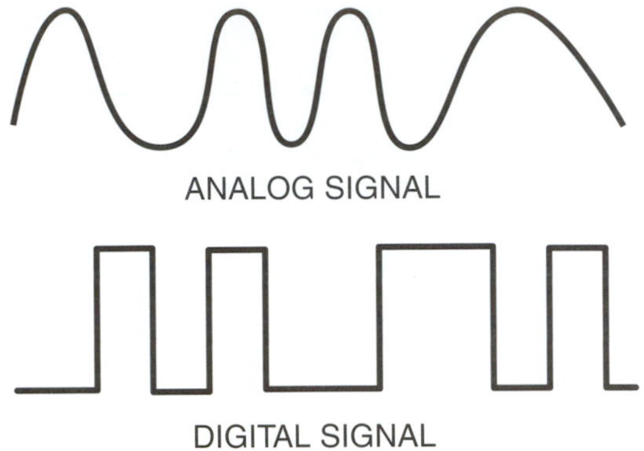

Figure 1–4 Analog signals can be constantly variable. Digital signals are either on/off or low/high.

Temperature-sensing devices usually give off an analog signal. Digital signals also vary but not continuously, and time is not needed for the change to occur. Turning a switch on and off creates a digital signal; voltage is either there or it is not. Digital signals are often referred to as square wave signals.

Binary Code. A computer converts a series of digital signals to a binary number made up of ones and zeros; voltage above a given threshold value converts to 1, and voltage below that converts to 0. Each 1 or 0 represents a **bit** of information. Eight bits equal a **byte** (sometimes referred to as a word). All communication between the microprocessor, the memories and the interfaces is in binary code, with each information exchange being in the form of a byte. If a series of digital signals converts to a binary number of 01111010, the numerical value (in base, 10, which we are most familiar with) can easily be derived, as shown in Figure 1–5A. A power is assigned to each place in the binary number, starting from the right and working to the left. The rightmost place is given a power of 1, with the power doubling with each successive place to the left. Each digit in the binary number is multiplied by its respective power. The products are then added together to yield the base-10 numerical value. As you will see in Figure 1–5B, the largest value that an 8-bit computer can communicate is 255. Automotive

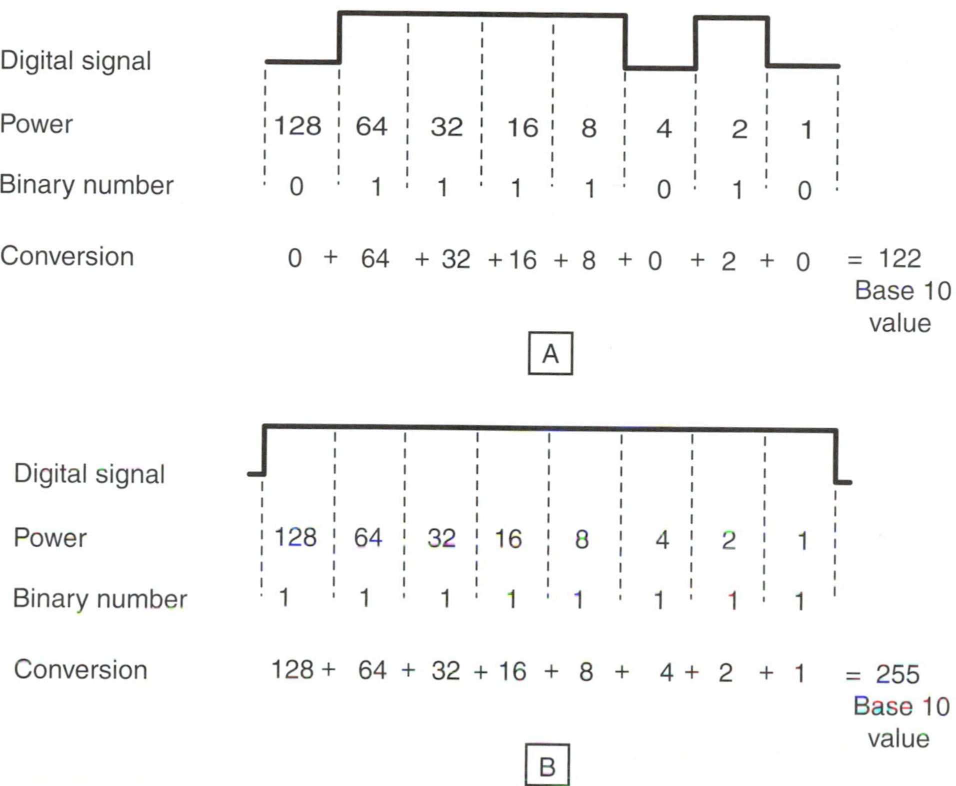

Figure 1–5 Binary numbers.

computers used 8-bit microprocessors for many years, but as these computers have taken on more responsibility and have had to communicate more rapidly, 16-bit and 32-bit computers have become commonplace.

Interface. The microprocessor is the heart of a computer, but it needs several support functions, one of which is the **interface**. A computer has an input and an output interface circuit, Figure 1–6. The interface has two functions: it protects the delicate electronics of the microprocessor from the higher voltages of the circuits attached to the computer, and it translates input and output signals. The input interface translates all analog input data to binary code; most sensors produce an analog signal. It is sometimes referred to as A/D, analog to digital. The output interface, D/A, translates digital signals to analog for any controlled functions that need an analog voltage.

Memories. The microprocessor of a small computer does the calculating and makes all of the decisions, or data processing, but it cannot store information. The computer is therefore equipped with information storage capability called **memory**. The computer actually has three memories: read-only memory (**ROM**), programmable read-only memory (**PROM**) and random-access memory (**RAM**), Figure 1–7.

The ROM contains permanently stored information that instructs (programs) the microprocessor on what to do in response to input data. The microprocessor can read information from the ROM but cannot put information into it. The ROM unit is soldered into the computer and is not easily removed.

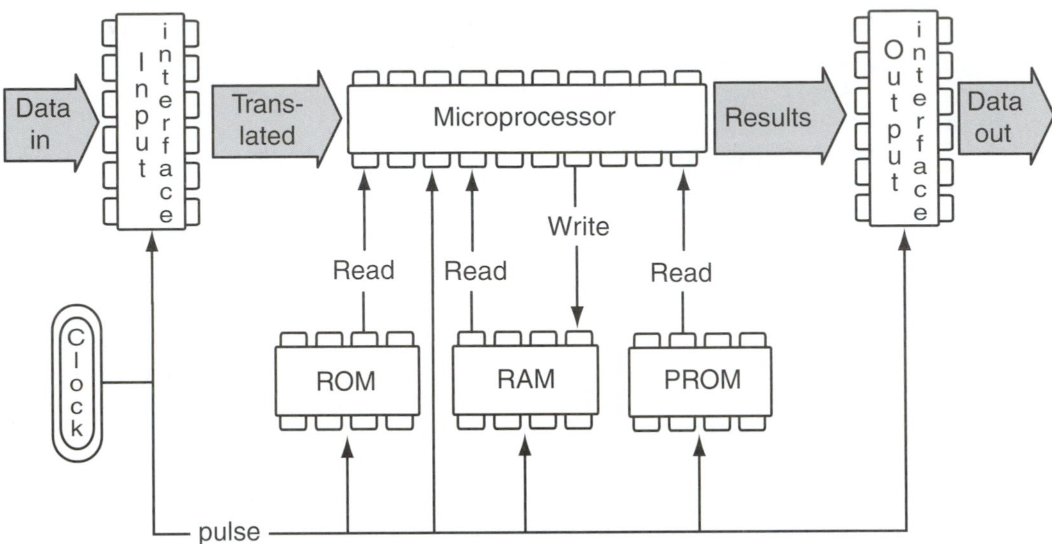

Figure 1–6 Interaction of the microprocessor and its support system.

The PROM differs from the ROM in that it plugs into the computer and is more easily reprogrammed or replaced with one containing a revised program. It contains program information specific to different vehicle model calibrations.

Some of the information stored in both the ROM and the PROM is stored in the form of look-up tables similar to the tables one might find in the back of a chemistry or math textbook. These tables enable the computer to interpret and make decisions in response to sensor data. For example, a piece of information (a voltage value of 2.4 volts, about half throttle) is received from the throttle position sensor. This information plus information from the engine speed sensor is compared to a table for spark advance. This preprogrammed table tells the computer what the spark advance should be for that throttle position and engine speed. The computer then modifies this spark advance value, by consulting other tables, with information concerning engine temperature and atmospheric pressure. These tables are sometimes referred to as maps.

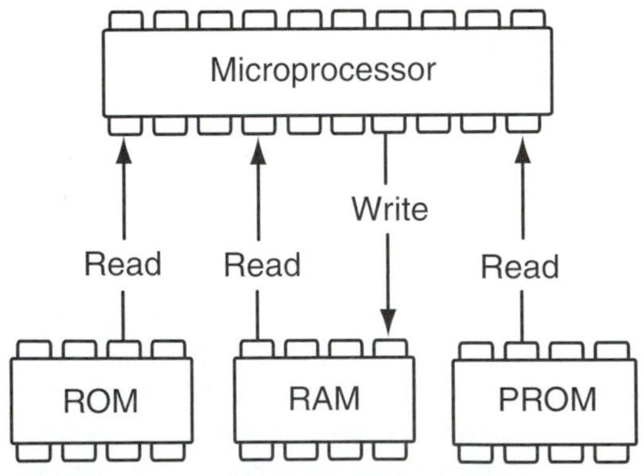

Figure 1–7 The three memories within a computer.

The RAM is where temporary information is stored. It can be both read from and written into. If, for example, data comes in from a sensor and will be used for several different decisions, such as manifold vacuum information, the microprocessor will record it in RAM until it is updated. In the example presented in the preceding paragraph, the information concerning throttle position and engine speed came from the RAM.

In most automotive computers, the computer tests at least some of the input and output signals to see whether the circuit that sent the signal or received the command is working properly. The computer knows what the voltage will be if the circuit were open and what the voltage will be if it were shorted. It watches to see that the voltage remains between those voltage values. In some circuits, the oxygen sensor circuit for instance, the computer expects to see the voltage go up and down, recrossing a certain value. If during its continuous testing it sees something wrong (a fault), it will record information in RAM concerning the fault. The RAM is also soldered in place and not easily removed.

There are two kinds of RAMS: volatile and nonvolatile. A volatile RAM (sometimes called keep-alive memory or **KAM**) must have a constant source of voltage to continue to exist and is erased when disconnected from its power source. In automotive applications, a volatile RAM is usually connected directly to the battery by way of a fuse or fusible link so that when the ignition is turned off, the RAM is still powered. A nonvolatile RAM does not lose its stored information if its power source is disconnected. Vehicles with digital display odometers usually store mileage information in a nonvolatile RAM.

The terms ROM, PROM and RAM are fairly standard throughout the computer industry; however, all car manufacturers do not use the same terms in reference to their computer's memory. For instance, instead of using the term PROM, Ford calls their equivalent unit an engine calibration assembly. Another variation of a PROM is an E-PROM. An E-PROM has a housing with a transparent top that will let light through. If ultraviolet light strikes the E-PROM, it reverts back to its unprogrammed state. To prevent deprogramming or memory loss, a cover must protect it from light. The cover is often a piece of tape. An E-PROM is sometimes used where ease of changing stored information is important. General Motors uses it as a PROM on some applications because it has more memory capacity.

Another variation of RAM is KAM (keep-alive memory). Often used as a subsection of the RAM, KAM is usually powered by a fuse straight from the battery rather than by the ignition or an ignition-controlled circuit. Therefore, when the ignition is turned off, the KAM remains powered and its stored information is retained. Other types of memory unique to specific manufacturers are introduced in appropriate chapters.

Clock Pulses. In order to maintain an orderly flow of information into, out of and within the computer, a quartz crystal is used to produce a continuous, consistent time pulse, Figure 1–6. The frequency of these clock pulses determines the computer's clock speed. Once the clock speed is established, the computer's **baud rate** can be established. The baud rate is the frequency at which bits of information are communicated, thereby establishing an orderly flow of information within, or from, the computer.

For example, a computer with a baud of 5,000 could transmit 5,000 bits of binary code information per second. Bauds in automotive computers have gone up considerably since those used in the early eighties. The computers used by General Motors in the early eighties had a baud of 160 (some of those computers were still used on some applications in the late eighties), while their P-4 electronic control module introduced in the late eighties has a baud of 8,192 with development underway to make them even faster. Ford introduced a computer in 1988 with a baud rate of 12,500. (See Chapter 7 for current baud rates.)

Data Links. When computers communicate with other electronic devices such as control panels, modules, some sensors or other computers in the form of digital signals, they communicate through circuits called data links. Some data links transmit data in only one direction, although others transmit in both directions. What makes a data link different from an ordinary circuit is the manner in which it is controlled. Some control must be used so that the devices at each end of the data link know when to transmit and when to receive. Several control methods can be used. Chapter 7, "Multiplexing Concepts," provides further discussion of this topic.

EXHAUST GASSES

The computer of a computerized engine control system has a mission: to reduce emissions, improve mileage and improve driveability. The priority varies, however, under some driving conditions. For example, during warmup, driveability has a higher priority than mileage; and at full throttle, some systems give the performance aspect of driveability a higher priority than emissions or mileage.

There are six major exhaust gases from an automobile: oxygen (O_2), hydrocarbons (HC), water vapor (H_2O), carbon dioxide (CO_2), carbon monoxide (CO) and oxides of nitrogen (NO_x).

1. *Oxygen.* Oxygen forms the basis for all combustion. It constitutes about 21 percent of our air and is the chemical that supports all flame in the atmosphere. The burning of oxygen and fuel releases the chemical energy that does all the work of the engine and the vehicle. A properly operating feedback control system results in a slightly fluctuating amount of residual oxygen in the exhaust. A very rich mixture would have so much fuel that all the oxygen would be consumed; a very lean mixture would leave too much oxygen unused in the exhaust.
2. *HC.* Gasoline is a hydrocarbon compound, as is engine oil. When hydrocarbons burn properly, the hydrogen and carbon atoms separate; each combines with oxygen to form either water (H_2O) or carbon dioxide (CO_2). If for any reason the gasoline in the cylinder fails to burn, it is pumped into the exhaust system as a raw HC molecule, a most undesirable emissions result. There can be several reasons for all or part of the HC in the cylinder not to burn:

   - Any electrical malfunction, which prevents the spark plug from igniting the fuel mixture.
   - A lack of mechanical integrity, which allows fuel to escape from the cylinder before the spark occurs or to fail to reach a sufficiently high compression pressure to ignite.
   - An overly lean mixture, which hampers flame propagation (the flame goes out too soon).
   - An exceedingly rich mixture, which uses up the oxygen before all of the HC has been consumed.
   - A low cylinder combustion chamber surface temperature, which rapidly draws heat from those HC molecules and the oxygen in contact with it, and thus prevents them from reaching or maintaining ignition temperature. A low cylinder surface temperature can be caused by a low coolant (low temperature or stuck-open thermostat) or an overadvanced ignition timing. When an engine is first started on a cold day, of course, the combustion chamber surfaces will all initially be lower than enough to sustain clean emissions performance.

   Any hydrocarbons not burned in the combustion chamber are burned in the catalytic converter.
3. *H_2O.* Water vapor forms the bulk of the exhaust product of most engines, even those running very poorly. The major source of energy in the fuel is the hydrogen, which readily combines chemically with the atmospheric oxygen drawn in with the intake air, releasing heat energy and forming water vapor. This is the most benign product of the engine's combustion, and is of concern only if the engine is run so little that condensed water is left on the interior surfaces, providing a favorable environment for rust.
4. *CO_2.* Carbon dioxide forms in the burning of the carbon portion of the fuel. Carbon dioxide is a harmless portion of the exhaust. It is the same gas used by plants in their respiration cycle, and it is the same gas used to carbonate beverages. While some carbon dioxide is harmless, environmental concerns have focused recently on the sheer amount of the gas produced by internal combustion engines. Some people have expressed the

fears that a significant increase in the level of atmospheric carbon dioxide could have a "greenhouse effect" on the earth's atmosphere, eventually changing the climate.

5. *CO.* Carbon monoxide is a very deadly poison gas, odorless and colorless, and lethal in almost undetectably small concentrations. During normal combustion, each carbon atom tries to take on two oxygen atoms (CO_2). If, however, there is a deficiency of oxygen in the combustion chamber, some carbon atoms are only able to combine with one oxygen atom and thus produce CO. The carbon monoxide actively tries to obtain another oxygen atom wherever it can find one, and this oxygen-scavenging property of CO is what makes it so lethal. If breathed into the lungs, it not only provides no oxygen, it removes what oxygen it finds there. An overly rich air/fuel ratio is the only cause of CO production.
6. *NO_x.* Air is made up of about 78 percent nitrogen and 21 percent oxygen. Under normal conditions nitrogen and oxygen do not chemically unite. When raised to a high enough temperature, however, they unite to form a nitrogen oxide compound, NO. In the presence of atmospheric oxygen, NO rapidly becomes nitrogen dioxide, NO_2. These compounds are grouped into a family of compounds referred to as oxides of nitrogen, NO_x. Oxide of nitrogen compounds start to form at about 2,500°F with production increasing as temperatures go higher.

Other Exhaust Emissions from Gasoline Engines

Aldehydes
Ammonia
Carboxylic acids
Inorganic solids
- lead (from leaded fuel)
- soot

Sulfur oxides (from fuel impurities)

The reduction beds of the catalytic converter help to reduce excessive NO_x gas into harmless constituents.

Catalytic Converter

Controlling the air/fuel ratio and the spark timing, keeping the coolant temperature near 93°C (200°F) or more and controlling intake air temperature and combustion temperature dramatically reduce exhaust emissions. Even these measures, however, do not eliminate them. The catalytic converter is the single most effective device for controlling exhaust emissions. Placed in the exhaust system, the **catalyst** agents cause low-temperature oxidation of HC and CO and reduction of NO_x, yielding H_2O, CO_2, O_2 and nitrogen. Placing the converter downstream in the exhaust system allows the exhaust gas temperature to drop significantly before the gas enters the converter, and thus prevents the production of more NO_x in the oxidation process. The catalysts that are most often used, **platinum** and **palladium** for oxidation and **rhodium** for reduction, are coated over a porous material, usually aluminum oxide in pellet form or aluminum oxide in a honeycomb structure. Aluminum oxide's porosity provides a tremendous amount of surface area on which the catalyst material can be applied to allow maximum exposure of the catalyst to the gases.

There have been three different types of catalytic converters used on automobiles: 1) a single-bed, two-way, introduced on cars in the mid-seventies; 2) a single-bed, three-way; and 3) a dual-bed, three-way. The three-way converters were introduced on cars in the late seventies.

A single-bed converter has all of the catalytic materials in one chamber, Figure 1–8. A dual-bed converter has a reducing chamber for breaking down NO_x and an oxidizing chamber for oxidizing HC and CO. The exhaust passes through the reducing chamber first, Figure 1–9. Additional oxygen is introduced into the oxidizing chamber to make it more efficient. A two-way converter oxidizes HC and CO; a three-way converter oxidizes HC and CO and reduces NO_x.

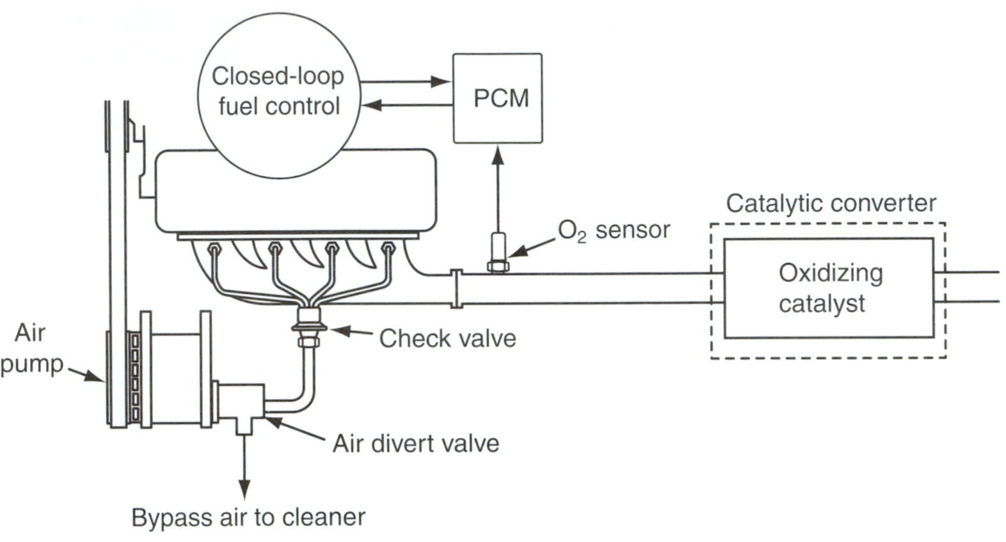

Figure 1–8 Single-bed, three-way catalytic converter.

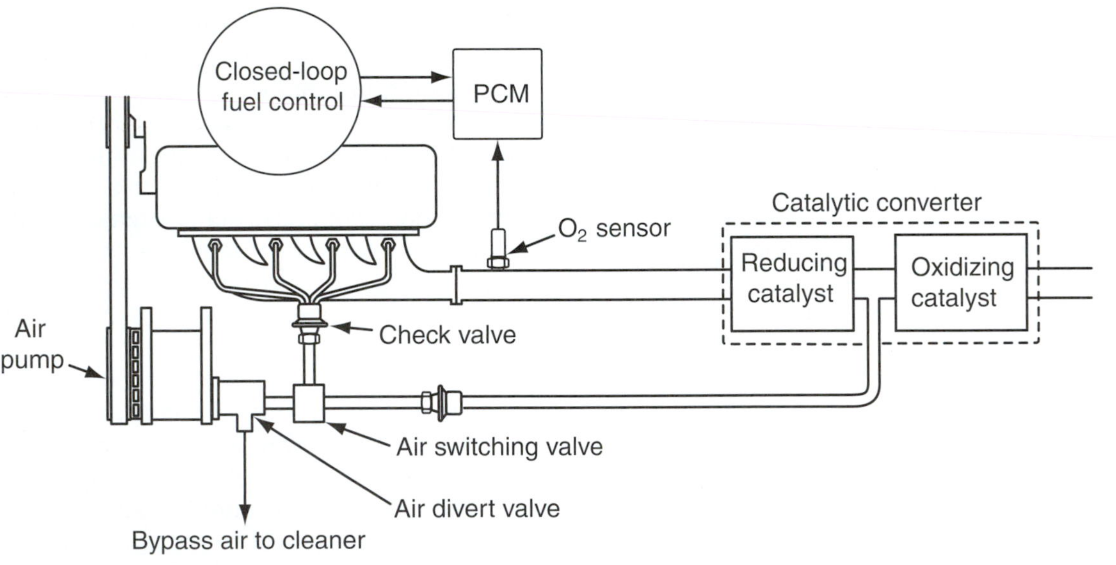

Figure 1–9 Dual-bed, three-way catalytic converter.

The rhodium-reducing catalyst of a catalytic converter causes the NO_x molecules to break down again into nitrogen and oxygen. It can only do so, however, if there is deficiency of oxygen in the exhaust stream. A rich air/fuel ratio is therefore needed, resulting in the production of a small amount of CO. The CO molecule is an oxygen-deficient molecule. When mixed with the NO_x, the

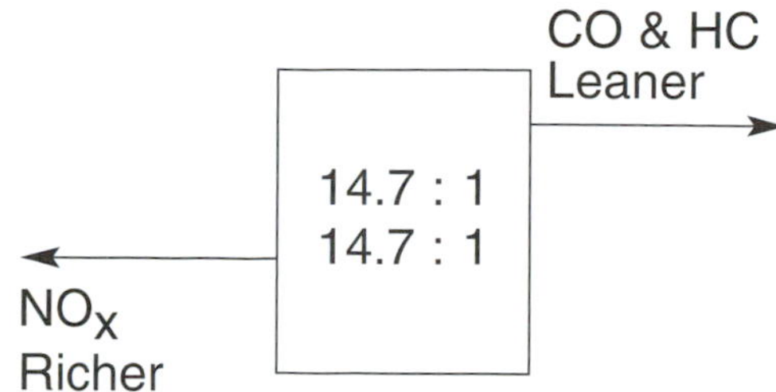

Figure 1–10 Three-way catalytic converter.

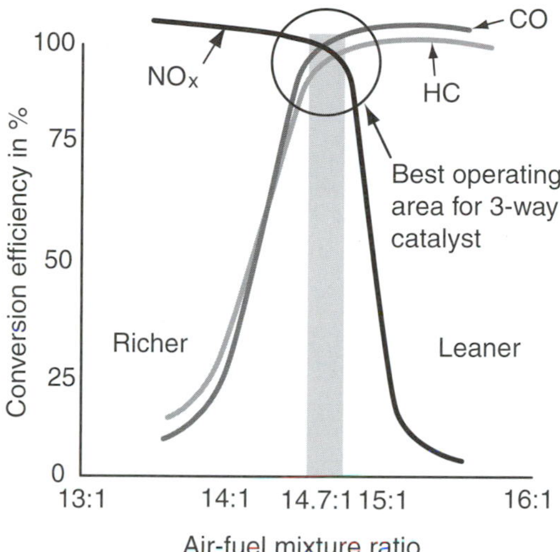

Figure 1–11 Converter efficiency.

CO molecule promotes the breakup of the NO_x molecule by attracting an oxygen atom from the NO_x molecule, resulting in the breakup of the entire molecule. Conversely, the platinum- and palladium-oxidizing catalysts need abundant available oxygen to oxidize HC and CO. A lean air/fuel mixture is needed to yield leftover oxygen, Figure 1–10. A **stoichiometric** air/fuel ratio (approximately 14.7 to 1) is essential to make the three-way converter work effectively, Figure 1–11. If the fuel mixture is leaner than 14.7 to 1, the reducing side of the converter is not effective.

Stoichiometry and 14.7 to 1

Keep in mind that a stoichiometric ratio is the one that most effectively allows all the components of the emissions control system, most notably the catalytic converter, to minimize exhaust emissions. When these systems were first invented, 14.7 to 1 was the ideal ratio for standard gasoline. Since that time there have been many changes to the fuels also, including different blends for different seasons of the year, different altitudes and even different areas depending on their compliance with federal air quality standards. But while *oxygenated* and other special fuels have changed the numbers of the ratio somewhat, the objective remains the same: to maintain the fuel/air ratio so that the resulting exhaust emissions are kept at their optimal state.

If the mixture is richer than 14.7 to 1, the additional CO and the oxygen liberated by the reducing catalyst, plus additional oxygen pumped into the oxidizing chamber in the case of a dual-bed converter, cause the converter to overheat as a result of the abundant oxidation process. One of the major reasons for creating a computerized engine control system was to have the ability to maintain a 14.7 to 1 air/fuel ratio and therefore enhance the effectiveness of the three-way catalytic converter.

Most manufacturers have experienced some problems with catalytic converters becoming restricted and have developed methods of testing for this condition. Depending on the severity of the restriction, the symptoms vary from slight loss of power after several minutes of driving to the engine not starting in the first place.

Other Emission-Control Devices

Other emission-control devices currently used on most light-duty vehicles and those that can be included as part of the computerized engine control system, or added to it, are discussed here.

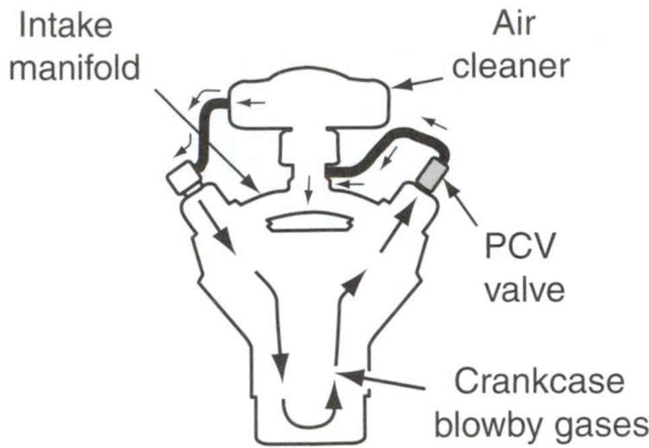

Figure 1–12 PCV system basic operation.

Positive Crankcase Ventilation. The positive crankcase ventilation (PCV) system is designed to recirculate crankcase gasses from the crankcase through the intake manifold and into the cylinders in order to burn them in the combustion chamber. This system draws filtered air in from the air cleaner, then through the crankcase in order to pick up any blow-by gasses, Figure 1–12. The combination of fresh air and blow-by gasses are then drawn through a metering device, either a PCV valve or a metering orifice, and into the intake manifold, thus allowing the crankcase hydrocarbons to be burned in the cylinders. This combination of fresh air and blow-by gasses tends to lean out the air/fuel mixture, so the PCV system is considered to be a calibrated vacuum leak. If, for any reason, the crankcase vapors exceed the PCV system's capacity, crankcase vapors are routed into the intake air stream through the intended crankcase inlet and will still be drawn through the throttle body into the intake manifold. The PCV system is designed to reduce HC emissions that were once vented to the atmosphere through a road draft tube.

Thermostatic Air Cleaner. The thermostatic air cleaner (TAC) system uses a heat stove wrapped around an exhaust manifold to heat intake air prior to entering the air cleaner during engine warmup, Figure 1–13. It is used on systems that have both air and fuel moving through the intake manifold and is used to reduce CO and HC emissions by reducing the tendency of the fuel to condense against the cold intake manifold walls and puddle on the floor of the intake manifold. It is usually not controlled by the engine computer and is not used on most port fuel injection systems.

Early Fuel Evaporation. Early Fuel Evaporation (EFE) systems are also designed to reduce the tendency of the fuel to puddle in a cold intake manifold during engine warmup in order to reduce CO and HC. The EFE system comes in two forms: a heat riser valve, Figure 1–14, is used on most V-type engines to direct exhaust gas from one exhaust manifold through an exhaust passage in the intake manifold to the exhaust mani-

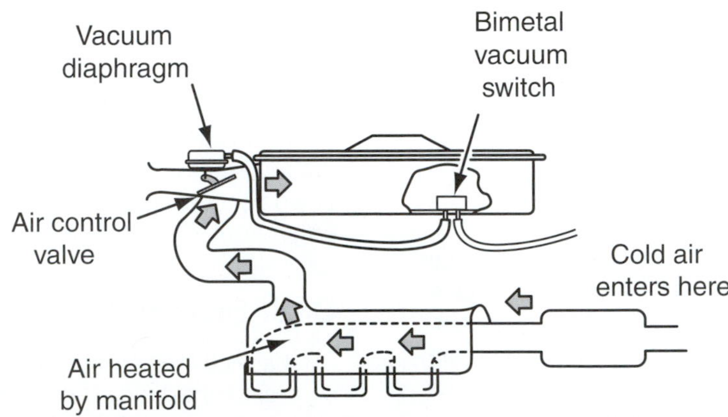

Figure 1–13 A typical thermostatic air cleaner design.

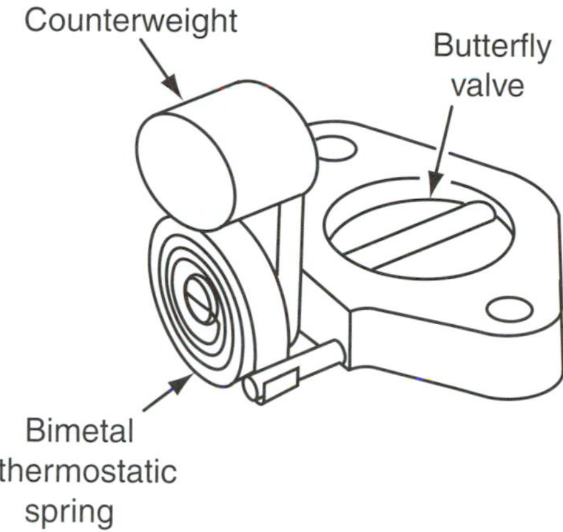

Figure 1–14 A typical mechanical heat riser valve used in an exhaust heat EFE system.

fold on the other side of the engine, and an electric heat grid, Figure 1–15, may be built into the base gasket of the carburetor or throttle body. These systems may be under the control of the engine computer and are not generally used on a port fuel injected engine.

Air Injection Reaction System. An air injection reaction (AIR) system, also commonly known as a secondary air injection system, uses an air pump to pump air into the exhaust manifold to help oxidize CO and HC coming out of the combustion chamber. (While most systems use a belt-driven air pump, some newer systems use an electrically driven pump.) A check valve is used to prevent hot exhaust gasses from reaching the rest of the AIR system while still allowing the air pump to push air into the exhaust system. Most systems also have a provision to dump the air to the atmosphere when rich air/fuel conditions are present in order to prevent backfire out of the exhaust system or overheating of the catalytic converters. Some versions use exhaust pressure pulses to pump air instead of an air pump and are known as *pulse AIR* (PAIR) systems. An exhaust system develops positive pressure pulses (pressure slightly above atmospheric pressure) as each puff of exhaust exits the cylinders, as well as negative pressure pulses (pressure slightly below atmospheric pressure) in between the positive pressure pulses. The negative pressure pulses are developed due to the fact that the exhaust is flowing (moving air tends to create a low pressure area). In a pulse air system, a check valve is used to allow the negative pressure pulses to bring air into the exhaust system while blocking the positive pressure pulses from reaching the rest of the AIR system. Typically, a pulse air system cannot flow as much volume as those systems that use an air pump and are therefore generally used on engines of smaller displacement.

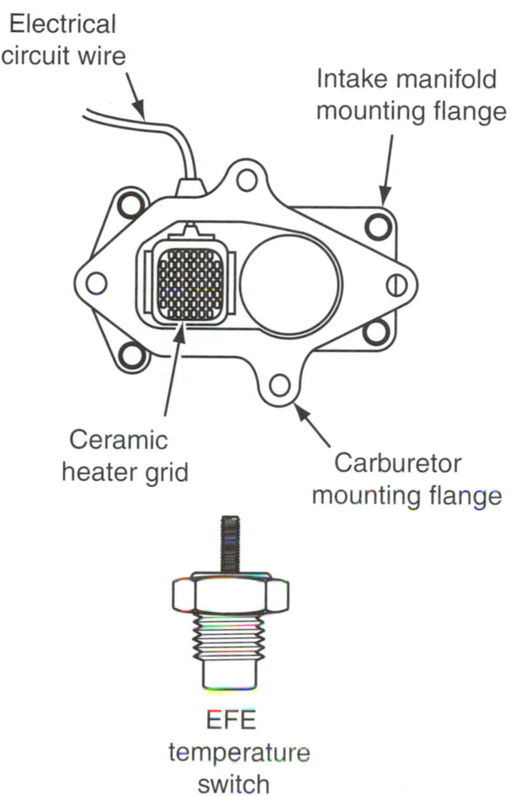

Figure 1–15 A typical EFE heat grid with its controlling temperature switch.

When used with a closed loop fuel control system, the only time that air is directed to the exhaust manifold is during engine warmup to help prewarm the oxygen sensor(s) and the catalytic converter(s). If this system is used with a dual bed catalytic converter (or multiple converters), air is then switched to a downstream exhaust port located after the NO_x reduction bed but before the CO/HC oxidation bed during closed loop operation, Figure 1–16. This is because directing the additional air ahead of either the oxygen sensor or the NO_x reduction converter would interfere with the proper function of either of these devices once the engine is warmed up and in closed loop. The air is then dumped to the atmosphere during rich engine operating conditions such as wide-open throttle (WOT), deceleration or extremely cold initial engine starts. This system is typically under the control of the engine computer.

Evaporative and Canister Purge Systems. These systems are designed to prevent the release of HC emissions to the atmosphere through evaporation. The fuel tank (and carburetor float bowl, if it applies) is vented to a charcoal canister, Figure 1–17. Then, when the engine is running, manifold vacuum purges the HC emissions from the canister to the intake manifold for combustion in the cylinders. On early systems, the manifold vacuum signal used to purge the canister was controlled by a ported vacuum signal to a purge demand valve. On newer systems, purging may be under the control of the engine computer, either by controlling a vacuum solenoid in conjunction with a purge demand valve or by controlling the purging directly with a purge solenoid. On Board Diagnostic (OBD) II applications (described in Chapter 6) use an evaporative system that has become increasingly more complex on late model vehicles due to the added ability of the engine computer to check the system for leaks to minimize the escape of HC emissions to the atmosphere through a missing gas cap or other fuel tank leak.

Exhaust Gas Recirculation. Exhaust gas recirculation (EGR) systems use an EGR valve, Figure 1–18, to allow recirculation of exhaust gasses back into the intake manifold in order to replace incoming air and oxygen with inert exhaust gas. This reduces the amount of air getting into the cylinder. The carburetor or the computerized fuel management system also reduces the amount of fuel in order to maintain the proper

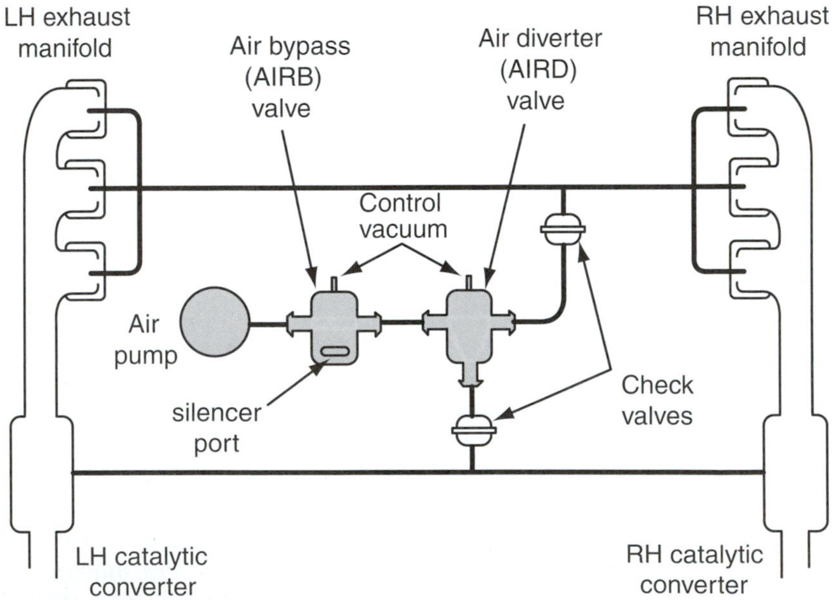

Figure 1–16 A typical secondary air injection system.

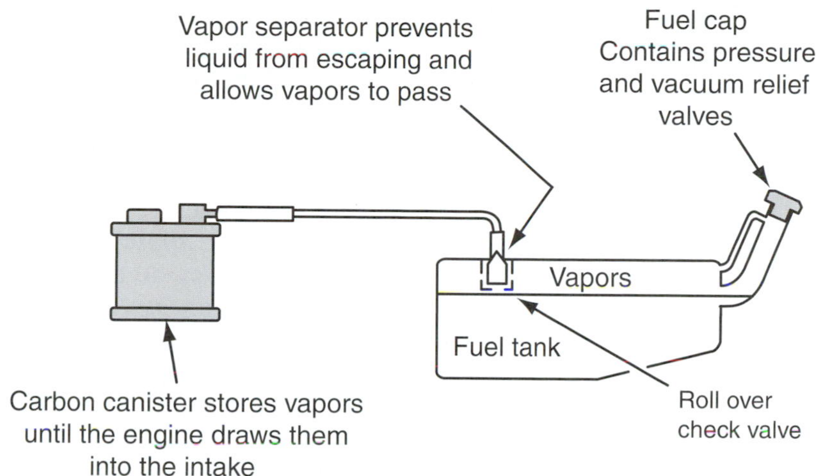

Figure 1–17 In a typical evaporation emission system, the fuel tank is vented to a charcoal canister.

air/fuel ratio. Ultimately, combustion chamber temperatures are reduced to reduce the amount of NO_x that is produced during combustion. Use of an EGR system allows the engineer to design an engine with a higher compression ratio or increased volumetric efficiency while still keeping NO_x production at a minimum. An added benefit is that a more aggressive spark advance program can be programmed into the middle rpm and load ranges, resulting in increased engine performance and fuel economy, while continuing to control spark knock. An EGR valve may be a vacuum-operated valve, which, in turn, controls a passage between the intake manifold and the exhaust manifold. It may be under the control of a ported vacuum signal or under the control of an engine computer via vacuum control and/or vacuum vent solenoids. The engine computer may **modulate** these solenoids or it may operate a solenoid on a duty cycle in order to control the percentage of time that a solenoid is applying vacuum versus the time that it is venting vacuum. Other EGR valves are electrically operated valves that control the exhaust gasses directly and without the use of vacuum. EGR valves are not opened at idle so as not to detract from the idle quality of the engine, nor are they opened at wide-open throttle so as not to interfere with all-out engine performance. On some systems, the engine computer monitors the EGR valve's position or flow rate, resulting in a feedback type of control of the EGR system.

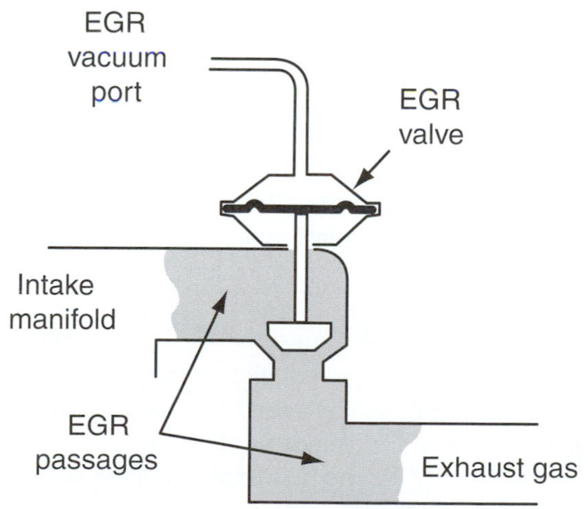

Figure 1–18 Opening of an EGR valve allows exhaust gasses to flow into the intake manifold, displacing incoming ambient air and oxygen.

Stoichiometric Air/Fuel Ratios

A stoichiometric air/fuel ratio provides just enough air and fuel for the most complete combustion resulting in the least possible total amount of leftover combustible materials: oxygen, carbon, and hydrogen. This maximizes the production of CO_2 and H_2O, and minimizes the potential for producing CO, HC and NO_x. It also provides good driveability and economy.

The most power is obtained from an air/fuel ratio of about 12.5 to 1, although the best economy is obtained at a ratio of about 16 to 1. Air/fuel ratios this far from stoichiometric, however, are not compatible with the three-way catalytic converter. The leaner mixtures also increase engine temperature as a result of slower burn rate.

ECONOMY AND DRIVEABILITY

Computerized engine control systems contribute to fuel economy and driveability in several ways.

Fuel Economy Controls

Air/Fuel Mixture. Maintaining a 14.7 to 1 air/fuel ratio promotes fuel economy by eliminating the fuel waste that occurs with richer fuel mixtures. Richer mixture usage can easily occur in an uncontrolled engine. It also provides an efficient mixture for combustion and thus enhances economy and driveability. Systems that use electronic fuel injection also contribute to economy with more efficient fuel control such as:

- shutting off fuel during deceleration.
- replacing the mechanical choke with a more articulated, electronically controlled cold engine enrichment.
- replacing the accelerator pump and vacuum-controlled full-power enrichment functions of a carburetor with an electronically controlled acceleration enrichment.

In many cases these features also help to improve driveability.

Electronic Ignition Timing. Ignition timing is well known to be a major factor in both fuel economy and driveability. It is also a critical factor in the control of exhaust emissions. With computer-controlled ignition timing, timing does not have to be dependent on just engine speed and load. It can be adjusted to meet the greatest need during any driving condition. For example, following a cold start, it can be advanced for driveability. During light load operation with a partially warmed-up engine, it can be retarded slightly to hasten engine warmup and reduce exhaust emissions. During acceleration or wide open throttle operation, it can be adjusted for maximum torque.

Idle Speed Control. Electronic idle speed control can more nearly maintain the best compromise between a speed at which the engine is prone to stall and one that wastes fuel and produces an annoying lurch as the automatic transmission is pulled into gear. It can keep idle speed from "yo-yoing" up and down as the air conditioning clutch cycles on and off, and it can be designed to prevent dieseling. It can also adjust idle speed in an effort to compensate for low charging system voltage, low engine temperature and engine overheating.

Torque Converter Clutch Control. Many engine control computer systems control a clutch in the torque converter. This function eliminates converter slippage and heat production and thus contributes to fuel economy. On some manual transmission applications, the computer activates an upshift light, which alerts the driver when to upshift to obtain maximum economy.

Air Conditioning Control. Some computerized engine control systems feature an air conditioning clutch cutout function that disengages the air conditioning clutch during heavy throttle operation. This function contributes somewhat to driveability.

Turbocharger Boost Control. To increase performance, each manufacturer is offering at least one turbocharged engine option. In most cases, the computer controls the amount of boost that the turbocharger can develop. It is worth noting that each of the manufacturers is using an air-to-air intercooler on selected turbo applications. This lowers manifold air temperature by 120° to 150°F and allows for even more aggressive turbo boost and spark advance due to the denser air charge.

Closed Loop, Open Loop

Closed Loop. Made possible by the development and use of the oxygen sensor, a feature of all comprehensive computerized engine control systems is the ability to operate the engine in a closed-loop mode. The term **closed loop** refers to the fact that the engine computer monitors the results of its own control through a **feedback** circuit. With the use of a feedback circuit, a computer is able to monitor what actually happened as a result of the last command it issued to an actuator. Without a feedback circuit, the computer must *assume* that it is controlling the actuator properly and that the actuator is responding properly. As computer technology moves forward, increasingly more output circuits are monitored with feedback circuits and sensors. With engine control systems, closed loop refers to the use of the oxygen sensor to monitor the results of the last commands issued to the fuel injector(s) (or mixture control solenoid in a feedback carburetor). Closed loop is a mode that allows the engine computer to maintain an air/fuel ratio that stays very close to stoichiometric. When operating in closed loop, the engine computer responds to the other critical sensors in bringing the air/fuel ratio close to the ideal. These sensors may include throttle position, engine load, barometric pressure, engine rpm, engine coolant temperature and intake air temperature. Then the engine computer uses the oxygen sensor to fine-tune the air/fuel ratio. The oxygen sensor reports to the computer a voltage representing the amount of oxygen that was left over after the burn during the most recent combustion. The computer sends a command to the fuel-metering control device to adjust the air/fuel ratio toward a stoichiometric air/fuel ratio (14.7 to 1). The adjustment usually causes the air/fuel ratio to cross stoichiometric, and it starts to move away from 14.7 in the opposite direction. The oxygen sensor sees this and reports it to the computer. The computer again issues a command to adjust back toward 14.7 to 1. With the speed at which this cycle occurs, the air/fuel ratio never gets very far from 14.7 to 1 in either direction. This cycle repeats continuously.

Closed loop is the most efficient operating mode, and the computer is programmed to keep the system in closed loop as much as possible. However, the following criteria must be met before any of the systems can go into closed loop:

- The oxygen sensor must reach operating temperature (around 315°C / 600°F).
- The engine coolant temperature must reach a criterion temperature (varies somewhat but tends to be around 65°C / 150°F).
- A predetermined amount of time must elapse from the time the engine was started (this varies from a few seconds to 1 or 2 minutes).

Other operating conditions, such as hard acceleration or, in some cases, prolonged idle, force the system out of closed loop.

Open Loop. Open loop is used during periods of time when a stoichiometric air/fuel ratio is not appropriate such as during engine warmup or at wide-open throttle (WOT). During this operational mode, the computer uses input information from several sensors to determine what the air/fuel ratio should be. These sensors may include throttle position, engine load, barometric pressure, engine rpm, engine coolant temperature and intake air temperature.

Once the necessary information is processed, the computer sends the appropriate command to the mixture control device. The command does not change until one of the inputs changes. In this mode, the computer does not use oxygen sensor input and therefore does not know if the command it sent actually achieved the most appropriate air/fuel ratio for the prevailing operating conditions. Certain failures within the system prevent it from going into closed loop or cause it to drop back out of closed loop.

ATTITUDE OF THE TECHNICIAN

Historically, many automotive technicians have had a negative attitude about manufacturers' continuous changes to cars. Changes mean always having to learn something new, having to cope with a new procedure and so on. There is another way to look at it, though: every time new knowledge or a new skill is required, it represents a new opportunity to get ahead of the pack. Most of us are in this business to make money; the more you know that other people have not bothered to learn yet, the more money you can make. Successful, highly paid automotive technicians *earn* the money they get because of what they *know*. Their knowledge enables them to produce more with the same effort. More important, it enables them to do things that other people cannot.

Computerized automotive control systems represent the newest and probably the most significant and most complicated development ever to occur in the history of automotive service. (It is complicated, not because it is more difficult, but because it involves more processes.) We have seen only the beginning. In recent years, every major component and function on the automobile has come under the control of an electronic control module, and engineers will undoubtedly continue to find new ways to use electronic control as we move forward. Those who appreciate the capability of these systems and learn everything they can about them will be able to service them and do very well while other people scratch their heads and complain.

✔ SYSTEM DIAGNOSIS AND SERVICE

Approaching Diagnosis

Diagnosing problems on complicated electronic systems takes a little different approach than many automotive technicians are used to. The flat-rate pay system has encouraged many of us to take shortcuts whenever possible; however, shortcuts on these systems get you into trouble. Use the service manual and follow procedures carefully. It is very useful to spend some time familiarizing yourself with the diagnostic guides and charts that manufacturers present in the service manual or manuals you will be using.

Prediagnostic Inspection. Although most computerized engine control systems feature some degree of self-diagnostic capability, none of them monitors spark plugs, spark plug wires, valves, vacuum hoses, PCV valves and other engine support and emission-control components that are not a part of the computer system. A seemingly unrelated part such as a spark plug wire can have a direct impact on the performance of the system. For example, a shorted spark plug wire prevents its cylinder from firing. The unburned oxygen coming out of that cylinder, if it is on the same side as the oxygen sensor on a V-type engine, causes the oxygen sensor to mistakenly see a lean condition. The computer responds by enriching the air/fuel mixture. Replacing the oxygen sensor, half or all of the other sensors, the computer, the headlights and the front bumper does not solve the problem. Replacing the shorted spark plug wire and possibly cleaning the spark plug *does* solve the problem.

The computer knows the engine operating parameters (throttle position, atmospheric pressure, engine speed, temperature, load etc.), and with this information it can calculate exactly what the spark timing should be. It continuously makes calculations and sends its spark timing commands. It does not, however, know what base timing is (except on the very latest models); and the command it sends out is added to base timing. If base timing is incorrect, a spark plug will fire at the wrong time in spite of the computer.

Having a closed thermostat stick or a restricted radiator causes the engine to overheat, possibly causing the computer to think that the coolant temperature-sensing device or its connecting circuit is shorted and setting a fault code in its diagnostic memory. The coolant sensing device, however, is only reporting what it sees; the fault is in the cooling system. Do not overlook the possibility that someone has replaced a 195-degree thermostat with one that opens at 160 degrees, or that the charcoal canister is saturated with gasoline or that it is being purged when it should not be. Do not overlook the possibility of loose or shorted electrical wires or of cracked or misrouted vacuum hoses.

Any such problem, although it may not be directly related to the computer system, can cause a driveability problem and can in some cases cause the computer to mistakenly report a problem in one of its circuits. The service manual for each vehicle with an engine control computer system should provide instructions for making a prediagnostic inspection, because the system's diagnostic procedures assume that all such unrelated, or perhaps it would be better to say *semi-related*, systems are functioning properly. All of the engine components, especially engine support components, and some nonengine components such as the transmission, are more related in their influence on each other than they have ever been before the development of computer control systems. It is extremely important to remember that for the computer control system to work properly, all other engine-related systems must be operating to manufacturer's specifications. If routine maintenance, such as changing the oil, PCV valve, spark plugs and spark plug wires, is due, it is advisable to do it before beginning diagnostic procedures.

SUMMARY

In this chapter, we have begun building our conceptual foundation for understanding why carmakers use computers in cars and what the computers are supposed to do. The principal design objectives are to control exhaust emissions to a legally acceptable level, to optimize the vehicle's driveability and to get the best fuel economy consistent with the driving conditions.

We have learned the major components of a microprocessor, the central element of the engine control computer. It employs a PROM (programmable, read-only memory) or hardwired memory to store general information about the specific vehicle. It maintains a KAM (keep-alive memory) to store adaptations to its control maps that it has learned in past driving trips. It includes interface connections for all its sensors and outputs as well as power to the computer itself.

We have learned the various elements in a car's exhaust that relate to the emissions laws: hydrocarbons or unburned fuels, carbon monoxide and carbon dioxide. We have seen how residual oxygen is used as a measure of the computer's mixture control success and how water vapor is produced as a result of the hydrogen in the fuel.

The concept of stoichiometry has been introduced, a concept that will play a role in each successive chapter. A stoichiometric ratio is the exact air/fuel mixture that allows the catalytic converter to keep emissions to a minimum. The only way to accurately maintain this ratio is through the precision control of a computer.

We have learned the concepts of open and closed loop. Open loop describes the conditions under which the computer determines air/fuel mixture and ignition timing based on information stored in its memory, about temperature, load, speed and other system parameters. Closed loop, in contrast, is the feedback mode in which the computer controls air/fuel mixture by adding in the output signal from the oxygen sensor.

Finally, we have learned the importance of following precise diagnostic steps if this type of

Excessive HC Production

Overadvancing ignition timing by 6 degrees can cause HC production to go up by as much as 25 percent and NO_x production to increase by as much as 20 to 30 percent. A thermostat that opens at 160°F can cause HC production to go up as much as 100 to 200 parts per million.

computer system is to be diagnosed. Without such a sequence, the technician can be misled by other problems, or by countermeasures the computer employs to solve some other problem that has affected the system.

▲ DIAGNOSTIC EXERCISE

In an uninformed and socially irresponsible attempt to improve his car's performance, a do-it-yourselfer disconnected, disabled or removed most of the emissions controls on his car. He slabbed off the EGR, disconnected most of the vacuum lines and ran the distributor advance vacuum line directly to manifold vacuum.

What will the consequences be for the car? Will its driveability, power or fuel economy improve? What should the repair/diagnostic technician's approach be?

REVIEW QUESTIONS

1. *Technician A* says that computerized engine control systems are intended to reduce exhaust emissions.
   Technician B says that computerized engine control systems are designed to improve fuel mileage and driveability.
   Who is correct?
   A. A only
   B. B only
   C. Both A and B
   D. Neither A nor B
2. An analog voltage is a voltage signal that does which of the following?
   A. It is continuously variable within a given range and requires time to change.
   B. It varies, but not continuously, and requires no time to change.
   C. It is often referred to as a square wave.
   D. Both B and C
3. A digital voltage is a voltage signal that does which of the following?
   A. It is continuously variable within a given range and requires time to change.
   B. It varies, but not continuously, and requires no time to change.
   C. It is often referred to as a square wave.
   D. Both B and C
4. What is the largest base-10 numerical value that can be communicated by an 8-bit computer?
   A. 80
   B. 128
   C. 255
   D. 1024
5. The binary code 01101110 equals what base-10 numerical value?
   A. 5
   B. 110
   C. 124
   D. 421
6. Which of the following components in an engine computer is a nonvolatile memory that contains information specific to a particular vehicle or model?
   A. A/D converter
   B. ROM
   C. RAM
   D. PROM
7. Which of the following components in an engine computer translates analog input signals to binary code?
   A. A/D converter
   B. ROM
   C. RAM
   D. PROM

8. Which of the following components in an engine computer is a volatile memory used by the microprocessor to store temporary information?
   A. A/D converter
   B. ROM
   C. RAM
   D. PROM
9. A cylinder misfire (due to a mechanical problem, an ignition problem or an overly lean air fuel ratio) will likely result in excessive production of which of the following?
   A. HC
   B. CO
   C. NO_x
   D. Both A and B
10. An overly rich air fuel ratio will likely result in excessive production of which of the following?
   A. HC
   B. CO
   C. NO_x
   D. Both A and B
11. High combustion chamber temperature can result in excessive production of which of the following?
   A. HC
   B. CO
   C. NO_x
   D. Both A and B
12. What air/fuel ratio is required to allow a three-way catalytic converter to work effectively?
   A. 12:1
   B. 13.7:1
   C. 14.7:1
   D. 17:1
13. The catalytic materials used most often inside a three-way catalytic converter include all except which of the following?
   A. Zirconium
   B. Platinum
   C. Palladium
   D. Rhodium
14. Which of the following is the single most effective device/system for controlling exhaust emissions?
   A. Exhaust gas recirculation valve
   B. Catalytic converter
   C. Positive crankcase ventilation system
   D. Charcoal canister
15. Which emission device/system is designed specifically to help control the production of NO_x?
   A. Exhaust gas recirculation valve
   B. Catalytic converter
   C. Positive crankcase ventilation system
   D. Charcoal canister
16. *Technician A* says that on a vehicle with a dual-bed, three-way catalytic converter, the NO_x reduction bed is located before the CO and HC oxidation bed.
   Technician B says that the rhodium in the NO_x-reducing bed needs to have plenty of oxygen present in the exhaust stream in order to break up the NO_x molecules.
   Who is correct?
   A. A only
   B. B only
   C. Both A and B
   D. Neither A nor B
17. *Technician A* says that closed loop operation is designed to allow the engine computer to monitor the results of its own fuel control through an oxygen sensor mounted in the exhaust stream.
   Technician B says that closed loop is the best engine-operating mode for wide-open throttle performance because of the increased precision that closed loop provides.
   Who is correct?
   A. A only
   B. B only
   C. Both A and B
   D. Neither A nor B
18. Closed loop operation requires all except which of the following?
   A. The oxygen sensor must reach operating temperature.
   B. The engine coolant temperature must reach a criterion temperature.
   C. The vehicle speed must be at least 20 mph.
   D. A predetermined amount of time must have elapsed since the engine was started

19. *Technician A* says that on a vehicle with a three-way catalytic converter, it is critical that the engine control system be capable of using input from an oxygen sensor in order to successfully hold the air/fuel ratio close to stoichiometric.
    Technician B says that if a vehicle has a three-way catalytic converter, the converter can either be a single-bed or a dual-bed design.
    Who is correct?
    A. A only
    B. B only
    C. Both A and B
    D. Neither A nor B

20. *Technician A* says that a fault code indicating a coolant temperature sensor problem could be the fault of a defective coolant temperature sensor or its circuit.
    Technician B says that a fault code indicating a coolant temperature sensor problem could be the result of a defective thermostat in the engine cooling system.
    Who is correct?
    A. A only
    B. B only
    C. Both A and B
    D. Neither A nor B

Chapter 2

Common Components for Computerized Engine Control Systems

OBJECTIVES

Upon completion and review of this chapter, you should be able to:

- ❑ Define the features common to most computerized engine control systems.
- ❑ Explain the functional concepts of each of the most common sensing devices.
- ❑ Understand the differences between a speed density system and a mass airflow system.
- ❑ Explain the operation of the most common types of actuators: solenoids, relays, motors and stepper motors.
- ❑ Define how a computer's control of an actuator's on-time through pulse width modulation differs from duty cycle control.

KEY TERMS

Actuator
Barometric Pressure
Clamping Diode
Duty Cycle
E-cell
Gallium Arsenate Crystal
Hall Effect Switch
Hz (Hertz)
Impedance
Piezoelectric
Piezoresistive
Potentiometer
Pulse Width
Schmitt Trigger
Sensor
Speed Density Formula
Stepper Motor
Thermistor
Volumetric Efficiency (VE)
Wide-Open Throttle (WOT)
Zirconium Dioxide (ZrO_2)

Among the different car manufacturers, vehicle models, and years, there are many different computerized engine control systems. Although there are significant differences between the various systems, when compared, they are actually more alike than they are different. Some of the more common components and circuits are discussed in this chapter to avoid needless duplication when discussing the specific systems. As you read the following chapters, you may find it useful to refer to this chapter to clarify how a particular component or circuit works.

COMMON FEATURES

Computers

The comprehensive automotive computer is a special-purpose, small, highly reliable, solid-state digital computer protected inside a metal box. It receives information in the form of voltage signals from several **sensors** and other input sources. With this information, which the computer rereads several thousand times per second, it can make "decisions" about engine calibration functions such as air/fuel mixture, spark timing, EGR application, and so forth.

These decisions appear as commands sent to the **actuators**—the solenoids, relays and motors that carry out the output commands of the computer. Commands usually amount simply to turning an actuator on or off. In most cases, the ignition switch provides voltage to the actuators either directly or through a relay. The computer controls the actuator by using one of its internal solid-state switches (power transistors and quad drivers are two types) to ground the actuator's circuit. While in most actuator circuits the computer completes the actuator's ground side, occasionally the computer may be circuited to complete the power side. If the wire that completes the circuit between the actuator and the computer were to short to the engine block or to chassis ground and the computer was providing power to the actuator, this short would likely destroy the computer, Figure 2–1. But if the computer uses this circuit to complete the actuator's ground path, a short to ground would not overload the computer's circuits. However, it would keep the actuator energized, Figure 2–2. In certain instances where safety would be compromised by a short that could energize the actuator, the computer may be circuited to provide power to the actuator.

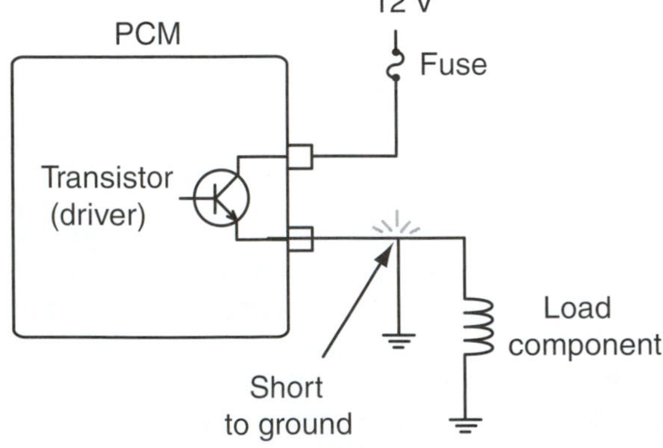

Figure 2–1 Computer providing power to an actuator. A short to ground can destroy the computer's driver.

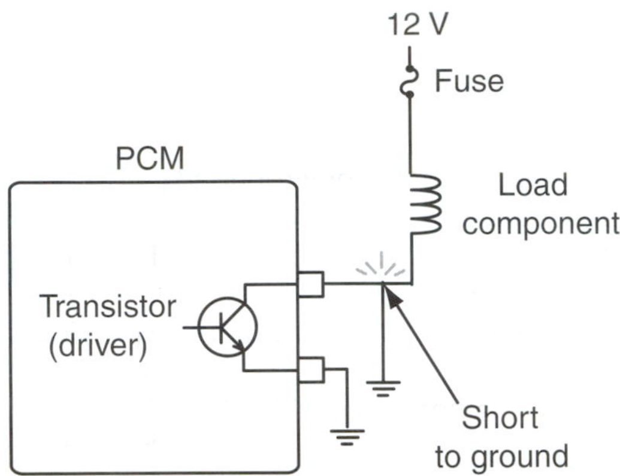

Figure 2–2 Computer completing an actuator's ground circuit. Short to ground keeps the actuator energized, but does not overload the computer.

CAUTION: No attempt should be made to open the computer's metal housing. The housing protects the computer from static electricity. Opening it or removing any circuit boards from it outside carefully controlled laboratory conditions will likely result in damage to some of its components. Certain late-model systems use computers with a special kind of memory that can be reprogrammed by exposing that component to light, usually by removing a piece of opaque tape. If you do not intend to do such reprogramming, or do not know what the effect will be, do not open a computer just to see what the internal parts look like. Over time, a technician can expect to find a completely failed computer that can be opened up without doing any additional harm.

Location. Over the years, most manufacturers located the engine computer within the passenger compartment rather than in the engine compartment. It is sometimes mounted under the instrument panel, behind either kick panel or under a front seat. This placement helped to protect it from the harsh environment of the engine compartment with its extreme temperature changes. Other manufacturers, such as Daimler-

Chrysler's systems or Ford's early Microprocessor Control Unit (MCU) system, located it in the engine compartment. These computer cases are designed to protect the computer from the engine compartment's environment. In fact, late model DaimlerChrysler engine computers have cases that also shield the computer's internal electronics from radio frequency interference (RFI) and electromagnetic frequency interference (EFI). Many engine computers are located on the passenger compartment side of the firewall under the instrument panel and attach to a harness connector on the engine compartment side of the firewall. Some manufacturers are now mounting the engine computer on the engine block, a concept that originated in the heavy-duty industry where the computer is matched to the engine rather than to the vehicle.

Engine Calibration. Because of the wide range of vehicle sizes and weights, engine and transmission options, axle ratios, and so forth, and because many of the decisions the computer makes must be adjusted for those variables, the manufacturers use some type of an engine calibration unit, Figure 2–3. The calibration unit is a chip that contains information that has to be specific to each vehicle. For example, the vehicle's weight affects the load on the engine, so to optimize ignition timing, ignition timing calibration must be programmed for that vehicle's weight. In a given model year, a manufacturer might have more than a hundred different vehicle models counting different engine and transmission options, but might use less than a dozen different computers. This is made possible by using an engine calibration unit specific to each vehicle.

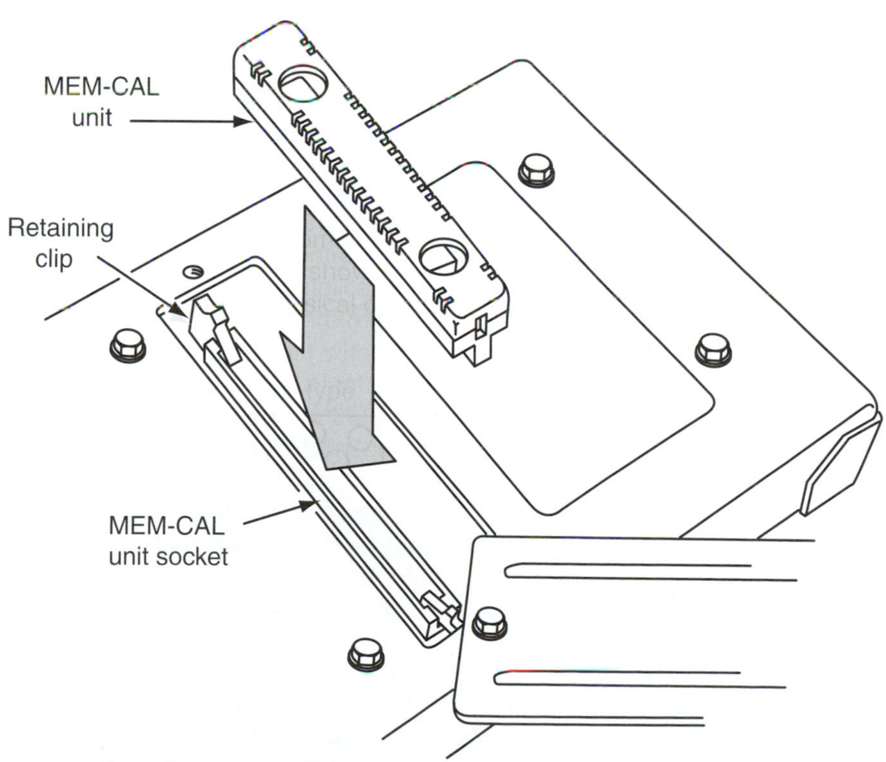

Figure 2–3 General Motors' electronic control module and PROM.

The engine calibration unit may be referred to as the "PROM," or some other term may be used. Some manufacturers make the calibration unit removable, and some make it a permanent part of the computer.

Five-Volt Reference

With few exceptions, all of the systems use 5 volts to operate their sensors. Within the electronics industry, 5 volts has been almost universally adopted as a standard for information transmitting circuits. This voltage value is high enough to provide reliable transmission and low enough not to damage the tiny circuits on the chips in the computer. Of course, use of a computer-industry standard voltage makes parts specification more economical for the car makers.

Fuel Injection

There are two types of fuel injection systems used with comprehensive computerized engine control systems: single point and multipoint. They each use an intermittent or timed spray to control fuel quantity (There is one exception, the Bosch K-Jet systems, described more fully in the European systems chapter of this book.)

Single-Point Injection. Single-point injection is often referred to as *throttle body injection*. Single point means that fuel is introduced into the engine from one location. This system uses an intake manifold similar to what would be used with a carbureted engine, but the carburetor is replaced with a throttle body unit, Figure 2–4. The throttle body unit contains one or two solenoid-operated injectors that spray fuel directly over the throttle blade (or blades). Fuel under pressure is supplied to the injector. The throttle blade is controlled by the throttle linkage just as in a carburetor. The computer controls voltage pulses to the solenoid-operated injector, which opens and sprays fuel into the throttle bore. The amount of fuel introduced is controlled by the length of time the solenoid is energized. This is referred to as ***pulse width***. The amount of air introduced is controlled by the opening of the throttle blade and is also affected by any air that is introduced by the idle air control valve (IACV).

The EFI system is characterized, especially on smaller engines, by excellent throttle response and good driveability. Experience has shown, however, that the system is best suited for engines with small cross-sectional area manifold runners that at low speeds will keep the fuel mixture moving at a higher velocity. This reduces the tendency for the heavier fuel particles to fall out of the airstream.

Multipoint Injection. Multipoint injection is often referred to as port injection and means that fuel is introduced into the engine from more than

Constant Spray and Timed Spray Injector

There are two basic methods of controlling the quantity of fuel introduced by a fuel injection system: the constant spray injection and the timed spray injection. Historically these have been port injection systems (placing an injector in the center of the intake manifold, above the throttle blades, is a comparatively new development). As the term *constant spray* implies, in this system each injector sprays a continuous stream of fuel into each intake port; and the quantity of fuel introduced is controlled by a variable fuel pressure. Fuel pressure is controlled by an airflow meter. The Bosch K-Jetronic is the most common example of this type of system.

In timed spray systems, the difference between fuel pressure and intake manifold pressure is held constant by a pressure regulator, and the injectors are turned on and off. Fuel quantity is controlled by how long the injectors are turned on, and the injectors are controlled by an electronic module (a single-function computer system). The Bosch D-Jetronic and L-Jetronic systems are the best known examples of this type of system.

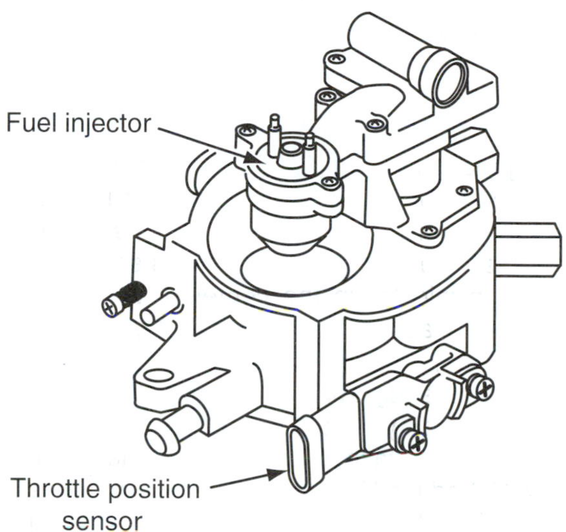

Figure 2–4 TBI Unit.

one location. This system uses an injector at each intake port. Fuel is sprayed directly into the port, just on the manifold side of the intake valve.

The multipoint injection system provides the most advanced form of fuel control yet developed. It offers the following advantages:

- Spraying precisely the same amount of fuel directly into the intake port of each cylinder eliminates the unequal fuel distribution inherent when already mixed air and fuel are passed through an intake manifold. There is simply no way to make certain that equal amounts of fuel and air get to each cylinder.
- Because there is no concern about fuel condensing as it passes through the intake manifold, there is less need to heat the air or the manifold.
- Because there is no concern about fuel molecules falling out of the airstream while moving through the manifold at low speeds, the cross-sectional area of the manifold runners can be larger and thus offer better cylinder-filling ability [**volumetric efficiency (VE)**] at higher speeds. Some manufacturers have even been able to design variable-geometry intake runners to optimize air ingestion at different engine speeds and loads.
- Most of the manifold-wetting process is avoided, though some wetting still occurs in the port areas and on the valve stems. If fuel is introduced into the intake manifold, some will remain on the manifold floor and walls, especially during cold engine operation and acceleration. Fuel metering has to allow for this fuel in order to avoid an overlean condition in the cylinders. It has to be accounted for again during high-vacuum conditions because it will then begin to evaporate and go into the cylinders.

In general, port fuel injection provides better engine performance and excellent driveability while maintaining or lowering exhaust emission levels and increasing fuel economy. The major disadvantages are somewhat greater cost and reduced serviceability because of the larger number of components and their relative inaccessibility.

Fuel Pressure Regulators. All fuel injection systems that are part of a comprehensive computerized engine control system use pulse width as the primary means of controlling fuel metering. A pressure regulator is used to provide a constant pressure to the injector. What is held constant is the difference between the pressure of the fuel in the injector and the air pressure in the intake manifold. This ensures a consistent flow of fuel for a given pulse width; there is no need for the computer to make allowances for flow differences based on pressure differences. Very late model systems include computers that can make these adjustments, and can run at an absolutely constant fuel pressure.

An electric fuel pump supplies fuel under pressure to the regulator. The fuel pressure regulator is located on the fuel rail or manifold and allows excess fuel to return to the tank. It has a diaphragm with fuel pressure on one side and atmospheric pressure on the other side, Figure 2–5. A diaphragm spring pushes the diaphragm against fuel pressure. As the diaphragm moves in

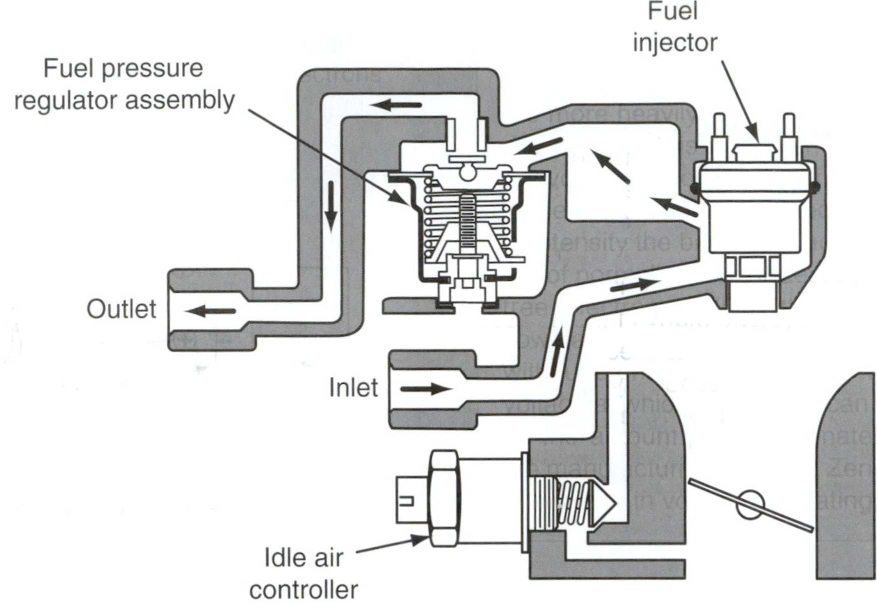

Figure 2–5 Fuel pressure regulator.

the direction the spring is pushing it, a flat disc in the center of the diaphragm closes off the fuel return passage. This allows less fuel to return to the tank and causes fuel pressure to go up, closer to fuel pump pressure. The increased fuel pressure moves the diaphragm against the spring. This allows more fuel to escape through the return line and thus causes fuel pressure to drop. The diaphragm will always find a balanced position that will keep fuel pressure at the desired value, maintaining an exact pressure differential.

On some single-point injection systems, the spring side of the diaphragm is exposed to atmospheric pressure. Changes in atmospheric pressure produce a slight change in fuel pressure. Some multipoint injection systems connect manifold vacuum to the spring side of the diaphragm.

Changes in atmospheric pressure or in manifold pressure change the total pressure on the spring side of the diaphragm and therefore the fuel pressure. The intent, however, is not to increase or decrease fuel delivery but to maintain a constant pressure drop across the injector. For example, when manifold pressure goes up, the injector has to spray fuel into a higher pressure environment. For the correct amount of fuel to be delivered during the pulse width, fuel pressure must be increased in proportion to the manifold pressure increase.

Fuel Injection History

Fuel injection for gasoline engines dates as far back as the middle 1930s. It has captured the interest of automotive engineers and performance enthusiasts for years by promising to be a superior method of fuel induction. Only recently, however, has its technology evolved far enough (see **Measuring Air Mass** in this chapter) to enable fuel injection to surpass the carburetor's ability to deliver performance, driveability, economy and emissions reduction under all driving conditions.

Returnless Fuel Systems

For years, automotive fuel systems were designed so that excessive fuel unneeded at the fuel injectors was returned to the fuel tank. When the fuel was in the engine compartment, it absorbed heat and carried that heat back into the fuel tank. Some vehicles were equipped with a fuel tank temperature sensor. This heated fuel increased the vapor collection requirements of the fuel containment system. Many of the newest vehicles are equipped with a returnless fuel supply system. In these returnless fuel systems, fuel sent to the engine compartment is prevented from returning to the tank. It simply waits in the fuel line until it is injected into the engine. The pressure regulator returns excess fuel from the high volume pump. It is located either inside the fuel tank or immediately outside of it.

The following examples demonstrate numerous ways the returnless fuel system is applied to various vehicles. The Corvette by General Motors incorporates the fuel filter and the pressure regulator in the same container located a short distance from the fuel tank, Figure 2–6. This container has three lines attached to it. One line runs from the tank; one line returns to the tank. The third line goes to the engine's fuel rail.

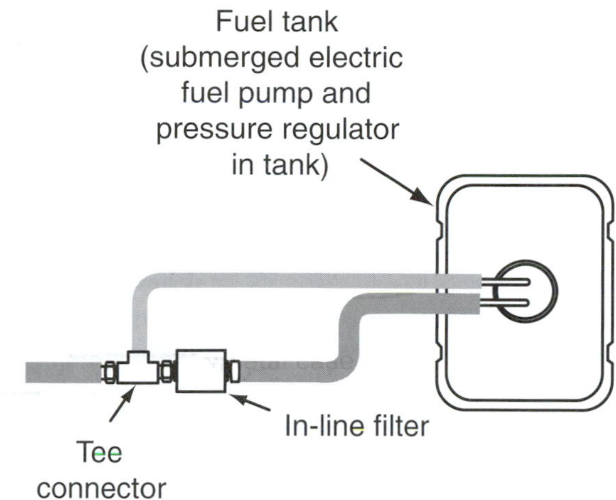

Figure 2–7 Camaro/Firebird fuel tank.

The Camaro/Firebird returns fuel to the fuel tank through a line that is teed in a short distance from the tank. Fuel returns to the tank through the fuel pressure regulator located on the fuel tank module inside the fuel tank, Figure 2–7.

Although there are still some traditional fuel systems in use for the current model year, many manufacturers use some variation of these two systems.

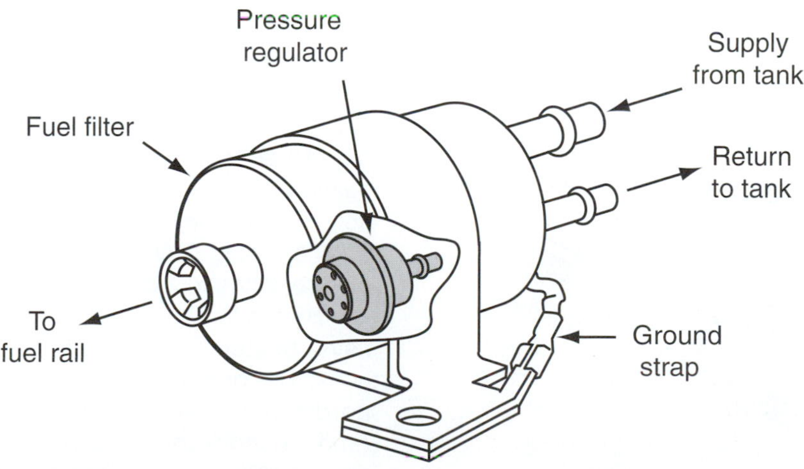

Figure 2–6 Corvette fuel filter/pressure regulator.

SENSING DEVICES

Exhaust Oxygen Sensors

One of the major components of all comprehensive computerized engine control systems is the oxygen sensor, Figure 2–8. The amount of free oxygen in the exhaust stream is a direct result of the air/fuel ratio. A rich mixture yields little free oxygen; most of it is consumed during combustion. A lean mixture, on the other hand, yields considerable free oxygen because not all of the oxygen is consumed during combustion. With an oxygen sensor in or near the exhaust manifold, measuring the amount of oxygen in the exhaust indirectly reveals the air/fuel ratio burned in the cylinder.

The heart of an oxygen sensor, Figure 2–9, is a hollow ceramic body that is closed at one end and contains within it a substance known as **zirconium dioxide**, a white crystalline compound that becomes oxygen-ion conductive at approximately 600°F. The inner and outer surfaces of the ceramic body are coated with separate, superthin, gas-permeable films of platinum. The platinum coatings serve as electrodes. The outer electrode surface is covered with a thin, porous, ceramic layer to protect against contamination from combustion residue. This body is placed in a metal shell similar to a spark plug shell except that the shell has a louvered nose that encloses the exhaust-side end of the ceramic body. When the shell is screwed into the exhaust pipe or manifold, the louvered end extends into the exhaust passage. The outer electrode surface contacts the shell; the inner electrode connects to a wire that goes to the computer. Ambient oxygen is allowed to flow into the hollow ceramic body and to contact the inner electrode.

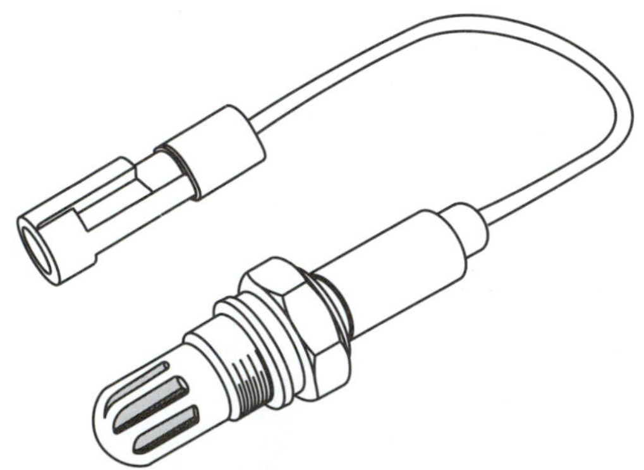

Figure 2–8 Oxygen sensor.

The oxygen sensor has an operating temperature range of 300°C to 850°C (572°F to 1,562°F). When the zirconium dioxide reaches 300°C it becomes oxygen-ion conductive. If the oxygen in the exhaust stream is less than that in the ambient air, a voltage is generated. The greater the difference, the higher the voltage signal generated, Figure 2–10. Voltage generated by the oxygen sensor will normally range from a minimum of 0.1 volt (100 millivolts) to a maximum of 0.9 volt (900 millivolts). The computer has a preprogrammed value, called a set-point, that it wants to see from the oxygen sensor during closed-loop operation. The set-point is usually between 0.45 volt and 0.5 volt and equates to the desired air/fuel ratio. Voltage values below the set-point are interpreted as lean (the percent of oxygen in the exhaust stream and in the ambient air becomes closer), and values above it are interpreted as rich (percentages of oxygen in the two areas move apart).

The engine computer has two terminals for the oxygen sensor circuit, one for the oxygen sensor's positive signal and one for the oxygen sensor's negative signal (signal ground). That is, the voltage sensing circuits within the computer must use two wires to measure the oxygen sensor's voltage, similar to how you would use a voltmeter to do the same thing. Although some oxygen sensors complete both circuits using insulated wiring, resulting in a two-wire oxygen sensor, many applications use the metal of the engine block to complete the signal ground circuit, resulting in a single-wire oxygen sensor.

By the late 1980s, most manufacturers were using heated oxygen sensors on many applications. The operation of the oxygen sensor itself is no different, but a heating element is added to (and placed within) the oxygen sensor to heat it up to operating temperature more quickly and to help it maintain proper operating temperature

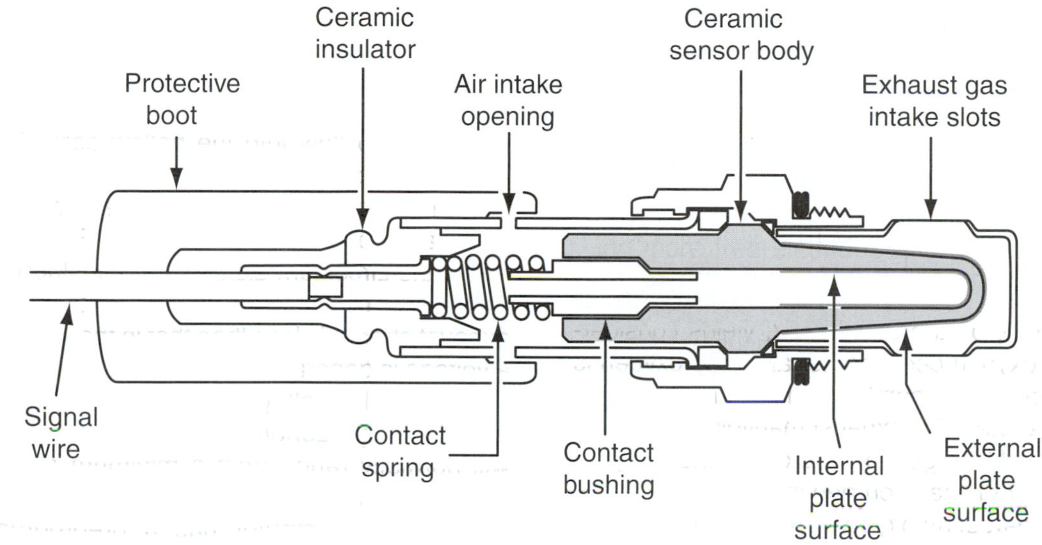

Figure 2–9 Exhaust gas oxygen sensor.

when contacted by cooler exhaust gasses, such as at idle. The addition of a heating element also allows it to be placed further downstream in the exhaust system so that it can sample a more accurate average of the exhaust gas produced by all cylinders. The addition of a heating element results in the addition of two current-carrying wires to the oxygen sensor, one that provides power for the heater when the ignition is switched on and one that is connected to a fixed ground. As a result, what was previously a single-wire oxygen sensor will now have three wires and what was previously a two-wire oxygen sensor will now have four wires. The heater circuit is not part of the computer control system, but on newer applications, the engine computer monitors the heater circuit for continuity and proper operation.

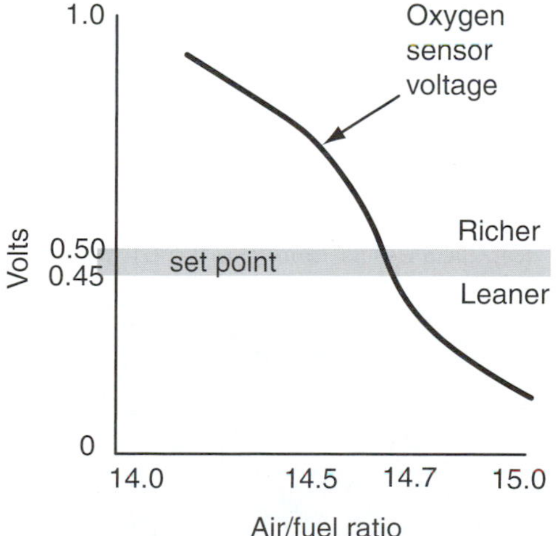

Figure 2–10 Relation of oxygen sensor signal voltage to air/fuel ratio (approximate).

Oxygen sensors are subject to contamination under certain conditions. Contaminating either electrode surface (the outside electrode is more vulnerable) will shield it from oxygen and adversely affect its performance. Prolonged exposure to rich fuel mixture exhaust can cause it to be carbon fouled. This can sometimes be burned off by operating in a lean condition for two or three minutes. The oxygen sensor will also be fouled by exposure to:

- exhaust from leaded fuel.
- the vapors from some silicone-based gasket-sealing compounds.
- the residues of coolant leaking into the combustion chamber.

Diagnostic & Service Tip

Oxygen sensors and mixture. Technicians should remember that, while the oxygen sensor's signal is central to the control of air/fuel mixture, it does not directly sense that mixture. All the oxygen sensor can respond to is the difference in oxygen in the two different gasses. Anything other than combustion that allows a change in the oxygen content of the exhaust will thus make the sensor's signal inaccurate. If there is an air leak into the exhaust, perhaps around the exhaust manifold, the sensor will report a lean mixture and the computer will attempt to correct for it by enrichening the mixture, exactly the wrong tactic. If a spark plug fails to deliver the ignition spark, the air in that cylinder's charge will go into the exhaust unburned. The sensor will again see an overlean condition and the computer will try to enrichen the mixture. For the computer's sensors, you must remember exactly what each one is measuring before you can know what the possibilities are for corrective measure.

Fouling by these materials will probably require that the oxygen sensor be replaced. Additionally, if the ambient air vent were to become plugged with mud or some other substance such as undercoating, ambient air would be restricted from reaching the zirconium dioxide element. This is also cause for replacement.

Thermistors

A **thermistor** is a resistor made from a semiconductor material. Its electrical resistance changes greatly and predictably as its temperature changes. At –40°C (–40°F), a typical thermistor can have a resistance of 100,000 ohms. At 100°C (212°F) the same thermistor can likely have a resistance between 100 and 200 ohms. Even small changes in temperature can be observed by monitoring the thermistor's resistance. This characteristic makes it an excellent means of measuring the temperature of such things as water (engine coolant) or air.

There are negative temperature coefficient (NTC) and positive temperature coefficient (PTC) thermistors. The resistance of the NTC thermistor goes down as its temperature goes up, while the resistance of the PTC type goes up as its temperature goes up. Most thermistors, including those used as temperature sensors on automotive computer systems, are the NTC type. A good example is an engine coolant temperature sensor, common to just about all computerized engine control systems. The coolant temperature sensor consists of a thermistor in the nose of a metal housing, Figure 2–11. The housing screws into the engine, usually in the head, with its nose extending into the water jacket so the thermistor element will be the same temperature as that of the coolant. The computer sends a regulated voltage signal (reference voltage, usually 5 volts) through a fixed resistance and then on to the sensor, Figure 2–12. A small amount of current flows through the thermistor (usually shown as a resistor symbol with an arrow diagonally across it) and returns to ground. This is a voltage divider circuit (current flows through the first resistance unit and then through the second resistance unit) and is commonly used as a temperature-sensing circuit.

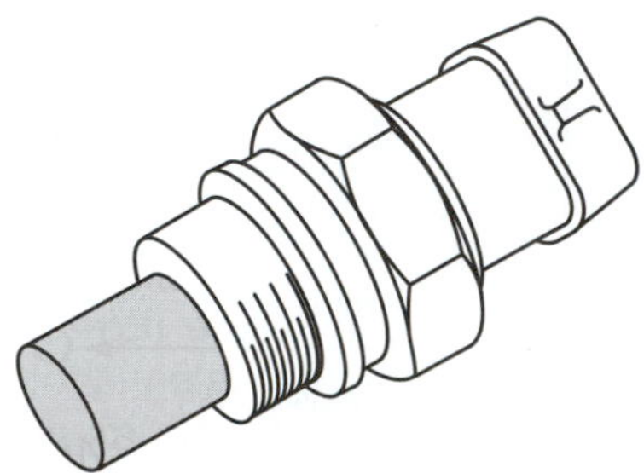

Figure 2–11 Coolant temperature sensor.

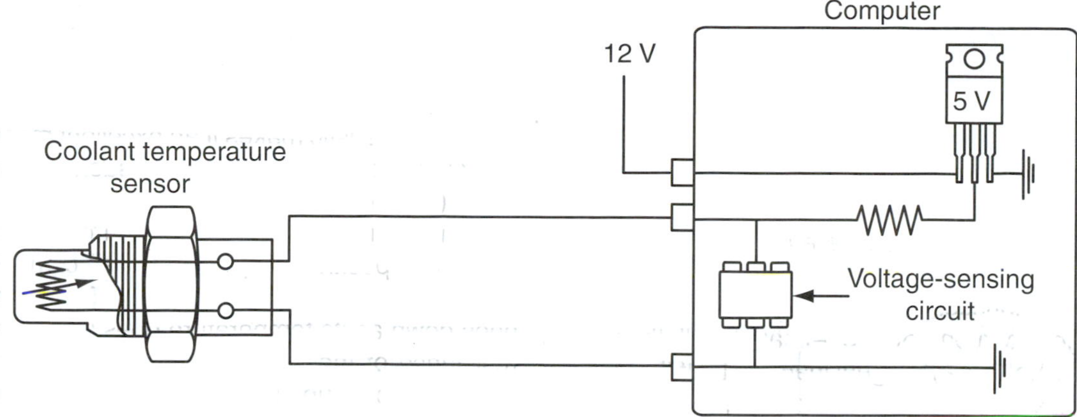

Figure 2–12 Temperature-sensing circuit.

Because the resistance of the thermistor changes with temperature, the voltage drop across it changes also. The computer monitors the voltage drop across the thermistor and, using preprogrammed values, converts the voltage drop to a corresponding temperature value.

Some newer systems use a more complex engine coolant temperature sensor called a *range-switching* or *dual range* sensor. As the engine begins to warm up, the thermistor's resistance decreases, causing the voltage drop across the sensor to decrease. At about 110°F to 120°F the computer lowers the effective resistance within the computer (by adding another resistor in parallel to it). This instantly increases the voltage drop across the sensor by effectively transferring it from the internal resistance. As the engine continues to warm up, the voltage drop across the sensor decreases a second time. This allows the computer to see more voltage change as the engine warms up than the supplied 5-volt reference signal would normally allow, thereby increasing the accuracy of the sensor input. See Chapter 17 for additional information on dual range temperature sensors.

If an open circuit develops in either of these temperature-sensing circuits (either by reason of the sensor or the circuitry itself), the computer sees full reference voltage and interprets this as a signal representing about –40°F. This is due to the fact that without any current flow in the circuit, the fixed resistor within the computer will not drop any voltage.

Potentiometers

A **potentiometer** is another application of a voltage divider circuit. It is formed from a carbon resistance material (or, in higher wattage potentiometers, a wound resistance wire) that has reference voltage supplied to one end and is grounded at the other end. It has a movable center contact (or wiper) that senses the voltage at a physical point along the resistance material, Figure 2–13. At any position of the wiper there is always some resistance both before and after the point of wiper contact. That is, some of the reference voltage is dropped before the point of wiper contact and the remainder of the reference voltage is dropped after the point of contact. As the wiper slides along the resistance material, the sensed voltage changes. This sensed voltage equates to the voltage drop that occurs after the point of wiper contact and is fed to the computer as the sensor's voltage signal. A potentiometer measures physical position and is used to sense either linear or rotary motion.

For example, when a potentiometer is used as a throttle position sensor as depicted in Figure 2–13, the potentiometer would typically be mounted

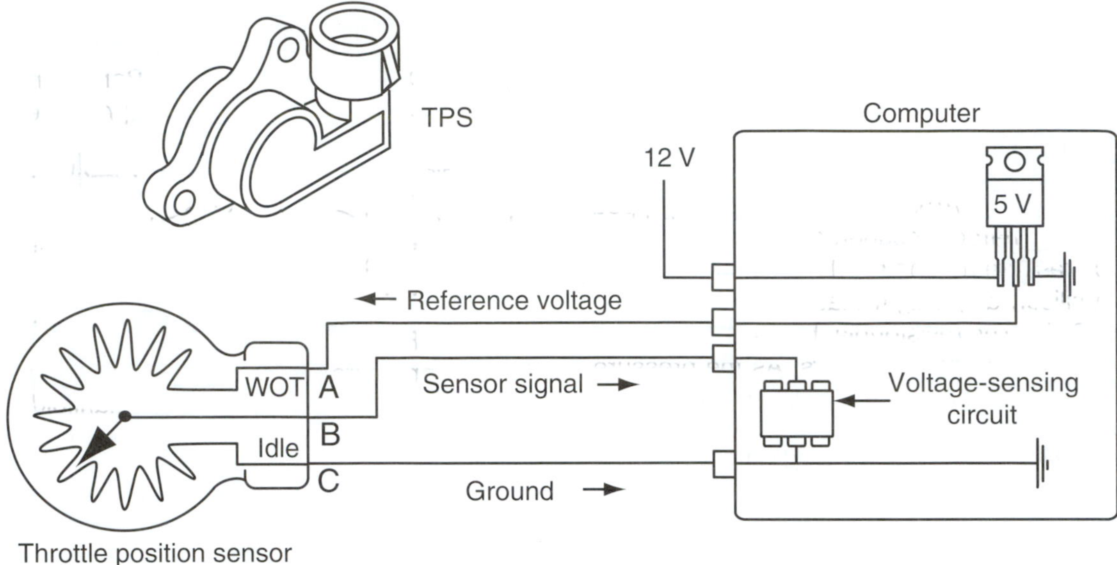

Figure 2–13 Potentiometer as throttle position sensor.

on one end of the throttle shaft. As the throttle is opened, the throttle shaft rotates and moves the wiper along the resistance material. The computer sends a 5-volt reference voltage to point A. If the wiper is positioned near point A (wide-open throttle on most applications), there will be a low voltage drop between points A and B (low resistance), and a high voltage drop will exist between points B and C. When the wiper is positioned near point C (idle position on most applications), there is a high voltage drop between points A and B and a low drop between points B and C. The computer monitors the voltage drop between points B and C and interprets a low voltage, 0.5 for example, as idle position. A high voltage, around 4.5, will be interpreted as **wide-open throttle (WOT)**. Voltages between these values will be interpreted as a proportionate throttle position.

A potentiometer has a higher wear factor than any other type of sensor due to the frequent physical movement of the wiper across the resistance material. As a result, it is quite common for a potentiometer to develop points along the wiper's travel that result in a loss of electrical contact. The term *potentiometer sweep test* refers to the testing of a potentiometer throughout its range of operation to determine if any electrical opens exist. Additionally, if a potentiometer loses its ground, the computer will sense full reference voltage in all positions of the wiper because the voltage-sensing circuits of the computer do not load the potentiometer circuit.

Pressure Sensors

Pressure sensors are commonly used to monitor intake manifold pressure and/or **barometric pressure**. Intake manifold pressure is a direct response to throttle position and engine speed, with throttle position being the most significant factor. The greater the throttle opening, the greater the manifold pressure becomes (lower vacuum). At WOT manifold pressure is nearly 100 percent of atmospheric pressure, reduced only by the slight energy consumed by friction with the intake channel walls. Intake manifold pressure can therefore be translated as close to engine load, and it is a critical factor in determining many engine calibrations. Barometric pressure is the actual ambient pressure of the air at the engine. It also impacts on manifold pressure and in most systems is considered when calculating engine calibrations such as air/fuel mixture and ignition timing.

Intake manifold pressure (usually a negative pressure or *vacuum* unless the vehicle is equipped with a turbocharger or supercharger) and/or barometric pressure are, on most systems, measured with a silicon diaphragm that acts as a resistor, Figure 2–14. The resistor/diaphragm (about 3 mm wide) separates two chambers. As the pressure on the two sides of the resistor/diaphragm varies, it flexes. The flexing causes the resistance of this semiconductor material to change. The computer applies a reference voltage to one side of the diaphragm. As the current crosses the resistor/diaphragm, the amount of voltage drop that occurs depends on how much the diaphragm is flexed. The signal is passed through a filtering circuit before it is sent to the computer as a DC analog signal. The computer monitors the voltage drop and, by using a look-up chart of stored pressure values, determines from the returned voltage the exact pressure to which the diaphragm is responding.

Two slightly different pressure sensor designs use the **piezoresistive** silicon diaphragm: the absolute pressure and the pressure differential sensors, Figure 2–15. In one design, the chamber under the diaphragm is sealed and contains a fixed reference pressure. The upper chamber is exposed to either intake manifold pressure or to atmospheric pressure. If the upper chamber is connected to the intake manifold, the sensor functions as a manifold absolute pressure sensor. Absolute, as used in manifold absolute pressure (MAP), refers to a sensor that compares a varying pressure to a fixed pressure. The output signal (return voltage) from this sensor

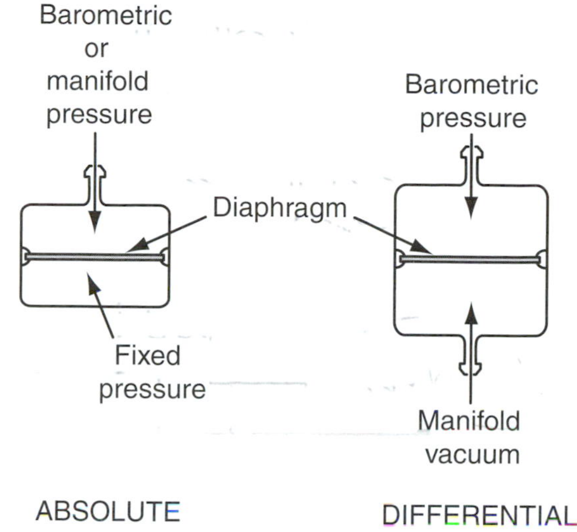

Figure 2–15 Two types of silicon diaphragm pressure sensors.

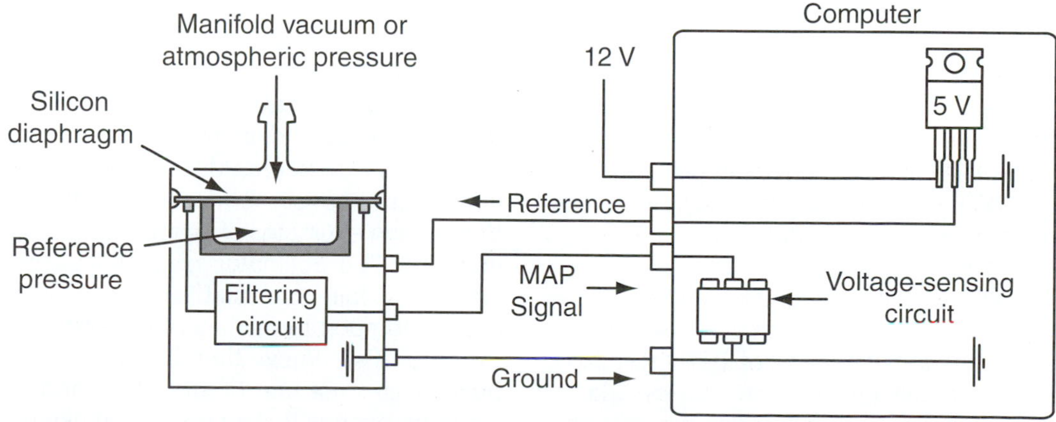

Figure 2–14 Silicon diaphragm pressure sensor.

increases as pressure on the variable side of the diaphragm increases (wider throttle opening for the engine speed). If the upper chamber is exposed to atmospheric pressure, the sensor functions as a barometric pressure sensor. Many systems take an initial MAP sensor reading as a reference barometric pressure reading for the next driving trip.

The differential pressure sensor combines the functions of both the MAP and the BARO sensors. Instead of using a fixed pressure, one side of the resistor/diaphragm is exposed to barometric pressure and the other side is connected to intake manifold pressure. The output signal is the result of subtracting manifold pressure from barometric pressure. The output signal is opposite to the MAP sensor, however, because as manifold pressure decreases (higher vacuum), output voltage increases. The pressure differential sensor had limited use on some early General Motors Computer Command Control systems.

Other types of pressure sensors are discussed in the appropriate chapters.

Speed Density Formula. To know how much fuel to meter into the cylinder, the microprocessor must know how much air, as measured by weight (more accurately, by mass), is in the cylinder. Since the microprocessor has no way to actually weigh the air going into a cylinder, some other method must be used to determine this value. Many systems use a mathematical calculation called the ***speed density formula***.

$$\frac{EP \times EGR \times VE \times MAP}{AT} = \text{air density in cylinder}$$

where: MAP = manifold absolute pressure
VE = volumetric efficiency
EP = engine parameters
EGR = EGR flow
AT = air temperature

To make this calculation, all of the factors that influence how much air gets into the cylinders must be considered. Throttle position and engine speed affect air intake, and engine temperature (measured as coolant temperature) affects how much heat the air gains as it passes through the induction system and thus affects the air's density in the cylinder. These factors are grouped together and called engine parameters. The microprocessor gets these pieces of information from some of its sensors that are common to all systems.

The amount of air getting into the cylinders is reduced by the amount of exhaust gas that the EGR valve meters into the induction system. Some systems use a sensor to determine EGR flow, and some use estimates (based on engine parameters) of EGR flow that are stored in computer memory.

Volumetric Efficiency. You know the diameter, length and shape of the intake manifold runners; the valve size, lift and timing; the combustion chamber design; and the cylinder size and compression ratio all affect how much air can get into the cylinders. This group of factors is called ***volumetric efficiency***. A volumetric efficiency value for any given set of engine parameters is stored in the computer's memory. In principle, volumetric efficiency is the amount of air that gets into the cylinder by the time the intake valve closes, divided by the theoretic amount displaced by the moving piston. Ordinarily this is less than 100 percent, but the volumetric efficiency varies with engine speed and load. With careful use of valve overlap, volumetric efficiency can sometimes go just above 100 percent for certain engine speeds. Maximum engine torque, of course, occurs just at the highest volumetric efficiency for a given engine.

Air Temperature. The air's temperature affects its density and, therefore, how much air can get into the cylinders. Most systems use an air temperature sensor to measure air temperature. Some have estimates of air temperature stored in memory. The estimates of air temperature are based on a fairly predictable relationship between coolant temperature and air temperature.

You also know that barometric pressure pushes air past the throttle valve and manifold pressure pushes that air past the intake valve into the cylinder. The last known values that must be determined are how much air the engine will inhale during any given set of operating conditions, the barometric pressure and the manifold pressure. The MAP sensor, as you will see in later chapters, almost always provides both pressure values. A MAP sensor identifies those systems that use the speed density formula.

Measuring Air Mass

Port fuel injection or multipoint injection has become the predominate fuel-metering system. Although a multipoint injection system offers several distinct advantages over other types of fuel-metering systems (discussed earlier in this chapter), it has one shortcoming. A multipoint injection system provides less opportunity (in time duration) for the fuel to evaporate than does a system where the fuel is introduced earlier in the intake channels, into the center of the intake manifold, and only the fuel that evaporates before combustion occurs is usable. In the past, this problem was probably most apparent as a lean stumble when accelerating from idle and was the result of the high manifold pressure that comes from the sudden increase in throttle opening. With the throttle open, atmospheric pressure forces air into the manifold faster than the engine can use it. The turbulence produced during the intake and compression strokes (and the increased oxygen density on turbocharged applications) helps to atomize the fuel and hasten its evaporation. This condition is only completely overcome, however, by spraying in additional fuel to compensate for the failure of the heavier hydrocarbon molecules to evaporate. The capability of the digital microprocessor provides the ability to calculate just the right amount of fuel to avoid a lean stumble without sacrificing economy or emissions.

For an engine at a given operating temperature and an identified set of atmospheric conditions (air temperature, pressure, and humidity), there is one precise amount of fuel that should be injected into the intake port for each manifold pressure value within the operating range. This amount will provide an evaporated 14.7 to 1 air/fuel ratio in the combustion chamber with the lowest possible left-over unevaporated, unburned HC.

Mass Airflow (MAF) Sensor. The MAP sensor used with the speed density formula has been pretty effective in enabling the microprocessor to accurately determine the mass of the air that goes into the cylinder. Keep in mind that the microprocessor quantifies both the air and the fuel by weight when calculating air/fuel ratios. The vane meter has also been fairly effective in providing the microprocessor with the needed information to calculate the air's mass. The vane meter is discussed in the **Inputs** section of the appropriate chapter. Each of these two methods have, however, been unable to measure one factor that affects the air's mass—humidity.

The MAF sensor, introduced in the mid-1980s, provides information to the microprocessor that accounts for the airflow's rate and density (including those things that affect the air's density: temperature, pressure and humidity). An MAF sensor can actually determine the airflow mass by *measuring* the quantity of air molecules that enter the engine (as opposed to a speed density application wherein the computer *calculates* how much air enters the engine). It does this by measuring the air's ability to cool. Air flowing over an object that is at a higher temperature than the air carries heat away from the object. The amount of heat carried away depends on several factors: the difference in temperature between the air and the object, the object's ability to conduct heat, the object's surface area, the air's mass and its flow rate. An MAF sensor tends to retain a higher level of accuracy as the engine wears than a speed density system. This is because as an engine wears, it tends to develop less manifold vacuum, causing the MAP sensor on a speed density application to interpret this (when compared to barometric pressure) as increased engine load.

The MAF sensor is placed in the duct that connects the air cleaner to the throttle body so that all of the air entering the induction system must pass through it, Figure 2–16. In the MAF sensor's main passageway is a smaller passage that a fixed percentage of the air passes through.

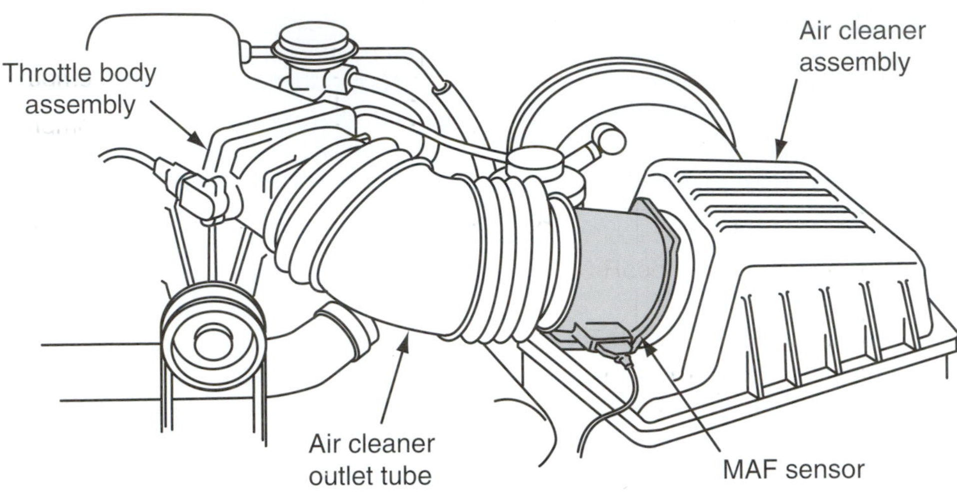

Figure 2–16 MAF sensor location.

In the smaller passage is a thermistor and a heated wire, Figure 2–17. The thermistor measures the incoming air temperature. The heated wire is maintained at a predetermined temperature above the air's temperature by a small electronic module mounted on the outside of the sensor's body.

The heated wire is actually one of the resistors of a balanced bridge circuit. Figure 2–18 represents a simplified MAF sensor circuit. The module supplies battery voltage to the balanced bridge. Resistors R1 and R3 form a series circuit in parallel to resistors R2 and R4. The voltage at the junction between R1 and R3 is equal to the voltage drop across R3. Because the values of R1 and R3 are fixed, the voltage at this junction is constant; it will vary, but only as battery voltage varies. Resistors R1 and R2 have equal values. With no airflow across R4, its resistance is equal to R3, and the voltage at the junction between R2 and R4 is equal to the voltage at the R1/R3 junction.

As airflow across R4 increases, its temperature drops; and because it is a positive temperature coefficient resistor, its resistance goes down with its temperature. As the air's mass increases, R4's resistance goes down even more. As the resistance of R4 goes down, a smaller portion of the voltage applied to the circuit formed by R2/R4 is dropped across R4. Therefore, a larger portion of the voltage is dropped across R2. So, as the resistance of R4 goes down, the voltage at the R2/R4 junction goes down. The computer reads the voltage difference between the R1/R3 junction and the R2/R4 junction as the MAF sensor value. By looking at a table in the computer's memory, the microprocessor converts the voltage value to a mass airflow rate.

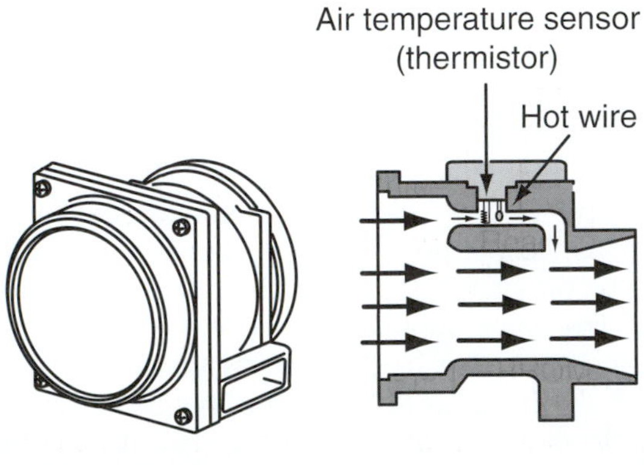

Air temperature sensor (thermistor)

Hot wire

Figure 2–17 Components of a hot wire-type mass airflow sensor.

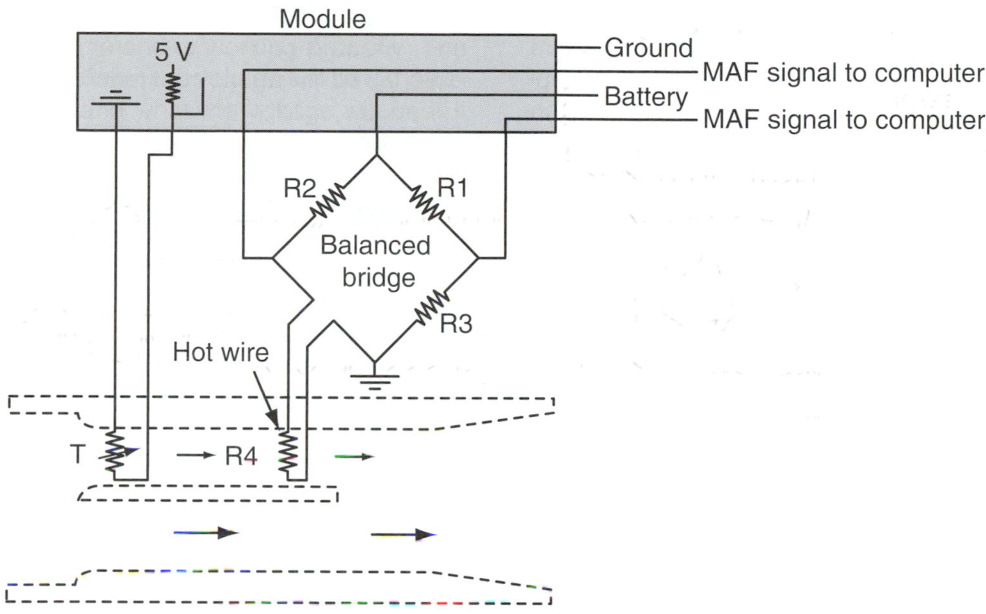

Figure 2–18 MAF sensor circuit.

Magnetic Sensors

Magnetic sensors, also sometimes known as permanent magnet sensors or variable reluctance sensors, consist of a coil of wire wrapped around a permanent magnet. Magnetic sensors are voltage-generating devices. A permanent magnet sensor operates on the concept that magnetic fields are able to cause electron flow as they expand or collapse across a coil of wire. A series of iron teeth and notches are passed near the magnet. As an iron tooth approaches the magnet, the magnet's magnetic field strength is increased and a voltage potential is generated within the coil of wire that is wrapped around the magnet. Then as the iron tooth pulls away from the magnet, the magnet's magnetic field strength is decreased and a voltage potential of the opposite polarity is generated within the coil of wire. As a result, a magnetic sensor produces an AC voltage. Generally, the computer is not so concerned with how much voltage is produced (as long as enough voltage is produced to make the signal recognizable to the computer), but it is instead more concerned with the timing of the AC voltage pulse and/or the frequency of the voltage pulses.

Testing of a magnetic sensor consists of using an ohmmeter to measure the resistance of the sensor's coil of wire and comparing the result to the sensor's specification. Then a voltmeter should be used (on an AC voltage scale) to test the sensor's ability to produce a voltage signal while the sensor is operating. Also, if the sensor has a metallic body, an ohmmeter should be used to verify that the sensor's coil of wire is not grounded (shorted) to the sensor's frame (this step is not important with plastic-bodied magnetic sensors).

Permanent magnet sensors are commonly used as ignition pickups and vehicle speed sensors, and are used in many other applications as well.

Hall Effect Switches

A **Hall effect switch** is frequently used to sense engine speed and crankshaft position. It provides a signal each time a piston reaches top

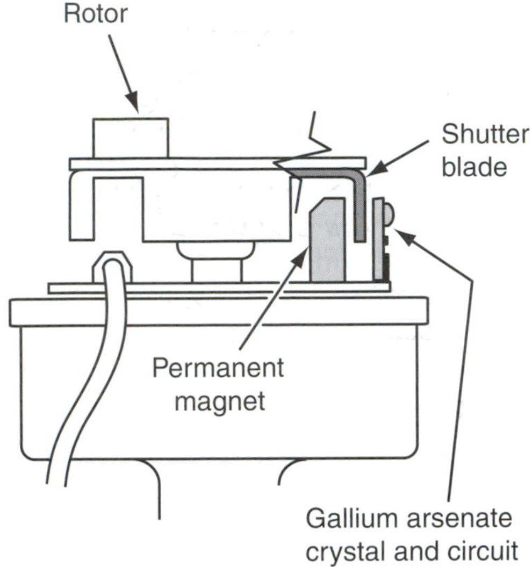

Figure 2–19 Hall effect switch.

dead center, and its signal serves as the primary input on which ignition timing is calculated. Because of the frequency with which this signal occurs and the critical need for accuracy, the Hall effect switch is a popular choice because it and its related circuitry provide a digital signal. On some applications, a separate Hall effect switch is used to monitor camshaft position as well.

A Hall effect switch consists of a permanent magnet, a **gallium arsenate crystal** with its related circuitry and a shutter wheel, Figure 2–19. The permanent magnet is mounted so that a small space is between it and the gallium arsenate crystal. The shutter wheel, rotated by a shaft, alternately passes its vanes through the narrow space between the magnet and the crystal, Figure 2–20. When a vane is between the magnet and the crystal, the vane intercepts the magnetic field and thus shields the crystal from it. When the vane moves out, the gallium arsenide crystal is invaded by the magnetic field. A steady current is passed through the crystal from end to end. When the magnetic lines of force from the permanent magnet invade the crystal, the current flow across the crystal is distorted. This results in a weak voltage potential being produced at the crystal's top and bottom surfaces—negative at one surface and positive at the other. As the shutter wheel turns, the crystal provides a weak high/low voltage signal.

This signal is usually modified before it is sent to the computer. For example, as shown in Figure 2–20, the signal is sent to an amplifier, which strengthens and inverts it. The inverted signal (high when the signal coming from the Hall effect switch is low) goes to a **Schmitt trigger** device,

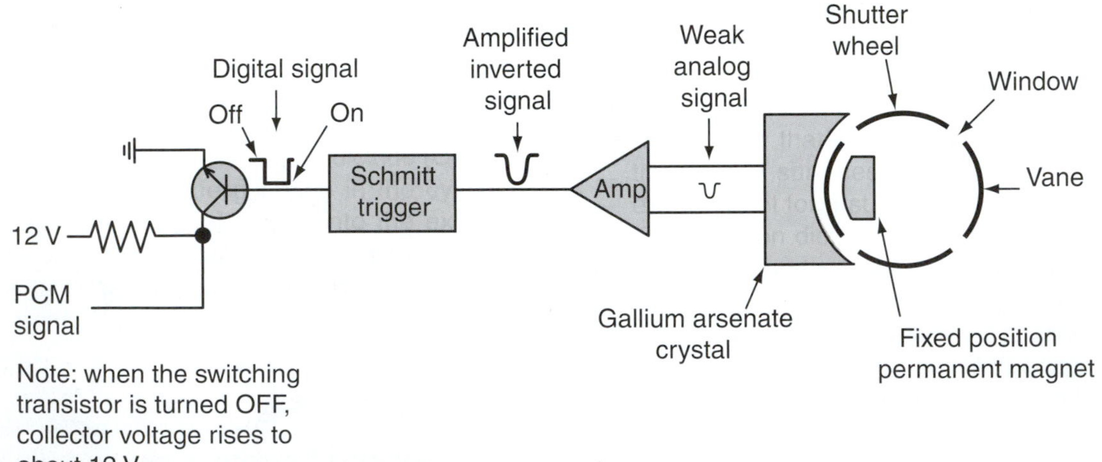

Figure 2–20 Hall effect switch circuit.

which converts the analog signal to a digital signal. The digital signal is fed to the base of a switching transistor. The transistor is switched on and off in response to the signals generated by the Hall effect switch assembly.

When the transistor is switched on, current flows through it from another circuit to ground. The other circuit can come from the ignition switch or from the computer, as shown in Figure 2–20. In either case it has a resistor between its voltage source and the transistor. A voltage sensing circuit in the computer connects to the circuit between the resistor and the transistor. When the transistor is on and the circuit is complete to ground, a volt drop occurs across the resistor. The voltage signal to the voltage sensing circuit is less than 1 volt. When the transistor is switched off, there is no drop across the resistor; and the voltage signal to the voltage-sensing circuit is near 12 volts. The computer monitors the voltage level and can determine by the frequency at which it rises and falls what engine speed is. Each time it rises, the computer knows that a piston is approaching top dead center.

Detonation Sensors

To optimize performance and fuel economy, ignition timing must be adjusted so that combustion will occur during a specific number of crankshaft degrees beginning at top dead center of the power stroke. If it occurs any later, less pressure will be produced in the cylinder, and less useful work will be done for the amount of fuel burned. If it occurs any sooner, detonation will occur. Due to variations in fuel quality, cylinder cooling efficiency, machining tolerances and their effect on compression ratios, a preprogrammed spark advance schedule can result in spark knock under certain driving conditions. To provide an aggressive spark advance and avoid spark knock, many systems use a detonation sensor, often referred to as a *knock sensor*. It alerts the computer when spark knock occurs so that the timing can be retarded.

The knock sensor, Figure 2–21, screws into some section of the engine where it will be subjected to the high-frequency vibration caused by the spark knock. Most knock sensors contain a **piezoelectric** crystal. Piezoelectric refers to a characteristic of certain materials whereby they produce an electrical signal in response to physical stress, or experience stress in response to an electrical signal. The most commonly used piezoelectric material produces an oscillating voltage signal of about 0.3 volt or more in response to pressure or vibration. The oscillations occur at the same frequency as the spark knock (5,000 to 6,000 **hertz [Hz]**, or cycles per second. Spark knock is essentially the ringing of the engine block when struck by the explosion of detonation). Although most vibrations within the engine and many elsewhere in the vehicle cause the knock sensor to produce a signal, the computer only responds to signals of the correct frequency.

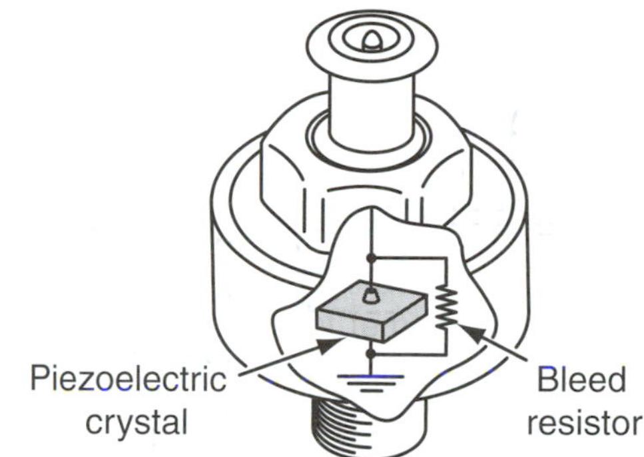

Figure 2–21 Knock sensor.

Switches

Most systems have several simple switches that are used to sense and communicate information to the computer. The information includes whether the transmission is in gear, the air-conditioning is on, the brakes are applied and so on.

Two types of circuits are used by a computer to sense switch inputs. In the most common, the switch completes the path to ground when closed and therefore pulls applied reference voltage low. Because closing the switch pulls the voltage low, this is known as a *pull-down* circuit, Figure

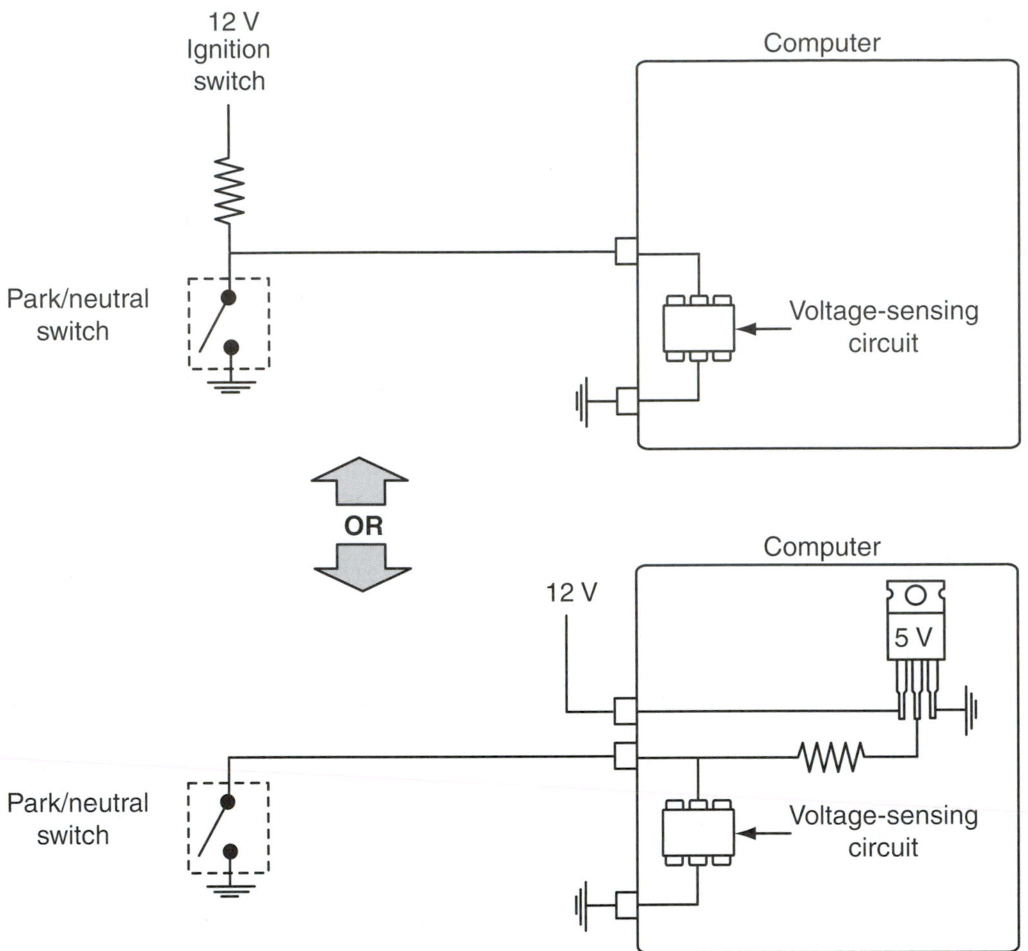

Figure 2–22 Pull-down circuit.

2–22. In the other, a switch applies voltage to a computer-monitored circuit. Because closing the switch pulls voltage high, this is known as a *pull-up* circuit, Figure 2-23.

E-Cells

Some selected systems use an **E-cell** to indicate to the computer when a specified amount of engine operation has occurred. The E-cell actually measures a specific amount of ignition on-time, which equates to an estimated number of vehicle miles. Its purpose is to alert the computer when the predetermined mileage has occurred so that the computer can adjust specific calibrations to compensate for predictable engine wear.

The E-cell contains a silver cathode and a gold anode. Ignition voltage is applied to the E-cell through a resistor. The passage of current through the cell controls a signal to the computer. As current passes through the cell, a chemical reaction takes place; the silver on the cathode is attracted to the gold anode. After a certain amount

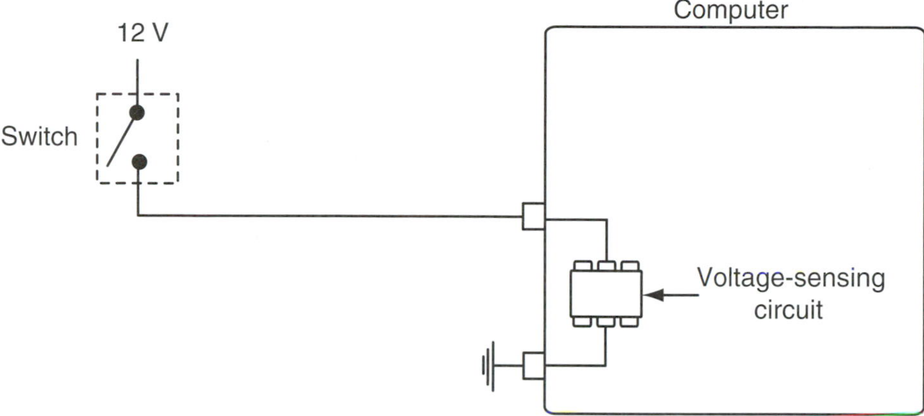

Figure 2–23 Pull-up circuit.

of on-time, the silver cathode is completely depleted and the circuit opens. How long this takes is controlled by the amount of current allowed to flow through the E-cell (how much resistance is in the circuit).

ACTUATORS

Solenoids

The most common actuator is a valve powered by a solenoid. The valve usually controls vacuum, but it can also be a valve designed to control fuel vapor, airflow, oil or water flow and so forth. A solenoid consists of a coil of wire with a movable iron core, Figure 2–24. When the solid-state switching circuit in the computer grounds the solenoid circuit, the coil becomes energized. The magnetic field developed by the current passing through the coil windings pulls the iron core so that it occupies the entire length of the coil. This movement of the core opens the valve attached to its other end.

Duty-cycled and Pulse Width Solenoids. In some cases, the switching circuit that drives the solenoid turns on and off rapidly. If the solenoid circuit is turned on and off (cycled) 10 times per second, then the solenoid is duty cycled. To complete a cycle, it must go from off to on, to off again. If it is turned on for 20 percent of each tenth of a second and off for 80 percent, it is said to be on a 20 percent **duty cycle**. In most cases where a solenoid is duty-cycled, the computer has the ability to change the duty-cycle to achieve a desired result. The best example of this is the mixture control device on many computer controlled carburetors, where the duty-cycle will frequently change to control the air/fuel mixture.

If a solenoid is turned on and off rapidly but there is no set number of cycles per second at which it is cycled, the on-time is referred to as pulse width. For example, let's assume that the computer has decided that the EGR valve should be open, but that it should be open to only 60 percent of its capacity. The computer can energize a solenoid valve such as the one in Figure 2–24 to allow vacuum to the EGR valve. To control or modulate the amount of vacuum to the EGR valve, the computer issues pulse width commands to a solenoid that when energized opens a port that allows atmospheric pressure to bleed into the EGR vacuum line, Figure 2–25. Because the computer wants the EGR valve open to 60 percent of its capacity, it selects pulse widths from a look-up chart to appropriately weaken the vacuum signal to the EGR valve.

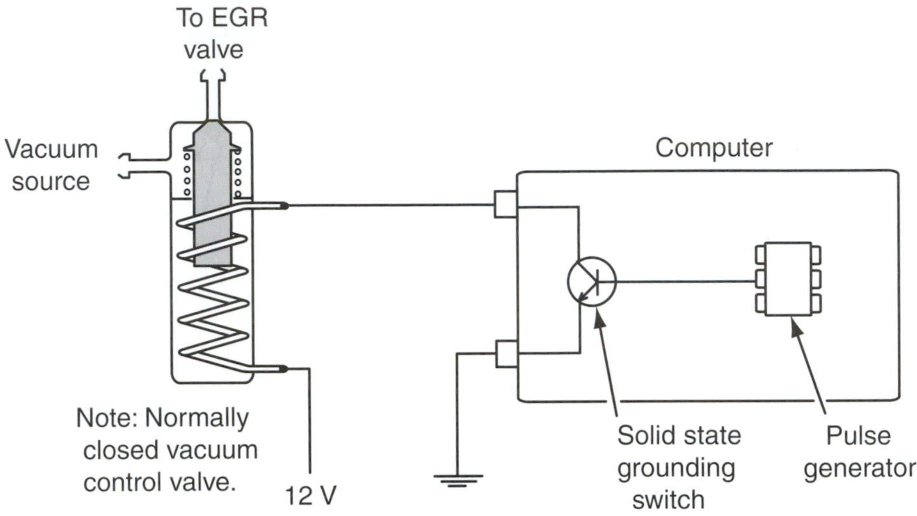

Figure 2–24 Typical solenoid-controlled vacuum valve.

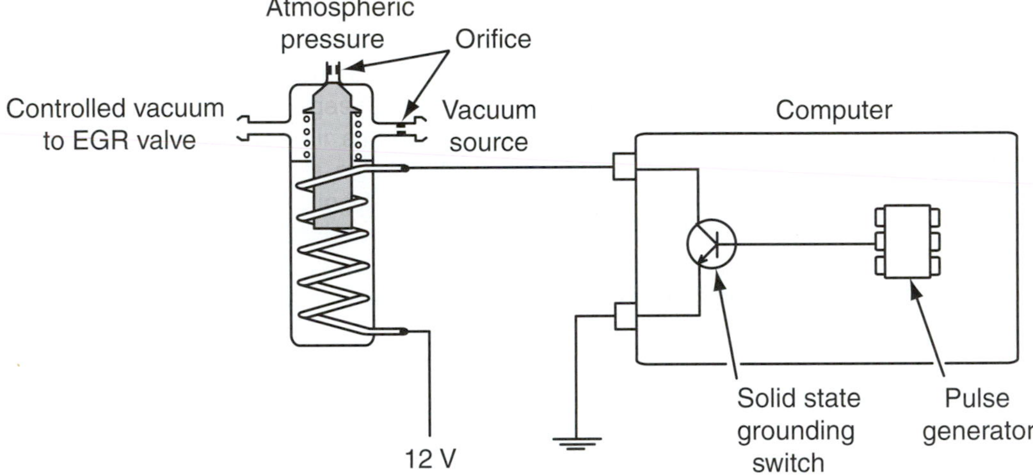

Figure 2–25 Typical pulse-width controlled vacuum valve.

Relays

A relay is a remote-control switch that allows a light-duty switch such as an ignition switch, a blower motor switch, a starter switch or a driver switch in a computer to control a device that draws a relatively heavy current load. The most common example in automotive computer application is a fuel pump relay, Figure 2–26. When the computer wants the fuel pump on, it turns on its driver switch (transistor). The transistor applies 12 volts to the relay coil, which develops a magnetic field in the core. The magnetic field pulls the armature down and closes the normally open contacts. The contacts complete the circuit from

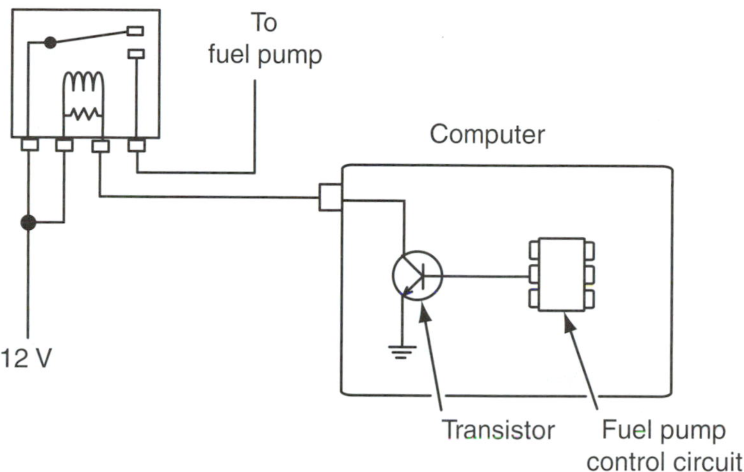

Figure 2–26 Typical fuel pump relay circuit.

the battery to the fuel pump. On some systems, the fuel pump relay is one of the few examples where the computer supplies the voltage to operate a device instead of the ground connection.

Electric Motors

When electric motors are used as actuators, two unique types are used.

Stepper Motors. A **stepper motor**, as used in these applications, is a small electric motor powered by voltage pulses. It contains a permanent magnet armature and usually either four or two field coils, Figure 2-27. In the design with four field coils, all coils receive battery voltage and the computer completes the ground paths, one at a time, to pulse current to each coil in sequence. In the design with two field coils, the computer not only pulses current to energize each coil in sequence, but it also can reverse the polarity applied to each coil as part of the sequencing. Each time the computer pulses current to a field coil, it causes the stepper motor to turn a specific number of degrees. In either design, reversing this sequence causes the stepper motor to turn in the opposite direction. Ultimately, a

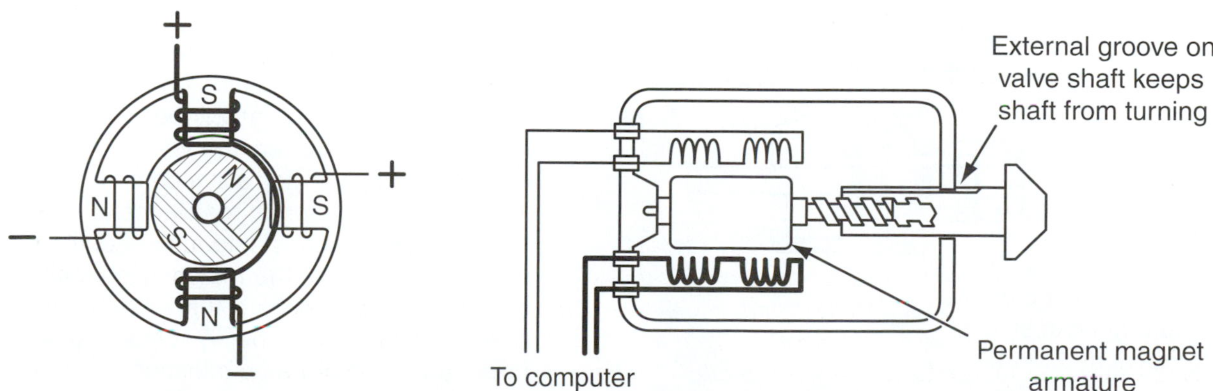

Figure 2–27 Typical stepper motor with two field coils.

stepper motor is controlled by the computer to turn a number of specific steps or increments, with the direction being controlled by the sequencing of the current pulses to the windings.

The armature shaft usually has a spiral on one end and this spiral connects to whatever the motor is supposed to control. As the motor turns one way, the controlled device, a pintle valve for instance, is extended. As it turns the other way, the valve is retracted. The computer can apply a series of pulses to the motor's coil windings in order to move the controlled device to whatever location is desired. The computer can also know exactly what position the valve is in by keeping count of the pulses applied. The stepper motor has been used to control air/fuel mixture and idle speed.

Permanent Magnet Field Motors. Some systems use this type of motor to control idle speed, Figure 2–28. It is a simple, reversible, DC motor with a wire-wound armature and a permanent magnet field. The polarity of the voltage applied to the armature winding determines the direction in which the motor spins. The armature shaft drives a tiny gear drive assembly that either extends or retracts a plunger. The plunger contacts the throttle linkage; and when it is extended, the throttle blade opening is increased. When it is retracted, the throttle blade opening is decreased. The computer has the ability to apply a continuous voltage to the armature until the desired idle speed is reached.

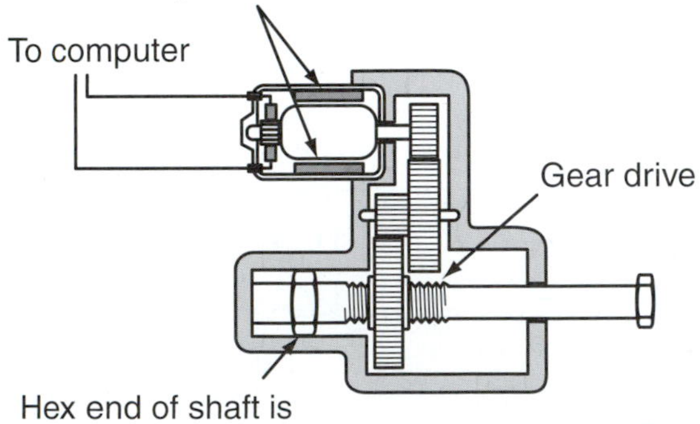

Figure 2–28 Permanent magnet field reversible DC motor.

✔ SYSTEM DIAGNOSIS AND SERVICE

Special Tools

A good service manual that applies specifically to the vehicle being serviced is essential. It should contain all of the necessary charts and diagnostic procedures and should list the tools necessary for diagnostic operations.

Most systems have a data link used in the factory to test the system's operation as it leaves the assembly line. Several companies have designed and are selling test equipment that can connect to the data link and get service code information from the computer. In most cases other information such as sensor voltage values and the state of position of some actuators can also be obtained. This information is either shown as a digital readout on the test instrument or is printed on a piece of paper. This type of tool is very helpful and in some cases is essential.

High-Impedance Digital Volt-Ohmmeter (DVOM). When diagnosing most systems, a DVOM is required for two reasons:

1. Many of the readings to be taken require a higher degree of accuracy than is obtained by reading a needle position on a scale.
2. Many of the circuits that will be tested have high resistance and very low voltage and amperage values. Applying a standard meter with its relatively low internal resistance to such a circuit can easily constitute a short to the circuit. In other words the meter's circuit offers an easier electrical path than the circuit or portion of the circuit being tested. A high-**impedance** meter has a minimum of 10 million ohms of internal resistance; a standard meter has only about 10 thousand.

Diagnostic & Service Tips

Oxygen Sensor. When handling or servicing the oxygen sensor, several precautions should be observed:

- A boot is often used to protect the end of the sensor where the wire connects. It is designed to allow air to flow between it and the sensor body. The vent in the boot and the passage leading into the chamber in the oxygen-sensing element must be open and free of grease or other contaminants.
- Those with soft boots must not have the boot pushed onto the sensor body so far as to cause it to overheat and melt.
- No attempt should be made to clean the sensor with any type of solvent.
- Do not short the sensor lead to ground, and do not put an ohmmeter across it. Any such attempt can permanently damage the sensor. The single exception is that a digital high-impedance (10 meg ohm minimum) voltmeter can be placed across the sensor lead to ground (the sensor body).
- If an oxygen sensor is reinstalled, its threads should be inspected and if necessary cleaned with an 18 mm spark plug thread chaser. Before installation, its threads should be coated with an antiseize compound, preferably liquid graphite and glass beads. Replacement sensors may already have the compound applied.

WARNING: It is often recommended that the oxygen sensor be removed while the engine is hot to reduce the possibility of thread damage. If doing so, wear leather gloves to prevent burns on the hand.

Vacuum Leaks. Vacuum leaks can have far-reaching effects. Examine Figure 2–29. Assume that the diaphragm ruptured in the primary vacuum break (or it could be a vacuum hose cracked or left off anywhere in that circuit). The results would be as follows:

- The primary vacuum break would not work.
- The secondary vacuum break would not work (rich condition, possibly stalling during warmup).
- The heated air system would not work (poor cold driveability).
- The vacuum sensor would report "no vacuum" to the computer, which would in turn issue erroneous spark and fuel mixture commands.
- A vacuum sensor code would probably be set.

Vacuum leaks downstream from the airflow sensor on a multipoint injection system cause poor idle if the engine idles at all.

Charcoal Canister. Charcoal canister saturation can often be the result of fuel tank overfilling. Slowly filling the tank right to the top of the filler neck causes the antioverfill chamber in the tank to fill and thus leaves no room for fuel expansion. When the fuel does expand, it is pushed up and into the charcoal canister. Canister purging then dumps so much hydrocarbon into the intake manifold that the system may not be able to compensate, even in closed loop. Many driveability problems can be linked to charcoal canister problems; do not overlook it in diagnostic procedures.

Avoid Damaging the Computer. When servicing a vehicle with a computerized engine control system, observe the following recommendations in order to reduce the potential for damage to the computer:

- Avoid starting the vehicle with the aid of a battery charger. Finish charging the battery first, then disconnect the charger before starting.
- Turn the ignition off before connecting jumper cables.

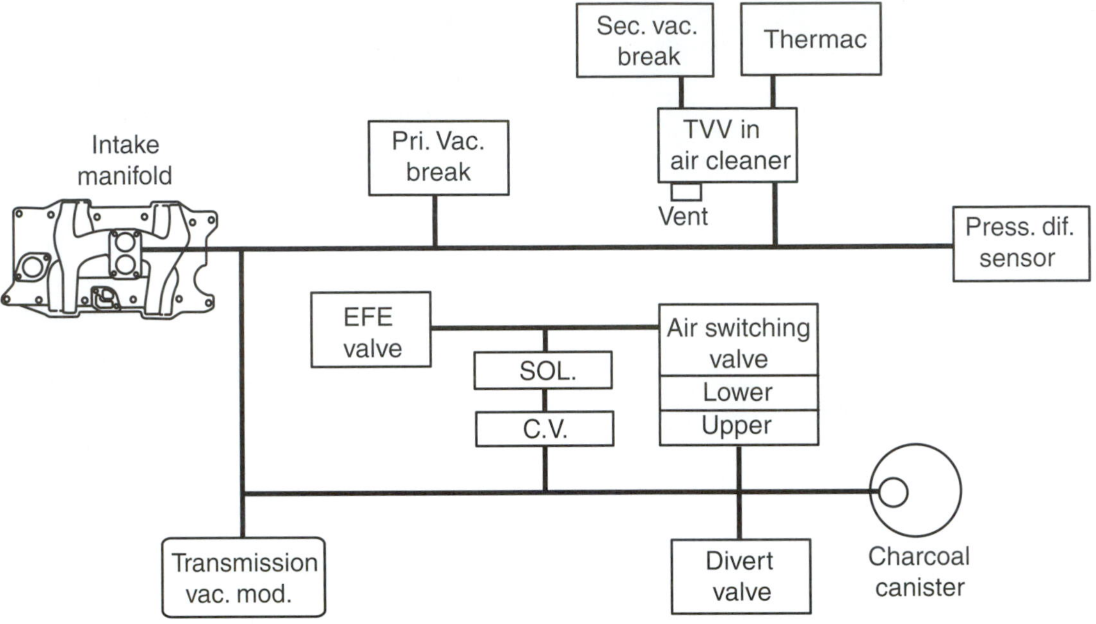

Figure 2–29 Typical vacuum circuit.

- Turn the ignition off before disconnecting the battery or computer connectors.
- Replace any system actuator with less than specified resistance or one that has a shorted clamping diode. A **clamping diode** is a diode placed across an electrical winding to suppress voltage spikes that are produced when the winding is turned on and off.
- Be sure that the winding of any electrical accessory added to the vehicle has sufficient voltage spike containment capability. A resistor of 200- to 300-watt capacity can be used instead of a diode in some cases.
- Use caution when testing the charging system, and never disconnect the battery cables with the engine running.

SUMMARY

This chapter has described single-point and multipoint fuel injection systems. The single-point systems use a single fuel injector, or often a pair of two injectors for a V-form engine, to deliver the fuel upstream of the throttle blade. These systems resemble an electronic carburetor. The multipoint systems use an injector for each cylinder, allowing somewhat more precise fuel mixture delivery. Each system varies the amount of fuel delivered by varying the pulse width of the injector, the time the injector solenoid is on and fuel sprays. Multipoint systems may spray individually or in cylinder banks in synchronization with the camshaft speed, varying quantity by pulse width modification.

We have learned how the computer system's most important sensors work, as variable resistors, thermistors varying their internal electrical resistance or as signal generators. The oxygen sensor produces a voltage corresponding to the amount of oxygen remaining in the exhaust gas; the crankshaft position sensor sends a signal corresponding to the position and speed of the crankshaft; the temperature sensors modify a reference voltage to match their temperature; and the throttle position sensor varies its return voltage signal through a variable potentiometer, so the computer knows exactly where the throttle is and how fast it has moved recently.

We reviewed both magnetic sensors and Hall effect sensors. The magnetic sensors use a magnet and a coil to produce an alternating current corresponding to the speed of a crankshaft or camshaft. The Hall effect sensor functions basically as an on-off switch, signaling the same sort of information. We have also seen how simple on-off switches can provide information on subjects like the engagement or disengagement of the air conditioner compressor or when the throttle closes completely, leaving the engine speed under the computer's control. Finally, we have seen how the computer operates the actuators not by powering them directly but by providing them an electric path to ground, thus protecting the computer itself from shorts in the circuit.

▲ DIAGNOSTIC EXERCISE

A car is brought into the shop with various running and driveability concerns. Among the things the technicians notice is a great deal of rust throughout the car, including on the connections for the electrical system. What kinds of problems would you anticipate might occur as a result of this kind of excessive rust formation? How would you check circuits to see how much, if at all, they were affected? How would you correct the electrical resistance problems that would occur from the rust?

REVIEW QUESTIONS

1. *Technician A* says that most often the computer is designed to control the ground side of an actuator.
   Technician B says that most often the engine computer is located in the trunk or luggage compartment of the vehicle.
   Who is correct?
   A. A only
   B. B only
   C. Both A and B
   D. Neither A nor B
2. With few exceptions, the reference voltage that engine computers send to the sensors is equal to which of the following?
   A. 1 volt
   B. 5 volts
   C. 12.6 volts
   D. 14.2 volts
3. Which of the following sensors produces or generates a voltage signal?
   A. Oxygen sensor
   B. Thermistor
   C. Potentiometer
   D. Piezoresistive silicon diaphragm
4. Zirconium dioxide is a substance that is found in which of the following?
   A. Oxygen sensor
   B. Throttle position sensor
   C. Engine coolant temperature sensor
   D. Manifold absolute pressure sensor
5. During closed loop operation, the engine computer attempts to cause the oxygen sensor to average which of the following?
   A. 100 millivolts
   B. 450 to 500 millivolts
   C. 900 millivolts
   D. 5 volts
6. *Technician A* says that as the temperature of an NTC thermistor increases, its resistance decreases.
   Technician B says that the engine computer monitors the voltage drop across an NTC thermistor in a temperature-sensing circuit in order to know the measured temperature.
   Who is correct?
   A. A only
   B. B only
   C. Both A and B
   D. Neither A nor B
7. *Technician A* says that a temperature-sensing circuit is a current-divider circuit and has an NTC thermistor in parallel to a fixed resistance within the computer.
   Technician B says that a temperature-sensing circuit is a voltage-divider circuit and

has an NTC thermistor in series to a fixed resistance within the computer.
Who is correct?
A. A only
B. B only
C. Both A and B
D. Neither A nor B

8. A sensor that varies resistance in order to measure the physical position and motion of a component would be which of the following?
A. Thermistor
B. Potentiometer
C. Piezoresistive silicon diaphragm
D. Piezoelectric crystal

9. If a temperature-sensing circuit with an NTC thermistor were to develop an electrical open, the computer would see which of the following?
A. About –40°F
B. About 0°F
C. About 150°F
D. About 212°F

10. If a circuit with a potentiometer were to develop an electrical open in the ground wire, which of the following would be true concerning the voltage that the computer sees?
A. It would be steady at 0 volts
B. It would be steady at 2.5 volts (1/2 of reference voltage)
C. It would be steady at 5 volts (reference voltage)
D. It would vary normally with the mechanical position of the sensor

11. The speed density formula is a method of which of the following?
A. Calculating how much air is entering the engine
B. Measuring how much air is entering the engine
C. Calculating how much fuel is entering the engine
D. Measuring how much fuel is entering the engine

12. *Technician A* says that exhaust gasses that are metered into the intake manifold by the EGR valve reduce the amount of ambient air that gets into the cylinders.
Technician B says that some engine computers use sensors to determine EGR flow rates, while others assume EGR flow rates by using estimates of EGR flow that are stored in a computer memory.
Who is correct?
A. A only
B. B only
C. Both A and B
D. Neither A nor B

13. Which of the following sensors is used to measure the amount of intake air by measuring the cooling effect of the air that is entering the engine?
A. Oxygen sensor
B. NTC thermistor
C. MAP sensor
D. MAF sensor

14. *Technician A* says that a Hall effect switch is frequently used to sense engine coolant temperature.
Technician B says that a Hall effect switch is frequently used to sense engine speed and crankshaft position.
Who is correct?
A. A only
B. B only
C. Both A and B
D. Neither A nor B

15. What is a detonation sensor (knock sensor)?
A. A thermistor
B. A potentiometer
C. A piezoresistive silicon diaphragm
D. A piezoelectric crystal

16. *Technician A* says that when a switch is closed in a pull-down circuit, it will apply battery voltage to a computer input.
Technician B says that when a switch is closed in a pull-up circuit, it will apply a ground to a computer input.
Who is correct?
A. A only
B. B only
C. Both A and B
D. Neither A nor B

17. An E-cell is an input that is designed to inform the computer of which of the following?
    A. Vehicle speed
    B. Engine speed and crankshaft position
    C. When a specific amount of ignition on-time has occurred
    D. When a specific engine temperature has been reached
18. If a solenoid is turned on and off rapidly but there is no set number of cycles per second, what is its on-time referred to as?
    A. Percent duty cycle
    B. Pulse width
    C. Frequency
    D. Capacitance
19. *Technician A* says that a solenoid-operated valve may control flow of vacuum, fuel vapors, air or liquids.
    Technician B says that a relay is an electrically operated switch that allows a light-duty switch to control a relatively heavy current load.
    Who is correct?
    A. A only
    B. B only
    C. Both A and B
    D. Neither A nor B
20. *Technician A* says that air/fuel mixture can be controlled with a stepper motor.
    Technician B says that idle speed can be controlled with either a stepper motor or a permanent magnet field motor.
    Who is correct?
    A. A only
    B. B only
    C. Both A and B
    D. Neither A nor B

Chapter 3

Diagnostic Concepts

OBJECTIVES

Upon completion and review of this chapter, you should be able to:

- ❑ Describe the two types of faults associated with automotive computer systems.
- ❑ Describe the two types of diagnostic trouble codes associated with automotive computer systems.
- ❑ Understand the function and purpose of a data stream.
- ❑ Understand the various types of functional tests that may be programmed into an engine computer.
- ❑ Recognize the function of a pinpoint test.
- ❑ Recognize the function of a flowchart.
- ❑ Understand the function and purpose of an electrical schematic.

KEY TERMS

Data Stream
Diagnostic Trouble Code (DTC)
Electrical Schematic
Flowcharts
Functional Test
Hard Fault
Memory DTC
On-Demand DTC
Pinpoint Test
Powertrain Control Module (PCM)
Self-Test
Soft Fault

Understanding the product-specific diagnostic methods in the manufacturer-specific chapters of this textbook may be a difficult task and may seemingly require a large amount of memorization. But all of these methods are based on basic diagnostic concepts. This chapter is designed to cover some of the basic concepts involved in diagnosing modern electronic engine control systems. Understanding the material in this chapter will promote increased understanding for the reader when the manufacturer-specific chapters are read.

TYPES OF FAULTS

Soft Faults

A **soft fault** is commonly known as an *intermittent fault.* As a technician, you should always attempt to verify the symptom as the first step of diagnosis. If the symptom cannot be recreated, then you are likely dealing with a soft fault. Following are some realistic tips for helping you diagnose a soft fault.

Diagnosing Soft Faults. Is it possible that the engine computer, also known as the **Powertrain Control Module** or **PCM**, has identified the area of the problem and has set a diagnostic trouble code (DTC) in memory? For example, a symptom of stalling on a hot day, and then restarting after a 20-minute cool down period can be the result of a faulty ignition module. But it can also be the result of a faulty electric fuel pump or even a faulty fuel pump relay. Beginning in the mid- to late 1980s, feedback circuits were typically added to the PCM in order to assist the technician in identifying the problem area. Look for a memory fault code that might direct you to one area or the other.

Normal procedures used to test load components and their circuits may be used to diagnose an intermittent fault even when the symptom does not appear to be present. If the symptom (or memory DTC) leads you to suspect an intermittent fault with a particular circuit (for example, the fuel pump circuit), try performing the following procedures.

If the intermittent problem is a fault with the circuitry, rather than a fault in the load component, you may be able to diagnose it through voltage drop testing. For example, if the switch (or relay-operated switch) that controls a circuit has built up excessive resistance in the contacts, this resistance will likely increase with an increase in temperature. The increased resistance reduces current flow in the circuit and may result in an apparent symptom. By the time you attempt to verify the symptom, the switch contacts may have cooled, reducing the effective resistance and increasing the current flow enough that the symptom appears to have disappeared. However, bear in mind that unwanted resistance is still present in the switch contacts and, while the current flow has increased enough to allow the load component to operate, it is still below the designed current value for this circuit. At this point, a voltage drop check of the circuitry identifies the unwanted resistance, even with no apparent symptom present. This procedure can also identify intermittent faults associated with a loose or corroded connection in the circuitry.

If the intermittent problem is with a load component, such as a solenoid, relay or motor, current ramping (explained in Chapter 4) can help you identify the fault. If the intermittent fault is with an electronic module, such as an ignition module, use a choke tester to heat (and cool) the module while watching for the symptom to resurface. You may also tap on electronic devices in an attempt to verify the problem.

Hard Faults

Hard faults are faults currently present at the time that you attempt to verify the symptom. These faults are normally easier to diagnose than a soft fault, and simple use of the concepts explained in this chapter will usually identify the fault.

DIAGNOSTIC TROUBLE CODES (DTCs)

While it is commonly known that most PCMs can communicate to the technician numeric readouts that represent fault codes, some additional explanation is required in terms of the types of fault codes that can be obtained from a PCM. All manufacturers now refer to these fault codes as **diagnostic trouble codes** or **DTCs**.

Memory DTCs

A **memory DTC** is a DTC that the PCM stored in its RAM IC chip at some point in the past. The technician may be able to bring up the memory fault code either through manual code-pulling methods or with a scan tool, depending on the make and year of the vehicle. At some point, most PCMs are programmed to erase the memory fault on its own if it does not see the fault reoccur for a preset number of restarts of the engine, or, in many instances, a preset number of warmup and cool down cycles of the engine. A memory fault code could be either a soft fault or a hard fault. Just because it is set in the PCM's memory does not really prove whether the fault is still present. If the particular vehicle application only has memory code capability and more than one DTC has set in memory, you should erase all stored memory codes and road test the vehicle. Any DTCs that reset during the road test are probably hard faults. After repair of a memory DTC, the technician should take the proper steps to erase the DTC from the PCM's memory in order to avoid inadvertently chasing an "already-repaired DTC" at a later time. Then the vehicle should be road tested in order to verify that the DTC does not reset.

On-Demand DTCs

An **on-demand DTC** is set as a result of the PCM running a **self-test** (performance test) of the system at the instruction of the technician. A self-test is defined as a test that the computer is programmed to perform on the system that it controls. The technician may be able to initiate a self-test either through manual methods or with a scan tool, depending on the make and year of the vehicle. After the technician initiates the self-test, the PCM puts the system through a variety of tests as it checks each circuit's ability to function. The resulting DTCs are hard faults in that the PCM has identified their existence and immediately reported the results to the technician. A PCM on a given application may be programmed to perform Key On, Engine Off (KOEO) and Key On, Engine Running (KOER) self-tests. In this event, it is important to perform both tests. The engine should be properly warmed up prior to running this type of test and all accessories should be turned off. Also, in some cases, the technician will be asked to cycle certain switches at some point during the self-test in order to allow the PCM to check its operation during the test. The PCM will likely perform electrical checks of the circuitry during the KOEO self-test and will likely perform functional tests during the KOER self-test. For example, if the PCM controls the air injection system (described briefly in Chapter 1) through vacuum solenoids, it may energize and de-energize these solenoids during the KOEO self-test while monitoring voltage drop in these circuits. Then, during the KOER self-test, the PCM will perform a functional test of this system by functioning the solenoids and watching the oxygen sensor(s) in order to know when the system has switched air to the exhaust manifold(s) (upstream).

An on-demand DTC does not set in the PCM's memory; therefore, you cannot really erase it from memory in the way that you would erase a memory DTC. Instead, you must make the indicated repair and rerun the self-test. If the repair was successful, the DTC will not reset. Also, if the application has both self-test and memory code abilities (as with the Ford EEC IV and EEC V systems described in Chapters 14 and 15), a simple comparison of the types of faults received will allow you to know which faults are hard and which faults are soft without having to erase the memory faults and road test the vehicle. Remember, if a DTC is set as a result of a self-test, it is a hard fault. If a DTC has been set in memory and does not set as a result of a self-test, then it is likely a soft fault.

DATA STREAMS

A **data stream** is information about a computer system that is delivered via serial data from the PCM (or other control module) to your scan tool (see scan tools in Chapter 4). The PCM believes this information to be true about the system that it controls. There is no manual method to access this type of information as it comes from the PCM's memories in the form of binary code. This information may include DTCs, input voltages and/or PCM-interpreted values, output commands, fuel trim numbers (see Fuel Trim in Chapter 10), PROM ID number and other types of information.

FUNCTIONAL TESTS

Functional tests are tests that the PCM has been programmed to perform, but the test results are usually not rendered by the PCM. Most often, it is up to the technician to determine pass/fail of each functional test. Functional tests are usually used in specific circumstances to aid in diagnosing specific problems and are not used on every vehicle that comes in the door. Many times, a flowchart will guide the technician to a specific functional test. Through the use of a functional test, the technician may be able to command the PCM to energize/de-energize certain actuators or command the PCM to perform other test procedures such as an engine power balance test.

PINPOINT TESTING

After a DTC has been retrieved in order to direct the technician to a general area or circuit, then the identified circuit and component must be **pinpoint tested** in order to determine the exact fault. Bear in mind that connectors and wires can set the same DTCs that a defective input sensor or output actuator can set. Some early systems did not have any self-diagnostic ability programmed into the PCM, therefore requiring that the entire computer system should be pinpoint tested by the technician when a symptom is encountered. While this can be a huge undertaking, fortunately many of these early systems were not as complex as the later systems. Still, this can be quite a daunting task.

Pinpoint testing may include voltage drop testing the PCM's power and ground circuits with the PCM powered up, testing the circuitry that connects the PCM to each of its sensors and actuators and testing each of the sensors and actuators for proper operation. The pinpoint test procedure should ultimately determine the exact repair needed so that the technician can identify for the customer what is required in order to properly repair the vehicle. In many cases, the pinpoint test procedure may identify a particular fault that the customer needs to be made aware of before overall testing is complete. Because identified faults should be repaired before continuing with the diagnostic procedure, the technician may have to contact the customer more than once when multiple faults exist. For example, all on-demand faults associated with Ford's KOEO self-test should be repaired before attempting to initiate a KOER self-test on the Ford EEC IV and EEC V systems discussed in Chapters 14 and 15. As a result, multiple phone calls to the customer may be necessary.

CAUTION: When using a DVOM (digital volt-Ohmmeter) or DSO (digital storage oscilloscope) to pinpoint test a circuit, *never* insert a probe into a female terminal. You will enlarge it and thus create a loose connection when it is reconnected.

Voltage Drop Testing the PCM's Power and Ground Circuits

All too often, technicians overlook the voltage drop testing of the PCM's power and ground circuits, Figures 3–1 and 3–2. As an electronic device, if either the power or ground circuit is faulty, the PCM cannot function properly. The maximum battery-to-PCM power or ground voltage drop should never exceed 100 millivolts.

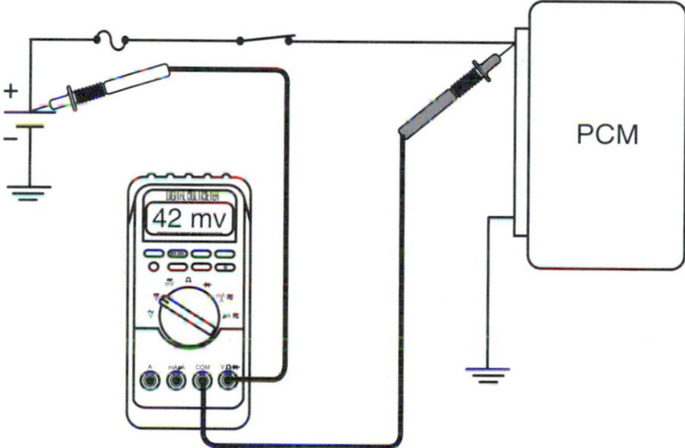

Figure 3–1 Performing a positive-side voltage drop test.

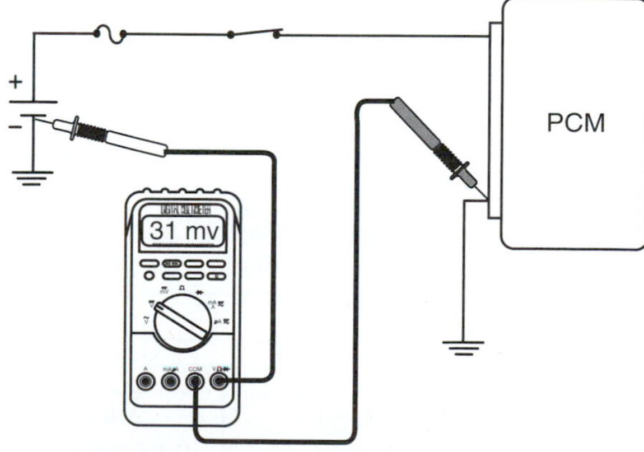

Figure 3–2 Performing a negative-side voltage drop test.

FLOWCHARTS

In most cases, either a symptom or a DTC can direct the technician to a **flowchart** in order to assist the technician in making pinpoint tests efficiently. A flowchart instructs the technician to perform a specific pinpoint test and note the results. Depending on the results, it may either direct the technician to another specific pinpoint test or, at some point, to make a specific repair. If all the steps in a flowchart are designed to fit on one page, it is commonly known as a *troubleshooting tree* Figure 3–3. Figure 3–4 shows a manufacturer-specific flowchart.

When following a flowchart, the technician should begin with the particular step as indicated by either the DTC or the symptom. It is critical at this point to read the flowchart's test instruction carefully and to perform the test exactly as indicated. It is equally important to be sure of the condition of any diagnostic tools used in each test step. If a required jumper wire has no continuity or if a DVOM lead has an intermittent connection, you may be falsely led to an incorrect test step, resulting in misdiagnosis of the system. Also, equally important, do not skip any test steps and assume their outcome. Some test steps may seem difficult to perform, but skipping a test step can also result in misdiagnosis of the system.

ELECTRICAL SCHEMATICS

When manufacturers design a flowchart, it is through their understanding of how the system is designed to operate that they are able to instruct the technician as to what tests should be performed for each of the possible symptoms and/or DTCs. Likewise, the technician should attempt to understand how a system with which he or she is unfamiliar is designed to operate. Part of your ability to do this lies in understanding each of the concepts outlined throughout this textbook, from simple inputs, processing and outputs to the more detailed descriptions in the various chapters of how each of the systems function. However, your diagnostic ability will be increased if you are also able to interpret the **electrical schematic** for the system you are attempting to diagnose. In fact, many flowcharts incorporate an electrical schematic for this purpose.

Many electrical schematics are simple, dedicated schematics. Others are complicated by the fact that an otherwise simple schematic is part of

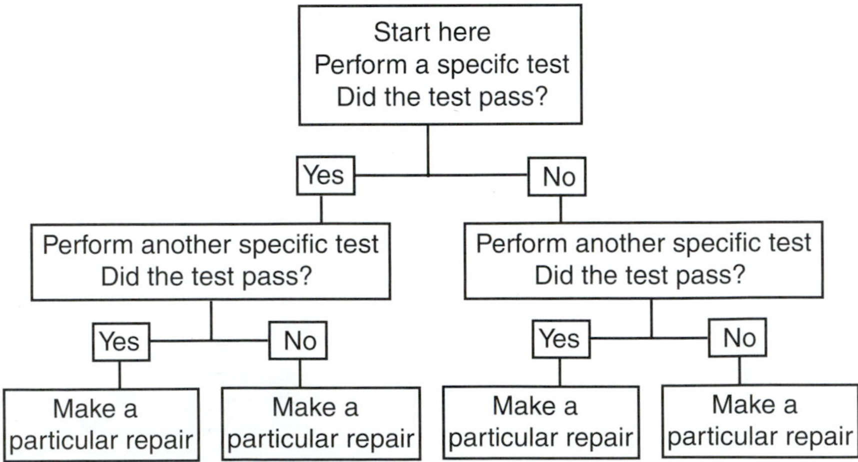

Figure 3–3 Typical form of a troubleshooting tree.

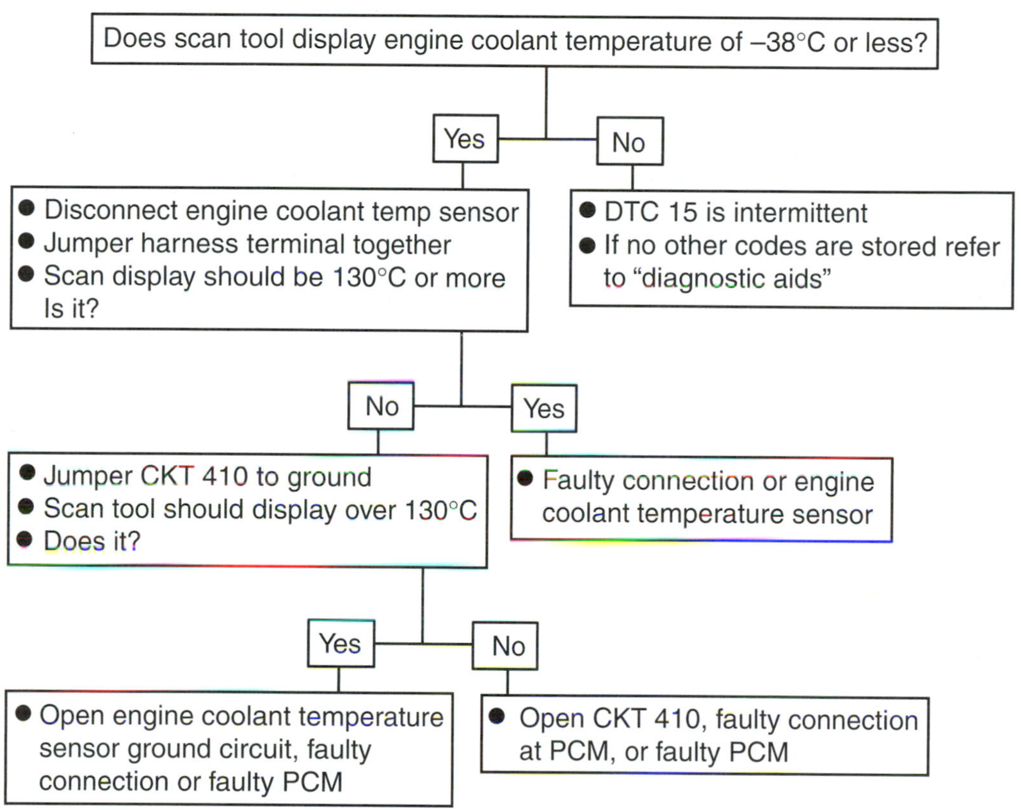

Diagnostic aid

Temperature vs. resistance values (Approximate)		
°C	°F	Ω
100	212	177
80	176	332
60	140	667
40	104	1459
20	68	3520
0	32	9420
–20	–4	28680
–40	–40	100700

Figure 3–4 Diagnostic flow chart for locating the cause of an ECT sensor DTC.

a larger schematic, such as one that shows the circuits for the entire vehicle. Even a dedicated schematic may be complicated simply by reason of the technology involved. As you practice interpreting electrical schematics, be guided by the following recommendations, which will simplify even the most complex schematics.

The first thing that is important in interpreting an electrical schematic is to identify the components. If you can, make a copy of the schematic on a copier, then you can use a highlighter to identify the components and specific circuits. Taking the time to identify the components not only gives you some immediate insight into the system, but it will also keep you from inadvertently attempting to interpret the outline of a component as a circuit. Second, identify sources of power and grounds. Then you should identify any input circuits to a control module, output circuits from a control module and data circuits between control modules on a network.

Certain complex electrical schematics will always be a challenge to any technician, no matter the level of experience. But as you practice interpreting electrical schematics, you will find that they become noticeably easier to read. If you take the time to compare the verified symptom to your understanding of system operation through the system's electrical schematic, you will be able to narrow down the area where you need to begin testing the system. This is essentially what a flowchart does for you. However, it is not recommended that you use an electrical schematic *in place of* a flowchart, but rather *with* a flowchart to enhance your diagnostic abilities.

OTHER GENERAL DIAGNOSTIC CONCEPTS

Not all faults (either soft or hard) will actually generate a DTC, particularly with the early engine computer systems. The PCM was programmed to monitor the system for total failures or those failures that resulted in a voltage value outside the parameters programmed into the PCM. For example, a throttle position sensor (TPS) could actually have an intermittent glitch that would not set a DTC in the PCM's memory because the voltage did not fall below the lower parameter programmed into the PCM. However, it would still result in a lean hesitation each time it was encountered. By the late 1980s, the PCM had the ability to compare certain sensors and determine if a fault existed even if the voltage was not outside of the parameter values. OBD II (discussed in Chapter 6) further enhanced this concept, and the ability of the PCM to monitor the system for efficiency as well as for failure became a standard that applies to all OBD II vehicles. As a result, the PCM is able to alert the technician through a DTC a higher percentage of the time on newer vehicles. Additionally, as technology progressed, the PCM was able to monitor more functions through additional feedback circuits. Therefore, on a newer vehicle, a DTC can direct the technician to a more specific area as opposed to some of the older systems. However, no matter how specific the area that a DTC is able to direct you to, you should always verify any fault through pinpoint testing. Also, bear in mind that multiple DTCs should be repaired in the order specified by the manufacturer.

SUMMARY

This chapter presented the basic concepts pertaining to diagnosing an engine computer system: hard faults and soft faults, memory code pulling and self-tests and several other concepts used by the manufacturer's systems and discussed throughout the remainder of this textbook. By presenting the concepts that constitute the building blocks of the various diagnostic programs, it should be easier for the reader to identify with the various concepts as the manufacturer-specific systems are covered.

▲ DIAGNOSTIC EXERCISE

A vehicle is towed into the shop because it "cranks, but won't start." The Malfunction Indicator Lamp (MIL) does not come on and all attempts to pull DTCs result in "no response." When a scan tool is connected, it cannot seem to communicate with the PCM. What tests should the technician do next?

REVIEW QUESTIONS

1. What is a soft fault?
   A. A fault that exists intermittently
   B. A fault that is always present
   C. A fault that can only be identified through the use of a scan tool
   D. A fault that cannot be properly diagnosed
2. What is a hard fault?
   A. A fault that exists intermittently
   B. A fault that is always present
   C. A fault that can only be identified through the use of a scan tool
   D. A fault that cannot be properly diagnosed
3. What is a memory DTC?
   A. A fault code that sets in the PCM's memory during normal driving conditions
   B. A fault code that must be erased from the PCM's memory after it has been properly repaired
   C. A fault code that sets as a result of a self-test initiated by the technician, but does not set in the PCM's memory
   D. Both A and B
4. What is an on-demand DTC?
   A. A fault code that sets in the PCM's memory during normal driving conditions
   B. A fault code that must be erased from the PCM's memory after it has been properly repaired
   C. A fault code that sets as a result of a self-test initiated by the technician but does not set in the PCM's memory
   D. Both A and B
5. *Technician A* says that some engine computer systems allow the technician to pull memory DTCs.
   Technician B says that some engine computer systems allow the technician to pull memory DTCs and initiate performance self-tests of the system.
   Who is correct?
   A. A only
   B. B only
   C. Both A and B
   D. Neither A nor B
6. Which of the following statements is false concerning a data stream?
   A. It consists of serial data (binary code) coming from the PCM.
   B. It includes information about input sensor values.
   C. It includes information about output actuator commands.
   D. It can be interpreted with manual methods, such as through using a voltmeter.
7. Which of the following statements is false concerning a functional test?
   A. It is a test that is programmed into the PCM.
   B. It should be used on every vehicle that comes in the shop for a tune-up.
   C. It is designed to aid the technician in identifying specific problems.
   D. A flowchart may direct the technician to perform a functional test.
8. Which of the following statements is false concerning a pinpoint test?
   A. It is designed to direct the technician to the general area of the problem.
   B. It is designed to help the technician identify the exact fault.
   C. It may include testing the sensors and actuators as well as the related circuitry.
   D. It may include performing voltage drop tests of the PCM's power and ground circuits.
9. Which of the following statements is false concerning a flowchart?
   A. It is designed to assist the technician in making pinpoint tests in order to identify the exact fault.
   B. The technician may be directed to a certain step of a flowchart by reason of a DTC.
   C. The technician may be directed to a certain step of a flowchart by reason of a symptom.

   D. It is okay to skip a test step in a flow chart and assume the test result if the test step is too difficult to perform.

10. Which of the following statements is false concerning an electrical schematic?
   A. It can aid a technician in understanding how a system is designed to operate.
   B. It may show components, power and ground circuits, input and output circuits and data stream circuits.
   C. It should be used with a flowchart for maximum diagnostic potential.
   D. There is nothing to be gained from taking the time to follow an electrical schematic because flowcharts are always 100 percent accurate and should therefore be used alone.

Chapter 4

Diagnostic Equipment

OBJECTIVES

Upon completion and review of this chapter, you should be able to:

- ❑ Describe the types of equipment that may be used in diagnosing symptoms associated with computerized engine control systems.
- ❑ Understand the functional concepts and procedures associated with each type of diagnostic tool.
- ❑ Interpret the readings associated with each type of diagnostic tool.
- ❑ Understand the advantages, disadvantages and limitations of each type of diagnostic tool.

KEY TERMS

Breakout Box
Current Ramping
DSO
DVOM
Logic Probe
Nonself-Powered Test Light
Scan Tool or Scanner
Trigger

Your noblest efforts to properly diagnose today's electronic systems may be limited by your knowledge of the diagnostic tools available to you as a technician. This chapter is designed to enhance your understanding of today's tools, their purpose and function and their limitations.

SCAN TOOLS

A **scan tool** (or **scanner**) is probably the most familiar tool to anyone working with today's computerized systems. Figures 4–1 and 4–2 show typical scan tools. However, it is important to recognize the distinct abilities of a scan tool. Let us begin by covering the purposes of a scan tool.

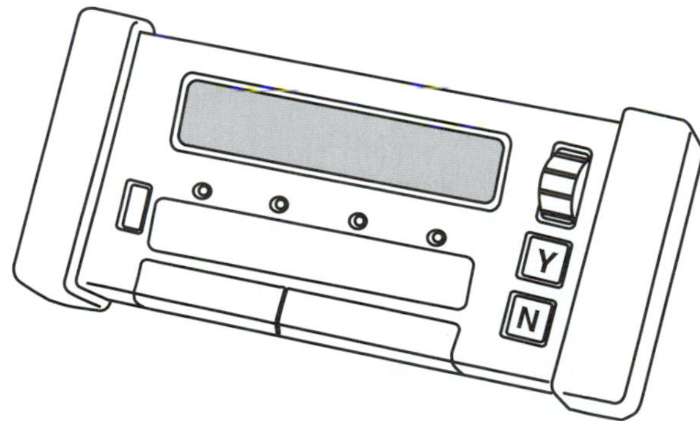

Figure 4–1 A typical scan tool that can be used to retrieve data stream information from the PCM.

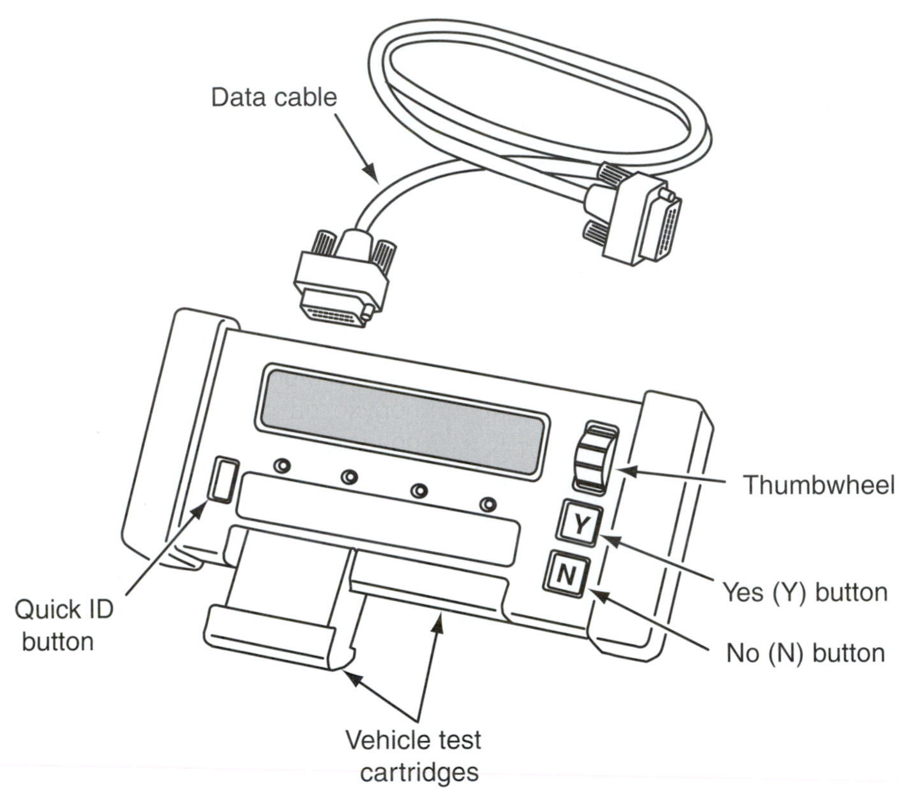

Purposes

Code Pulling. A scanner's ability to pull codes includes both pulling DTCs from a computer's memory and causing the computer to perform a self-test of the system and display the resulting DTCs. In certain applications in which code pulling cannot be done with a scan tool but must be accomplished through manual methods, the scan tool may still have the ability to instruct the technician as to the proper procedure, even though it cannot be connected to the system.

Functional Tests. Scan tools can also be used to perform functional tests according to those tests that the vehicle manufacturer has programmed into the on-board computer.

Data Stream. Scan tools can read data stream information from the on-board computers. This information includes the *computer's interpretation* of input sensor values, the computer's *intended* output commands and certain pieces of information from within the computer's memory (such as fuel trim values and PROM ID numbers).

Additional Features

Many scan tools provide additional features such as a data capture mode (sometimes called a *snapshot* or *movie mode*) for use in identifying causes of intermittent problems. When attempting to verify an intermittent problem, this mode may be used to capture data while road testing the vehicle. When the symptom is felt, this data can be frozen within the scan tool. Then it can be reviewed later, frame by frame while watching the input/output values closely at the point where the symptom occurred.

Other additional features may include idle reset and/or idle rpm control through the scanner, octane adjustment, tire size input for vehicle speed sensor (VSS) calibration and the ability to disable or enable certain features for various systems. Additionally, some scan tools also provide an internal source of information (electronic shop manual) including DTC meanings, module pinouts and term definitions.

Purpose and Functional Concept

Scan tools are designed to communicate with the on-board computers using serial data (binary code). When connected to a diagnostic link connector (DLC), the scanner actually becomes a node on the data bus and can typically communicate with any other node (or module) on the bus. How much information that can be accessed through a scanner varies among scan tool manufacturers. The vehicle manufacturer's scanner is, of course, the most capable unit. Generally, aftermarket scan tool manufacturers are more limited in their capabilities. However, a technician working in an independent repair shop will usually find that the aftermarket scan tool is the best investment.

Functional Procedure

A scan tool simply plugs into the DLC after it has been equipped with the correct adaptor. Once powered up, it requires vehicle identification number (VIN) and data input from the technician. This automatically sets the scan tool to the baud rate of the modules on the data bus. Then the technician must select the module to be diagnosed as well as the desired menu item, such as memory DTCs, KOEO self-test, KOER self-test, functional tests and data stream.

Functional Interpretation

When looking at scan tool data, bear in mind that the scanner only offers the computer's interpretation of the data. For example, input values are measured at the computer and are subject to errors induced by faulty connections as well as poor computer power and ground connections. Output commands indicate the computer's intentions and, most often, cannot compensate for defective drivers, poor connections and deficient actuators.

Advantages

A scan tool has the advantage that it is easy to connect and can very quickly give the technician a very large amount of information. Also, the display typically provides a written menu and instructions that make this tool very easy to use.

Disadvantages and Limitations

Although a scan tool may offer a very quick approach to diagnostics, it can only direct you to a problem *area* or circuit. It cannot tell you the exact cause of a symptom. For example, if it shows that a temperature circuit is sending a 5-volt signal to the computer (equating to approximately –40°F), through your understanding of an NTC thermistor circuit you know that the circuit is open. This open could be in the sensor, a wire, a connector or the PCM itself. To identify the exact cause of the open circuit, further pinpoint testing is needed. Also, bear in mind that when a computer tells you via a scanner that an action has been commanded, this only means that the computer is attempting to forward bias the output driver (transistor) that controls this action. It does not really know (in most cases) whether the action has truly taken place. The exception would be on those systems that monitor the output function with a feedback signal. This becomes more common on newer systems, particularly those that are OBD II compliant, or safety systems such as antilock braking systems (ABS). Certain output circuits have been monitored on Ford computerized systems since 1980, such as EGR operation or those circuits that were tested during a Key On, Engine Off (KOEO) or Key On, Engine Running (KOER) self-test.

Scan tools are also limited in their ability to identify quick intermittent problems. Scan tool data is only updated according to the speed programmed into the PCM, and then the data must be updated, each in turn. This increases the probability that a glitch in a TPS circuit, for example, may happen between scan tool updates and therefore may never be seen by the technician who is using a scan tool.

BREAKOUT BOXES

In the 1980s, a **breakout box** was a diagnostic tool used primarily to diagnose Ford computerized engine control systems. Many technicians equated it to a scan tool, but a breakout box has a totally different function and purpose than a scan tool, and the technician is best served by access to both diagnostic methods. As a result, today other manufacturers and systems have breakout boxes designed to help you in diagnosis, and aftermarket universal breakout boxes have become widely available. Figure 4–3 shows a typical breakout box.

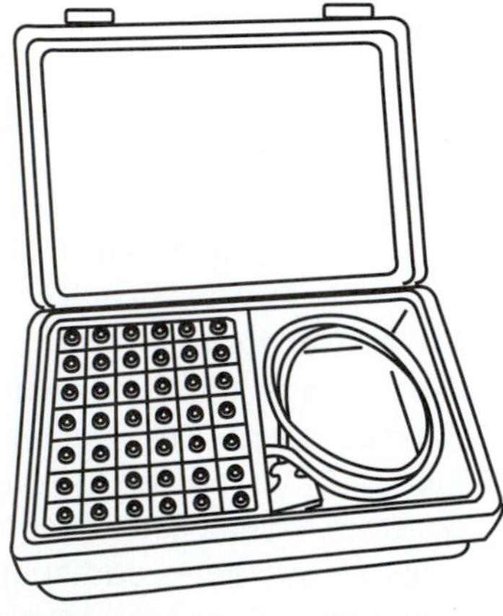

Figure 4–3 A typical breakout box that can be used to pinpoint test the computer's circuits.

Purpose and Functional Concept

A breakout box gives the technician a central testing point to make pinpoint tests of input sensors, output actuators and the associated wiring as well as computer power and ground circuits. It connects into the harness between the PCM and the harness connector. As a result, it makes all computer circuits entering or exiting the PCM accessible to your DVOM or DSO while maintaining the operating ability of the computer system.

Functional Procedure

Carefully disconnect the PCM from the harness connector. Then connect the harness connector to the proper breakout box connector. Make only this connection and leave the PCM disconnected if you need to make resistance checks of the circuits or if you will be using jumper wires to test actuators from the box. (Of course, due to their low resistance and high current potential, do not ever use jumper wires to manually energize high-energy ignition coils or fuel injectors.) Connect the PCM to the breakout box for the system to be fully operational for testing voltage, frequency or duty cycle on-time. Then use a logic probe, DVOM or DSO to probe the designated pins on the breakout box, depending on which circuits you need to test.

Functional Interpretation

Tests made with a breakout box indicate raw data (voltage, voltage drop, resistance, frequency, duty cycle or pulse width). Therefore they verify not only the input sensors or output actuators, but also the circuitry that connects these components to the PCM.

Advantages

When used in conjunction with a voltmeter or oscilloscope, a breakout box provides the user with a single location to access all circuits for testing. Test results are actual values, not computer-interpreted values.

Disadvantages and Limitations

Depending on the particular vehicle application, breakout boxes can be quite time consuming to connect into the system. Also, a breakout box cannot be used to identify things in the computer's memory such as a PROM ID number, fuel trim values or how an octane adjustment has been set with a scan tool.

NONSELF-POWERED TEST LIGHTS

Nonself-powered test lights, Figure 4–4, continue to have a place in your toolbox for certain quick checks. However, many technicians continue to use these devices in circuits where they can be very dangerous. Therefore it is important to understand a few facts about them.

Purpose and Functional Concept

The test light's only purpose is to serve as a quick-check tool, designed to show if a voltage potential is present. A voltage differential across the test light causes the bulb to glow.

Functional Procedure

Begin by connecting the test light across the battery's terminals to verify the integrity of the test light. Then, while grounded, touch the circuit in question and note the light. A test light makes a great tool for testing all of the vehicle's fuses electrically in a short amount of time. It also can be used to test switching of circuits such as ignition coils and fuel injectors as long as the resistance of the test light equals or exceeds the resistance of the intended load. A grounded test light can also be used to provide a path of low resistance for identifying poor insulation in secondary ignition wiring. Secondary ignition wiring that is arcing to ground can be difficult to identify in the shop when cylinder pressures are low and resistance across the plug gap is also low. But many times, even with the engine idling, a secondary wire will arc to a grounded test light to help identify the specific insulation breakdown. For this test, pass the tip of the grounded test light along each of the secondary wires with the engine running.

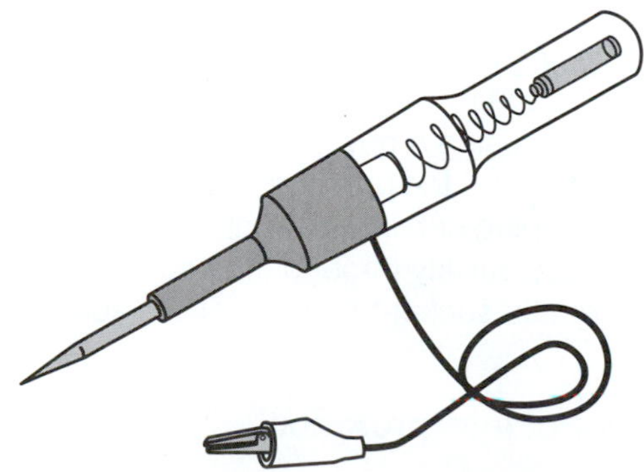

Figure 4–4 A typical nonself-powered test light.

Functional Interpretation

The fact that a test light illuminates only proves that a voltage potential is present. In situations in which accurate voltage readings are required, you should never try to guess how much voltage is present by the bulb's brightness.

Advantages

The advantages of a test light are that it is easy to use, it is inexpensive and it takes up very little room in your toolbox. Although it has several limitations, it makes a great quick-check tool for several initial tests as you begin diagnosing certain types of faults.

Disadvantages and Limitations

A test light cannot indicate voltage values; therefore there are many tests that it cannot perform. For example, it would not be a good tool to use in performing voltage drop tests. Additionally, it has a great potential for damaging electronic circuits. If the test light's resistance is less than the intended load, it will draw more current than the circuit was designed to carry. This can destroy a module's output driver. Many computers have been destroyed because someone used a test light to read code output. These circuits are generally designed to let you count codes manually using an analog volt-ohmmeter and were not designed to carry the current that a test light draws and are thus destroyed by the use of a test light. Also remember that use of a test light on an air bag system can result in accidental deployment of the air bag(s). For all practical purposes, think of a test light as a jumper wire.

LOGIC PROBES

A **logic probe**, Figure 4–5, is basically a test light that is safe for use on today's vehicles because it has high resistance designed into its internal circuitry. In fact, the term *logic probe* implies that it was designed for use within today's computers. Logic probes are generally available from electronics suppliers at a reasonable price. You can also purchase similar devices known as *high-impedance test lights* from the tool suppliers.

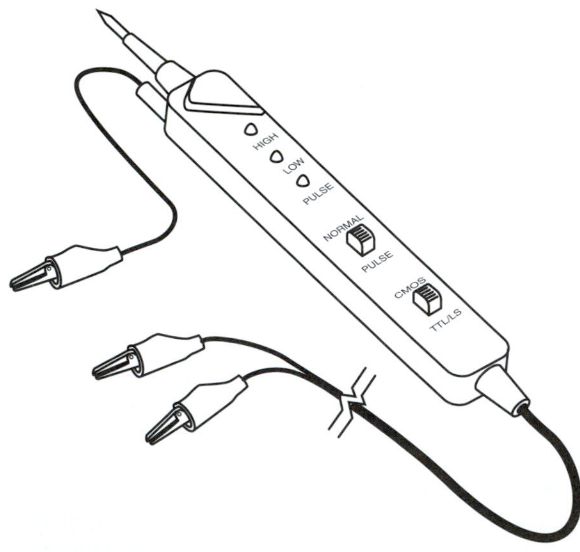

Figure 4–5 A logic probe is an excellent tool for testing for the presence of electrical signals.

Purpose and Functional Concept

Like a test light, a logic probe provides the technician with a quick-check tool. Once powered up, a logic probe can be used to check for voltage and ground potential. It is also excellent for testing for the presence of switched signals within the primary ignition system or at the fuel injectors.

Functional Procedure

As an electronic device, a logic probe must first be powered. To do this, connect the power leads to battery positive and negative. Then, for best results, connect the tip's negative lead to a clean unpainted ground. (The ground lead in the power lead set contains a protective diode that influences the ground level if it is the only ground used.) If the logic probe has a CMOS/TTL switch, set it as follows: set it to CMOS for working with voltages from 0 to 12 volts and set it to TTL for working with voltages from 0 to 5 volts. Then set the NORMAL/PULSE switch as follows: when it is set it to NORMAL the red and green light emitting diodes (LEDs) will indicate high or low voltage levels within the selected voltage range and a tone-generator emits a high or low tone, and when it is set to PULSE, a yellow LED is also used to indicate a *pulsed* voltage signal. Unfortunately, the yellow LED is also lit by induced voltage from the secondary ignition wires if you are in close proximity to them on a running engine.

Also, the alligator clips may be cut off and replaced by soldering in a cigarette lighter adaptor (while paying attention to the required polarity). In this configuration, a logic probe can be quickly powered up using a power point in the passenger compartment for the purpose of electrically testing fuses. The plug can still be connected to the vehicle's battery quickly through the use of a cigarette lighter female socket-to-alligator clips adaptor.

Functional Interpretation

With the tip of the probe connected to the circuit in question, note the LEDs. The red LED (and high tone) responds to a higher voltage and the green LED (and low tone) responds to a lower voltage. Brightness of one LED versus the other can also give you a rough indication of duty cycle on-time versus off-time. If you have it connected to the ground side of a ground-side switched actuator, increasing the on-time increases the brightness of the green LED.

Advantages

A logic probe is an easy-to-use tool best suited for determining the presence (or loss) of primary ignition and fuel injection signals. It also has high internal resistance and is therefore safe for use in diagnosing electronic circuits.

Disadvantages and Limitations

A logic probe, like a standard test light, cannot inform you of exact voltage values or voltage drop readings. Therefore it is limited to certain quick checks.

DIGITAL VOLT-OHMMETERS (DVOMs)

A **DVOM** is probably the choice of most technicians for most pinpoint testing of the computer and its associated circuits and components. A typical manual-scaling DVOM is shown in Figure 4–6 and a typical autoscaling DVOM is shown in Figure 4–7.

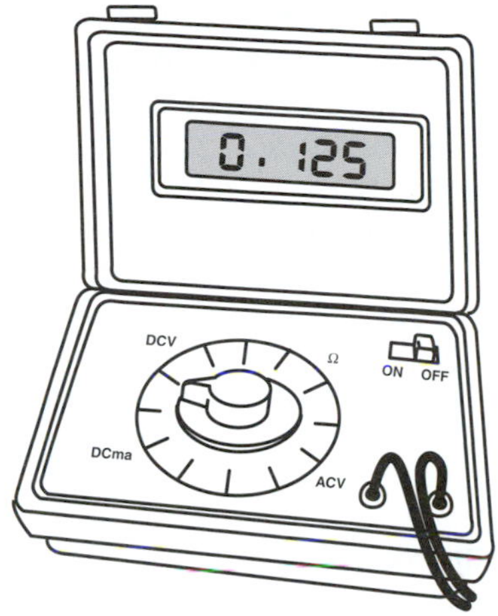

Figure 4–6 A typical manually scaled DVOM.

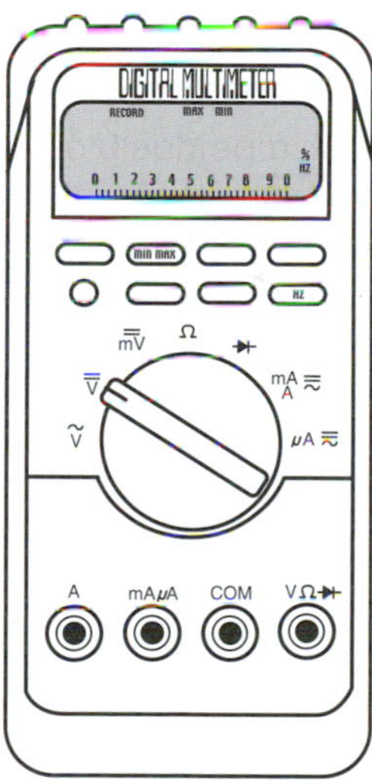

Figure 4–7 A typical automatic-scaling DVOM.

Purpose and Functional Concept

A DVOM is, most often, the tool of choice for making accurate pinpoint tests of the electrical characteristics of a circuit. When purchasing a DVOM, look for the following features: resistance, AC and DC voltage, AC and DC current flow, frequency, duty cycle on-time, pulse width on-time, temperature and rpm. Also, you might want a bar graph, a Min and Max function, a Hold function and an Average function. If the DVOM is self-scaling, also look at the unit's response time (in the specification sheet) or how long it takes for the DVOM to sense the signal, rescale and display the reading. This time can range from 2 seconds on some low-end units to 100 milliseconds (ms) or better on high-end units. The DVOM should also be rated at no less than 10 megaohm impedance when scaled to a DC voltage scale. This high resistance through the meter ensures that it does not damage electronic circuits, plus meter accuracy is increased when measuring DC voltage. Bear in mind that when the DVOM is functioning as an ammeter, it effectively acts as a jumper wire.

A DVOM is designed to provide in-depth information about the electrical characteristics of a circuit. However, all too often, technicians get careless in their interpretation of what they see on the DVOM's display screen and misdiagnose a problem.

Functional Procedure

A DVOM has an easy two-lead connection. Connect the black lead to the jack marked Common and the red lead to the jack marked for the type of measurement needed. Bear in mind, leads that are connected into the amp jacks effectively turn the DVOM into a jumper wire. Now set the dial to the scale needed. Connect the red and black leads to the circuit/component that you need to test according to the type of measurement being made (voltage, voltage drop, amperage or resistance). The red lead should connect to the point that is most positive in the circuit and the black lead to the point that is most negative. If the DVOM is a manual-scaling unit, for most accuracy be sure to ultimately scale down to the lowest scale that can read the value being measured. Then note the reading on the display and interpret the reading as per the type of scale to which the DVOM is currently set.

Functional Interpretation

A DVOM provides very exact measurements of raw data, and therefore it makes an excellent pinpoint test tool. But particular attention needs to be paid to the scale's factor when interpreting the reading on the display. This factor may be shown on either the scale selection dial or on the display depending on the particular DVOM. A K (kilo) means that whole numbers represent thousands and a capital M (mega) means that whole numbers represent millions. The decimal point should be moved 3 or 6 places to the *right*, respectively, to convert these values to numbers that represent single whole units. Likewise, a small m (milli) means that whole numbers represent thousandths and a μ (micro) means that whole numbers represent millionths. The decimal point should be moved 3 or 6 places to the *left*, respectively, to convert these values to numbers that represent single whole units. If a factor value is not shown on the scale selection or the display, then read the display as it appears. The ability to use a DVOM on any type of scale (ohms, amps, volts, Hertz) depends on the technician's ability to interpret these factors. Figure 4–8 shows the four most common factors used with DVOMs. Figure 4–9 shows four sample interpretations of these factors.

Advantages

A DVOM is very easy to connect, read and interpret. It provides an easy method for accurate pinpoint testing of most components or circuits. Again, you need to take care in the interpretation of the displayed reading.

PREFIX (Factor)	SYMBOL	RELATIONSHIP TO BASIC UNIT
Mega	M	1,000,000
Kilo	K	1,000
Milli	m	1/1,000
Micro	μ	1/1,000,000

Figure 4–8 The four most common DVOM scale factors.

1.234 MΩ = 1,234,000 Ohms
1.234 KΩ = 1,234 Ohms
1234 mA = 1.234 Amps
1234 μA = .001234 Amps

Figure 4–9 Examples of converting each of the factors to the base unit.

Disadvantages and Limitations

When connected to a voltage that switches quickly, a DVOM reads the average voltage, which indicates duty cycle on-time. Also, a DVOM is unable to read "time" and may be ineffective in diagnosing certain component failures in which this is important, for example, an oxygen sensor's response time.

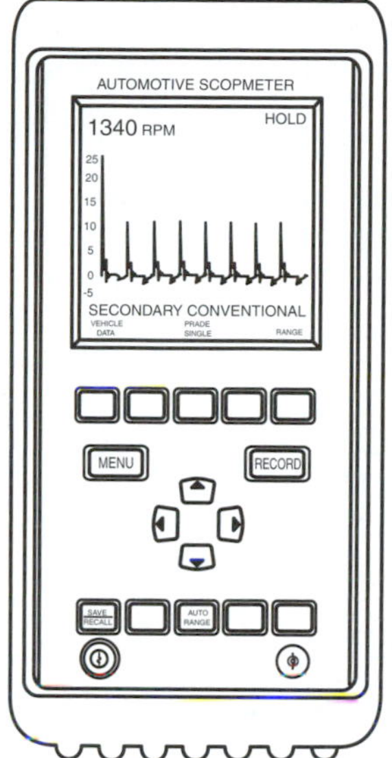

Figure 4–11 A typical digital storage oscilloscope.

DIGITAL STORAGE OSCILLOSCOPES (DSOs)

A **DSO** (also sometimes referred to as a *lab scope*) is one of the great forward leaps in diagnosing automotive computer systems. Designed similarly to the analog bench-top oscilloscopes of the past that were used in the electronics field, these units also utilize computer-generated waveforms that can be frozen in time, measured, downloaded to a PC and printed out. Figures 4–10, 4–11, 4–12 and 4–13 show typical DSOs or lab scopes. Also, graphing multimeters can perform much of the same function as a DSO.

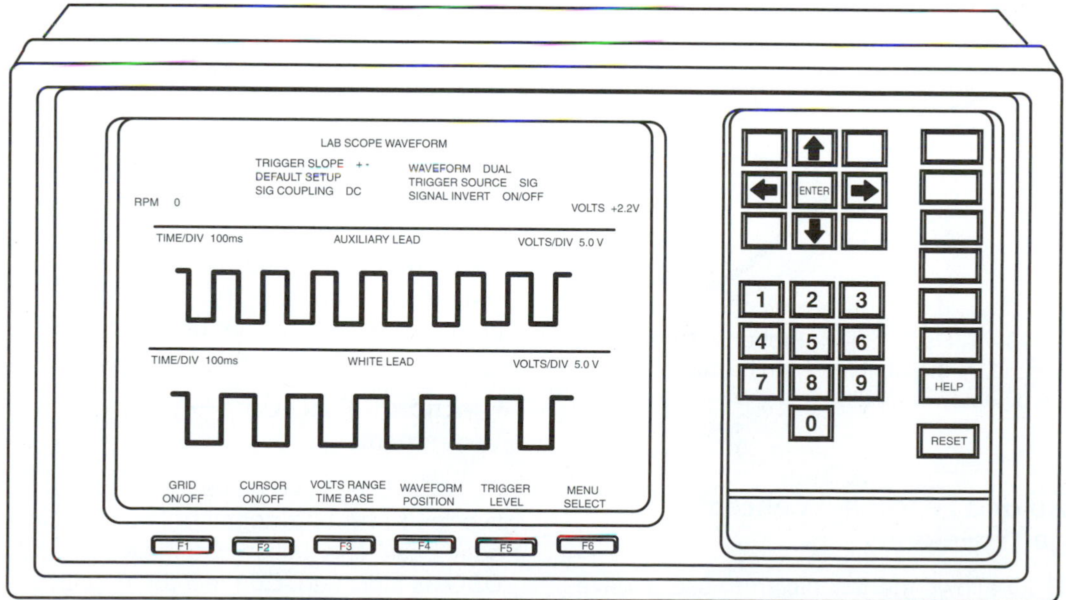

Figure 4–10 A multifunction, dual-trace lab scope.

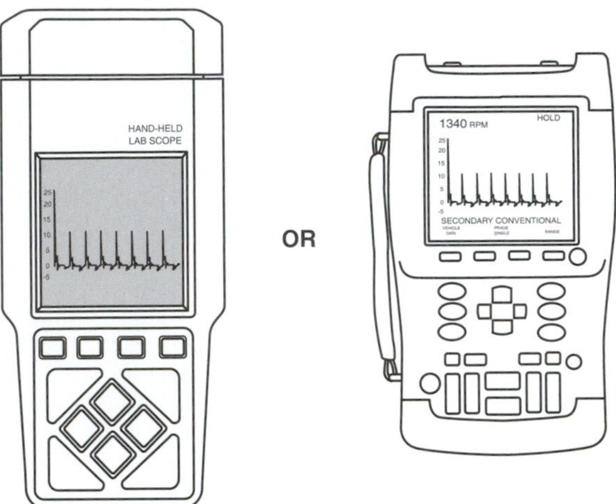

Figure 4–12 The modern hand-held lab scope is powerful, yet convenient to use.

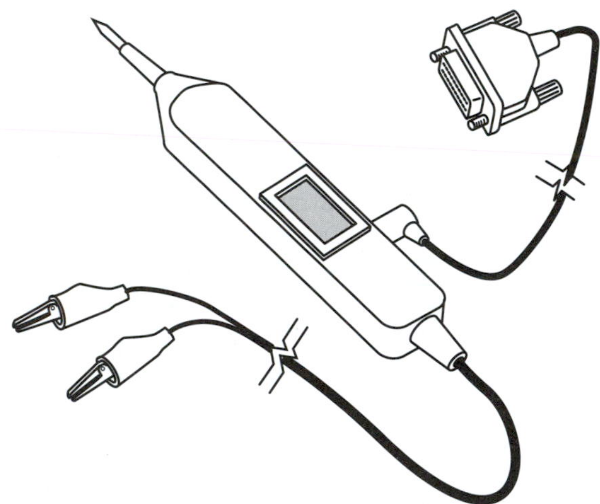

Figure 4–13 A Radio Shack Probe Scope that can be used alone, or the waveform can be downloaded to a desktop or laptop computer.

Purpose and Functional Concept—Voltage Waveforms

A DSO allows the technician to see a waveform of *voltage over time*, Figure 4–14, in order to allow accurate pinpoint testing of system faults as-

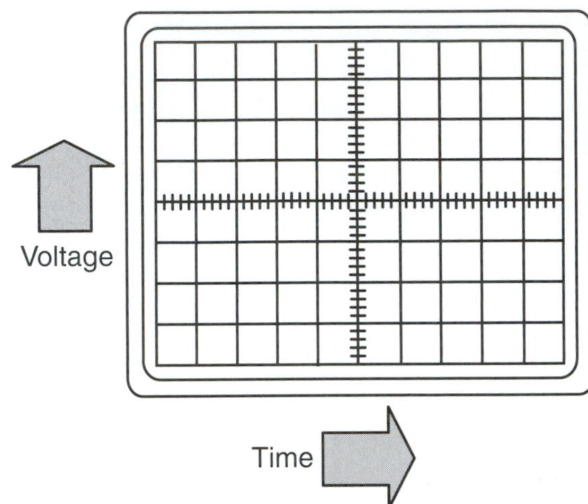

Figure 4–14 Typical DSO screen.

sociated with these voltage signals. It also allows evaluation of the characteristics associated with these voltage signals, such as frequency, duty cycle on-time and pulse width on-time. A DSO can also be adapted to display a waveform of *current over time* in order to allow the technician to use a load component's current draw in diagnosis.

A DSO can provide in-depth information about the electrical characteristics of a circuit. Some DSOs provide multiple channel (trace) capability, allowing the technician to capture more than one waveform simultaneously, Figure 4–15. A DSO may have one-, two-, three- or four-channel capability. It has one lead set for each channel. Multiple channels may share a common ground lead. The DSO screen is typically divided into 8 voltage divisions and 10 time divisions, although this can vary among manufacturers.

Functional Procedure—Voltage Waveforms

A DSO has two leads per channel, one positive lead and one negative lead, making for an easy basic connection. Connect the leads to the DSO as indicated. Connect the negative probe to a clean, unpainted ground. Connect the positive probe to any voltage you wish to measure.

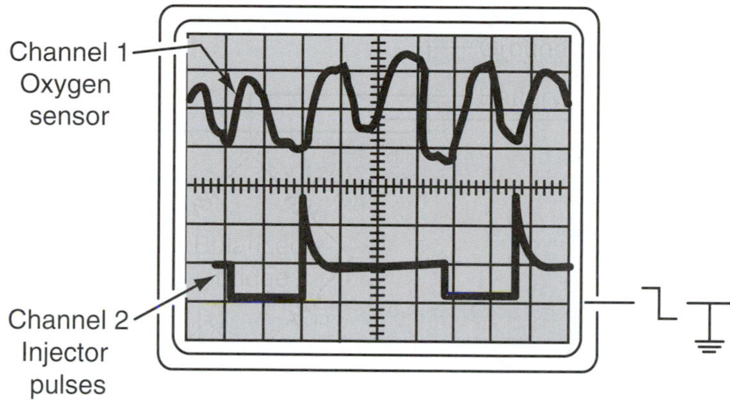

Figure 4–15 Lab scopes have two or more traces, so you can compare signals, for example, that of an oxygen sensor to injector pulse width.

To adjust the DSO's characteristics to see a proper waveform, make the following adjustments. Adjust the Coupling to the type of voltage signal you will be connected to—AC, DC or Ground. The Ground selection allows you to see waveform detail of a higher voltage while on a sensitive voltage scale by indexing it to the average voltage. This avoids the voltage being "off of the screen."

Adjust the ground level (sometimes known as the vertical waveform position) to the middle of the screen to read an AC voltage and toward the bottom of the screen (ideally, one division from the bottom) to read a DC voltage.

Adjust the Volts/Division (or on some units, volts-per-screen) to allow you to see the entire amplitude (or range) of the voltage. Adjusting the Volts/Division to a value that is too small may cause your waveform to move off the visible screen and seemingly disappear. Adjusting it to a value that is too large will make the detail of the waveform more difficult to see.

Adjust the Time/Division (or on some units, time-per-screen) to allow you to see two or three complete patterns of the waveform. Adjust this *shorter* to see more *detail* of the waveform and *longer* to see more of the waveform's *trend*. For example, if you were testing an oxygen sensor's ability to produce proper voltage values on an operating engine for the sake of testing the sensor itself, you would ideally adjust this short enough to see more detail—maybe 100 milliseconds per division. But if you then wanted to use the oxygen sensor's waveform to verify the ability of the engine computer to properly control the air/fuel ratio while driving the vehicle, you would ideally adjust this to see more of a trend—maybe 1 second per division. Examples of both an oxygen sensor waveform and a mass air flow (MAF) sensor waveform are shown in Figures 4–16 and 4–17 with the time adjusted to 200 ms/div in both figures.

You also need to adjust the **trigger** characteristics as follows. Trigger is designed to determine the instant in time that the DSO begins to display a waveform. If the trigger characteristics are misadjusted, the DSO randomly guesses as to where to begin a waveform, resulting in a waveform that will likely appear to walk across the screen. For example, when using an ignition oscilloscope to capture the secondary waveform of an ignition system, the secondary coil lead captures the waveform for all cylinders. But the number one cylinder trigger lead instructs the scope as to which of the cylinders' waveforms to begin with on the display screen. Trigger characteristics are divided into four adjustments, each of which is used by the DSO in determining when to begin

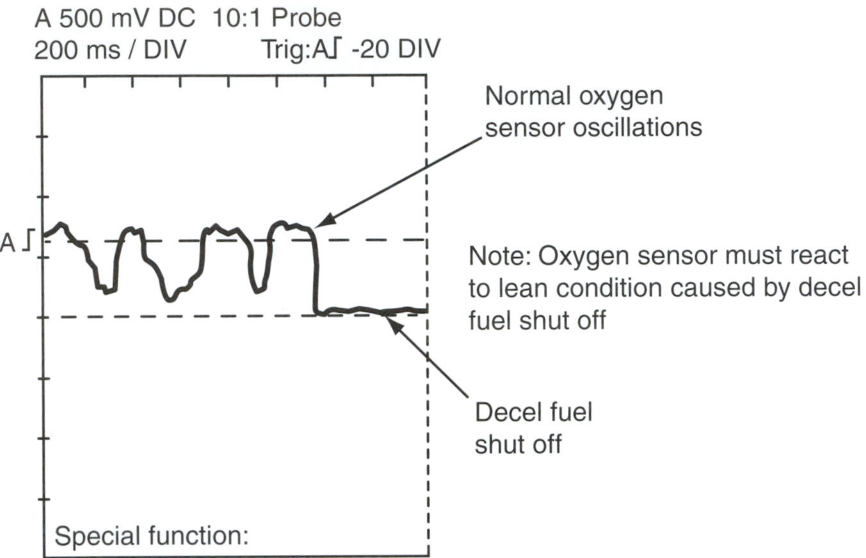

Figure 4–16 A lab scope allows you to see both the quantity and the quality of the signal, as with this O_2 sensor pattern.

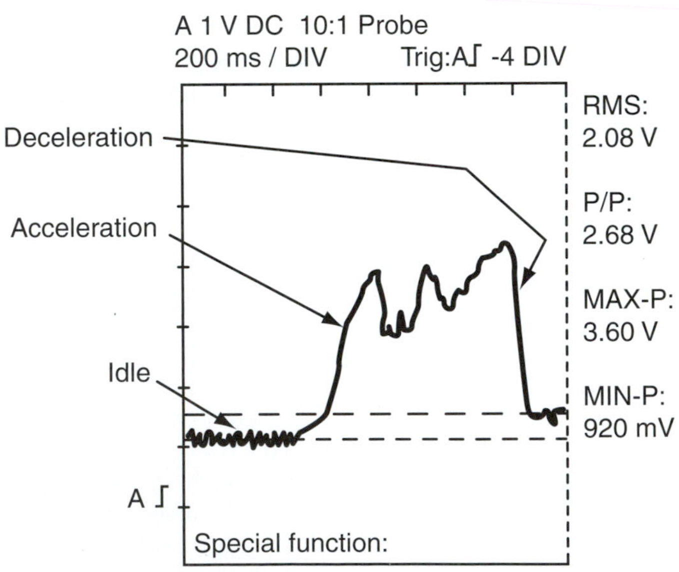

Figure 4–17 All the detail necessary to catch intermittent glitches and fast-occurring faults is present and controllable with a modern lab scope.

drawing the waveform: trigger delay, trigger level, trigger slope and trigger source.

Trigger Delay. Trigger delay is also sometimes known as the horizontal waveform position. Adjust this to begin the trace at a particular point on the screen—typically one division from the left edge of the screen. If you adjust this too far to the left edge of the screen, you may not see the initial event. This is why many ignition scopes do not show the initial firing line for cylinder number one at the left edge of the screen.

Trigger Level. Trigger level is a selected voltage value. The DSO begins the trace as it sees the voltage waveform cross this value. Adjust the trigger level to a value that is more than the lowest voltage of the waveform and less than the highest voltage of the waveform. Other factors may help you to determine the proper trigger level as well; for example, you would not want to trigger off a fuel injector's voltage spike, but rather off the turn-on signal, Figure 4–18.

Trigger Slope. Trigger slope is used to determine whether the DSO will begin the trace on a voltage shift from low-to-high (upslope) or high-to-low (downslope) as the waveform crosses the trigger level. Adjust this to begin the trace on either an upslope or a downslope according to what you want to see in the waveform. For example, on a fuel injector waveform you would normally want the trace to begin with the turning on of the injector, not the turning off. This is accomplished by selecting a downslope in order to begin the trace as the voltage is pulled low on a ground-side switched injector. See Figure 4–18.

Trigger Source. Trigger source is used to determine which lead is used to signal the DSO that the trace should begin. In most cases the trigger source is set to the lead you are using to capture the waveform (also known as *internal*). This keeps the DSO hookup simple; just one positive lead and one negative lead need to be connected to the circuit. In certain cases you may want to use another lead to trigger the waveform other than the one being used to capture the waveform (also known as *external*). Secondary ignition scopes use an external trigger in that they use the number one trigger lead to trigger the waveform, even though it is the secondary coil lead

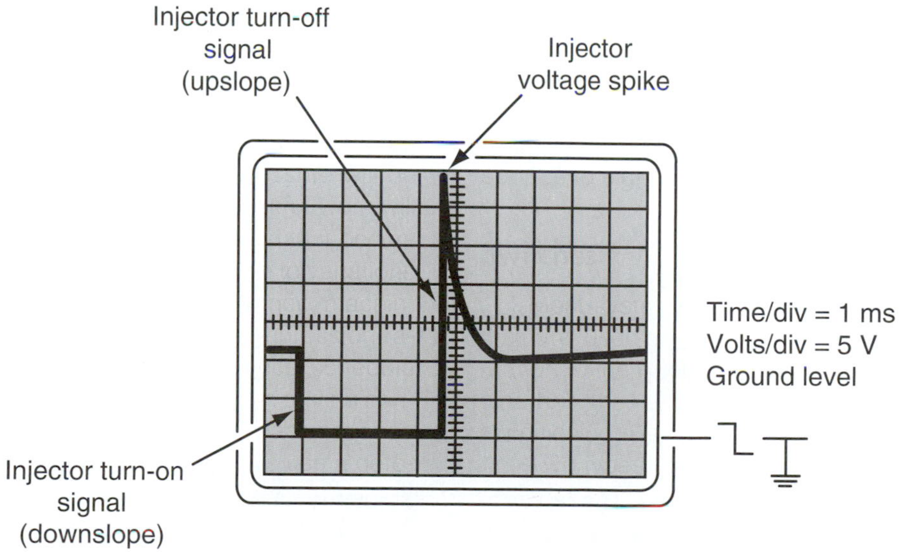

Figure 4–18 A typical fuel injector voltage waveform.

that is capturing the secondary waveform for all cylinders.

Functional Interpretation—Voltage Waveforms

When interpreting a voltage waveform, you should look at the following characteristics. Depending upon what type of component or circuit that you are testing, some of these characteristics may be more important than others.

Amplitude. The amplitude is how high and low the voltage goes. For example, a voltage that does not fall all the way to ground when a ground-side switched fuel injector is turned on indicates that there is still resistance in that circuit after the point where the DSO lead is connected to it. Or, if when scoping the waveform of a magnetic pickup coil in the distributor of a running engine, you notice that the amplitude varies with each rotation of the distributor at a *steady* engine rpm, worn bushings or a bent distributor shaft is indicated, due to a reluctor-to-pickup air gap that varies with each rotation.

Time. Time in the waveform is important with such things as digital pulse trains (variable frequency, duty cycle and pulse width), O_2 sensor testing and signals used to identify speed or rpm. For example, you can see on a DSO's waveform how long it takes for the O_2 sensor to respond to an air/fuel mixture change. This type of failure would indicate a "lazy" O_2 sensor and is difficult to identify without the use of a DSO.

Shape/Symmetry/Sequence of Events. Inspect the waveform for glitches that could cause an improper signal to the PCM, for example, a TPS signal that intermittently drops out. A DSO makes these types of faults much easier to identify.

Cursors. Horizontal and vertical cursors are provided on most DSOs to allow precise measurements of both voltage and time. The horizontal cursors are used to measure a voltage difference and the vertical cursors are used to measure a time difference. These differences may be indicated by the delta symbol (Δ). For example, ΔV equals delta volts or the voltage difference between the positions of the two horizontal cursors, and ΔT equals delta time or the time difference between the positions of the two vertical cursors.

Current Ramping

Another diagnostic approach gaining popularity is the use of a DSO to look at a current waveform, rather than a voltage waveform. This procedure is known as ***current ramping***. It allows the technician to identify problems with the load component that cannot be identified through the use of voltage measurements or waveforms. To do this, you must couple the DSO with a *quality*, low-current, inductive current (amp) probe. Inductive current probes are often used with a DVOM to measure current nonintrusively in that they simply clamp around a wire. They are self-powered and convert any measured current to a DC voltage (usually millivolts). The DVOM is then set on a DC voltage scale to read the current. An adjustment is provided on the current probe to zero out the value on the DVOM when the probe is not clamped around a wire.

A current probe works much the same way when used with a DSO. After using the current probe's adjustment to zero out the waveform to the ground level position, clamp the probe around the wire of a load component's circuit. It should be clamped as close to the load component as possible to avoid picking up current for multiple load components. Current ramping is not recommended for input sensor circuits as these circuits usually have high resistance resulting in very low current levels. Input sensors rely primarily on voltage values to communicate their information to the PCM.

Once the current probe is clamped around a wire, the circuit should be energized. The same adjustments already discussed under Digital Storage Oscilloscopes (DSOs) should then be performed. Also, most current probes have an adjustment that determines the millivolts-to-amps relationship. For example, you may be able to adjust the current probe so that 10 millivolts (mV) equals one amp or so that 100 mV equals one amp. Then you should set your volts per division on the DSO. Now you can calculate how many amps are represented by the height of the waveform on the DSO screen.

When a circuit is first switched on, the closing of the switch does not force the electrons to move. Rather, it simply allows the voltage pressure present in the circuit to begin pushing the electrons. Therefore it takes a few milliseconds for the current to fully saturate. Of course, Ohm's law dictates the level of full saturation. Therefore the current waveform *ramps upward* until it is fully saturated, thus the term *current ramping*. However, when the switch is opened, this action forces all electron flow in the circuit to stop almost immediately. Therefore the current returns to zero much faster than it saturated. For this reason voltage spikes are never produced by expanding electromagnetic fields when a circuit is switched on, but instead are produced by collapsing electromagnetic fields when a circuit is switched off. See Figure 4–19 for a typical current waveform of a solenoid such as a standard fuel injector.

When using current ramping to test a winding of a solenoid (or primary winding of an ignition coil), current that builds too quickly indicates that the windings are shorted (the electrons did not have as far to travel as intended). When using current ramping to test a motor, such as a cooling fan motor or a fuel pump motor, the current waveform can tell you quite a lot about the motor's condition. Each time that a pair of commutator segments line up with the brushes, the current momentarily rises. As the commutator segments slip away from the brushes, the current momentarily falls until the next pair of commutator segments begins to line up again. Most electric fuel pumps contain either four or five windings resulting in either 8 or 10 commutator segments (respectively) per each full rotation. You can even figure the motor's rpm by looking at the frequency of the commutator pulses. See Figure 4–20 for a typical current waveform of a fuel pump motor.

When capturing the current waveform of a motor, consider the following types of faults. A trace that drops to zero twice per rotation indicates an open winding. A trace that seems to increase, then decrease, in overall current with each rotation indicates physical binding, possibly due to worn bushings. Noise on the waveform indicates poor electrical contact between the brushes and the commutator segments. Each of

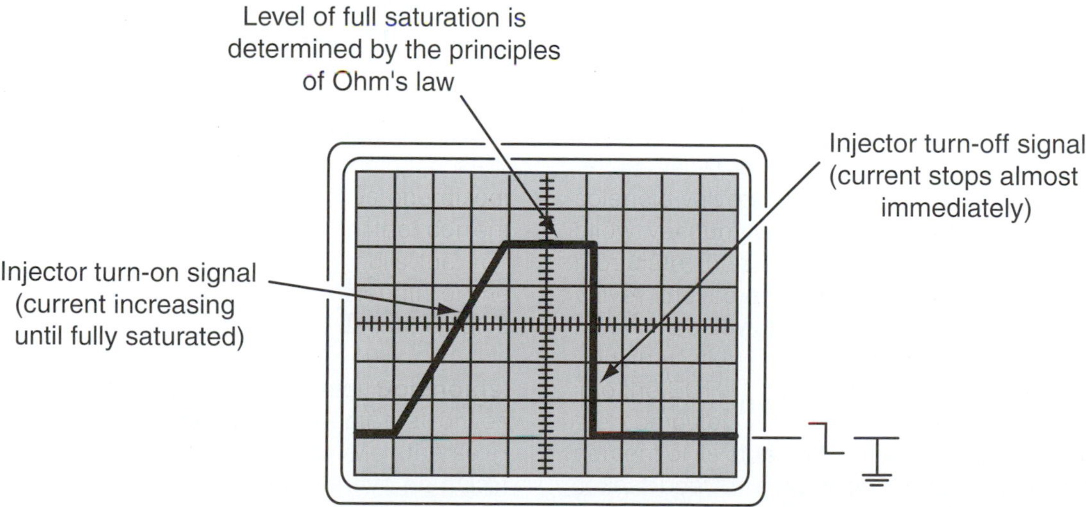

Figure 4–19 A typical fuel injector current waveform.

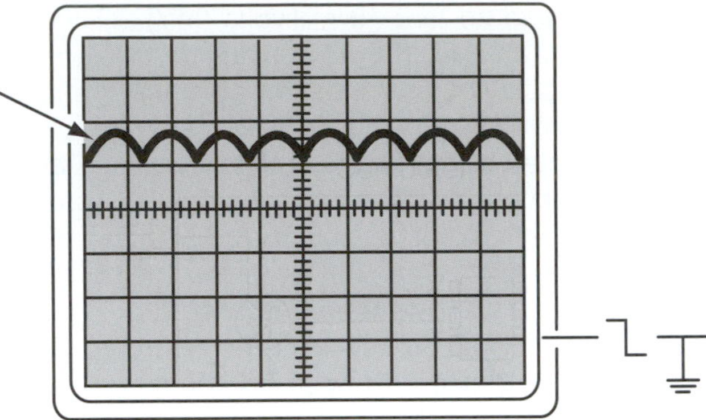
A motor's current pulses are created by brush-to-commutator segment contact

Figure 4–20 A typical DC motor current waveform.

these conditions can be responsible for *intermittent* symptoms. Therefore, current ramping can allow you to diagnose *intermittent* problems with load components more effectively than any other method.

Advantages

The advantages of a DSO are that it can allow you to perform pinpoint tests of the components and circuits in a computer system more effectively than any other means available. Voltage waveform databases are becoming more readily available. Unfortunately, current waveform databases are virtually nonexistent at this time, but you can build your own database by recording current waveforms of both good and defective load components (before and after repair waveforms) and then downloading these waveforms to a personal computer. As with any other testing technique, as you practice using a DSO, you will find that you become extremely proficient with it and will wonder how you ever did without it.

Disadvantages and Limitations

When a lab scope function is built into a "big box" type of ignition scope, it does not lend itself to procedures such as verifying oxygen sensor waveforms while driving the vehicle under realistic load conditions unless you also have access to a dynamometer in the shop. Even the more portable versions are typically larger than a DVOM, so you may want to use a standard DVOM for many tests due to its smaller size, even though most DSOs have a DVOM function built into them.

GAS ANALYZERS

Gas analysis and gas analyzers are more fully explained in Chapter 5 because a full chapter needs to be devoted to this topic. But in the context of the current chapter, the use of a gas analyzer is designed to quickly direct the technician to a problem area. That is, gas analysis is not in itself a pinpoint test procedure, but instead can quickly direct the technician to the problem areas where pinpoint testing should begin.

SUMMARY

This chapter outlined many pieces of diagnostic equipment along with the advantages and limitations of each. It is designed to give the technician an overall view of when to use each piece of equipment, what types of tests can be performed with each piece of equipment and what the test results mean to the technician. This chapter also differentiated between diagnostic equipment designed to lead the technician to the general area of a fault quickly, such as a scan tool or gas analyzer, and equipment suitable for making pinpoint tests of a circuit, such as a DVOM or DSO, whether used alone or in conjunction with a breakout box.

▲ DIAGNOSTIC EXERCISES

1. A fuel-injected engine does start and run. But after connecting a scan tool to the DLC of this vehicle and entering the requested data, a technician finds that after the ignition is switched on, the scan tool displays a message indicating "No communication." What types of problems could cause this?
2. After connecting a DSO to an electrical circuit in the engine compartment of a vehicle, a technician discovers that there is no visible trace on the screen. What types of problems could cause this?

REVIEW QUESTIONS

1. Which of the following diagnostic tools is designed to help the technician quickly identify the area of a fault?
   A. A scan tool
   B. A DVOM
   C. A DSO
   D. Both B and C
2. Which of the following diagnostic tools is designed to help the technician pinpoint test the exact fault?
   A. A scan tool
   B. A DVOM
   C. A DSO
   D. Both B and C
3. Which of the following diagnostic tools is designed to allow the technician to access the PCM's data stream through serial data (binary code) communication with the PCM?
   A. A scan tool
   B. A DVOM
   C. A DSO
   D. Both B and C
4. Which of the following diagnostic tools is designed to allow the technician to measure time, such as the response time of an oxygen sensor?
   A. A scan tool
   B. A DVOM
   C. A DSO
   D. Both B and C
5. Which of the following diagnostic tools is the best tool for helping the technician identify a quick, intermittent glitch such as what might occur in a Throttle Position Sensor?
   A. A scan tool
   B. A DVOM
   C. A DSO
   D. A nonself-powered test light
6. *Technician A* says that if a scan tool says that the EGR valve is being commanded to open 50 percent, then you can be assured that the valve is working properly and no further testing is necessary regarding the EGR valve.
   Technician B says that if a scan tool shows that the voltage for the ECT sensor is out of range, then you can be assured that the ECT sensor itself is at fault and no further pinpoint testing is necessary regarding the ECT sensor circuit.
   Who is correct?
   A. A only
   B. B only
   C. Both A and B
   D. Neither A nor B
7. Which of the following statements is false regarding the use of a breakout box?
   A. It connects in between the PCM and the computer harness.
   B. Once properly connected, it allows you to access all computer circuits with a DVOM or DSO.

C. It can be used to access information in the PCM's memory such as fuel trim data.
D. It allows you to make pinpoint tests of all computer circuits from one central location.

8. Which of the following pieces of diagnostic equipment is best suited for making exact voltage measurements when pinpoint testing an electronic circuit?
   A. Nonself-powered test light
   B. Logic probe
   C. High-impedance DVOM on a DC volts scale
   D. Scan tool
9. Which of the following pieces of diagnostic equipment can damage computer circuits if not used properly?
   A. Nonself-powered test light
   B. Logic probe
   C. High-impedance DVOM on a DC volts scale
   D. Scan tool
10. Which of the following pieces of diagnostic equipment is a computer-safe tool that can be used to quickly test for the presence of switched signals within the primary ignition system or at the fuel injectors, although it cannot read exact voltages?
   A. Nonself-powered test light
   B. Logic probe
   C. DVOM
   D. Scan tool
11. A technician is using a DVOM to measure the resistance of a secondary ignition wire. The DVOM is set to a 20 K ohm scale. The display shows 7.92. To which of the following does this number equate?
   A. 7.92 Ω
   B. 792 Ω
   C. 7,920 Ω
   D. 158,400 Ω (7.92 × 20,000)
12. A technician is using a DVOM to measure the parasitic draw of a vehicle's electrical system on the battery. The DVOM is set to a 200 mA scale. The display shows 32. To which of the following does this number equate?
   A. 32 amps
   B. 3.2 amps
   C. .32 amps
   D. .032 amps
13. A customer brings a vehicle into the shop and complains of intermittent stalling requiring a cool-down period before it will restart. During a road test the engine stalls. Testing shows that the fuel system has lost pressure and the fuel pump does not seem to run. After you get the vehicle back to the shop, you make voltage drop tests of the fuel pump's power and ground circuits to test for the intermittent condition and these tests prove that the circuitry is okay. What is the best test that can be performed in order to verify whether the fuel pump itself is responsible for the intermittent failure?
   A. Perform a voltage drop test across the fuel pump motor.
   B. Use a DVOM to measure the average current that the fuel pump motor draws.
   C. Use a DSO to look at the fuel pump circuit's voltage waveform.
   D. Use a DSO and an inductive current probe to look at the fuel pump motor's current waveform.
14. Which of the following can a scan tool not provide?
   A. Diagnostic Trouble Codes
   B. Data stream information
   C. Functional testing of the computer system
   D. Pinpoint testing of the computer system
15. *Technician A* says that when adjusting the time-per-division on a DSO, adjusting this shorter allows the technician to see more detail of the waveform.
   Technician B says that when adjusting the time-per-division on a DSO, adjusting this longer allows the technician to see more of the waveform's trend.
   Who is correct?
   A. A only
   B. B only
   C. Both A and B
   D. Neither A nor B

16. A DSO is properly connected to a circuit, but the waveform appears to be randomly "walking across the screen." Which of the following adjustments will most likely correct this condition?
   A. Ground level
   B. Volts-per-division
   C. Time-per-division
   D. Trigger characteristics
17. *Technician A* says that exact time measurements can be made on a DSO screen with the horizontal cursors.
   Technician B says that exact voltage measurements can be made on a DSO screen with the vertical cursors.
   Who is correct?
   A. A only
   B. B only
   C. Both A and B
   D. Neither A nor B
18. When scoping a voltage waveform with a DSO, which of the following characteristics may be important to proper diagnosis?
   A. Amplitude of the voltage
   B. Time
   C. Shape, symmetry and sequence of events
   D. All of the above
19. *Technician A* says that an "internal trigger" means that the DSO is triggering the waveform off the same lead being used to capture the waveform.
   Technician B says that an ignition oscilloscope triggers the secondary waveform off an external trigger lead known as the "number one trigger lead."
   Who is correct?
   A. A only
   B. B only
   C. Both A and B
   D. Neither A nor B
20. When looking at the current waveform of a fuel injector solenoid or an ignition coil primary winding, current that builds too quickly to the saturation level suggests what type of fault?
   A. Open winding
   B. Shorted winding
   C. Circuit has excessive resistance on the insulated side
   D. Circuit has excessive resistance on the ground side

Chapter 5

Exhaust Gas Analysis

OBJECTIVES

Upon completion and review of this chapter, you should be able to:

- ❑ Understand the theory of gas analysis.
- ❑ Describe the exhaust gasses measured by a gas analyzer.
- ❑ Recognize the methods used by gas analyzers to sample emission gasses.
- ❑ Understand how to use emission gas levels in diagnosing engine performance, fuel economy complaints and emission failures.

KEY TERMS

Carbon Dioxide (CO_2)
Carbon Monoxide (CO)
Concentration Sampling
Constant Volume Sampling (CVS)
Exhaust Gas Analyzer
Grams Per Mile (gpm)
Hydrocarbons (HC)
Nitrogen (N_2)
Oxides of Nitrogen (NO_x)
Oxygen (O_2)
Water (H_2O)

Understanding what produces each of the gasses that exit the tailpipe of an automobile is the basis for using these gasses to help in diagnosing symptoms of poor engine performance, poor fuel economy and failure of emission tests. However, bear in mind that gas analysis is used in diagnosis to quickly narrow down the possible areas where pinpoint testing needs to be done. Gas analysis is not a pinpoint test, but rather, needs to be followed up by the proper pinpoint tests. Ultimately, gas analysis can be a very effective diagnostic tool when used properly. It is designed to complement the other diagnostic tools available to today's technician.

THEORY OF GAS ANALYSIS

What Goes In Must Come Out

When the engine is running, it is actually operating as an air pump. Everything that comes into the combustion chamber through the intake valve is ultimately expelled in some form through the exhaust valve into the exhaust system. By measuring the gasses that exit the tailpipe, we can get a good idea of what happened in combustion. But first, a good understanding of how each of the gasses is formed is in order. Although several exhaust gasses are produced during

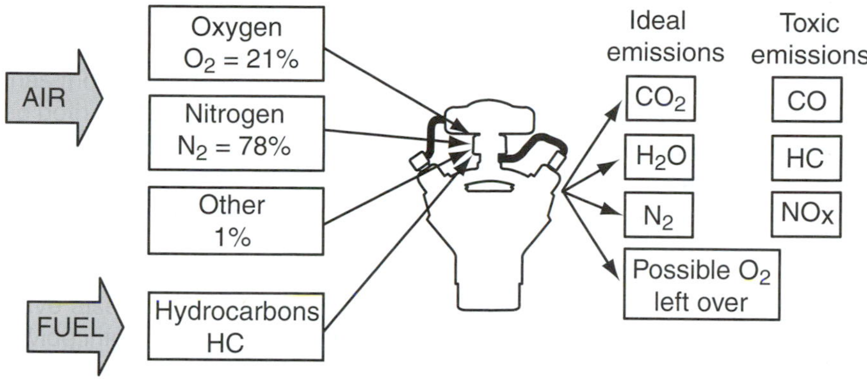

Figure 5–1 Exhaust gasses that result from combustion.

combustion in minute amounts, we present the production of the primary exhaust gasses formed and their concern to the technician, Figure 5–1. An initial description of these gasses is also found in Chapter 1, but they are discussed in greater detail in this chapter.

To begin, both atmospheric air and fuel are introduced into the engine. Atmospheric air consists of about 78 percent **nitrogen** (N_2) and between 20 percent and 21 percent **oxygen** (O_2). Fuel is made up primarily of **hydrocarbons**, each molecule containing one atom of hydrogen and one atom of carbon (HC). As the combustion within the cylinder takes place, these components combine in a chemical reaction and then exit the cylinder.

Nontoxic Gasses

If an engine has good compression, ignition and a stoichiometric air/fuel ratio, the combustion process will produce nontoxic gasses as follows. During combustion of the HCs, the carbon separates from the hydrogen. Both of these elements then recombine with oxygen. The hydrogen atoms combine with oxygen atoms to form molecules of **water** vapor or H_2O, with each molecule consisting of two hydrogen atoms and one oxygen atom. The carbon atoms combine with oxygen atoms to form molecules of **carbon dioxide** or CO_2, with each molecule containing one carbon atom and two oxygen atoms. In order to form CO_2, combustion must take place. Any nitrogen will, for the most part, go through the engine unchanged. And if the air/fuel ratio is slightly lean, a small amount of oxygen may be left over after combustion has taken place.

Toxic Gasses

If some of the fuel does not burn during combustion, it exits the engine as unburned hydrocarbons (or HC molecules). If a *total misfire* occurs within the cylinder for any reason, HC levels will be *dramatically high*. This includes any misfire due to compression problems, ignition misfire or a lean misfire, which occurs on a spark-ignition engine when a lean air/fuel ratio causes the fuel molecules to be spread so far apart that the flame front cannot propagate (about a 17:1 air/fuel ratio). If the air/fuel ratio is too rich, there is not enough oxygen to burn all of the fuel completely. This is sometimes known as a *partial misfire* and results in *moderately high* HC levels.

Carbon monoxide (CO) is *always* produced as a result of combustion taking place with a rich air/fuel ratio resulting in a shortage of oxygen. Misfires do not produce CO. For example, replacing the spark plugs or spark plug wires would not be an effective repair to correct high CO levels. Think of a CO molecule as a molecule that is starved for oxygen. Eventually it will seek out another oxygen atom and be converted to CO_2. In the meantime, carbon monoxide can cause dizziness, nausea and even death if you breathe it in. You cannot detect the presence of CO in the air through sight, taste or smell.

Oxides of nitrogen (NO_x) are produced as a result of combining nitrogen and oxygen (atmospheric air) with temperatures in excess of about 2,500°F. Nitrogen combines easily with other elements in the presence of high heat and/or pressure. In this case, if the combustion chamber temperature begins to exceed 2,500°F, NO_x is formed within the cylinder. NO_x, when mixed with hydrocarbons in direct sunlight, forms photochemical smog, which is an irritant to the eyes and lungs.

MEASURED GASSES

Exhaust Gas Analyzers

Of the seven gasses just defined (four nontoxic gasses [N_2, O_2, H_2O and CO_2] and three toxic gasses [CO, HC and NO_x]), up to five of them are generally measured depending on the particular **exhaust gas analyzer**. Early gas analyzers only measured HC and CO emissions. Unfortunately, on a vehicle that has a catalytic converter in good working order and up to proper operating temperature, HC and CO emissions may both be effectively oxidized within the converter. If the purpose is to perform an emission test, this does not present a problem. But if the purpose is to analyze the exhaust gasses in an effort to help diagnose engine performance problems, this false reduction of HC and CO can give the impression that the engine is performing more efficiently than it really is.

Today's four-gas analyzers not only measure HC and CO but also measure CO_2 and O_2. Both CO_2 and O_2 are, of course, nontoxic gasses, but they also are a valuable aid in using exhaust gasses for analytical purposes if properly understood.

Modern five-gas analyzers also measure NO_x in addition to HC, CO, CO_2 and O_2. The additional capability to measure NO_x adds very little to our analytical abilities, but is a necessary addition if you work in a geographical area where local emission laws require testing for NO_x.

How the Air/Fuel Ratio Affects the Performance of the Gasses

Hydrocarbon levels will increase by reason of any misfire occurring within the cylinder for any reason. If the engine compression and ignition systems are in good working order and valve and spark timing are correct, HC levels will then be low at a stoichiometric air/fuel ratio (14.7:1). As the air/fuel ratio moves richer from 14.7:1, HC levels increase moderately as the air needed for complete combustion is reduced, Figure 5–2. As the air/fuel ratio moves leaner from 14.7:1, HC levels remain low until the point of lean misfire, about 17:1. Beyond this point, they tend to increase dramatically.

Carbon monoxide levels are not increased by reason of engine misfire (in fact, they are slightly reduced). They are also low at a stoichiometric air/fuel ratio. As the air/fuel ratio moves richer from 14.7:1, CO levels increase as the air needed for complete combustion is reduced, Figure 5–3. As the air/fuel ratio moves leaner from 14.7:1, CO levels remain low. As a result, you can determine

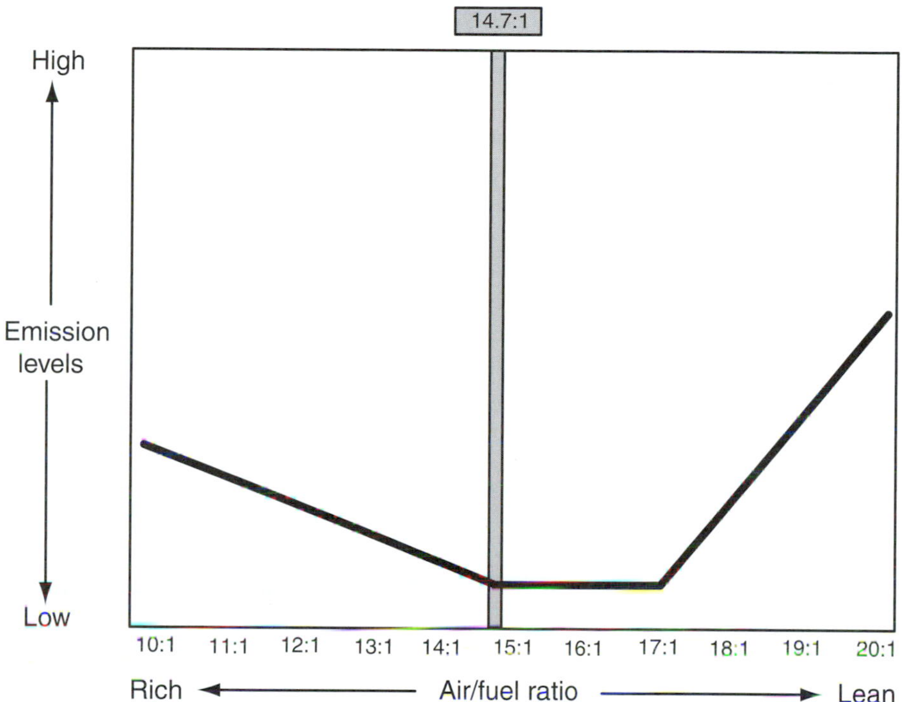

Figure 5–2 Effect of air/fuel ratio on hydrocarbons.

from watching CO levels how far rich of stoichiometric the air/ratio is (assuming the catalytic converter is not yet up to operating temperature), but CO levels cannot be used to determine how lean from stoichiometric the air/fuel ratio is.

Oxygen levels are slightly increased by reason of misfire (misfires result in failure to combust both the fuel and the oxygen). When misfire is not present, O_2 levels are low at stoichiometric but increase as the air/fuel ratio moves leaner from 14.7:1 due to increased levels of oxygen being "left over" after combustion, Figure 5–4. As the air/fuel ratio moves richer from 14.7:1, O_2 levels stay low. As a result, you can determine from watching O_2 levels how far lean of stoichiometric the air/ratio is, but O_2 levels cannot be used to determine how rich from stoichiometric the air/fuel ratio is. Of course, all of this assumes that no false atmospheric air is getting into the exhaust system. Therefore, unlike when performing an emission test, it is important for gas analysis purposes that any existing air injection system has been disabled. Also, the exhaust system must be tight. Any false air getting into the ex-

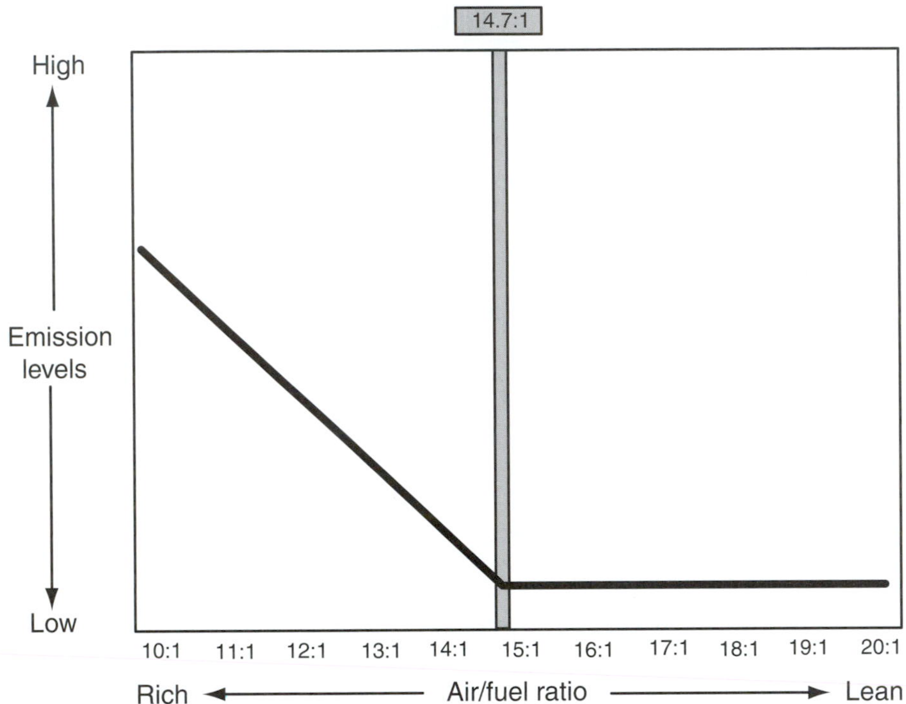

Figure 5–3 Effect of air/fuel ratio on carbon monoxide.

haust system falsely increases the O_2 levels and falsely minimizes the readings of all other gasses.

Carbon dioxide levels will be at their highest point when the engine is running at its utmost efficiency. Anything that reduces this level of efficiency also reduces the CO_2 levels. Mechanical or ignition misfire, misadjusted spark timing and improper air/fuel ratios all have the potential to reduce CO_2 levels. If the air/fuel ratio is either too lean or too rich, CO_2 is reduced, Figure 5–5. CO_2 will be at its highest level at stoichiometric air/fuel mixtures.

Oxides of nitrogen levels may not be much help in diagnosing engine performance problems because there are other possible causes for high NO_x, such as an inoperative electric cooling fan. But it is important to understand what causes high NO_x emissions in order to diagnose and correct failures concerning high NO_x. NO_x is produced as a result of high temperatures. This usually takes

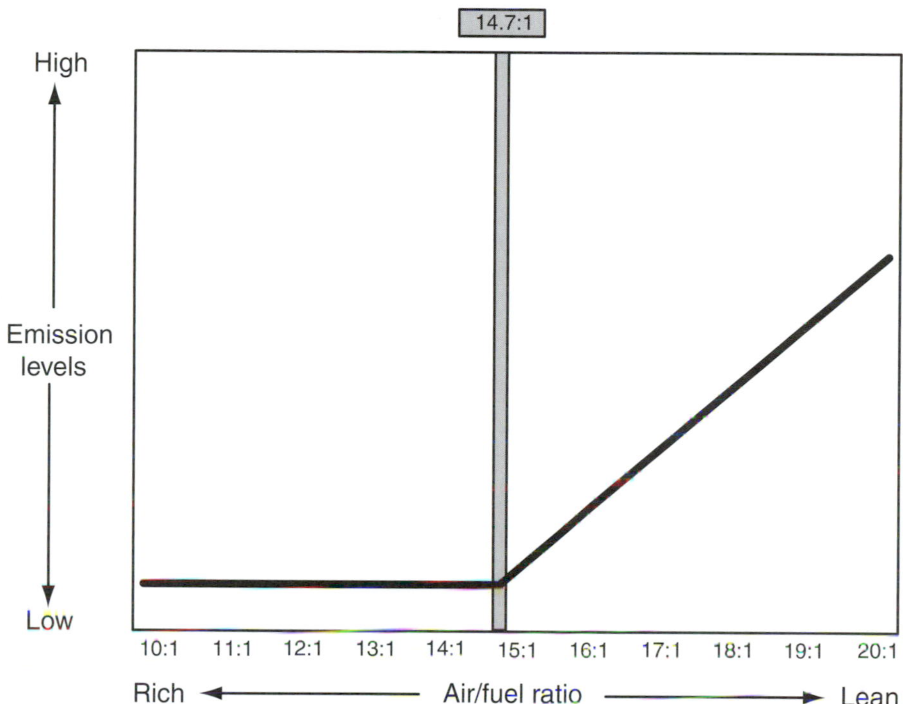

Figure 5–4 Effect of air/fuel ratio on oxygen.

place within the combustion chamber, but can also happen when overheating occurs in the exhaust manifold due to excessive CO and HC being present in the manifold and atmospheric air being added by the air injection system. Many reasons exist for high temperatures occurring in the combustion chamber that can result in production of excessive NO_x. These include the buildup of carbon within the combustion chamber (which increases the compression ratio), failure of the exhaust gas recirculation (EGR) system, engine cooling system problems, overadvanced spark timing, spark knock or ping, thermostatic air cleaner (TAC) systems stuck in the hot air position after the engine is fully warmed up, or early fuel evaporation (EFE) system problems such as a heat grid that is stuck turned on or a heat riser that is stuck closed after the engine is fully warmed up. The effect of air/fuel ratio on NO_x is that leaner mixtures run hotter and therefore

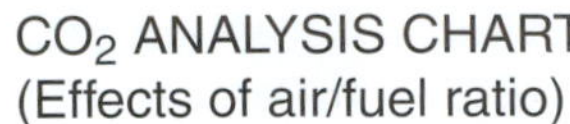

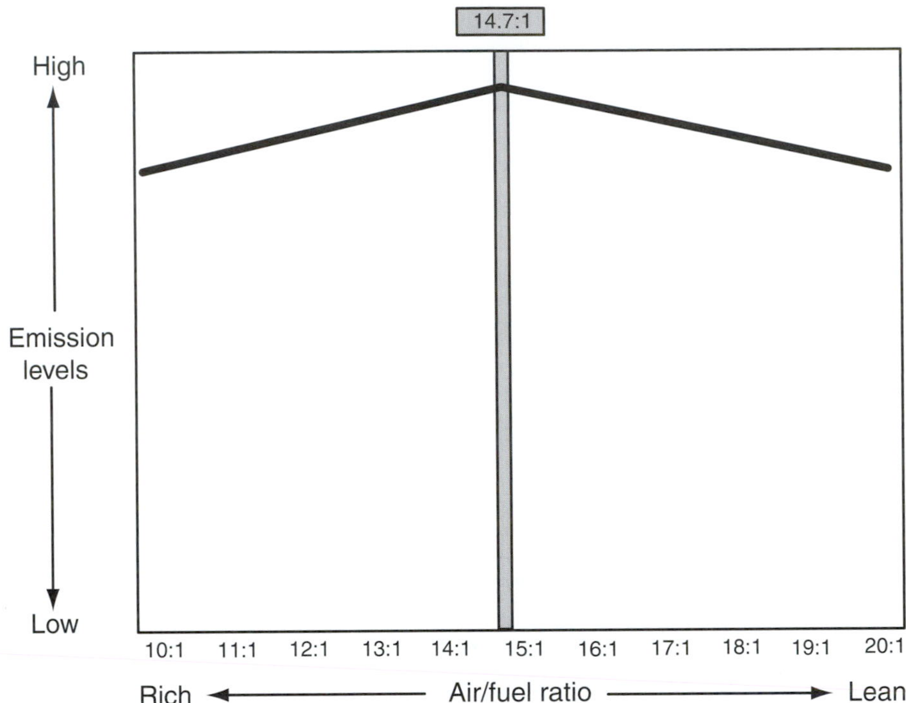

Figure 5–5 Effect of air/fuel ratio on carbon dioxide.

create more NO_x until the point of lean misfire, Figure 5–6.

GAS ANALYZERS

Concentration Sampling

The concentration sampling gas analysis machine is the analyzer found in most automotive shops today. Its design may be a big box type, Figure 5–7, or it may be portable, Figure 5–8. A probe, which is simply inserted into the exhaust pipe, allows the analyzer to sample only a small percentage of the total exhaust gas volume when the engine is running. In turn, each of the gasses is measured as a percentage, or concentration, of the total sample, thus the term ***concentration sampling***. All of the measured gasses are either measured as percent (parts per hundred), or as parts per million (ppm), with ppm simply being a smaller percentage. Because concentration sampling is the method used in most shops, as a technician you need to have a thorough understanding of the levels to expect of each of the gasses of a well-tuned running engine. Keep in mind that these figures are guidelines only and the results will vary with the year and model of each vehicle. Also, many states issue emission test cut-points for the various years and classifications of vehicles. Such cut-points do not reflect the ideal emission levels for these gasses, but only reflect the point at which the particular state has chosen to fail the vehicle during an emission

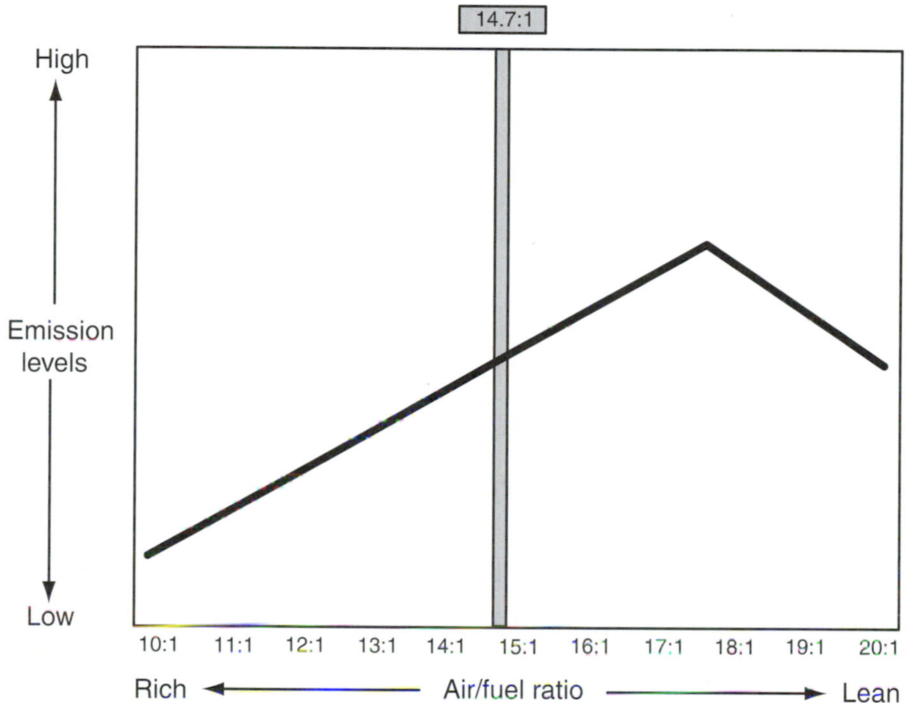

Figure 5–6 Effect of air/fuel ratio on oxides of nitrogen.

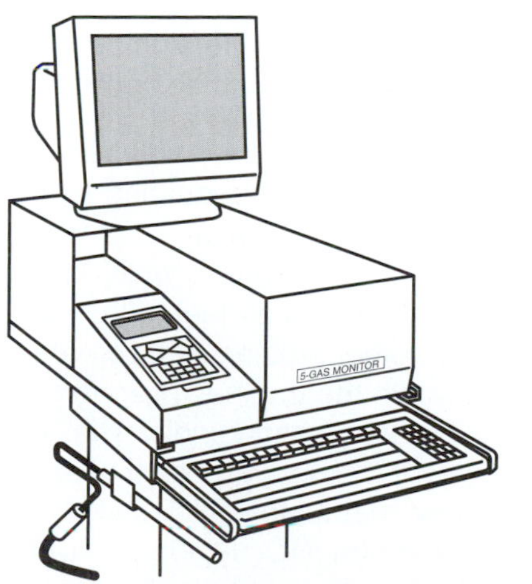

Figure 5–7 Five-gas emissions analyzer. *(Courtesy of OTC Division, SPX Corporation.)*

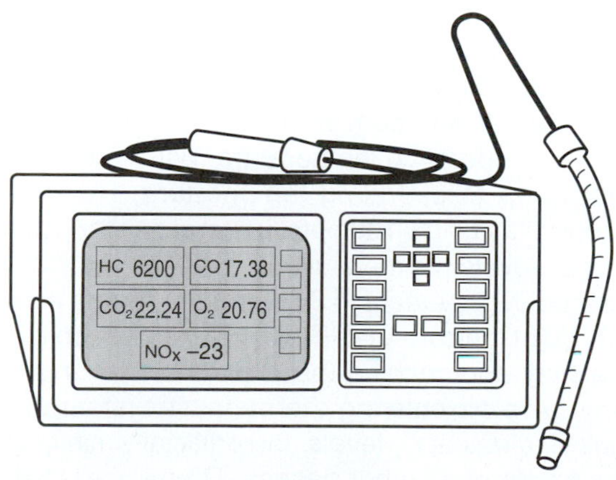

Figure 5–8 A portable exhaust gas analyzer.

test. However, these cut-points can give you some indication of what emission levels to expect if you remember that these cut-points are usually quite generous.

This type of emission testing is usually done within the shop with the vehicle at idle or at 2,500 rpm, but does not require the vehicle to be under any operating load. Therefore this type of test is also referred to as a *no-load emission test.* If you have access to either a dynamometer or a portable gas analyzer that can monitor emission levels while driving the vehicle on public roads, you will be able to monitor emissions with the engine under load, which increases the ability of the analyzer to identify problem areas. But, in most shops, the test is still done without the vehicle being under any engine load.

WARNING: Looking away from the road in order to monitor a portable gas analyzer while driving is extremely dangerous. You should have another technician drive the vehicle for you, allowing you to concentrate on the gas analyzer readings.

CO levels are measured in percent and should stay below 2.5 percent to 3 percent if the vehicle does not have a catalytic converter or if the converter is not up to operating temperature. With a converter that is up to operating temperature, CO levels should stay below 0.5 percent, preferably near 0 percent.

HC levels are measured in ppm and should stay below 300 to 400 ppm if the vehicle does not have a catalytic converter or if the converter is not up to operating temperature. With a converter that is up to operating temperature, HC levels should stay below 100 ppm, preferably near 0 ppm.

O_2 levels are measured in percent and should stay below 2 percent to 3 percent. O_2 levels above this point indicate that the air/fuel ratio is too lean, possibly approaching the area of a lean misfire. Any false air entering the exhaust system artificially increases O_2 levels and artificially minimizes the levels of all other gasses. Remember to disable any air injection systems by pinching off the air hoses and to check the exhaust system for leaks.

CO_2 levels are measured in percent, will generally be above 12 percent to 13 percent and may reach as high as 16 percent. Because engine design parameters (such as valve timing and lift) affect CO_2 levels, it is not possible to set a hard specification for CO_2. Therefore, if the CO_2 is reading close to 16 percent, the engine is likely performing well. If the CO_2 is reading at 11 percent or less, the engine likely has performance problems that need to be diagnosed. But if the CO_2 is between 12 percent and 14 percent, it could mean that the engine is performing well, but it could also mean that there are performance problems with the engine. For example, an engine that emits 15.5 percent CO_2 when running at its peak efficiency might only emit 13 percent CO_2 when a problem is present. The technician must therefore evaluate all of the other gasses when making a determination.

NO_x levels are measured in ppm and should preferably stay quite close to 0 ppm. Unfortunately, while NO_x problems can show up on a no-load test (such as engine overheating), other problems that can create NO_x only show up under engine load (such as spark knock or ping). And although it is important to monitor NO_x in emission program areas that test for it, it is not otherwise as critical a measurement for diagnostic purposes as the other gasses are. For example, a lean air/fuel mixture causes an increase in NO_x levels, but you can determine from the other gasses (leftover O_2) whether the engine is running lean. You are not dependent on NO_x levels to tell you when the engine is lean. Furthermore, while high NO_x levels can indicate a lean air/fuel ratio, they can also be caused by other problems that create high temperatures within the engine. Therefore, high NO_x levels do not necessarily indicate a lean condition. However, if you are within an emission program area that tests vehicles for NO_x, it is important to monitor NO_x levels both before and after repairs are made in order to know if NO_x levels have been reduced on those vehicles that failed for high NO_x.

Some states use a concentration method of exhaust sampling to sample emissions under no load at idle, then 2,500 rpm and then again at idle. This type of test may be referred to as a *no-load test* or an *idle test*. The measured emissions at 2,500 rpm may or may not be used as a pass/failure standard. This type of testing is used most often with older vehicles. Other states have incorporated an emission program referred to as an *Acceleration Simulation Mode* (ASM) test. The ASM test uses a concentration method of exhaust sampling combined with the use of a dynamometer to test CO, HC and NO_x levels under vehicle load. The ASM test is divided into two parts. During the ASM 5015 part of the test, the vehicle is driven on a dynamometer at 15 mph. During the ASM 2525 part of the test, the vehicle is driven on a dynamometer at 25 mph. Both ASM tests require that the vehicle must pass concentration type emission standards for at least 10 seconds of a 90-second test, beginning no sooner than 25 seconds into the test. The results of either the idle test or the ASM test are measured in percent and parts per million.

Constant Volume Sampling

Several states have now adopted an emission-testing program that imitates some characteristics of the federal test procedure (FTP) used to certify new vehicles before they are legal for sale in the United States. This test, called an *Inspection/Maintenance test* (I/M test), places the vehicle under realistic load on a dynamometer designed to imitate the vehicle's curb weight, wind resistance and other rolling resistance. If the vehicle is run for 240 seconds on the dynamometer during the test, then the test is referred to as an *IM240 test*, Figure 5–9. An IM240 emission test uses a process called ***constant volume sampling*** (CVS), which requires that all of the exhaust be captured and measured along with enough ambient air to create a known volume of gasses. Then the density of this known volume is calculated. Finally, the concentration (percentage) of each gas is calculated in order to figure how many grams of each gas are being expelled from the exhaust system. The analytical computer is also able to use the dynamometer to determine the distance the vehicle has traveled in any instant in time or over a longer time period. Ultimately, each gas is calculated in **grams per mile** (gpm).

Constant volume sampling generally tests for four gasses: the three primary toxic gasses (CO, HC and NO_x) and also one nontoxic gas (CO_2). The primary advantage of this type of test is that it tests the vehicle under realistic load conditions. Sometimes the engine computer's strategies are different when the engine is not under load as opposed to when the vehicle is driven down the road. Or a secondary spark plug wire that is developing poor insulation may only arc to the engine block when under realistic load due to the

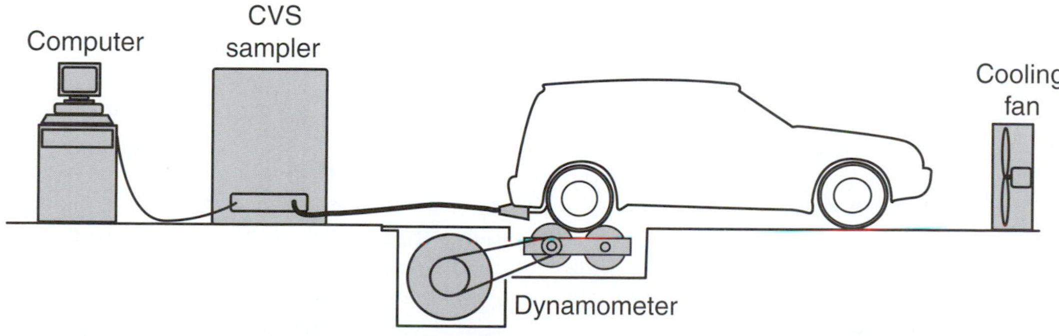

Figure 5–9 Components of a typical IM240 test station.

increased cylinder pressures when under load. This same problem may not show up at all during a no-load emission test due to the fact that the spark plug gap may continue to be the path of least resistance when cylinder pressures are low. Ultimately, the best way to test how many emission gasses exit the tailpipe when the vehicle is traveling on a road is to test the vehicle under the same conditions.

DIAGNOSING WITH THE GASSES

Concentration Sampling

Because most automotive shops have the ability to perform concentration sampling of exhaust gasses either with a big box type scope or with a portable gas analyzer, the emphasis of this book is on this type of testing.

When using gas analysis to aid in diagnosing engine performance problems, fuel economy problems or emission test failures, the technician should take and record the readings of all gasses at both idle and 2,500 rpm. It is critical to be able to test CO, HC, O_2 and CO_2 levels for best engine performance evaluation. If you are working in an emission program area that tests for NO_x, you should have a gas analyzer that also tests for NO_x.

If the reason that you are testing for emission levels is because the vehicle failed an emission test, you should record both before and after repair readings. Although this is always a good idea, it is especially important if the emission test performed on the vehicle produced exhaust gas readings in gpm, as there is no way to reliably convert gpm into percent or ppm. Taking exhaust gas readings before you make any repairs (even before you perform a visual inspection or wiggle any electrical connectors) is known as *baselining*. Baselining the vehicle before you make any repairs allows you to take exhaust gas readings after the repairs have been made and compare them to the baseline readings to determine if your repairs have been effective.

Preparing the Gas Analyzer

Before testing the vehicle, the gas analyzer should be properly warmed up. The analyzer should also be calibrated weekly using a calibration gas and following the manufacturer's specified procedures for the particular analyzer. This test usually includes a leak test of the probe and sample hose.

Before each test, the probe should be left open to sample ambient air while the readings are evaluated. HC, CO and CO_2 should read very close to zero. O_2 levels should read between 20 percent and 21 percent. If not, the O_2 sensor in the analyzer should be replaced. (It is also a good idea to watch the O_2 reading when the gas calibration is performed. It should show very close to zero during the calibration.)

Evaluating the Gasses

This book recommends that the technician first evaluate CO_2 levels, both at idle and 2,500 rpm, in order to make a quick determination concerning the efficiency of the engine at both engine speeds. If the CO_2 is low at either measurement, then you know a problem exists at that engine speed. However, even if the CO_2 is high, be sure to evaluate the other gasses as well. Do the other gasses also support the high level of CO_2 or should the CO_2 be even higher still?

Once the CO_2 has been evaluated, then look at leftover O_2 to check for a lean engine operating condition at either test. (The air injection system must be disabled.) If the engine is running lean at idle but is performing well at 2,500 rpm, check for vacuum leaks. Conversely, if the engine is running lean at 2,500 rpm but is performing well at idle, check for a partially plugged fuel filter.

Look at CO levels to check for a rich engine operating condition at either rpm, while keeping in mind that if the catalytic converter is hot, this value may be falsely minimized. (If you are evaluating emission levels to perform an emission test, you want the converter to be up to full operating temperature. But if you are evaluating emission levels for diagnostic purposes, it is best if you avoid getting the converter up to full operating temperature if possible.) If the engine is running rich at idle but is performing well at 2,500 rpm, check for misadjusted idle mixture screws (carbureted engines) or a small diaphragm tear in the fuel pressure regulator (fuel-injected engines). Of course, larger tears in the regulator's diaphragm will cause overly rich conditions at both engine speeds. Conversely, if the engine is running rich at 2,500 rpm but is performing well at idle, check for a partially plugged air filter.

Look at HC levels to check for indications of a misfire condition at either rpm, while keeping in mind that if the catalytic converter is hot, this value may be falsely minimized. If HC levels are moderately high and CO levels are also high, the HC levels are high because the air/fuel ratio is rich. When you correct the overly rich condition so as to decrease CO levels, HC levels also decrease. If HC levels are dramatically high, then a total misfire is present (although it may be intermittent), either due to an ignition misfire, a lean misfire or mechanical problems. If HC levels are high at 2,500 rpm only or are high at both engine speeds, you need to scope the secondary ignition system. Ignition misfire problems do not decrease with increased rpm. That is, if HC levels are high at idle but are normal at 2,500 rpm, the problem is not ignition related, but instead is due to either a lean misfire or mechanical problems. At this point you may add propane (using a propane enrichment tool) to artificially enrich the air/fuel ratio. If the HC levels now decrease, the problem is air/fuel ratio related. If the HC levels stay high, you should run engine compression tests.

CAUTION: Never spray carburetor cleaner into the throttle body on a running engine as a method of artificial enrichment while a gas analyzer's probe is in the tailpipe. You may damage internal components of the gas analyzer.

Other Tests

A gas analyzer may also be used to perform certain other tests. For example, you may remove the radiator cap from the radiator while the engine is cold, and then use the analyzer probe to check for CO or CO_2 emission gasses escaping from the radiator. Do not evaluate the HC levels for this test, as they are increased by the antifreeze. However, because CO and CO_2 are only produced as the result of combustion taking place within the engine, any CO or CO_2 present in the radiator is a good clue that a head gasket leak is present.

An additional test that may be performed with a gas analyzer is a test for the ability of the ignition and fuel management systems to deliver spark and fuel to each of the cylinders. For example, if a cylinder power balance test shows that one of the cylinders is performing poorly but emission levels seem to be normal, use the scope to perform the cylinder power balance test again while watching the emission gas levels. When the scope disables the spark for each cylinder, if fuel is being delivered, the HC levels should increase dramatically. When the spark is re-enabled the HC levels should decrease again. This test may also be run manually by placing a small wire under the plug wire boot (at either end of the plug wire), then using a grounded test light to short the spark to ground. If the HC levels remain very low when the spark is removed, then fuel is not being delivered to that cylinder. Conversely, if HC levels are always high and removing the spark makes little difference, then spark is not likely being delivered to that cylinder to begin with.

Also keep in mind that the ability to use a gas analyzer to evaluate HC levels can allow you to diagnose other sources of unburned fuel, such as leaking mechanical fuel pumps or leaking fuel injectors.

CAUTION: Never allow the analyzer probe to suck in liquid coolant, because it will damage the gas analyzer.

SUMMARY

This chapter presented how each of the gasses are produced and what emission levels mean in terms of diagnostic purposes. We reviewed some basic information on the types of analyzers and what this means to the technician and the importance of baselining a vehicle before making any repairs. And last we discussed how gas analysis can be used to quickly isolate the general problem area to reduce the amount of pinpoint testing that must be done during diagnosis.

▲ DIAGNOSTIC EXERCISE

A fuel-injected vehicle is pulled into the shop with a complaint of poor gas mileage and poor performance. The engine is properly warmed up and in closed loop for all of the following tests. A scan tool shows that the oxygen sensor is stuck lean and the PCM is commanding a wide pulse width to the fuel injectors. Further testing proves that the oxygen sensor is okay. A gas analyzer probe is inserted in the tailpipe. In order to accurately measure the emission gasses, the technician disables the air injection system. All emission gasses are normal at both idle and 2,500 rpm. Additionally, the technician notices that the scan tool is now showing that the O_2 sensor is now cross-counting and the injector pulse width is normal. At this point the technician re-enables the air injection system. The scan tool again shows that the O_2 sensor is stuck lean and the PCM is commanding a wide pulse width to the injectors. But the gas analyzer is showing a high level of CO at both idle and 2,500 rpm. What apparently is wrong?

REVIEW QUESTIONS

1. Which of the following best describes the primary makeup of atmospheric air?
   A. 21 percent CO and 78 percent O_2
   B. 21 percent O_2 and 78 percent N_2
   C. 21 percent N_2 and 78 percent O_2
   D. 21 percent O_2 and 78 percent H_2O
2. Which of the following emissions of a gasoline engine does a five-gas analyzer not measure?
   A. H_2O
   B. CO
   C. CO_2
   D. O_2
3. Which of the following emission gasses should be as high as possible and is an efficiency indicator?
   A. O_2
   B. HC
   C. CO
   D. CO_2
4. Higher than normal HC emission levels may be the result of which of the following?
   A. An ignition misfire, a lean misfire or a mechanical misfire
   B. An overly rich air/fuel ratio
   C. High combustion chamber temperatures
   D. Both A and B
5. Higher than normal CO emission levels may be the result of which of the following?
   A. An ignition misfire, a lean misfire or a mechanical misfire
   B. An overly rich air/fuel ratio
   C. High combustion chamber temperatures
   D. Both A and B
6. Higher than normal NO_x emission levels may be the result of which of the following?
   A. An ignition misfire, a lean misfire or a mechanical misfire
   B. An overly rich air/fuel ratio
   C. High combustion chamber temperatures
   D. Both A and B
7. All of the following problems can result in the production of excessive NO_x within the engine *except* which of the following?
   A. Failure of the exhaust gas recirculation (EGR) system
   B. A thermostatic air cleaner (TAC) that is stuck in the cold air position
   C. Engine cooling system problems that result in engine overheating
   D. Spark knock or ping under load

8. Which of the following describes concentration sampling?
   A. Uses a dynamometer to load the engine to realistic loads
   B. Captures all of the exhaust and displays each gas as gpm
   C. Samples only a small portion of the total exhaust sample, then displays each gas as either percent or ppm
   D. Both A and B
9. Which of the following describes constant volume sampling?
   A. Uses a dynamometer to load the engine to realistic loads
   B. Captures all of the exhaust and displays each gas as gpm
   C. Samples only a small portion of the total exhaust sample, then display each gas as either percent or ppm
   D. Both A and B
10. When diagnosing engine performance problems, it is important to disable the air injection system and to be sure that the exhaust system does not leak when using a gas analyzer that uses concentration sampling. This is because any false air getting into the exhaust system will do which of the following?
    A. Falsely decrease O_2 levels
    B. Falsely increase O_2 levels
    C. Falsely decrease the levels of all of the other measured gasses (except O_2)
    D. Both B and C
11. If a gas analyzer's probe is placed near a fuel line leak, the analyzer readings will show an increase in which of the following?
    A. H_2O levels
    B. CO levels
    C. HC levels
    D. Both B and C
12. When a gas analyzer is properly warmed up and the probe is sampling ambient air, what should the O_2 level read?
    A. 0 percent
    B. Between 1 percent and 5 percent
    C. Between 20 percent and 21 percent
    D. About 78 percent
13. How would an ignition misfire affect the following gasses?
    A. O_2 levels would be lower; CO and CO_2 levels would be higher.
    B. CO_2 levels would be lower; O_2 and CO levels would be higher.
    C. CO_2 and CO levels would be lower; O_2 levels would be higher.
    D. CO_2 and O_2 levels would be lower; CO levels would be higher.
14. *Technician A* says that you take exhaust gas readings before performing any repair work on a vehicle so that when the repair work is complete you have some initial gas readings with which to compare your final gas readings.
    Technician B says that in order to convert gpm emission values into percentages (or concentrations), you must multiply the gpm by the atmospheric air pressure and then divide by 78 percent.
    Who is correct?
    A. A only
    B. B only
    C. Both A and B
    D. Neither A nor B
15. The following readings are obtained on a gas analyzer:

	Idle	2500 rpm
HC	1,335 ppm	112 ppm
CO	0.00 percent	0.48 percent
CO_2	8.90 percent	14.80 percent
O_2	6.80 percent	0.73 percent

    What type of problem is most likely indicated?
    A. Insulation breakdown on a secondary ignition wire
    B. A vacuum leak resulting in a lean condition at idle
    C. A plugged fuel filter resulting in a lean condition at 2,500 rpm
    D. A plugged air filter resulting in a rich condition at 2,500 rpm
16. The following readings are obtained on a gas analyzer:

	Idle	2500 rpm
HC	1,585 ppm	1,980 ppm
CO	0.00 percent	0.00 percent
CO_2	9.12 percent	7.98 percent
O_2	0.80 percent	0.96 percent

What type of problem is most likely indicated?

A. Insulation breakdown on a secondary ignition wire
B. A vacuum leak resulting in a lean condition at idle
C. A plugged fuel filter resulting in a lean condition at 2,500 rpm
D. A plugged air filter resulting in a rich condition at 2,500 rpm

17. The following readings are obtained on a gas analyzer:

	Idle	2500 rpm
HC	10 ppm	560 ppm
CO	0.02 percent	6.70 percent
CO_2	15.12 percent	10.79 percent
O_2	0.70 percent	0.00 percent

What type of problem is most likely indicated?

A. Insulation breakdown on a secondary ignition wire
B. A vacuum leak resulting in a lean condition at idle
C. A plugged fuel filter resulting in a lean condition at 2,500 rpm
D. A plugged air filter resulting in a rich condition at 2,500 rpm

18. A vehicle comes into the shop that has failed an emission test for high CO.
Technician A says that the spark plugs and spark plug wires may be at fault and should be replaced in order to correct this problem.
Technician B says that the first thing that should be done is to test the vehicle on the shop's gas analyzer in order to verify the readings before any repairs are made.
Who is correct?
A. A only
B. B only
C. Both A and B
D. Neither A nor B

19. An engine is suspected of having a head gasket leak. With the engine cold, the radiator cap is removed and then the engine is started. A gas analyzer probe is then placed in the vicinity of the radiator neck, taking care not to let it come in contact with liquid coolant.
Technician A says an increase in HC levels would indicate a head gasket leak.
Technician B says that an increase in CO or CO_2 levels would indicate a head gasket leak.
Who is correct?
A. A only
B. B only
C. Both A and B
D. Neither A nor B

20. A cylinder power balance test is being run on a vehicle with a V6 engine with port fuel injection while a gas analyzer is sampling the emission gasses. With all cylinders enabled, HC levels are low. When the spark for cylinders 1, 2, 4 and 6 is turned off individually, HC levels increase dramatically. When the spark for cylinders 3 and 5 is turned off individually, HC levels stay low.
Technician A says that this is evidence that cylinders 1, 2, 4 and 6 are getting too much fuel.
Technician B says that the fuel injectors at cylinders 3 and 5 may be plugged.
Who is correct?
A. A only
B. B only
C. Both A and B
D. Neither A nor B

Chapter 6

OBD II Self-Diagnostics

OBJECTIVES

Upon completion and review of this chapter, you should be able to:

- ❑ Define the reasons for the OBD II program.
- ❑ Explain the major aspects of the OBD II program.
- ❑ Describe the features standardized for all manufacturers within the OBD II program.
- ❑ Understand the monitoring conditions that the OBD II PCM requires.
- ❑ Describe the conditions that will cause an OBD II PCM to set a diagnostic trouble code and turn on the Malfunction Indicator Lamp.
- ❑ Explain the strategies of the diagnostic management software.
- ❑ Understand the monitoring sequences required on all vehicles, both domestic and imported, by the OBD II program.
- ❑ Know how to approach an OBD II-equipped vehicle in terms of diagnostics.

KEY TERMS

Drive Cycle
Freeze Frame
Malfunction Indicator Light (MIL)
Misfire Detection
Monitor
OBD II
Snapshot
Standardized 16-Pin Diagnostic Link Connector (DLC)
Trip

Through the years, the federal government has been a driving force for change in the automotive industry. Federal regulations have changed not only for the domestic product but also for all vehicles marketed in the United States. The second update of On Board Diagnostic (**OBD II**) standards has made both foreign and domestic vehicles more similar than dissimilar. This chapter demonstrates those similarities in a broad overview of new vehicles.

With the 1996 models, many car manufacturers built their first OBD II-compliant vehicles. The OBD II program is intended to standardize the diagnosis of emissions and driveability-related problems on all new cars sold in the United States. The new standardized self-diagnostic systems are on most cars for the 1996 model year. (Temporary waivers were allowed initially for some manufacturers with unique technical difficulties, but by the 1998 model year all manufacturers were required to be in compliance.)

WHY OBD II?

The OBD II system (and its predictable successors) should be the basis for driveability and emissions diagnosis for many years to come. It makes diagnostic tools, codes and procedures similar, regardless of manufacturer or country of origin. The system was originally crafted to allow plenty of room for growth and the incorporation of many additional subsystems. While there are some differences among carmakers and among models (because of different system control components), the purpose of the program is to make the diagnosis of emissions and driveability problems simple and uniform; it will no longer be necessary to learn entirely new systems for each manufacturer.

The chemistry of gasoline combustion, the mechanics of a four-cycle engine and the emissions control strategies that have proved successful are the same for all carmakers. These facts plus federal law should make emissions and driveability diagnosis both more successful and easier to learn in the future.

Much of this information involves technical changes introduced gradually, since emissions concerns first began to shape combustion control measures and since in any given year most carmakers' changes are largely enhancements or refocusings of systems introduced previously. Some subsystems described here started as early as 1963 (PCVs); others appear on only a few 1996 models; some will remain unique to specific vehicles. The OBD II system affords the options to do all of this.

WHAT DOES OBD II DO?

The idea behind OBD II is that any properly trained automotive service technician can effectively diagnose any engine performance, fuel economy or emissions concern on any vehicle built according to the OBD II standard using the same diagnostic tools, regardless of vehicle make. As a result, vehicles will be better maintained by the repair industry, improving the overall air quality. OBD II standards have since been enhanced to provide availability of additional manufacturer service information to the repair industry by 2003 to help aftermarket technicians achieve more effective repairs. Meanwhile, carmakers can introduce special diagnostic tools or capacities for their own systems, so long as standard scan tools, along with digital volt/ohmmeters and oscilloscopes can analyze the system. These dealer tools can, of course, have additional capacities beyond the designated OBD II functions. One of the mandated capacities of OBD II systems, for example, is **freeze frame**, the ability of the system to record data from its sensors and actuators at a time when the system turns on the **Malfunction Indicator Light** (MIL). General Motors expands this capacity to include "failure records," which does the same thing as freeze frame, but includes any fault stored in the computer's memory, not just those related to emissions component circuit failures.

The goal of OBD II is to monitor the effectiveness of the major emission controls and to turn on the MIL and store a DTC whenever the effectiveness deteriorates to a point where research indicates the emission level reaches 1.5 times the allowable standard for that gas and that vehicle, based on the FTP.

Besides enhancements to the computer's capacities, the program requires some additional sensor hardware to monitor the emissions performance closely enough to fulfill the tighter constraints and beyond merely keeping track of component failures. In most cases this hardware consists of an additional heated oxygen sensor down the exhaust stream from the catalytic converter, upgrading specific connectors and components to last the mandated 100,000 miles or 10 years, in some cases a more precise crankshaft or camshaft position sensor (to detect misfires more accurately) and the new standardized 16-pin data link connector (DLC), Figure 6–1.

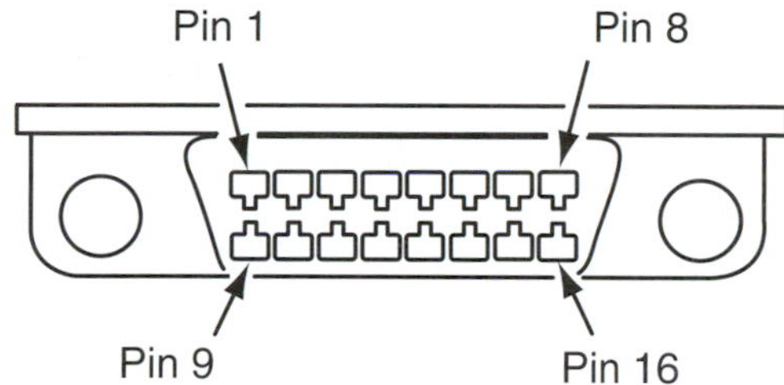

Pin1: Manufacturer discretionary
Pin 2: J1850 bus positive
Pin 3: Manufacturer discretionary
Pin 4: Chassis ground
Pin 5: Signal ground
Pin6: Manufacturer discretionary
Pin 7: ISO 1941-2 "K" line
Pin 8: Manufacturer discretionary

Pin 9: Manufacturer discretionary
Pin 10: J1850 bus negative
Pin 11: Manufacturer discretionary
Pin 12: Manufacturer discretionary
Pin 13: Manufacturer discretionary
Pin 14: Manufacturer discretionary
Pin 15: ISO 9141-2 "L" line
Pin 16: Battery power

Figure 6–1 Standardized OBD II Diagnostic Link Connector (DLC).

STANDARDIZATION

Besides the closer monitoring of emissions performance, the other major change of OBD II is the standardization of diagnosis. While not all vehicles use identical systems, there is much overlap in the types of systems used (catalytic converters, oxygen sensor feedback), and the OBD II program is designed to reduce the confusion between one system and another by mandating not only the standard diagnostic link, but also the specific codes and the descriptions of components in manufacturers' literature. These standards and standard descriptions and trouble codes were all prepared by the Society of Automotive Engineers (SAE) to achieve the following:

- common terms and acronyms (SAE standard J1930)
- common DLC and location (SAE standard J1962)
- common diagnostic test modes (SAE standard J2190)
- common scan tools (SAE standard J1979)
- common diagnostic trouble codes (SAE standard J2012)
- common protocol standard (SAE standard J1850)

Common Terms

All vehicle manufacturers will have to employ common names and abbreviations for components serving similar purposes. For example, the sensor reporting crankshaft position and speed information to the computer will be called a crankshaft position sensor by each manufacturer, and it will be abbreviated CKP, Figure 6–2. The computers will

Old Acronyms			New Technology, All Manufacturers	
Chrysler	Ford	GM	Acronyms	Terms
SMEC/SBEC	ECA	ECM	PCM	Powertrain Control Module
Diag. test Connector	Self-test connector	ALDL	DLC	Data Link Connector
DIS —	DIS/ EDIS	DIS/IDI/ C3I	EIS	Electronic Ignition System
CTS	ECT	CTS	ECT	Engine Coolant Temperature sensor
EVAP	CANP	CANP	EEC	Evap emission control solenoid
—	SPOUT	EST	IC	Ignition control
—	Dis module	C3I module	ICM	Ignition control module
CTS	ACT	MAT	IAT	Intake air temperature sensor
KS	KS	DS	KS	Knock sensor
—	PSPS	PS switch	PSP	Power steering pressure switch
PLL/Check engine light	MIL	Check engine/ Service engine	MIL	Malfunction indicator lamp
TPS	TPS	TPS	TP	Throttle position sensor
—	BP	BARO	BARO	Barometric pressure
Brake switch	BOO	Brake switch	BOO	Brake On/Off switch
Sync	CID	Sync	CMP	Camshaft position sensor
REF pickup	CPS	REF	CKP	Crankshaft position sensor
—	PFE		DPFE	Differential Pres. Feedback EGR
—	TFI-IV	HEI	DI	Distributor ignition
HO_2	HEGO	HO_2	H_2OS	Heated oxygen sensor

Figure 6–2 A partial list of J1930-standardized terminology and acronyms.

be described as PCMs. Most manufacturers began using these terms for their 1993 model year vehicles.

Common Data Link Connector (DLC)

Each vehicle will have a **standardized 16-pin diagnostic connector**, which will have a standard shape and size and will use the same pins for the same information. It will be located somewhere between the left end of the instrument panel and a position 300 millimeters to the right of the center, Figure 6–3. The connector contains seven pins that have been standardized as to what they are to be used for, Figure 6–1. The other nine pins are discretionary pins and may be used by the manufacturer for any purpose. The DLC is for use with a scan tool only. No jumper connections are to be made at the DLC (as was done in the past with some pre-OBD II systems). Also, note that the DLC contains both the power and ground circuits required for powering up a scan tool. Therefore, an additional power supply is no longer needed to power the scan tool as with many pre-OBD II systems.

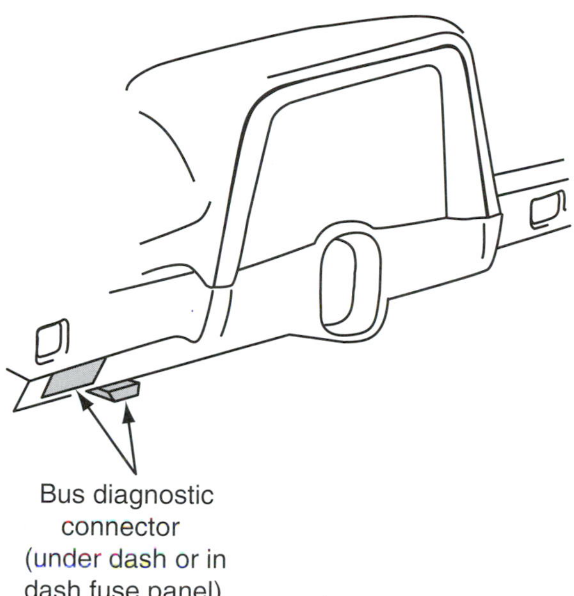

Figure 6–3 Not only is the DLC's configuration standardized, but so is its position—it must be between the left end of the instrument panel and a position no more than 300 millimeters to the right of center. This makes finding it easy for the technician.

Common Diagnostic Test Modes

These test modes are common to all OBD II vehicles and all can be accessed using an OBD II scan tool. Each mode is described in the following paragraphs.

Mode 1. Parameter Identification (PID) mode allows access to certain data values, analog and digital inputs and outputs, calculated values and system status information. Throughout the service literature, there are references to PID values. Some of the PID references are from a generic OBD II PID list that all scan tools must be able to reference. If a referenced nongeneric PID is not on this list, it can be accessed with the manufacturer-specific scan tool or equivalent. If a generic scan tool is used for nongeneric OBD II PIDs, a string of characters (a hexadecimal number in some cases) may have to be entered. The necessary numbers will be supplied by the scan tool maker. Later model and updated generic scan tools will have the capacity to directly access all or selected manufacturers' codes.

Mode 2. Freeze Frame Data Access mode permits access to emission-related data values from specific generic PIDs. These values represent the operating conditions at the time the fault was recognized and logged into memory as a DTC. Once a DTC and set of freeze frame data are stored in the computer's memory, they will stay in memory even when additional emission-related DTCs are stored. The number of such sets of freeze frame data that can be stored is, of course, limited. On General Motors vehicles, for example, the number of such sets is five for the 1996 model year.

One type of failure is an exception to that rule: misfire. Fuel system misfires will overwrite any other type of data and are not themselves overwritten. They can be removed only with the scan tool. When the scan tool is used to erase a DTC, it automatically erases all the freeze frame data associated with that DTC event.

Mode 3. This mode permits scan tools to obtain stored DTCs. The information is transmitted from the car computer to the scan tool following an OBD II Mode 3 request. Either the DTC, its descriptive text or both will display on the scanner. The specific menu access techniques to emission-related DTCs is left up to the scan tool manufacturer, but such data should be relatively simple to extract. As the number and sophistication of scan tools increases, independently supplied scanners will be able to access and interpret the manufacturer-specific codes as well. Complete printed lists of DTCs for all manufacturers are available.

Mode 4. The PCM (Powertrain Control Module) reset mode allows the scan tool to clear all emission-related diagnostic information from its memory. Once the PCM has been reset, the PCM stores an inspection maintenance readiness code until all the OBD II system monitors or components have been tested to satisfy an OBD II drive cycle without any other faults occurring. Quite specific conditions must be met before a given engine start and vehicle movement constitutes a "drive cycle" for the OBD II system, which is described later in this chapter.

Mode 5. The oxygen sensor monitoring test result indicates the on-board sensor fault limits and the actual oxygen sensor voltage outputs during the test cycle. The test cycle includes specific operating conditions that must be met (engine temperature, load, speed) to complete the test. This information helps determine the effectiveness of the exhaust catalytic converter. Here are some, but not all, of the available tests and test identification numbers:

Test ID	Test Description	Units
01	Rich to lean sensor threshold voltage for test cycle	VOLTS
02	Lean to rich sensor threshold voltage for test cycle	VOLTS
07	Minimum sensor voltage for test cycle	VOLTS
08	Maximum sensor voltage for test cycle	VOLTS

Mode 6. The output state mode (OSM) allows a technician to activate and deactivate the system's actuators on command and through the scan tool. When the output state mode is engaged, the actuators can be controlled without affecting the radiator cooling fans. The low- and high-speed radiator cooling fans are turned on separately, without energizing other output components.

Common Scan Tools

For OBD II, a scan tool must access and interpret emission-related diagnostic trouble codes regardless of the vehicle make or model. Most brand-specific and better quality aftermarket scan tools can also access additional information regarding driveability problems and other systems controlled or monitored by the PCM.

The scan tool includes a harness that mates with the standardized 16-pin connector, Figure 6–4.

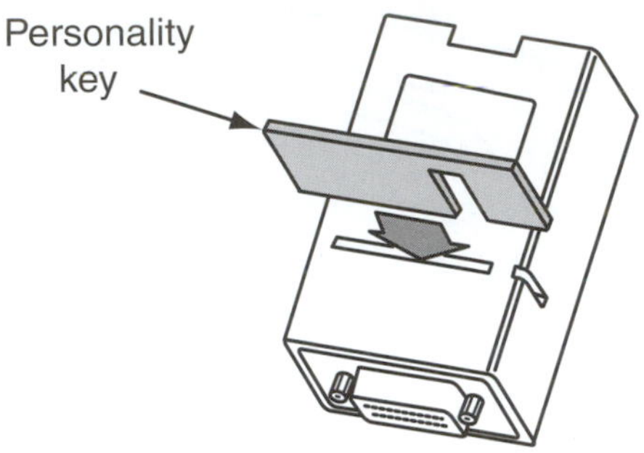

Figure 6–4 The connector of a generic scan tool fits the universal DLC. This aftermarket unit uses plug-in "personality keys" to help take advantage of each car maker's unique, proprietary diagnostics, those that go beyond OBD II.

Common Diagnostic Trouble Codes

SAE J2012 determines a five-character alphanumeric code in which each character has a specific meaning. The first character is the prefix letter indicating the range of the function:

P = Powertrain
B = Body
C = Chassis
U = Network or Data Link

The second character (the first number) indicates whether the DTC to follow is a standard SAE code or one specific to the manufacturer:

0 = SAE
1 = Manufacturer
2 = SAE
3 = Manufacturer

The difference is that a DTC with 0 as the second character means the same thing regardless of the make or model of the vehicle, while a DTC with 1 as the second character has its meaning defined by the manufacturer of that vehicle, so its meaning may be different from exactly the same DTC on another make and model of car.

The third character of a powertrain DTC (one beginning with P) indicates the system subgroup:

0 = total system
1 and 2 = fuel/air control
3 = ignition system/misfire
4 = auxiliary emission controls
5 = idle/speed control
6 = PCM and inputs/outputs
7 = transmission
8 = non-EEC powertrain

The fourth and fifth characters identify the specific fault detected. Some OBD II vehicles still flash a two-digit code through the malfunction indicator lamp, as previously. Typically these codes are not numerically identical with the last two characters of the OBD II (SAE J2012) code, but correspond to the manufacturer's older code tables.

Standard Protocol

A "protocol" in computer language is merely an agreed-upon digital code (or language) the computer uses to communicate with the scan tool. Compliance with OBD II means each manufacturer uses a standardized multiplexing language between the PCM and its sensors and actuators and with the diagnostic information sent to and received from the scan tool through the DLC.

MONITORING CONDITIONS

OBD II standards require the engine management system to detect faults, set DTCs, turn on or off the malfunction indicator light, or erase the DTCs for each monitored circuit according to very specific sets of operating conditions. The following definitions help identify the conditions determined so far:

Warm-up Cycle

OBD II standards define a warm-up cycle as a period of vehicle operation, after the engine was turned off, in which coolant temperature rises by at least 40°F and reaches at least 160°F. Most OBD II DTCs are automatically erased after 40 warm-up cycles if the failure is not detected again after the MIL is turned off. See the chart, Figure 6–5. Some manufacturers retain erased DTCs in a 'flagged' condition; forgiven, as it were, but not forgotten. This can be useful if the technician notices a pattern of component failure, all of which might be related to a single intermittent cause like low fuel pressure.

Trip

A **trip** is a key cycle that consists of Ignition On, Engine Run, Ignition Off and PCM Power Down, during which the enable criteria for a particular diagnostic test are met and the diagnostic test is run by the PCM, Figure 6–6. A trip is used by the PCM to perform a diagnostic test to confirm a symptom or its repair. If a trip is used by the PCM to confirm a symptom, the PCM responds by setting a DTC in memory if the symptom is still present. A trip can also be used by the PCM to confirm a repair after a DTC has been cleared from the PCM's memory with a scan tool. Because the enable criteria for each DTC are different, the required trip used to confirm a symptom or repair for each DTC is also different.

Drive Cycle

A **drive cycle** is a series of operating conditions that allows the PCM to complete all of the OBD II emission-related monitors, Figure 6–7. When all of the driving conditions have been met and, therefore, all of the diagnostics have been run, the system is said to be Inspection/Maintenance (I/M) ready and all of the electronic I/M

MONITOR	MONITOR TYPE (when it completes)	NUMBER OF MALFUNCTIONS (on separate drive cycles to set DTC)	NUMBER OF SEPARATE CONSECUTIVE DRIVE CYCLES (to light MIL and store DLC)	NUMBER AND TYPE OF DRIVE CYCLES (with no malfunction to erase pending DTC)	NUMBER AND TYPE OF DRIVE CYCLES (with no malfunction to turn MIL off)	NUMBER OF WARM-UPS TO ERASE DTC (after MIL is extinguished)
Catalyst efficiency	Once per drive cycle	1	3	1	3 OBD II drive cycles	40
Misfire type A	Continuous	1	1		3 similar conditions	40
Misfire type B/C	Continuous	1	2	1	3 similar conditions	40
Fuel system	Continuous	1	2	1	3 similar conditions	40
Oxygen sensor	Once per trip	1	2	1 trip	3 trips	40
EGR	Once per trip	1	2	1 trip	3 trips	40
Compre-hensive component	Continuous when conditions allow	1	2	1 trip	3 trips	40

Figure 6–5 OBD II DTC/MIL function chart.

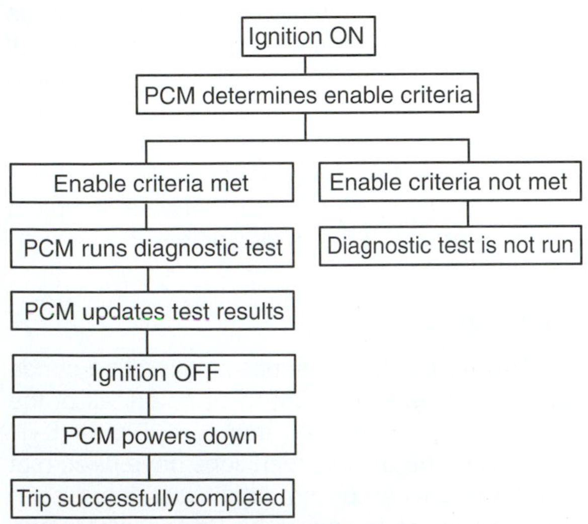

Figure 6–6 An OBD II trip.

"flags" in the PCM are set to Yes. An OBD II scan tool may be used to identify whether system diagnosis is complete, and, if not, which I/M "flags" are not yet set to Yes. The scan tool is only able to determine which diagnostics have not yet been run and are still needed to complete the full drive cycle. A scan tool is not able to identify specifically what the outcome of each test is, except, of course, for DTCs that have been set in the PCM's memory. The diagnostic conditions that make up a drive cycle may be run in any order. As the PCM sees each diagnostic condition achieved, it performs the associated monitors. Drive cycle diagnostic conditions and requirements vary among manufacturers. A pre-1998 General Motors drive cycle is shown in Figure 6–7. It requires 12 minutes to run the entire drive cycle. In 1998, General Motors lengthened its drive cycle to a total of 19 minutes.

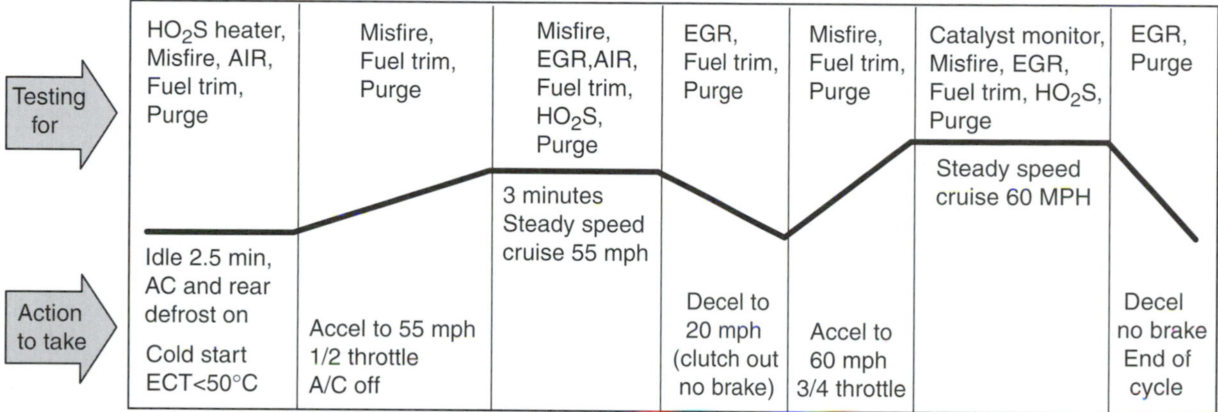

Figure 6–7 General Motors pre-1998 OBD II drive cycle.

Similar Conditions

Once the MIL has been turned on for misfire or a fuel system fault, the vehicle must go through three consecutive drive cycles including operating conditions similar to those existing at the time the fault was first detected to turn the MIL off. Similar conditions mean:

- engine speed within 375 rpm of the DTC flagged condition
- engine load within 10 percent of the same
- engine temperature the same (either cold or warmed up)

Note that with OBD II it is necessary for a circuit to fail twice (or three times for a catalytic converter fault) to set a DTC and turn on the MIL. Notice also that achieving operating conditions similar to those when the DTC was first set could take some time, particularly if the conditions are unusual. If the problem initially appeared only at WOT and high rpm, these conditions may not be met again until the next time Granny borrows the car.

SETTING DTCs AND TURNING ON THE MIL

When an emissions-related fault is detected for the first time, a DTC relative to that fault is stored as a pending or intermittent code. Different manufacturers describe this preliminary code differently. Ford calls it a "pending" code; DaimlerChrysler calls it a "maturing" code; and General Motors calls it a "failed last time" code. Under whatever name, during the next trip or drive cycle the pending fault will be erased if the monitoring sequence that first detected the fault is repeated and the same fault does not recur. If the fault does recur on the second trip or drive cycle, then the DTC is stored (now as a "history" code in General Motors language), and the computer turns on the MIL.

There are two current exceptions to this arrangement. In the first, the misfire monitor can store a DTC and start flashing the MIL in response to its first detection of a type A misfire (a type A misfire could overheat and damage the three-way catalytic converter). In the second exception, the catalyst monitoring sequence must detect a fault in three OBD II drive cycles before storing a DTC and turning the MIL on.

Different faults must fit different criteria for turning the MIL on or off. If the MIL was turned on by a monitoring sequence for the heated oxygen sensor, the EGR or the total system, it will turn off after three consecutive OBD II trips during which the problem does not recur under the expected circumstances. If a misfire or fuel mixture monitoring sequence detected the fault, it will be turned off after three repeated OBD II trips with the right conditions met but the problem not appearing. Similarly, a MIL illumination that resulted from a fault detected in the catalytic converter effectiveness monitoring sequence will also be automatically turned off after three OBD II trips with the proper conditions met but no recurrence of the fault. Turning the MIL off, however, does not automatically erase the stored DTC.

Erasing a DTC requires 40 warm-up cycles without the problem recurring. These 40 cycles begin only after the MIL is turned off. DTCs can also be erased by a technician using a scan tool.

DIAGNOSTIC MANAGEMENT SOFTWARE

Each PCM includes diagnostic management software to organize the complex testing procedures. The terms used for this diagnostic management software varies by manufacturer. Ford and General Motors call theirs the "diagnostic executive," while DaimlerChrysler calls the same thing the "task manager."

Each monitoring sequence performs its tests under a unique set of operating conditions, involving specific temperature, engine speed, load, throttle position and time duration from startup. These conditions must be met for the test to be completed. The diagnostic management software determines the sequence in which the tests will be run, whether the proper conditions have been met for each test and whether the duration of the test was long enough. If not, the test is aborted, and that trip is not counted relative to that test. The managing software waits for the next opportunity to run the appropriate monitoring sequence.

For an example, let's say an EGR monitoring sequence has detected a fault on the previous trip. A pending DTC has been set, and the diagnostic management software is waiting for the next trip to confirm the fault (and store the DTC and light the MIL) or to confirm correct functioning of the system and erase the pending DTC. During the next period of vehicle operation, the software waits for engine data indicating the correct temperature and idle speed, with over 4 minutes since startup. If these conditions have been met, it begins the EGR monitoring sequence, testing the EGR system. Suppose there has been enough idle time and the correct amount of acceleration, followed by a steady throttle position at 34 mph for well over a minute. Then let's assume 5 minutes of driving at speeds from 23 to 44 mph, but no WOT operation, followed by unbraked deceleration and idle for 15 seconds. Then the ignition is turned off.

But a trip (the minimum for an EGR monitoring sequence test) has not occurred yet because there was no 1/2 throttle acceleration to 55 mph. The EGR system has neither passed nor failed the second test; the diagnostic management software is still waiting for the next trip so the test can be run again complete. This sequence was enough of a drive cycle, however, to enable the misfire and fuel system monitoring sequences to complete their tests.

Sometimes the diagnostic management software may cancel a test because of an additional problem with significant connections to the test circuit. For example, if the computer already knows the oxygen sensor is not working, it will not conduct a monitoring sequence test of the catalytic converter because the results would be inconclusive. That test will be postponed until the oxygen sensor circuit problem is corrected and the correction is recognized by the computer.

Sometimes the computer will not run a given test because it conflicts with some other test currently underway. For example, if the computer is running the EGR monitoring sequence, it will not run the catalytic converter monitoring sequence at the same time. When the EGR test is underway, the variations in EGR gas temporarily cause conditions in the catalytic converter that are not representative of normal converter operation.

Sometimes the diagnostic management software will run a test, but suspend it until another monitoring sequence has been completed and its results have been evaluated. For another example, the catalytic converter monitoring sequence might be delayed until the oxygen sensor test sequence has been successfully completed.

Freeze Frame Data

Besides storing detected DTCs, the diagnostic management software keeps a running track record of all the relevant engine parameters for a given circuit. If a fault is detected and recorded, that information is stored as a **snapshot**. This data is used by the diagnostic management software for comparison and identification of similar operating conditions when they recur. The data is also available to the diagnostic technician for further information about what might be amiss in the system. The technician may also wish to use the freeze frame data to aid in duplicating the symptom during a road test. This information can be accessed with the scan tool. Freeze frame information typically includes:

- the DTC involved
- engine rpm
- engine load
- fuel trim (short and long term)
- engine coolant temperature
- MAP and/or MAF values
- operating mode (open or closed loop)
- vehicle speed.

The freeze frame data storage capacity for OBD II is only required to store one freeze frame for a DTC, though some manufacturers have chosen to install greater capacity (five in GM's case). On the basic system, freeze frame data is stored only for the one that occurred first, unless the later DTC is a misfire or fuel system fault. In that case, the diagnostic management software replaces the stored data from the lower priority DTC with the freeze frame data relative to the misfire or fuel system DTC.

MONITORING SEQUENCES

As mentioned previously, a monitoring sequence (sometimes called simply a **monitor**) is an operating strategy the computer uses to check the operation of a specific circuit, system, function or component. Some of the specific monitoring sequences commonly used are described in this section.

Catalyst Efficiency Monitoring Sequence

As the engine's fuel system cycles from slightly lean to slightly rich in response to the oxygen sensor signal, an effective three-way catalytic converter (TWC) stores some oxygen during the lean cycles and uses that oxygen to oxidize (burn) the excess hydrocarbons during rich periods. (Of course, an air injection system, if present, supplies additional oxygen.) Because of this phenomenon, the percentage of oxygen in exhaust discharged from the converter tends to be almost constant despite variations in the fluctuating amount entering. An effective converter will even this out, but one that has begun to lose its capacity to store oxygen will not be able to do so; therefore its downstream oxygen signal will begin to fluctuate with the oxygen content entering, just as the upstream sensor does (Figure 6–8).

Catalytic converter effectiveness is tested once per OBD II drive cycle, and this test is conducted by comparing the readings of the downstream (rear), secondary oxygen sensor with the signals from those in front of the converter (pre-cat sensors). The downstream, secondary oxygen sensor is sometimes called the *catalyst monitor sensor* (CMS). As the PCM commands the fuel mixture to be rich and lean across the stoichiometric notch, the upstream oxygen sensors

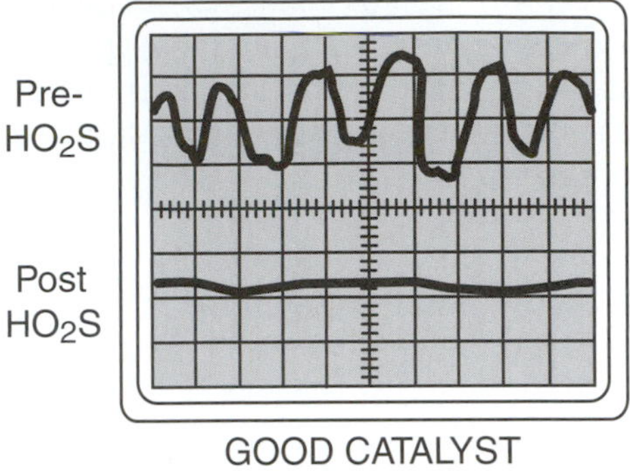

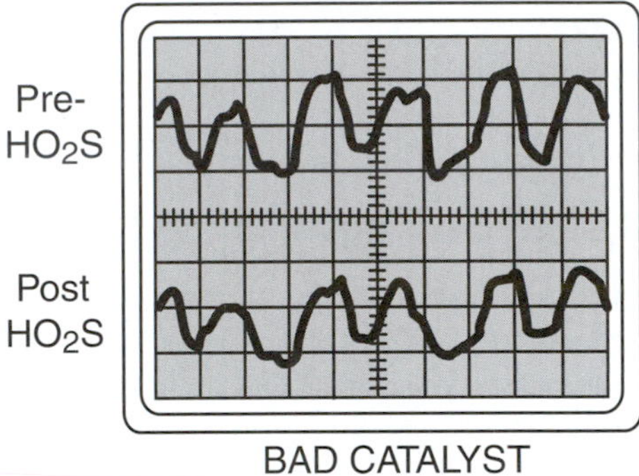

Figure 6–8 There is a great difference between the voltage signals of the two oxygen sensors if the catalytic converter is operating efficiently, but the signal from the rear O_2 sensor starts to fluctuate along with the front sensor if the catalyst is going bad.

should produce a rising and falling voltage signal in response to the residual oxygen after the combustion chambers. The downstream sensor (CMS) should produce a signal of much lower frequency and amplitude, as shown in Figure 6–8.

To prevent crack damage to the ceramic element of the CMS, some manufacturers delay turning on the heater for a period of time, usually until the engine coolant temperature sensor indicates a warmed-up engine. This allows the water condensation in the exhaust system to evaporate. To prevent mixups, the downstream sensor has a different harness connector than the pre-cat sensor, though operation of each sensor is identical.

Misfire Monitoring Sequence

Any time a cylinder misfires, the raw fuel and air are pumped from that cylinder into the exhaust and through the catalytic converter. With this much oxygen and fuel dumped into and burned in the catalytic converter, it gets very hot. The ceramic or aluminum-oxide honeycomb begins to melt into a solid mass, ceasing emissions functions and plugging the exhaust with a molten lump. Emissions performance and driveability both suffer dramatically.

Misfires are therefore monitored continually by measuring the contribution of each cylinder to engine speed. This is referred to as ***misfire detection***. As the engine runs, the crankshaft speed is not actually constant, but changes slightly as each cylinder delivers torque during its power stroke. In between power strokes, the crankshaft actually slows down by a few rpm, only to accelerate backup with the next cylinder's power stroke. By using a very high data rate crankshaft position sensor, manufacturers can provide the computer with a means to track these slight variations in engine rpm. When it does not detect the appropriate acceleration for a given cylinder (identified by the crankshaft position sensor, compared in some cases to data from the camshaft position sensor), the computer knows this cylinder has misfired.

Even on engines that are running perfectly, combustion is not perfect, and there are occasional misfires. Most OBD II systems allow a random misfire rate of about 2 percent before a misfire is flagged as a fault.

The OBD II misfire monitoring sequence includes an adaptive feature compensating for variations in engine characteristics caused by manufacturing tolerances and component wear. It also has the adaptive capacity to allow for vibration at different engine speeds and loads. When an individual cylinder's contribution to engine speed falls below a certain threshold, however, the misfire monitoring sequence calculates the vibration, tolerance and load factors before setting a misfire DTC. Also, on some applications, the

PCM receives input from the antilock brake system's wheel speed sensors to determine when perceived cylinder misfire is simply the result of drivetrain influences due to a rough road surface.

There are two different misfire thresholds: type A misfires and type B/C misfires (Figure 6–9).

Type A Misfire Monitoring Sequence. If during 200 revolutions of the crankshaft, the misfire monitoring sequence detects a misfire rate that would cause the catalytic converter temperature to reach 1,600°F or above, the MIL begins to flash, and the system defaults to open loop to prevent the fuel control system from commanding a greater pulse width (because of the extra oxygen in the exhaust stream from the misfired cylinder). Once the engine is out of the operating range where the high misfire occurs, the MIL stops flashing, but stays on constantly. A DTC is immediately set. Many manufacturers whose products use sequenced fuel injection have programmed the PCM to turn off one or two injectors in the faulty cylinders, to keep from pumping fuel into the exhaust. As a safety exception, if the engine is under load, as when passing or climbing a hill, the PCM does not deny fuel to the misfiring cylinder or cylinders when that capacity is available.

Type B/C Misfire Monitoring Sequence. If during 1,000 revolutions of the crankshaft the misfire monitoring sequence detects a misfire rate of 2 to 3 percent, it sets a pending DTC. At the same time, all the operating conditions at the time are recorded as a freeze frame. If the same pattern is repeated on the next drive cycle, the diagnostic management software turns on the MIL.

Fuel System Monitoring Sequence

Though this is one of the highest priority monitoring sequences, it is also one of the simplest. Whenever the system is running in closed loop, the fuel system monitoring sequence continuously watches short-term fuel trim and long-term fuel trim (as GM's former integrator and block learn are now called).

If a problem such as a vacuum leak, air flow restriction or incorrect fuel pressure causes the

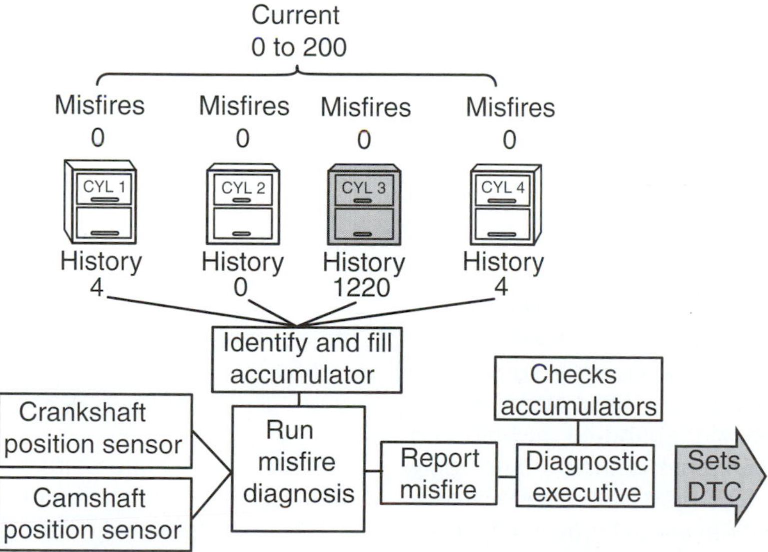

Figure 6–9 Misfire counters are similar to files kept on each cylinder. Current and historical misfire counters are maintained, and the diagnostic executive reviews this information before setting a DTC.

adaptive fuel control to make changes exceeding a predetermined limit in the short- or long-term fuel trim, the fuel system monitoring sequence reports a failure, and a pending DTC is set. On the next drive cycle, if the failure does not reappear, the pending code is erased; if it does appear again, a DTC is set and the MIL is turned on.

Oxygen Sensor Monitoring Sequence

The monitoring sequence tests upstream and downstream oxygen sensors separately, testing each once per drive cycle. Once the diagnostic management software identifies the correct engine operating conditions, the computer pulses the injectors at a fixed duty-cycle rate. The oxygen sensor monitoring sequence checks the frequency of the oxygen sensor signal to see that it produces a signal corresponding to the cycle rate of the injectors. The frequency is high enough that a slow-responding oxygen sensor is not able to keep up and exhibits a reduced amplitude signal as well. To pass the oxygen sensor monitoring sequence, the oxygen sensor must generate a voltage output greater than 0.67 volts, switch across 0.45 volts a minimum number of times during a 120 second period and demonstrate a rapid voltage rise and fall.

Because of the way the catalytic converter works, the rear oxygen sensor, sometimes called the catalyst monitor sensor (CMS), should normally produce a low amplitude and fairly low frequency signal. During CMS testing, the computer forces the air/fuel ratio from rich to lean to force the CMS to produce higher amplitudes. If a rich air/fuel condition is momentarily sustained going into the combustion chambers, available stored oxygen in the catalytic converter is consumed, and the converter will contain less oxygen. In consequence, the CMS signal should go higher in voltage. If a lean condition is then momentarily sustained, the converter gets saturated with oxygen and the CMS signal should drop in voltage.

Certain automotive manufacturers use gold-plated pins and sockets for the oxygen sensor harness connector to obtain increased reliability, to meet the extended period of emissions warranty required by the law.

EGR System Monitor

Different manufacturers use different methods to obtain EGR system feedback, to confirm that the system is working under engine conditions in which it is needed. One of the simplest methods is the one used by DaimlerChrysler.

In normal operation, when the EGR valve opens, the pulse width to the injectors is somewhat reduced, to compensate for the oxygen displaced by the inert exhaust gas. DaimlerChrysler's strategy is to select an operating condition that meets the test criteria, including the criterion that the EGR valve should open. Then, without any other change in the pulse width command to the injectors, the EGR valve is disabled and closed. This should cause the air/fuel ratio to go lean, since there is in effect more oxygen in the mixture. Monitoring the oxygen sensor feedback signal thus indicates the proper (or improper) function of the EGR valve.

Comprehensive Component Monitoring Sequence (CCM)

Remaining inputs and outputs affecting emissions may not be individually tested by a monitoring sequence. They are instead checked by the CCM. In many cases, monitoring these components is done in the same way it was on earlier OBD I systems (Figure 6–10).

Analog inputs are checked for open circuits, shorts and information signals that are out of range by monitoring the analog to digital converter input voltages. Typical examples are:

- intake air temperature (IAT)
- engine coolant temperature (ECT)
- throttle position (TP)
- mass air flow (MAF) and manifold absolute pressure (MAP).

Important: Not all vehicles have these components.

Components Intended to Illuminate MIL
Automatic transmission temperature sensor
Engine coolant temperature (ECT) sensor
Evaporative emission canister purge
Evaporative emission purge vacuum switch
Idle air control (IAC) coil
Ignition control (IC) system
Ignition sensor (cam sync, Diag)
Ignition sensor Hi res (7X)
Intake air temperature (IAT) sensor
Knock sensor (KS)
Manifold absolute pressure (MAP) sensor
Mass air flow (MAF) sensor
Throttle position (TP) sensor A and B
Transmission 3/2 shift solenoid
Transmission range (TR) mode pressure switch
Transmission shift solenoid A
Transmission shift solenoid B
Transmission TCC enable solenoid
Transmission torque converter clutch (TCC) control solenoid
Transmission turbine speed sensor (HI/LO)
Transmission vehicle speed sensor (HI/LO)

Figure 6–10 By California Air Resources Board (CARB) regulations, Comprehensive Component Monitoring will illuminate the MIL if there is a failure in any of these components.

Digital and frequency input signals are checked by plausibility. This is done by using other sensor values and calculations to see whether a given sensor's reading is approximately what would be expected for the existing conditions. For example, the diagnostic management software compares the CKP signal to the CMP signal. Some examples of these interrelated signals are:

- crankshaft position (CKP)
- camshaft position (CMP)
- ignition diagnostic monitor (IDM)
- vehicle speed (VSS)
- output shaft speed (OSS)

All PCM outputs are tested while the vehicle is operating. Some of the resulting input signals are monitored continuously. Others require output actuation so that the resulting change in conditions can be observed. This is commonly known as an *intrusive test*. An intrusive test is a system/component test initiated by the PCM, although, in concept, it is similar to a self-test (described in Chapter 3) that may be initiated by a technician. The driver may even feel an intrusive test as the PCM performs it.

A defective idle air control (IAC) system can cause either incorrect idle speed or a rough idle, either of which can also produce increased or fluctuating emissions performance. The IAC is a closed-loop feedback to the IAC solenoid if the idle speed reported to the computer by the CKP sensor is not correct. The comprehensive component monitoring sequence checks to see whether the corrections made exceed predetermined limits.

Other outputs and actuators are checked for open and short circuits by monitoring the voltage in the actuator's driver circuit. The actuators are energized in almost every case by switching the driver circuit to ground, which reduces the voltage to nearly zero. Whenever the actuator is not energized, the voltage in the circuit should be at the charging system voltage. The following are actuators typically monitored by OBD II systems:

- idle air control coil
- EVAP canister purge vacuum switch
- fan control (high speed)
- heated oxygen sensor heater (HO_2S)
- catalytic converter monitoring oxygen sensor (CMS)
- wide-open throttle air conditioning cutout (WAC)
- electronic pressure control solenoid (EPC)
- shift solenoid 1 (SS1)
- shift solenoid 2 (SS2)
- torque converter clutch (TCC)

Faults in the last four of these actuator circuits most frequently result in the computer's turning on the transmission control indicator light (TCIL) rather than the MIL, if there is such a light on the vehicle.

Control of the two oxygen sensor circuits, the primary upstream mixture control sensor and the downstream catalytic converter monitoring sensor, is different for different manufacturers. Ford, for example, uses the PCM to actuate the heater circuits. In that case it is easy to monitor whether the circuit is working by monitoring the circuit voltage.

General Motors handles the testing differently. In these vehicles, the PCM feeds a bias voltage of approximately 0.45 volts to the HO_2S signal terminal. When the oxygen sensor reaches operating temperature and starts to generate a signal, the bias voltage is turned off, and the system works normally. When the ignition is turned on, it feeds battery voltage to the front oxygen sensor heater circuit, which has a fixed ground. After a cold start the PCM measures how long it takes for the forward oxygen sensor to start generating signals. The sensor, of course, reaches operating temperature faster with the heater circuit energized. If the PCM determines from its memory that it took too long to start generating mixture feedback signals, a pending DTC is set. The amount of time allowed for the sensor to start producing signals depends on the ECT and IAT temperature signals.

DaimlerChrysler uses yet another strategy to test the feedback control front oxygen sensor heater circuit. The heater is powered directly by the automatic shutdown relay (ASD), controlled by the PCM. After the ignition is shut off, the PCM uses the battery temperature sensor to sense ambient temperature. It then waits for a specific time, based on the ambient temperature, for the oxygen sensor to cool down long enough to stop generating any signal. After that time, the PCM energizes the ASD relay. If the heater brings the oxygen sensor back to operating temperature, it resumes generating a signal, even though the engine is not running. If the sensor does not produce a signal within a predetermined amount of time, a pending DTC is set.

Evaporative Fuel System Integrity

To further guard against the possibility of volatile hydrocarbons leaking from the evaporative emissions system, OBD II requirements call for a detection system that can detect a leak equal to a 0.040-inch opening in the system (Figure 6–11). While manufacturers use a variety of means to check this, the most prevalent system seems to be to equip the system with a vacuum solenoid that can, as the test is run between ignition on and engine cranking, use the MAP sensor to check for residual vapor pressure caused by the fuel vapors. Note that such a system, and other proposed systems, will be thrown off if the fuel tank cap is loose or leaks pressure when it is latched. A DTC is set in the usual way.

Secondary Air (AIR) Monitoring Sequence

Manufacturers are permitted to use a variety of methods to monitor the air injection system, provided it detects failures of the designated 1.5 times increase in emissions. The most common means is to use the upstream or downstream oxygen sensor to check on the amount of oxygen in the exhaust while commanding the air into the different parts of the exhaust system.

While most vehicles built to the OBD II standards in 1996 do not employ an air injection system, those that do are able to monitor its function with the downstream oxygen sensor.

Chlorofluorocarbon (CFC) Monitoring

While the OBD II standards required the monitoring of CFCs, other federal laws prohibit the

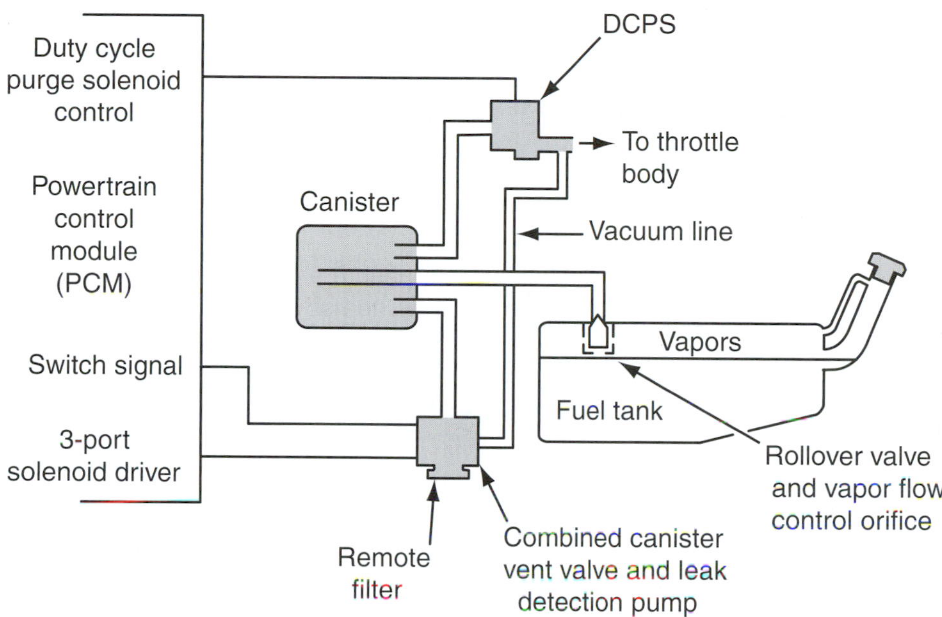

Figure 6–11 A complex network of components and monitor sequence is required to assure the integrity of the evaporative system.

production of the product, so virtually no vehicles will have that type of refrigerant. Without the CFCs, there is no requirement to monitor them.

DOMESTIC CAR MAKER CODES

Ever since computerized engine management systems became popular in the early 1980s, the domestic carmakers have been using their own on-board diagnostics, which includes proprietary fault or trouble codes. These can be made to flash out on the MIL, then one can look up the code number on a chart for the cause of the problem. Since the 1980s, with a proper scan tool, technicians have been able to access the data stream of both GM and DaimlerChrysler vehicles (Ford started making this information available to the technician much later).

With the advent of OBD II, DTCs have become standardized, but there are also make-specific codes (designated by a 1 after the first letter; generic codes have a 0 there).

It is still possible, in some cases, to access the old MIL codes as well. On a Chrysler product, for instance, you turn the ignition key on-off-on-off-on within a span of five seconds, then count the flashes of the MIL. A code 14, for example, tells you that the MAP sensor voltage is too low, which corresponds to generic/universal OBD II code P0107.

See the table (Figure 6–12) for sample DaimlerChrysler DTCs and what they mean, both those available through a generic scan tool and those that can be flashed out on the MIL.

MIL Code	Generic Scan Tool Code	HEX Code	DRB Scan Tool Display	Description
11	P1391	9D	Intermittent loss of CMP or CKP	Intermittent loss of either camshaft of crankshaft position sensor
	or			
		28	No crank reference signal at PCM	No crank reference signal detected during engine cranking
	or			
	P1398	BA	Misfire adaptive numerator at limit	CKP sensor target windows have too much variation
12			Battery disconnect	Direct battery input to PCM was disconnected within the last 50 key-on cycles
13	P1297	27	No change in MAP from start to run	No difference recognized between the engine MAP reading and the barometric (atmosphereric) pressure reading from start-up
14	P0107	24	MAP sensor voltage too low	MAP sensor input below minimum acceptable voltage
	or			
	P0108	25	MAP sensor voltage too high	MAP sensor input above maximum acceptable voltage
	or			
15	P0500	23	No vehicle speed sensor signal	No vehicle speed sensor signal detected during road load conditions

Figure 6–12 A comparison of DaimlerChrysler two-digit flash out codes versus OBD generic five-digit codes.

EUROPEAN APPROACH

Robert Bosch is the major European authority on the subject of computerized engine management systems, and its most highly evolved systems are in the Motronic family, used by Mercedes-Benz, Volkswagen, Volvo and others. Motronic has always meant a combination of fuel and spark management since it was first introduced in 1979. Now it has additional functions and numerical series designations like software, Motronic 4.3 is upgrading to 4.4 as of this writing.

The way the European carmakers adapt Motronic to satisfy OBD II regulations is interesting. In Mercedes-Benz C-Class cars, for instance, the Motronic control module (M-B calls the system HFM-SFI for Hot Film Engine Management-Sequential Fuel Injection) is next to a proprietary diagnostic connector that is to be used with the company's own scan tool, which is used by dealership technicians. OBD II regulations are satisfied by a separate dedicated diagnostic module connected to the Motronic PCM through a CAN (Controller Area Network) bus (see Chapter 7). This fulfills the legal requirements and sends data to a universal/generic DLC (Data Link Connector) under the dash. The OBD II module's only outputs are to the canister purge switchover valve and the MIL. The CAN bus also allows information to flow to and from the electronic accelerator or cruise/idle speed control module.

In Motronic, just as in every other computerized engine management system regardless of country of origin, there is a comprehensive list of sensors, or operating-data acquisition devices, including a chassis accelerometer to differentiate potholes from misfiring. This is different from the approach of other makers, such as GM, which uses signals from the ABS wheel speed sensors to indicate rough road surfaces.

The plausibility check of the air mass meter is a good example of the self-tests. The computer makes continuous calculations from throttle angle and rpm to arrive at a probable injection duration (similar to the operation of a speed-density EFI system). Then, it compares this value to that being obtained from the MAF sensor, and stores a DTC if there is a large enough discrepancy.

A complex test procedure is required for the evaporative emissions system. The following is Volvo's means of performing it:

1. The canister valve is closed and the evaporative system (EVAP) valve is inhibited, which should mean the tank is sealed and the pressure inside is stable. If pressure falls, a DTC for the EVAP valve is stored.
2. The canister valve is opened, venting the system. The EVAP valve starts to cycle and fresh air is drawn through the canister. Pressure in the tank should begin to fall slowly. A rapid drop will set a DTC for the canister shutoff valve.
3. With the canister valve closed, the EVAP continues to cycle, which should make tank pressure fall quickly. If not, the DTC will be for substantial leakage.
4. The EVAP valve is now closed, which should leave a vacuum in the tank. If this falls slowly, a small leak DTC is recorded.

To speed up the completion of trips during diagnosis and service, use this driving schedule from Volvo:

- With the coolant below 77°F and the A/C off, start the engine and put the transmission in gear.
- Accelerate gently to 1,500–2,000 rpm, then drive for 5 minutes at this speed.
- Idle the engine in gear for 70 seconds.
- Drive for 6 minutes at that speed.
- Idle the engine in gear for 40 seconds
- Drive for 5 minutes at 1,500–2,000 rpm again.

ASIAN EXAMPLE: TOYOTA

The first car maker on the street with a fully OBD II-compliant engine control system was Toyota, as early as 1994. This was a considerable engineering accomplishment, especially in light of the fact that some other import carmakers had stated early on in discussions about the OBD II situation that they would have to use add-on modules to meet the standards (similar to the way Mercedes-Benz/Robert Bosch have handled it).

Toyota's OBD II strategies closely resemble those of our domestic carmakers. Just as with every other vehicle manufacturer, however, Toyota uses some proprietary diagnostics connected with its own DTCs, as shown in Figure 6–13.

DTC	Detection Item	Trouble Area	MIL
P1300	Igniter circuit malfunction	• Open or short in IGF or IGT circuit from igniter to PCM • Igniter • PCM	YES
P1335	Crankshaft position sensor circuit (run) malfunction	• Open or short in crankshaft position sensor circuit • Crankshaft position sensor • PCM	NO
P1500	Starter signal circuit malfunction	• Open or short in starter signal circuit • Open or short in ignition switch or starter relay circuit • PCM	NO
P1600	PCM battery malfunction	• Open in back up power source circuit • PCM	YES
P1780	Park/neutral position switch malfunction	• Short in park/neutral position switch circuit • Park/neutral position switch • PCM	YES

Figure 6–13 This chart shows some of the DTCs specific to Toyota.

ENHANCED OBD II MONITORING

The Environmental Protection Agency (EPA) required a modification of the OBD II requirements (sometimes known as *Enhanced OBD II*) that applied to vehicles produced in the 1999 model year. This requirement has led to increasingly advanced OBD II systems. For example, this newer system alerts the vehicle operator if it detects fuel vapor leaks to the atmosphere in excess of the 0.040-inch pinhole requirement already set in place. But the method in which it does, it is more accurate, as opposed to the initial prevalent method that used the engine's MAP sensor to check for fuel tank pressurization from fuel vapors as the engine is started. Let us look at how this system works so that we can understand later modifications more easily.

Enhanced Evaporative Vapor Recovery System (EVAP)

The Enhanced Evaporative Vapor Recovery system consists of two computer-operated solenoids, a pressure sensor, a test port, a charcoal canister and the connecting hoses (Figure 6–14).

Testing the OBD II system begins when the charcoal canister is charged. The charcoal canister is charged when the vehicle is parked. The concrete, asphalt, gravel or dirt the vehicle is parked on has absorbed heat from the sun's rays. The warm surface of the parking pad causes heat to rise to the bottom of the fuel tank and warms the fuel inside. The close proximity of the exhaust pipe also aids this heat radiation. Fuel absorbs heat during the evaporation process through a process similar to air-conditioning heat transfer. That is, when heat is transferred into a liquid, it

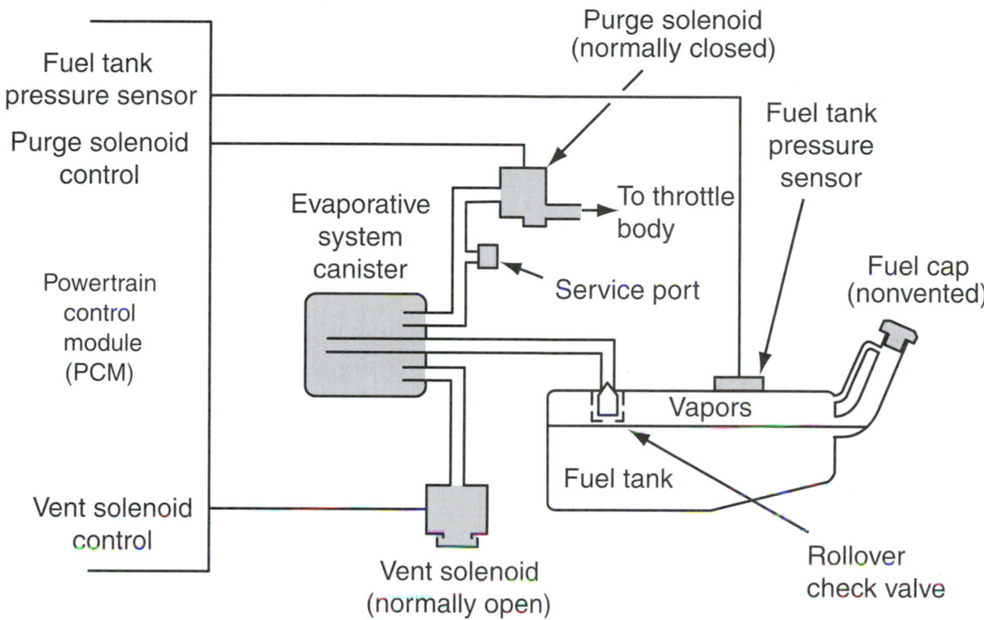

Figure 6–14 EVAP system components.

changes into a gas. The gas then contains the heat. There is only one way the expanded gas can leave the tank. It must go through the charcoal canister and the normally open computer-controlled vent valve. Fuel vapors travel through the activated charcoal of the canister where they are absorbed. This means that only heated air is permitted to vent. This cycle completes the charging process.

Charging does not occur on extremely cold days. The computer knows the outside temperature and compensates for it. The test schedule is modified by the computer if the fuel tank is over 80 percent or under 15 percent full. The tank sender unit provides fuel level information to the computer. In earlier systems, it was merely a sensor for the instrument cluster.

To begin the test, turn on the key but do not start the engine. This puts the normally open vent valve in the rest position. The computer should be able to sense atmospheric pressure in the fuel tank. The computer checks fuel tank pressure sensor against intake manifold pressure sensor. Both should read the same atmospheric value. If the values differ, the computer must decide which sensor has an incorrect reading. It does this through comparison. The two readings are compared against the computer's memory of the last 50 startups. Deviations from these memory parameters cause the computer to set a code. The computer can install a default value for intake manifold pressure sensor failure.

The computer opens the normally closed EVAP solenoid when the engine reaches normal operating temperature. The opened EVAP solenoid allows engine vacuum to pull fuel vapors from the charcoal. The computer uses the oxygen sensor to monitor the exhaust. *Cross counts* are the number of times the oxygen sensor switches from a rich air/fuel mixture to a lean air/fuel mixture within a specific length of time. The cross counts should indicate a sudden richness in the air/fuel mixture. If this richness is not indicated, the computer assumes that either the charcoal was not charged properly or that the charcoal lost its charge. Charcoal charge may be lost when the fuel tank cap is left off. It may also be lost if there is a leak in some of the fuel hoses. The computer does not run further tests unless it senses that the charcoal canister was not charged.

The next test has a relatively short cycle. It is called a *gross* or *large leak test*. The EVAP solenoid is open and allows intake manifold vacuum to flow into the charcoal canister. The computer simply activates the normally open vent valve. When this valve is closed, atmospheric pressure cannot enter the system. If the computer detects engine vacuum inside the fuel tank, it must act rapidly because if this vacuum draw continues for too long, it will suck the fuel tank flat. If the computer does not detect a vacuum, it suspects a gross leak or a pinched hose leading to the tank. If the computer detects a vacuum, it assumes the charcoal canister charge must have leaked out through a very small hole. The computer then runs the next test. This test is called the *small leak test*.

The vent solenoid is activated in a closed position; the EVAP solenoid is activated in an open position, which allows vacuum into the fuel tank. For the small leak test, the computer deactivates the EVAP solenoid. This traps intake manifold vacuum inside the fuel tank. The computer monitors how long the tank holds this vacuum. If the vacuum leaks out within a predetermined time, the computer has discovered a small leak. If the vacuum does not leak out and the ambient air temperature is insufficient to charge the canister, the computer finds no leak and decides the system is operating properly.

The most common problem found with this system is caused by the owner of the vehicle. He or she will fail to tighten the fuel filler cap properly. Although the computer does not detect this at the time of refueling, when the vehicle is started the next morning, the computer discovers a small leak, turns on the check engine light in the dashboard instrument cluster and stores a code. The owner then returns the vehicle to the dealership for service.

On-Board Refueling Vapor Recovery (ORVR) System

Another federal mandate requires vehicles to capture any vapors that occur during refueling. The refueling process begins with a spring-loaded, closed, antispitback valve, which is located at the bottom of the fuel tank filler tube. The filler tube is designed like the venturi area of a carburetor. The filler pipe diameter is reduced to 24 millimeters, or approximately one inch. When fuel travels down this small pipe, it produces a liquid seal that prevents the escape of fuel vapors from the tank during refueling. Vented fuel vapors in the vapor recirculation line are drawn back down the filler pipe by the venturi action of the pipe (Figure 6–15).

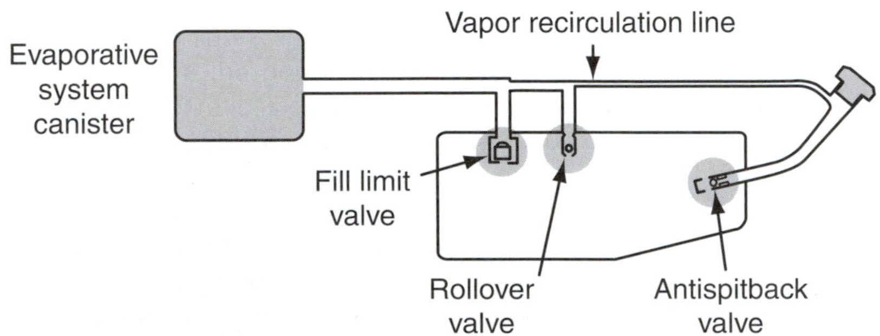

Figure 6–15 ORVR system components.

As the fuel tank fills, the fill limit valve rises. This causes the passageway to the charcoal canister to close and prevents liquid from overcharging the canister. Liquid in the tank rises up the filler neck and shuts off the automatic gas pump nozzle.

DIAGNOSTIC EQUIPMENT

Scan Tools

OBD II performs continuous tests to make sure everything that affects emissions is working properly and provides universal DTCs to help you in troubleshooting.

But you still need diagnostic equipment. The most prominent example of this is the scan tool, a microprocessor-based hand-held test instrument that gives you the capability of accessing the engine management data stream (see Chapter 4). You can then carefully consider this information to see if it makes sense when compared to the symptoms that caused the car to be brought in for service.

OBD II regulations state that DTCs and a large number of engine management sensor signals, computer commands and so forth must be readable on a universal generic scan tool and that all cars must carry a universal 16-pin DLC under the dash.

A scan tool will give you a great deal of important troubleshooting information, but many technicians do not make use of its full capabilities, using it only to pull codes.

That is unfortunate because in most cases the data a scan tool provides is exactly what is needed to find the cause of the trouble. Here is a partial list of the information available on GM cars through a typical scan tool:

- DTCs, both current and historical
- Oxygen sensor signal in millivolts
- Rear oxygen sensor signal in millivolts
- Loop status
- Rich/lean flag
- Power enrichment on/off
- Oxygen sensor cross-counts
- Injector pulse width in milliseconds
- Fuel trim cell
- Fuel trim index
- Short-term fuel trim and average
- Long-term fuel trim and average
- Engine rpm
- Desired idle rpm
- Coolant temperature
- Intake air temperature
- MAP signal
- Barometric pressure signal
- Throttle position volts
- Throttle angle as a percentage
- Calculated air flow
- Low octane fuel spark modifier
- Spark advance in degrees
- Knock retard in degrees
- Knock signal (yes/no)
- Open/closed loop indication
- Converter high-temperature condition (yes/no)
- Air control solenoid (port/atmosphere)
- EGR desired position as a percentage
- EGR actual position as a percentage
- EGR pintle position in volts
- EGR duty cycle percentage
- EGR auto-zero (inactive/active)
- Idle air control position
- Wastegate position
- Park/neutral position
- PRNDL switch position
- Commanded gear
- Brake switch on/off
- Mph/kph
- TCC/shift light on/off
- 4th gear switch on/off
- A/C request (yes/no)
- A/C clutch on/off
- Battery voltage
- Fuel pump volts
- Intake tune valve on/off
- Purge duty cycle percentage
- Purge learn memory
- PROM ID
- Time from start

This kind of information will help you cure the majority of driveability/emissions problems, regardless of whether they have set a code.

Another extremely important feature is that you can read all that important data simply by plugging the scan tool into the DLC and pushing buttons or scrolling through menus. In fact, most technicians who have taken the time to learn the simple basics of scan tool operation typically use it before they even consider hooking up any other type of diagnostic equipment. It is usually the first tool deployed in driveability/emissions diagnosis.

A scan tool may also allow you to activate components normally controlled by the PCM. In DaimlerChrysler's Circuit Actuation Test Mode, for instance, the company's DRB (Diagnostic Readout Box) scan tool (or aftermarket equivalent) can actuate:

- All ignition coils individually
- All fuel injectors individually
- Idle air control motor
- Radiator fan control module
- A/C clutch relay
- Auto shutdown relay
- Duty cycle EVAP purge solenoid
- Generator field
- Torque converter clutch solenoid
- EGR solenoid
- Fuel system test
- Speed control vacuum solenoid
- Speed control vent solenoid
- Fuel pump relay
- All other solenoids/relays
- Set rpm in 100 rpm increments from 800 to 2,000

This test procedure can be very helpful because if a component functions as it is supposed to (you may hear or feel a click or see fuel spray), you can be fairly sure its wiring and driver circuit are okay.

Certain computer system functions are only accessible by the carmakers' own scan tools; for example, bidirectional control of the braking system and the ability to force transmission shifts. Forcing abnormal system operation can cause system damage or even uncontrolled braking. The manufacturers have taken the position that the ability to control these functions should be left to dealership personnel.

Of course, scan tools designed for use on 1996 and newer cars also fulfill the requirements for OBD II tests. Besides engine data such as listed above, there is a category called Specific Engine Data. In GM, for example, this includes:

- information specific to EGR and EVAP system diagnosis
- data required to verify that the EGR and EVAP systems are operating properly
- information specific to the diagnosis of misfire
- information specific to the HO_2S 1 and HO_2S 2 sensors
- data required to verify the proper operation of both oxygen sensors

Another category is DTC Data. This includes freeze frame data, which is information gathered at the moment a DTC is set. From this, the technician can recreate the conditions that were present at the time. Also in DTC Data are failure records, which contain the information present at the time a diagnostic test was failed, but this data is not necessarily associated with MIL activity.

An especially powerful troubleshooting combination is the use of a scan tool along with a four-gas (HC, CO, O_2 and CO_2) or five-gas (add NO_x) infrared exhaust analyzer. This allows you to compare sensor signal or computer command information with actual tailpipe emissions to see if the logical result of these readings is indeed present.

Lab Scopes

No matter how well your scan tool works and how well you understand what it is telling you, there will be occasions when it simply will not let you "see" the problem. Perhaps the glitch occurs so quickly that the scan tool's display cannot show it, or maybe the OBD system simply is not programmed to recognize this discrepancy.

In these cases you may find the lab scope or DSO (covered in Chapter 4) to be very helpful. Because of the way in which a lab scope acquires and displays signals, it can add immense troubleshooting power to your ability to diagnose an OBD II system. Because the potential sampling rate is typically millions of samples per second (dependent upon the time-per-division adjustment), every important detail of a signal or event can be displayed. Think of your lab scope as a high-speed visual voltmeter with which you can test anything that you can test with a DVOM. Nothing about OBD II changes the way in which you use a lab scope, and no updated cartridges are needed for this piece of equipment. Simply use your scan tool with the power of an OBD II PCM to narrow down the area that you need to diagnose, then use your DSO for the pinpoint testing required.

SUMMARY

In this chapter we've covered the whys and the hows of OBD II, and shown that it will make diagnosis much more similar for all vehicles. We have seen how the program is standardized for all manufacturers and have reviewed how carmakers have similar emissions control strategies. Finally, we looked at the main tools for OBD II analysis: the scan tool and the oscilloscope.

▲ DIAGNOSTIC EXERCISES

1. An OBD II-equipped car has a burned valve. How will the system prevent fuel from passing through the cylinder and overheating the catalytic converter?
2. What advantage comes from reading sensors' information and testing actuators through the OBD II connector rather than directly testing them on the car with the harness disconnected?

REVIEW QUESTIONS

1. What advantage for the independent technician is provided by OBD II standards?
   A. It is designed to standardize diagnosis of vehicle emissions and driveability-related problems.
   B. It is designed to force all vehicle manufacturers to use exactly the same engine computer system, including sensors and actuators.
   C. It is designed to force vehicle manufacturers to use new, totally redesigned computer systems in place of the systems already in use.
   D. It is designed to standardize the wiring colors used by all manufacturers.
2. The new standardized self-diagnostic computer systems known as OBD II are required on all cars as of what year?
   A. 1986
   B. 1991
   C. 1996
   D. 2000
3. OBD II standardizes all *except* which of the following?
   A. DLC shape, size and location
   B. Terms and acronyms
   C. Wiring colors used throughout the vehicle
   D. DTC format
4. *Technician A* says that OBD II represents a revolutionary change in the way that computers control engines to maximize performance, economy and emissions.
   Technician B says that the snapshot feature of the OBD II self-diagnostics program is the ability of the PCM to record relevant engine parameters when a fault is recorded.
   Who is correct?
   A. A only
   B. B only
   C. Both A and B
   D. Neither A nor B
5. How does an OBD II PCM detect cylinder misfire?

A. Through input from the vehicle speed sensor (VSS)
B. Through input from a high data rate crankshaft position (CKP) sensor
C. Through input from the knock sensor
D. Through input from the manifold absolute pressure (MAP) sensor

6. If an OBD II PCM detects a type A misfire, it will default to open loop operation and shut down the fuel injectors on the misfiring cylinders in order to protect which of the following?
   A. Engine block
   B. Ignition system
   C. Oxygen sensors
   D. Catalytic converter
7. What is the purpose for placing a heated oxygen sensor after the catalytic converter?
   A. To help the PCM control the air/fuel mixture while in open loop
   B. To help the PCM control the air/fuel mixture while in closed loop
   C. To allow the PCM to monitor the catalytic converter's ability to store and use oxygen
   D. To allow the PCM to test the oxygen sensor placed before the catalytic converter
8. *Technician A* says that the reason that OBD II was adopted was to make diagnosis of a driveability complaint more uniform.
   Technician B says that the reason that OBD II was adopted was due to concern about automotive emissions and air pollution.
   Who is correct?
   A. A only
   B. B only
   C. Both A and B
   D. Neither A nor B
9. In order to evaluate the EGR system, DaimlerChrysler's OBD II PCM selects an operating mode in which the EGR is held open. Then it closes the EGR valve and maintains the current pulse width command to the fuel injectors while monitoring which of the following?
   A. MAP sensor
   B. Oxygen sensor
   C. Crankshaft rpm
   D. Vehicle speed sensor
10. *Technician A* says that an OBD II DTC that begins with P1 is for the powertrain and is specific to the vehicle manufacturer.
    Technician B says that an OBD II DTC that begins with P0 is for the powertrain and is an SAE standardized code.
    Who is correct?
    A. A only
    B. B only
    C. Both A and B
    D. Neither A nor B
11. *Technician A* says that an OBD II drive cycle is a specific set of driving conditions designed to allow the PCM to perform all of the required monitoring sequences.
    Technician B says that the PCM performs all of the tests required by OBD II instantaneously as soon as the engine is started.
    Who is correct?
    A. A only
    B. B only
    C. Both A and B
    D. Neither A nor B
12. What is indicated if, while in closed loop, the waveform from a catalyst monitor sensor (CMS) looks like the waveform from the precat sensor?
    A. The engine is producing a type A misfire.
    B. The engine is producing a type B or type C misfire.
    C. The catalytic converter is in good condition.
    D. The catalytic converter has lost its ability to store and use oxygen.
13. *Technician A* says that, in most cases, the first time that an emissions-related fault is detected, a "pending," "maturing," or "failed last time" code will be set.
    Technician B says that, in most cases, the second time that an emissions-related fault is detected, the PCM will set a DTC in memory and turn on the MIL.
    Who is correct?
    A. A only
    B. B only
    C. Both A and B
    D. Neither A nor B

14. *Technician A* says that if a misfire of sufficient severity to cause the catalytic converter to reach 1,600°F is detected, the PCM will flash the MIL.
    Technician B says that the PCM will not turn on the MIL until the third detection of a Type A misfire.
    Who is correct?
    A. A only
    B. B only
    C. Both A and B
    D. Neither A nor B
15. *Technician A* says that the Bosch Motronic OBD II system satisfies OBD II requirements by using a separate, dedicated OBD II diagnostic module connected to the Motronic PCM through a data bus.
    Technician B says that European cars are not required to meet OBD II requirements because their engine management systems are so sophisticated.
    Who is correct?
    A. A only
    B. B only
    C. Both A and B
    D. Neither A nor B
16. On 1999 and newer enhanced OBD II systems, how does the PCM test the fuel tank for leaks?
    A. By applying manifold vacuum to the fuel tank
    B. By using a pressure sensor to monitor the pressure/vacuum within the fuel tank
    C. By pressurizing the fuel tank with nitrogen, then monitoring for the presence of nitrogen in the air around the fuel tank
    D. Both A and B
17. What input values may be monitored by an OBD II PCM in order to determine if the secondary air injection system is operating properly?
    A. The manifold absolute pressure (MAP) sensor
    B. The mass air flow (MAF) sensor
    C. The oxygen (O_2) sensors
    D. The crankshaft position (CKP) sensor
18. *Technician A* says that an OBD II PCM will turn on the MIL if both the short-term and long-term fuel trims reach predetermined limits in two consecutive drive cycles.
    Technician B says that virtually all vehicles will be exempt from the OBD II standards that require the PCM to monitor for chlorofluorocarbons.
    Who is correct?
    A. A only
    B. B only
    C. Both A and B
    D. Neither A nor B
19. *Technician A* says that OBD II requirements are a "step backward" for some vehicles because it requires less data be made available to an aftermarket scan tool than some manufacturers were already providing with the manufacturer scan tool.
    Technician B says that both OBD II requirements and specific manufacturer diagnostic information coexist, therefore OBD II requirements are likely to provide an advantage to a diagnostic technician.
    Who is correct?
    A. A only
    B. B only
    C. Both A and B
    D. Neither A nor B
20. *Technician A* says that a scan tool can give the technician a large amount of information when connected to an OBD II DLC.
    Technician B says that a lab scope (DSO) may be needed to pinpoint test certain types of problems on an OBD II system.
    Who is correct?
    A. A only
    B. B only
    C. Both A and B
    D. Neither A nor B

Chapter 7

Multiplexing Concepts

OBJECTIVES

Upon completion and review of this chapter, you should be able to:

- ❑ Recognize why multiplexing is reducing hard wiring on many electronic systems, both automotive and nonautomotive.
- ❑ Describe the differences between the two types of circuitry commonly used to form a data bus.
- ❑ Understand the methods used to minimize the effects of an induced voltage on the data bus.
- ❑ Describe the electrical configurations of a data bus and their advantage and disadvantages.
- ❑ Describe how a control module communicates on an SAE-standardized J1850 data bus.
- ❑ Understand the abilities of a scan tool and a lab scope when using them to perform any diagnosis of computers that communicate on a data bus, or problems with the data bus itself.

KEY TERMS

Active Voltage
Arbitration
Asynchronous
Binary Code
Bit
Byte
Data Bus Network
Demultiplexor
Hard Wiring
Kilobyte
Loop Configuration
Master Slave
Megabyte
Multiplexing (MUX)
Multiplexor
Node
Passive Voltage
Protocol
Pulse Width Modulated (PWM)
Serial Data
Slave
Star Configuration
Variable Pulse Width (VPW)

For many years, when an automotive manufacturer decided to add another feature, it involved simply adding more wiring in order to complete the needed circuits. Even with computer-controlled systems, adding more features meant adding more computers, more actuators and more wiring to connect them. And many of these computers required inputs from the same sensors or from multiple similar sensors. For example, many older vehicles had three coolant temperature sensors or switches: one to control the operation of the electric cooling fan, one to control the coolant temperature gauge or warning light and one that acted as an input to the PCM. What if we could provide sensor information to only one computer and then have that computer simply communicate what it knew about these sensors to the other computers that needed the same information?

This concept is known as ***multiplexing (MUX)***. Ultimately, multiplexing eliminates wiring, therefore reducing vehicle weight, while improving dependability, increasing computer diagnostic ability and allowing the addition of many more features. In fact, on a fully multiplexed vehicle, the manufacturer could conceivably add a feature to an existing vehicle by simply providing a software update over the multiplex network to a particular control module. Doing so would install the added feature without having to add additional control modules or wiring.

MULTIPLEXING OVERVIEW

Hard Wiring

Before the utilization of multiplexing, each sensor was wired to each of the computers that needed that sensor's input with **hard wiring**. Hard wiring is any copper wire that performs only one function 100 percent of the time. This includes simple electrical circuits that operate electrical devices, input circuits from sensors to a computer, output circuits through which a computer controls an actuator and the circuits that provide power and ground to a computer.

Elimination of Hard Wiring

A multiplexing circuit is a circuit that can be used to communicate more than one message across it, thereby eliminating the need for multiple wires in order to perform the same work. By this simple definition, Ford Motor Company has been using multiplexing since the mid-1960s in their speed control systems. The Speed Control Command Switches (SCCS) were mounted in the steering wheel and consisted of either four or five switches. As the driver operated each switch, various voltage or ground signals, modified by resistances, were communicated to the transducer or control module over a single circuit. This was important because any additional wires would have created the need for additional slip rings in the steering column and steering wheel assembly (prior to the use of clock springs that are used with air bags).

Of course, today's modern multiplexing systems, as they are used with our digital computers, do not simply use resistances to modify the message, but instead communicate with digitized code known as ***binary code*** or ***serial data***. Circuits over which serial data are communicated are known as a ***data bus network***, Figure 7–1. Any computer that communicates on a data bus is called a ***node***. If a node contains a **multiplexor**, it then has the ability to send messages on the data bus. If a node contains a **demultiplexor**, it then has the ability to receive and decipher messages on the data bus. Many nodes contain both a multiplexor and demultiplexor, therefore having the ability to both send and receive messages on the bus. The data bus circuits do not fall into the category of hard wiring because they are not limited to one job 100 percent of the time, but instead may communicate one message in one instant and a totally different message in the next.

Some of the computers that may constitute nodes on the data bus include the PCM, the Body Control Module (BCM), the Instrument Panel Controller (IPC), the Electronic Brake Control Module (EBCM), the Automatic Transfer Case Module (ATCM), and several other computers including trip computers, message centers, antitheft systems computers, electronic suspension system computers, and the computer that controls the heating, ventilation and air-conditioning system. Instead of running hard wiring from each of the sensors to each of the computers, sensor information is shared on the bus. Additionally, the diagnostic link connector (DLC) may be connected to the data bus. This allows a scan tool to become a node on the bus for diagnostic purposes.

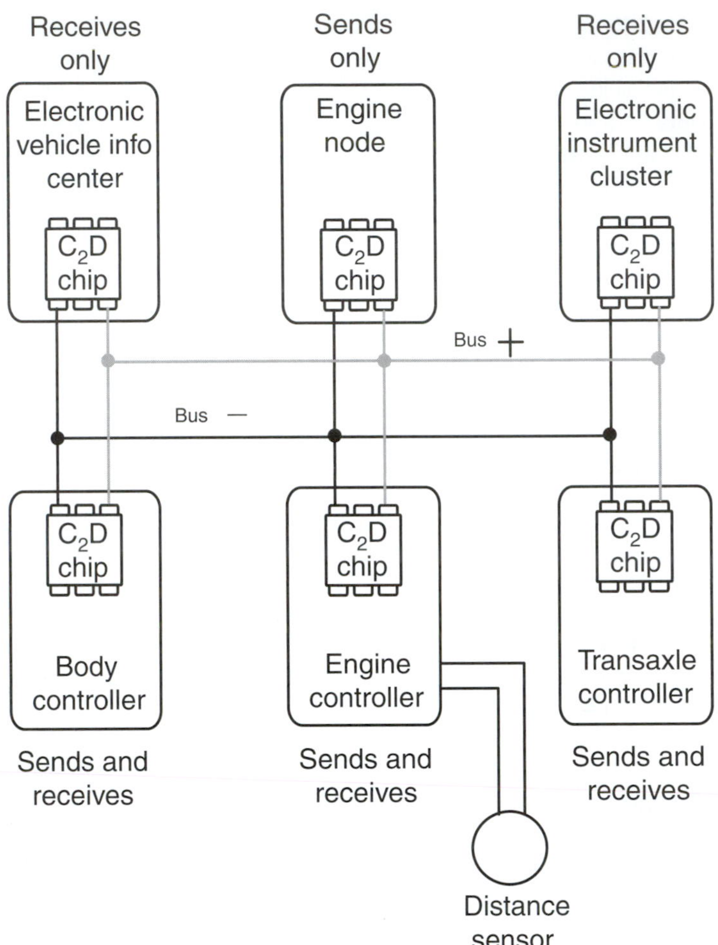

Figure 7–1 Data bus network schematic.

Typically, one computer, (often the BCM), known as the ***master slave***, controls the data bus. The other computers that are dependent on the data bus are known as ***slave*** computers.

The Popularity of Multiplexing

Multiplexing is employed in much of today's technologically advanced components and systems, both automotive and nonautomotive. Telephone systems employ multiplexing extensively. And your personal computer uses the same principles in communicating to and from the monitor, keyboard, printer, scanner and other computers on a network. For years popular video cameras have been equipped with a multiplex port that can be linked with a patch cable to a video editor. The patch cable provides a two-way data bus for communication purposes between the components that it connects together. And home security systems use a data bus to connect the keypads to the primary controller, which typically has the ability to communicate serial data over the phone lines to a central monitoring station. As you can see, learning about how automotive multiplexing systems work will enhance your understanding about how many of today's modern nonautomotive systems also work.

MULTIPLEX SYSTEM DESIGNS

Multiplex Communication

In certain instances, automotive multiplexing circuits may communicate combinations of light on and off signals that are sent over fiber-optic material (similar to how some CD/DVD players are connected to an audio receiver/amplifier), but most automotive applications communicate combinations of voltage on and off signals (or high-voltage and low-voltage signals) that are sent over copper wire. These signals are referred to as "ones" and "zeros." Each one or zero is called a ***bit*** of information and, when strung together, eventually forms a complete word or **byte**. (See Chapter 1 for a more detailed explanation of bits and bytes.) "Bit" is short for "binary digit" and "byte" is short for "binary term." The prefix "kilo" (normally used to indicate a multiplier value of 1,000), when used with "byte" to form **kilobyte**, indicates a value of 1,024 bytes of information. (1,024 is equal to the number of *combinations* of zeros and ones available when a base two numbering system is carried out to ten places. It is also the base ten value of the eleventh column in a base two numbering system.) Similarly, a **megabyte** indicates a value of 1,048,576 (1,024 squared) bytes of information.

Two-Wire Data Bus

Many multiplexed circuits use two wires to complete the data bus circuitry between all nodes, one for bus positive and the other for bus negative. The negative wire connected between all nodes on the bus makes for a cleaner signal on the bus positive wire (as is also done with most ignition systems between the ignition module and the PCM). The two wires are then twisted together in order to minimize the effects of an induced voltage, Figure 7–2. This data bus is known as a *twisted pair*. Comparison of the wires to each other, not to chassis ground, is used to decipher the binary code on these wires. By twisting the wires together, a voltage that might be induced on either wire by a nearby circuit is likely to be equally induced on both wires, therefore not affecting the voltage difference between the two wires and leaving the binary code unaffected.

Typically, two-wire busses use binary code that has a fixed pulse width. That is, all bits (zeros and ones) will be the same length, Figure 7–3. Therefore the serial data on this data bus is said to be **pulse width modulated (PWM)**.

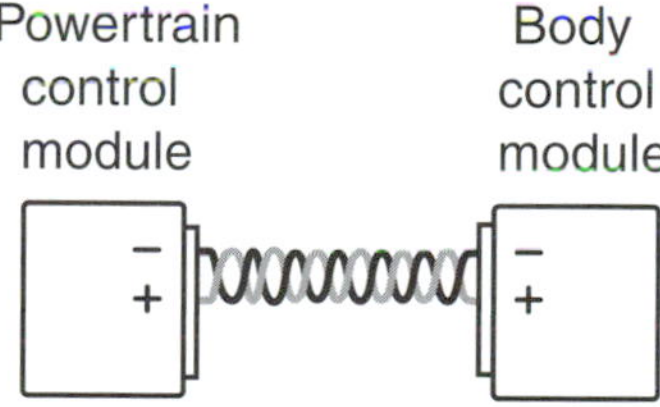

Figure 7–2 A twisted pair of wires forms the data bus that connects between nodes.

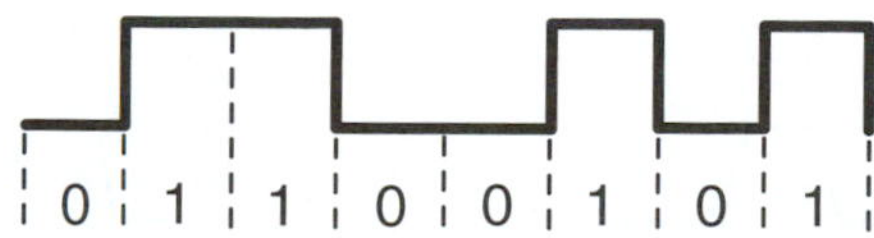

Figure 7–3 Pulse width modulated serial data is formed from binary code in which all bits of information are equal in length.

Single-Wire Data Bus

In order to reduce wiring complexity even further, many newer data busses use a single-wire bus to connect all of the nodes together, Figure 7–4. In this design, the serial data is typically a **variable pulse width (VPW)**, meaning that not all bits of binary code are the same length. A number one may be represented as a short high-voltage pulse, but it may also be represented as a long low-voltage pulse. And, conversely, a number zero may be represented as a short low-voltage pulse or a long high-voltage pulse. This allows the voltage to switch between high and low with every bit of information, Figure 7–5.

If you were to look at this serial data with a lab scope, you would see that the waveform is not truly vertical between bits of information. Each bit is typically slightly trapezoidal in shape, Figure 7–6. This allows the nodes on the data bus to distinguish between serial data and an induced voltage from a nearby circuit, which would tend to be more vertical at the edges.

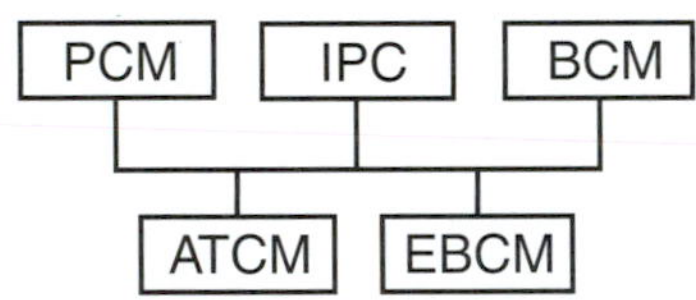

Figure 7–4 A single-wire data bus.

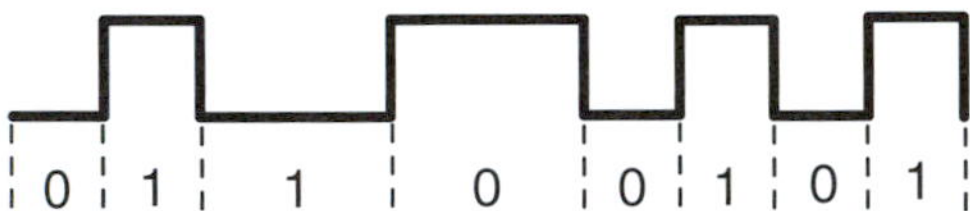

Figure 7–5 Variable pulse width serial data is formed from binary code in which bits of information are of different lengths.

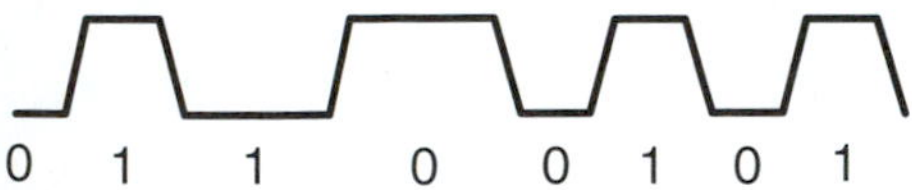

Figure 7–6 VPW serial data on a single-wire data bus has a slightly trapezoidal shape to its waveform.

Data Bus Configuration

Early automotive data busses were wired from one node to the next in a series circuit or **loop configuration**, Figure 7–7. Unfortunately, if one node on the bus were to lose power or ground, it could result in a total bus failure, affecting all the nodes on the bus. Newer configurations tend to wire the nodes to the data bus in a parallel fashion, known as a ***star configuration***, Figure 7–8. While the data bus itself could be shorted to power or ground, if a node on the bus were to lose power or ground, the other nodes on the bus would still be able to communicate. Another advantage of this design is that the system can be designed to allow scan tool connectivity to only one node, or to a selected few. A bus bar may be used to complete the bus connection between each node, Figure 7–9. If the bus bar is re-

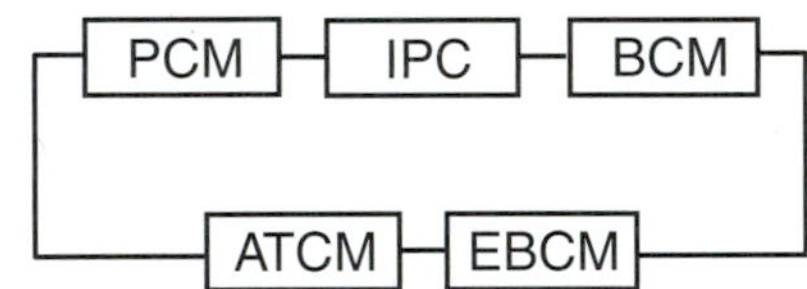

Figure 7–7 Loop design data bus with all nodes wired in series to one another.

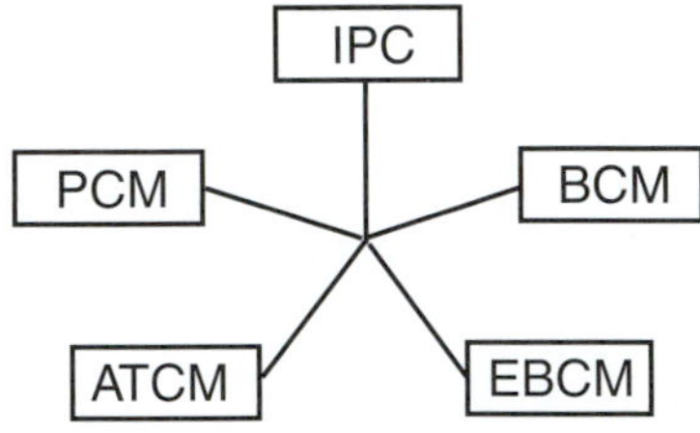

Figure 7–8 Star design data bus with all nodes wired in parallel to one another.

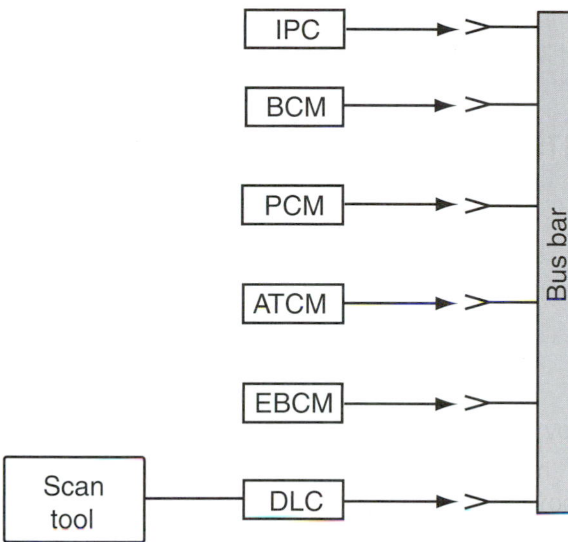

Figure 7–9 Star design data bus with a bus bar present for diagnostic purposes.

moved, then all nodes are isolated from one another. A jumper wire may then be used to connect the bus circuit from the DLC to a selected bus circuit. In this method, scan tool diagnosis can be done with the scan tool connected to one or more selected nodes.

MULTIPLEXING PROTOCOLS

Protocols: Language of the Computer

A **protocol** is simply a language in which computers communicate with one another over a data bus. Rephrased, it is a standardized binary code. Different protocols may vary in baud rate and may also vary as to whether they are pulse width modulated or a variable pulse width. Protocols may also differ according to what voltage level equals a one or a zero. And, while nodes on some data busses communicate continuously in a nonending, steady stream of information as a method of continually updating the other computers on the bus, others communicate intermittently only as needed. The latter is known as an ***asynchronous*** protocol.

Protocol Classes

SAE has defined three different classes of protocols according to their speed of communication or baud rate. A Class A protocol is a low-speed protocol that has a baud rate of up to 10,000 bits per second (10Kb/s) and is generally used for controlling such things as lighting, power windows, power seats and power door locks. A Class B protocol is a medium-speed protocol that has a baud rate of up to 100,000 bits per second (100Kb/s) and is typically used for sharing information between computers. A Class C protocol is a high-speed protocol with a baud rate of up to 1,000,000 bits per second (1Mb/s) and is typically used for select functions that require real-time control (such as drive-by-wire). As performance needs continue to increase, a D classification should be expected in the future that will be able to support baud rates in excess of 1Mb/s.

Common Protocols

Early protocols used on automotive applications include GM's UART (Universal Asynchronous Receiver-Transmitter) protocol and DaimlerChrysler C2D (Chrysler Collision Detection) protocol. These are class A protocols and handle PWM data. Chrysler's term *collision detection* refers to the process of preventing two or more messages from colliding on the data bus.

Most newer automotive protocols fall under OBD II standards such as the J1850 protocol. SAE has adopted this protocol as the class B standard protocol. This protocol may either use PWM data on a two-wire bus (Ford's Standard Corporate Protocol [SCP] and Audio Corporate Protocol [ACP]), or may use VPW data on a single-wire bus (GM's Class 2 protocol and DaimlerChrysler's

Programmable Communication Interface [PCI] protocol).

Another automotive class B protocol that falls under OBD II standards is also standardized by the International Standards Organization (ISO) and is known as the ISO9141 protocol. This is a European standard protocol, but is also used on some of the domestic applications as well. In fact, the California Air Resource Board (CARB) has adopted the ISO9141 protocol for all vehicles sold in California with feedback fuel management systems. ISO9141 standardizes a protocol to be used between the nodes on the bus and an OBD II standardized scan tool (as per SAE J1978 standards) for diagnostic purposes. It is a single-wire bus, which is called the *K-line*. The K-line allows for bidirectional communication between the scan tool and the nodes on the bus. An optional L-line is sometimes used to allow one-way communication from the scan tool to the nodes on the bus, Figure 7–10.

The principal class C protocol in use is called *Controller Area Network* (CAN) and was developed by Robert Bosch in the early 1980s. This protocol supports up to 1 Mb/s and is designed for real-time control of specific systems. The CAN protocol has been accepted in Europe as an ISO protocol. It is also gaining acceptance in the United States.

A point to note: Intel has been one of the primary companies involved in the design of automotive multiplexing networks and has designed products that incorporate both J1850 and CAN functionality into them.

COMMUNICATION ON A J1850 VPW DATA BUS

Overview of the J1850 Protocol

Because SAE has standardized the J1850 protocol as the class B standard protocol, it is presented in more detail here. The J1850 standard supports two versions, a PWM data bus that uses a twisted pair and supports a baud rate of 41.6 kilobytes per second (Ford's SCP and ACP systems) and a VPW data bus that uses a single-wire bus and supports a baud rate of 10.4 kilobytes per second (GM's Class 2 and DaimlerChrysler's PCI systems). Our emphasis here is on the VPW version in order to demonstrate the detail of such a protocol.

Passive versus Active Bits

A **passive voltage** is a voltage that is at rest. An **active voltage** is a voltage that is either actively pulled high or low by a computer circuit. On a J1850 data bus, a passive voltage is a low voltage that is close to ground (between 0 volts and 3.5 volts). The J1850 standard defines an active

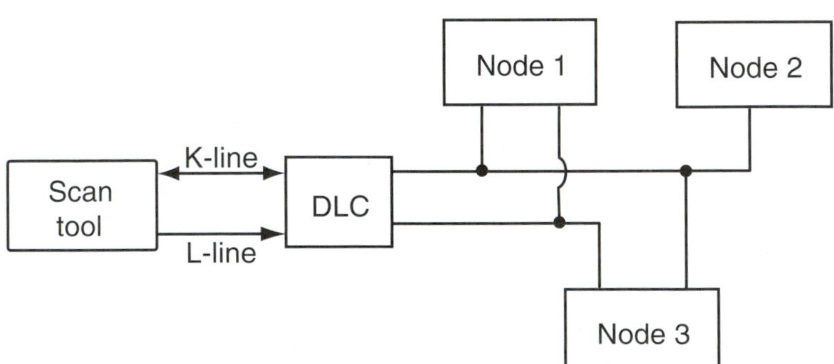

Figure 7–10 ISO9141 data bus.

voltage as one that is pulled high (between 4.25 volts and 20 volts). In order to allow the voltage to switch with each bit of binary code, a one may either be a short active voltage pulse (a high-voltage pulse that lasts 64μs) or a long passive voltage pulse (a low-voltage pulse that lasts 128μs). And, conversely, a zero may either be a short passive voltage pulse (a low-voltage pulse that lasts 64μs) or a long active voltage pulse (a high-voltage pulse that lasts 128μs), Figure 7–11.

The J1850 Message

The J1850 message begins with a Start of Frame (SOF) signal, an active voltage pulse that lasts for 200μs, Figure 7–12. This is immediately followed by the next portion of the message, called the *header*. The header's first byte designates message priority, whether the header is one or three bytes long, whether an In Frame Response (IFR) is required from the receiving node, an address identification that identifies both the sender and the intended receiver, and the intended message type.

The next portion of the message is the Data Field, which contains the "meat" of the message and may be up to 11 bits in length. This is immediately followed by a Cyclical Redundancy Check (CRC) byte. The CRC and the data field are, together, considered one "word" and do not exceed 12 bits in length. For example, if the data field is 11 bits, the CRC is 1 bit. But if the data field is less than 11 bits, then the CRC can be more than 1 bit as long as the total does not exceed 12 bits when added to the data field. The purpose of the CRC bit is to guard against message error. The receiving nodes add the CRC value to the value of the data field in order to calculate whether message errors have occurred.

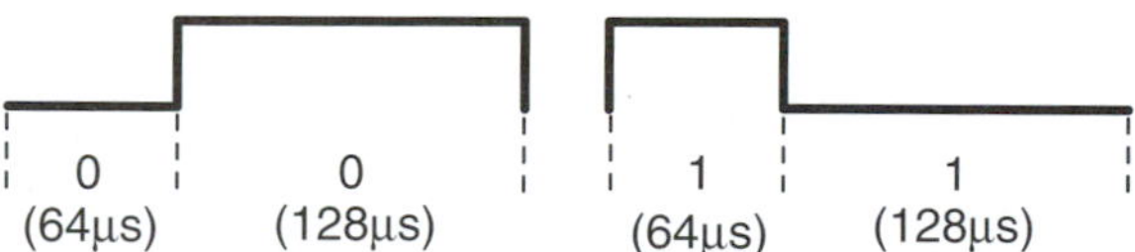

Figure 7–11 SAE J1850 protocol bit length.

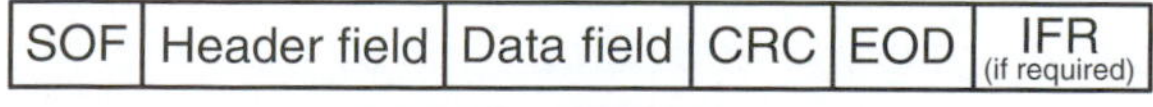

SOF	Header field	Data field	CRC	EOD	IFR (if required)

Figure 7–12 The SAE J1850 message.

The CRC is then followed by an End Of Data (EOD) pulse. This is a passive voltage that lasts 200μs. If requested in the header field, the receiving node must issue an IFR within the next 80μs. This allows the transmitting node to be certain that the intended receiver received the message. Think of it as a handshake between the transmitting node and the receiving node, similar in concept to the "handshake" that occurs between fax machines when the tones from the sending unit are responded to by a tone from the receiving unit.

Arbitration

A node on the data bus will "listen" to make sure that the bus is "quiet" before it begins transmitting data on the bus. But because the J1850 protocol supports peer-to-peer networking that allows equal access to all nodes on the data bus, occasionally more that one node may begin to transmit data simultaneously. **Arbitration** is the process of determining which node can continue to transmit data on the bus when two or more nodes begin transmitting data at the same time. The SAE J1850 protocol uses bit-by-bit arbitration to compare active and passive bits on the bus. When a node that is transmitting a passive, low-voltage pulse sees the data bus circuit voltage pulled actively high by another node on the bus, it quits transmitting. That is, when a node transmits a passive voltage and sees another node's active voltage, it is said to lose arbitration. The nodes with the active voltage continue to arbitrate until only one node is left transmitting data. This design is such that the node that initially transmits the most zeros wins arbitration and can continue to transmit data on the bus, Figure 7–13.

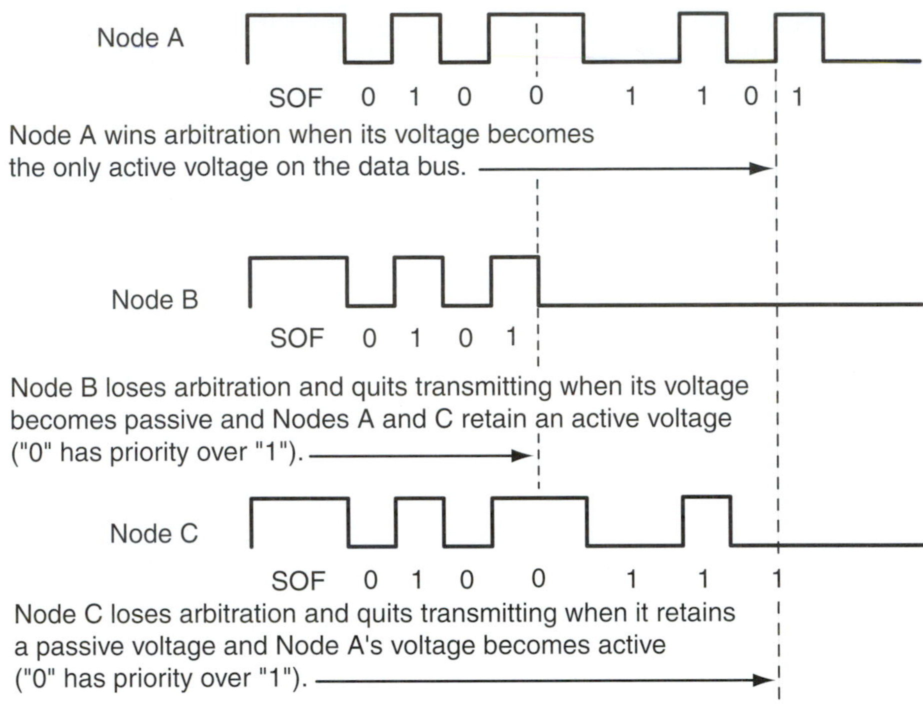
Node A
SOF 0 1 0 0 1 1 0 1
Node A wins arbitration when its voltage becomes the only active voltage on the data bus.
Node B
SOF 0 1 0 1
Node B loses arbitration and quits transmitting when its voltage becomes passive and Nodes A and C retain an active voltage ("0" has priority over "1").
Node C
SOF 0 1 0 0 1 1 1
Node C loses arbitration and quits transmitting when it retains a passive voltage and Node A's voltage becomes active ("0" has priority over "1").
An active voltage has priority over a passive voltage. As a result, an "0" has priority over a "1" due to the fact that a "0" either stays active longer or results in an earlier switch to an active voltage.

Figure 7–13 SAE J1850 arbitration.

The nodes that lost arbitration for the current message can try transmitting again after this message is complete.

MULTIPLEXING VARIATIONS

Smart Devices

A sensor or switch can be electronically equipped to transmit binary code that communicates to other components on a data bus what input values it is sensing. Conversely, an actuator can also be electronically equipped to receive, translate and react to binary code that it receives from other components on a data bus. These components are said to be electronically "smart."

For example, many of today's luxury vehicles have driver control switches mounted on the steering wheel assembly (either front side or back side) that are used to control the heating, ventilation and air conditioning system as well as the sound system. These switches send serial data over a data bus that connects both to the climate control module and to the radio. If a command is sent to increase the radio's volume, the binary code is received and acted upon by the radio, but is ignored by the climate control module.

DIAGNOSIS OF MULTIPLEXED CIRCUITS

Scan Tools

Multiplexing allows several computers to be accessed through the DLC for diagnostic purposes. Simply connect an OBD II standardized scan tool (SAE J1978 standard) to the DLC and then, after the VIN information has been entered, you can choose the computer you want to communicate with from the scanner's menu. Once you have selected a computer, you will be able to pull DTCs, access the computer's data stream, perform functional tests and perform other functions as the particular computer has been programmed to allow. On some systems, the data bus may connect to the climate control head and/or the audio system control head for diagnosis without the use of a scan tool. Follow the manufacturer's instructions as to what tests may be done by entering information on the buttons on the control head.

Lab Scopes

If a computer on the network does not seem capable of communicating with your scan tool, a DSO may be required in order to evaluate whether the computer is capable of transmitting the binary code that the scan tool needs to see. However, do not forget that common reasons for a scanner's failure to communicate is improper entry of the VIN information or if the scan tool has failed to power up properly. Also, in certain cases, another scan tool may communicate where one scan tool has failed.

If you cannot get a scan tool to communicate on the data bus, then use a lab scope to determine whether there is serial data present on the bus. With a single-wire data bus, connect the lab scope between the bus wire and a clean, unpainted ground. With a two-wire data bus, be sure to connect the lab scope between the two wires (bus positive and bus negative), rather than between bus positive and chassis ground to ensure that your waveform is not affected by voltage drop problems between bus negative and chassis ground.

Once you have connected the lab scope and turned on the ignition switch, you are simply looking for the presence (or lack thereof) of serial data. You should not expect to be able to interpret the serial data with a lab scope, as a scan tool does. In many systems, it is a good idea to have the scan tool connected to the bus while you are attempting to verify the presence of serial data with a lab scope.

SUMMARY

This chapter presented revolutionary electrical systems that had a limited introduction into the automotive world in the 1980s, but by the end of the 1990s, these systems were being used on most makes of vehicles to a very large extent. And you should expect the use of such systems to continue to expand. These electrical systems will be a part of what the technician needs to be able to diagnose when a vehicle comes into the shop for anything from a minor problem such as an inoperative trip computer to something as major as a no start condition. We discussed how multiplexing is used to reduce hard wiring, thereby reducing weight and increasing dependability. We also saw how multiplexing is able to allow additional features to be added without increasing the number of control modules or wires on the vehicle. Then we looked at how the protocols have been standardized and we saw the detail of one of the most popular standardized protocols currently in use. Finally, we discussed the abilities of both a scan tool and a lab scope to aid the technician in diagnosing problems with a multiplexed network.

▲ DIAGNOSTIC EXERCISE

A technician is diagnosing a vehicle that was towed into the shop with a "cranks, but won't start" complaint. After verifying the symptom, he connects a scan tool to the DLC and turns the ignition switch on. The scan tool displays a "no communication" message. At that point, he disconnects the bus bar in order to isolate all of the nodes on the data bus (star configuration). He then uses a jumper wire to connect just the PCM to the DLC. With the ignition switched on, the scan tool begins communicating with the PCM. What type of problem is indicated and how can the technician isolate the problem?

REVIEW QUESTIONS

1. *Technician A* says that a multiplexed circuit is a circuit that can be used to communicate multiple messages.
   Technician B says that modern multiplexed circuits transmit serial data in order to allow computers to communicate with each other.
   Who is correct?
   A. A only
   B. B only
   C. Both A and B
   D. Neither A nor B
2. Advantages of multiplexing include all *except* which of the following?
   A. Dedicated hard wiring and vehicle weight are both reduced.
   B. All functions on the vehicle are under the control of one physical computer.
   C. Dependability is increased.
   D. Computer diagnostic ability is increased.
3. A megabyte of information is equal to which of the following?
   A. 1,000 bytes of information
   B. 1,024 bytes of information
   C. 1,000,000 bytes of information
   D. 1,048,576 bytes of information
4. *Technician A* says that some multiplexing systems use a data bus that consists of two wires that are twisted together, known as a *twisted pair.*
   Technician B says that some multiplexing systems use a data bus that consists of only one wire.
   Who is correct?
   A. A only
   B. B only
   C. Both A and B
   D. Neither A nor B
5. Why are the wires twisted together on a data bus that uses two wires to communicate?
   A. In order to provide greater physical strength of both wires
   B. In order to increase the likelihood that if one wire was severed, they both would be severed (for safety purposes)
   C. Because the data bus circuitry flows high levels of current that could affect other nearby circuits if the data bus wires were not twisted together
   D. In order to minimize the effects of an induced voltage on the data bus
6. *Technician A* says that a data bus that connects the computers in a series circuit is known as a *star configuration.*
   Technician B says that a data bus that connects the computers in a parallel circuit is known as a *loop configuration.*
   Who is correct?
   A. A only
   B. B only
   C. Both A and B
   D. Neither A nor B
7. What is a standardized binary code (or computer language) known as?
   A. Protocol
   B. Node
   C. Byte
   D. Data bus
8. *Technician A* says that multiplexing is unique to the automotive industry and is only used on automotive applications.
   Technician B says that multiplexing allows features to be added to a vehicle while at the same time reducing wiring.
   Who is correct?
   A. A only
   B. B only
   C. Both A and B
   D. Neither A nor B

9. How are the effects of an induced voltage minimized on a single-wire data bus?
   A. The wire is routed inside of a grounded shield.
   B. The edges of each bit of information are slightly slanted, resulting in a slightly trapezoidal waveform shape.
   C. Multiple messages are transmitted simultaneously in order to ensure that the other computers on the data bus understand the message.
   D. Capacitors are connected to the data bus in order to minimize any voltage change.
10. *Technician A* says that PWM serial data has a fixed pulse width with all zeros and ones being the same length.
    Technician B says that serial data that has a VPW will have bits of different lengths.
    Who is correct?
    A. A only
    B. B only
    C. Both A and B
    D. Neither A nor B
11. What is a computer that can communicate on a data bus known as?
    A. Protocol
    B. Node
    C. Byte
    D. Control module
12. What is the term for the process of determining which of two computers on a data bus can continue to transmit data when they begin transmitting at the same time?
    A. Arbitration
    B. Qualification
    C. Annotation
    D. Defragmentation
13. On an SAE J1850 VPW data bus, if multiple nodes on the bus attempted to transmit data simultaneously, which of the following messages would win arbitration?
    A. 00101001
    B. 00100111
    C. 00100100
    D. 00100011
14. *Technician A* says that the computer that controls a data bus is known as the *master slave*.
    Technician B says that a dependent computer on a data bus is known as a "slave".
    Who is correct?
    A. A only
    B. B only
    C. Both A and B
    D. Neither A nor B
15. How is collision detection on a multiplexing network defined?
    A. Alerting all of the computers on the data bus when the potential for a vehicle collision is sensed
    B. The process of trying to keep the "twisted pair" insulated from each other
    C. The process of preventing two or more messages from colliding on the data bus
    D. The process of preventing a short from the data bus to chassis ground
16. *Technician A* says that some multiplexing networks may allow you to access the computers for diagnostic purposes by connecting a scan tool to the DLC.
    Technician B says that some multiplexing networks may allow you to access the computers for diagnostic purposes through the climate control and/or audio system control head.
    Who is correct?
    A. A only
    B. B only
    C. Both A and B
    D. Neither A nor B
17. *Technician A* says that when you connect a scan tool to the DLC, it becomes a node on the network to which it is connected.
    Technician B says that either a scan tool or a lab scope can be used by the technician to interpret the serial data on a data bus.
    Who is correct?
    A. A only
    B. B only
    C. Both A and B
    D. Neither A nor B

18. Which of the following reasons can keep a scan tool from communicating on a data bus when it is connected to the DLC?
    A. The scan tool is not properly powered up.
    B. The VIN is improperly entered into the scan tool.
    C. Serial data does not exist on the data bus.
    D. All of the above
19. Which of the following is the best tool to check for the presence of serial data on a data bus?
    A. Scan tool
    B. Lab scope
    C. Test light
    D. Short finder
20. Which of the following is the best tool to retrieve information from or issue commands to computers on a data bus?
    A. Scan tool
    B. Lab scope
    C. Test light
    D. Short finder

Chapter 8

General Motors' Computer Command Control

OBJECTIVES

Upon completion and review of this chapter, you should be able to:

- ❑ Understand some of the specific features of GMs' early PCMs associated with the CCC system.
- ❑ Describe the operating modes known as closed loop and open loop.
- ❑ Describe the inputs to the PCM that are used with a CCC system.
- ❑ Describe the systems controlled through output actuators by the PCM in the CCC system.
- ❑ Understand how the CCC system is designed to help the technician in diagnosing the system.

KEY TERMS

Aspirator
Bleed
Detonation
Divert Mode
Dualjet
Fail-Safe
Light-Emitting Diode
Maximum Authority
Milliamp
Millivolt
Minimum Authority
Pulsair System
Purge Valve
Vacuum Control Valve

The Computer Command Control (CCC) system was General Motors' first widely used comprehensive computerized engine control system. It was first introduced in mid-1980. Beginning with the 1981 model year, all of General Motors' carbureted passenger cars used the CCC system. It should be noted that the CCC system varies slightly from one engine application to another and from one model year to another. In 1982 four different CCC systems were used. Most 1982 engines used the full-function system, which was a slightly updated version of the 1981 system. Beginning in 1982, Oldsmobile 5-liter engines were equipped with a limited control system; the T-car (Chevrolet Chevette and Pontiac T-1000) used a minimum-function system. The Chevrolet and GMC S truck (S-10 and S-15, introduced in 1982) with the California emissions package and a four-cylinder gasoline engine used an imported Powertrain Control Module (PCM) slightly different from the other CCC systems. There are also slight variations within each of these systems.

Although all of this seems a bit bewildering, it need not be. All versions of the CCC system are much more alike than different. Once the basic system and function are understood and a good service manual is at hand for the specific directions and specifications of the vehicle, you will find the CCC systems are very manageable.

From EFC to CCC

General Motors introduced its first electronic closed-loop fuel control system in 1978. This single function system, called Electronic Fuel Control (EFC), was limited to a few four-cylinder engines. In 1979 GM introduced its first comprehensive engine control system on selected applications: Computer-Controlled Catalytic Converter (C-4). The system was improved for model year 1981 and was renamed the Computer Command Control system. Apparently someone at GM noticed the name and decided they did not want their engine control system named after a Ford automatic transmission.

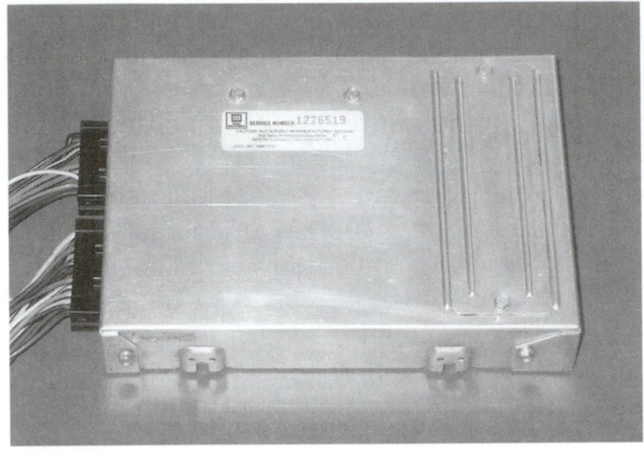

Figure 8–1 General Motor's Computer Command Control system PCM.

POWERTRAIN CONTROL MODULE (PCM)

The engine control computer used in the CCC system and in all other General Motors' systems was originally called an Electronic Control Module or ECM. Today, under OBD II standards for all manufacturers, it is referred to as the Powertrain Control Module or PCM, Figures 8–1 and 8–2. The inputs it receives and the actuators it controls are shown in Figure 8–3. The PCM is located in the passenger compartment, usually near the glove compartment or behind the passenger kick panel. The PROM is located under an access cover, Figure 8–4. It is responsible for fine-tuning engine calibrations for each specific vehicle and when plugged in becomes a functioning (in fact, essential) part of the PCM.

Figure 8–2 General Motors' CCC system PCM circuit board.

Keep-Alive Memory (KAM)

The PCM monitors its most important sensors and selected actuator circuits. If a fault such as an open or short circuit occurs during normal operation, a fault code is stored in the KAM. This code can be retrieved later to aid the technician in

Diagnostic & Service Tip

In earlier PCMs, the PROM can easily be installed backwards, which results in irreversible electrical damage to it. Consult a service manual or the System Diagnosis and Service section of this chapter before removing and replacing any GM PROM.

Inputs	Codes
Coolant Temperature Sensor	14, 15
Vacuum Sensor	34
Barometric Pressure Sensor (if used)	32
Throttle Position Sensor	21
Distributor Reference (crank position—engine speed)	12, 41
Oxygen Sensor	13, 44, 45
Vehicle Speed Sensor	24
Ignition On	
Air Conditioner On/Off	
Park/Neutral Switch	
System Voltage	
Transmission Gear Position Switch(es)	
Spark Knock Sensor (if used)	43
EGR Vacuum Indicator Switch (only on 1984 and later models)	53
Idle Speed Control Switch	35

PCM
ROM and PROM process information (inputs) and issue commands (outputs).
RAM monitors indicated circuits, sets codes and reads codes out when put in diagnostics.

Codes	Outputs
23	Fuel Mixture
42	Electronic Spark Timing
	Electronic Spark Retard (if used)
	Idle Speed
	AIR Management
	EGR
	Canister Purge
	Torque Converter Clutch (or shift light on manual transmission)
	Air Conditioner
	Early Fuel Evaporation
	Diagnosis (check engine light) (test terminal) (serial data)

Figure 8–3 Overview of Computer Command Control. Inputs and outputs with code numbers are monitored for faults. Those without are not.

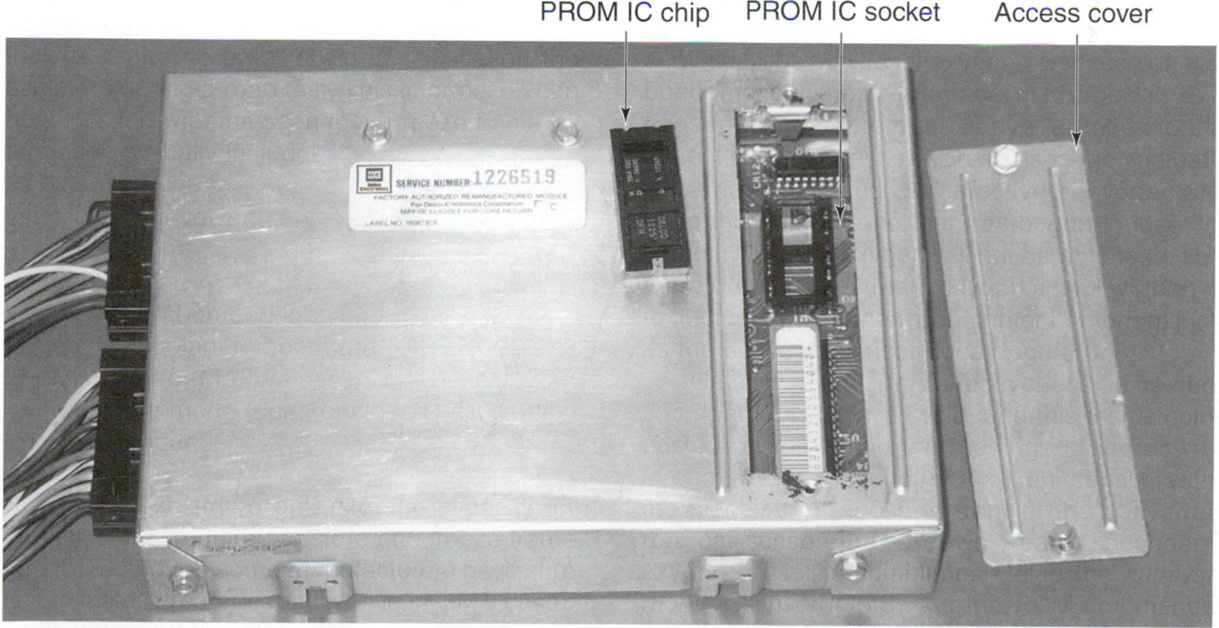

Figure 8–4 PCM and PROM IC chip.

locating the problem. This feature is especially useful because it enables the PCM to report a fault that occurred recently but is no longer present, as well as those that are currently in the system. The KAM receives battery power at all times so the codes can be retained in memory. This battery draw is very low, much less than the electrical power loss internal to the battery itself.

OPERATING MODES

There are basically two operating modes: closed loop and open loop.

Closed Loop

Before the system can go into closed loop, the coolant must reach a temperature of approximately 65°C (150°F), the oxygen sensor must reach at least 300°C (570°F), and a predetermined amount of time must have passed following engine startup. This time is programmed into the PROM and varies with the engine application, model year and so on. It can be as little as a few seconds on some engines, or as much as a couple of minutes on others.

The system drops out of closed loop if the vehicle is driven at or near wide-open throttle (WOT), if the coolant temperature drops below the criterion value or if certain crucial system component failures occur.

Diagnostic & Service Tip

1982 and later GM T-cars (Chevrolet Chevette and Pontiac T-1000) used a minimum-function CCC system. These systems use a coolant temperature switch instead of a coolant temperature sensor. As the engine reaches operating temperature, the switch closes and puts the system into closed loop (assuming the oxygen sensor is hot and the proper time has elapsed). If the engine cools down and the switch opens, the system stays in closed loop. Try to avoid inadvertent confusion by thinking that the temperature switch is shorted or open, when it is actually functioning normally.

Open Loop

This operating mode includes several suboperational modes.

Startup Enrichment. This mode provides a rich mixture command to the mixture control solenoid for a short time after each engine startup. The duration of the mode depends on engine temperature, as reported by the coolant temperature sensor at the time the engine starts. If the engine is started warm, this mode will be shorter than if it is started cold.

Blended Enrichment. This mode occurs during engine warmup (engine coolant and/or the oxygen sensor are not up to operating temperatures). In this mode, the PCM controls the air/fuel ratio using information from sensors concerning coolant temperature, throttle position, manifold pressure, barometric pressure and engine speed. As engine temperature warms, the air/fuel ratio is adjusted leaner by the PCM. In this open-loop mode, the PCM attempts to adjust the air/fuel ratio using the fixed, preprogrammed instructions in its read-only memory. If the air/fuel ratio is not correct because of a carburetor problem, vacuum leak and so forth, the PCM cannot recognize the problem and cannot take any countermeasures in response to it. In other words, in open loop the PCM does what it is programmed to do, but the engine does not run correctly unless other engine components function properly according to its expectations.

Power Enrichment. This mode occurs when the vehicle is operated at or near wide-open throttle. The PCM sends a steady power enrichment signal to the mixture control solenoid, which provides the rich mixture required by an engine when manifold vacuum is near zero.

Limp-In. This mode allows the engine to continue operating in spite of most major failures, such as a PCM failure, that can occur in the system. This is a **fail-safe** mode that provides no fuel mixture control (full-rich or full-metered) and no spark advance. The engine will run, but driveability, fuel economy and emissions quality performance will be compromised.

INPUTS

Engine Coolant Temperature Sensor (ECT)

With the exception of some T-car applications (Chevrolet Chevette and Pontiac T-1000), all CCC systems use an Engine Coolant Temperature sensor, the ECT, Figure 8–5, as described in Chapter 2. Most often it is near the thermostat housing. It is the single most important sensor in the system. Its input affects just about every command the computer sends (this statement is generally true of all computerized engine control systems, regardless of make, model or year).

Some 1982 and later GM T-cars are equipped with a minimum-function system and use a coolant temperature switch instead of an ECT sensor. The switch is open while the coolant is below a specified temperature and closes as the coolant approaches the normal operating temperature. In this system, the PCM only knows whether the coolant is below or above a specific temperature.

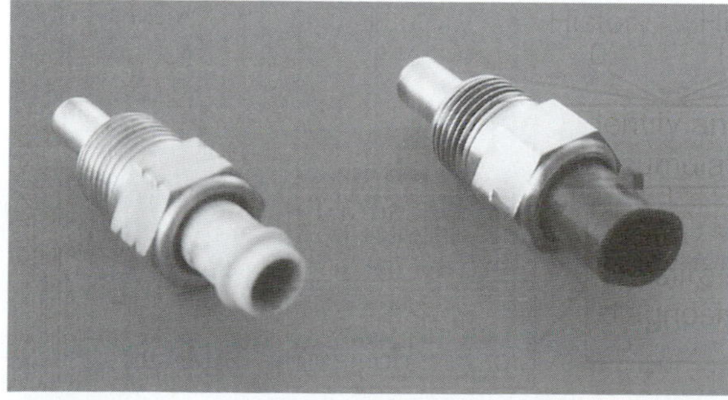

Figure 8–5 Early and newer styles of General Motors ECT sensors.

Pressure Sensors

Three different kinds of pressure sensors were used on various CCC systems, all of the piezoresistive silicon diaphragm type and all nearly identical in appearance, Figure 8–6. Each

BOTTOM VIEW

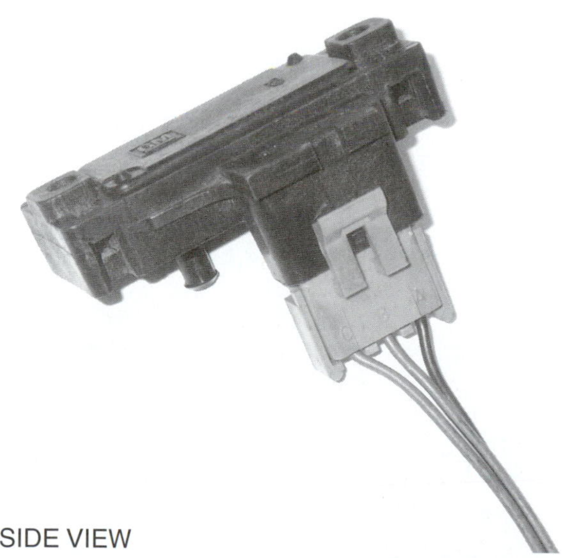

SIDE VIEW

Figure 8–6 MAP sensor. *(Courtesy of General Motors Corporation, Service Technology Group.)*

can, however, be distinguished by a colored plastic insert at its electrical harness connector.

Manifold Absolute Pressure (MAP) Sensor. The MAP sensor has either a red or orange connector (orange is for turbocharged engines). The MAP sensor is often used in combination with a BARO sensor.

Barometric Pressure (BARO) Sensor. The BARO sensor uses a blue connector. On some vehicles it is under the hood and near the MAP sensor. On others, it is inside the passenger compartment under the instrument panel.

The MAP and BARO sensors for 1980 and early 1981 vehicles were round metal units, both mounted on the same bracket under the hood.

Pressure Differential (VAC—for vacuum) Sensor. The VAC sensor uses a grey or black connector.

Oxygen Sensor

The oxygen sensor used on 1981 and earlier models used two wires with a vented silicone boot covering its open end, Figure 8–7. The second wire served as a backup ground circuit; the sensor shell is already grounded to the manifold. This backup is used because it is not uncommon for the heat fluctuations of the exhaust manifold to break a ground circuit through it. Beginning in 1982, the ground wire was not used and a metal boot replaced the silicone boot. The silicone boot melted if it was pushed too far over the oxygen sensor toward the exhaust manifold. These systems use an oxygen sensor ground wire running from the PCM to the engine block, not directly connected to the oxygen sensor.

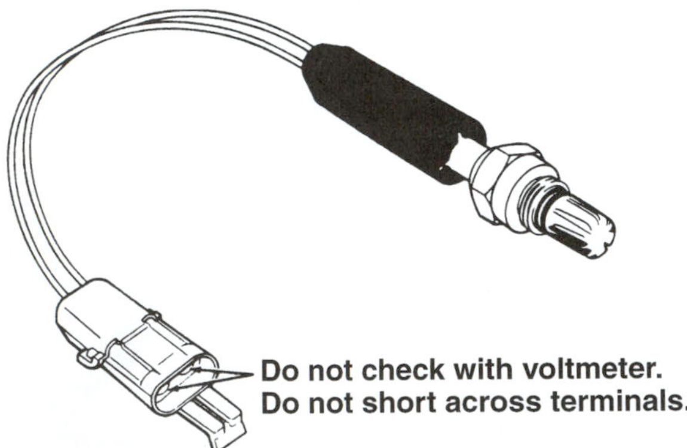
Do not check with voltmeter.
Do not short across terminals.

Figure 8–7 Oxygen sensor, early style (two wire). *(Courtesy of General Motors Corporation, Service Technology Group.)*

Service Interval for Oxygen Sensors

Prior to model year 1981 GM recommended a service interval of 15,000 or 30,000 miles for oxygen sensors. No replacement interval is set for vehicles built after model year 1981; replacement is called for instead only when the sensor is found to be defective.

Throttle Position Sensor (TPS)

The TPS is a linear potentiometer mounted in the bowl section of the carburetor, Figure 8–8. When the throttle is opened, the accelerator pump lever forces the TPS plunger down. This action moves the potentiometer's wiper. The TPS is the only adjustable sensor in the CCC system. Because engine vacuum and throttle position are so closely related, the PCM expects to see the readings from these two sensors closely synchronized.

Distributor Reference Pulse (REF)

The REF is also abbreviated as the DIST or REF pulse. This signal provides the PCM with information about engine speed and crankshaft position. The REF is obtained by tapping into the pickup coil circuit inside the high energy ignition (HEI) module, Figure 8–9.

The pickup coil signal is generated as the distributor shaft and timer core turn, moving the

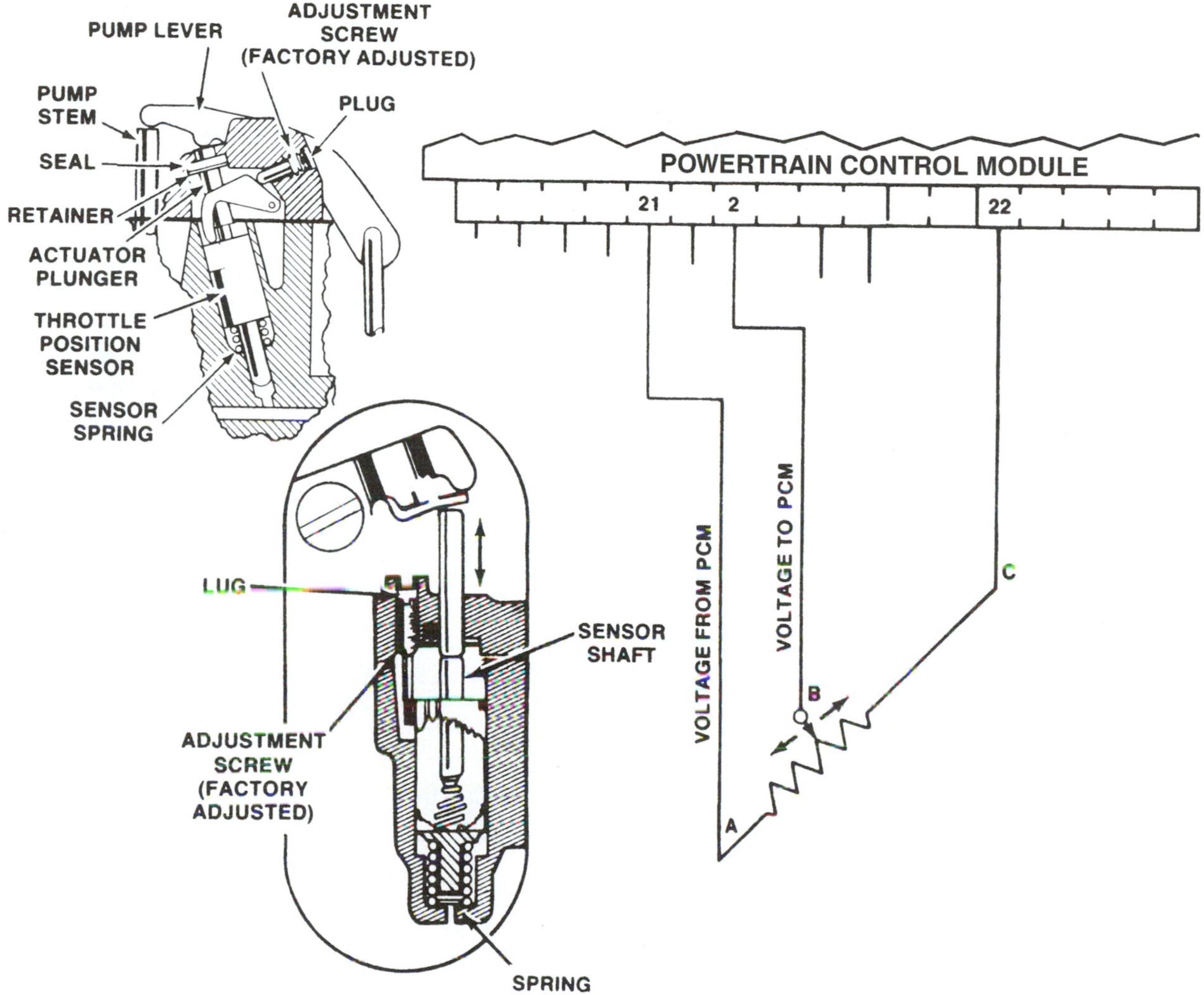

Figure 8–8 Throttle position sensor. *(Courtesy of General Motors Corporation, Service Technology Group.)*

points of the timer core (or reluctor) past the points of the magnetic pickup unit. Each time the magnetic field strengthens or weakens as a result of the points aligning and misaligning, a voltage signal is produced in the pickup coil. This signal occurs just as the circuit reverses polarity and the points of the reluctor and the pickup cross. In the HEI module, a signal converter changes the analog signal produced by the pickup coil to a digital signal. On HEI systems prior to computer controls, the pickup coil signal indicated to the module when to turn off the main switching transistor, which turned off the primary ignition circuit and fired the spark. On CCC systems, the only time the pickup coil signal is used by the module is during startup cranking and in some system failures, in fail-safe or limp-home mode. During normal operation the PCM uses the pickup coil signal only as a REF signal, a concept explained further in the Outputs section of this chapter.

From 1982 to 1985, a Hall effect switch, Figure 8–10, was positioned between the pickup coil and

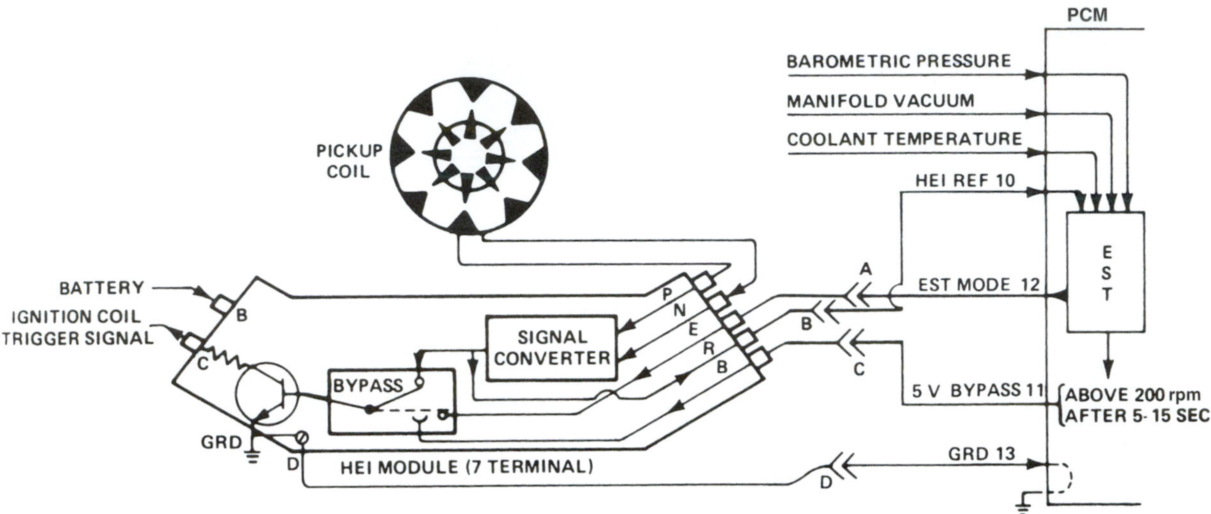

Figure 8–9 Electronic spark control circuit. *(Courtesy of General Motors Corporation, Service Technology Group.)*

the distributor rotor on the Chevrolet 3.8-liter (229 cid) odd-firing V6. This sensor was used to produce greater ignition timing accuracy. The odd-firing engine is especially susceptible to signal converter error. Notice in Figure 8–10 the pickup is still there. However, it is used only to start the engine. Notice that terminal R of the HEI module is not used (on applications without a Hall effect switch, terminal R feeds the REF signal to the PCM), and that the REF comes directly to the PCM from the Hall effect switch. Once the engine starts, the solid-state switching device, illustrated in

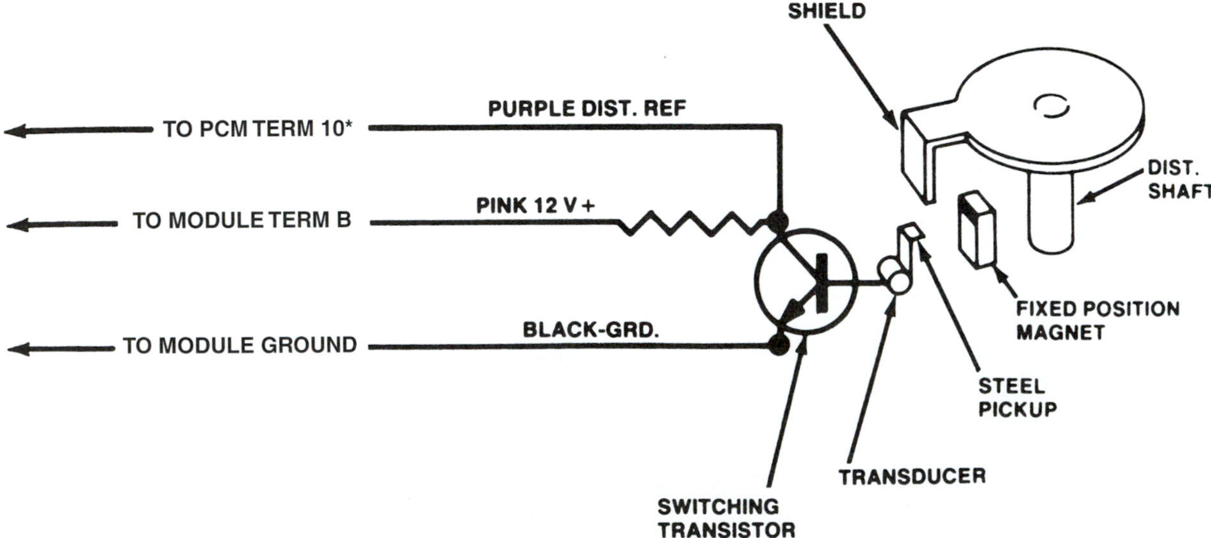

Figure 8–10 Hall effect switch. *(Courtesy of General Motors Corporation, Service Technology Group.)*

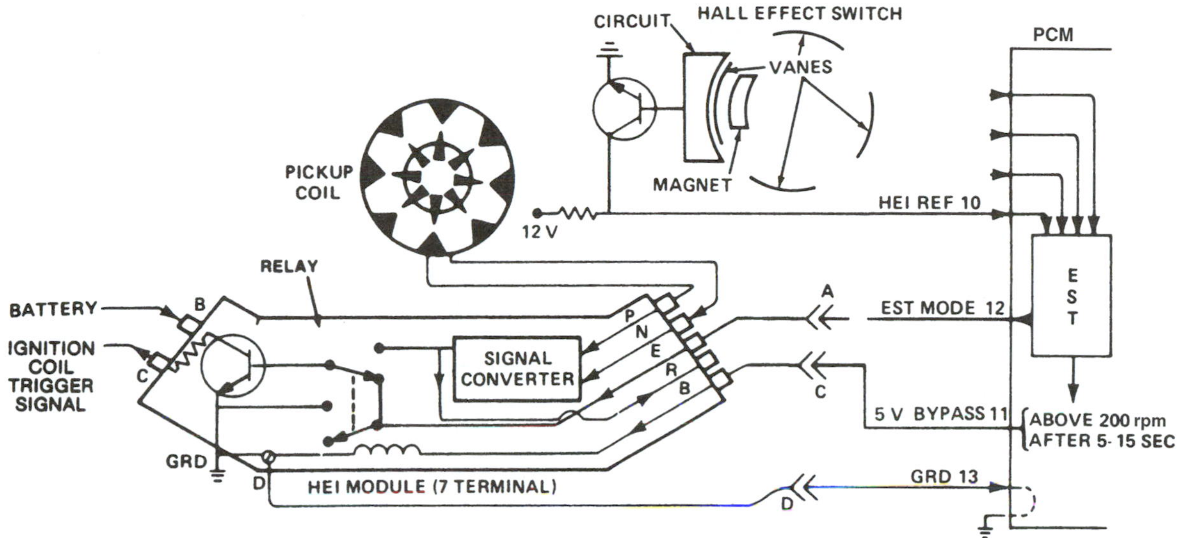

Figure 8–11 Electronic spark control system with Hall effect switch. *(Courtesy of General Motors Corporation, Service Technology Group.)*

Figure 8–11 as a relay, has disconnected the pickup coil from the main switching transistor and connected in its place the EST lead from the HEI terminal E. The PCM is now in charge of spark timing.

Detonation Sensor

The **detonation** or knock sensor, Figure 8–12, is part of the electronic spark control (ESC) system, a subsystem of CCC on many applications. A similar spark control system, also called ESC, has been an independent spark retard system for several years on non-CCC vehicles.

Figure 8–12 Detonation sensor.

On many General Motors engines, the knock sensor is screwed into the block in the crankcase area to reduce its sensitivity. On those applications, it was found that mounting it on the upper part of the engine allowed a clicking valve lifter to trigger the knock signal and erroneously retard the ignition timing.

Vehicle Speed Sensor (VSS)

Two kinds of vehicle speed sensors are used on CCC systems. The earlier type uses a **light-emitting diode (LED)** and a phototransistor. Both are in a plastic connector plugged into the back of the speedometer housing near the speedometer cable attachment point, Figure 8–13. On early models, the LED is powered through the PCM; on later versions, power comes directly from the ignition switch. With the ignition on, the LED directs its invisible infrared light beam to the back of the speedometer cup, which is painted black. The

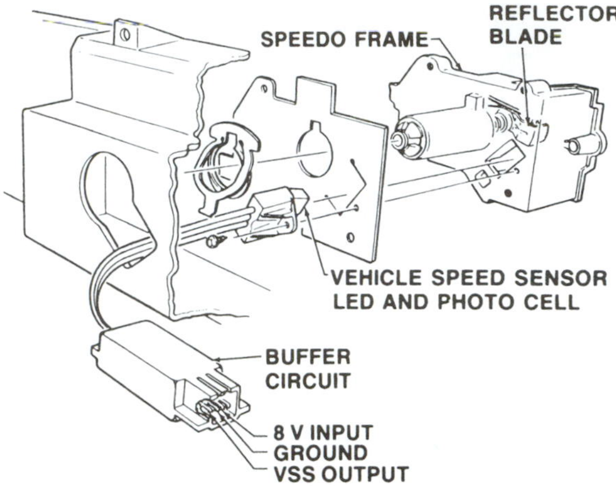

Figure 8–13 Vehicle speed sensor. *(Courtesy of General Motors Corporation, Service Technology Group.)*

drive magnet, spinning with the speedometer cable, has a reflective surface at one point. As the drive magnet moves into the light from the LED, the light reflects back to the receptive phototransistor. Each time the light strikes the phototransistor, it produces a voltage pulse. These pulses are fed to a buffer switch.

The VSS signal is an analog signal varying in amplitude as well as frequency as the vehicle speed changes. The buffer switch modifies the raw VSS signal to 2,002 digital pulses per mile (different pulse numbers for certain later models). The PCM uses this information to determine when to apply the torque converter clutch.

Diagnostic & Service Tip

The LED-generated VSS signal can be blocked by anything that can block the infrared light. Either a dustball or, not infrequently, a ball of grease pushed through the speedometer cable can shut off the signal. Many experienced technicians check this first when there is indication of a VSS signal failure.

Some later model vehicles use an electronically operated speedometer with no speedometer cable. These vehicles use a pulse generator device as a VSS. This device bolts to the transmission where the speedometer cable drive would be if it were used. This transmission-driven signal generator rotates a magnet near a coil. As each pole of the magnet swings by the coil, a weak voltage signal, similar to what a magnetic pickup in an ignition distributor generates, is produced in the coil. This signal produces an analog voltage signal proportional to the speed of the car. These signals are fed to a buffer, which converts them to 4,004 digital pulses per mile. These signals then go to the PCM, which calculates vehicle speed from the signal frequency and its own internal clock. Four-wheel-drive vehicles often have two similar sensors, one for transmission output and one for vehicle speed to enable the computer to distinguish when the vehicle is in the low transfer case range.

Ignition Switch

The ignition switch is one of the two power sources for the PCM. When the ignition turns on, the PCM initializes (starts its program) and gets ready to function. The ignition switch also powers most of the actuators the PCM will control in operation (by grounding and completing their circuits). The other PCM power supply comes from the battery through a fuse. This circuit powers the diagnostic memory. The PCM also monitors system voltage through these two inputs.

Park/Neutral (P/N) Switch

The P/N switch consists of another pair of contacts added to the neutral safety switch that were long a part of vehicles with automatic transmissions (used primarily to prevent the starter from engaging the flywheel in any gears but park or neutral). The computer's P/N switch signals the PCM whether the transmission is in gear or not, and this information is used to help control engine idle speed. On some vehicles the PCM also does not activate any spark advance or EGR commands unless the transmission is in some gear.

Air-Conditioning (A/C) Switch

On CCC cars with air conditioning, the A/C switch connects through a wire to the PCM. When the A/C is turned on or off, the PCM receives that information, which it uses in controlling idle speed.

Idle Speed Control (ISC) Switch

At closed throttle, the throttle lever, pulled by the return spring, presses against the ISC plunger. This pressure closes a set of contact points at the base of the plunger, Figure 8–14. This circuit signals the PCM that it is now in charge of idle speed. Opening the throttle releases the pressure on the plunger and opens the contact points, signaling the PCM not to control idle speed, Figure 8–15.

Transmission Switches

Most transmissions in CCC-equipped vehicles include one or more hydraulically operated electric switches in the valve body. These switches provide the PCM with signals indicating what gear the transmission is in. This information enables it to control the lockup torque converter clutch operation more effectively.

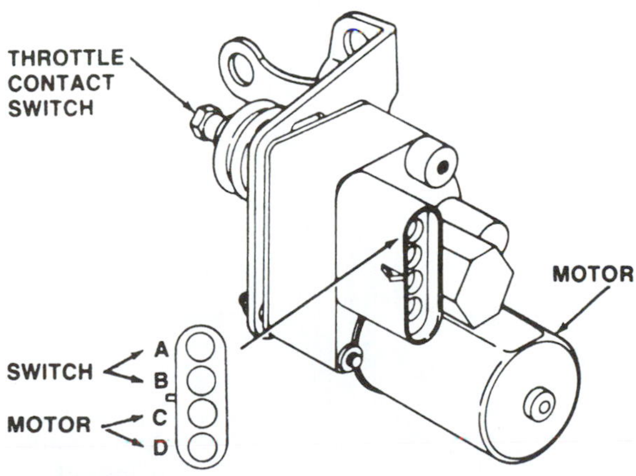

Figure 8–14 Idle speed control motor. *(Courtesy of General Motors Corporation, Service Technology Group.)*

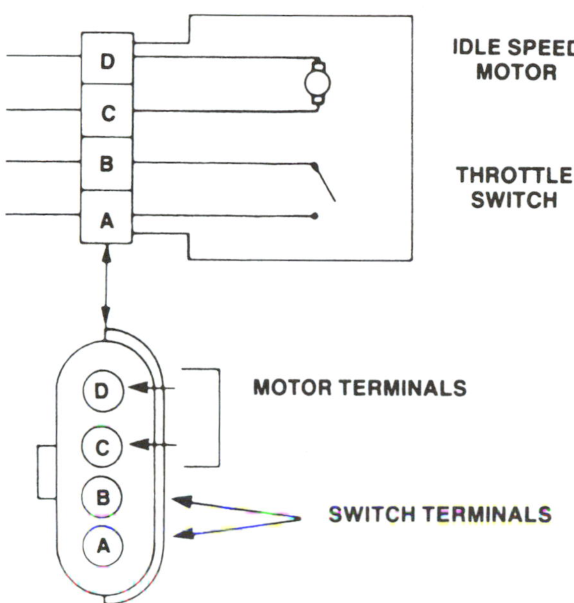

Figure 8–15 Idle speed control motor connector. *(Courtesy of General Motors Corporation, Service Technology Group.)*

EGR Vacuum Diagnostic Control Switch

Beginning in 1984, most CCC systems included a vacuum-operated switch tied into the vacuum hose between the EGR valve and the EGR control solenoid, Figure 8–16. When vacuum is applied to the EGR valve, the switch closes. If the PCM detects a closed EGR diagnostic switch during starting, idle or at any other time it has not commanded EGR flow, then it will turn on the check engine light and set a fault code in diagnostic memory. If it sees an open switch during any time it has commanded the EGR to recirculate exhaust gas, it will also turn on the check engine light and set the same code.

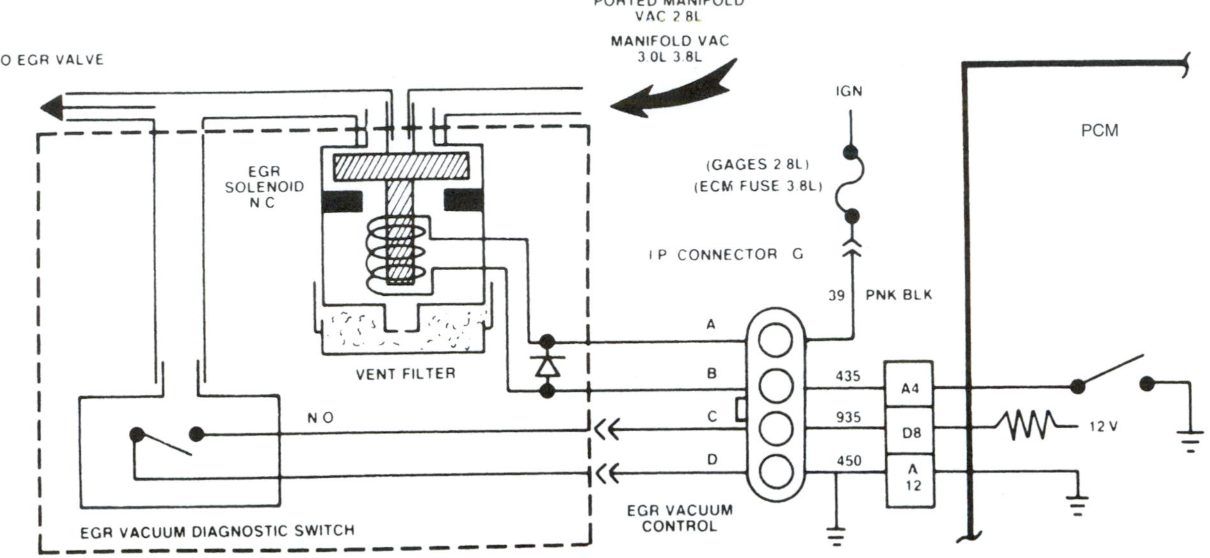

Figure 8–16 EGR diagnostic switch circuit. *(Courtesy of General Motors Corporation, Service Technology Group.)*

OUTPUTS

Mixture Control (M/C) Solenoid

In a CCC system, the mixture control solenoid effectively replaces the full power enrichment system that would otherwise be used in the carburetor of a non-CCC vehicle. This solenoid, however, has much greater detailed control of the air/fuel ratio over a much wider range of engine operating conditions than the conventional full power enrichment system does, Figures 8–17 and 8–18. The mixture control solenoid gets power directly through the ignition switch. The PCM duty cycles it on and off through a solid-state grounding switch (a power transistor), Figure 8–19. When the PCM grounds the M/C solenoid circuit, the solenoid turns on and drives the solenoid plunger down. When the plunger is driven down, it drives a metering rod (or rods) down into the main metering jet (or jets) to reduce fuel flow. On some carburetors, the tip of the solenoid plunger itself serves as the metering rod. When driven down, the solenoid provides a lean mixture; when

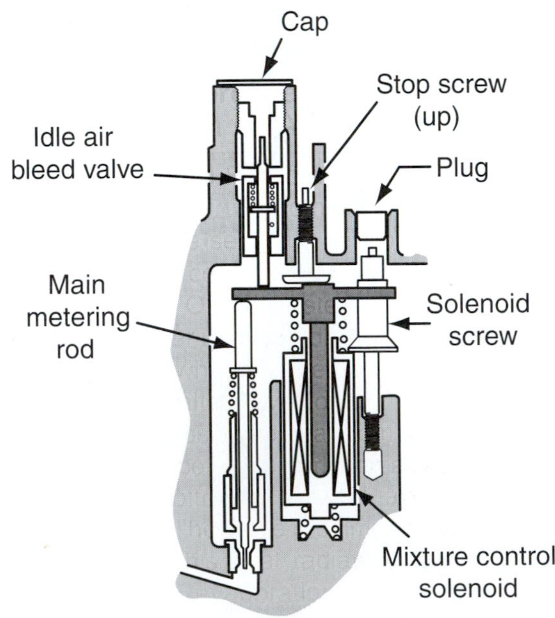

Figure 8–17 Mixture control solenoid, Rochester Dualjet or Quadrajet.

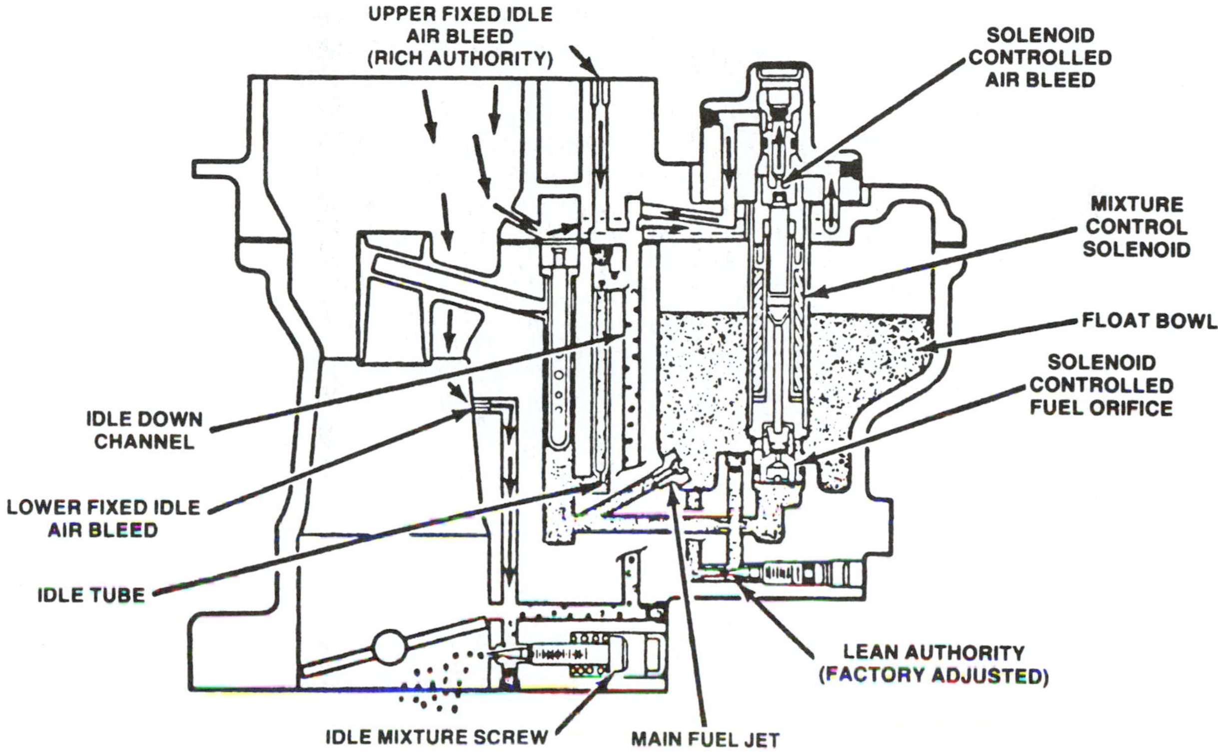

Figure 8–18 Mixture control solenoid, Holley 6510 C. *(Courtesy of General Motors Corporation, Service Technology Group.)*

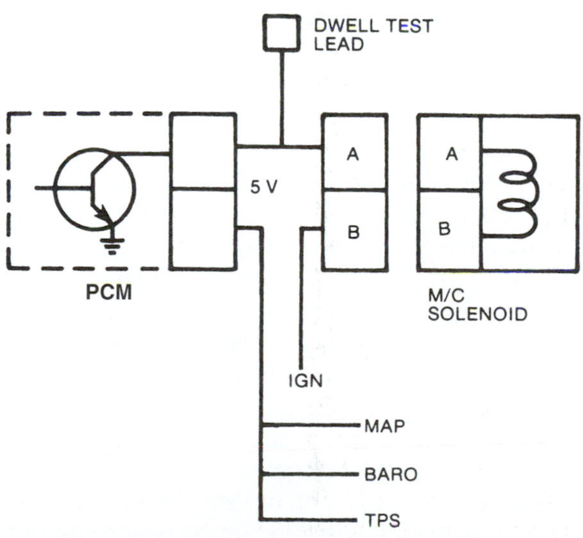

Figure 8–19 Mixture control solenoid circuit.

turned off and lifted by the spring, it provides a full rich mixture.

The PCM cycles the mixture control solenoid 10 times per second regardless of engine speed or load (except during shutdown, when the solenoid turns off). The PCM can vary the duty cycle anywhere from 90 percent to 10 percent (that is, it can hold the solenoid down leaning the mixture for 90 percent of the time; or it can let the solenoid up, richening the mixture for 90 percent of the time, depending on what mixture it calculates is required from the signals from the other sensors), Figure 8–20.

Measuring the Duty Cycle. You can measure the relative time the mixture solenoid is on or off with a dwell meter, Figure 8–21. The wire between the mixture control solenoid and the PCM has a special connector, the dwell test connector,

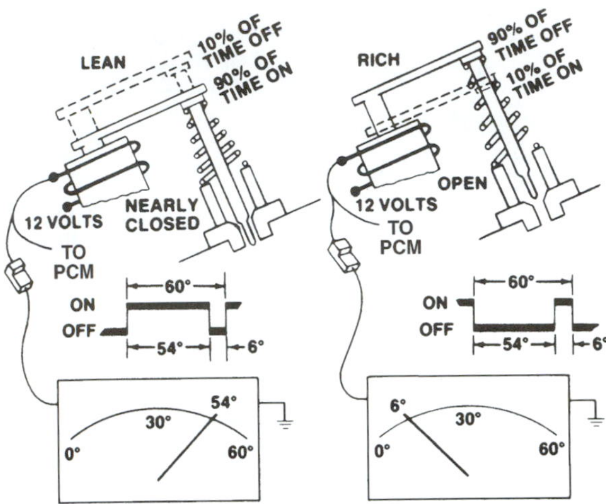

Figure 8–20 Mixture control solenoid duty cycle. *(Courtesy of General Motors Corporation, Service Technology Group.)*

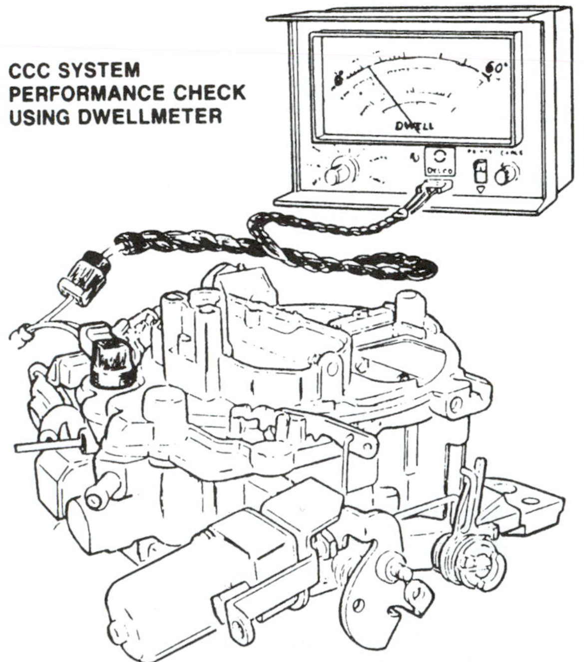

Figure 8–21 Dwell test lead. *(Courtesy of General Motors Corporation, Service Technology Group.)*

for this purpose. A dwell meter, once connected between the dwell test connector and to ground, and set on the six-cylinder (60 percent) scale, indicates in degrees the relative amount of on-time the M/C solenoid is energized, Figure 8–22. Many newer DVOMs include a duty-cycle measure that yields a reading directly in percent of on-time.

Dwell Readings. During engine cranking enrichment or wide-open throttle operation, a dwell reading of 6 degrees is expected (10 percent duty-cycle or less), for full rich. During engine warmup (open loop) an initial reading of about 10 degrees (about 15 percent) is considered normal, depending on how cold the engine is to begin with. As the engine warms up, the dwell reading should steadily increase. In open loop, the PCM selects an air/fuel ratio based on engine temperature, load, barometric pressure, throttle position and engine speed. This produces the best results in terms of exhaust emissions, fuel mileage and driveability. The fuel mixture, as reflected in the dwell reading, remains constant until some input information changes or until the system goes into closed loop. When coolant reaches a temperature of about 150°F, the oxygen sensor reaches about 600°F and the required amount of time has passed, the PCM puts the system into closed loop. At that moment, the dwell meter needle begins wagging. It usually wags, or varies, over a range of 5 to 10 degrees. The dwell reading is now taken by selecting the center or average of the range over which it sweeps.

The dwell varies because of the way the PCM responds to the oxygen sensor's feedback signals. When the oxygen sensor sees a lean condition (anything leaner than 14.7 to 1), it reports it to the PCM. The PCM reacts by sending a rich command to the mixture control solenoid. If everything is working properly, this will soon appear as "rich" at the oxygen sensor, and its signal will reflect that condition. The oxygen sensor's rich signal then produces a lean command from the PCM, and so on. In this way, the air/fuel ratio constantly fluctuates back and forth across the 14.7 to 1 stoichiometric ratio—the mixture at which the cat-

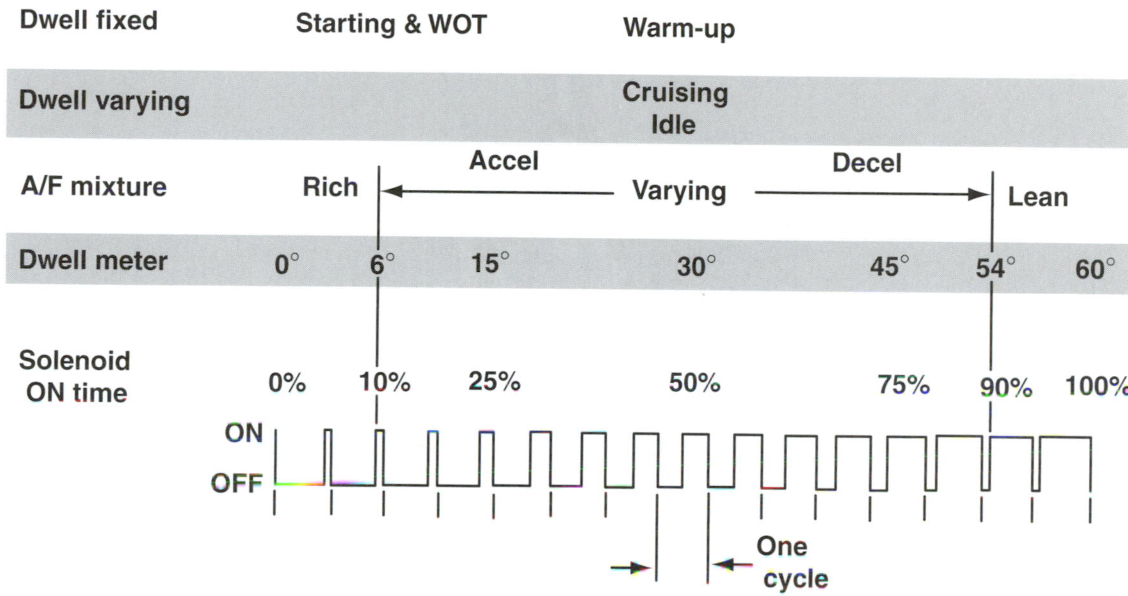

Figure 8–22 Dwell readings.

alytic converter can most effectively counteract any residual HC, CO and NO_x in the exhaust. The actual delivered mixture usually remains within a range of about 14.5 to 14.9 to 1.

This ratio, however, should not be taken as invariable. The original definition of *stoichiometric* (the Greek word for *ideal*) meant the combustion state of an engine whose exhaust contained no unburned fuel and no oxygen—a combustion, in short, that perfectly and completely burned all the fuel using all the air the throttle admitted, and no more. This earlier notion focused on power output and fuel economy, whereas the current concept of stoichiometry is that air/fuel ratio that leaves the exhaust exiting the engine at a chemical and thermal state optimal for reduction and oxidation of the emissions gases by the catalytic converter. The traditional way of describing this is 14.7 to 1, but with oxygenated fuels and various other modifications for seasons and altitudes, this may not be the actual stoichiometric ratio for a given engine under given driving conditions.

At the same time the mixture control solenoid cycles the metering system on the carburetor's main fuel delivery system, it also cycles the idle air bleed valve in the idle speed/low speed system (**bleed** is a controlled leak). The idle air bleed valve, Figure 8–23, is designed as a variable restriction—a sort of unfixed orifice—to determine the amount of air entering the idle/low speed circuit. When the mixture control solenoid metering valves are down, the idle air bleed valve plunger drops and allows maximum air into the idle speed/low speed circuits, driving the mixture lean. When the mixture control solenoid is up, the idle air bleed is also up and allows minimum air into the circuit, driving the mixture rich. With this arrangement, the same mixture control solenoid controls the air/fuel mixture using whichever fuel metering circuit is in use.

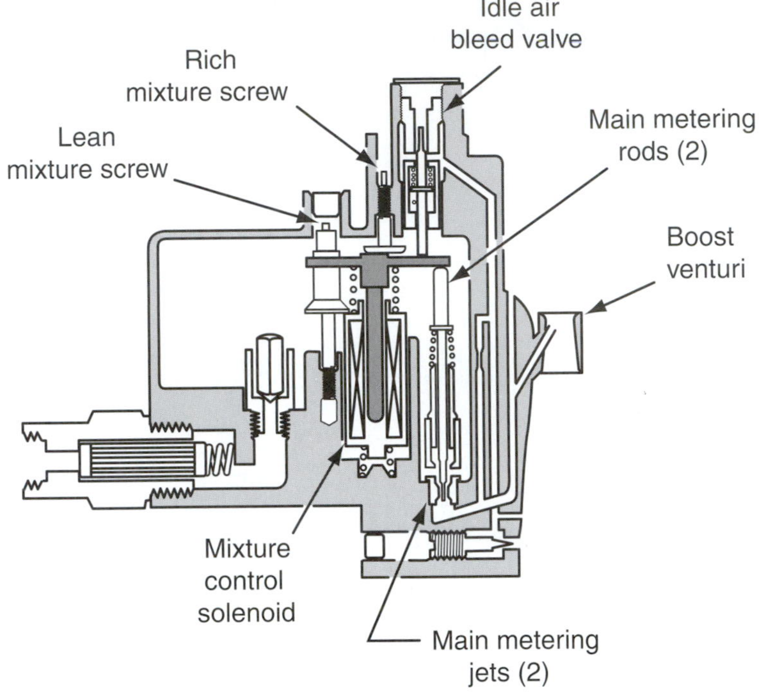

Figure 8–23 Rochester Dualjet or Quadrajet.

Electronic Spark Timing (EST)

With the CCC system, the PCM eliminates the vacuum and centrifugal advance mechanisms of the earlier HEI distributors, Figure 8–24 (and along with it most of the failed pickup coils, whose connections often broke from the frequent twisting back and forth). To achieve this, the HEI module was modified to work directly with the PCM. The PCM in the updated system controls timing by signaling the HEI module when to open the primary ignition circuit, therefore turning it off and firing the spark. The HEI module responds only to these spark timing commands from the PCM once the engine has started.

In the following discussion and in the accompanying illustrations, the HEI module is treated as though it has a mechanical bypass relay with a double set of contact points. This is done to make its operation easier to understand. Actually, solid-state components are used to achieve the functions discussed.

During cranking, the bypass relay is in the de-energized or module mode. In this mode, the main switching transistor of the HEI module is connected to the pickup coil, Figure 8–25. The pickup coil signal, after conversion from an analog to a digital signal, turns the switching transistor on and off. This same signal is sent to the PCM as the HEI reference pulse.

When the PCM sees a REF pulse signal of about 200 rpm (calculated from the signal and from its own internal clock), it decides the engine is running. Then it sends a 5-volt signal through the bypass wire to the bypass relay, causing the double contacts to move. The upper contact disconnects the base of the main switching transistor

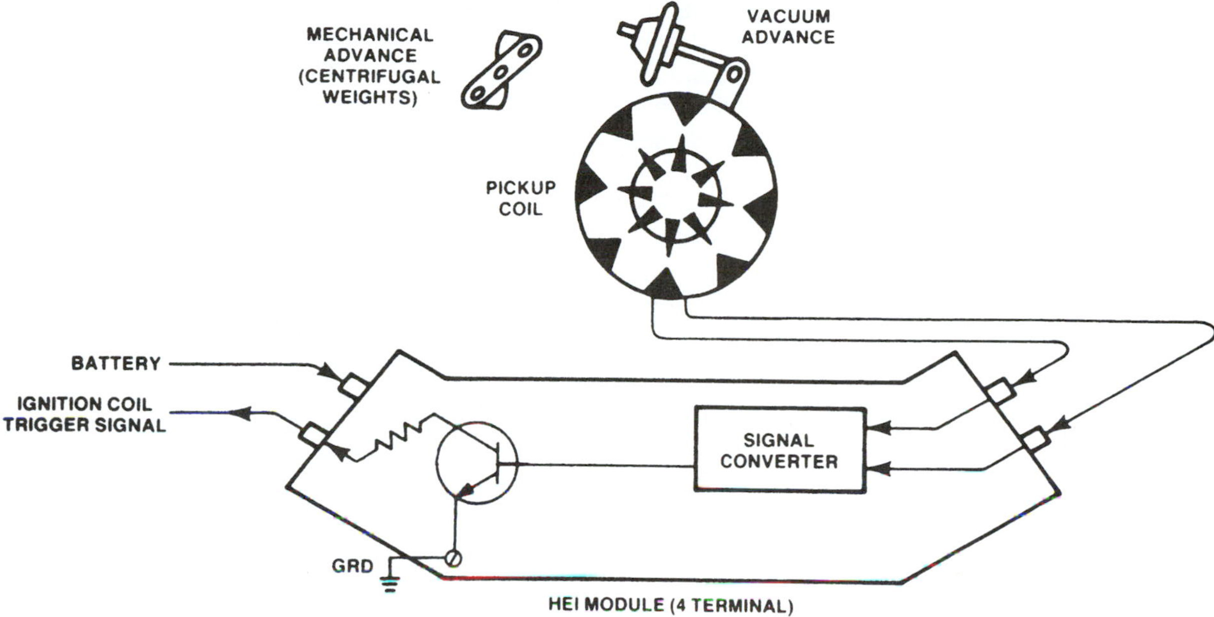

Figure 8–24 HEI circuit with centrifugal and vacuum advances. *(Courtesy of General Motors Corporation, Service Technology Group.)*

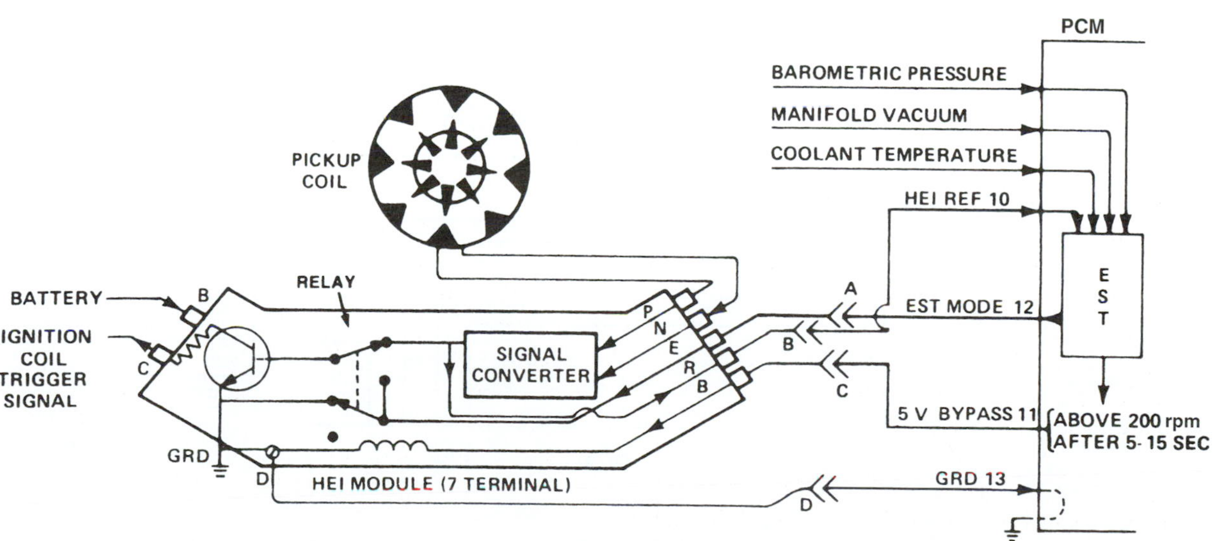

Figure 8–25 HEI circuit with EST. *(Courtesy of General Motors Corporation, Service Technology Group.)*

from the pickup coil and connects it to the EST wire. The lower contact simultaneously disconnects the EST wire from ground. The system is now in EST mode, and timing is controlled by the PCM. The PCM considers barometric pressure, manifold pressure, coolant temperature, engine speed, and crankshaft position and then sends the HEI module the calculated optimum spark timing command.

Some Chevrolet Chevette and Pontiac T-1000 CCC systems do not use an EST subsystem.

Electronic Spark Control (ESC)

Engines with a strong potential for spark knock (particularly higher performance engines) are equipped with ESC. The ESC system becomes a subsystem of the CCC system on the vehicle. It has its own electronic module working in conjunction with the PCM. Its function is to retard ignition timing when detonation (uncontrolled, rapid burning of the fuel charge) occurs. Two slightly different versions of ESC systems have been used.

Of the two ESC systems, the 1981 system, Figure 8–26, passes the EST and the bypass commands from the PCM through the ESC module on their way to the HEI module. If there is spark knock (as reported by the knock sensor), the detonation sensor informs the ESC module that it is occurring. The ESC module then modifies the EST command. Spark timing is retarded about 4 degrees per second until the detonation clears up. When the detonation sensor no longer "hears" knock, the ESC module begins restoring the spark advance it took away. The spark advance is restored at a slower rate than it was removed, approximately 2 degrees per second.

In the second version, used on model year 1982 and later vehicles, Figure 8–27, the EST and bypass commands do not pass through the ESC module. Instead, when the knock sensor produces a voltage indicating knock, the ESC module sends a request to the PCM for it to retard timing. Here's how it works: Ignition voltage is applied to terminal F of the ESC module. The module transfers this voltage, reduced to about 10

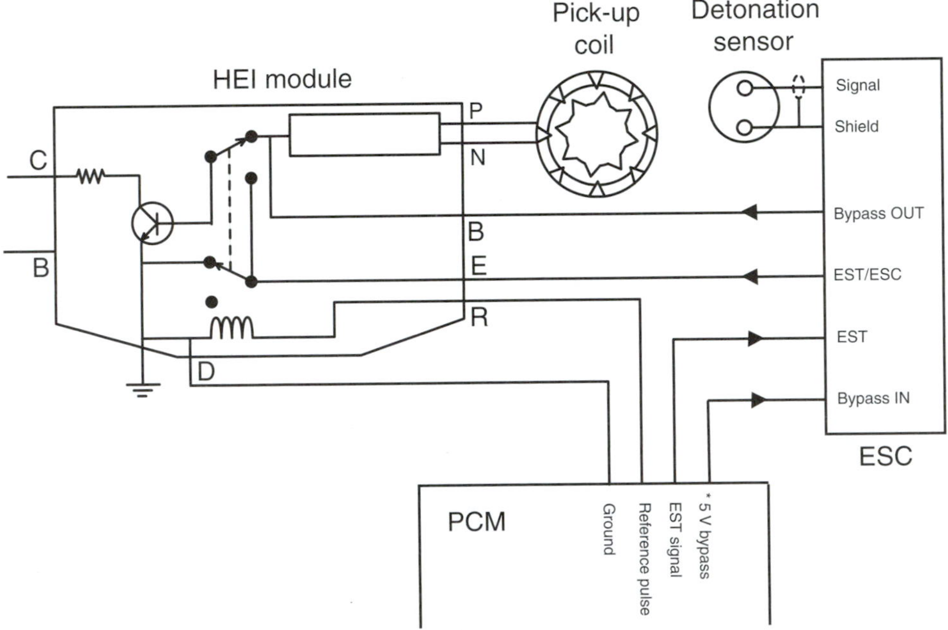

Figure 8–26 ESC circuit, pre-1982.

Knock, Detonation and Preignition

Detonation, knock and preignition are sometimes confused. Each of them constitutes effectively too-advanced timing and detracts from engine performance, sometimes also causing damage if the condition persists. Ordinarily the air/fuel mixture burns, rapidly but gradually (over about three thousandths of a second) proceeding from the spark plug and consuming the mixture in the combustion chamber. Detonation occurs if conditions in the combustion chamber reach such a combination of temperature and pressure that the entire charge explodes almost simultaneously. In this case, there is an extremely rapid buildup and loss of combustion pressure doing little torque-generating work. Detonation can easily occur if the spark advance is too great, and the explosion tries to force the piston back down the cylinder at the end of the compression stroke. A high compression engine is more inclined to detonation than a low compression engine, because it has much more heat and compression at the top of its compression stroke. An engine with combustion chamber deposits—which displace empty space and have the effect of raising the compression ratio—will have the same problem. These carbon deposits can also glow red hot and become a source of mistimed ignition themselves. In fact, since detonation also builds up deposits (because the fuel is incompletely burned), it can accelerate this problem.

Preignition occurs when a hot spot in the combustion chamber, either a hot exhaust valve or a hot carbon deposit, ignite the air/fuel mixture before the proper time. A car that diesels after shutdown is suffering from preignition. The effect is the same, but retarding spark on an engine with a combustion chamber hot spot obviously will not change the ignition point. The only repair for that condition is removal of the deposit or repair or replacement of the defective valve.

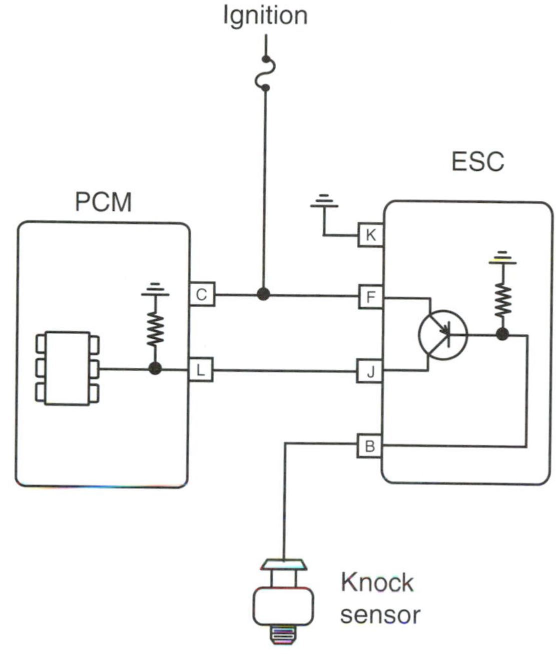

Figure 8–27 ESC circuit.

volts, to terminal J. It is then carried to terminal L of the PCM. When a detonation signal is sensed on terminal B of the ESC module, the module reduces the voltage at terminal J to below 1 volt. The PCM then retards timing until the signal voltage reappears at terminal L. Either of the ESC systems can be easily tested by tapping on the intake manifold adjacent to the knock sensor with a metal tool, while watching the timing with a timing light or meter. Be sure to consult the vehicle's appropriate service manual for specific directions.

Air Management Valve

As mentioned often before, one of the major goals of the CCC system is making the three-way catalytic converter work at maximum effectiveness. The system achieves this primarily by controlling the air/fuel ratio. On vehicles with a dual-bed, three-way catalytic converter, the system also works to optimize the effectiveness of the converter by introducing air from the air pump into the oxidizing side of the converter. The PCM-controlled air management system is responsible for controlling air injection into either the exhaust manifold or the catalytic converter.

CCC vehicles with dual-bed, three-way catalytic converters also have the appropriate dual air management valve. Those with a single-bed, three-way converter use a single air management valve, Figures 8–28 and 8–29.

Single Valve. The single valve system directs air to one of two places. During engine warmup, air is directed into the exhaust manifold. This:

- reduces HC and CO by oxidizing them in the exhaust system.
- raises oxygen sensor temperature more quickly from the heat produced by the above process.
- raises the catalytic converter temperature more quickly; the converter must be above 205°–260°C (410–500°F) to operate effectively.

As the engine coolant temperature climbs to 65°C (150°F), the temperature at which the CCC system goes into closed loop, the PCM commands the divert valve into **divert mode**. Depending on the particular application, the air is diverted either to the air cleaner or to the atmosphere. Had the air continued into the exhaust manifold, the signal from the oxygen sensor would have been inaccurate, not reflecting the results of the combustion mixture's burning. CCC systems with a single-bed converter are always in divert mode except during warmup.

Dual Valve. On vehicles with a dual-bed converter, air can be directed to one of three places. During engine warmup, the divert valve

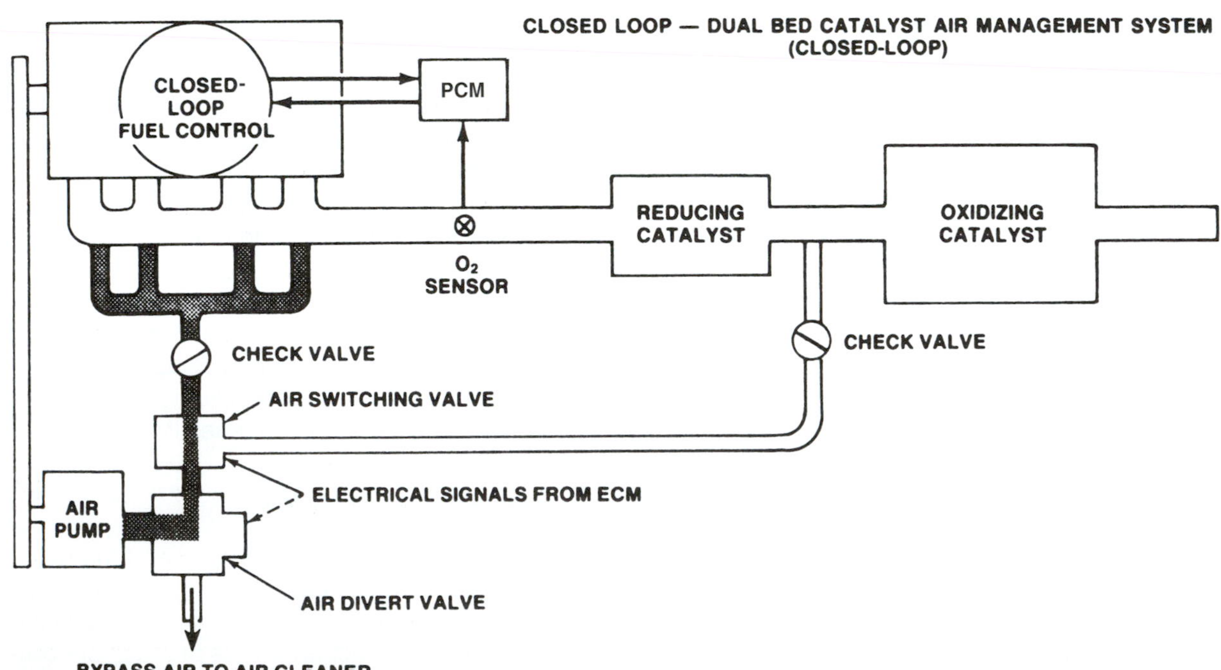

Figure 8–28 Air management with dual-valve and dual-bed catalyst. *(Courtesy of General Motors Corporation, Service Technology Group.)*

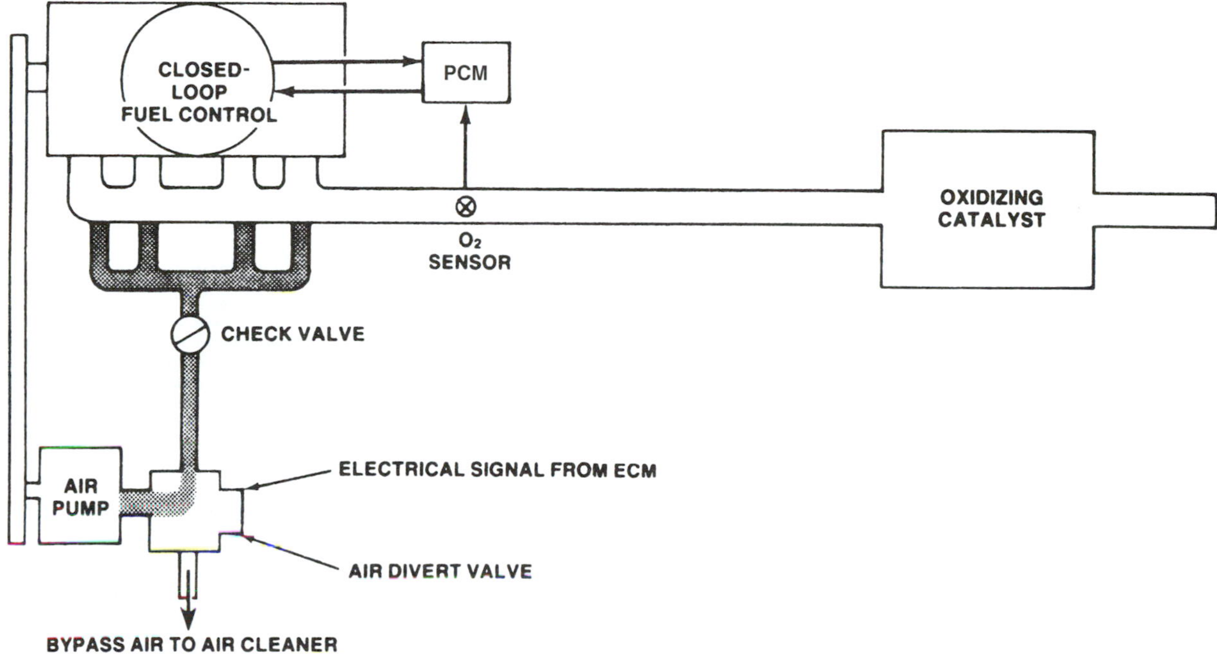

Figure 8–29 Air management with single valve and catalyst. *(Courtesy of General Motors Corporation, Service Technology Group.)*

directs air to the switching valve, Figure 8–28. The switching valve directs the air to the exhaust manifold. As the engine approaches closed-loop temperature, the PCM commands the switching valve to direct air to the oxidizing bed of the catalytic converter. Air pumped into the oxidizing bed of the converter has the following effects:

- It prevents the additional oxygen from flowing across the oxygen sensor, which would cause it to send an incorrect signal to the PCM.
- It lowers the temperature in the exhaust manifold. Continued pumping of oxygen into the exhaust manifold after the engine has reached normal operating temperature could produce additional NO_x.
- It makes the oxidizing bed of the catalytic converter operate at maximum efficiency without interfering with the efficiency of the reducing bed (which is upstream of the introduced oxygen and requires some of the residual carbon monoxide to work).

During WOT operation and during deceleration, the PCM commands the divert valve to dump the air, using the divert valve, because there is no other safe place to send it. Either of these driving conditions results in increased amounts of HC and CO in the exhaust system. If during these conditions, air were to continue to be pumped into any part of the exhaust system, the converter would quickly overheat. Certain failures within the CCC system, such as in the mixture control solenoid circuit or in the EST system, also cause the PCM to put the air management system into the divert mode.

Air Management Valve Designs. Many different valve designs have been used on various CCC applications, so a discussion of each different valve is impractical. Most of them, however, are more alike than they are different. They

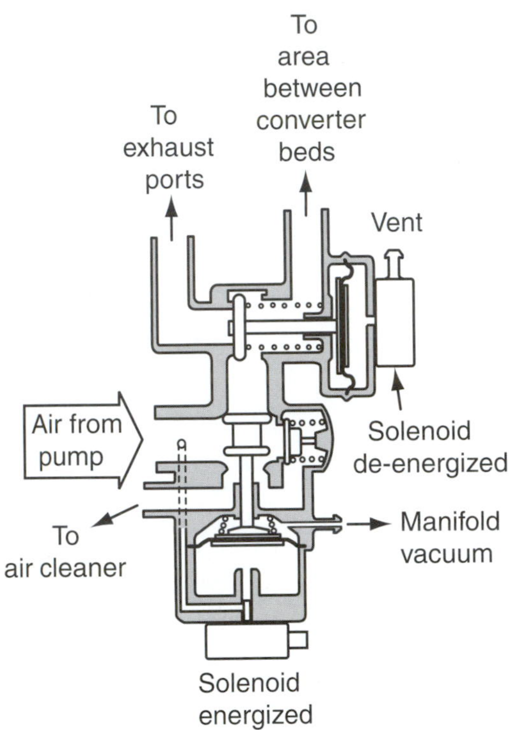

Figure 8–30 Divert mode (deceleration).

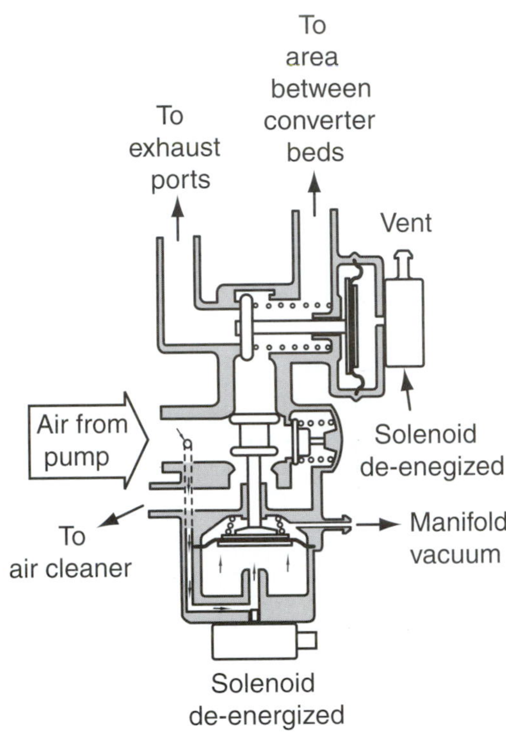

Figure 8–31 Divert mode (PCM commanded).

all direct airflow by moving an internal valve to open or block a passage. The valves can be moved by a vacuum diaphragm, an electric solenoid or a combination of the two. Figures 8–30, 8–31 and 8–32 show a typical dual valve in different modes of operation.

In Figure 8–30 the vehicle is decelerating. The high vacuum signal has pulled the diaphragm of the divert valve (lower valve) up. The passage to the switching valve (upper valve) is blocked, and air is diverted to the air cleaner. The air cleaner serves as a pump silencer. Diverting the injection pump air to it has no effect on the air/fuel ratio since the injection air enters the intake airstream before any metering of air by the throttle.

In Figure 8–31 the PCM has put the air management system into divert mode by de-energizing the solenoid. This allows pump air pressure to be routed to the decel timing chamber. With the help of the manifold vacuum on the other side of the diaphragm, the pressure forces the diaphragm and divert valve up.

Figure 8–32 shows the air management valve directing air to the exhaust manifold during engine warmup. Manifold vacuum is normally not strong enough to pull the divert valve diaphragm up, and the PCM keeps the divert solenoid energized, blocking pump air pressure from entering the decel timing chamber. The spring holds the divert valve down and allows air to flow up to the switching valve. The switching valve solenoid, energized by the PCM, allows vacuum to apply to the upper diaphragm. This pulls it and the attached switching valve to the right to block the converter air passage, thus opening the exhaust port passage.

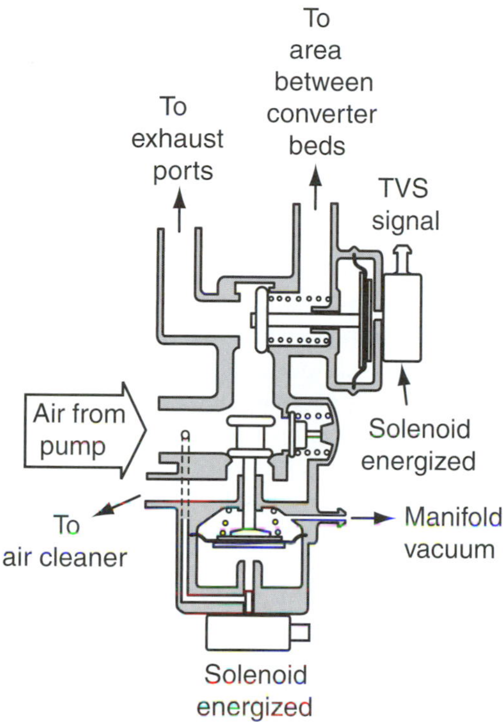

Figure 8–32 Air directed to exhaust ports.

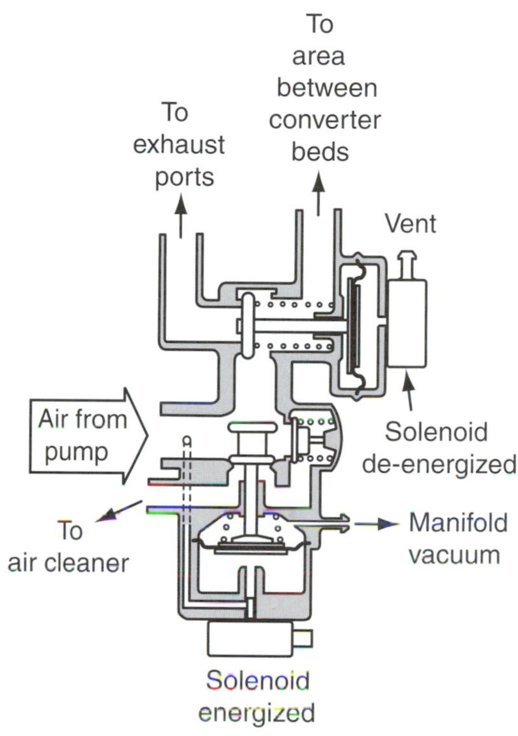

Figure 8–33 Warm engine mode (closed loop).

In Figure 8–33, the engine has warmed up, and the CCC system is ready to go into closed loop. The PCM has de-energized the switching valve solenoid, which blocks vacuum to the diaphragm. The spring moves the valve to the left, opening the converter passage and blocking the exhaust port passage.

All air management valves used with CCC systems use a pressure relief valve. At high engine speeds, the pressure relief valve exhausts excess air to reduce system pressure. This can easily be mistaken for a divert mode function.

Pulsair System. Some vehicles use a **Pulsair system** instead of an air injection system, especially on four-cylinder engines. Between each exhaust pulse in the exhaust manifold, a low pressure pulse develops. This low pressure can be used to syphon air into the exhaust ports. This air is drawn through a tube with a check valve (usually a reed-type flap). The check valve blocks reverse airflow during the exhaust pulse.

Pulsair Shutoff Valve. When a Pulsair system is combined with the CCC system, the PCM uses a Pulsair shutoff valve to control air flowing into the exhaust system, Figure 8–34.

During engine warmup, the solenoid is energized and vacuum applied to the diaphragm. The valve is pulled down, and air is allowed to flow into the exhaust system. With the exception of how air is made to flow, this system is very similar to the single-valve air management system.

Idle Speed Control (ISC)

Idle Speed Control Motor. To achieve the established goals for the exhaust emission quality,

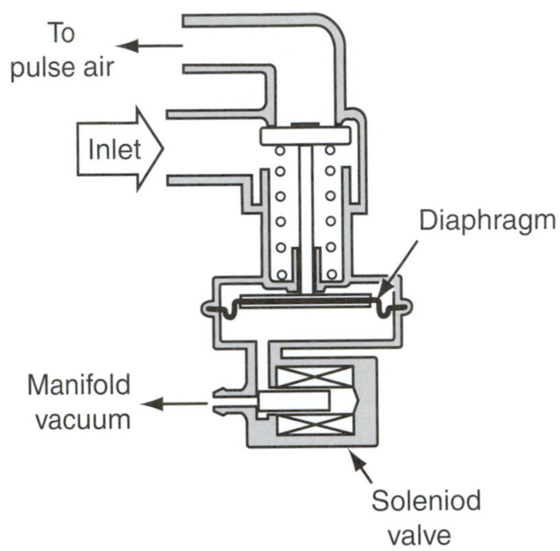

Figure 8–34 Pulsair shutoff valve.

fuel economy and driveability, an articulated idle speed control is needed. This is achieved by the use of a permanent magnet field reversible electric motor attached to the side of the carburetor, Figure 8–35.

The PCM normally has a fixed idle speed that it wants to maintain during idle. It moves the throttle as needed with the ISC motor. However, it commands a higher idle speed in response to any one of the following three conditions:

- During closed choke idle, the fast idle cam holds the throttle blade open enough to lift the throttle linkage off the ISC plunger. This allows the ISC switch to open so the PCM does not monitor idle speed. As the choke spring allows the fast idle cam to fall away and the throttle returns to warm idle position, the PCM notes the still low coolant temperature and commands a slightly higher idle speed.
- If the engine starts to overheat, the PCM commands a higher idle speed to increase coolant flow by turning the water pump faster.
- If system voltage falls below a predetermined value, the PCM commands a higher idle speed to increase generator speed and output.

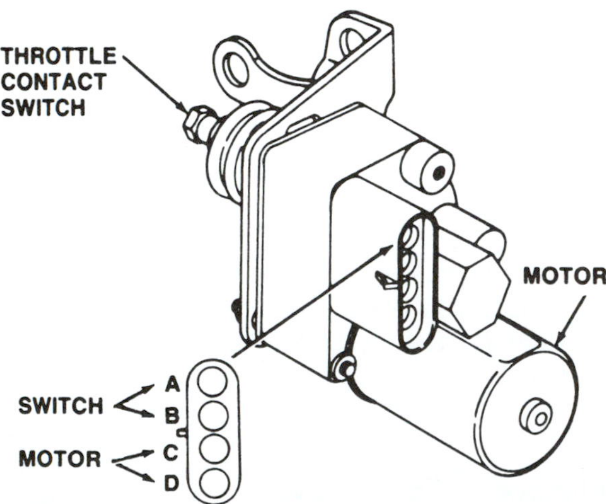

Figure 8–35 ISC motor assembly. *(Courtesy of General Motors Corporation, Service Technology Group.)*

During warm idle, if the automatic transmission is put in gear (forward or reverse), the park/neutral switch signals the PCM of the impending load on the engine. The PCM then commands the ISC motor to extend the plunger. The throttle blade opens at about the same time the transmission engages, and no appreciable change in engine speed occurs. If the air conditioning is turned on, the PCM receives a signal from the air-conditioning switch. The PCM commands a wider throttle blade opening to compensate for the additional load of the air-conditioning compressor.

Starting in 1982, models with smaller engines came with an ISC relay. The ISC relay uses a solid-state fixed time delay device to keep the ISC motor activated in a retract mode after the key is turned off. This allows the throttle to completely close each time the engine is turned off and thus helps prevent dieseling.

There is no way to adjust idle speed on CCC systems using an ISC motor. Attempting to adjust the idle speed by adjusting the plunger screw merely results in the PCM moving the plunger away to neutralize the adjustment. Idle speed does not change until the ISC plunger motor uses up all its travel trying to compensate for the adjustment, at which point the system is completely out of calibration. Proper calibration can be restored, however by following the **minimum** and **maximum authority** adjustment procedures outlined in the service manual for the specific vehicle. When idle speed driveability problems occur, the ISC system is usually responding to or being affected by the problem, not causing it.

> **CAUTION:** Be sure the ignition is turned off before connecting or disconnecting the ISC motor, or damage can result to the PCM.

Idle Stop Solenoid. Some engines do not use an ISC motor. Instead they use an idle stop solenoid. The idle stop solenoid is essentially the same as an antidiesel solenoid or dashpot, used for years and dating back to the first emission control systems. When the ignition is turned on, the solenoid is energized and extends its plunger to hold the throttle lever off the idle stop screw. The extended plunger provides the curb idle position for the throttle blade. When the ignition is turned off, the plunger retracts and allows the throttle blade to move fully closed to prevent dieseling. The same device can be connected to the A/C switch. In this case, the idle stop screw provides curb idle, and the solenoid increases idle speed slightly when the A/C is turned on.

Idle Load Compensator (ILC). Beginning in 1982, the Oldsmobile limited control CCC system used an ILC instead of an ISC motor, Figure 8–36. The ILC is not controlled by the PCM; it is controlled by engine vacuum. The ILC is very similar in construction to a vacuum advance unit on an older distributor. It contains a diaphragm with a spring pushing against one side. The other side attaches to a plunger that extends from the unit in the form of an adjustable screw. The head of the plunger screw acts as the throttle stop to control idle speed. Manifold vacuum is applied to the spring side of the diaphragm. When the automatic transmission is put into gear, the load on the engine causes vacuum to go down slightly. This allows the spring to push the diaphragm and plunger a little farther, thus opening the throttle a little. The same thing occurs for any load or condition that reduces engine vacuum.

Idle speed adjustments can be made to the ILC; however, the adjustment is made at the vacuum side of the diaphragm, not at the plunger screw. The idle adjusting screw in the stem on the back of the unit changes the preload on the

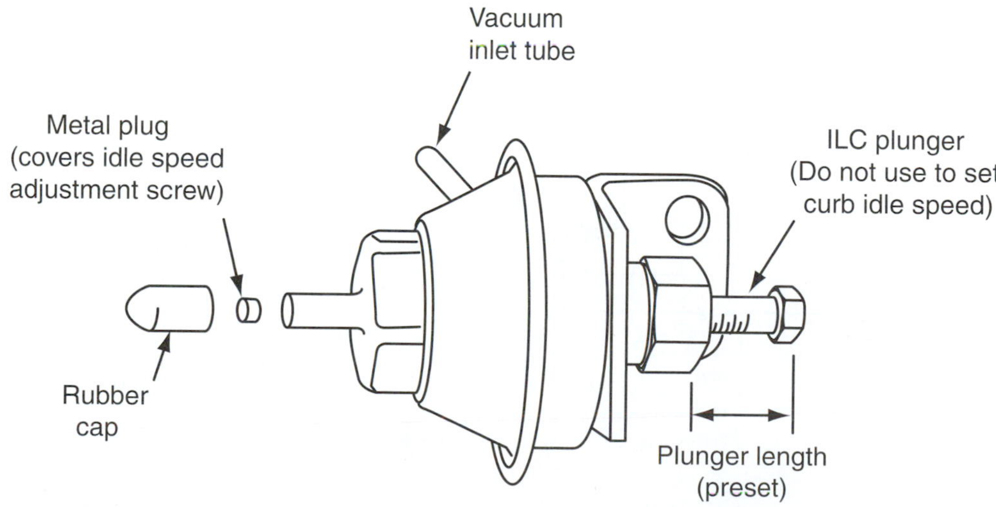

Figure 8–36 Idle load compensator.

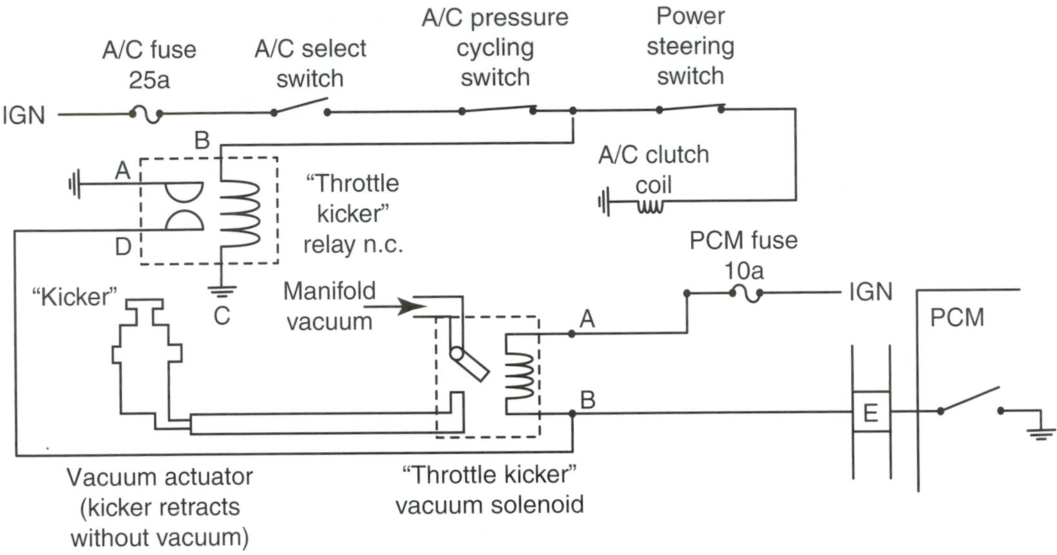

Figure 8–37 Vacuum-operated throttle kicker circuit (2.8-liter).

diaphragm spring. Refer to an appropriate service manual for adjustment procedures.

Throttle Kicker. Some later CCC vehicles use either a vacuum or electronic throttle kicker. The functions of a throttle kicker are explained in Figures 8–37 and 8–38.

Torque Converter Clutch (TCC)

The TCC improves fuel economy by eliminating the hydraulic slippage and heat production in the torque converter once a cruise speed is achieved.

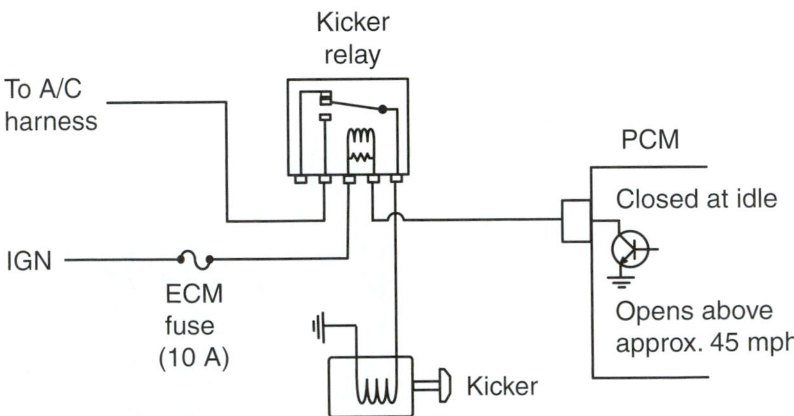

Figure 8–38 Electrically operated throttle kicker circuit.

Clutch. The major component of the TCC system is the lockup clutch itself. It becomes a fourth element added to a conventional torque converter, Figure 8–39. The clutch plate, splined to the turbine, has friction material bonded to its engine side and near its outer circumference. The converter cover has a machined surface just inside, where the converter drive lugs attach. This machined surface mates with the friction material on the disc when the clutch is applied. When hydraulic pressure is applied to the turbine side of the disc, the disc is forced against the converter cover. The friction between the clutch's friction material and the cover locks up the unit and causes the torque converter to rotate as a solid unit. The pressure source is converter feed oil coming from the pressure regulator valve in the transmission valve body. In non-TCC applications, converter feed oil is used to charge and cool the converter by circulating through the converter and then the transmission cooler in the radiator. On TCC applications, however, converter feed oil must pass through a converter clutch apply valve on its way to and as it returns from the converter. The apply valve, a small valve in the transmission, controls the direction of oil flow through the torque converter, Figure 8–40.

With the apply valve in its at-rest position and the transmission in neutral, or with the car moving at low speed, converter feed oil is directed into the converter through the release passage by way of the hollow turbine shaft. This oil is fed into the converter between the converter cover and the clutch disc. The oil forces the disc away from the cover and thus releases the clutch. Converter feed oil flows over the circumference of the disc and circulates through the converter. It exits through the apply passage, a passage between the pump drive hub and the stator support shaft.

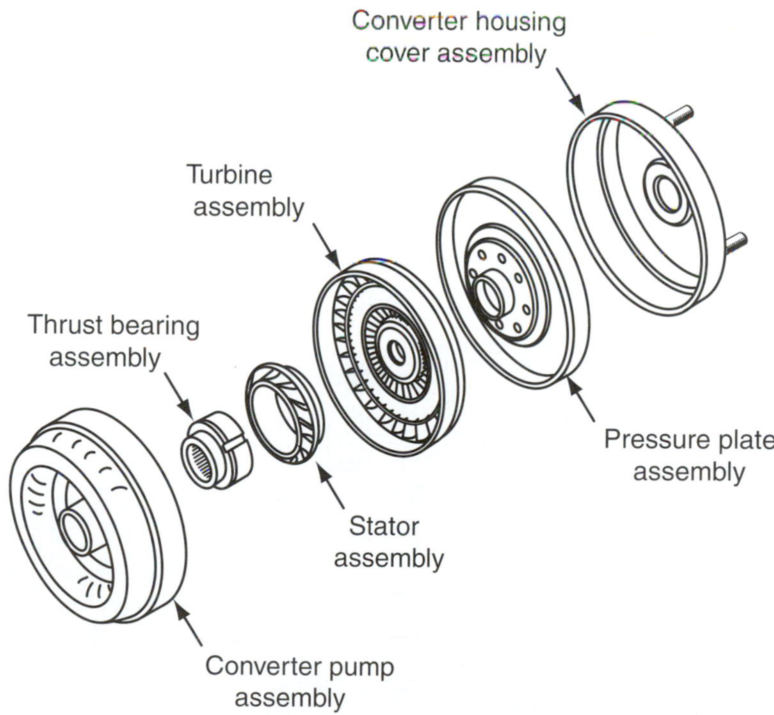

Figure 8–39 Torque converter with TCC.

Two criteria must be met before the clutch applies. The transmission must be ready hydraulically, and the PCM must be satisfied with engine temperature, throttle position, engine load and vehicle speed. When the transmission is in the right gear (this varies from transmission to transmission), hydraulic pressure is supplied at one end of the clutch apply valve, Figure 8–41. This hydraulic pressure (converter apply signal) has the potential to move the apply valve into the apply position. The apply valve is moved, however, only if the PCM is ready for the lockup clutch to engage.

The PCM controls the position of the clutch apply valve with a solenoid that opens or closes an exhaust port at the converter apply signal end of the apply valve. If the solenoid is not energized, the exhaust port is open and the converter apply signal oil exhausts from the signal end of the apply valve as fast as it arrives. Sufficient pressure does not develop to move the apply valve. Converter feed oil continues to flow into the converter through the release passage, Figure 8–40. When the solenoid is energized, the exhaust port is blocked. The converter apply signal oil develops pressure, and the apply valve moves and is held in the apply position. Converter feed oil now flows into the converter through the apply passage, Figure 8–41. The clutch is applied and pre-

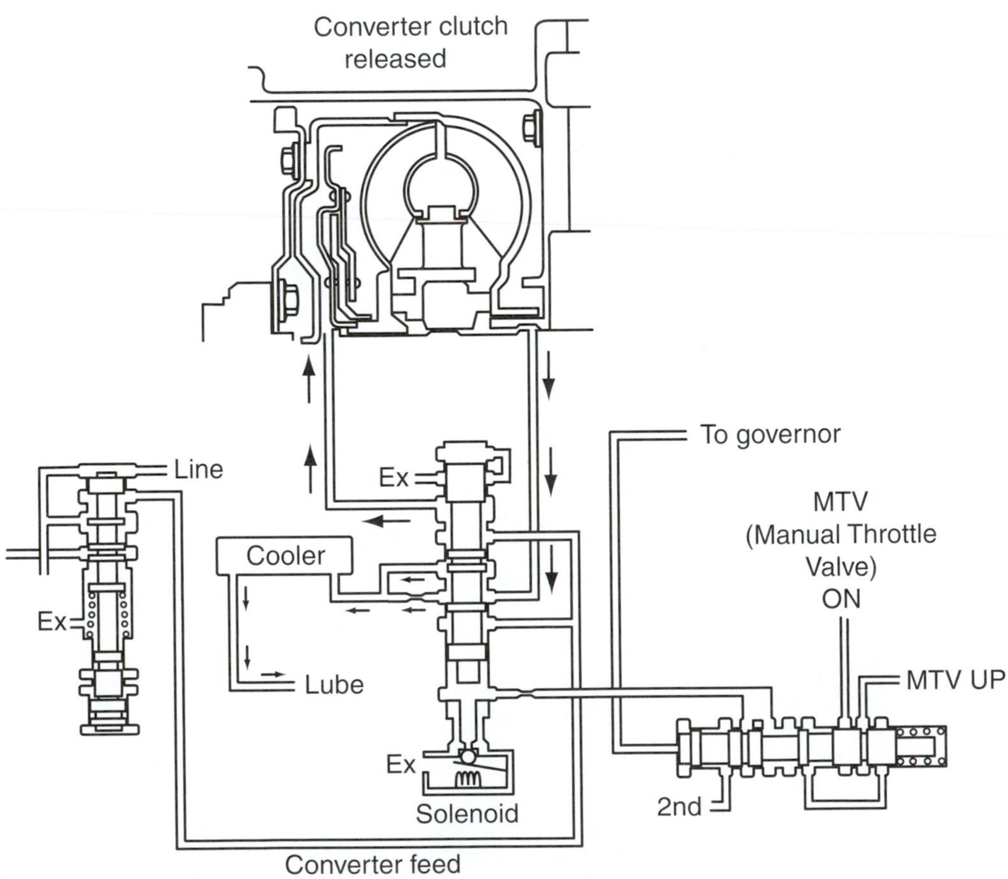

Figure 8–40 TCC apply circuit in release position.

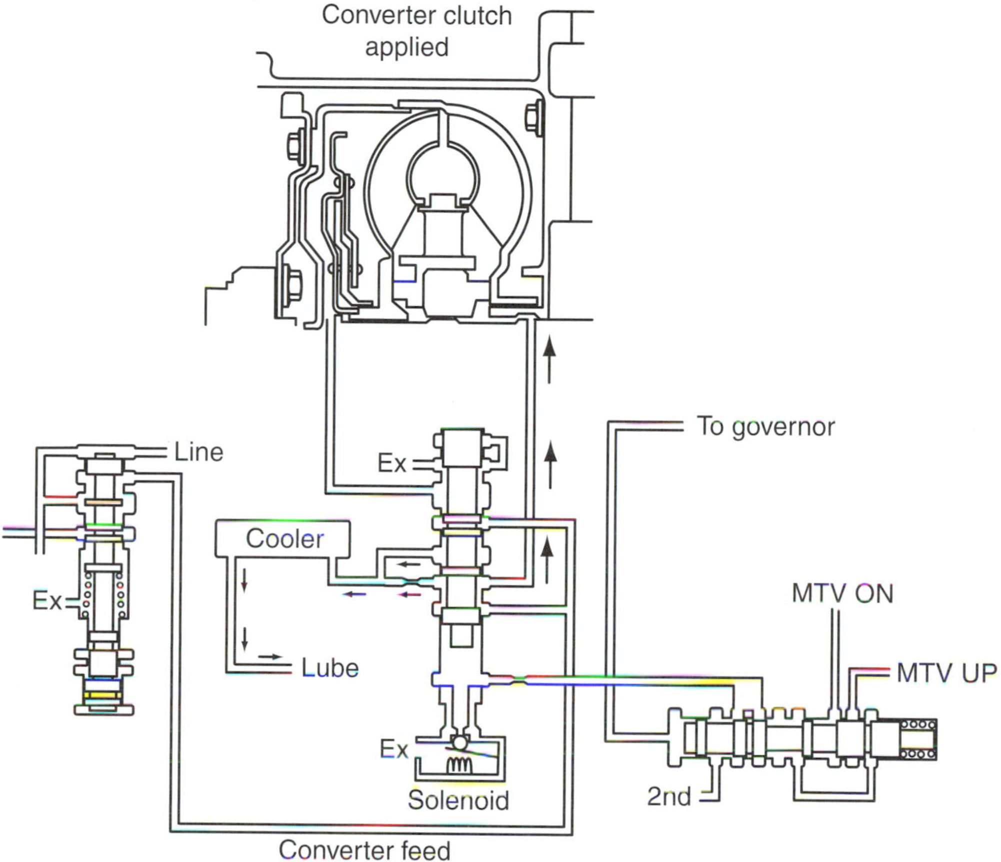

Figure 8–41 TCC apply circuit in apply position.

vents oil from exhausting through the release pressure, holding static pressure.

Figure 8–42 shows a typical TCC solenoid control circuit. The power source is the ignition switch, through the gauge fuse. The brake switch is in series on the high-voltage side of the circuit. Anytime the brakes are applied, the solenoid is de-energized. After the current passes through the apply solenoid, it arrives at the 4-3 pulse switch, used only on automatic transmissions with a fourth-speed overdrive. The 4-3 pulse switch momentarily disengages the TCC to allow a smooth 4-3 downshift. It is closed at all other times. From there, the current goes to the PCM, where the circuit is grounded when the PCM is ready to apply the TCC. When the transmission goes into fourth gear, the fourth-gear switch opens to alert the PCM that the transmission is in overdrive. In overdrive the PCM holds the TCC on through a much wider range of throttle position than it does in second or third gear. To put it in engineering language, the PCM holds the TCC on throughout a wider throttle position window. The term *window* refers to a range within a range. For instance, if the TPS signal to the PCM were 0.5 volt at idle and 4.5 volts at WOT, the PCM might select 2.0 to 3.5 as the window in which it would keep the TCC applied.

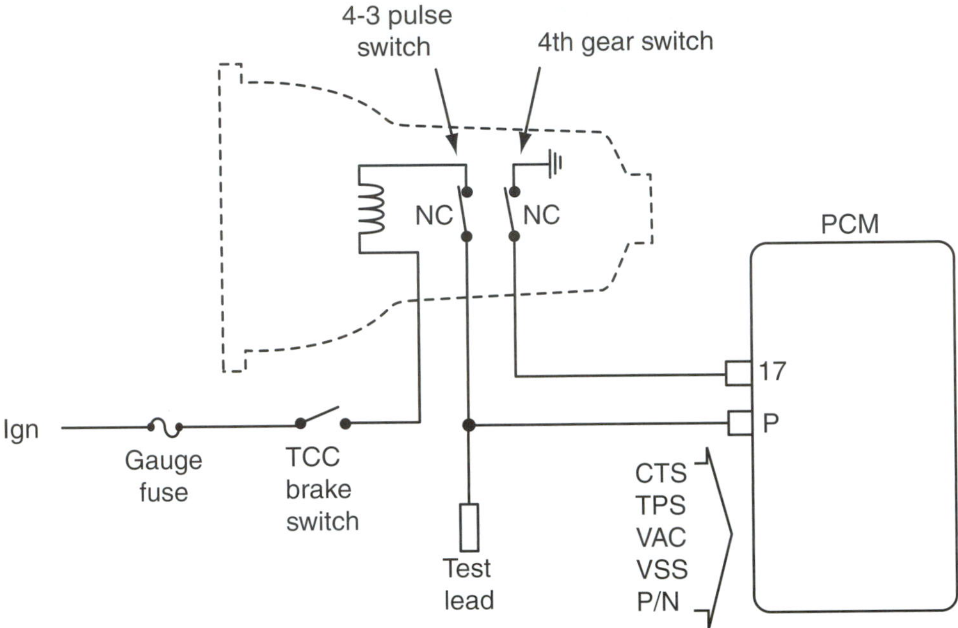

Figure 8–42 TCC electronic control circuit.

The TCC test lead, which comes from the low-voltage side of the circuit, can be used to tell when the PCM has electrically applied the TCC circuit or to ground the circuit and thus override the PCM. Its use is covered in more detail under System Diagnosis and Service later in this chapter.

TCCs and Four-Speed Automatic Transmissions

On vehicles using a four-speed automatic transmission, the TCC must be applied while the transmission is in fourth gear. Any failure resulting in the clutch not applying during prolonged fourth gear operation can result in transmission damage as a result of oil overheating. This can occur in a relatively short period of steady-state fourth-gear operation.

Exhaust Gas Recirculation (EGR) Valve

The primary purpose of the EGR valve is to control the production of oxides of nitrogen (NO_x) by lowering combustion temperatures. It does this by metering a certain amount of inert exhaust gas into the incoming air/fuel mixture, slowing its burn rate. The EGR also serves somewhat to control detonation. The pressing need for better fuel economy and driveability, combined with the development of computer-controlled ignition timing has pushed ignition timing to the ragged edge of detonation. Without the cooling effect of EGR on combustion temperature, detonation would occur in almost all engines.

Even so, excessive EGR gas causes serious driveability problems, especially when engines are cold. It makes sense to have the PCM in charge of EGR operation to closely monitor and control its application. On CCC engines, the EGR is turned on or off or on some later models is modulated (the amount of opening is controlled) by one or more PCM-controlled solenoids. The exact method and amount of EGR control varies considerably with engine and model year application. Generally, however, EGR systems controlled by CCC can be summarized by one of the following statements:

- The PCM operates a solenoid that blocks vacuum when energized. When not energized, it passes ported vacuum to the EGR valve, Figure 8–43.
- The PCM operates a solenoid that bleeds atmospheric pressure into the ported vacuum signal to the EGR valve when activated. A bleed solenoid can be used to turn the EGR valve off or to partially close (modulate) it, Figure 8–44.
- The PCM operates two solenoids, one to block or pass ported vacuum to the EGR valve and the other as a bleed solenoid to modulate it, Figure 8–45.
- The PCM operates two blocking solenoids in series. Each one is operated according to different sets of engine calibration criteria, Figure 8–46.
- A pulse-width modulating solenoid (rapidly turned on and off with the on-time variable) precisely controls the amount of vacuum allowed to the EGR valve, Figure 8–47.

One other EGR control system that can be encountered on a CCC vehicle should be discussed at least briefly, the **aspirator**-controlled EGR, Figure 8–48. The aspirator, mounted in the

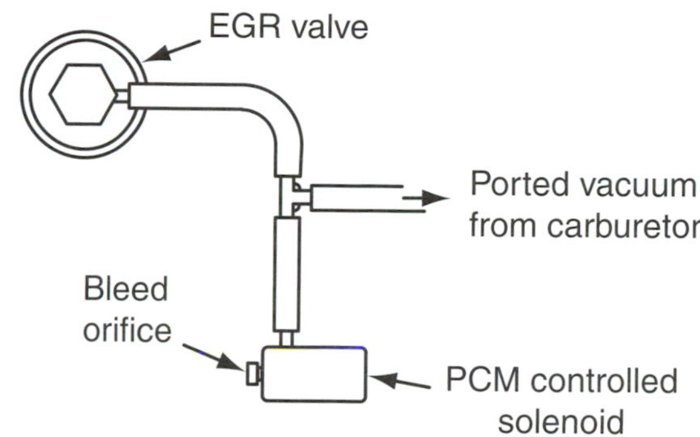

Figure 8–44 EGR bleed solenoid.

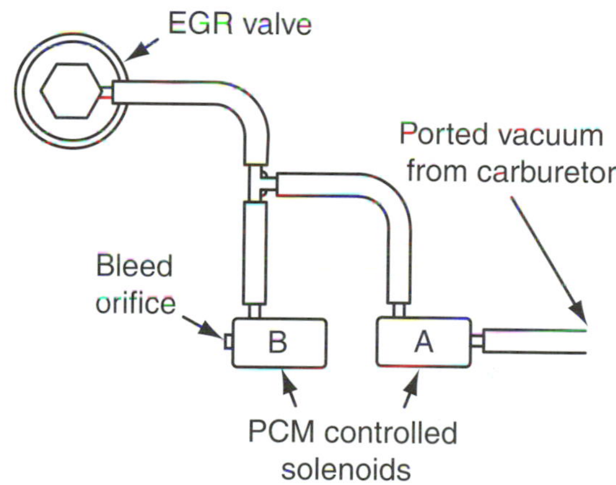

Figure 8–45 EGR bleed and control solenoids.

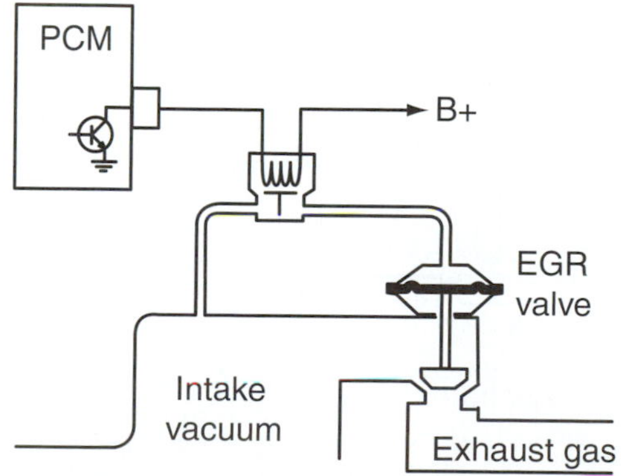

Figure 8–43 Typical EGR valve.

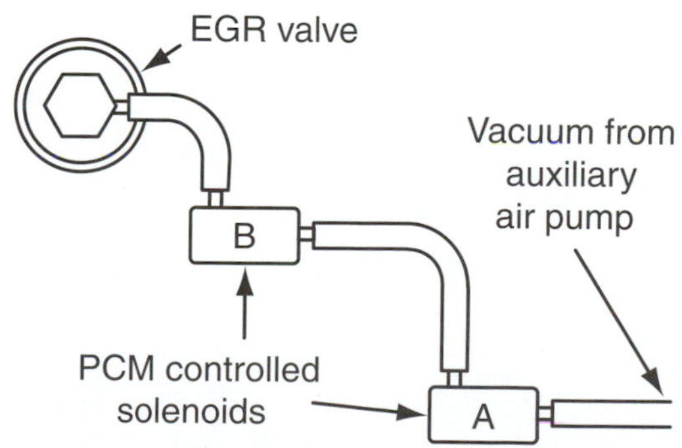

Figure 8–46 Two EGR control solenoids.

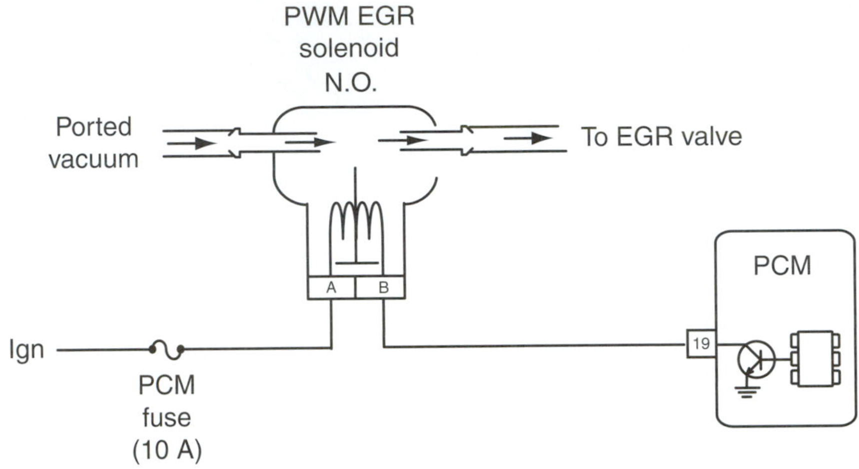

Figure 8–47 Pulse-width modulated EGR control solenoid.

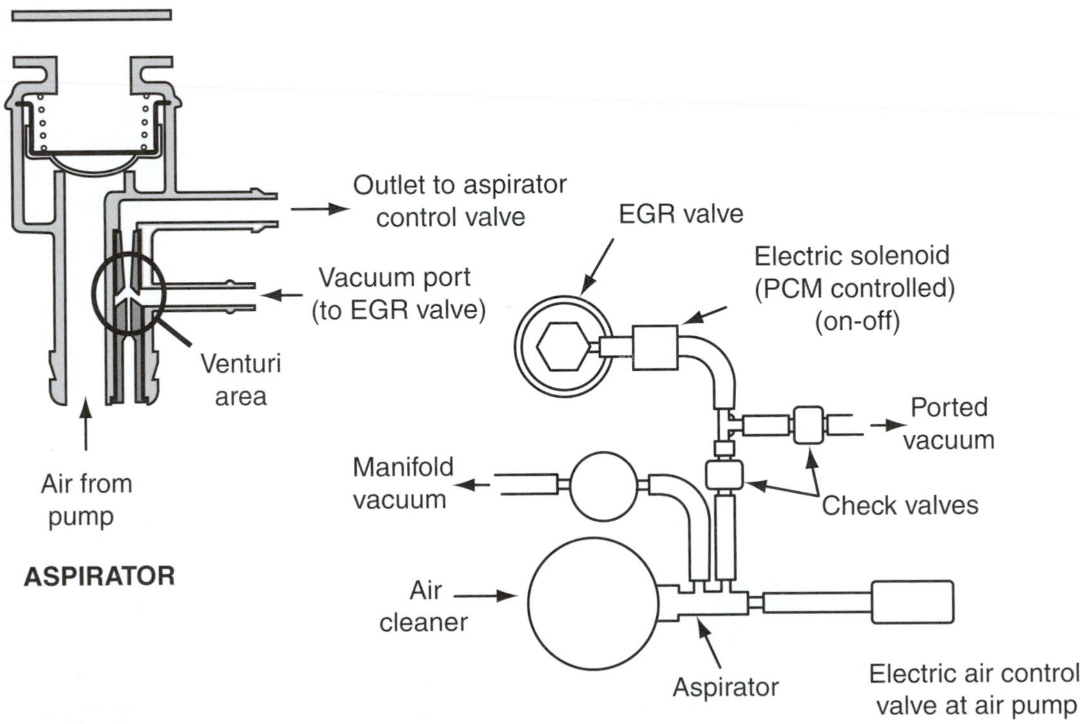

Figure 8–48 Aspirator-assisted EGR valve.

air cleaner, contains a small venturi. Air from the air pump is directed to the aspirator. During periods of low engine vacuum, the aspirator control valve allows air to pass through the venturi portion of the aspirator to produce a vacuum that keeps the EGR valve open. When engine vacuum is high, the aspirator control valve closes to prevent air from passing through the aspirator venturi. The EGR valve is now operated by ported vacuum.

Generally, if a failure occurs within the EGR electrical control system or if the PCM goes into limp-in mode, the EGR valve is simply controlled according to a ported vacuum signal.

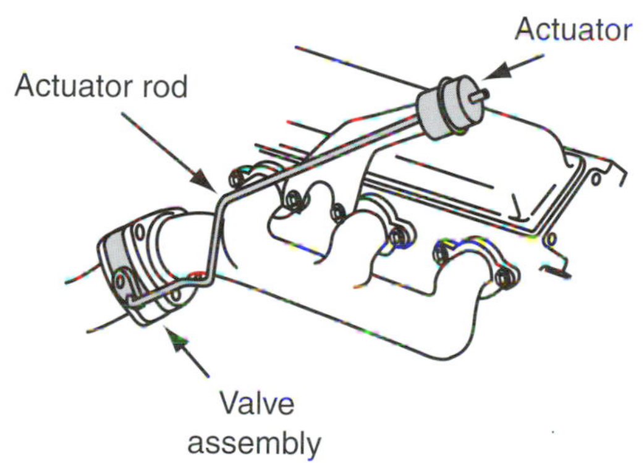

Figure 8–49 Vacuum-actuated EFE valve.

Early Fuel Evaporation (EFE)

The EFE system applies heat to the intake manifold area beneath the carburetor to help evaporate the fuel and to keep it in a vapor state. CCC systems use one of two different types of EFE systems.

Exhaust Heat Type. An EFE valve (known earlier as a heat riser valve, located where the exhaust pipe connects to the exhaust manifold on one side of the engine) is closed by a vacuum motor, Figure 8–49. This forces exhaust gases through a passage in the intake manifold beneath the space over which the carburetor mounts (the intake manifold plenum). Vacuum to the vacuum motor is controlled by a solenoid under the control of the PCM, Figure 8–50. This system varies from earlier, non-CCC, EFE systems in that they used a **vacuum control valve (VCV)** to control vacuum to the EFE actuator.

Electric Grid Type. The electric EFE uses a ceramic-encased heating element under the carburetor, Figure 8–51. The heater element is powered by a relay, Figure 8–52. When the ignition is on, voltage is available at one of the normally open relay contacts. Ignition voltage is also applied to the relay coil waiting for ground by the PCM. When coolant temperature is low, the PCM grounds the coil and the relay contacts close. This powers the heater element. Electric EFE circuits vary slightly with engine application and model year.

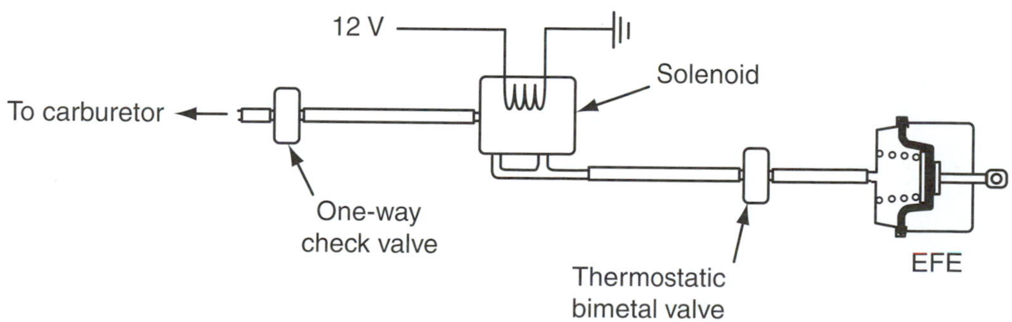

Figure 8–50 EFE vacuum control circuit.

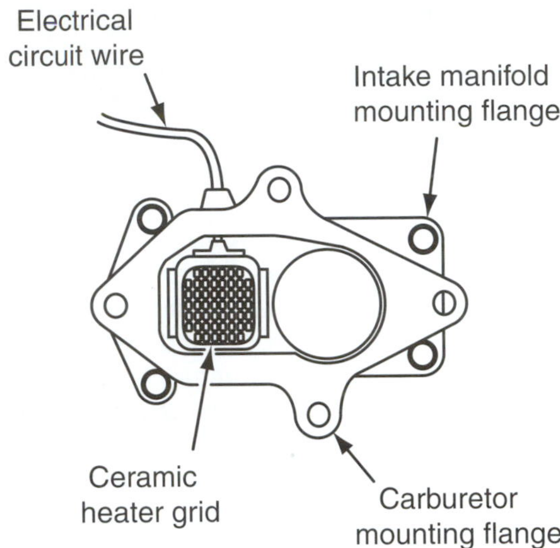

Figure 8–51 Electrically heated EFE valve.

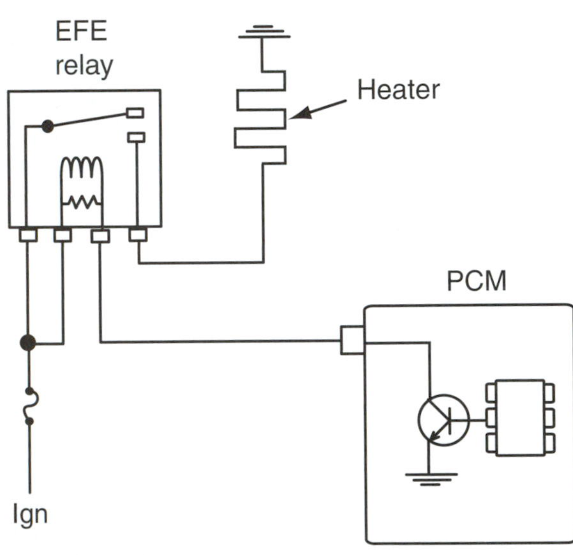

Figure 8–52 Electric EFE control circuit.

Controlled Canister Purge (CCP)

On many CCC vehicles, the PCM also operates a solenoid to control purge vacuum to the **purge valve** on the charcoal canister. When the CCC system is in open loop, or before a predetermined time period has elapsed since engine startup, or below a specific rpm, the purge solenoid is energized and blocks purge vacuum. In closed loop, after a specified elapsed time and above a specific rpm, the purge solenoid is de-energized and canister purging occurs.

Malfunction Indicator Lamp and Lamp Driver

All CCC systems use a Malfunction Indicator Lamp (MIL), known by GM as a Check Engine Light prior to OBD II. If the PCM sees a fault in one of the circuits that it monitors for malfunctions, it turns on the MIL located on the instrument panel. This warns the driver that a malfunction exists. By grounding the test terminal in the diagnostic link connector (DLC) located under the instrument panel, the technician can instruct the PCM to flash the MIL in order to count out DTCs. These flashes of the MIL represent one or more DTCs that identify the circuit or circuits in which the fault exists. This is explained more fully under System Diagnosis and Service in this chapter.

In 1982, a remote lamp driver was added to the system, Figure 8–53. It is a separate module, located under the instrument panel in most cases. Its purpose is to light the MIL as a default until instructed by the PCM to turn it off. A feedback-carbureted engine will actually run if the PCM fails to power up, although the engine will be at base timing and the carburetor will be full rich. But, because the failed PCM will not likely be able to inform the driver of the fault through the MIL, the remote lamp driver automatically lights the MIL so that the driver is made aware that a problem exists.

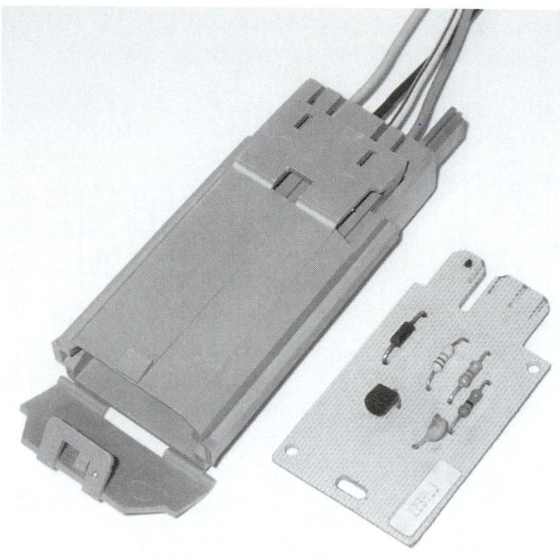

Figure 8–53 Remote lamp driver removed from its housing.

E-cell

Some Oldsmobile and Chevrolet eight-cylinder engine CCC applications incorporate an E-cell, often referred to as green engine calibration unit. The E-cell slightly modifies some engine calibrations during engine break in. It is not necessary to replace the E-cell even if it fails prematurely.

✔ SYSTEM DIAGNOSIS AND SERVICE

Self-Diagnosis

The PCM monitors the major input sensors, the mixture control solenoid, the EST and their respective circuits for proper operation. If the PCM sees a fault, such as an open, a short or a voltage value that stays too high or too low for too long, in any of the circuits it monitors, it turns on the MIL on the instrument panel and records a code number in its diagnostic memory. Some later models have a service engine soon light and a service engine now light, either of which can operate as the MIL depending on the severity of the problem. The code number identifies the circuit in which the fault exists. The MIL warns the driver that a fault exists. The MIL should come on, however, anytime the ignition is on without the engine running, as a bulb check.

Diagnostic Memory. A portion of the PCM's random-access memory is devoted to diagnostic memory, sometimes referred to as long-term memory. The diagnostic memory enables the PCM to store code numbers, referred to as diagnostic trouble codes, or DTCs (known by GM as "trouble codes" prior to OBD II), that identify the type of fault and the circuit in which the fault exists. A technician can obtain the DTCs by either of two methods. Grounding the test terminal in the DLC (known by GM as an Assembly Line Communications Link or ALCL, or Assembly Line Diagnostic Link or ALDL prior to OBD II) puts the system into diagnostics, Figure 8–54. (The DLC is most often located under the instrument panel and between the steering column and the radio.) The PCM pulses out its stored DTCs by flashing the MIL. All DTCs are stored as two-digit numbers. Two quick flashes followed by a pause of about two seconds represents a two; three more quick flashes represents the second digit, a three. Thus, the DTC would be a code 23. Figure 8–55 demonstrates a code 12 being pulsed. Once the two-digit number completes, a slightly longer pause separates it from the next two-digit DTC. Once a DTC is displayed, it repeats twice more (for a total of three times) before moving on to the next code.

The second method of obtaining stored DTCs is to connect a scan tool to the DLC and, with the ignition turned on, use it to pull the DTCs from the PCM's memory. The scan tool may also be used (with most CCC systems) to look at data stream information from the PCM including input sensor values as interpreted by the PCM as well as output commands from the PCM.

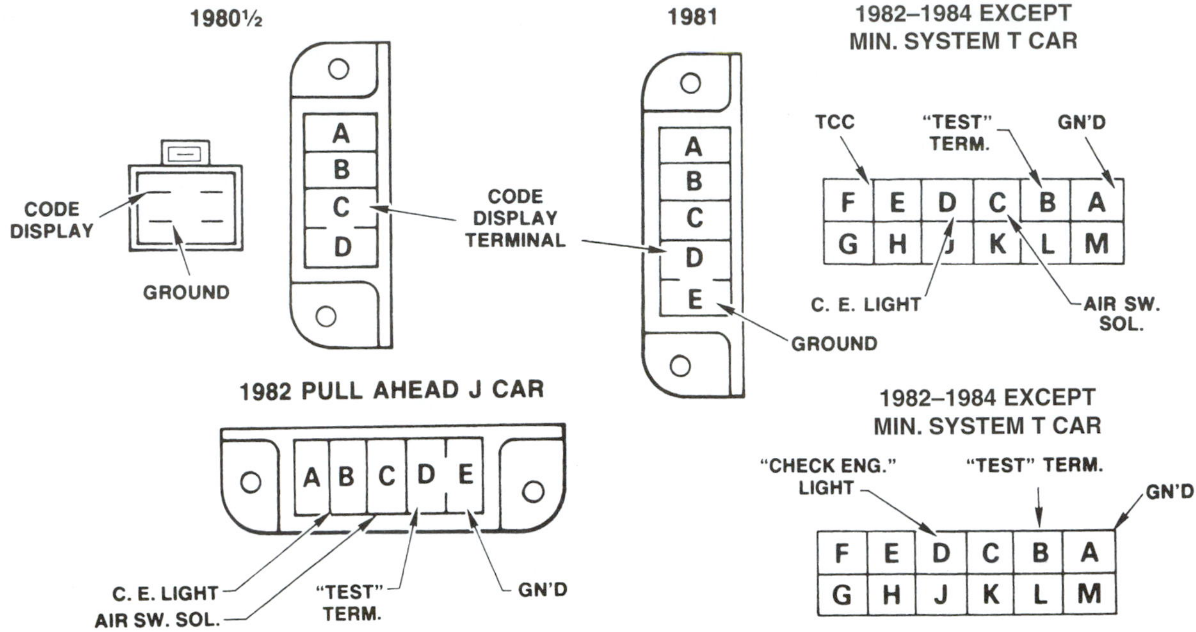

Figure 8–54 DCL connectors. *(Courtesy of General Motors Corporation, Service Technology Group.)*

Intermittent Faults. If the PCM sees a fault, it turns on the MIL and sets a DTC in memory. It keeps the MIL on until the ignition is turned off, or until the PCM no longer senses the fault. If the perceived fault clears up, the MIL goes off, but the DTC remains in memory. If the fault does not repeat again within the next fifty ignition cycles (starting the engine and turning it off constitutes one ignition cycle), the DTC is automatically erased.

CHECK ENGINE Pause CHECK ENGINE CHECK ENGINE Long Pause

Flash Flash Flash

Figure 8–55 Check engine light in diagnostics.

Clearing the Memory. The DTC can also be erased by disconnecting the power supply from terminal R of the PCM for 10 seconds. This can be done most easily either by pulling the fuse marked PCM from the fuse panel, disconnecting the negative battery lead or on some later models, disconnecting a fusible link near the positive battery terminal.

Behavior in Diagnostics

When the test terminal in the DLC is grounded and the engine is not running, the MIL should begin flashing code 12. Code 12 indicates no reference pulse is coming from the distributor, which is what you would expect since the engine is not running. Any additional stored codes display in numerical sequence after code 12.

The ignition should always be on before grounding the test terminal; otherwise the MIL does not function properly.

There are some other noteworthy points about code 12. It is not a storable code (it is not stored in memory). It is most often used as an indication that the self-diagnostic function of the PCM is working properly. If you put the system into diagnostics with the engine off and do not see a code 12, the self-diagnostic function is not working and must be repaired before you can continue. The only time the presence of code 12 indicates a fault is if it displays when the engine is running. Then it means the REF pulse is not coming in from the distributor to the PCM.

All DTCs, beginning with code 12, will be seen a total of three times each. When the final DTC has flashed out on the MIL, the entire set of DTCs begins to flash out over again, beginning with code 12. Therefore, the technician should recognize that when the fourth code 12 is seen, all DTCs have been displayed. If code 12 is displayed a fourth time without any other DTCs being flashed out, then a "pass" is indicated. Code 12 does not mean "system pass." In fact, GM does not use a DTC to mean "system pass," but rather the absence of fault codes between sets of code 12.

While the codes are displayed, the PCM also:

- sends a steady 30-degree dwell command to the mixture control solenoid.
- energizes all PCM-controlled solenoids.
- pulses the ISC motor in and out.

If the test terminal is grounded while the engine is running, the PCM will:

- disable the open-loop timer, which maintains the time lapse before the system can go into closed loop.
- take out some of the adaptive enrichment modes.
- send a 30-degree dwell command to the mixture control solenoid if the system is in open loop and is not in an enrichment mode.
- set the EST at a fixed spark advance position.

While in diagnostics (test terminal grounded), the PCM will not store new codes that might be encountered.

The 1982 and later T-car (Chevrolet Chevette and Pontiac T-1000) has no diagnostic memory. It can only display codes concerning faults it currently perceives. To receive fault codes from the minimum function system used on these cars, the engine must be running when the test terminal is grounded, and of course, the fault must be present.

Diagnostic Procedures

The diagnostic procedures are contained in several different charts or groups of charts. They are as follows:

- diagnostic circuit check
- customer complaint (or in newer manuals, driveability symptoms)
- system performance check
- diagnostic charts without trouble codes
- diagnostic charts with trouble codes
- diagnostic charts on related components

It is important to follow the directions very carefully when any of the diagnostic procedures or flow charts are used. Failing to do so or taking other shortcuts usually results in inaccurate diagnostic conclusions.

When using diagnostic procedures, be sure to use procedures or charts that apply to the specific engine or application and model year. Procedures vary for different engines and are often changed by model year.

Diagnostic Circuit Check. (Figure 8–56) This chart of procedures should always be used before attempting to diagnose any PCM-controlled system. It is primarily responsible for sorting out why a code is stored. It also aids in discovering why the check engine light is not

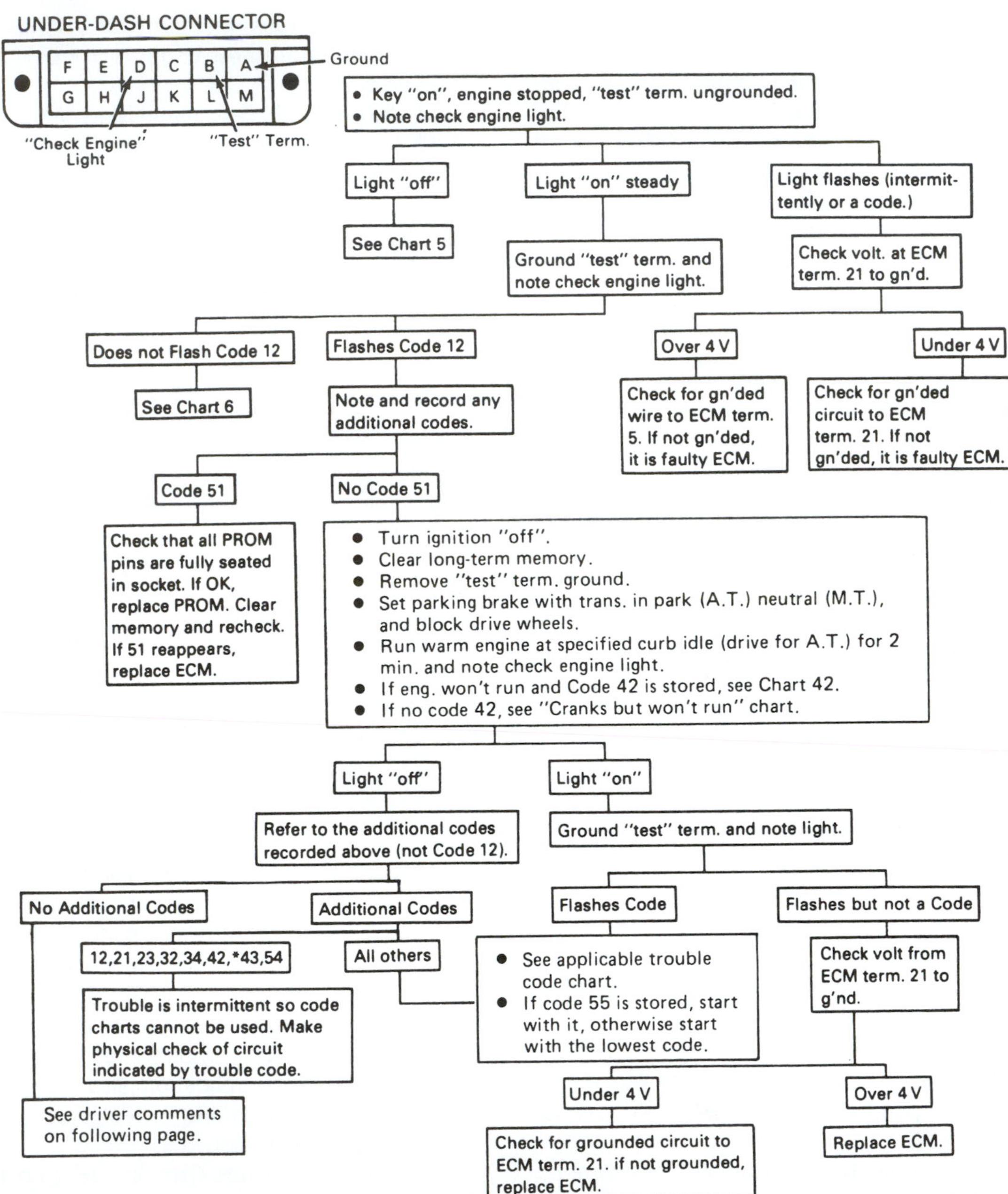

Before replacing an ECM, always check PROM for correct application and installation. Also, remove terminal(s) from ECM connector for circuit involved, clean terminal contact and expand it slightly to increase contact pressure and recheck to see if problem is corrected. In case of repeat ECM failure, check for a shorted solenoid or relay controlled by the ECM.

The system performance check should be performed after any repairs to the system have been made.

*It is possible to set a false Code 42 on starting, but the "Check Engine" light will not be "on". No corrective action is necessary.

Figure 8–56 Diagnostic circuit check. Charts 5 and 6, referred to in the diagram, are found in the manufacturer's service manual. *(Courtesy of General Motors Corporation, Service Technology Group.)*

working properly if it is not. There can be several reasons for a particular code's being stored:

- *Existing problem.* A fault can exist, in which case its corresponding code is called a *hard code*.
- *Intermittent problem.* A fault can develop and clear up by itself, in which case the resulting code is intermittent.
- *No real problem.* Someone previously working on the car can open a CCC system circuit that the PCM monitors while the ignition is on. This can set a code even though there is nothing wrong with the circuit. Sometimes a strong radio signal, such as can be encountered near an airport, can set a DTC. This type of code is often referred to as a *phantom code*.

It is important that this type of code identification be made. The DTC charts are written assuming a real fault exists. If one of the DTC charts is used to pursue a phantom or intermittent code, the chart more often leads to an invalid conclusion because it is beginning with an invalid assumption—that a fault currently exists when in fact none does.

Driver Comment Chart/Driveability Symptom Section. (Figure 8–57) This chart or section, depending on the model year, acts as a guide to aid the technician in determining the cause of problems that either do not set DTCs or that set intermittent DTCs.

System Performance Check. (Figure 8–58) This procedure should always be used after repairing any part of the CCC system or any component the PCM controls. It verifies that the heart of the system, the fuel mixture control, is working properly. If the fuel mixture is not controlled properly, it refers to another chart for further diagnosis.

A DTC found in memory and shown by the diagnostic circuit check to be intermittent should be dismissed as a phantom code when no driver complaint or performance problem was identified and the system performance check shows satisfactory results.

Diagnostic Charts without Trouble Codes. This is a series of charts contained in the service manual. Any of the charts within the series can assist in finding the cause of any one of several specific fault conditions that do not have a corresponding DTC. These charts should only be used when referred to by one of the preceding charts (diagnostic circuit check, driver comment or system performance check) or when a specific condition addressed by one of them has been identified. As mentioned previously, improper use of the charts most often leads to mistakes, sometimes costly ones.

Diagnostic Charts with DTCs. This series of charts contained in the service manual can be used to find the cause of a fault once the fault has been identified by a code stored in the PCM's memory, and after the diagnostic circuit check has been performed to verify that the fault currently exists.

Diagnostic Charts on Related Components. This series of charts, contained in the service manual, helps with the diagnosis of components and related circuits controlled by the PCM but not monitored by it for proper operation. These components and circuits include the air management valve and the torque converter clutch.

Dwell Diagnosis

If the system is in closed loop and the dwell varies between 10 and 50 degrees on the six-cylinder scale, we know the system is able to maintain the desired air/fuel ratio, Figure 8–59. If the dwell is not varying and is between 10 and 50 degrees, the system is not in closed loop. If the dwell stays below 10 or above 50 degrees, some condition is causing the air/fuel mixture to run either too lean or too rich and is beyond the capacity of the system to compensate. It is possible, more remotely, that the system electronics have malfunctioned and are unable to recognize the delivered air/fuel ratio.

Let's assume, for example, that we are getting a steady dwell of 54 degrees, which indicates a rich condition. To determine whether an electronic malfunction has occurred or whether the engine is actually running rich, run the engine for two minutes at a fast idle to make sure it is in

DRIVER COMMENTS
(Stalling, detonation, surge, fuel economy, etc.)

IF THE "CHECK ENGINE" LIGHT IS NOT ON, NORMAL CHECKS THAT WOULD BE PERFORMED ON ENGINES WITHOUT COMPUTER SYSTEMS SHOULD BE DONE FIRST.
IF THE GENERATOR OR COOLANT LIGHT IS ON WITH THE CHECK ENGINE LIGHT, THEY SHOULD BE DIAGNOSED FIRST. INSPECT FOR POOR ELECTRICAL CONNECTIONS AT THE COOLANT SENSOR, M/C SOLENOID, ETC., AND DAMAGED OR LOOSE VACUUM HOSES. REPAIR AS NECESSARY.

Intermittent check engine light but no codes are stored.
    Check for a loose connection in the circuit from:
        Ignition coil to ground, and arcing at plugs or plug wires
        Bat. to PCM terminals. C and R
        PCM terminals. A and U to ground
        EST wires should be kept away from spark plug wires, distributor and generator
        PCM term 13 wire to distributor should have a good ground for its shield
        Open diode across A/C compressor clutch

Loss of long term memory
    Grounding the dwell lead for 10 seconds with the test terminal ungrounded, engine running, should result in a code 23. This code should be retained in long term memory after the engine is stopped and restarted. If not retained the PCM is faulty.

Stalling, rough idle, or improper idle speed
    See idle speed control on page 4A-60

Detonation (spark knock)
    Check: ESC performance on page 4A-55
        MAP or vacuum sensor on pages 2A-20 and 2A-21
        EGR operation on pages 4A-56 and 4A-57
        TPS operation on chart 4
        HEI operation on page 4A-54

Poor performance and/or fuel economy and surging
    See: Carb. on car service - Service Manual
        EFE operation on pages 4A-58 thru 4A-60
        TCC operation check on page 4A-63
        EST diagnosis on page 4A-54
        ESC diagnosis on page 4A-55

Poor full-throttle performance
    See chart 4 if equipped with TPS

Intermittent no-start
    Incorrect pick-up coil or ignition coil. See "Cranks but won't start" chart
    Intermittent ground connections on PCM

All other comments
    Make system performance check on warm engine. System performance check should be performed after any repairs to the CCC system.

Figure 8–57 Driver complaint chart.

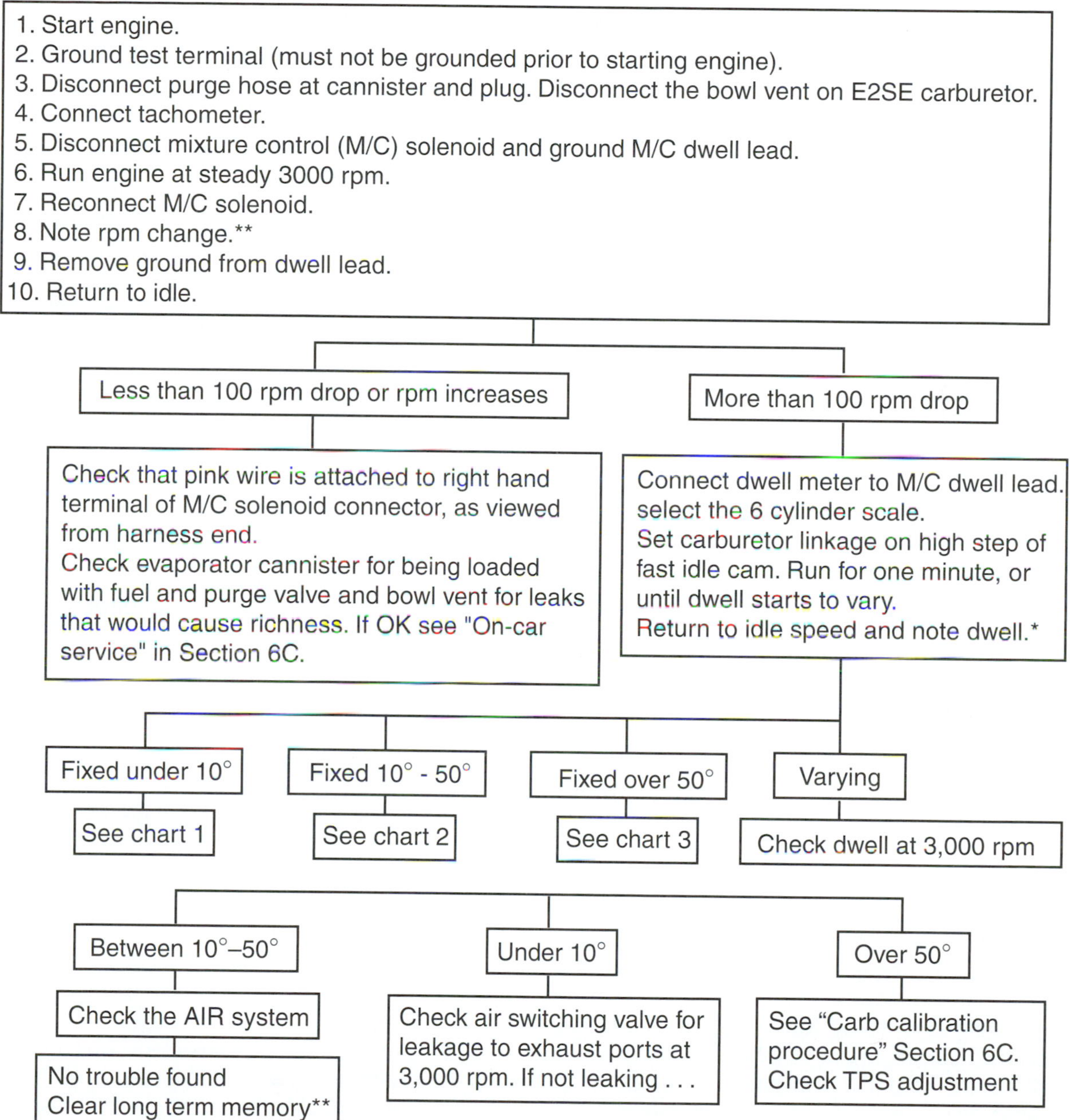

1. Start engine.
2. Ground test terminal (must not be grounded prior to starting engine).
3. Disconnect purge hose at cannister and plug. Disconnect the bowl vent on E2SE carburetor.
4. Connect tachometer.
5. Disconnect mixture control (M/C) solenoid and ground M/C dwell lead.
6. Run engine at steady 3000 rpm.
7. Reconnect M/C solenoid.
8. Note rpm change.**
9. Remove ground from dwell lead.
10. Return to idle.
Less than 100 rpm drop or rpm increases
More than 100 rpm drop
Check that pink wire is attached to right hand terminal of M/C solenoid connector, as viewed from harness end.
Check evaporator cannister for being loaded with fuel and purge valve and bowl vent for leaks that would cause richness. If OK see "On-car service" in Section 6C.
Connect dwell meter to M/C dwell lead. select the 6 cylinder scale.
Set carburetor linkage on high step of fast idle cam. Run for one minute, or until dwell starts to vary.
Return to idle speed and note dwell.*
Fixed under 10°
See chart 1
Fixed 10° - 50°
See chart 2
Fixed over 50°
See chart 3
Varying
Check dwell at 3,000 rpm
Between 10°–50°
Check the AIR system
No trouble found
Clear long term memory**
Under 10°
Check air switching valve for leakage to exhaust ports at 3,000 rpm. If not leaking . . .
Over 50°
See "Carb calibration procedure" Section 6C.
Check TPS adjustment

* Oxygen sensors may cool off at idle and the dwell change from varying to fixed. If this happens, running the engine at fast idle will warm it up again.

** If the vehicle is equipped with an electric cooling fan, it may lower the rpm when it engages.

Figure 8–58 System performance check.

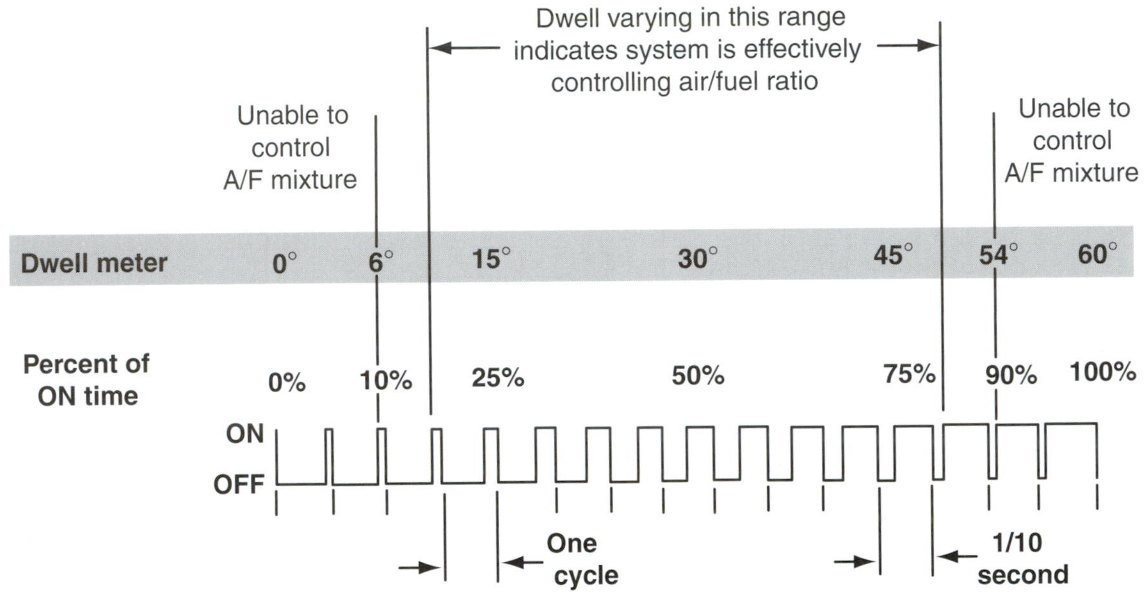

Figure 8–59 Dwell reading and system performance.

closed loop. Next pull a large vacuum hose (just short of what would kill the engine), and observe the dwell. It should begin to go down. If it does not respond, there is an electronic problem; the system did not recognize the change in the air/fuel mixture.

Some dwell meters, particularly swinging needle analog types originally designed to test dwell on contact point ignition, sometimes do not work well on CCC systems because they draw too much current and affect the signal to the mixture control solenoid. If a particular dwell meter causes any change in engine performance when it is connected, do not use it to test CCC mixture delivery.

You do not necessarily need a so-called dwell meter to test the mixture. Duty-cycle is the same measurement as dwell, but in terms of percent of on-time instead of dwell. On a duty-cycle reading, 100 percent corresponds to 60 degrees on a dwell meter set to six cylinders; 50 percent corresponds to 30 degrees, and so on. The additional advantage to a duty-cycle measurement is that the tool ordinarily has enough electrical impedance (low draw) that it cannot affect the delivery of the signal in any significant way.

Carburetor and M/C Solenoid Adjustment

Varying degrees of adjustment capacity exist on the four different carburetors used on CCC vehicles. The 6510-C Holley, used on the T-car, has only its idle mixture screw for an external mixture adjustment. Mixture adjustments are made at idle only using a dwell meter. The E2SE Varajet has an idle mixture screw, and inside the bowl is a lean mixture screw, Figure 8–60. The lean mixture screw is a main metering adjustment and is made at 3,000 rpm using a dwell meter. The idle mixture screw is

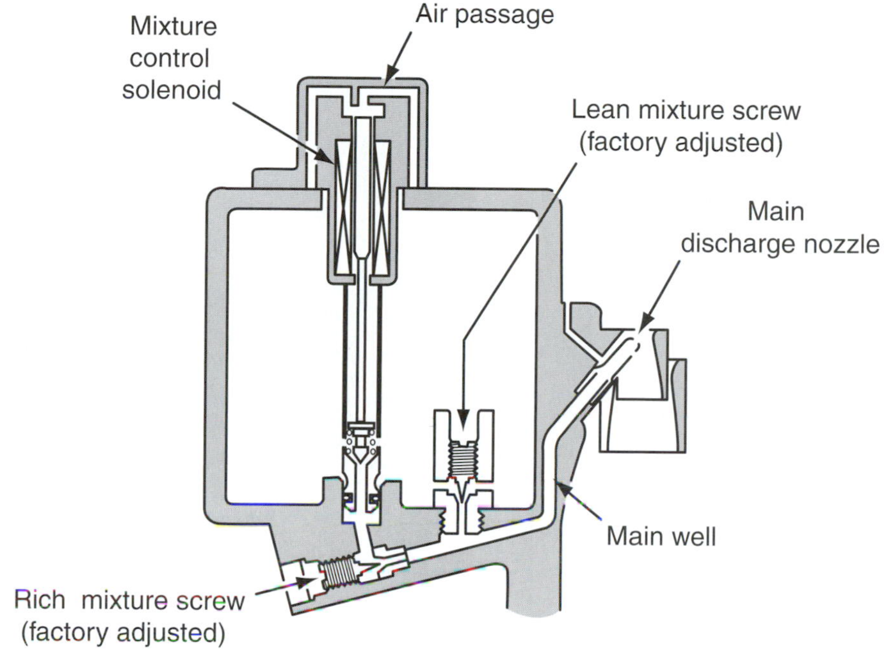

Figure 8–60 Rochester Varajet (E2SE).

adjusted at idle using a dwell meter. The E4ME Quadrajet and the E2ME **Dualjet** carburetors have much more adjustment capacity. Their adjustments are identical and include idle mixture screws, an idle air bleed valve and adjustments on the mixture control solenoid itself.

WARNING: When performing mixture adjustments when the engine is running, either at idle or at higher speeds, be certain that all instrument leads are clear of the coolant fan and the accessory drive belts. Also be sure that you have no loose clothing such as shirt or coat sleeves that can be caught in the machinery, and that the wheels are securely chocked so vehicle movement is impossible.

Since the CCC system's introduction, several revisions have been made in adjustment specifications and procedures. One significant procedure revision was made in 1983 for the E4ME and E2ME carburetors, used from 1981 to 1984. The E4ME and E2ME carburetors included on 1985 and later models have internal changes that prevent the use of earlier adjustment procedures. Although such carburetor adjustments should be performed only when clearly indicated, many driveability problems can be eliminated by merely restoring proper float level and carefully making updated adjustments.

WARNING: Make no attempt to adjust the rich mixture screw on the E2SE carburetor shown in Figure 8–60. Removing and replacing the covering plug produces a potential fuel leak and fire hazard.

PCM and PROM Service

Take considerable care when replacing a PROM. A finger touching one of the PROM's pins can discharge static electricity into it and damage it, Figure 8–61. To insert the PROM into the PCM,

Figure 8–61 PROM IC chip and carrier.

position the PROM in the carrier. The carrier only fits in the PROM cavity of the PCM in one position. The PROM, however, fits in the carrier in either of two ways. If the PROM is installed in the PCM incorrectly, it will be damaged electrically. To avoid this, identify the small half-circle notch on one end of the PROM. Now look in the PROM cavity and identify a similar notch on one end of the PROM seat, Figure 8–62. To avoid bending

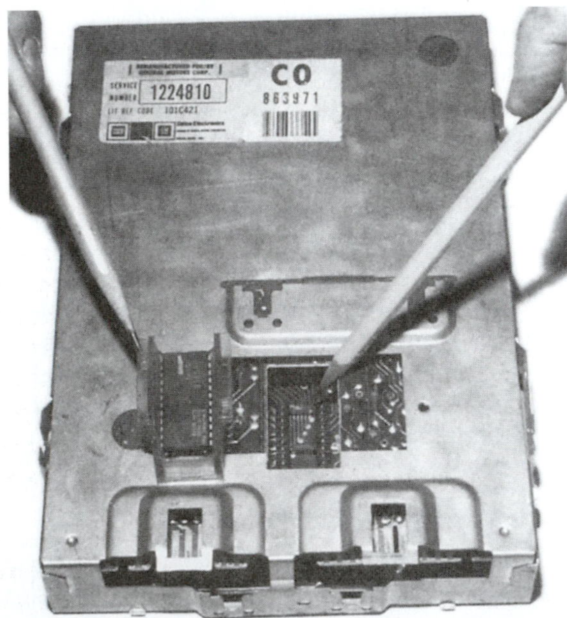

Figure 8–62 PROM, PROM carrier and PROM socket.

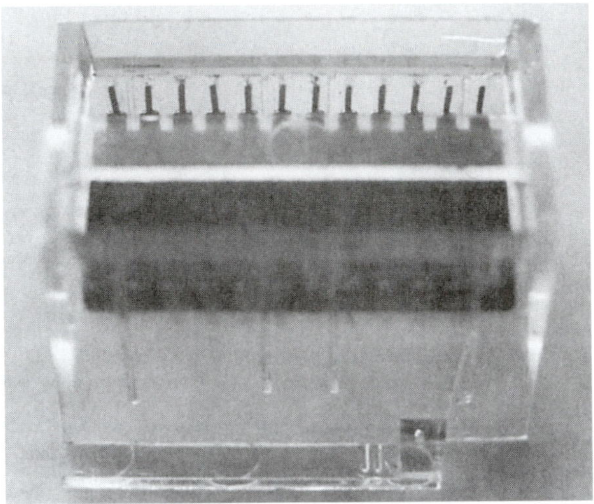

Figure 8–63 PROM pins/PROM carrier shoulder.

the pins when installing the PROM, be sure the PROM is positioned in the carrier with the tips of the pins above the shoulder of the carrier, as shown in Figure 8–63, before placing the PROM and carrier in the PROM cavity. This allows the carrier to guide each pin into place as the PROM is pressed down.

The PROM and PCM are serviced separately. Remove the PROM before exchanging the PCM for a new one. Each PROM encodes a large amount of information for each specific vehicle, including its unique set of accessories. Occasionally, GM issues revised, updated PROMs for earlier vehicles. The PROM number, available through the scan tool, identifies the unit installed. Some driveability problems that do not yield to other diagnostic measures can be corrected with an updated PROM.

Weather-Pack Connectors

All CCC system harness connections under the hood are made with weather-pack connectors, Figure 8–64, designed to provide environmental protection for sensitive electrical connections so corrosion and contamination buildup is held to a minimum. Many of the circuits in the CCC system do not carry more than 1 or 2 **milliamps** at about 100 **millivolts** (0.1 volt). Any resistance caused by oxidation or contamination of an electrical connection can have a significant impact on the performance of the system. It is important that the integrity of the weather-pack connectors be maintained. Don't break the seal by inserting a probe into the connector. Voltage readings can be obtained by opening the connector and temporarily installing short jumper wires made for this purpose (many voltage checks require that the circuit be electrically connected during the test). If the weather-pack connector has been damaged, replace it as shown in Figure 8–64 or solder and tape the connection as shown in Figure 8–65.

Diagnostic & Service Tip

TPS. The TPS is the one sensor that can be adjusted. Sometimes driveability complaints can be cured by adjusting the TPS to exact specifications, which often vary with engine and model year application. Be sure to check the applicable service manual for correct procedures and specifications.

Diagnostic & Service Tip

TCC Test Terminal. When using the TCC test terminal, remember the following points:

- Voltage is present when the TCC is not applied and is near zero when the TCC is applied, Figure 8–42, because applying the TCC causes the voltage to drop across the apply solenoid.
- The TCC can be applied by grounding the TCC test terminal.
- Applying voltage to the TCC test terminal damages the PCM if the PCM tries to apply the TCC while the voltage is applied. (On earlier models with five-terminal DLCs, the TCC test terminal is in the fuse panel, and thus increases vulnerability to this mistake).

CCC System Behavior

Some of the behaviors of the CCC system can be amusing or frustrating, depending on whether they are understood. For instance, driving up a long hill in the mountains at near full throttle can set a code and turn on the check engine light. Changing altitude rapidly or driving backward for about a quarter of a mile can have the same kinds of effects. There is nothing necessarily wrong: the system was just not programmed for those kinds of driving conditions.

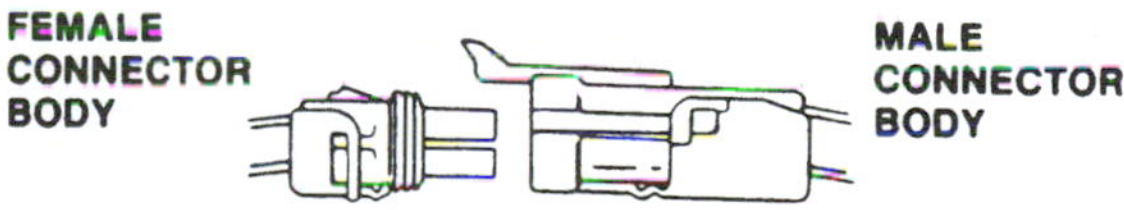
FEMALE CONNECTOR BODY
MALE CONNECTOR BODY
1. OPEN SECONDARY LOCK HINGE ON CONNECTOR

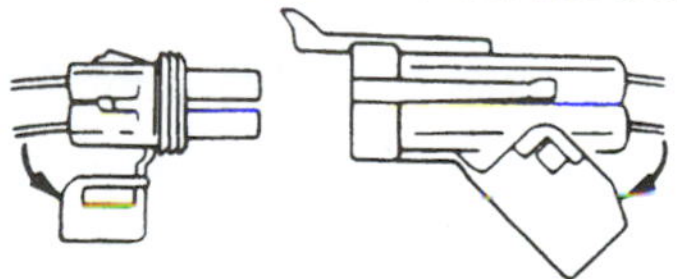
2. REMOVE TERMINALS USING SPECIAL TOOL
J-28742

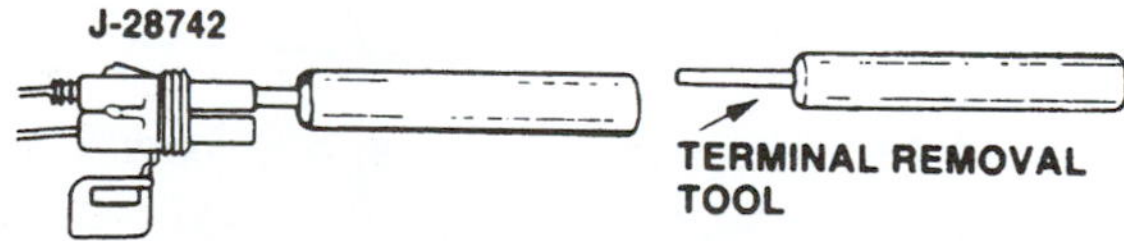
TERMINAL REMOVAL TOOL
3. CUT WIRE IMMEDIATELY BEHIND CABLE SEAL
4. SLIP NEW CABLE SEAL ONTO WIRE
(IN DIRECTION SHOWN) AND STRIP 5.00mm (.2")
OF INSULATION FROM WIRE. POSITION CABLE
SEAL AS SHOWN.

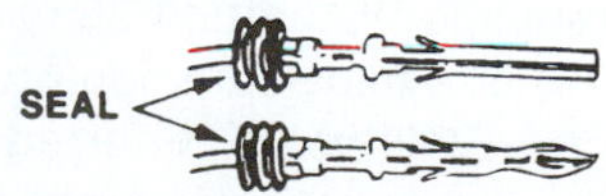
SEAL

Figure 8–64 Weather-pack connector. *(Courtesy of General Motors Corporation, Service Technology Group.)*

TWISTED/SHIELDED CABLE

1. **Remove outer jacket.**
2. **Unwrap aluminum/mylar tape. Do not remove mylar.**

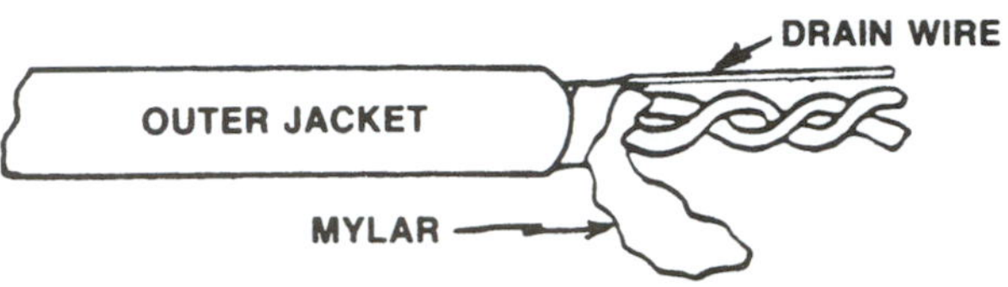

3. **Untwist conductors. Strip insulation as necessary.**

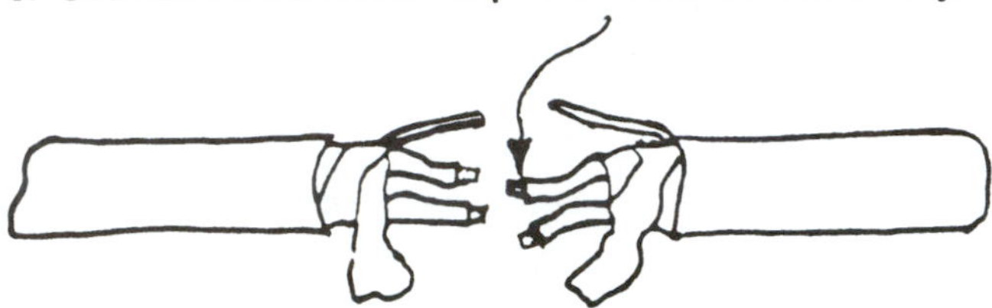

4. **Splice wires using splice clips and rosin core solder. Wrap each splice to insulate.**

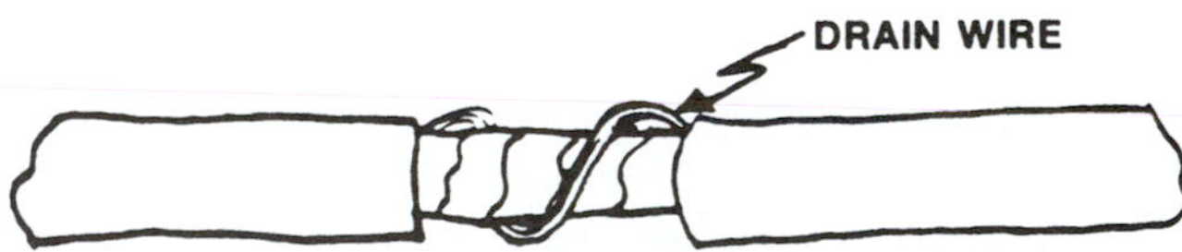

5. **Wrap with mylar and drain (uninsulated) wire.**
6. **Tape over whole bundle to secure as before.**

TWISTED LEADS

1. **Locate damaged wire.**
2. **Remove insulation as required.**

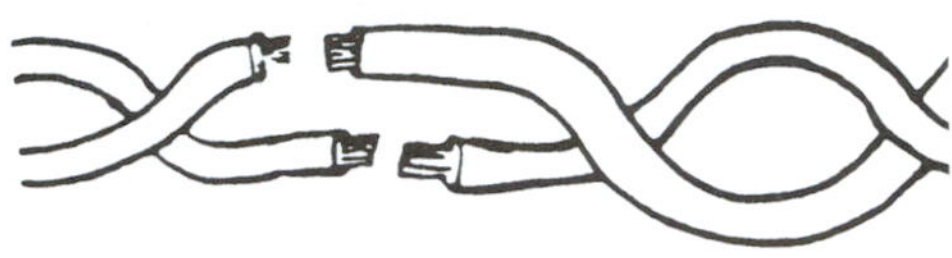

3. **Splice two wires together using splice clips and rosin core solder.**

4. **Cover splice with tape to insulate from other wires.**
5. **Retwist as before and tape with electrical tape to hold in place.**

Figure 8–65 Wire repair. *(Courtesy of General Motors Corporation, Service Technology Group.)*

Surging Complaints. Surging complaints are occasionally heard concerning CCC vehicles at steady-state speeds around 35 miles per hour or higher. This is a fairly common problem, often caused by the EGR valve, referred to as EGR chuggle. The inert gases introduced in the intake manifold by the EGR valve cause an increased frequency of cylinder misfires. Although this has been occurring since the EGR valve's introduction in the early 1970s, it was not noticeable to the driver until a clutch appeared in the torque converter, making it a solid coupling. Combined with lower axle numerical ratios, this means the chuggle is now more perceptible. This complaint is sometimes misdiagnosed as a malfunction of the lockup torque converter. To isolate the problem, temporarily disconnect and plug the vacuum line from the EGR valve. Drive to see whether the surge still occurs. If it does, the problem is not EGR chuggle.

CAUTION: Do not leave the EGR valve disconnected. To do so is not only emissions tampering and against the law, it can also result in engine damage from detonation.

SUMMARY

We have seen how the torque converter lockup clutch in an automatic transmission or transaxle works to eliminate the slip that would otherwise occur and improve fuel economy. We reviewed several methods used to control the air/fuel mixture with a mixture solenoid on a feedback carburetor, fine-tuning the delivered mixture by adjusting the amount of air and fuel in the main metering circuits of the carburetor; and we have seen how a technician can diagnose this component using a dwell meter.

This chapter covered the method used by the CCC system to control the engine's idle speed under various temperature and load conditions. We also learned how the air injection system works, directing injected air to the exhaust manifold, the catalytic converter, or diverting it to the atmosphere when not needed.

Finally, we began considering the computer's ability to test itself and its circuits—its self-diagnostic capacities. We have seen the gradual development of these capacities from the early flashing LEDs to more informative types of code. This capacity will, as we will see in later chapters, grow in complexity as the computers gain control responsibilities.

▲ DIAGNOSTIC EXERCISE

A vehicle is brought in with surging and other driveability complaints. When put into diagnostic mode, the computer displays DTCs pointing to the TPS circuit. Describe the steps a technician should follow between finding that DTC and deciding to replace the TPS.

REVIEW QUESTIONS

1. Requirements for the CCC system to enter closed loop include all *except* which of the following?
   A. The engine coolant temperature must reach about 150°F.
   B. The temperature of the catalytic converter must be above 1,400°F.
   C. The temperature of the oxygen sensor must reach at least 570°F.
   D. A predetermined amount of time must have passed since the engine was started.
2. What mode does the PCM operate in during engine warmup?
   A. Closed loop
   B. Blended enrichment
   C. Power enrichment
   D. Limp-in
3. Which of the following sensors is the only adjustable sensor in the CCC system and may be adjusted by the technician?
   A. ECT
   B. MAP
   C. BARO
   D. TPS
4. What is the service replacement interval recommended by GM for an oxygen sensor on vehicles built after 1981?
   A. 15,000 to 30,000 miles
   B. 30,000 to 45,000 miles
   C. Every 60,000 miles
   D. GM does not recommend a replacement interval, but the sensor should be replaced if it is found to be defective.
5. The ESC system is a subsystem of the CCC system.
   Technician A says that the ESC system uses input from a MAP sensor.
   Technician B says that the ESC system is used to control detonation.

Who is correct?
A. A only
B. B only
C. Both A and B
D. Neither A nor B

6. Idle speed control by the PCM is influenced by all *except* which of the following input signals?
A. Detonation sensor
B. P/N switch
C. A/C switch
D. ISC switch

7. *Technician A* says that most CCC-equipped vehicles use one or more hydraulically operated electric switches in the transmission's valve body that communicate to the PCM in which gear the transmission is operating.
Technician B says that the hydraulically operated switches in the transmission are used by the PCM control the lockup torque converter clutch more effectively.
Who is correct?
A. A only
B. B only
C. Both A and B
D. Neither A nor B

8. In order to measure the duty cycle of a mixture control solenoid, a technician may use which of the following?
A. A Hertz meter
B. A dwell meter
C. The duty cycle scale on a DVOM
D. Either B or C

9. *Technician A* says that when the mixture control solenoid's duty cycle begins to vary, it indicates that the PCM has entered closed loop.
Technician B says that, while in closed loop, if the O_2 sensor reports a lean exhaust condition, the PCM will react by increasing the amount of duty cycle on-time.
Who is correct?
A. A only
B. B only
C. Both A and B
D. Neither A nor B

10. In EST mode, the spark timing is affected by all *except* which of the following?
A. Vacuum and centrifugal advance mechanisms
B. Barometric pressure
C. Manifold vacuum
D. Coolant temperature

11. *Technician A* says that while the engine is cranking, spark timing is controlled directly by the pickup coil in the distributor.
Technician B says that once the engine is running, the PCM delivers a 5-volt signal to the bypass relay within the ignition module, causing spark timing to be put under the control of the PCM's EST signal.
Who is correct?
A. A only
B. B only
C. Both A and B
D. Neither A nor B

12. In order for the catalytic converter to operate properly, its internal temperature must be above at least which of the following?
A. 150°F
B. 410° to 500°F
C. 1,400°F
D. 2,500°F

13. On an AIR system with a dual-bed, three-way catalytic converter, during closed loop operation the PCM causes the AIR management valve to switch air pump air flow to which of the following?
A. The exhaust manifold
B. The NO_x reducing bed of the converter
C. The oxidizing bed of the converter
D. The atmosphere

14. *Technician A* says that the primary purpose of an EGR valve is to control the production of carbon monoxide on a running engine.
Technician B says that an EGR valve can help to control detonation on a running engine.
Who is correct?
A. A only
B. B only
C. Both A and B
D. Neither A nor B

15. *Technician A* says that the PCM controls a vacuum solenoid, which, in turn, controls a vacuum signal to the canister purge valve.
Technician B says that canister purging is allowed only on an engine operating in open loop.
Who is correct?
    A. A only
    B. B only
    C. Both A and B
    D. Neither A nor B
16. On a four-speed automatic transmission, what can happen if the TCC fails to apply during fourth gear operation?
    A. The transmission oil may overheat.
    B. The transmission may be damaged.
    C. The vehicle will not accelerate properly when the throttle is opened further.
    D. Both A and B
17. A vehicle with a TCC is being driven with a grounded test light connected to the TCC test terminal. At idle, the test light is illuminated. As the vehicle reaches cruising speed, the test light goes out. What does this condition indicate?
    A. The fuse for the TCC solenoid control circuit has probably blown.
    B. The PCM has energized the TCC solenoid control circuit.
    C. The PCM has de-energized the TCC solenoid control circuit.
    D. The PCM has determined that a fault exists and has turned off the voltage to the test light in order to alert the technician.
18. A GM vehicle with a CCC system is being put into self-diagnostics by grounding the test terminal in the DLC.
Technician A says that a code 12 means "system pass."
Technician B says that a code 12 indicates that no reference pulse is being received from the distributor, and that a "system pass" is indicated by the absence of DTCs between sets of code 12.
Who is correct?
    A. A only
    B. B only
    C. Both A and B
    D. Neither A nor B
19. *Technician A* says that when replacing a PROM in the PCM, care must be taken to avoid touching the PROM's pins in order to avoid damage from a static electrical discharge.
Technician B says that if the PROM is installed in the carrier backward and then inserted in the PCM, the PROM will be damaged electrically.
Who is correct?
    A. A only
    B. B only
    C. Both A and B
    D. Neither A nor B
20. *Technician A* says that when testing CCC system circuits that use weather-pack connectors, jumper wires should be installed temporarily in order to gain access to these circuits.
Technician B says that weather-pack connectors should not be back-probed with a DVOM probe or other probe.
Who is correct?
    A. A only
    B. B only
    C. Both A and B
    D. Neither A nor B

Chapter 9

General Motors' Electronic Fuel Injection

OBJECTIVES

Upon completion and review of this chapter, you should be able to:

- ❑ Describe the operating modes of a GM EFI (TBI) system.
- ❑ Describe the inputs associated with a GM EFI (TBI) system.
- ❑ Describe the operation of a GM EFI (TBI) fuel system.
- ❑ Describe the operation of the idle speed control system associated with a GM EFI (TBI) system.
- ❑ Describe the operation of the spark management systems associated with a GM EFI (TBI) system.
- ❑ Describe the operation of the emission control systems associated with a GM EFI (TBI) system.

KEY TERMS

Block Learn
Calculation Packet (CALPAK)
Fuel Trim
Pickup Coil
REF Pulse
Throttle Body Injection (TBI)

General Motors' Electronic Fuel Injection (EFI)—not to be confused with the Digital Fuel Injection (DFI) system used by Cadillac beginning in the late 1970s and sometimes also called EFI—is a single-point fuel injection system. It is often called a **Throttle Body Injection (TBI)** system. That name, however, more precisely described a major component of the system, Figure 9–1, the throttle body with its single or double fuel injector. This system was introduced by General Motors in 1982.

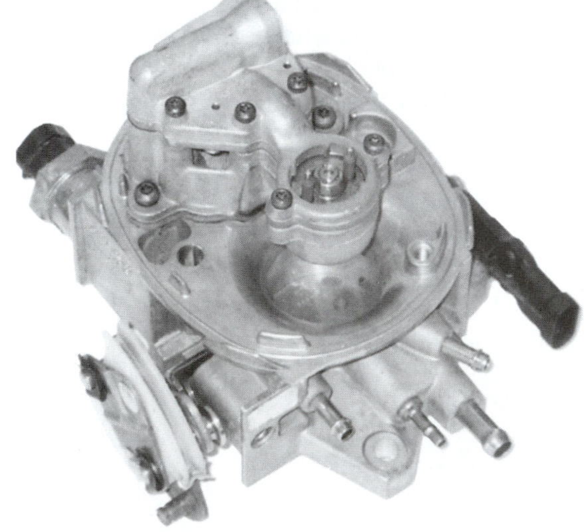

Figure 9–1 GM TBI unit.

There were two versions of the EFI system when it first appeared in 1982: one used a single TBI unit for four-cylinder engines and a second, higher performance version (the Crossfire system) had two separate TBI units for eight-cylinder engines. This latter is not to be confused with TBI systems using two fuel injectors in a single throttle body. The two-injector TBI unit was introduced on some V6 engines beginning in 1985.

POWERTRAIN CONTROL MODULE

The PCM is usually above or near the glove compartment. On some models, notably the Pontiac Fiero, it is in the console. On some Corvettes, it is in the battery compartment behind the driver's seat. Its inputs and the functions it controls (its outputs) are shown in Figure 9–2.

PCM for 2.5-Liter, EFI Operates Cruise Control

The PCM for selected engines such as the 2.5-liter four-cylinder engine was given more extensive and powerful internal circuitry to enable it to perform additional functions, such as control of the cruise control in 1988.

Throttle Body Backup (TBB)

The throttle body backup (TBB) is a fuel backup circuit within the PCM. It is primarily responsible for providing fuel pulses to the injector solenoid in the event of a general PCM failure, a failure that prevents it from running its program. The PCM used on the 2.0-liter engine uses a removable **calculation packet** (**CALPAK**) that provides fuel backup. The CALPAK plugs into the PCM the same way the PROM does.

Fuel Trim

The EFI PCM has a learning ability called **Fuel Trim**, originally called **Block Learn**. It is designed to correct for gradually changing input values that result from component wear that may affect air/fuel ratios. It also can compensate for driving conditions such as changing altitude, and even learns to modify output commands to

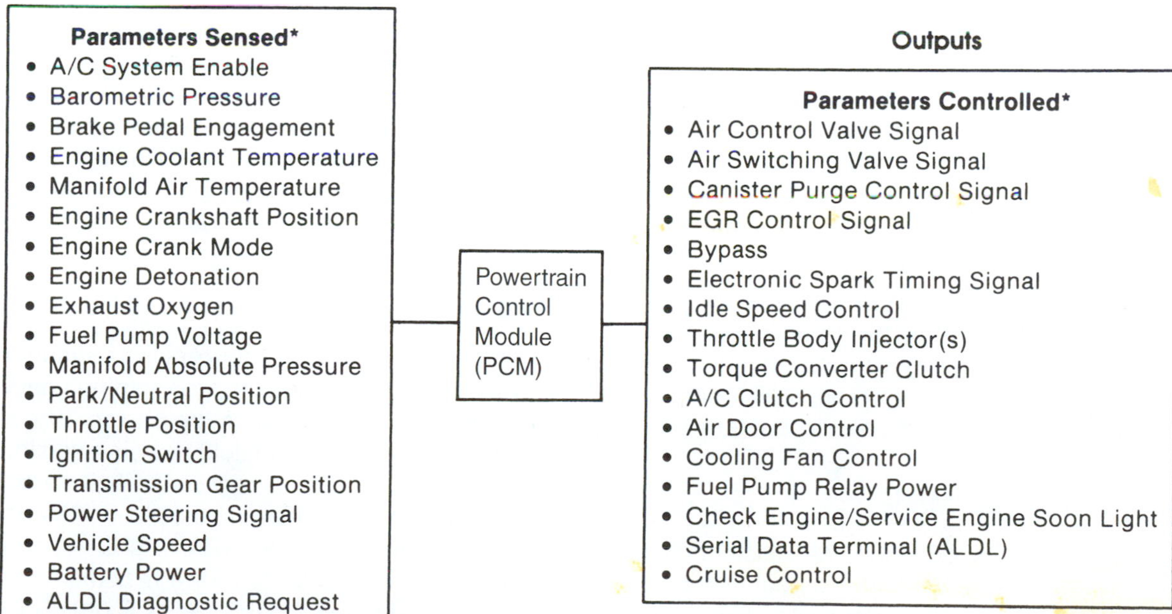

Inputs
Parameters Sensed*
• A/C System Enable
• Barometric Pressure
• Brake Pedal Engagement
• Engine Coolant Temperature
• Manifold Air Temperature
• Engine Crankshaft Position
• Engine Crank Mode
• Engine Detonation
• Exhaust Oxygen
• Fuel Pump Voltage
• Manifold Absolute Pressure
• Park/Neutral Position
• Throttle Position
• Ignition Switch
• Transmission Gear Position
• Power Steering Signal
• Vehicle Speed
• Battery Power
• ALDL Diagnostic Request
Powertrain
Control
Module
(PCM)
Outputs
Parameters Controlled*
• Air Control Valve Signal
• Air Switching Valve Signal
• Canister Purge Control Signal
• EGR Control Signal
• Bypass
• Electronic Spark Timing Signal
• Idle Speed Control
• Throttle Body Injector(s)
• Torque Converter Clutch
• A/C Clutch Control
• Air Door Control
• Cooling Fan Control
• Fuel Pump Relay Power
• Check Engine/Service Engine Soon Light
• Serial Data Terminal (ALDL)
• Cruise Control
*Some features not used on all engines.

Figure 9–2 EFI system overview.

complement a particular driver's driving habits. For an in-depth discussion of Fuel Trim, see System Diagnosis and Service in Chapter 10.

Keep-Alive Memory (KAM)

The PCM has a KAM like that discussed in earlier chapters. As long as it remains connected to a source of electrical power (the battery), it will retain the information it has accumulated in the previous driving intervals. The amount of power required to maintain this stored information is very small, much less than the battery drain caused by internal forces in the battery itself.

OPERATING MODES

The EFI system features the typical closed-loop and open-loop modes. During open loop, the PCM is programmed to provide an air/fuel ratio most suited to driveability as well as economy and emissions concerns. Of course, like all computerized engine control systems, emissions concerns are primary except in certain safety-related circumstances.

Synchronized Mode

In synchronized mode, the injector is pulsed once for each reference pulse from the distributor. To describe it another way, the injector, operating in synchronized mode, sprays fuel once for each time a cylinder fires. On dual-TBI units, the injectors are pulsed alternately to avoid fuel pressure pulse problems. All closed-loop operation is in synchronized mode as is most open-loop operation.

Nonsynchronized Mode

In nonsynchronized mode, the injector pulses every 12.5 milliseconds, regardless of distributor reference pulses. On dual-TBI systems, each injector pulses every 12.5 milliseconds, but because the injectors pulse alternately, the engine actually "inhales" a fuel spray pulse every 6.25 milliseconds. Nonsynchronized pulses occur only in response to one or more of the following operating conditions:

- The injector is on or off (open or closed) and time becomes too small for accurate duty-cycle control (about 1.5 milliseconds), as occurs near full throttle or during deceleration. Even though the injector only opens a few thousandths of an inch, it is a mechanical device opened by a magnetic field and closed by a spring. It has a definite, finite maximum speed in contrast to its virtually instantaneous electric current flow. When the injector is required to open or close at a rate that approaches its mechanical response capacity limits, the PCM stops making the injector keep up with the reference pulse and employs the slower mode of operation.
- During *prime* pulses. On the 1982 Crossfire EFI system, once the ignition was turned on and if the coolant temperature was low enough, the PCM commanded the injectors to deliver a spray or two into the manifold to prime the engine for starting. This action was similar to pumping the throttle before starting a carbureted engine and squirting a small amount of fuel into the manifold by way of the accelerator pump. The Crossfire system was, however, discontinued in 1983.

Cranking Mode

The fuel injector(s) delivers an enriched air/fuel ratio particularly suited for starting conditions when the starter cranks the engine. These conditions include high manifold pressure and low airflow velocity, both of which make fuel vaporization more difficult. The actual pulse widths depend on coolant temperature. At –36°C (–32°F), the pulse width is calibrated for a maximum rich air/fuel, approximately 1.5 to 1. At 94°C (201.5°F), the pulse width is calibrated for a maximum lean air/fuel ratio, approximately 14.7 to 1. The higher

Diagnostic & Service Tip

While the purpose of the clear flood mode is to allow the engine to start when flooded, a malfunctioning or completely misadjusted throttle position sensor can indicate an 80 percent or wider open throttle even when the driver has not depressed the pedal at all. This condition, of course, prevents starting. The most common cause of this kind of problem occurs if the TPS loses its ground and cannot reduce the reference signal back to the computer.

the temperature, the shorter will be the pulse width or injector on-time, unless the engine is overheated. In the case of overheating, pulse width is widened somewhat to provide a slightly richer mixture, required for starting in that condition.

Clear Flood

If an engine floods, that is, if the spark plug electrodes are damp enough with fuel that the electric pulse grounds through the fuel rather than jumping the spark plug gap, thus misfiring, the PCM adopts the clear flood strategy. Depressing the throttle to 80 percent of wide-open throttle or more during crank indicates to the computer to issue an air/fuel ratio of about 20 to 1. It stays in this mixture mode until the engine starts or until the throttle is closed to below 80 percent WOT. If the engine is not flooded and the throttle is held at 80 percent WOT or more, it is unlikely the engine will start with the superlean mixture.

Run

As soon as the PCM sees a reference pulse from the distributor indicating 600 rpm or more, it puts the system into open loop. In open loop the PCM does not use information from the oxygen sensor to determine air/fuel mixture. It monitors the oxygen sensor signal to see whether the system is ready to go into closed loop, but its delivered mixture commands are determined from engine coolant, speed and load information. The following criteria must be satisfied before the engine can go into closed loop:

- The oxygen sensor must produce voltage signals crossing 0.45 volt (450 millivolts). Before it can do this, the oxygen sensor must have reached a temperature of at least 300°C (570°F).
- The engine coolant temperature must reach a specified temperature, about 65.5°C (150°F). This temperature varies depending on model and build year.
- A specified amount of time must elapse since the engine was started, regardless of component temperatures. This factor also varies by model.

The values for each of these criteria are encoded in the PROM. When all of the criteria are met, the PCM puts the system into closed loop. At this point, the PCM uses the input from the oxygen sensor to calculate fine adjustments to the air/fuel mixture commands. Of course, if one of the other sensors indicates closed-loop criteria have failed, the PCM will return to open-loop calculations.

Semiclosed Loop

To improve fuel economy under certain cruise conditions, the PCM for some throttle body injected engines is programmed to go out of closed loop during some sustained cruise conditions at highway speeds. During these periods, the air/fuel mixture may go as lean as 16.5 to 1. The PCM monitors the following information parameters and puts the system into this fuel control mode only when they are within predetermined values (varying somewhat by model and year):

- engine temperature
- spark timing
- canister purge
- constant (sustained) average vehicle speed

The PCM periodically switches back to closed loop to check on engine operating parameters. If the parameters are still within the required limits, it goes back to the lean calibration, open-loop mode. Operation in the lean calibration, open-loop mode, and switching back and forth between closed loop and open loop are smooth and should not be detectable to the driver.

INPUTS

Engine Coolant Temperature (ECT) Sensor

This is the same sensor used in the computer command control system and on the other General Motors systems. On later models it has a slightly different appearance, Figure 9–3.

Manifold Absolute Pressure (MAP) Sensor

The MAP sensor on EFI applications, Figure 9–4, is similar to that used on CCC applications, but the PCM is programmed to use the sensor as a barometric pressure sensor also. When the ignition is first turned to run, but before it is turned to start, the PCM takes a reading from the MAP sensor. Since the engine is not running, the manifold pressure is identical to atmospheric pressure. The PCM records this reading as the base barometric pressure and uses it for calculations of fuel mixture while in open loop and for calculating spark timing. This BARO reading is retained and used until the engine is restarted or until the throttle is opened to WOT. At that time the engine vacuum goes to nearly atmospheric pressure, and an updated BARO reading is taken.

Figure 9–3 ECT sensor.

Figure 9–4 Manifold Absolute Pressure (MAP) sensor.

Oxygen Sensor

This sensor is the same single-wire sensor used on other earlier General Motors systems. Once it reaches operating temperature, the oxygen sensor provides a low-voltage signal inversely varying with the amount of residual oxygen remaining in the exhaust system after combustion and before the catalytic converter.

Throttle Position Sensor (TPS)

The throttle position sensor (TPS) is a variable resistor (potentiometer) mounted on the TBI unit, connected to the end of the throttle shaft,

Figure 9–5 Throttle Position Sensor (TPS).

Figure 9–5. On some EFI applications the TPS is adjustable. On others, there is no way to adjust the TPS.

Distributor Reference Pulse (REF)

The REF, sometimes called the **REF pulse**, tells the PCM what the engine speed is (when factored against the PCM's internal clock) and what the crankshaft position is. On most EFI applications, the distributor **pickup coil** provides the REF signal, as described in earlier chapters. It is important to note that if the REF signal fails to arrive at the PCM while the engine is running, fuel injection shuts off.

The earlier 2.5-liter EFI engines used a Hall effect switch to supply the REF signal, Figure 9–6. The pickup coil at the distributor was retained for starting, but that was its only function.

Denotation (Knock) Sensor

The knock sensor used on the EFI system is similar to that used on other GM systems. It is a crystal piezoelectric sensor that generates a signal when excited by vibrations characteristic of knock frequencies.

Vehicle Speed Sensor (VSS)

Two kinds of vehicle speed sensors have been used on EFI vehicles, similar to those described in

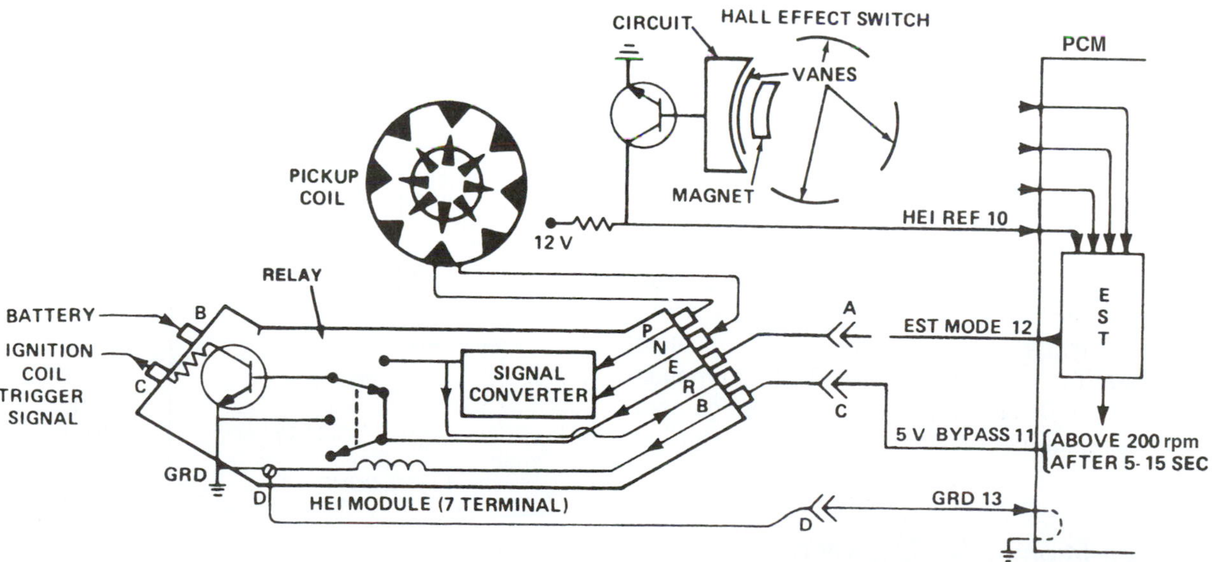

Figure 9–6 HEI module with EST and Hall effect switch. *(Courtesy of General Motors Corporation, Service Technology Group.)*

the previous chapter, and similar to those used by other manufacturers as well. On EFI vehicles the VSS input is also used by the PCM to help identify a deceleration condition. This is discussed further under Injector Assembly in the Outputs section of this chapter.

Ignition Switch

The ignition switch is one of the two power sources for the PCM. When the ignition is first turned on, the PCM initializes (starts its program bootup sequence) and is quickly ready to function. The ignition switch also powers most of the actuators that the PCM controls.

Park/Neutral (P/N) Switch

The park/neutral switch, usually a part of the shifter assembly, closes when the vehicle is shifted into park or neutral. This provides the PCM with information about whether the vehicle is in an active gear. The PCM uses this information to control engine idle speed. If the P/N switch is disconnected or significantly out of adjustment (enough to signal one gear when it is in another), idle quality can suffer while in park or neutral. An out of adjustment P/N switch can also prevent engagement of the starter, as has been the case for many years before computer controls.

Transmission Switches

Most transmissions used in CCC and EFI vehicles employ one or more hydraulically operated electric switches screwed into the valve body. These switches send the PCM signals indicating what gear the transmission is in. This information allows the PCM to properly control engagement and disengagement of the torque converter lockup clutch.

Air-Conditioning (A/C) Switch

On EFI cars with air conditioning, the A/C switch includes a wire to the PCM. When the A/C is turned on or off, this wire conveys this information to the PCM, which uses the signal in its determination of the proper idle speed.

Ignition Crank Position

The circuit that energizes the starter solenoid also signals the PCM through a wire. This informs the PCM that the engine is cranking, which the PCM uses in determining fuel injection programs for startup.

Fuel Pump Voltage Signal

Some EFI systems have a wire from the power side of the fuel pump (positive) to the PCM. This wire indicates to the PCM that the fuel pump is on, Figure 9–7. Not all vehicles include this feature.

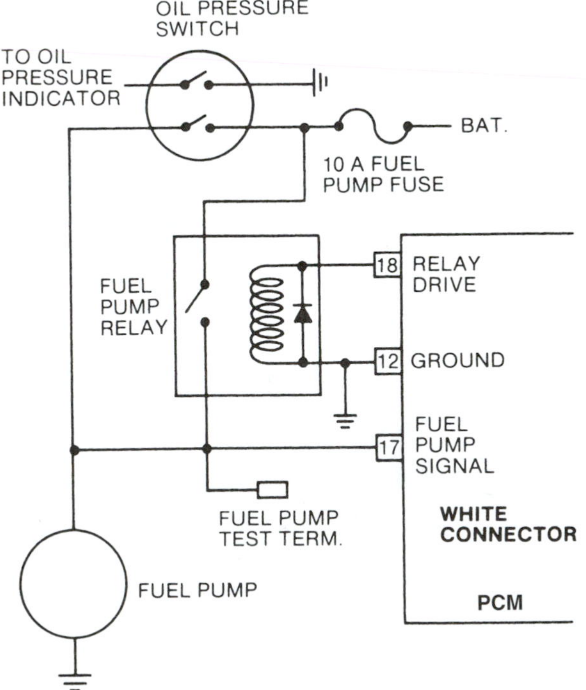

Figure 9–7 Fuel pump control circuit.

OUTPUTS

Throttle Body Injection (TBI) Unit

The throttle body injection (TBI) unit is made of three castings: the throttle body, the fuel meter body and the fuel meter cover, Figure 9–8. The throttle body contains the throttle bore and valve. It provides the vacuum ports for EGR, canister purge and so on, similar to what is found on the base of a carburetor. Mounted to it are the idle air control motor and the TPS. The fuel meter body contains the fuel injector and the fuel pressure regulator, Figure 9–9.

Injector Assembly

The injector is a solenoid-operated assembly. When energized by the PCM (pulsed by grounding), a spring-loaded metering valve lifts off its seat a few thousandths of an inch. Fuel under pressure passes through a fine screen filter fitting around the tip of the injector and sprays in a conical pattern toward the walls of the throttle bore, just above the throttle blade. This angle and position was chosen to maximize the exposure of fuel droplets to onrushing air, thus optimizing vaporization. The angle of injection sprays the fuel droplets across the incoming air as well as into the throttle body, maximizing vaporization by maximizing exposure to passing air.

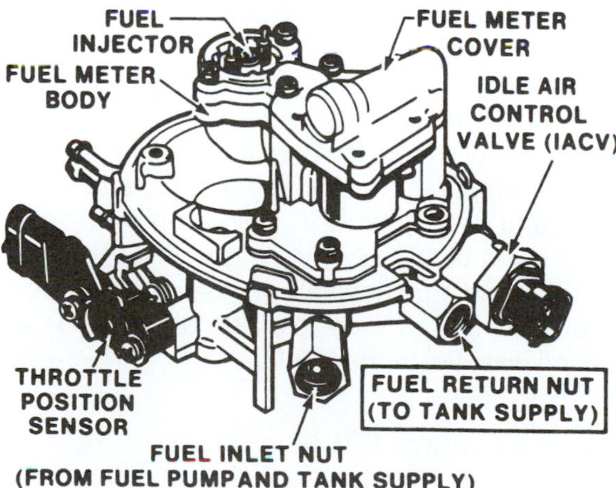

Figure 9–8 TBI unit. *(Courtesy of General Motors Corporation, Service Technology Group.)*

The TBI

The fascinating thing about the EFI system is that the TBI unit performs every function a carburetor does, but with greatly increased control. When the TPS moves rapidly, the injector momentarily sprays more fuel (increased pulse width), just as with an accelerator pump. On cold starts, the injector sprays additional fuel to duplicate the function of a choke. With the idle speed controller, the throttle body achieves fast idle by simply opening the idle control motor. The TBI requires no mechanical choke, fast idle cam, accelerator pump, and piston and so on. Everything is done electronically. Even better than with a carburetor, on deceleration the TBI unit can lean the fuel mixture maximally, even shutting fuel off altogether in some cases. This can compensate for any fuel that evaporates from manifold walls as a result of the momentarily higher vacuum under deceleration conditions. Hard deceleration—a closed throttle, a sharp drop in manifold pressure and a decrease in vehicle speed—determines the shutoff of fuel.

The major difference is in the control circuits: the carburetor functions as a pneumatic/mechanical/hydraulic computer to meter the air/fuel mixture. The TBI unit, however, is a dumb actuator, responding entirely to the computer's control commands. It does no controlling itself.

Fuel Pressure Regulator

Inside the TBI assembly is a fuel pressure regulator, Figure 9–9. The purpose of this regulator is to maintain the fuel pressure within a range

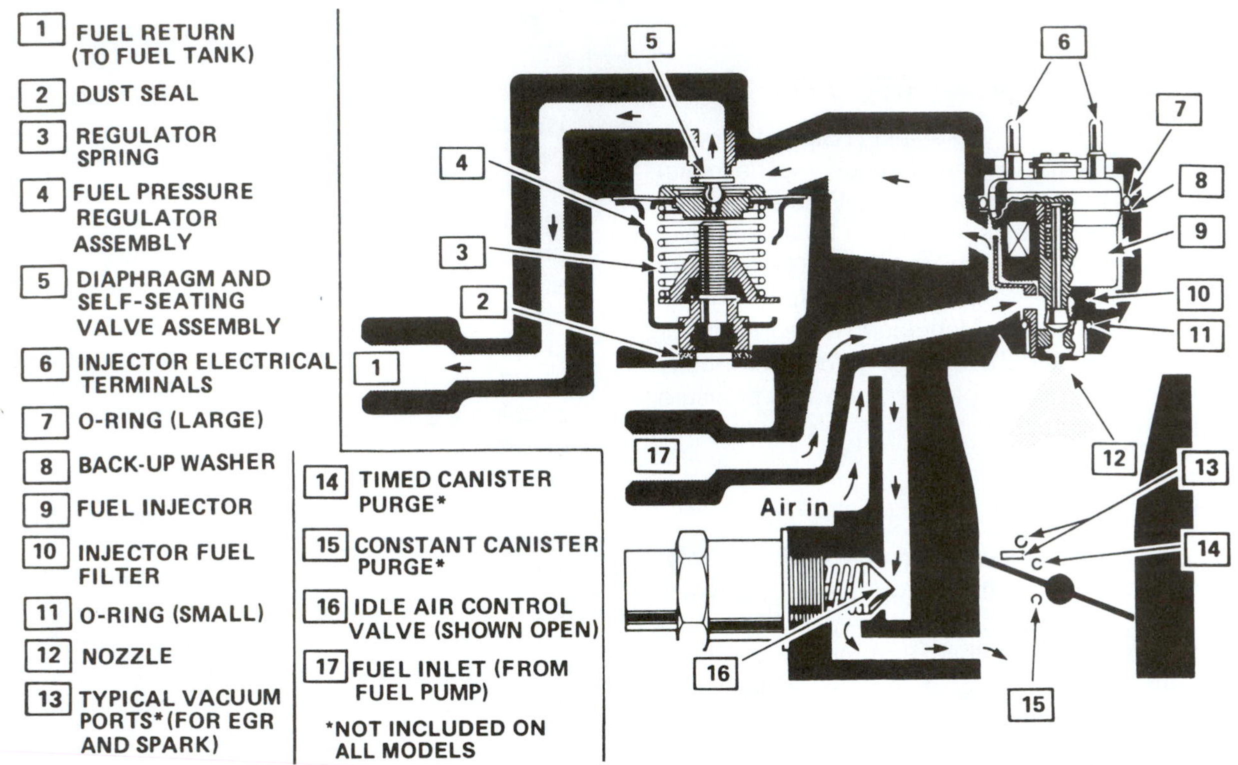

Figure 9–9 Idle air bypass circuit. *(Courtesy of General Motors Corporation, Service Technology Group.)*

so that the PCM's pulse-width injection commands will translate into the proper stoichiometric ratio in the combustion chambers. The spring side of the diaphragm is exposed to atmospheric pressure, so fuel pressure varies slightly with changes in atmospheric pressure. At lower atmospheric pressures, the fuel pressure applied to the inlet of the fuel injector is reduced equally to how much the atmospheric pressure at the outlet tip of the injector is reduced. This keeps the pressure differential across the fuel injector(s) consistent regardless of changes in altitude or weather. This in turn ensures that the flow rate will be consistent anytime an injector is energized. Therefore, the only variable used to control how much fuel is introduced into the engine is the injector's on-time or pulse width. This gives the PCM absolute control of the air/fuel ratio. On most models, fuel pressure should be maintained between 9 and 13 psi by the fuel pressure regulator.

On Crossfire systems, the rear TBI unit has the pressure regulator, while the front TBI uses a pressure compensator, Figure 9–10. It works like the regulator except that its function is to make up for the temporary drop in fuel pressure between front and rear units and to keep the pressure identical at both. Such drops are usually caused by the diaphragm and valve in the pressure regulator moving to a more closed position, thus allowing less fuel to the front unit.

Idle Air Control (IAC)

The IAC assembly controls idle speed by controlling the amount of air bypassing the throttle. It consists of a small, reversible electric step-

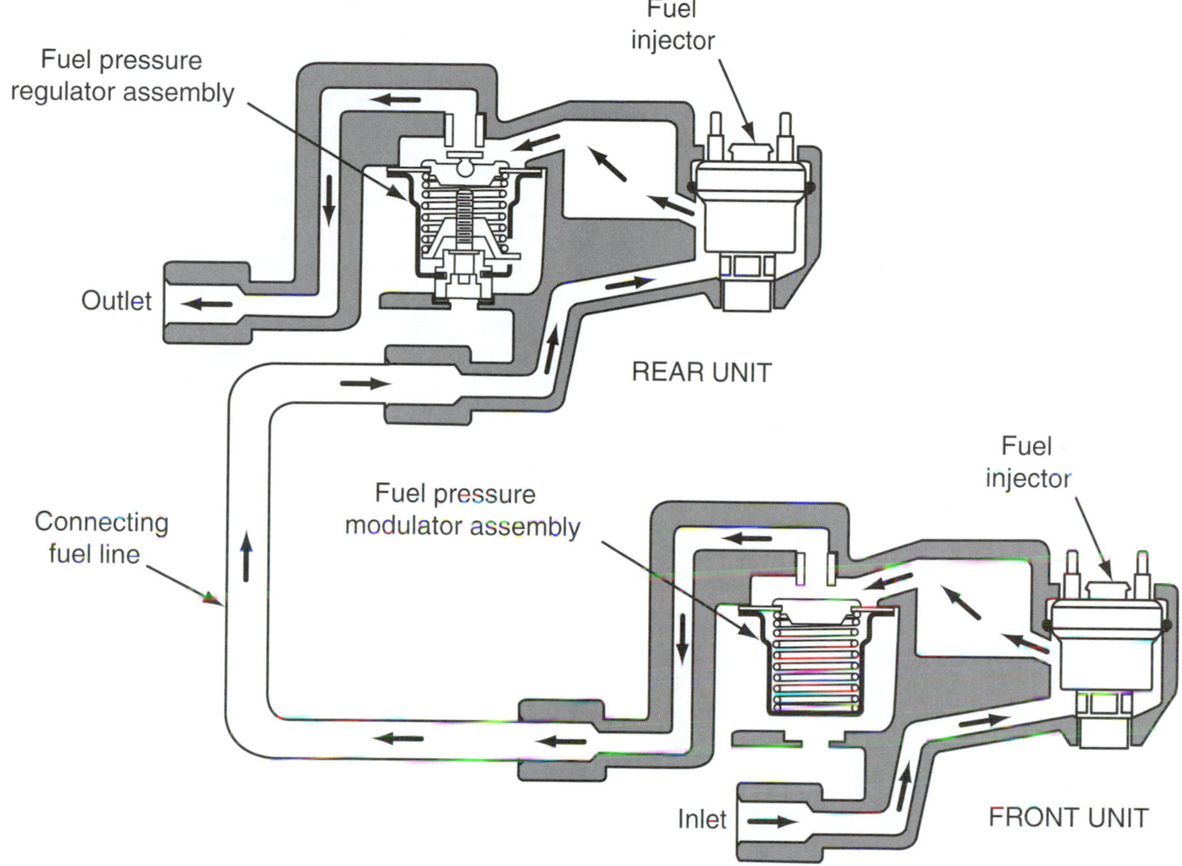

Figure 9–10 Crossfire TBI units.

per motor and a pintle valve, Figure 9–9. As the motor's armature turns, the pintle valve extends or retracts depending on the direction the motor is turning. During idle, the throttle blade is in a fixed, nearly closed position, allowing a constant amount of idle air into the intake manifold. The pintle valve allows additional air through the bypass passage. The PCM uses the stepper motor to position the pintle valve for desired idle speed. During warm engine operation, the PCM attempts to maintain a fixed idle speed by adjusting the IAC valve position for load variations (transmission in or out of gear, air conditioner on or off, and so on).

If during idle or at low vehicle speed (below 10 mph) engine speed drops below a specified rpm, the PCM puts the IAC into antistall mode. The IAC motor retracts the pintle to allow additional air into the intake manifold to raise engine speed. It momentarily raises engine speed above base idle.

The stepper motor is unique: its armature has two separate windings. The direction in which the armature turns depends on which winding is powered. Power is applied by the PCM in short pulses. Each pulse rotates the armature about 30 degrees and extends or retracts the pintle valve a corresponding amount. The PCM applies as many pulses as necessary to set the valve where it will achieve the desired idle rpm.

The PCM counts the pulses applied and thus knows where the pintle valve is at all times. There

Diagnostic & Service Tip

It is not uncommon to find a car in a shop that does not idle properly, but it would not be a natural check to see whether the vehicle speed sensor is working properly if the technician did not know the connection between the two elements of the computer control system. An experienced technician makes sure to check that the speedometer is working on any vehicle that does not idle properly. Note that failure to idle properly can mean idling not at all, too slowly or too fast.

are 12 pulses per revolution and 255 total positions. Fully extended (closed bypass passage) is the reference position: zero. Fully retracted (wide-open bypass passage) is step 255.

To keep an accurate track of the IAC assembly's position, the PCM references itself frequently. When the engine comes off idle and the car starts to move, the PCM begins looking for a reading from the VSS representing 30 mph. The first time it sees a 30-mph signal, it moves the pintle to the park (fully closed) position to reorient the zero reference. It then moves the pintle to a preprogrammed distance from closed. If for any reason the VSS is disabled, the PCM will not be able to reestablish the correct idle speed.

Fuel Pump Control

Fuel and fuel pressure are supplied by an electric pump inside the fuel tank. The fuel pump assembly contains a check valve preventing the fuel from bleeding back from the fuel line into the tank. If the check valve leaks, the engine cranks longer before starting while the fuel line is purged of vapor.

The fuel pump turns on and off through a fuel pump relay, usually mounted in the engine compartment. The fuel pump relay is controlled by the PCM. Be aware that the fuel pump relay is identical in appearance and usually mounted beside the air-conditioning relay, also controlled by the PCM, Figure 9–11. On some models, the cooling fan relay is also mounted in the same location and looks just like the others.

When the ignition is turned on, the PCM activates the fuel pump relay. If the PCM does not get an ignition REF signal within 2 seconds, it turns off the fuel pump through the relay and does not turn it back on until it does receive the REF signal. Once the engine is running, the PCM deactivates the relay when the ignition is turned off or anytime the REF signal disappears from the PCM. Note also that the oil pressure switch, which operates the oil pressure indicator on the instrument panel, is electrically in parallel with the fuel pump relay. This is done to provide a backup to the fuel pump relay. The engine starts after a fuel pump relay failure if cranked long enough to build about 4 psi oil pressure.

Fuel Pump Test Terminal. A wire connects to the circuit powering the fuel pump. The open end of this wire has a terminal to which a jumper lead can be connected to power the fuel pump. A voltmeter connected to the test terminal quickly shows if either the fuel pump relay or oil pressure switch is supplying power to the fuel

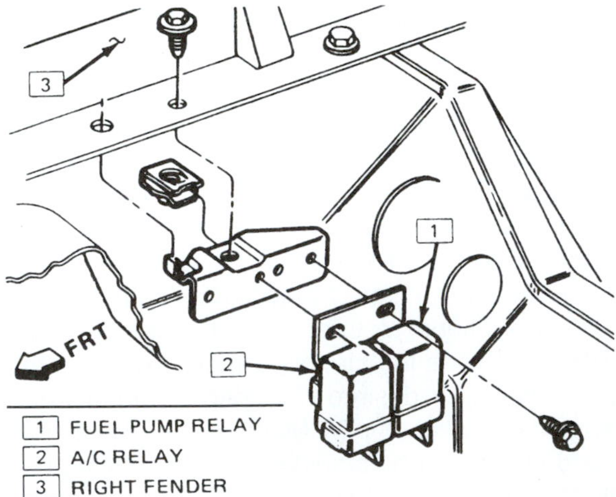

Figure 9–11 Fuel pump and A/C relay. *(Courtesy of General Motors Corporation, Service Technology Group.)*

pump circuit. This terminal is located in the engine compartment or in the ALDL under the instrument panel, depending on the model and year.

Voltage Correction. During vehicle operation, system voltage can vary considerably as a result of the various electrical accessories sometimes used and sometimes off, and the state of the battery's charge. Variations in the system's voltage can cause differences in the amount of fuel delivered at the injectors because of differences in the amount of time the injectors are open (a higher voltage opens the injector more quickly and vice versa). This affects the amount of fuel injected during a pulse width. Because of this problem, the PCM monitors system voltage and multiplies the pulse width by a voltage correction factor, Figure 9–12, programmed into its permanent memory. As system voltage goes down, the pulse width increases. As system voltage goes up, the pulse width decreases. If voltage goes down to a criterion value, the PCM increases dwell to maintain good ignition performance by keeping the spark hot and by increasing the idle speed.

Electronic Spark Timing (EST)

EST maintains the optimum spark timing under all conditions of engine load, speed, temperature and barometric pressure. Two basic functions are incorporated in the system. Dwell control is provided to allow sufficient energy to

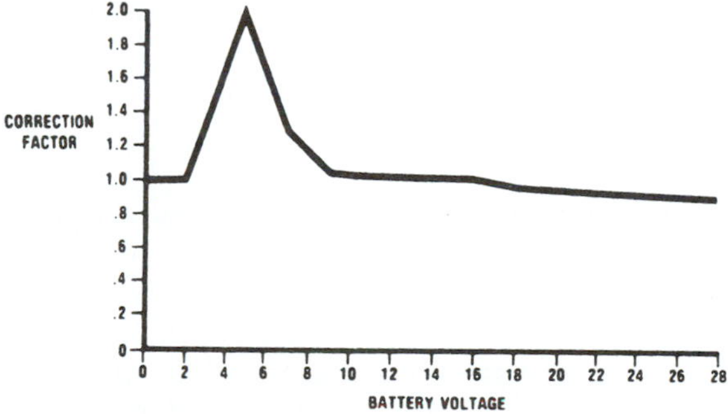

Figure 9–12 System voltage correction graph. *(Courtesy of General Motors Corporation, Service Technology Group.)*

build the magnetic field in the ignition coil for proper ignition voltage secondary output. Spark is provided at just the right time to start combustion at the moment of peak pressure in the compression stroke. The EST function of the PCM eliminates the vacuum and centrifugal advance mechanisms in the classic distributor. The HEI module accepts spark timing commands from the PCM once the engine is started.

During cranking, Figure 9–13, the solid-state switching circuit connects the main switching transistor of the HEI module to the pickup coil. The pickup coil signal turns on and off the main switching transistor. This state is called module mode and provides no spark advance; the timing is fixed at the reference timing point. On applications that do not use a Hall effect switch, this same signal, after it is converted from an analog signal to a digital signal by a signal converter circuit in the module, is sent to the PCM as the REF signal. On applications that do use the Hall effect switch, it supplies the REF signal. Note in Figure 9–14 that terminal R is not used.

When the PCM sees a REF pulse signal corresponding to an engine rpm of about 600 or more, it decides the engine is running. It then sends a 5-volt signal through the bypass wire to the solid-state switching circuit. The switching circuit in turn disconnects the pickup coil from the base of the main switching transistor. The EST wire is simultaneously disconnected from ground and connected to the main switching transistor's base. The system is now in EST mode, and timing is controlled by the PCM. The PCM considers barometric pressure, manifold pressure, coolant temperature, engine speed and crankshaft position. It then sends the HEI module the optimum spark timing command it has calculated.

Electronic Spark Control (ESC)

The Crossfire EFI system is equipped with ESC. Its function is to retard ignition timing when detonation occurs. The system is effectively the same as the later version of ESC described in the Outputs section of Chapter 8.

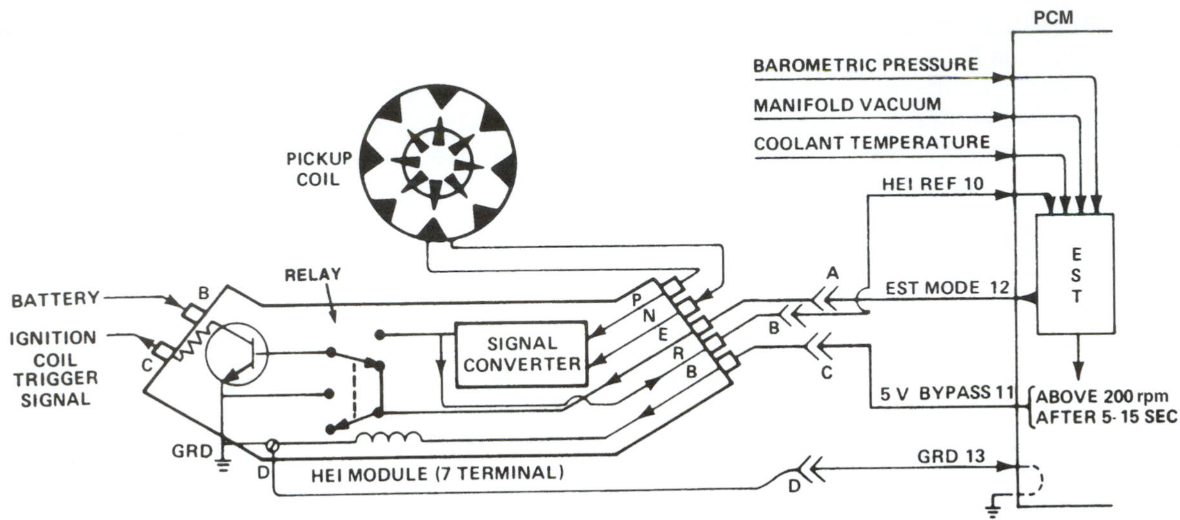

Figure 9–13 HEI module with EST. *(Courtesy of General Motors Corporation, Service Technology Group.)*

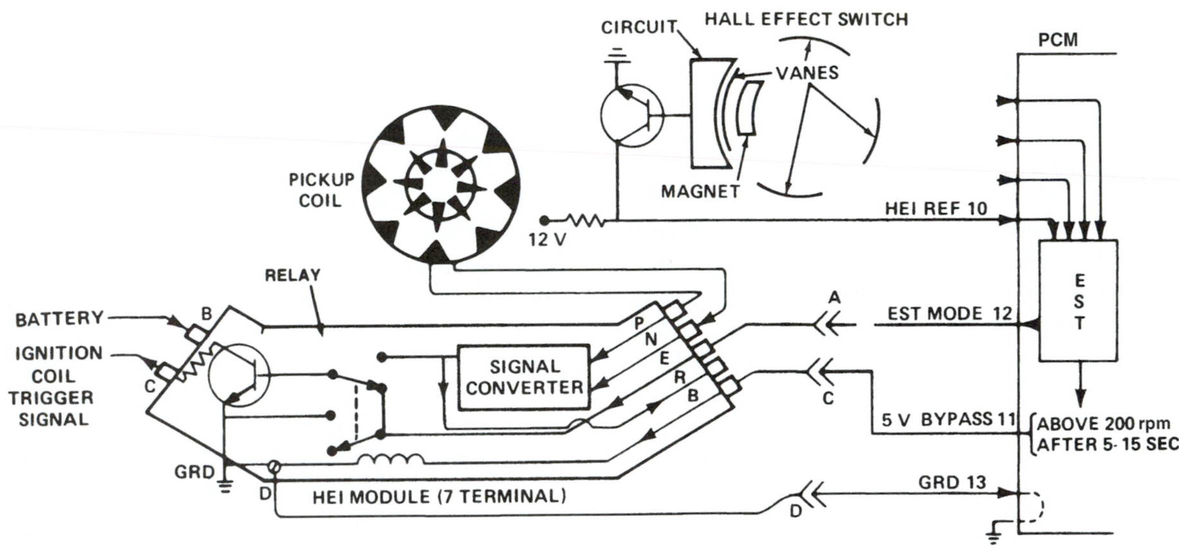

Figure 9–14 HEI module with EST and Hall effect switch. *(Courtesy of General Motors Corporation, Service Technology Group.)*

Torque Converter Clutch (TCC)

The purpose of the torque converter lockup clutch is to improve fuel mileage. It does this by eliminating the hydraulic slippage and heat production in the torque converter once the vehicle has attained a steady cruise speed. Its operation is essentially the same as the one described in the Outputs section of Chapter 8.

Manual Transmissions. On manual transmissions, the TCC output terminal on the PCM can

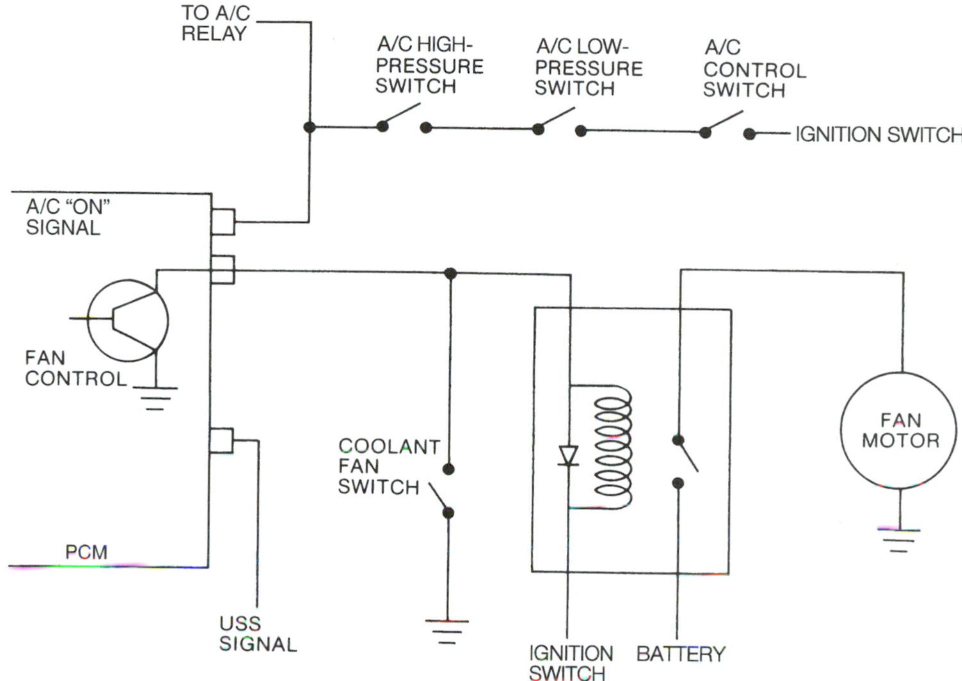

Figure 9–15 Cooling fan control circuit (typical).

be used to operate the upshift light on the instrument panel. The PCM looks at engine speed, engine load, throttle position and vehicle speed. Then it alerts the driver when to shift into the next gear for optimum fuel economy. Note that this shift point is strictly for fuel economy, not for maximum engine life. Most experienced mechanics as well as experienced drivers will let the engine turn somewhat faster than the upshift light would indicate.

Cooling Fan Control

Transverse engines almost universally use an electrically powered radiator cooling fan to pull air through the radiator and across the fins when they are moving at speeds too low for ram air to provide sufficient cooling. On most EFI systems, the PCM has some control over the cooling fan operation. The cooling fan can be turned on either by the PCM itself or by a coolant temperature switch. On some applications, the PCM has full control. The fan is turned on and off by a relay operated by the PCM, Figure 9–15. On most vehicles the fan turns on when the coolant temperature is high, when the vehicle is traveling below 30 mph with the air conditioner on and in some cases when the air-conditioning high side pressure reaches a certain criterion point.

Hood Louver Control

The Crossfire EFI system includes an electrically operated air door the PCM can open using a relay and solenoid. When the hood louver door opens, it allows fresh air from above the engine hood through a hole in the top of the air cleaner housing to the air cleaner elements. The PCM opens the hood louver door above a specified coolant temperature or under WOT conditions.

Air-Conditioning (A/C) Relay

The A/C relay is what actually turns the compressor clutch on and off, Figure 9–16. When the

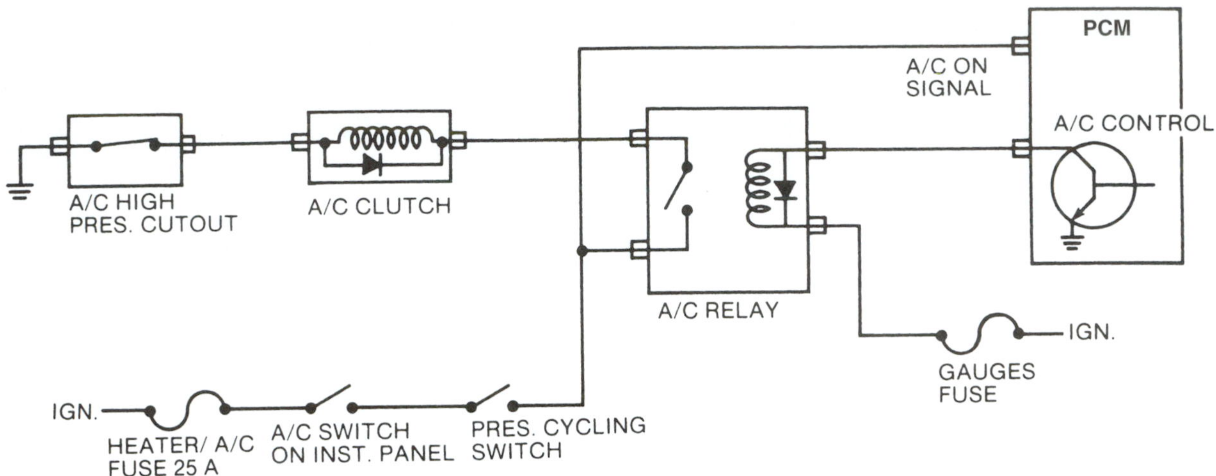

Figure 9–16 A/C relay circuit (typical).

driver turns on the A/C switch, a signal goes to the PCM and power is made available at terminal B of the relay. The PCM waits a half-second and then grounds terminal C of the relay. This energizes the relay and engages the compressor clutch (if the engine is at idle, the half-second allows the PCM time to adjust the IAC position and the engine idle speed before the compressor clutch engages). The PCM deactivates the relay at WOT, during heavy engine load, or when the IAC is in power steering antistall mode. As seen in Figure 9–16, the compressor clutch circuit can be opened also by the high pressure switch or the pressure cycling switch, both of which are standard features of the air-conditioning system.

Related Emission Controls

On all but the Crossfire EFI system, EGR, canister purge and air management control are not controlled by the PCM. The EGR, canister purge and air management controls on the Crossfire system are essentially the same as those on V8 engines with computer command control (CCC) discussed in Chapter 8.

Catalytic Converter. Most EFI engine systems use single-bed, three-way catalytic converters. No supplemental air is introduced into the converter, so no air injection is used. In some cases a Pulsair system is used and works independently of the computer.

EGR. Many four-cylinder EFI systems use a self-modulating EGR valve with no PCM control (most vehicles have a back-pressure EGR).

Under the main diaphragm found in a conventional EGR valve is a second diaphragm. A small spring pushes downward on the second diaphragm. A passage in the EGR valve stem allows exhaust manifold pressure to enter into a chamber under the second diaphragm. When exhaust pressure is low, the small spring can hold the second diaphragm down in spite of the exhaust pressure pushing it up. Atmospheric pressure flows through a vent in the lower part of the EGR valve housing and through a port in the second diaphragm, into the space between the two diaphragms. If the second diaphragm is down, the atmospheric pressure can flow through another port in the main diaphragm into the vacuum chamber. With atmospheric pressure finding its way into the vacuum chamber, not enough vacuum is developed to lift the main diaphragm and open the EGR valve. As the throttle opening increases, exhaust manifold pressure increases. When exhaust pressure is high enough, the second diaphragm is forced up and closes off the port through the main diaphragm into the vacuum chamber. The vacuum chamber is no longer vented to atmospheric pressure, and sufficient vacuum can now develop to open the EGR valve.

Canister Purge. On four-cylinder EFI engines, canister purge control is achieved with a thermal vacuum switch blocking purge vacuum until the engine is in closed loop. In closed loop, any additional fuel from the canister can be compensated for by varying the injection on-time, while still maintaining the stoichiometric ratio of 14.7 to 1 air to fuel.

Service Engine Soon Light

All EFI systems use a service engine soon light. On early 1980s vehicles, this was called the check engine light. If the PCM discovers a fault in one of the circuits it monitors for malfunctions, it turns on the service engine soon light on the instrument panel. This warns the driver that a malfunction exists. By grounding the test terminal in the Diagnostic Link Connector (DLC), located under the dash, the PCM flashes the service engine soon light. The sequence of these flashes indicates one or more trouble codes identifying the circuit or circuits in which the fault was detected. This is explained more fully under System Diagnosis and Service in this chapter.

✔ SYSTEM DIAGNOSIS AND SERVICE

Self-Diagnosis

The self-diagnostic capacity and procedures for the EFI system are essentially the same as for the CCC system described in the System Diagnosis and Service section of Chapter 8. An abbreviated discussion pointing out just the variations from the CCC system is what is presented here.

Diagnostic procedures for 1987 and later are written to include the use of a scanner or scan tool. Previously, diagnostic procedures were written around the use of a voltmeter to obtain such measurements as sensor readings.

Approach to Diagnosis

When any of the diagnostic procedures are used, it is important to follow the directions carefully. Failing to do so or taking shortcuts usually results in inaccurate conclusions and leads to unneeded and ineffective repairs.

Diagnostic Circuit Check. When diagnosing a problem on an EFI system, after verifying that all non-EFI engine support systems are working properly, start with a diagnostic circuit check, Figure 9–17. The diagnostic circuit check either verifies that the system is working properly or refers to another chart for further diagnosis. Be sure the diagnostic circuit check and trouble code charts come from the same model year manual as the vehicle you are working on. While some techniques and codes are retained over several years and for various vehicles, others are changed with some frequency. On earlier systems, the diagnostic circuit check is designed to sort out whether the fault currently exists or if it is intermittent, present once but not now. On those systems, the trouble code charts are written assuming the fault exists currently. If one of them is used to pursue an intermittent code, the chart most often leads to an invalid diagnosis because it is beginning with an invalid assumption—that a fault currently exists when in fact none does.

For later systems, the diagnostic circuit check and many of the trouble code charts are written to allow the trouble code charts to determine whether the code obtained from the diagnostic memory identifies a current fault (a hard fault) or an intermittent fault.

Field Service Mode. Grounding the test terminal with the engine running puts the system into field service mode. In this mode the PCM uses the check engine light to show whether the system is in open or closed loop and whether the system is running rich or lean. While in this mode, if the check engine light flashes at a rate of two times per second, the system is in open loop. If it flashes at a rate of one time per second, the system is in closed loop. If the light is turned *off* most or all of the time, the system is in closed loop and

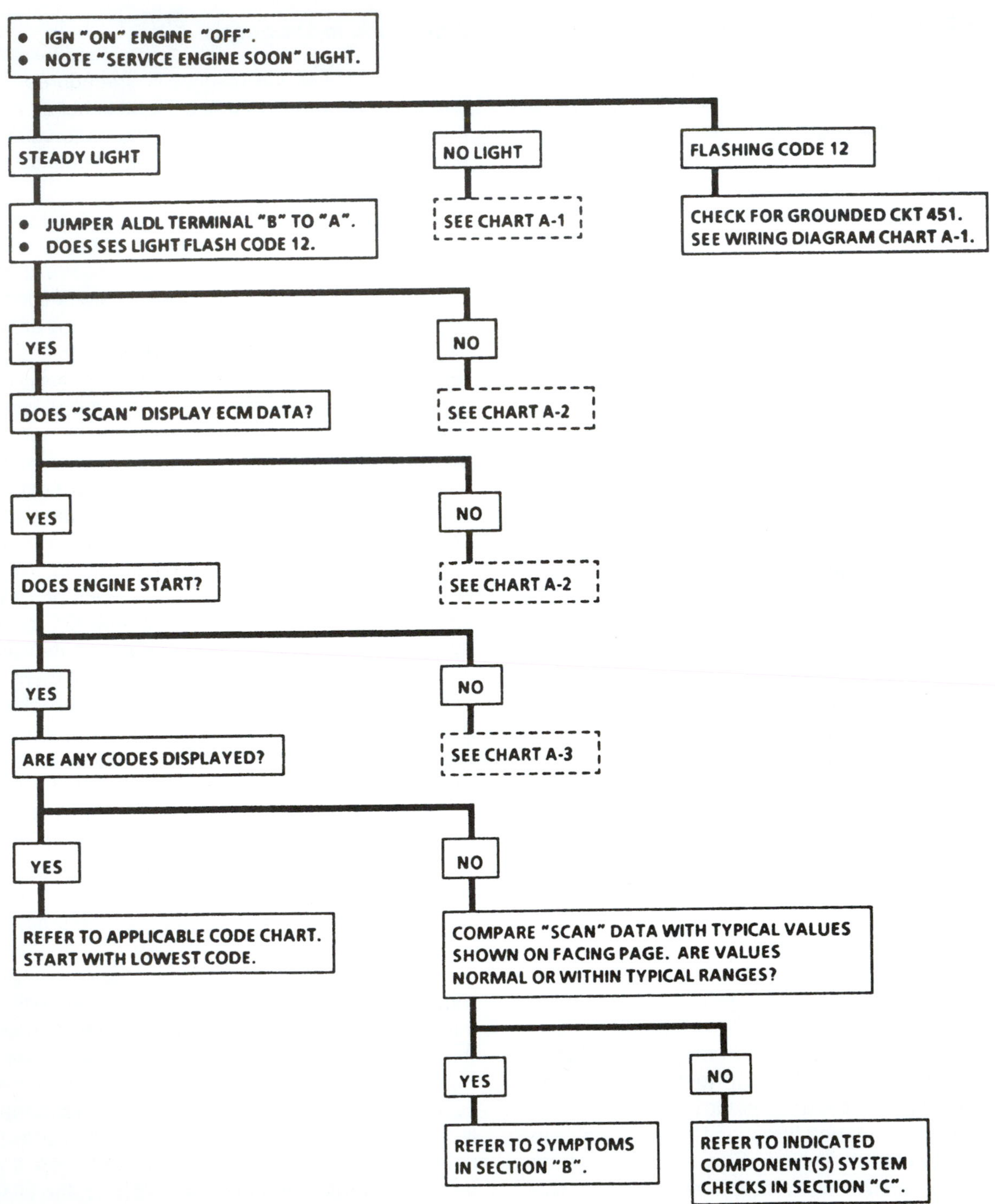

Figure 9–17 Diagnostic circuit check. *(Courtesy of General Motors Corporation, Service Technology Group.)*

a lean exhaust condition is indicated (O_2 sensor stays *below* 450 millivolts most or all of the time). If the light is turned *on* most or all of the time, the system is in closed loop and a rich exhaust condition is indicated (O_2 sensor stays *above* 450 millivolts most or all of the time).

The field service mode was incorporated into the diagnostic circuit check in 1984 and was discontinued when the diagnostic procedures were written around the use of a scanner. The scan tool provides the information the field service mode was designed to provide.

PCM and PROM Service

Considerable care should be taken when working on a PROM or CALPAK (used only on 2.0-liter engine PCMs). A finger touching one of the pins can discharge enough static electricity into it to damage its memory data, Figure 9–18. The PROM carrier fits into the PROM cavity of the PCM in only one position. The PROM, however, fits into the carrier in either of two ways. If the PROM is installed in the PCM reversed end to end from its correct position, it will be damaged electrically and may also damage the computer. Consult the service manual for detailed installation instructions.

The PROM, CALPAK and PCM are serviced separately. Therefore, the technician should remove the PROM and CALPAK before exchanging the PCM for a new one. The same PCMs are used on a variety of vehicles; PROMs and CALPAKs are specific to that vehicle model, year and accessory package. If the CALPAK or PROM is missing or damaged, the engine may not start.

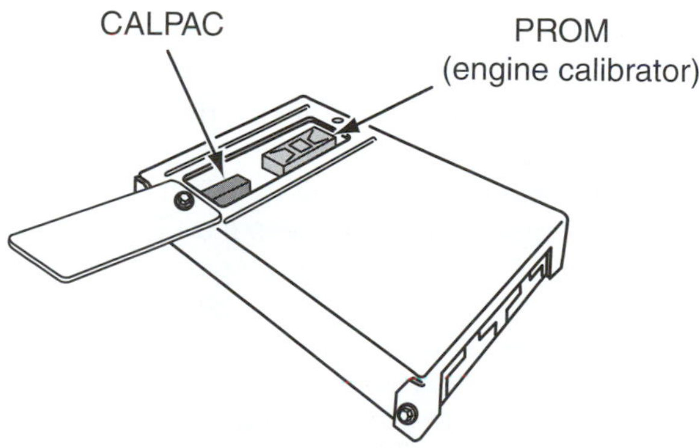

Figure 9–18 PROM and CALPAK in PCM.

Weather-Pack Connectors

All EFI system harness connections under the hood in the engine compartment are made through weather-pack connectors, as discussed in the System Diagnosis and Service section of Chapter 8.

Vacuum Leaks. A vacuum leak on an EFI system usually results in increased idle speed while in closed loop.

TBI. The TBI unit has two O-rings, Figure 9–19, between the injector solenoid and the fuel meter body. The O-rings prevent pressurized fuel from leaking past the injector solenoid. A leak at either of these O-rings can result in dieseling and/or an overly rich mixture. A suspected leak can be easily checked. With the engine off and a jumper lead supplying battery voltage to the fuel pump test terminal, inspect the injector for signs of leaks.

WARNING: Fuel Pressure Test. Some driveability complaints require a fuel pump pressure test. Before the fuel system is opened, pressure in the line should be relieved. The system is designed to retain operating pressure in the line after the engine is shut off. Opening the line under pressure causes a considerable spray of fuel, resulting in a fire and personal safety hazard. To relieve the pressure, remove the fuel pump fuse and crank the engine for several seconds. See the service manual for further test procedures.

Erasing Learned Ability. Anytime battery power is removed from the PCM, its learned memory is lost. This can produce a noticeable change in performance. Learning can be restored

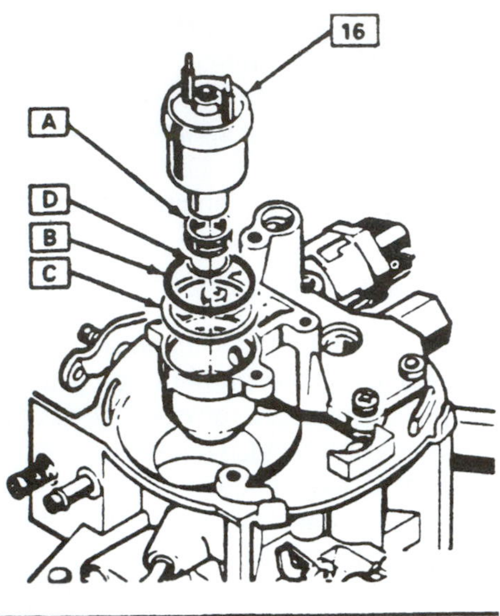

16 FUEL INJECTOR
A FILTER
B LARGE O-RING
C STEEL BACK-UP WASHER
D SMALL O-RING

Figure 9–19 EFI injector assembly. *(Courtesy of General Motors Corporation, Service Technology Group.)*

by driving the car for usually not more than half an hour at normal operating temperature, part throttle and idle, and at moderate acceleration.

SUMMARY

We have covered the most important information inputs the computer uses to calibrate the engine's delivered fuel mixture and spark advance, factors such as engine and air temperatures, engine speed and load, throttle position and vehicle speed. The various ways these sensors work and what use the computer makes of each kind of signal were reviewed. We have learned how fuel pump and ignition circuits work in the system as well as diagnostic precautions and procedures.

The chapter has explained the way the computer makes use of a hard-wired memory, but also builds a body of learned information to correct the original computer instructions and enable the various mixture and advance maps to most effectively optimize exhaust emissions, performance and fuel economy.

▲ DIAGNOSTIC EXERCISES

1. A car is brought into the shop and will not start. Spark and fuel pressure are good, while there are no codes, hard or intermittent. Spark plug tips are black, and replacing them allows the engine to start with difficulty. Soon, however, it dies again. Fuel pressure, it turns out, drops rapidly after shutdown even with feed and return lines clamped off. What component should be checked first?
2. A car with the EFI system will not hold a consistent idle. It idles very poorly when cold, idles slowly when warm and stalls when the air conditioner is turned on. What is the most fruitful sequence of tests to determine the cause of the problem?

REVIEW QUESTIONS

1. What does the PCM do in synchronized mode?
   A. Pulses both TBI fuel injectors simultaneously.
   B. Pulses the TBI fuel injector(s) once for each tach reference signal from the distributor.
   C. Pulses the TBI fuel injector(s) once for each pulse from the VSS.
   D. Pulses the TBI fuel injector(s) every 12.5 milliseconds.
2. *Technician A* says that during cranking mode, if the throttle is opened less than 80 percent, the PCM determines the proper fuel injector pulse width from the ECT sensor.
   Technician B says that during cranking mode, if the throttle is opened wider than 80 percent, the PCM leans out the air/fuel ratio to about 20:1 to help the driver relieve a flooded condition.

Who is correct?
A. A only
B. B only
C. Both A and B
D. Neither A nor B

3. *Technician A* says that if the ground-side wire of the TPS open-circuited resulting in a 5-volt signal to the PCM, this would cause the PCM to enter Clear Flood Mode when the engine is cranking.
Technician B says that if the ground-side wire of the TPS were to open-circuit, resulting in a 5-volt signal to the PCM, the engine could be very difficult to start under normal conditions when it is not flooded.
Who is correct?
A. A only
B. B only
C. Both A and B
D. Neither A nor B

4. What other function does the MAP sensor serve beyond informing the PCM of manifold absolute pressure?
A. It informs the PCM of intake air temperature.
B. It informs the PCM of the atmosphere's relative humidity.
C. It informs the PCM of barometric pressure.
D. It serves no other function.

5. Which of the following inputs affect the PCM's control of the engine's idle speed?
A. Park/neutral (P/N) switch
B. Air-conditioning (A/C) switch
C. Fuel pump voltage signal
D. Both A and B

6. *Technician A* says that the fuel pressure regulator is used to maintain fuel pressure at the fuel injector(s) within a range of 9 to 13 psi.
Technician B says that the spring-side of the fuel pressure regulator's diaphragm is exposed to atmospheric pressure to allow slight fuel pressure adjustments so as to maintain a consistent pressure differential across the injector(s) regardless of altitude or weather conditions. This results in a consistent flow rate anytime an injector is energized.
Who is correct?
A. A only
B. B only
C. Both A and B
D. Neither A nor B

7. The idle air control assembly used by the PCM to control idle speed is which of the following?
A. Vacuum-operated diaphragm
B. Reversible DC motor
C. Stepper motor
D. Duty cycle solenoid

8. What type of device is used by the PCM to energize the fuel pump, the cooling fan and the A/C clutch?
A. Solenoid
B. Diode
C. Potentiometer
D. Relay

9. What function(s) does the fuel pump test terminal serve?
A. It allows a technician to use a jumper wire to manually power the fuel pump motor.
B. It allows a technician to use a voltmeter to see if the fuel pump relay or oil pressure switch is providing power to the fuel pump motor.
C. It allows a technician to test the resistance of the fuel pump relay winding.
D. Both A and B

10. What are the first two steps when pursuing a driveability complaint on an EFI-equipped vehicle?
A. Replace the ignition module, then scope the primary ignition system to be sure that a tach reference signal is present.
B. Replace the PCM, then test the fuse labeled "PCM."
C. Replace the oxygen sensor, then replace the EGR valve.
D. Verify all non-EFI engine support systems are working properly, then perform a diagnostic circuit check.

11. A vehicle equipped with EFI is brought into the shop with a complaint of poor fuel mileage and is also emitting black smoke out the tailpipe during some driving conditions. A fuel

pressure check shows that the fuel pressure is too high. What is the most likely cause?
   A. A defective fuel pump
   B. A defective fuel pressure regulator
   C. High resistance on the ground side of the fuel pump
   D. A defective PCM
12. When the engine is cranking, the EST system operates the switching transistor according to the pickup coil signal (base timing). What mode is this known as?
   A. Module mode
   B. REF mode
   C. EST mode
   D. HEI mode
13. Once the engine is running, the PCM begins to control the switching transistor in the EST system (computed timing). What mode is this known as?
   A. Module mode
   B. REF mode
   C. EST mode
   D. HEI mode
14. When determining computed timing on an EFI system, the PCM considers all of the following inputs *except* which of the following?
   A. Barometric and manifold pressure
   B. Engine coolant temperature
   C. Engine oil pressure
   D. Crankshaft position and rpm
15. What is the purpose of the ESC system?
   A. To determine how far computed timing should be advanced
   B. To retard ignition timing when detonation occurs
   C. To alert the PCM to a cylinder that is misfiring
   D. To sense how much spark advance is being added in by the vacuum and centrifugal advance mechanisms
16. On most EFI vehicles, the PCM will turn on the electric cooling fan when the A/C is turned on and vehicle speed drops below which of the following?
   A. 20 mph
   B. 30 mph
   C. 40 mph
   D. 50 mph
17. Most EFI applications use which of the following?
   A. Single-bed, two-way catalytic converter with no air injection into the converter
   B. Single-bed, three-way catalytic converter with no air injection into the converter
   C. Dual-bed, three-way catalytic converter with no air injection into the converter
   D. Dual-bed, three-way catalytic converter with an air injection system
18. *Technician A* says that a GM EFI system uses weather-pack connectors at all computer system connections in the engine compartment.
   Technician B says that a vacuum leak on an EFI application can result in an idle speed that is too fast.
   Who is correct?
   A. A only
   B. B only
   C. Both A and B
   D. Neither A nor B
19. When replacing the PCM on an EFI application, what component(s) should be transferred from the old PCM to the new PCM?
   A. RAM
   B. PROM
   C. CALPAK
   D. Both B and C
20. *Technician A* says that anytime that battery power has been disconnected from the PCM, the learned memory can be lost, resulting in a noticeable change in engine performance.
   Technician B says that it is a good idea to road test the vehicle after the battery has been disconnected in order to begin the PCM's relearning process before returning the vehicle to the customer.
   Who is correct?
   A. A only
   B. B only
   C. Both A and B
   D. Neither A nor B

Chapter 10

General Motors' Port Fuel Injection

OBJECTIVES

Upon completion and review of this chapter, you should be able to:

- ❑ Describe the operating modes of a GM PFI system.
- ❑ Describe the inputs associated with a GM PFI system.
- ❑ Describe the operation of a GM PFI fuel system.
- ❑ Describe the operation of the idle speed control system associated with a GM PFI system.
- ❑ Describe the operation of the spark management systems associated with a GM PFI system.
- ❑ Describe the operation of the emission control systems associated with a GM PFI system.
- ❑ Describe the operation of a turbocharger and the turbocharger wastegate.
- ❑ Describe how the Fuel Trim program is designed to help a PCM compensate for wear and aging of parts on any fuel injected system.

KEY TERMS

AC
Three-Coil Ignition (C3I)
Fuel Trim
High-pressure Cutout Switch
Integrator
Long-Term Fuel Trim
Longitudinal
Normally Aspirated
Pressure Cycling Switch
Short-Term Fuel Trim
Thermac
Thermo-time Switch
Wastegate

Excluding the Cadillac Digital Fuel Injection system discussed in Chapter 11, Port Fuel Injection (PFI) represents the fourth generation of General Motors' comprehensive computerized engine control systems. PFI was introduced with limited applications on 1984 models. It followed General Motors' first closed-loop fuel system of 1978, the CCC system introduced in 1980, and the EFI system introduced in 1982. In 1985 the PFI system was expanded to several engine applications. Some General Motors car divisions have chosen special names for the system because of some special feature designated for a given engine or body application. These special features include:

- Sequential Fuel Injection (SFI) on the turbocharged 3.8-liter engine.
- Multiport Fuel Injection (MFI) on most other engines.
- Tuned Port Injection (TPI) on the Corvette 5.7-liter engine. The TPI version includes tuned intake manifold runners, matched in shape, length, and cross-sectional area to help maximize volumetric efficiency.

Because only air moves through the intake manifold, there is no concern about holding fuel in a vapor state as it passes through the runners. Consequently the heated intake air (**Thermac**) and the early fuel evaporation (EFE) systems are eliminated. Most of the PFI applications do run engine coolant through a passage in the throttle body to prevent the formation of ice around the throttle blade.

POWERTRAIN CONTROL MODULE (PCM)

The PCM is usually located under the instrument panel or behind the passenger kick panel. On the P car (Pontiac Fiero) it is in the console between the seats. It contains a removable PROM and CALPAK. The CALPAK provides calibration for cold-start cranking and for fuel backup in the event of a PCM failure. The fuel backup circuit (FBC) provides an operating mode when any of the following conditions occur:

- PCM voltage falls below 9 volts (most likely during cold cranking).
- The PROM is missing or not functioning.
- The PCM is unable to provide its normal computer-operated pulses.

The FBC makes use of the throttle position sensor, the coolant temperature sensor signal, and the engine rpm signal. It is powered through the ignition switch. During FBC mode the engine runs erratically and sets a code 62.

Keep-Alive Memory (KAM)

The PCM has basically the same self-diagnostic capacity and KAM as discussed in earlier chapters. The inputs fed into the PCM and the functions controlled for most systems are shown in Figure 10–1.

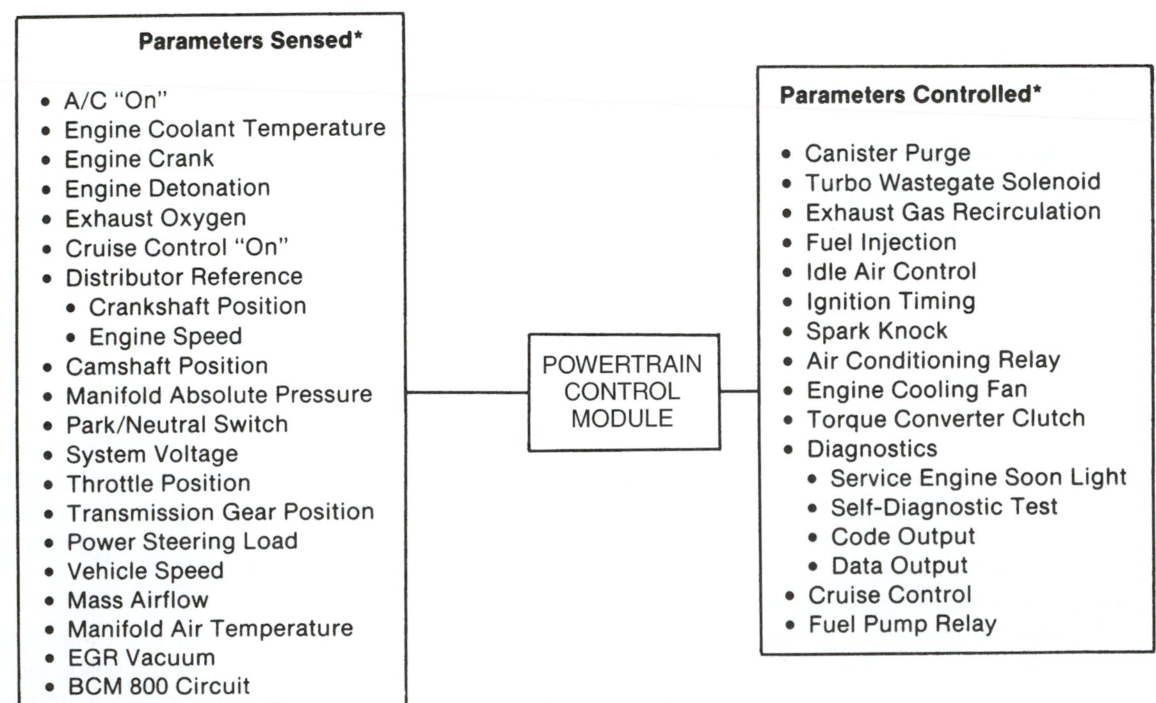

Figure 10–1 PFI system overview.

OPERATING MODES

The different operating modes of the PFI system control how much fuel to introduce into the intake manifold for the various possible operating conditions.

Starting Mode

When the ignition is turned on, the PCM turns the fuel pump relay on. If it does not see an ignition signal within 2 seconds telling it that the engine is cranking, it turns the fuel pump off again. The fuel pump provides fuel pressure to the injectors. As the system goes into cranking mode, the PCM checks coolant temperature and throttle position to calculate what the air/fuel ratio should be for this startup. The air/fuel ratio ranges from maximally rich, 1.5 to 1, at –36°C (–32°F), to maximally lean, 14.7 to 1 at 94°C (201°F). The PCM varies the air/fuel ratio by controlling the time the injectors are turned on, what is referred to as the pulse width.

The 2.8-, 5.0- and 5.7-liter engine PFI applications also use a cold-start injector. This is an additional fuel injector in the intake manifold that improves cold-starting by spraying additional fuel into the intake manifold during cranking. It is not PCM-controlled. Spraying the additional fuel for cold-start enrichment into the manifold gives the fuel more opportunity to evaporate in the cold, dense air than if it were sprayed directly into the intake ports. The injector is activated by the cranking system and is controlled by a timing mechanism. The subsystem is described more fully in the Outputs section of this chapter.

Clear Flood Mode

The clear flood mode works essentially the same as it does on the EFI system. As long as the engine is turning below 600 rpm and the throttle is held to 80 percent WOT or more, the PCM will keep the air/fuel ratio at 20 to 1, except on the 2.8-liter engine, which cuts off fuel completely in clear flood. The same troubleshooting considerations apply to this system's clear flood mode as to the system in Chapter 9.

Run Mode

The run mode consists of the open- and closed-loop operating conditions. When the engine starts and the rpm rises above 400, the system goes into open loop. In open loop, the PCM ignores information from the oxygen sensor and determines air/fuel ratio commands based on input from other sensors (coolant temperature, vacuum or mass airflow and throttle position, as well as engine speed if the vacuum sensor is used). The system stays in open loop until:

- the oxygen sensor produces a varying voltage showing that it is hot enough to work properly.
- the coolant is above a specified temperature.
- a specified amount of time has elapsed since the engine last started.

These values vary with engine application and are hardwired into the PROM. When all three conditions are met, the PCM puts the system into closed loop. In closed loop, the PCM uses oxygen sensor input to calculate air/fuel ratio commands and keeps the air/fuel ratio at a near perfect 14.7 to 1. During heavy acceleration, wide-open throttle or hard deceleration, the system temporarily drops out of closed loop.

Acceleration Mode

Rapid increases in throttle opening and manifold pressure or airflow signal the PCM to enrich the air/fuel mixture. This compensates for the reduced evaporation rate of the gasoline resulting from the higher manifold pressure.

Deceleration Mode

Rapid decreases in throttle opening and manifold pressure or airflow cause the PCM to lean the air/fuel mixture. If the changes are severe enough, fuel is momentarily cut off completely.

Battery (Charging System) Voltage Correction Mode

The battery voltage correction feature on the PFI system is different from that on the EFI system. On the PFI system, when battery voltage drops below a specified value, the PCM:

- enrichens the air/fuel mixture according to a preset formula.
- increases the throttle opening if the engine is idling.
- increases ignition dwell to compensate for a weakened ignition spark should the coil's magnetic field not build sufficiently to generate a hot secondary spark.

Fuel Cutoff Mode

When the ignition is turned off, the PCM immediately stops pulsing the injectors, to prevent dieseling as well as to consume the fuel mixture remaining in the ports. Injection is also stopped anytime the distributor reference pulse stops coming to the PCM.

FUEL SUPPLY SYSTEM

Fuel pressure is supplied by an electric pump in the fuel tank, Figure 10–2. The pump is turned on and off by a fuel pump relay controlled by the PCM. A pressure relief valve in the pump limits

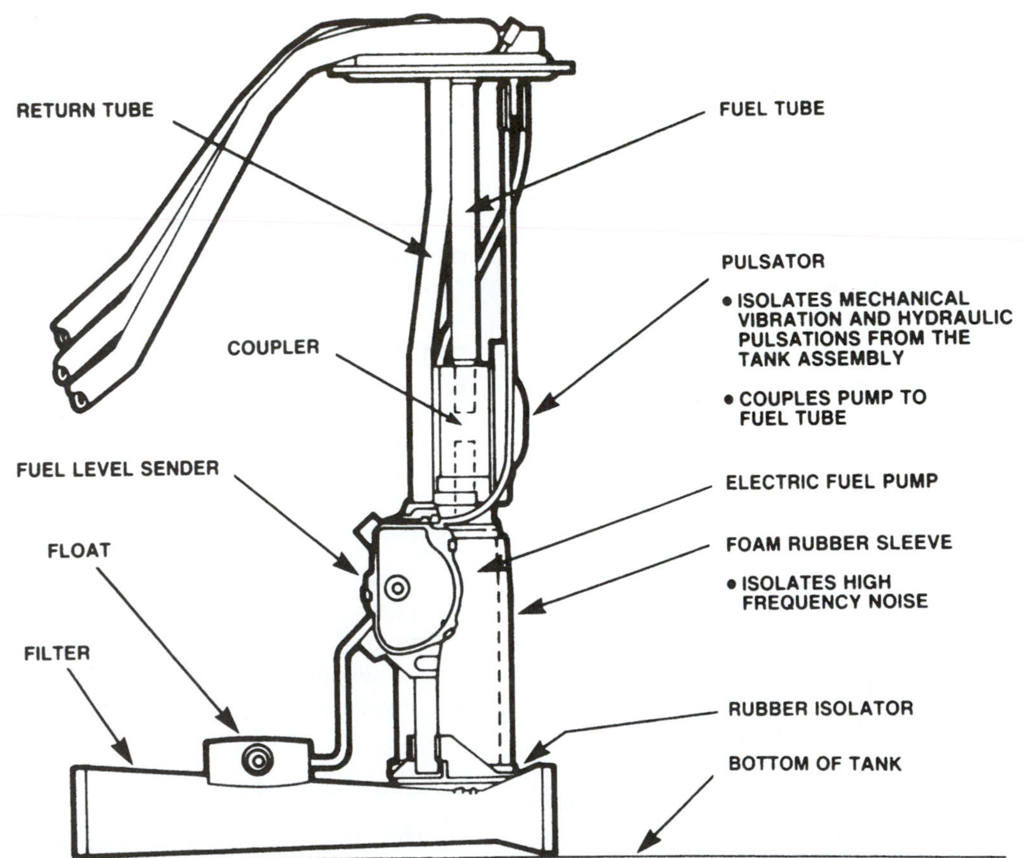

Figure 10–2 Fuel pump and sending unit assembly. *(Courtesy of General Motors Corporation, Service Technology Group.)*

maximum pump pressure to between 80 and 90 psi. This pressure is realized only if flow stops and the pump is working against static pressure. The filter mounted to the bottom of the pump assembly is a 50-micron filter; downstream is another filter, a 10- to 20-micron in-line filter (one micron is one millionth of a meter). Because of the high operating pressure, an O-ring is used at all threaded connections, and all flex hoses have internal steel reinforcement.

Fuel pressure is controlled by the fuel pressure regulator, Figure 10–3, which is usually mounted on the fuel rail in the engine compartment, Figure 10–4. Manifold vacuum is supplied to the spring side of the diaphragm. At light throttle, the high vacuum helps pull the diaphragm up and against the spring. This allows more fuel to return to the tank, thus reducing output pressure. Pressure at the injectors ranges from 34 psi at idle to 44 psi at full throttle.

At idle, when pressure within the intake manifold is *low* (high vacuum), fuel rail pressure is also *low*. At WOT, when pressure within the intake manifold is *high* (low vacuum or no vacuum), fuel rail pressure is also *high*. If the intake manifold pressure is increased beyond atmospheric pressure through turbo charging, fuel rail pressure is also equally increased. By indexing the fuel rail pressure to a manifold vacuum (actually, manifold absolute pressure) signal, the fuel pressure varies equally with the pressure in the intake manifold. This keeps the pressure differential across the fuel injectors (between the inlet and outlet ends) constant under all engine operating conditions. Since the injector orifice size is fixed on a given engine, if the pressure differential is held constant, then the flow rate will also be constant anytime the injector is open. This leaves only one variable in controlling the air/fuel ratio: how long the PCM keeps the injector held open each time that it energizes it. Many technicians falsely believe that the regulator is indexed to manifold vacuum in order to increase the *flow rate* at WOT and decrease the flow rate at idle. In reality, the regulator is indexed to manifold vacuum so that the flow rate will be held *constant* regardless of throttle position. This constant flow rate is programmed into the PCM's PROM chip. Then the base fuel calculations in PROM only have to calculate injector on-time (pulse width) when controlling the air/fuel ratio.

The manifold vacuum signal to the pressure regulator is only necessary on a PFI application

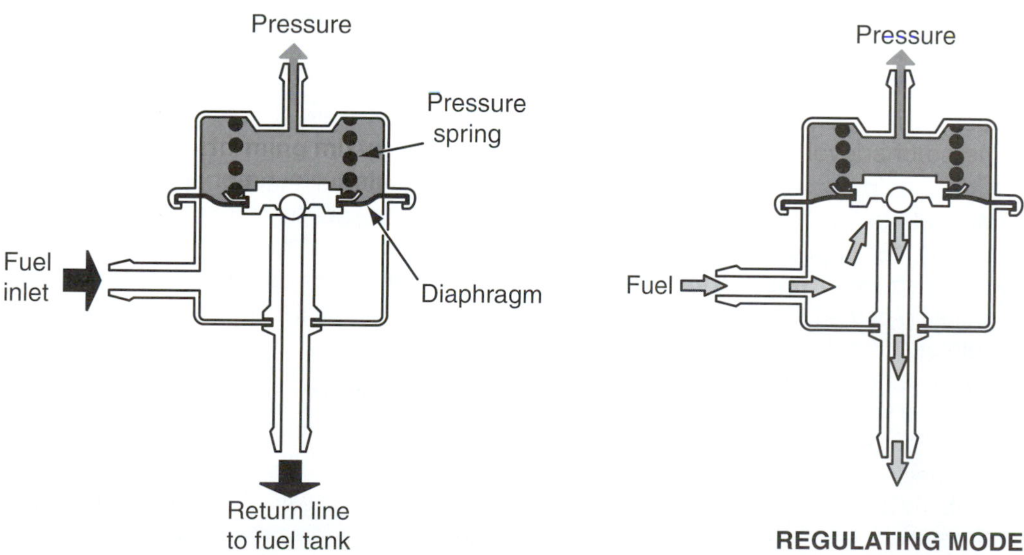

Figure 10–3 Operation of a diaphragm-operated fuel pressure regulator.

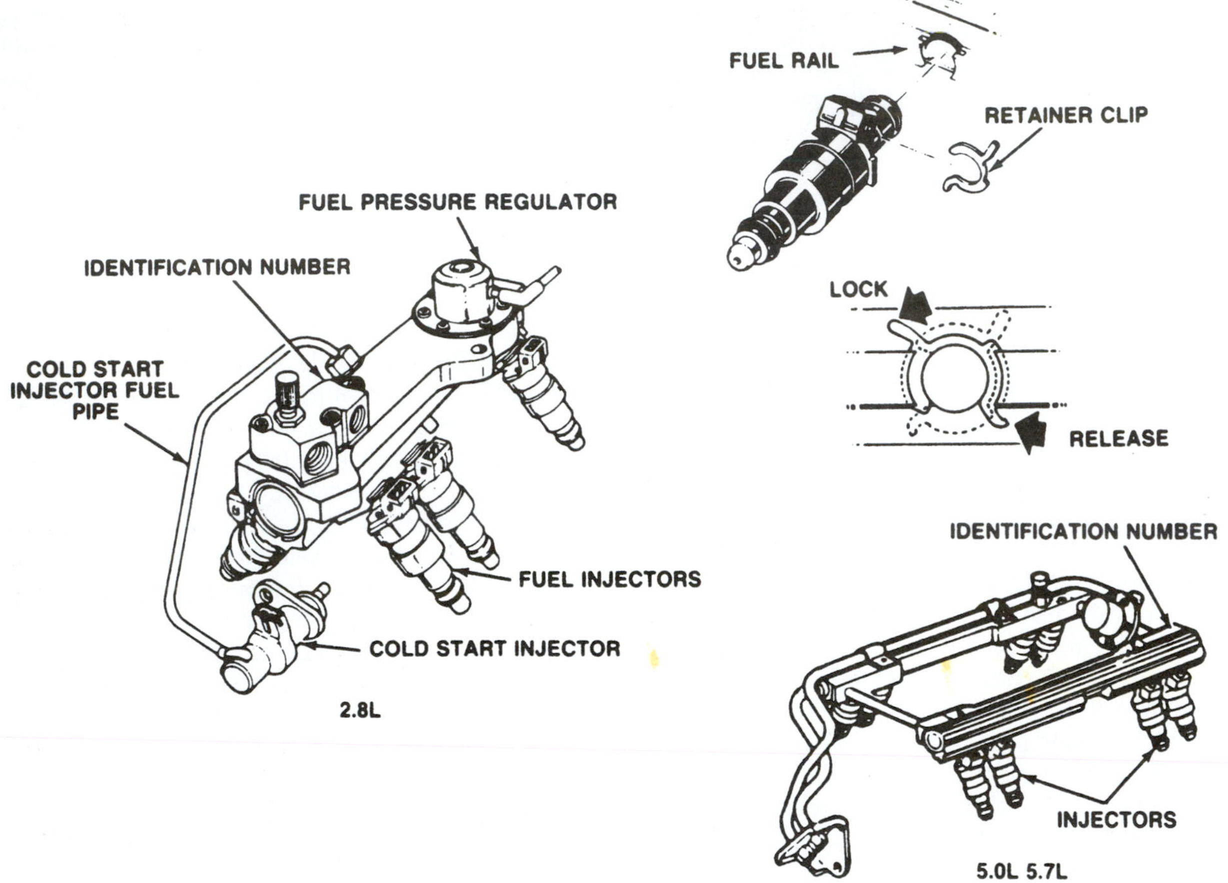

Figure 10–4 Fuel rail: 2.8-liter, 5.0-liter and 5.7-liter engines. *(Courtesy of General Motors Corporation, Service Technology Group.)*

because the outlet tip of the injector is located in the intake manifold *below the throttle plate(s)*. GM (and most other) TBI applications do not require referencing the regulator to a vacuum signal because the outlet tip of the injector is located above the throttle plate(s) in a throttle body designed to reduce or eliminate the effects of engine load in the throttle bore area. Thereby, the fuel pressure on a TBI application does not need to vary with engine load because the pressure at the outlet tip of the injector is fairly consistent resulting in a steady pressure differential.

Additionally, by releasing excess fuel into a return line, the fuel pressure regulator keeps a constant stream of fuel flowing back to the tank to ensure that cool fuel from the tank is always supplied to the fuel rail and injectors. This, combined with the higher operating pressures, helps to reduce or eliminate the chance of vapor lock.

INJECTORS

An injector is installed in the intake port of each cylinder with 70 to 100 millimeters between the tip of the injector and the center of the valve on V-type engines, Figure 10–5. The nozzles spray fuel in a 25-degree conical pattern. O-rings are

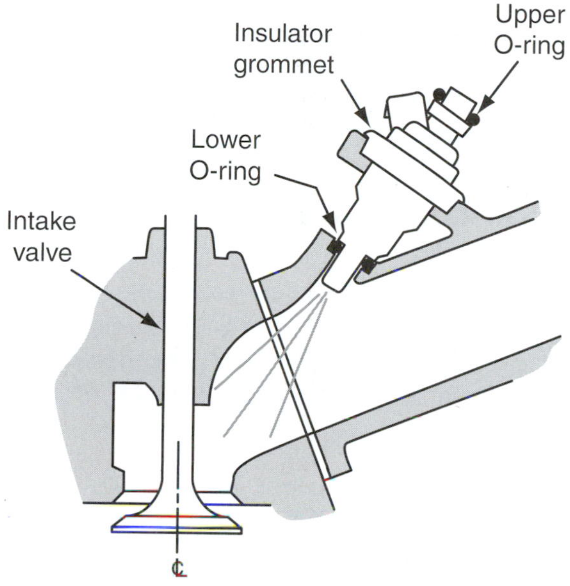

Figure 10–5 Port injection sprays fuel into the intake port and fills the port with fuel vapor before the valve opens.

used to seal between the nozzle and the fuel rail and between the nozzle and the intake manifold. The O-rings also serve to retard the heat transfer from the engine to the nozzle of the injector and to reduce nozzle vibration. The O-rings should be lubricated or replaced as necessary whenever the injectors are removed. A damaged or hardened O-ring allows a fuel or vacuum leak. PFI systems are very sensitive to vacuum leaks.

THROTTLE BODY

The throttle body on a PFI system controls the amount of air allowed into the engine's induction system with a throttle blade and shaft controlled by the accelerator pedal, just as in a carburetor or TBI unit, Figure 10–6. In this case, however, only air passes the throttle blade. An idle stop screw determines how nearly closed the throttle blade is at idle. A throttle position sensor keeps the PCM informed as to the throttle valve's position. Idle

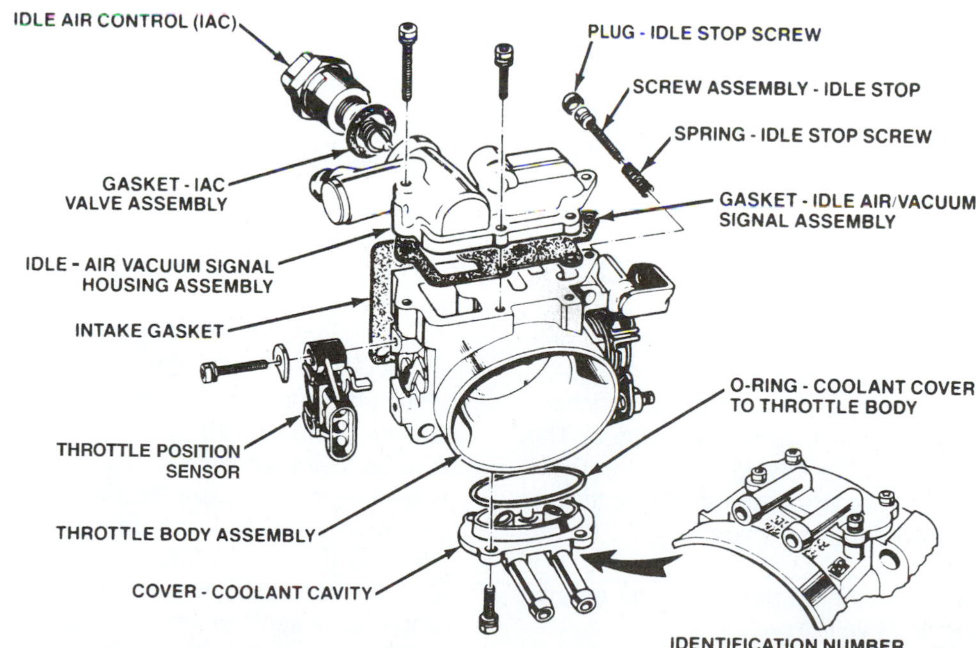

Figure 10–6 Throttle body assembly, 2.8-liter engine. *(Courtesy of General Motors Corporation, Service Technology Group.)*

Diagnostic & Service Tip

One of the unexpected problems with port fuel injection throttle bodies has proved to be fuel deposits on the throttle blades. The source of these fuel vapors is what drifts back from the intake ports when the engine is shut down hot. The heat vaporizes most or all of the residual fuel, and the coolest surface it encounters in the intake path is the throttle blade itself. The problem is that these deposits change the shape and size of the opening in the smallest throttle opening angles. This can adversely affect driveability in unexpected ways, such as hesitation, erratic idle, hard starting and surging. The throttle blades can be visually inspected for such deposits and cleaned with solvent. This inspection should be a routine maintenance task on cars with this fuel injection system.

speed is regulated by an idle air control motor and valve assembly. It varies the amount of air allowed to bypass the throttle blade through a bypass passage and is controlled by the PCM. Most throttle body units have a cavity for engine coolant to flow through to prevent throttle blade icing.

NON-PCM EMISSION CONTROLS

A trend is developing in the automotive industry where as computerized engine control systems become more efficient, fewer traditional mechanical emissions control devices are needed. This is evidenced by the following list of non-PCM-controlled emission control devices during the first or second year of PFI production:

- 1.8-liter—canister purge, PCV and internal EGR (no EGR valve)
- 2.8-liter—PCV and deceleration valve (admits additional air during deceleration on manual transmission vehicles)
- 3.0- and 3.8-liter—PCV
- 5.0- and 5.7-liter—PCV

INPUTS

Engine Coolant Temperature (ECT) Sensor

This is the same sensor used in the Computer Command Control system, the Electronic Fuel Injection System, and other GM electronic engine management systems. The ECT input affects control of:

- air/fuel mixture (in open loop).
- spark timing.
- spark knock control (on some engines).
- engine idle.
- torque converter lockup clutch actuation.
- canister purge (except on the 1.8-liter engine up to 1985).
- EGR (except for the 1.8-liter engine up to 1985).
- cooling fan (on some vehicles, especially with transverse-mount engines).

Mass Airflow (MAF) Sensor

Many General Motors PFI systems either use or have used a MAF sensor. In the mid-1980s, two MAF sensors were used: one on Buick engines and one on Chevrolet engines. Buick engines used a MAF manufactured by AC, a division of General Motors. Chevrolet engines used a Bosch unit. In the late 1980s, the AC unit was discontinued and a unit built by Hitachi of Japan was used on Buick engines. Chevrolet engines were by then no longer equipped with a MAF sensor.

AC. The **AC** unit contains a screen to break up the airflow, Figure 10–7. After the air passes through the screen, it flows over an air temperature-sensing resistor. A sample tube then directs some of the air to flow over a heated foil sensing element. The power to heat the sensing element comes from a fuse (a relay as on some

C3I System

For many years, automotive engineers generally felt that it did not make much difference *when* the fuel was sprayed into the intake port as long as precisely the *right amount* of fuel was sprayed. When you consider how fast all of this happens (the intake valve opens and closes 25 times per second at 3,000 rpm), it would not seem to make much difference.

With the introduction of the **three-coil ignition** system (**C3I**) by Buick in 1984—and since extended to most other divisions—it became necessary to employ a sequential fuel injection system. This is because the C3I fires two plugs at one time; one at the regular spark advance position at the end of the compression stroke, the other at the cylinder 360 degrees out of sequence with the first, the cylinder just completing its exhaust stroke. This is the waste spark system explained in the Outputs section of this chapter. The second "waste" spark serves no purpose in combustion; it is just an unavoidable by-product of the ignition system's electrical design. If, however, a pressurized combustible mixture (the first applications of C3I were on turbocharged engines) was just on the other side of the open intake valve, a backfire could occur in the intake manifold. To avoid this, each injector pulse is timed so the fuel is immediately purged from the manifold by its respective cylinder's intake stroke.

However, the additional circuitry and related expense seems to have been proved worthwhile, as GM has rapidly expanded the application of sequential fuel injection to other engines. Ford introduced sequential fuel injection in 1986 on some 5-liter engines and has expanded its use of sequential injection to many other engines since then.

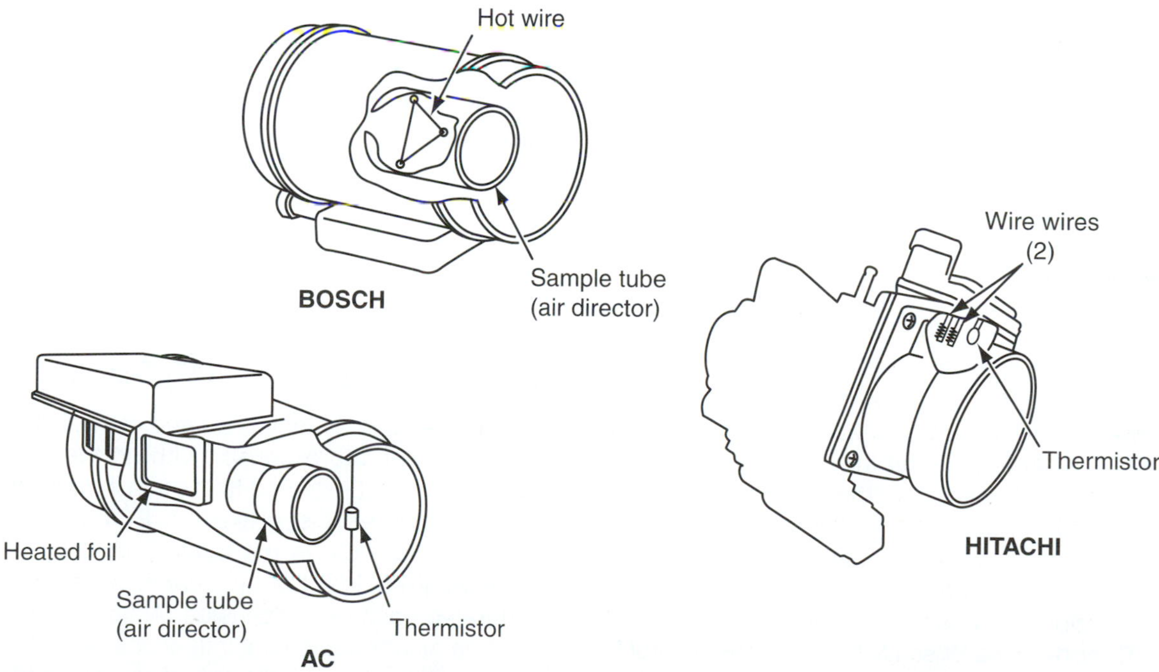

Figure 10–7 Mass airflow sensor (AC type).

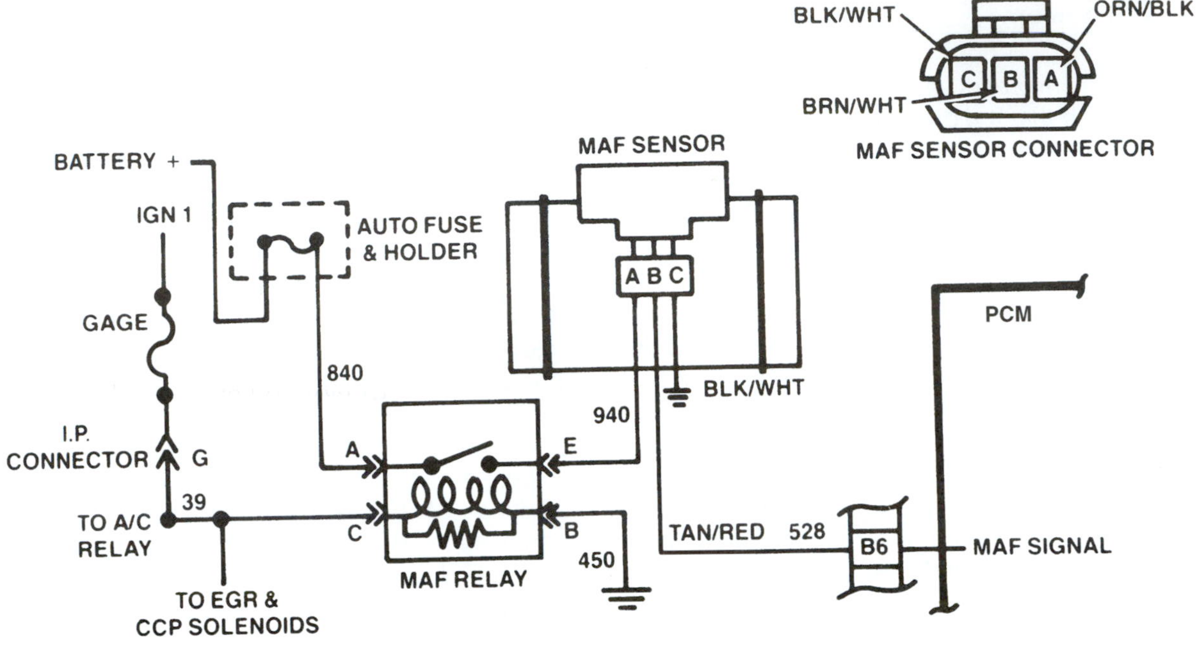

Figure 10–8 MAF relay circuit. *(Courtesy of General Motors Corporation, Service Technology Group.)*

applications), Figure 10–8. The circuitry of the MAF sensor controls current flow through the foil sensing element to maintain it at 75°C above the incoming air temperature, as measured by the temperature-sensing resistor. The power required to keep the sensing foil element 75°C above incoming air temperature is a measure of mass airflow. This value is sent to the PCM as a digital signal ranging in frequency from 30 cycles per second (30 hertz or 30 hz) to 150 hz (150 hz represents the highest mass airflow rate). The PCM compares the MAF sensor information to a preprogrammed look-up chart and finds airflow in grams per second. Using this value, engine temperature, and engine rpm, the PCM can calculate exactly how much fuel is required to achieve the desired air/fuel ratio. Mass airflow readings are taken and air/fuel mixture calculations are revised about 160 times per second.

Bosch. The Bosch MAF sensor works much the same way as the AC unit except that a wire heat element is used instead of a foil sensing element. Also, each time the ignition is turned off after the system has been in closed loop, a separate burn-off module momentarily puts enough current through the wire heat element to make it red hot, Figure 10–9. This burns off any residual accumulation to keep the sensor accurate. It is critical that the sensor wire's surface be clean because any accumulation of deposits retards its ability to transfer heat.

Intake Air Temperature (IAT) Sensor

The IAT sensor looks like and essentially is a coolant temperature sensor with its thermistor-sensing element exposed to the intake air instead of being fully enclosed like the ECT sensor. It screws into a hole in the intake manifold or on some vehicles, into the air cleaner housing. The perforated nose extends inside. Although it is not used on all applications, it influences air/fuel mixture, spark timing and idle speed control on those engines that use it.

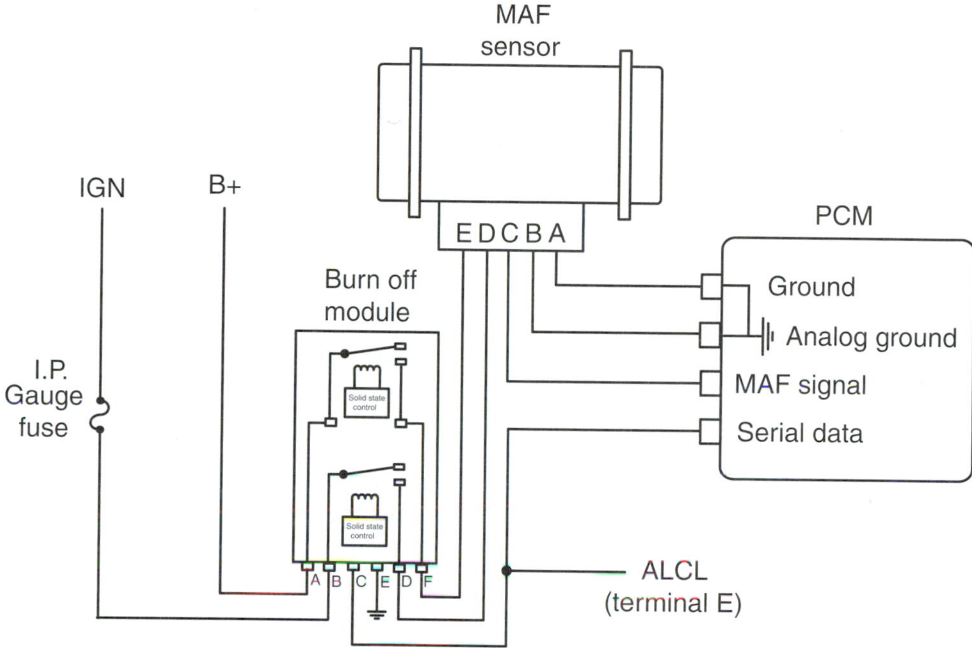

Figure 10–9 Bosch MAF circuit and burn-off module.

Throttle Position Sensor (TPS)

The TPS is a potentiometer mounted on the side of the throttle body and attached to the end of the throttle shaft, Figure 10–6. Its input affects most PCM-controlled functions including fuel control, spark timing, idle air control, torque converter lockup clutch engagement and air conditioning control.

Reference Pulse (REF)

On all but those engines equipped with a distributorless ignition system, the REF signal comes from the distributor pickup coil. Without this signal the engine does not start or run because without it the PCM will not pulse the injectors. A more complete discussion of the REF signal is found in the Inputs section of Chapter 8.

The distributorless ignition systems use sensors to provide information to the ignition module relative to crankshaft and camshaft position and speed. The C3I (Computer-Controlled Coil Ignition) systems use Hall effect switches while both the Direct Ignition System (DIS) and Integrated Direct Ignition (IDI) systems use a permanent magnet pulse generator, Figure 10–10.

C3I. There are three different versions of the C3I system: Type I, Type II and Type III. Each of the three types uses a Hall effect switch mounted behind the crankshaft balancer, Figure 10–11. The balancer has three vanes, which are positioned to form a circle, and they extend back toward the engine. As the crankshaft turns, the vanes alternately pass through the Hall effect switch, producing a signal that is fed to the PCM and the ignition module. When used on the 3.0-liter engine, the Type I system uses a second Hall effect switch mounted behind the crankshaft balancer with a single vane extending back, nearer the outer circumference of the balancer. This vane acts as a camshaft position sensor.

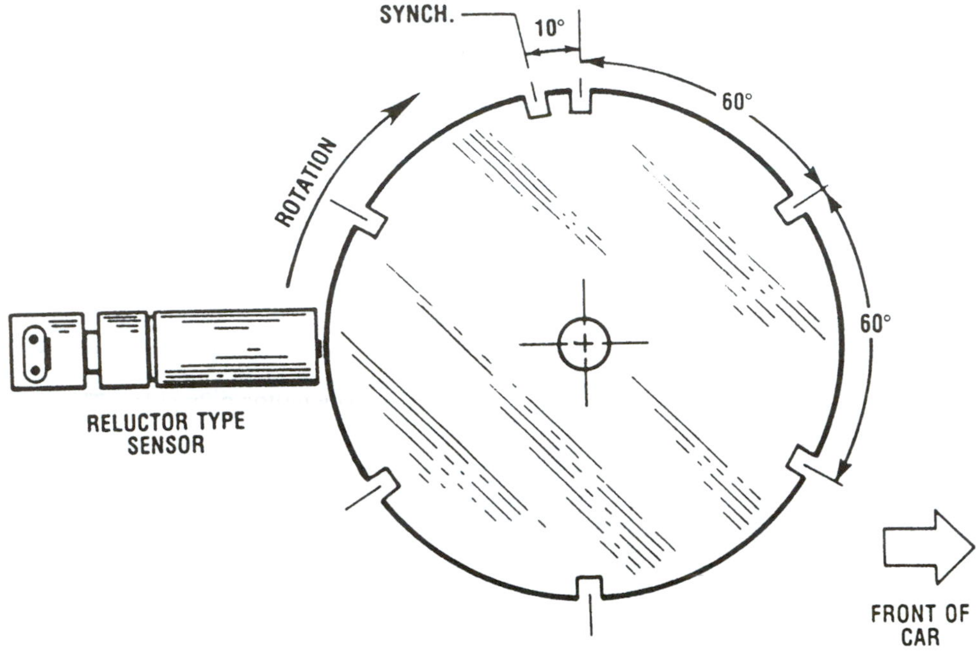

Figure 10–10 Crankshaft position sensor.

Figure 10–11 Hall effect crankshaft sensor and harmonic balancer with three vanes and three windows.

The Type I or Type II systems, when used with the 3.8-liter turbo engine, include a Hall effect switch mounted in what used to be the distributor hole to serve as a camshaft position sensor, Figure 10–12. This sensor is driven by the camshaft at the same speed (unlike the harmonic-balancer mounted sensor, which signals at crankshaft speed). When used on the 3.8-liter nonturbocharged SFI engine, these same two systems use a Hall effect cam sensor mounted in the timing cover and are triggered by a magnet attached to the camshaft timing gear. The cam sensor signal alerts the PCM when the number one cylinder is at TDC so it can start the firing order over again and time the injector pulses. Loss of this signal results in a code 41 being set. If the signal is absent or lost during cranking, the engine will not start; if the signal is lost while the engine is running, the engine will continue running, but will not restart. In more recent General Motors literature, the camshaft position sensor is referred to as a "sync sensor."

The Type III system, sometimes referred to as the "Fast Start," also has its sync sensor located in the timing cover. On this system, a third sensor is used. It is a Hall effect switch and is located behind the harmonic balancer along with the crankshaft sensor. This sensor, referred to as the "Crank 18x," is triggered by 18 evenly spaced interrupter vanes protruding rearward from near the outer circumference of the balancer. The Crank 18x sensor provides 18 signals per crankshaft revolution to the ignition module, which, in turn, sends on to the PCM, Figure 10–13.

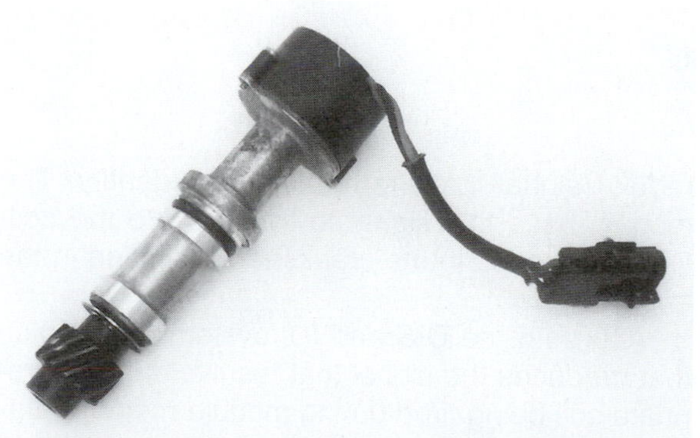

Figure 10–12 Hall effect camshaft position sensor that mounts in the distributor hole.

On this system, the crank sensor is called the *Crank 3x* sensor. Its three interrupter vanes, nearer the center on the balancer, are not symmetrical as they are on the other C3I systems. The windows between the vanes are 10, 20 and 30 degrees wide as shown in Figure 10–13. The signal produced by the leading and trailing edge of each window constitutes a 3x pulse. The number of 18x signals that occurs during each 3x pulse enables the PCM to determine which 3x pulse it is reading. There is one 18x signal during the 10-degree window pulse, two during the 20-degree window pulse and three during the 30-degree window pulse. This enables the system to fire a coil for the appropriate cylinders within the first 120 degrees of crankshaft rotation. This system provides the following advantages:

- faster starts, because the PCM can identify cylinder position more quickly
- more accurate REF signals to the PCM, especially at low speeds
- increased run reliability, as the engine can continue to operate without the cam sensor
- the potential for the PCM to read crankshaft acceleration and deceleration rates

DIS and IDI. Both the IDI and the DIS use a magnetic pulse generator on the crankshaft instead of the Hall effect switches for crankshaft reference (position and speed). The crankshaft for those engines which use either the DIS or IDI system has a round steel disc machined into its center. This disc is concentric to the crankshaft's centerline at its center counterweight, Figure 10–14. This disc has seven notches cut into it and is called a *reluctor*. The term *reluctor* refers to any device that changes the reluctance of a material to conduct magnetic lines of force. The same reluctor is used on both four- and six-cylinder engines. The crankshaft sensor is mounted to the engine block

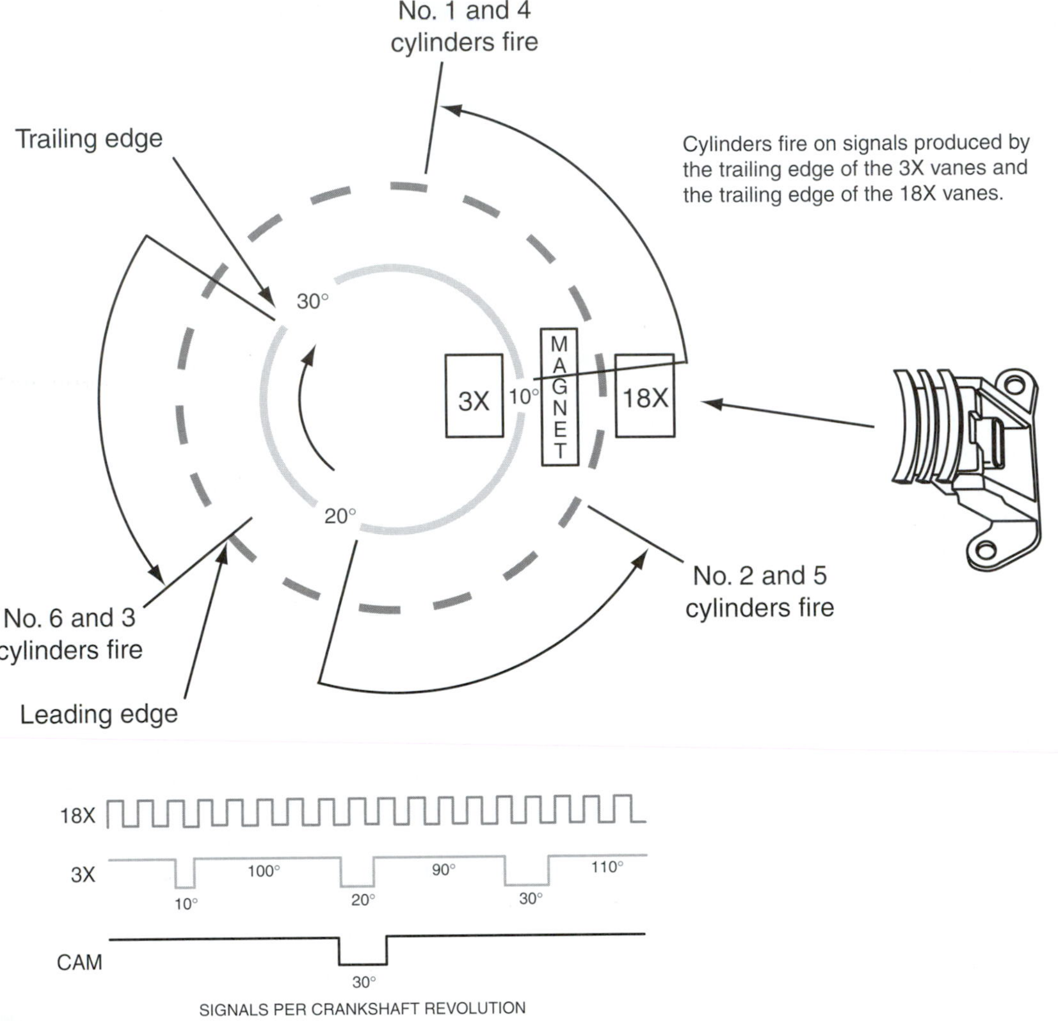

Figure 10–13 Combination sensor for C3I (Type III) Fast Start.

and extends through a hole in the crankcase. Its tip, containing a permanent magnet and a wire coil, is spaced 0.050 inch from the reluctor.

Six of the reluctor's seven notches are spaced evenly around it at 60-degree intervals, Figure 10–14. The seventh notch is called the *sync* notch and is located between notches 6 and 1, 10 degrees from notch 6 and 50 degrees from notch 1. By being placed at an odd position in relation to the other notches, the sync notch provides an irregular signal that the module can identify. The module uses this signal to synchronize the coil-firing sequence to the crankshaft's position in the engine cycle.

On both the DIS and IDI systems, the notch that produces the signal that results in the appropriate coil being fired during module mode operation is often referred to as the *cylinder event* notch. The crank sensor signal that occurs 60 degrees before the cylinder event notch causes voltage in

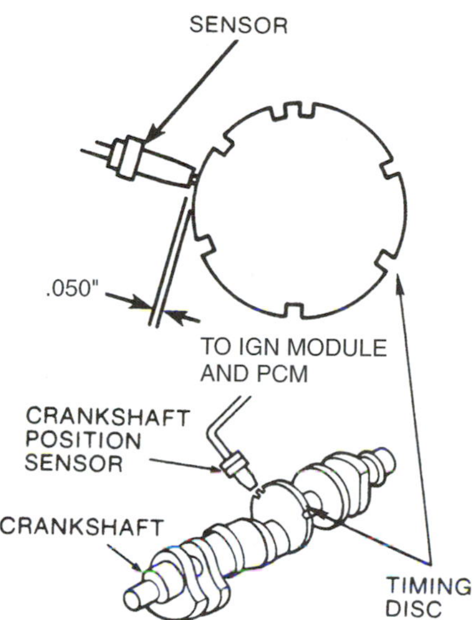

Figure 10–14 Crankshaft timing disc.

the REF wire to go low. The cylinder event notch causes voltage in the REF wire to go high. When the engine operates in the electronic spark timing mode, the PCM uses this signal as an input for calculating:

- ignition timing while operating in EST mode.
- fuel injector pulses.

GM's Opti-Spark Distributor

In 1992, GM introduced an optical distributor on the Corvette, similar to those already in use by Chrysler on their Mitsubishi engines, as well as Nissan vehicles. The GM opti-spark distributor is located at the front of the engine just behind the water pump, thus explaining its overall flat design, Figure 10–15. Like their counterparts, this sensor has two photo diodes placed opposite two LEDs that monitor two sets of slots in a slotted wheel, Figure 10–16. When light is allowed through a slot to strike the photo diode, a voltage pulse is produced.

The outer row has 360 slots, one for every degree of distributor rotation, which produces the high-resolution signal. The inner row has one slot for every cylinder—eight slots for the Corvette application. This produces the low-resolution signal. This is the concept behind the Chrysler and Nissan optical distributors as well. Unlike Chrysler, whose PCM figures out where the number one cylinder is

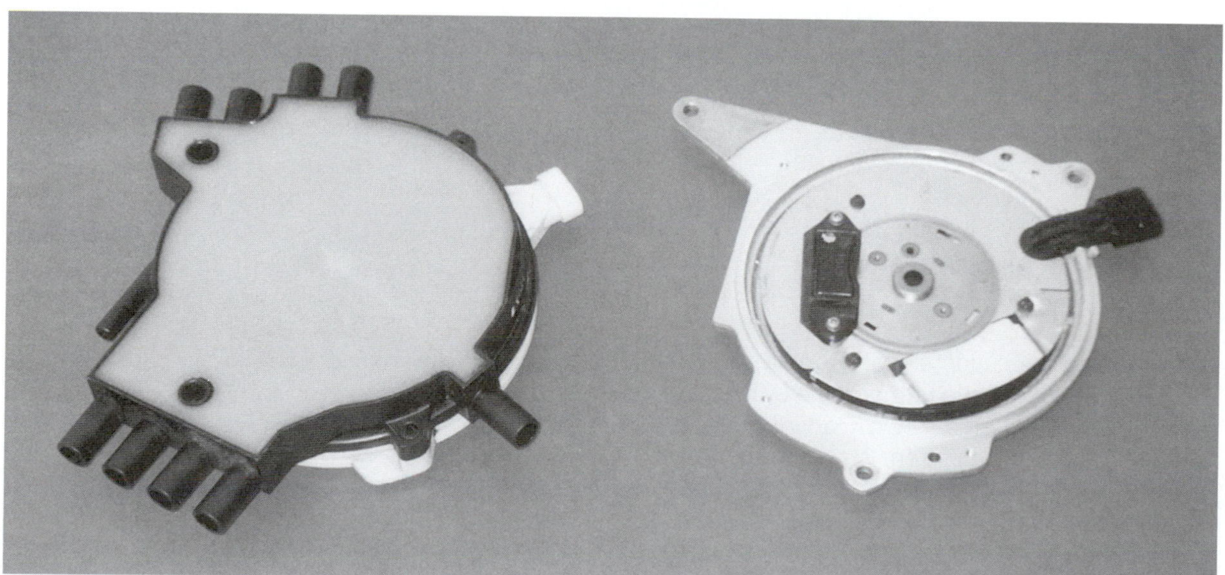

Figure 10–15 Opti-spark distributor.

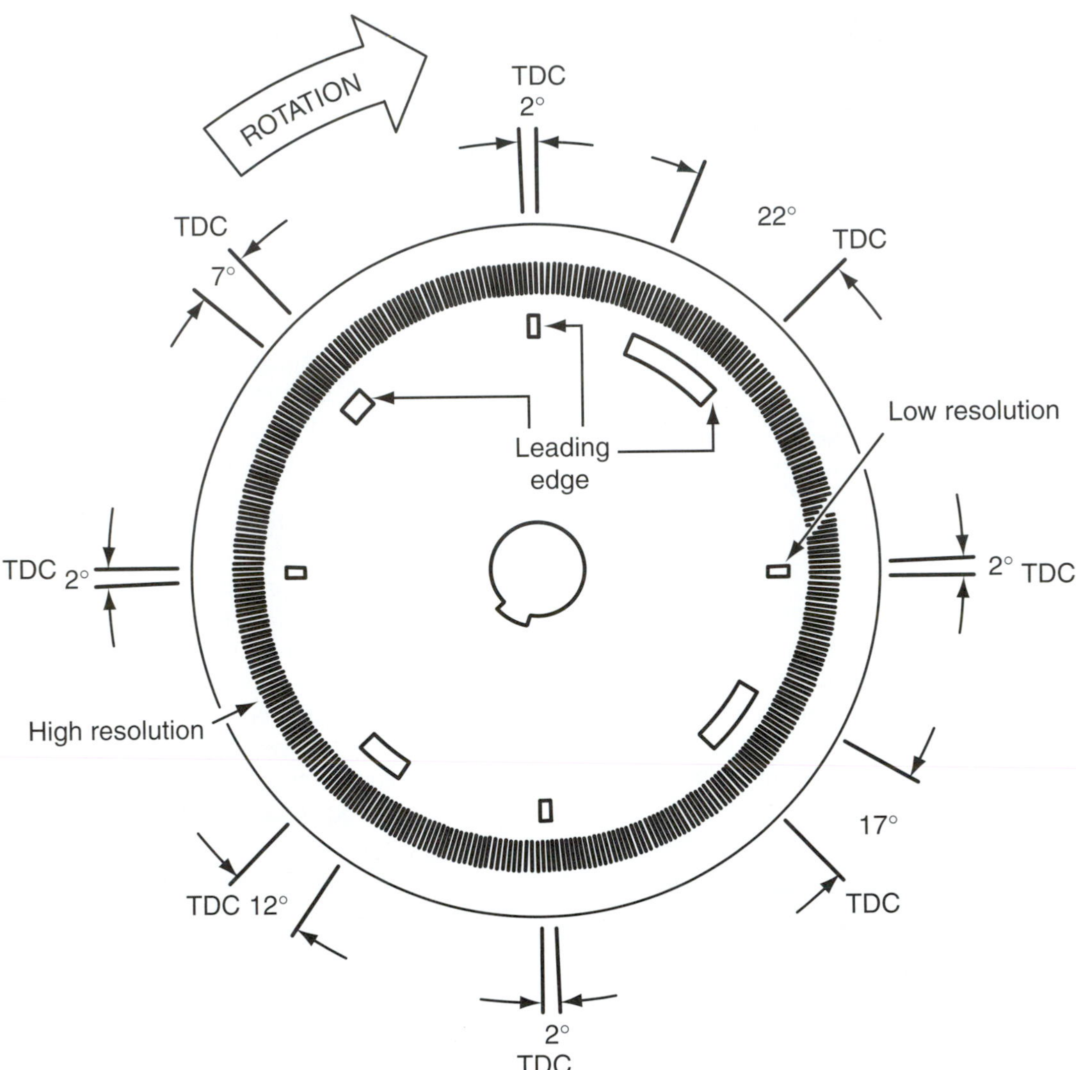

Figure 10–16 Opti-spark sensor disk.

from the high-resolution signal, Corvette and Nissan systems identify the number one cylinder from the low-resolution signal. On Corvette, the low-resolution slots are different sizes in degrees of rotation, Figure 10–16. By comparing these slots to the high-resolution slots, the PCM identifies the location of each of the cylinders. Therefore, this design provides the information that the PCM needs to control the fuel injectors sequentially.

Oxygen Sensor

This oxygen sensor is the same single-wire sensor used on other General Motors vehicles. In some vehicles in later model years, multiwire sensors are used, both to provide independent ground circuits and to provide electric resistance heat to the sensor to put the system into closed loop more quickly.

Vehicle Speed Sensor (VSS)

The two types of VSS discussed in Chapter 8 are also used on this system.

EGR Diagnostic Switch

On PFI systems that control EGR operation, an EGR diagnostic switch is used to tell the PCM whether the EGR is actually being applied when it is commanded to apply. Two types are used. The four- and six-cylinder engines use the same vacuum-operated switch discussed in Chapter 8.

The 5.0- and 5.7-liter engines use a thermal switch for exactly the same purpose as the vacuum switch used on the smaller engines. It, however, screws into the base of the EGR valve, Figure 10–17. If the EGR valve is open, the heat of the exhaust gasses flowing past the heat-sensing tip of the thermal switch, causes the switch to close. The PCM wants to see 12 volts on circuit 935 when the EGR is supposed to be off and less than 1 volt when the EGR is supposed to be on.

Detonation Sensor

Most of General Motors' PFI systems incorporate an electronic spark control (ESC) system using a piezoelectric knock sensor, as described in Chapter 8.

Park/Neutral (P/N) Switch

The park/neutral switch, which can be located near the shifter assembly or in the transmission itself, is normally open while driving. When the vehicle is shifted into park or neutral, it closes. The PCM monitors the P/N switch and thus knows whether the transmission is in gear. This information is used to help control engine idle and EGR control.

Air-Conditioning (A/C) Switch

On PFI cars equipped with air conditioning, the A/C control switch on the instrument panel connects through a wire to the PCM. When the A/C is turned on, the voltage signal travels from the A/C switch to the **pressure cycling switch** to the PCM. Some systems also put the **high-pressure cutout switch** in series in the same circuit. This signal tells the PCM that the A/C switch is on and the pressure cycling switch is closed. The PCM uses this information to turn on the A/C relay and/or adjust the IAC valve.

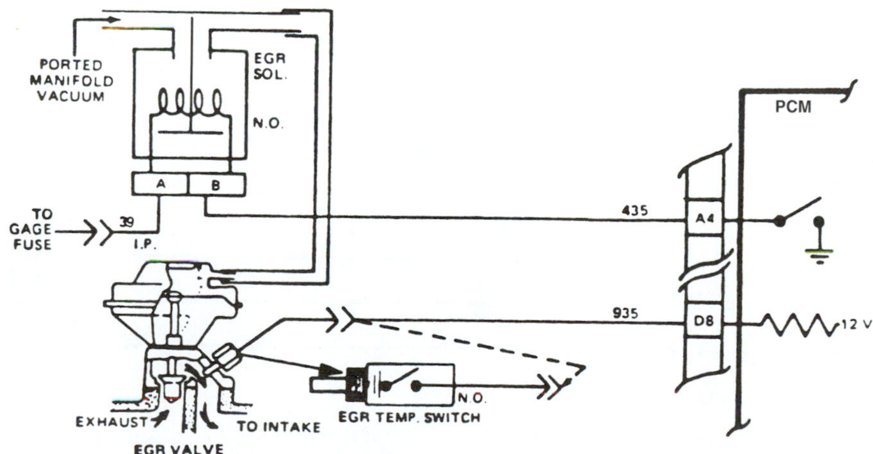

Figure 10–17 EGR valve with thermal diagnostic switch. *(Courtesy of General Motors Corporation, Service Technology Group.)*

Transmission Switches

Transmissions used in PFI-equipped vehicles have one or more hydraulically operated electric switches screwed into the valve body. These switches provide the PCM with signals indicating what gear the transmission is in. This information enables it to more effectively control torque converter lockup clutch operation.

Cranking Signal

On some engines, a wire from the starter solenoid circuit feeds a cranking signal to the PCM to alert it that the engine is being cranked. This puts the system in cranking mode and leads to an enriched air/fuel mixture for easier starting. On other engines, the cranking signal is fed to the cold-start injector in the intake manifold and provides the necessary enrichment through that mechanism. The cold-start injector, on vehicles where it is found, is not controlled by the PCM.

System Voltage/Fuel Pump Voltage

The PCM monitors system voltage through one of its battery voltage inputs. If voltage drops below a preprogrammed value, the system goes into battery voltage correction mode. Some PFI systems have a wire from the positive side of the fuel pump power circuit to the PCM, Figure 10–18. This signal is used to make fuel delivery compensations based on system voltage and causes a code 54 to be stored if pump voltage is lost while the engine is running.

Power Steering (P/S) Switch

A power steering switch is used on PFI systems to alert the PCM when power steering pressure is high enough to significantly affect engine idle performance (because of the momentary high load), Figure 10–19. The P/S switch is normally open and closes in response to high P/S pressure. When the switch is open, feedback voltage to the PCM is about 12 volts; when the switch closes, voltage falls below 1 volt. Also, when the switch

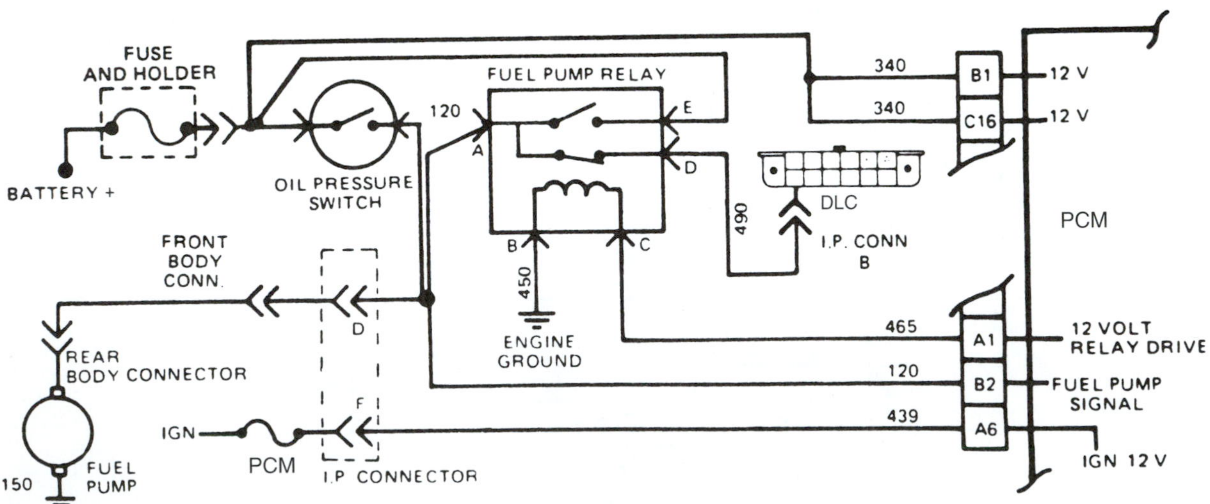

Figure 10–18 Fuel pump control circuit (typical). *(Courtesy of General Motors Corporation, Service Technology Group.)*

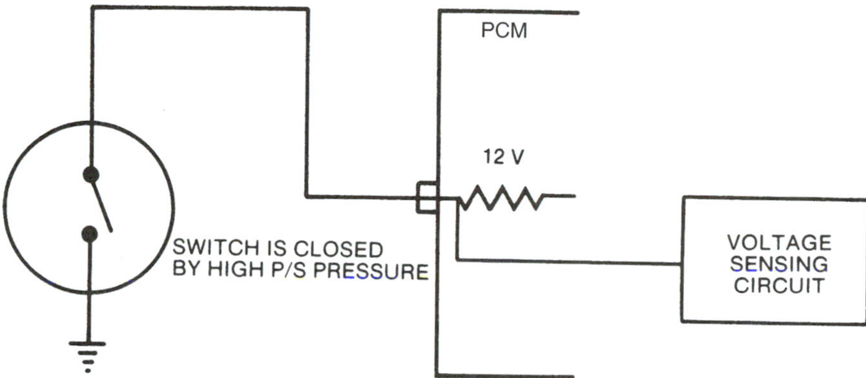

Figure 10–19 Power steering switch (typical).

closes, the PCM increases idle air bypass and retards ignition timing. On some systems, the A/C compressor clutch is also disengaged (also to reduce idle load).

Body Computer Module (BCM) 800 Circuit

Some late-model PFI applications also feature a body computer module, not unlike the Cadillac DFI system. The BCM controls air conditioning and other systems such as a driver information center. The driver information center displays information about fuel economy, time elapsed, distance to destination, time, date and so forth. The BCM shares information with the PCM. For instance, the PCM may send the BCM information about vehicle speed, coolant temperature or indicate when the vehicle is operating at WOT. The BCM can send the PCM such information as outside air temperature, the temperature in the high-pressure side of the A/C system, or indicate when the rear window defog system is on.

This information is transmitted over a wire between the two computers. The wire is called a *data link*. If information can be transmitted both ways over the same wire, it is called a *bidirectional data link*, or it may be called a *universal asynchronous receiver/transmitter*. General Motors refers to this data link on later model applications as the BCM 800 Circuit. The BCM 800 Circuit also extends to modules that the BCM or PCM communicates with, Figure 10–20. See the BCM sections of Chapter 12 for more information about body computers.

Cruise Control

When cruise control is engaged, the PCM is notified by the cruise control module. The PCM uses this information to modify its control of the torque converter lockup clutch.

OUTPUTS

Injectors

The fuel injector is a solenoid-operated nozzle controlled by the PCM, Figure 10–21. The PCM pulses it by grounding the return side (low voltage side) of the injector circuit, Figure 10–22. Fuel under nearly constant controlled pressure is sprayed past the open needle and seat assembly into the intake port of each cylinder. On all except SFI applications, all injectors are simultaneously

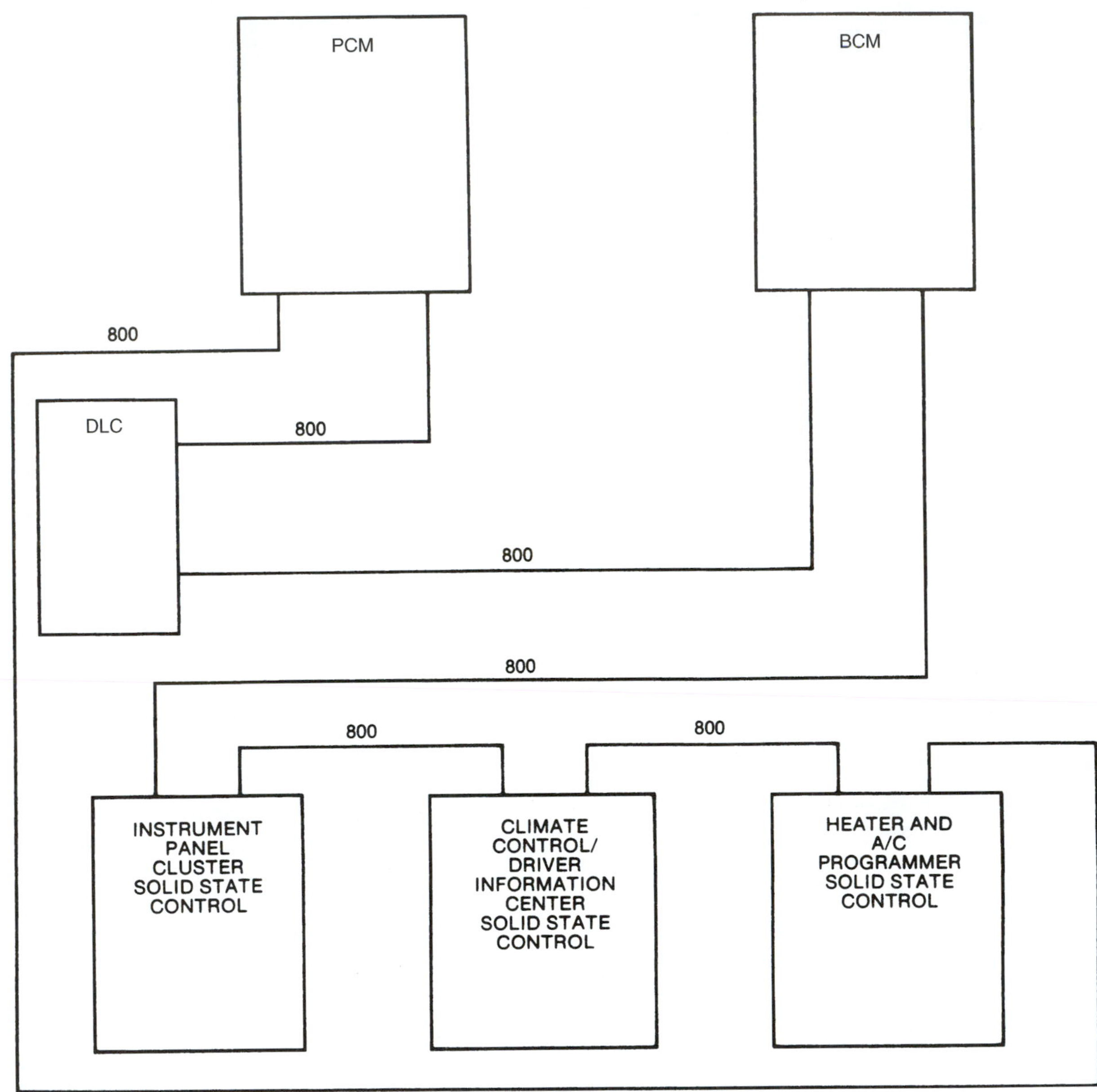

Figure 10–20 BCM 800 circuit.

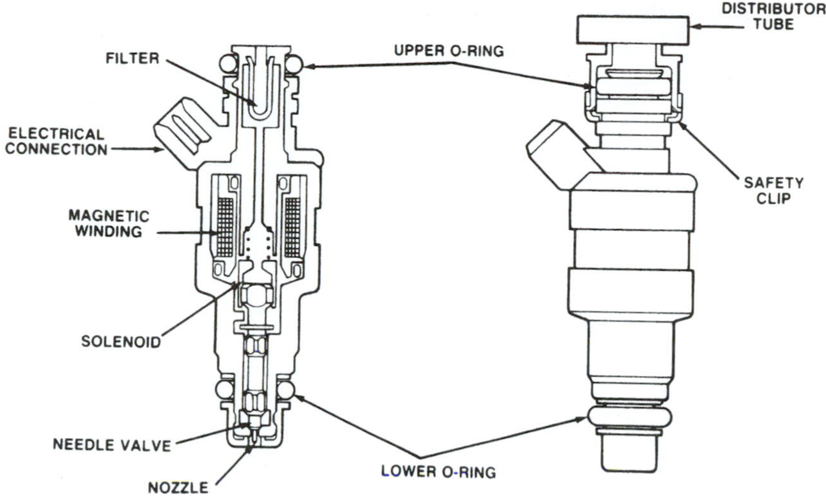

Figure 10–21 Injector (cutaway schematic). *(Courtesy of General Motors Corporation, Service Technology Group.)*

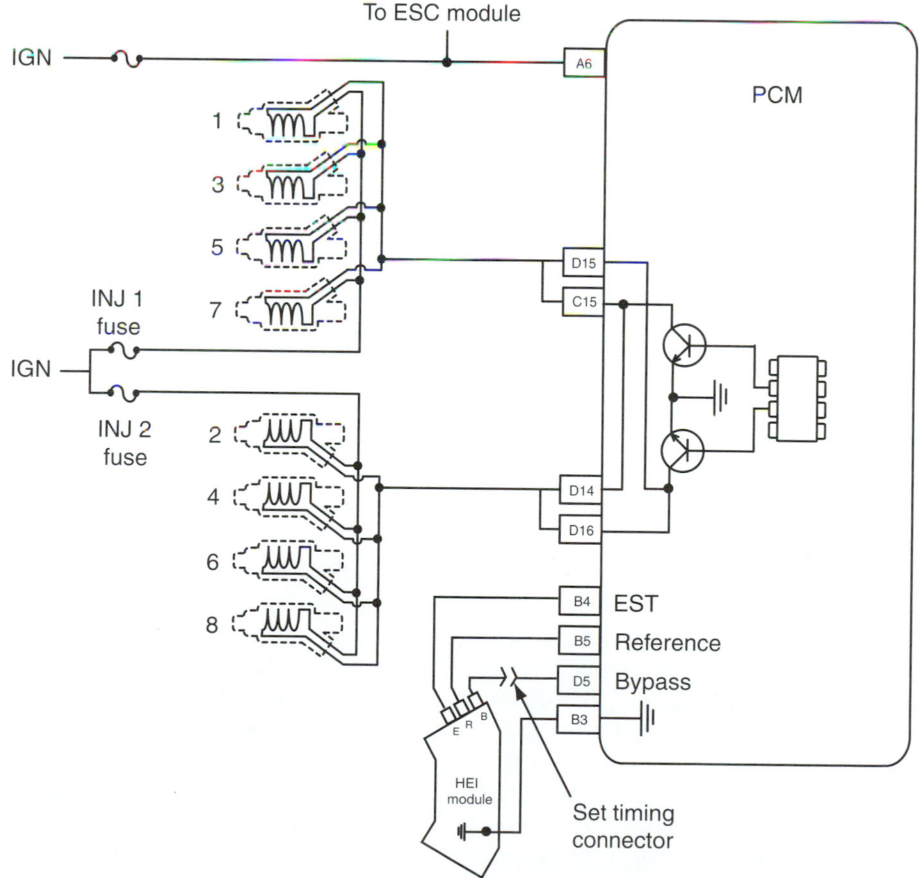

Figure 10–22 Injector control circuit.

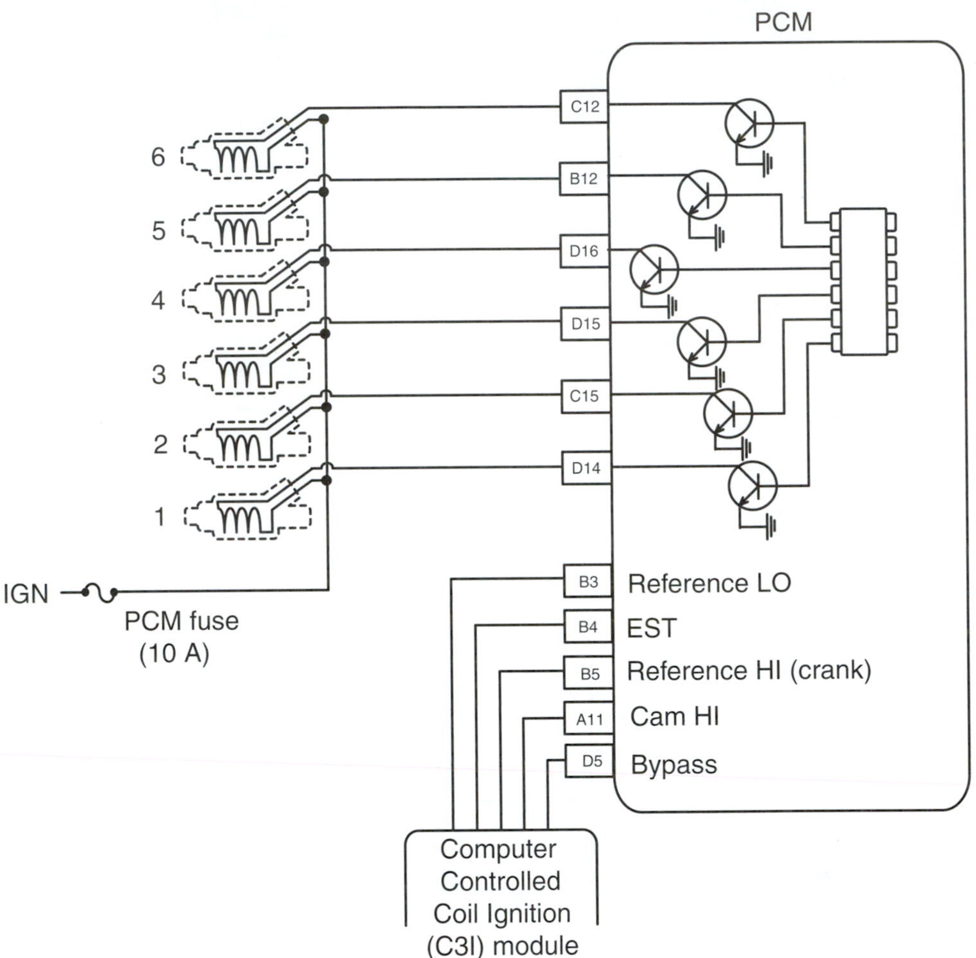

Figure 10–23 Electrical schematic of a typical SFI circuit.

pulsed once per crankshaft revolution. The design of the intake manifold is such that any fuel sprayed while the intake valve is closed simply sits and vaporizes until the intake valve opens. The SFI system pulses each injector individually, once per engine cycle in the firing order, Figure 10–23. During each pulse the injectors are held open for about one or two milliseconds at idle (depending on coolant temperature), to perhaps 6 or 7 milliseconds at WOT (1 millisecond equals 1/1,000 of a second).

Multec Injector. Rochester Products, a division of General Motors, has developed a new style injector for multipoint injection systems, Figure 10–24. It is referred to as a *Multec*, which is short for Multiple Technology Injector. This injector features a lower opening voltage requirement (important during cold weather cranking), fast response time, better fuel atomization, improved spray pattern control and less susceptibility to fouling due to fuel blends or contamination. The traditional pintle, as shown in Figure 10–21, is replaced

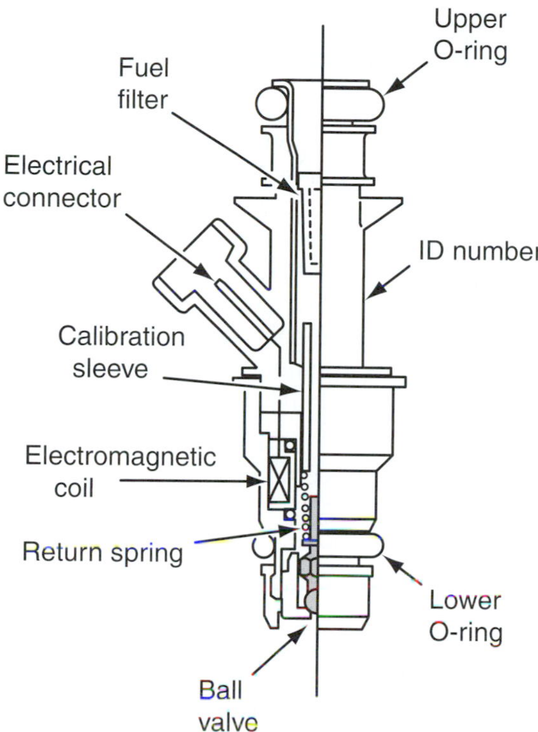

Figure 10–24 Multec injector schematic.

with a ball shape on the solenoid core. When the ball is lifted off its seat, fuel sprays through the seat opening and past the director plate into the intake port.

Cold-Start Injector. The cold-start injector is controlled by a thermal-time switch screwed into the engine water jacket, Figure 10–25. When the starter solenoid is engaged, voltage is applied to circuit 806 by way of the crank fuse. Circuit 806 powers a heat element in the **thermo-time switch**. Circuit 806 also supplies voltage to the cold-start injector. Ground for the cold-start injector is provided by a bimetallic switch in the thermo-time switch. If the bimetallic switch is cool and closed when voltage is applied to the injector, the injector is activated. As soon as voltage is applied to the injector, however, it is also applied to the heat element in the thermo-time switch. If the coolant is at 20°C (68°F) or below, the bimetallic switch stays closed for a maximum of 8 seconds before the heat from the heat element opens the control switch. So the cold-start injector turns on for a maximum of 8 seconds or less, depending on coolant temperature, while and only while the engine is cranking. If the engine is started hot, of

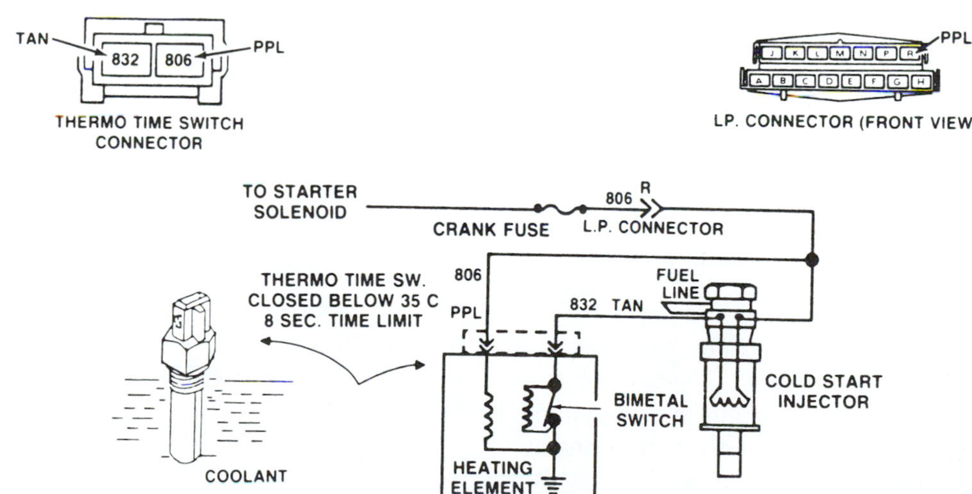

Figure 10–25 Cold-start injector control circuit (typical). *(Courtesy of General Motors Corporation, Service Technology Group.)*

Diagnostic & Service Tip

Although fuel injectors have no difficulty spraying fuel when they are clean and open, experience has shown that certain brands and blends of fuel are not entirely compatible with the conditions under which the injectors operate. In particular, the lower volatile compounds in the fuel tend to remain on the tip of the injector and in the pintle and seat opening as the engine is shut off. The heat of the engine then hardens the residue into deposits that can either limit the spray quantity or even degrade the spray pattern so as to prevent proper vaporization. Since this problem is usually not identical at each cylinder of the engine, the quantity sprayed and the quality of the spray patterns gradually begin to vary across the engine's injectors. The system, of course, does not include measures to identify which cylinder is spraying rich or which is spraying lean. So the computer has to work with the average mixture feedback signal it gets from the oxygen sensor. Once the fuel injectors' delivery volume varies by more than about 3 percent, however, there is no pulse-width setting that will deliver the correct mixture. In these cases, only fuel injector cleaning with special chemicals or fuel injector replacement will clear up the delivery problem.

course, the cold-start injector does not come on at all.

Electric Fuel Pump

Other than providing a higher fuel pressure, the fuel pump for the PFI system operates and is controlled just like the pump for the EFI system. Greater fuel pressure allows finer atomization of the fuel and resists vapor lock problems more effectively.

Electronic Spark Timing (EST)/Distributorless Ignition Systems

EST maintains optimum spark timing under all conditions of engine load, speed, temperature, and air density or mass. The distributor type ignition systems used on PFI engines operate the same as those discussed previously, except none of them use Hall effect switches.

Given the developments we have seen in computerized engine controls, there is no need to continue using a distributor with its tendencies toward wear and mechanical failure because:

- reference signals can be taken directly from the crankshaft and camshaft.
- an ignition module can be mounted almost anywhere in the vehicle.
- spark advance can be more effectively optimized by a fast microprocessor.
- secondary spark distribution can be managed by separate coils and a module.

This should not be very surprising. After all, even the old points and condenser distributors were essentially nothing more than camshaft position sensors, combined with a simple mechanical/vacuum spark advance calculator and an on/off switch.

The term *distributorless ignition system* has come to be used as a generic label for an ignition system that does not use a distributor and may be abbreviated in some literature as DIS. However, General Motors most often uses the same abbreviation for Direct Ignition System, one of their specific distributorless ignition systems. General Motors has introduced several different distributorless ignition systems on various engine applications:

- the C3I (Computer-Controlled Coil Ignition, of which there are three subtypes)
- the DIS (Direct Ignition System)
- the IDI (Integrated Direct Ignition)

Each system has its own unique module receiving information about engine speed and crankshaft/camshaft positions from the system sensors. The module monitors this information and passes it on to the PCM. A coil pack, with one ignition coil for every pair of firing-order-opposite cylinders, is mounted on and controlled by the module, Figure 10–26. The module on these systems performs the same function as an HEI module in a distributor type ignition system, with the same REF, bypass, EST and ground circuits connecting the module to the PCM, Figure 10–27. In this case the module must select and fire each coil in the correct sequence. The module and coil pack may be mounted anywhere on the engine. The end of each coil's secondary winding is connected to a spark plug by way of a spark plug cable. Each time a coil's primary circuit opens, that coil fires both of its spark plugs simultaneously.

The spark plugs the coil fires are in companion cylinders; those cylinders arrive at TDC at the same time but opposite each other in firing order. For instance, if number 1 cylinder is being fired at the end of its compression stroke (assume the firing order is 1-6-5-4-3-2), number 4 will be at the top of its exhaust stroke and will be fired also. It only takes about 2 to 3 kilovolts (kV) to fire a spark plug during the exhaust stroke, about the same additional voltage required to jump the rotor gap in a distributor.

The C3I system was introduced in 1984 on the 3.8-liter turbocharged Buick engine. The C3I II system soon followed, and C3I III was introduced in 1988 on the 3800 (simply another name for the same displacement, but a new V6 engine, introduced that year). The ignition module is connected to the PCM by a 14-pin connector, Figure 10–28. Ignition switch voltage at terminal N powers the module; ignition switch voltage at terminal P powers the ignition coils. If the ignition is turned on without a cranking signal appearing within 1 to 2 seconds, the module will shut off primary circuit current to prevent coil overheating. The PCM has a spark timing range capacity of zero degrees to 70 degrees and will provide dwell times between 3 milliseconds at high rpm and 15 milliseconds at low rpm.

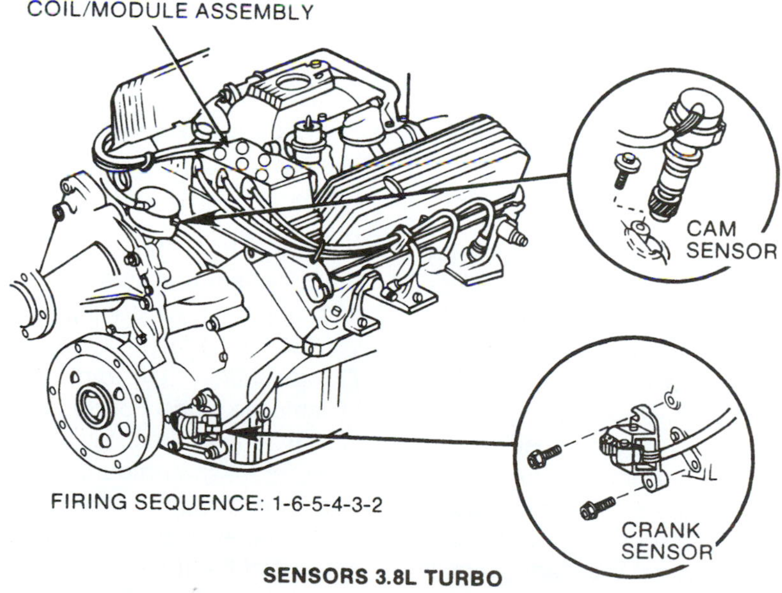

Figure 10–26 C3I system components. *(Courtesy of General Motors Corporation, Service Technology Group.)*

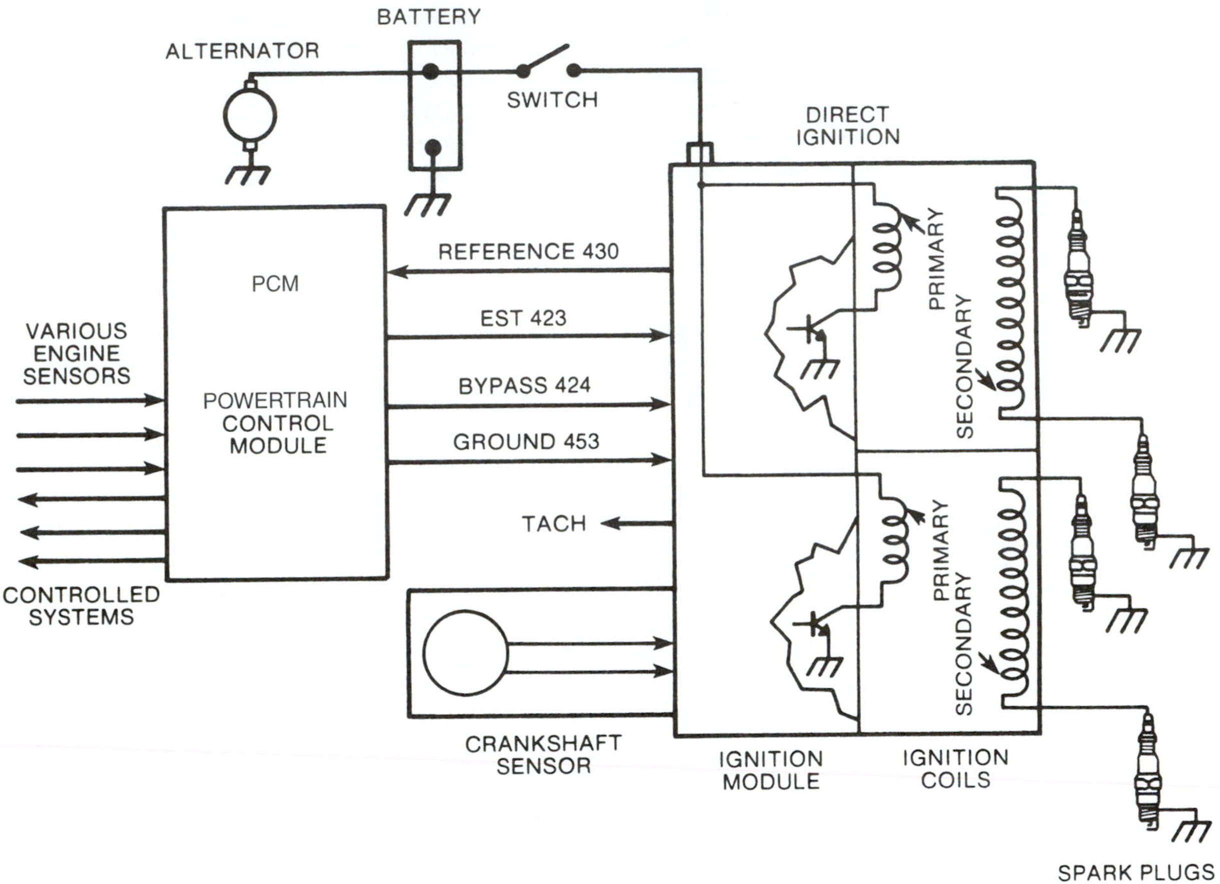

Figure 10–27 Four-cylinder DIS systems schematic. *(Courtesy of General Motors Corporation, Service Technology Group.)*

On Type I and Type II C3I systems and on DIS and IDI systems, the module has to see a signal from the cam or sync sensor during cranking before it knows what position the crank is in and which coil to fire. Thus no spark occurs until the module sees the sync signal. Once that signal has occurred, the module knows what position the crankshaft is in and fires a coil for the next appropriate cylinder event signal, which is the second cylinder in the firing order. Depending on what position the crankshaft is in when the starting crank begins, the crankshaft may spin more than a full revolution before a spark occurs. On these systems the first cylinder to fire is always the second cylinder in the firing order. The C3I Type III (Fast Start) system provides slightly faster starting due to its improved sensor input.

IDI. The latest system is called *Integrated Direct Ignition* (IDI) and is used on the 2.3-liter Oldsmobile Quad Four engine. The IDI system is similar to the Direct Ignition System. It uses the same crankshaft-mounted, seven notch reluctor and crankcase-mounted magnetic sensor, but uses a different module and harness connectors. No part of the IDI system is visible from the exterior of the engine. The coil, module, spark plugs and wires are mounted under the topmost part on the engine, Figure 10–29.

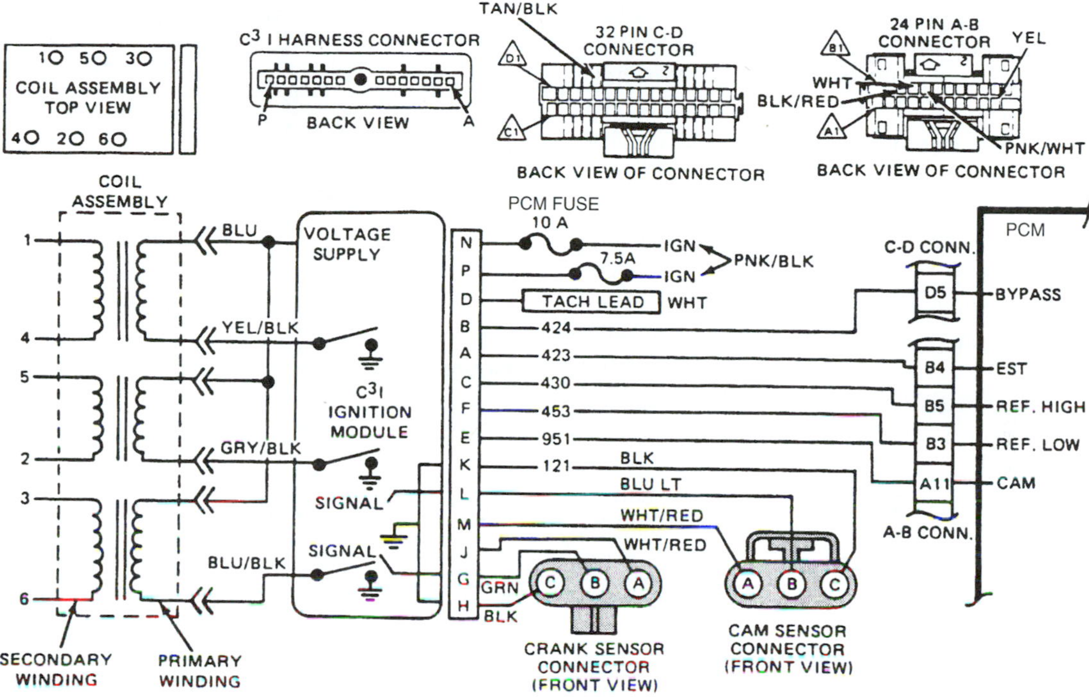

Figure 10–28 Wiring diagram for a C3I system 3.8-L turbocharged SFI V6 engine. *(Courtesy of Buick Motor Division, General Motors Corporation.)*

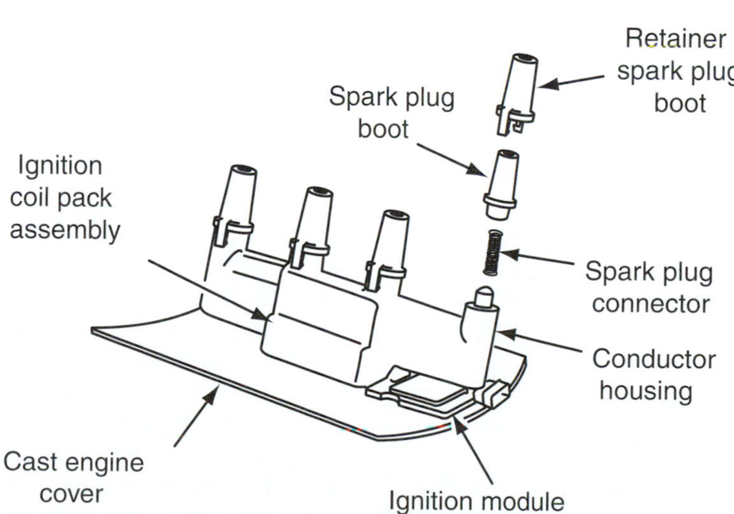

Figure 10–29 A Quad Four ignition system that does not use spark plug wires.

Note that the Quad Four engine and other four-cylinder engines using this system nonetheless use the seven notch plate on the crankshaft. Software hard-wired into the module allows the system to make allowances for the correct number of cylinders.

DIS. During module mode operation (see Chapter 9 Electronic Spark Timing) on four-cylinder applications, the signal produced by the sync notch tells the ignition module to skip the next signal (produced by the number 1 notch) and to fire the 2-3 coil on the signal produced by the number 2 notch, Figure 10–30. The module then skips the signals produced by notches 3 and 4 and fires the 1-4 coil in response to the signal produced by notch number 5. Notches 6 and 7 pass the sensor without the module responding to them and the process begins again. Notice the module is not concerned with cylinder firing order; it is concerned with coil firing order.

On the six-cylinder engine, the signal produced by notch number 7 references the ignition module to skip the signal from notch 1 and respond to the signal from notch number 2 to fire the 2-5 coil, Figure 10–31. The module then skips the signal from notch number 3 and fires the 3-6 coil on the signal from notch number 4. The signal from notch number 5 is skipped and the 1-4 coil is fired on the signal from notch number 6.

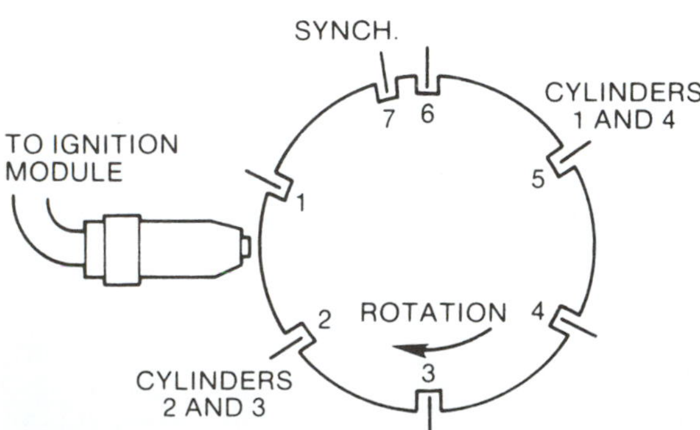

Figure 10–30 Four-cylinder coil firing sequence.

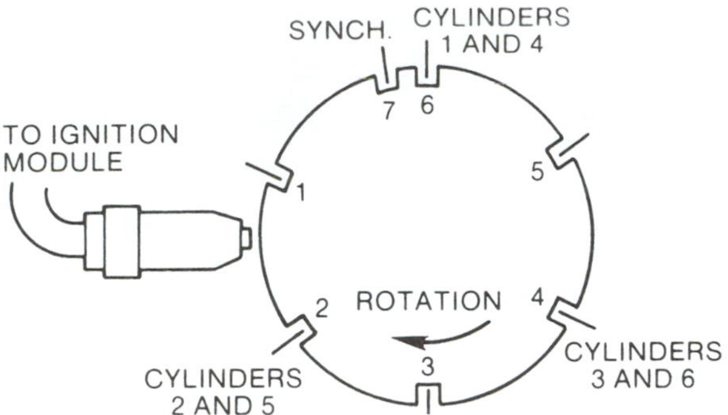

Figure 10–31 Six-cylinder coil firing sequence.

Electronic Spark Control (ESC)

All PFI systems except those on the 2.8-liter engines are equipped with ESC. Its function is to retard ignition timing if detonation occurs. Since modern engines are designed to run at the ragged edge of knock, this requires precise controls. The ESC has its own hybrid module that works in conjunction with the PCM, Figure 10–32. Its operation is essentially the same as the later style ESC system. The conditions under which ignition timing are retarded in response to spark knock are:

- engine speed is above 1,100 rpm.
- ignition voltage to the PCM is above 9 volts.
- no code 43 is present.

Idle Air Control (IAC)

The IAC assembly is a stepper motor and pintle valve just as on the EFI system. It is screwed into the throttle body and controls idle speed, Figure 10–33. With one exception, it works like the IAC used on the EFI system. The exception is that the PCM rereferences its count of the IAC's pintle valve position by moving the valve to its zero position (closed) each time the ignition is turned off, instead of when the PCM gets a 30 mph signal from the VSS as the EFI system does. The PCM then issues a preprogrammed number of pulses to open the valve to a ready position for the next

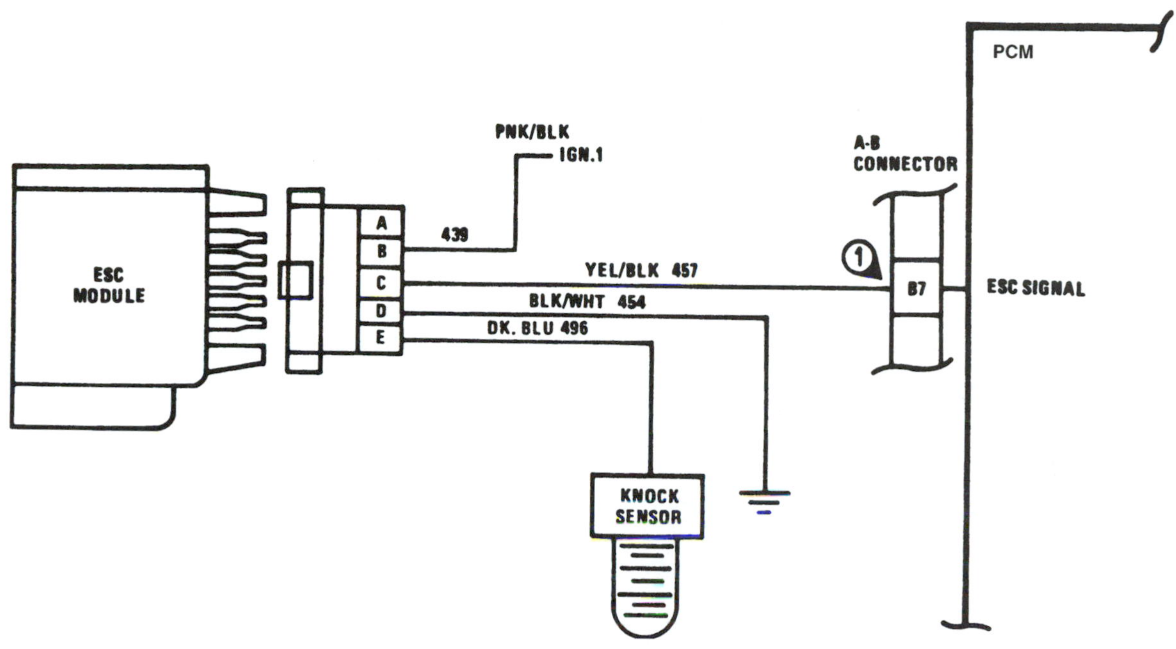

Figure 10–32 ESC circuit. *(Courtesy of General Motors Corporation, Service Technology Group.)*

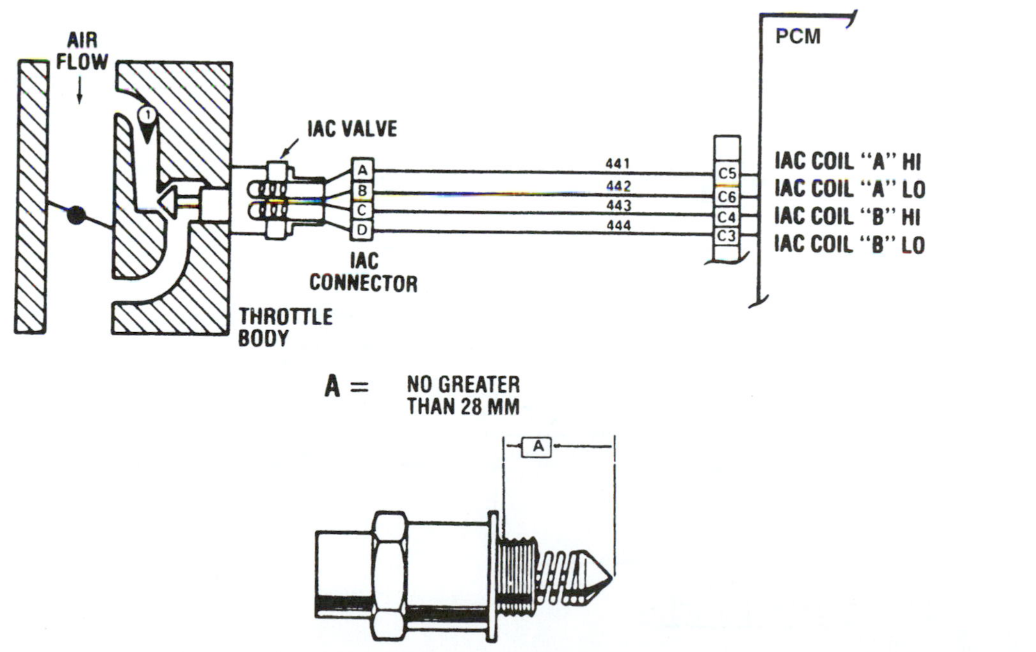

Figure 10–33 IAC motor control circuit. *(Courtesy of General Motors Corporation, Service Technology Group.)*

startup. The following inputs can affect idle speed performance:

- battery voltage
- coolant temperature
- throttle position
- vehicle speed
- mass airflow
- engine speed
- A/C clutch signal
- power steering pressure signal
- P/N switch signal

Air Injection Reaction (AIR) System

Only some of the PFI systems require an AIR system.

2.8-Liter Engines. The 1985 2.8-liter engine equipped with manual transmission uses an AIR system with a single-bed, three-way converter. During warmup the PCM energizes the divert valve solenoid, Figure 10–34, which then allows vacuum to be applied to the divert valve. The pump's air is directed to the exhaust ports. As the engine approaches normal temperature (and just before going into closed loop), the PCM turns off the solenoid, which in turn blocks vacuum from the divert valve. The divert valve then directs the air to the air cleaner (divert mode).

5.0- and 5.7-Liter Engines. These engines use dual-bed, three-way catalytic converters. The air pump pumps air to the system control valve; the control valve, which contains two valves in one housing, directs air to any one of three places, Figure 10–35:

- to the exhaust ports during engine warmup.
- to the oxidizing chamber of the catalytic converter during closed-loop operation.
- to atmosphere (divert mode—during WOT, deceleration, when any failure occurs that causes the PCM to turn on the check engine light or when rpm is high and causes the pressure relief valve inside the control valve to open, though pressure relief is not really a divert mode). Divert mode is intended to prevent converter overheating.

Exhaust Gas Recirculation (EGR)

The EGR valve lets small amounts of inert exhaust gas into the intake manifold to lower combustion temperature by slowing combustion, thereby reducing the production of oxides of nitro-

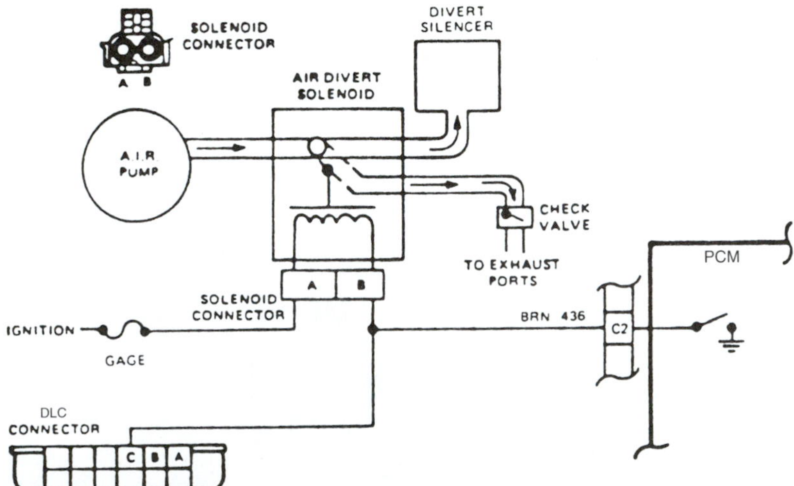

Figure 10–34 Divert valve control circuit (2.8-liter engine). *(Courtesy of General Motors Corporation, Service Technology Group.)*

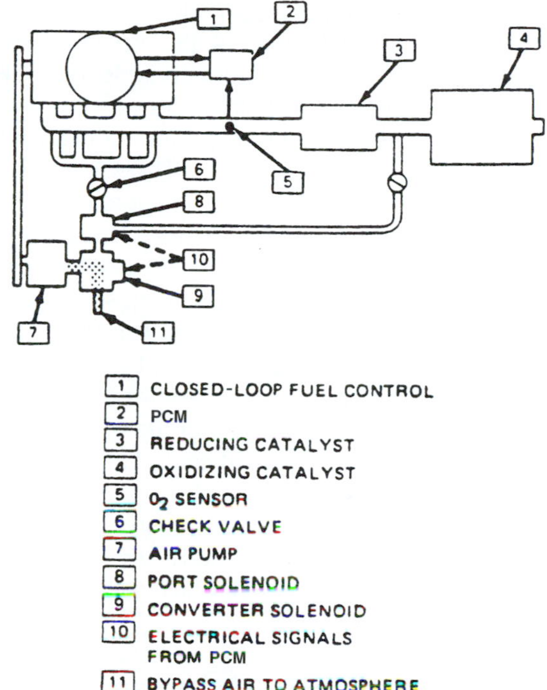

Figure 10–35 Air management system (5.0-liter and 5.7-liter engines). *(Courtesy of General Motors Corporation, Service Technology Group.)*

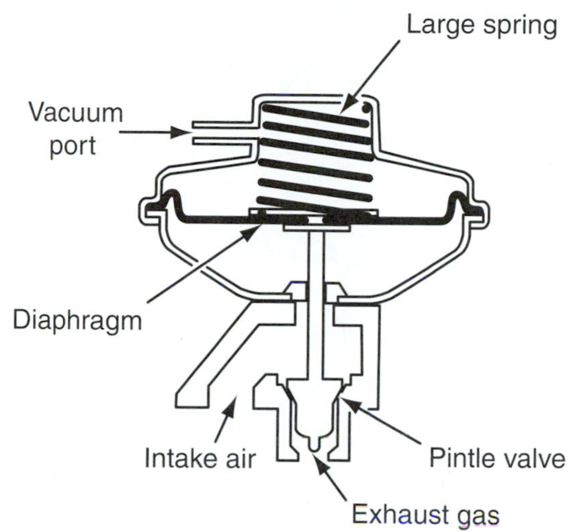

Figure 10–36 Typical design of an EGR valve.

gen (NO_x). The EGR valve is opened by ported vacuum (manifold vacuum on some applications), which pulls the diaphragm up against the diaphragm spring, Figure 10–36. Three types of vacuum-operated EGR valves are used: the standard ported valve as shown in Figure 10–36, a positive backpressure valve and a negative backpressure valve. The other two valves are very similar in design to the standard valve, except that they are sensitive to and modulated by exhaust backpressure or a combination of exhaust backpressure and manifold vacuum, respectively. They can be identified by:

- a "P" stamped on the top of the positive backpressure valve, Figure 10–37.
- an "N" stamped on the top of the negative backpressure valve.
- a blank space on top of the port valve.

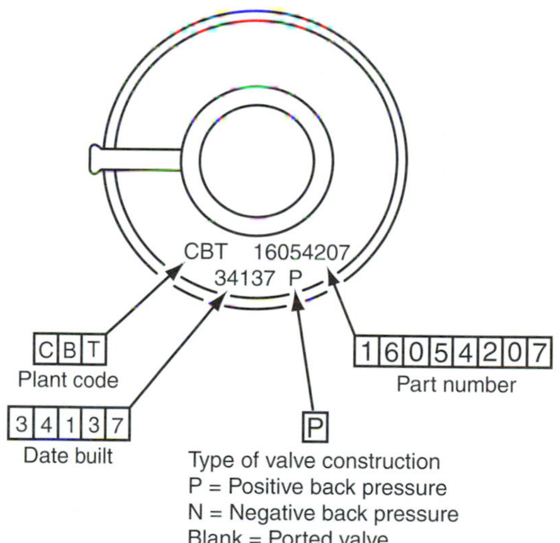

Figure 10–37 EGR valve identification.

Regardless of the valve used, vacuum to it is controlled by a vacuum control assembly, Figure 10–38. The control assembly solenoid is pulsed many times per second by the PCM. When the

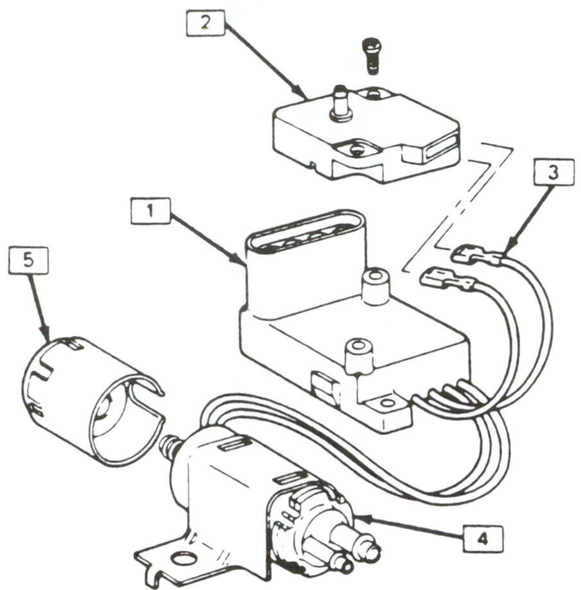

Figure 10–38 EGR vacuum control assembly. *(Courtesy of General Motors Corporation, Service Technology Group.)*

solenoid is turned on, vacuum to the EGR port is blocked. The strength of the vacuum signal, which controls how far the EGR valve opens, is controlled by varying the pulse width. The solenoid is fitted with a vent filter that must be replaced periodically. The PCM controls the solenoid using input concerning coolant temperature, throttle position and mass airflow. The EGR valve is not opened unless the engine is warm and above idle.

Integrated EGR Valve. In 1987, GM introduced an integrated EGR valve on the 2.8 L V6. This valve is vacuum-operated by a ported vacuum signal. It contains two additional components, a normally closed vent solenoid and a potentiometer. If the PCM leaves the solenoid de-energized, then ported vacuum controls the EGR valve. If the PCM wishes to interfere with this control, it energizes the solenoid, opening the vent. The potentiometer is used by the PCM to monitor the position of the EGR pintle, similar to the EGR Valve Position sensors that Ford introduced in 1980 on their EEC systems. The GM valve also includes a replaceable filter on the vent solenoid. If it is determined that a fault exists with either the vent solenoid or the potentiometer, the entire assembly must be replaced. The integrated EGR valve is easily recognized by a group of five wires connected to the valve (two to control the solenoid and three for the potentiometer).

Digital EGR Valve. A new concept in EGR valves is employed on the 3800 V6 engine for 1988 and beyond. The digital EGR valve provides the following advantages:

- vacuum is not required for its operation.
- increased control of EGR flow.
- faster response to PCM commands.
- greater PCM diagnostic capacity.

The digital EGR valve is directly operated by solenoids rather than by a vacuum diaphragm. Instead of trying to regulate EGR flow by controlling how far the valve opens or by duty cycling it, three different valves are used to open or close three different-sized orifices. Each valve can be opened or closed independently of the others, allowing any of seven different increments of exhaust gas flow into the induction system, Figure 10–39. Orifice number one, when open, flows 14 percent of the maximum EGR. Orifice number two, when open flows 29 percent of the maximum. Number three flows 57 percent of maximum when open. By manipulating which valve or combination of valves is open at any given time, EGR flow can be easily and accurately controlled.

The digital EGR assembly consists of the EGR base, the EGR base plate and the solenoid and mounting plate assembly, Figure 10–40.

Increment	Orifice #1 (14%)	Orifice #2 (29%)	Orifice #3 (57%)	EGR Flow (%)
0	closed	closed	closed	0
1	open	closed	closed	14
2	closed	open	closed	29
3	open	open	closed	43
4	closed	closed	open	57
5	open	closed	open	71
6	closed	open	open	86
7	open	open	open	100

Figure 10–39 Increments of EGR flow with digital EGR valve system.

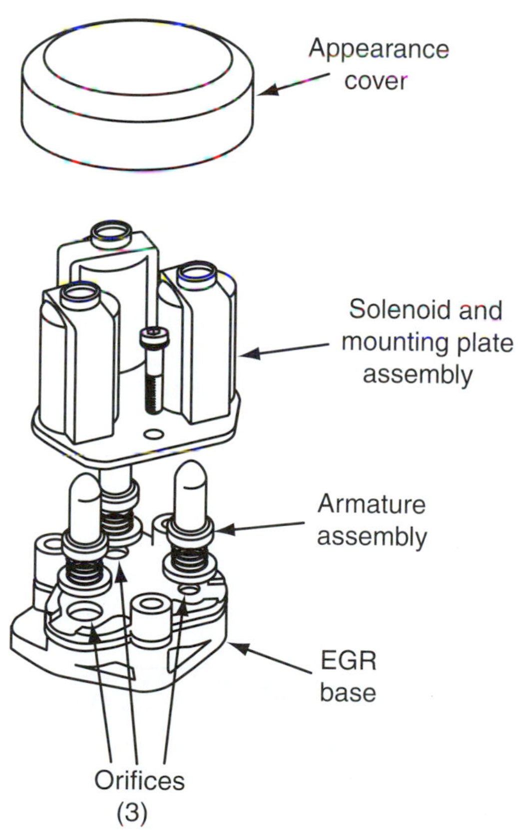

Figure 10–40 A digital EGR valve with three solenoids.

When the base and base plate are fitted together, there is a sealed cavity between them with the base forming the floor of the cavity and the base plate forming the ceiling of the cavity. There are four holes in the base (floor). The center hole is always open and allows exhaust gas to enter the cavity. The other three holes are closed by the three EGR valve pintles, Figure 10–41. The pintles are attached to their pintle shafts by a ball-joint type connection so they can more readily align with the seats they rest on and maintain a more positive seal.

As shown in Figure 10–41, the pintle and its shaft are part of the armature assembly. Each pintle-shaft-armature assembly is held down by its armature return spring, which is attached to the lower portion of the shaft. When one of the solenoid windings in the solenoid assembly is turned on by the PCM, the resulting magnetic field lifts the corresponding armature assembly. This opens the valve and compresses the return spring. When the PCM turns the solenoid off, the spring drives the armature assembly back down and closes the valve.

There are two seals on the pintle shaft. The lower seal is pushed down against the top of the base plate. It prevents exhaust gasses in the base cavity from leaking out between the stem and the base plate. On this EGR valve design,

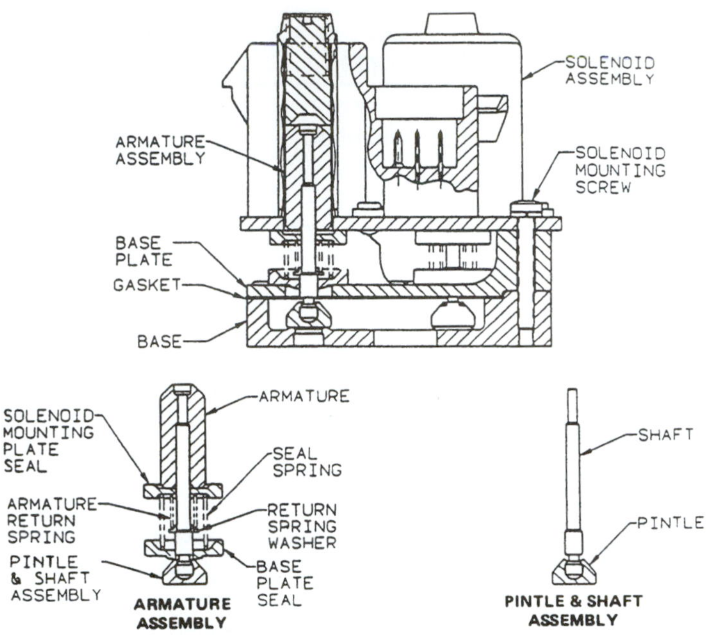

Figure 10–41 Digital EGR assembly. *(Courtesy of General Motors Corporation, Service Technology Group.)*

ambient air has much less tendency to leak into the induction system by way of the EGR valve stem seal than it does on the traditional EGR valve design, because exhaust pressure in the base cavity prevents a low pressure from developing under the seal.

The upper seal is pressed against the bottom of the solenoid assembly. It keeps dirt and dust from entering the armature cavity of the solenoid assembly to reduce wear and provide greater reliability. A spring fitted between the seals holds them in place. The return spring also helps hold the upper seal in place.

Some engines use a two-solenoid digital EGR valve, which works the same way except that it only has three different EGR flow rates. When the first valve is open, about 33 percent of the total EGR will flow. When the second valve is open, about 66 percent of the total will flow. When both valves are open, 100 percent of EGR flows.

By not relying on vacuum to operate the EGR valve assembly, the PCM can be made to monitor EGR system operation for fault diagnosis without any additional external hardware. If a short or open occurs in any one of the solenoid circuits, a code will be set in the PCM's keep-alive memory. Each solenoid has its own assigned code number.

Linear EGR Valve. In 1994, GM introduced the linear EGR valve, Figure 10–42. This valve, like the digital EGR valve, controls exhaust gas recirculation directly without the use of vacuum as

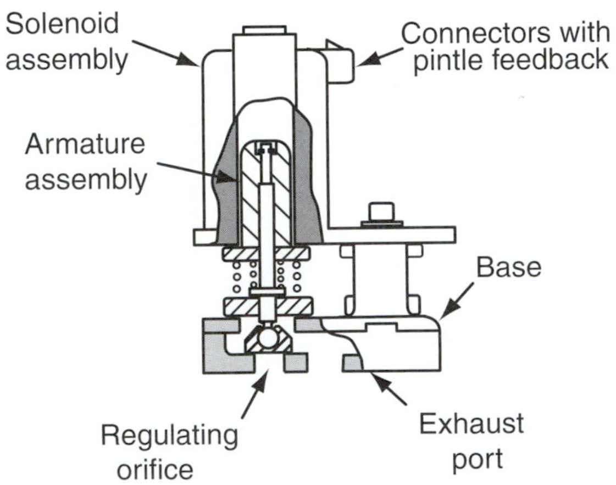

Figure 10–42 Linear EGR valve design.

a control medium. The linear EGR valve uses a single, normally closed, solenoid-operated pintle valve operated by the PCM using a pulse width modulated signal in order to control the volume of exhaust gasses allowed to recirculate into the intake manifold. The movement of the pintle is monitored by a potentiometer, which produces a feedback signal for the PCM, ranging between about 1 volt when the valve is closed and about 4.5 volts when the valve is open.

Charcoal Canister Purge

The PCM operates a solenoid that controls vacuum to the purge valve on the charcoal canister, Figure 10–43. The solenoid is turned on and blocks vacuum to the purge valve when the engine is cold or at idle, Figure 10–44. The PCM turns the solenoid off and thus allows any stored hydrocarbons (fuel) to be purged through the purge valve when:

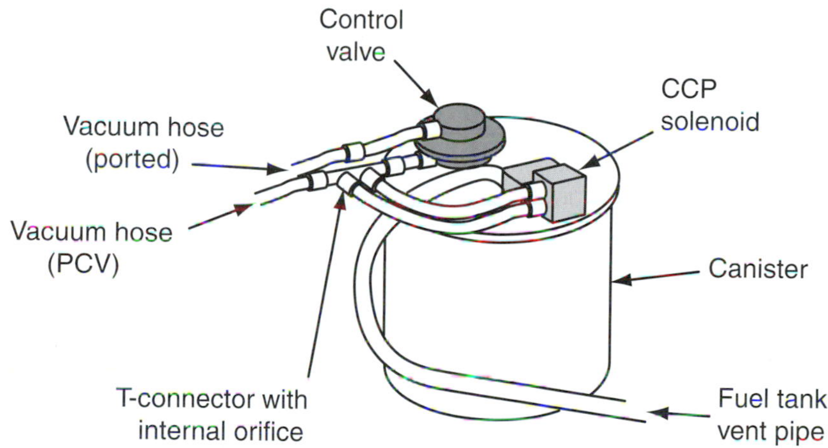

Figure 10–43 Charcoal canister.

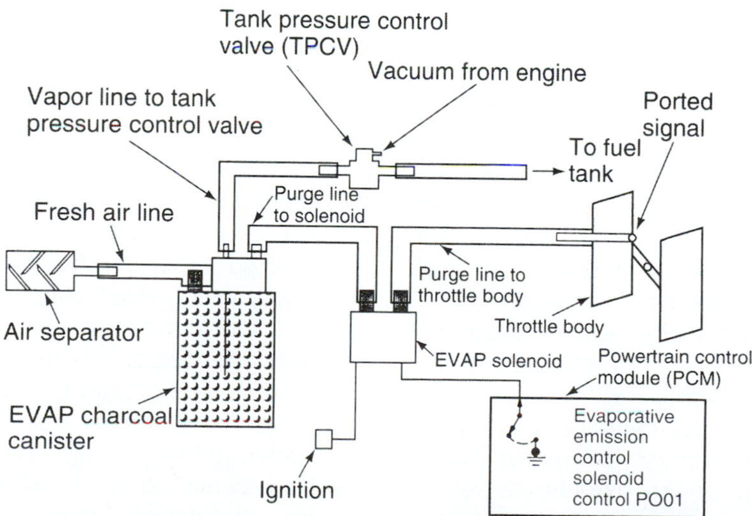

Figure 10–44 EVAP system components. *(Courtesy of Cadillac Motor Car Division, General Motors Corporation.)*

- the engine is warm.
- the engine has run for a preprogrammed amount of time since it was started.
- a preprogrammed road speed is reached.
- a preprogrammed throttle opening is achieved.

In addition to the purge valve, some engine application canisters also have a non-PCM-controlled control valve. It is controlled by ported vacuum. When vacuum is applied to it, it allows canister purging to occur through another port connected to manifold vacuum.

For 1987, the Chevrolet 5.0- and 5.7-liter engines have a revised canister purge strategy. When the engine is operating in the closed-loop mode, the PCM will duty cycle the purge solenoid to control the amount of vapors admitted into the engine's induction system, rather than just turning the purge solenoid on to stop purging and off to allow purging. The PCM uses input from the oxygen sensor to determine the volume of vapors to purge from the canister. If a rich condition is indicated by the oxygen sensor, purge volume is reduced. This strategy is intended to provide improved driveability.

Turbocharger

One of the major factors that controls power is how much air and fuel are put into the cylinders. Putting more fuel in is a fairly simple matter. Putting more air in is not so simple (the proper air/fuel ratio must be maintained). The amount of air that can be put into the cylinder is limited by atmospheric pressure unless some means is used to force air in at greater than atmospheric pressure. This is accomplished with some type of supercharger of which the turbocharger is the most popular and generally the most efficient.

A turbocharger is a centrifugal, variable-displacement air pump driven by otherwise wasted heat energy in the exhaust stream. It pumps air into the intake manifold at pressures that are limited only by the pressure that can be developed without significantly increasing the air's temperature.

Most **normally aspirated** engines (those that rely on atmospheric pressure to fill the cylinder) only fill the cylinder to about 85 percent of atmospheric pressure (they actually achieve 85 percent volumetric efficiency, or VE, at full throttle with the engine at its maximum torque speed). Turbocharging produces VE values in excess of 100 percent.

Turbochargers

Because a turbocharger puts more air into the cylinder, it raises the engine's compression pressure. Because it does not produce significant boost pressure at low exhaust flow rates, it allows an engine the benefits of low compression during light throttle operation and high compression during heavy throttle operation. Low compression offers lower combustion chamber temperatures, which result in less wear and lower NO_x emissions. High compression provides higher combustion chamber temperatures, which result in more complete combustion and more power for the amount of fuel consumed.

Another way to look at it is in terms of displacement. If an engine is operating at an atmospheric pressure of 14 psi, and the turbocharger is producing a boost of another 14 psi, the cylinders are charged at a pressure of 28 psi. The engine is consuming approximately twice the amount of air it could without the turbocharger. We have effectively doubled the engine's displacement. The engine certainly does consume twice as much fuel, too, but not as much as it would with twice the displacement, because the higher compression pressure and temperature produces more complete combustion. The turbocharged engine also weighs less than an equivalently powerful naturally-aspirated engine, thus saving on the amount of work done.

Turbocharger Operation. After leaving the manifold, the hot exhaust gasses flow through the vanes of the exhaust turbine wheel and thus spin it, Figure 10–45. The more exhaust volume coming out of the engine, the faster the turbine spins. It can achieve speeds in excess of 130,000 rpm. The exhaust turbine drives a short shaft that drives a compressor turbine. The high speed produces centrifugal force to move the air from between the vanes and thus causes it to flow out radially. The air is then channeled into the intake manifold. As air is thrown from the turbine vanes, a low pressure develops in its place. Atmospheric pressure pushes more air through the air cleaner, the mass airflow sensor (if used) and the throttle body.

A criterion turbine speed must be reached before a pressure boost is realized; this speed varies with turbine size and turbocharger design. If turbine speed goes too high, boost pressure (and charge temperature) will also go too high and can cause preignition and engine damage. To limit boost pressure, a **wastegate** is used to divert exhaust gasses away from the exhaust turbine and thus limit its speed.

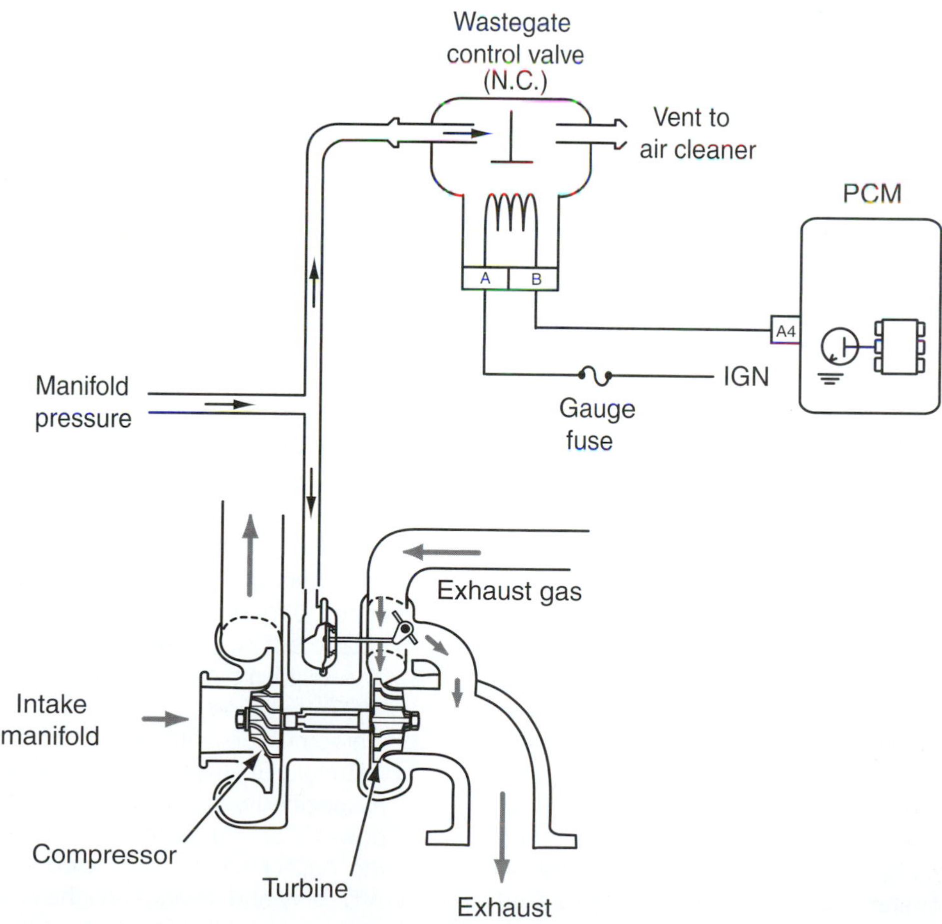

Figure 10–45 Turbocharger and wastegate control (typical).

Wastegate. The wastegate opens by a wastegate actuator. A spring in the actuator holds the wastegate closed. When manifold pressure (turbo boost) reaches approximately 8 psi, it overcomes the actuator spring and opens the wastegate. If however, engine operating parameters are favorable (coolant temperature, incoming air temperature, for example), the PCM pulses a wastegate solenoid, which in turn bleeds off some of the pressure acting on the actuator. When this occurs, boost pressure is allowed to rise to 10 psi before the wastegate opens.

A code 31 is set if:

- an overboost is sensed by the MAP sensor on the 1.8-liter engine.
- the PCM, monitoring the wastegate solenoid circuit operation, sees a malfunction in the circuit while the solenoid is operated between a 5 percent and a 95 percent duty cycle on the 3.8-liter engine.

Transmission Converter Clutch (TCC)

Notice that the name has changed slightly from this unit's application on CCC and EFI systems, where TCC meant "torque converter clutch." It is still the same part, however. The purpose of the TCC is still to increase fuel economy. It does so by eliminating hydraulic slippage during cruise conditions and by eliminating heat production in the torque converter, especially during overdrive operation.

Electric Cooling Fan

All General Motors vehicles with transverse mounted engines and a few with **longitudinal** engines (parallel to the center line of the car) are equipped with an electric cooling fan to pull air through the radiator and A/C condenser. Control of the fan varies somewhat with engine application. In all cases, however, the fan is turned on when:

- coolant temperature exceeds a specified value. This can be done by either the PCM or a coolant temperature override switch on some applications and only by the PCM on others.
- when A/C compressor output pressure (head pressure) exceeds a specified value. This is done by a switch in the high pressure side of the A/C system on some engine applications. It is done by the PCM on other applications (on those applications, an A/C head pressure switch feeds head pressure information to the PCM).

On most applications, the PCM turns on the fan anytime the A/C is on and vehicle speed is less than a specified value. Others turn it on under a specified speed whether the A/C is on or off. Once the vehicle reaches a criterion speed, enough air is pushed through the radiator without the aid of the fan, unless overheating or high A/C head pressure conditions exist. This is a fuel economy feature. Study Figure 10–46 as a typical example.

Air-Conditioning (A/C) Control

The PCM controls the relay that turns the A/C clutch on and off for two and in some cases three reasons:

1. When the A/C control switch (on the instrument panel) is turned on, the PCM delays A/C clutch engagement for 0.4 second to allow time to adjust the IAC valve.
2. The PCM disengages the A/C clutch during WOT operation.
3. On some applications, the PCM turns off the A/C clutch if power steering pressure exceeds a specified value during idle. On others the power steering switch is in series with either the A/C clutch or the control winding of the A/C relay and disengages the A/C clutch without relying on the PCM.

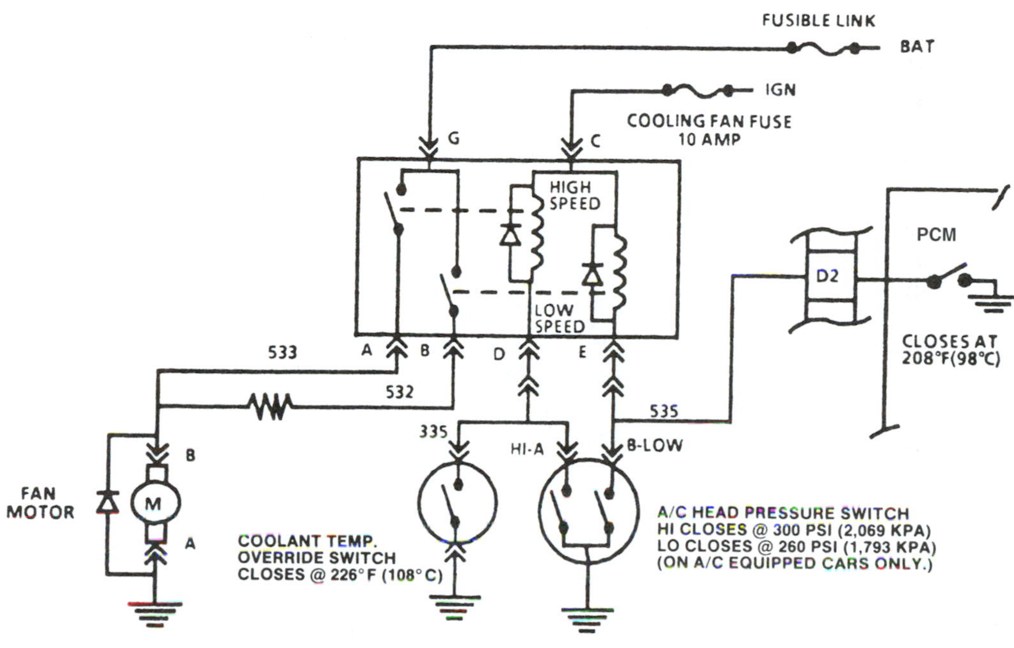

COOLANT FAN CONTROL OPERATING CONDITIONS (WITH A/C)					
A/C SW.	HEAD PRESS.	ROAD SPEED	ENG. TEMP.	FAN SPEED	FAN ON BECAUSE:
OFF/ON	UNDER 260 PSI	UNDER 45 MPH	OVER 98°C	LOW	ECM TURNED ON
OFF/ON	UNDER 260 PSI	UNDER 45 MPH	UNDER 95°C	OFF	
OFF/ON	UNDER 260 PSI	OVER 45 MPH	UNDER 106°C	OFF	
ON	OVER 260 PSI	N/A	N/A	LOW	A/C HEAD PRESS. SW. (LOW) TURNED ON
ON	OVER 300 PSI	N/A	N/A	HIGH	A/C HEAD PRESS. SW. (HIGH) TURNED ON
OFF/ON	N/A	N/A	OVER 106°C	HIGH	COOLANT TEMP. OVERRIDE SW. TURNED ON
WITHOUT A/C					
IGN. SW.		ROAD SPEED	ENG. TEMP.	FAN SPEED	FAN ON BECAUSE:
ON		N/A	OVER 98°C	LOW	ECM TURNED ON
ON		OVER 45 MPH	UNDER 108°C	OFF	
ON		OVER 45 MPH	OVER 108°C	HIGH	COOLANT TEMP. OVERRIDE SW. TURNED ON

Figure 10–46 Coolant fan control circuit (typical). *(Courtesy of General Motors Corporation, Service Technology Group.)*

✔ SYSTEM DIAGNOSIS AND SERVICE

Self-Diagnosis

The self-diagnostic capacities and procedures of the PFI systems are essentially the same as those of the CCC system discussed in the System Diagnosis and Service section of Chapter 8. In 1988 General Motors made a significant increase in the self-diagnostic capacity of most PFI engine applications. The PCM was designed to monitor more circuits, both input and output. New diagnostic code numbers were also assigned. It is important to note that from year to year, and in some cases even from engine to engine within a single model year, code numbers may have different meanings. Be sure the service literature you use is applicable to the engine being serviced and to the model year of the vehicle.

TCC Test Lead. The TCC test lead is in cavity F of the Diagnostic Link Converter (DLC). It can be used to monitor the TCC circuit operation with a voltmeter or test light or it can be used to ground the TCC test lead, overriding the PCM.

Fuel Pump Test Lead. On some vehicles, the fuel pump test lead is found in cavity G of the DLC. On others, it is found on the left side of the engine compartment. A voltmeter or test light can be connected to this lead to determine whether the fuel pump relay or oil pressure switch has supplied power to the fuel pump power lead. Or, a jumper wire can be connected from the test lead to a 12-volt source. This powers the fuel pump for fuel pressure or injector tests.

WARNING: When powering the fuel pump by the above method, be sure that there are no fuel leaks and avoid causing any sparks. Otherwise there is a serious fire hazard.

Diagnostic procedures for 1987 and later are written to include the use of a scanner or scan tool. Previously, diagnostic procedures were written around the use of a voltmeter to obtain such measurements as sensor readings.

Diagnostic Procedure

Diagnostic Circuit Check. After making a thorough visual inspection, paying particular attention to possible causes of vacuum leaks and correcting any problems found, perform the diagnostic circuit check. This procedure is outlined in the service manual. It is designed to identify the type of problem causing the complaint and it refers to the next step or chart to use.

Fuel Pressure Test. Some diagnostic charts call for a fuel pressure test. Fuel pressure is critical to the performance of a fuel injection system. The fuel rail has a fuel pressure test fitting that contains a Schraeder valve like that used in an A/C pressure test fitting or in the valve stem of a tire.

WARNING: While the fuel pressure gauge hose is screwed to the test fitting, a shop towel should be wrapped around the fitting to prevent gasoline from being sprayed on the engine. Remember, the fuel is under high pressure and the residual check valve in the fuel pump can hold pressure for some time after the engine is shut off. Spraying fuel can be not only a fire danger, but also a risk to the eyes and nose.

Injector Balance Test

To realize some of the advantages of PFI, the injectors must all deliver the same amount of fuel to each cylinder. The injector balance test tests injector performance and uniform fuel delivery. Check the applicable service manual for specific directions.

Fuel Trim

Fuel trim is a function programmed into the PCM on fuel-injected applications and is designed to affect the PCM's control of the air/fuel ratio. It is designed to allow the PCM to adjust the fuel injector's pulse width in order to compensate for the wear and aging of components as they affect air/fuel ratios. This is similar to the manner in which a PCM on a feedback-carbureted system is able to adjust the duty cycle to maintain a stoichiometric air/fuel ratio.

Fuel trim is divided into two parts: **short-term fuel trim** (STFT) and **long-term fuel trim** (LTFT). Prior to the OBD II standardization of terms and acronyms, GM referred to LTFT as *Block Learn* and STFT as ***Integrator***. STFT is the immediate reaction to the oxygen sensor. LTFT is an adjustment designed to keep the oxygen sensor averaging the ideal 450 millivolts (mV). Most applications today allow the technician to look at both of these on a scan tool to aid in diagnostics. Therefore, it is important that the technician be able to understand how these functions affect the PCM's control of air/fuel ratio.

Both fuel trims consist of electronic look-up tables (electronic charts) in the PCM's memory chip known as RAM (the only memory chip into which the microprocessor can record information). These look-up tables are divided into 16 cells (or blocks) spread across the possible combinations of engine load and rpm, Figure 10–47. Prior to OBD II standardization, GM used the number 128 as the neutral number. This is the middle value of the numeric range available to an 8-bit computer. (The binary value of 11111111 equals the base ten value of 255. If you include the binary value of 00000000, there are actually 256 different combinations of zeroes and ones available to an 8-bit computer.) When the number in a cell is 128, the base fuel calculation will not be modified. If the oxygen sensor reports that a lean condition exists, the microprocessor (central processing unit or CPU) records this information in the fuel trim by increasing the value in the cell. If the oxygen sensor reports that a rich condition exists, the CPU records this information in the fuel trim by decreasing the value in the cell. When the CPU is again ready to energize a fuel injector, it looks at the fuel trim numbers. Numbers higher than the neutral number tell the CPU that it needs to add fuel to the base fuel calculation by increasing

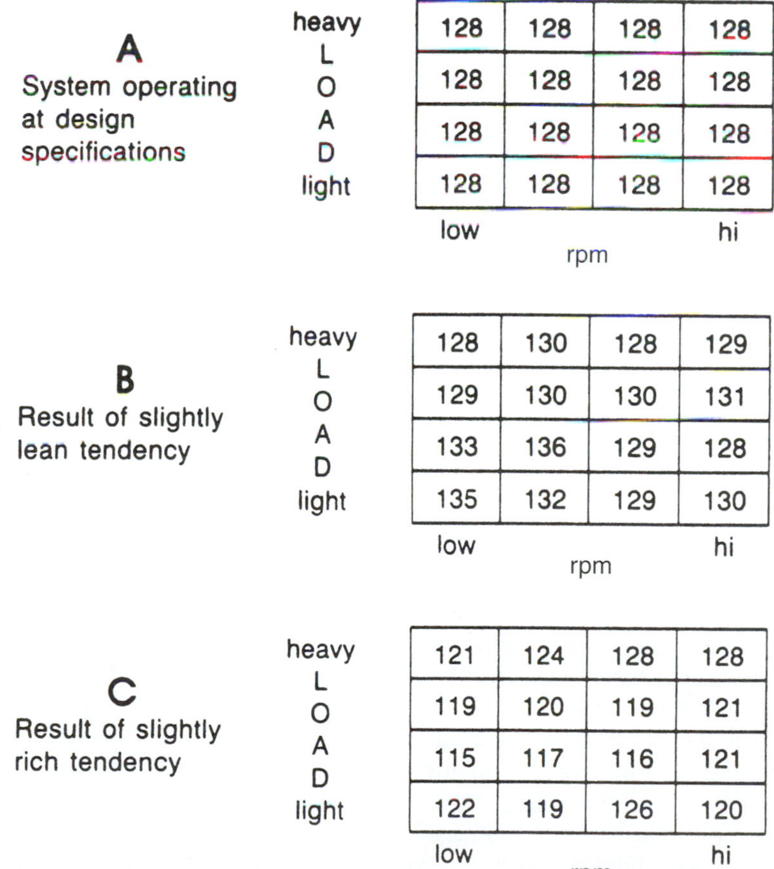

A System operating at design specifications

LOAD \ rpm	low			hi
heavy	128	128	128	128
	128	128	128	128
	128	128	128	128
light	128	128	128	128

B Result of slightly lean tendency

LOAD \ rpm	low			hi
heavy	128	130	128	129
	129	130	130	131
	133	136	129	128
light	135	132	129	130

C Result of slightly rich tendency

LOAD \ rpm	low			hi
heavy	121	124	128	128
	119	120	119	121
	115	117	116	121
light	122	119	126	120

Figure 10–47 Block learn information.

injector on-time; numbers lower than the neutral number tell the CPU that it needs to subtract fuel from the base fuel calculation by decreasing injector on-time. Note the results of slightly lean tendencies and slight rich tendencies of the engine in Figure 10–47. A technician can retrieve this information with a scan tool in order to know where the rich or lean tendencies are. Under the standardization of OBD II, all manufacturers are now using the whole number one as the neutral number. The advantage of this is that a scan tool now shows the fuel that the PCM has learned to add or subtract as a *percentage* of the base fuel calculation, making it easier for a technician to interpret the severity of a problem.

Both fuel trims affect the next decision that the PCM makes concerning a fuel injector's pulse width. When the CPU is preparing to turn a fuel injector on, it looks at three pieces of information to determine the required pulse width.

First, the CPU looks at the base fuel calculation (stored in the PROM chip). The base fuel calculation tells the CPU what a fuel injector's ideal pulse width should be under the engine's current operating conditions (engine load, coolant temperature and so forth). All possible combinations of engine operating conditions are stored in the PROM chip for the particular engine application. Let's create an example and suggest that under the engine's current operating conditions, the base fuel calculation in the PROM chip is 1.78 milliseconds (ms).

Second, the CPU looks at the LTFT (block learn) in order to know how to compensate for the engine's wear and aging of parts as it affects air/fuel ratios. Maybe the engine in the example has developed a vacuum leak. As the CPU has been watching the oxygen sensor cross-count in closed loop recently, it noticed that the oxygen sensor was only averaging 300 millivolts instead of 450 millivolts. The CPU then increased the value in the LTFT. Now the LTFT issues an instruction to increase the pulse width of the base fuel calculation (1.78 milliseconds) by another .13 milliseconds for a total of 1.91 milliseconds of injector on-time.

Third, the CPU looks at the STFT (integrator) to know what the oxygen sensor was last reporting the last time that the CPU looked at it. The CPU has no memory of its own and must record the value from each one of the input sensors in the RAM chip to use this information in making output decisions. The latest oxygen sensor reading is stored as the STFT value. As the oxygen sensor cross-counts in closed loop, this value will change. In fact, the changing of this value causes the oxygen sensor to cross count. If the oxygen sensor last reported a lean condition, the STFT number will be high, and if the oxygen sensor last reported a rich condition, the STFT number will be low. In our example, this might cause our calculated injector on-time to vary by +/– .04 milliseconds, from 1.87 milliseconds to 1.95 milliseconds.

Finally, if an engine problem becomes severe enough, the LTFT may reach a threshold limit, even though the STFT may still be able to move far enough to get the oxygen sensor to cross-count. The result would be an oxygen sensor that is no longer averaging 450 millivolts. If the problem becomes more severe, the CPU may not be able to adjust the pulse width far enough to get the oxygen sensor to cross to the other side of the stoichiometric value, in which case the oxygen sensor voltage is no longer cross-counting (even though the PCM may still legally be in closed loop). At this point, both the LTFT and STFT are said to have reached their threshold limits. If the vehicle is an OBD II application, the PCM will set a DTC and turn on the MIL.

PCM, PROM and CALPAK Service

If correctly used diagnostic procedures call for a PCM to be replaced, check the PCM and PROM service and part numbers to be sure they are the right units for that vehicle. If a PROM is replaced, check that it has not been superseded by an updated replacement PROM. This can be done with the aid of the dealer's parts department or with service bulletins. Some aftermarket service manuals contain service bulletins. Carefully remove the PROM and CALPAK from the PCM to be replaced. The service (replacement) PCM does not include a new PROM or CALPAK. Be careful not to touch the pins of either unit with your fingers because static electricity applied to the pins can damage them. Store them in a safe place while they are out of the PCM.

When the PROM is reinstalled, be sure to orient it properly to the PROM carrier, Figure 10–48. While it can be inserted backward, doing so will destroy it when the ignition is turned on.

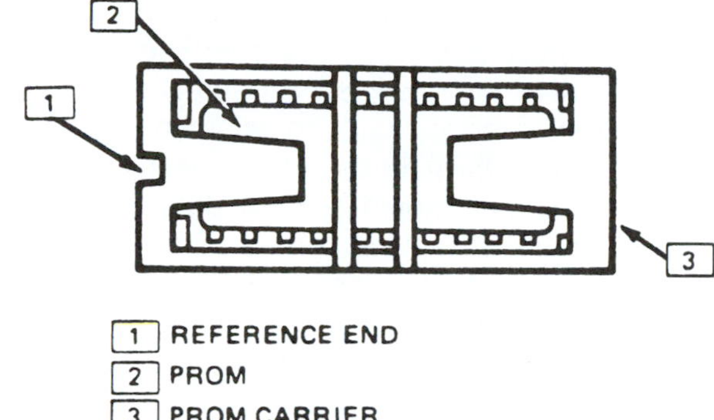

Figure 10–48 PROM and PROM carrier. *(Courtesy of General Motors Corporation, Service Technology Group.)*

Weather-Pack Connectors

All PFI system harness connections under the hood are made with weather-pack connectors.

Diagnostic & Service Tips

IAC Assembly. If the engine is run with the IAC harness disconnected or if the IAC assembly is removed and reinstalled, idle speed will probably be incorrect when it is reconnected. To rereference idle speed, simply start the engine with the IAC harness connected. Allow the engine to warm up and then turn the ignition off. The 3.0- and 3.8-liter engines rereference simply by cycling the ignition on and off.

If the IAC assembly is removed, be sure that the distance from the tip of the pintle valve to the motor housing does not exceed 28 millimeters ($1^1/_8$ inch), Figure 10–49. Otherwise it will be damaged when it is installed. See the service manual for detailed instructions.

When it is installed, use the correct torque specifications. Overtorquing can distort the IAC housing and cause the motor to stick. Complaints of engine stalling as the car comes to a stop can be a result of this condition. Also, if the IAC assembly is replaced, be sure to get the right part. There are three different pintle shapes, and the correct one must be used.

Minimum Idle Speed Adjustment. Minimum idle speed is the speed achieved when the IAC valve is in its closed position. The only air getting into the intake manifold is what goes by the throttle blade (unless there is a vacuum leak). An idle stop screw is

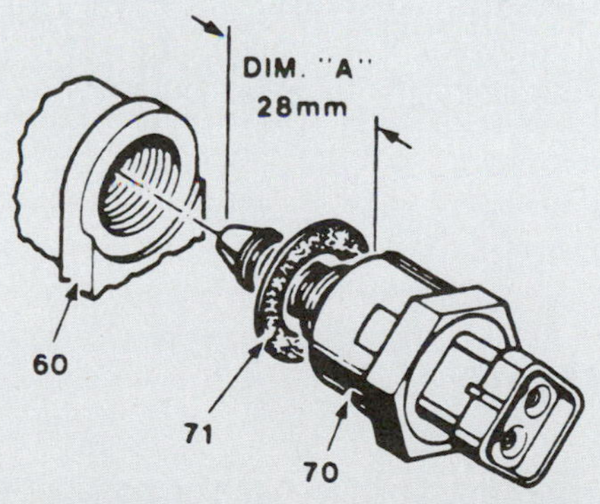

Figure 10–49 IAC motor. *(Courtesy of General Motors Corporation, Service Technology Group.)*

(*continued*)

used to position the throttle blade. The head of the idle stop screw is recessed and covered with a metal plug. The plug can be removed and the idle stop screw adjusted. However, this should only be done if the throttle body is replaced or a service procedure calls for it. This is not a normal service routine adjustment.

To set the minimum idle speed adjustment, the engine should be warm and the IAC valve fully extended (closed). Consult the appropriate service manual for specific directions and specifications.

TPS Adjustment. The TPS on some PFI systems is adjustable. Adjustment is checked by connecting a voltmeter between terminals A and B at the sensor connection with the ignition on. Gain voltmeter access to the terminals by disconnecting the three-wire weather-pack connector and inserting three short jumper leads to temporarily reconnect the circuit. Performing a minimum idle speed adjustment will probably cause the TPS to need adjustment too.

Injector O-Rings. The O-ring that seals between the bottom of the injector and the manifold is a potential source of vacuum leaks. Carefully examine the O-rings and/or replace them when the injectors are serviced.

3.8-Liter Turbo (SFI) Injector Leads. Because the injectors are pulsed individually and in firing order on the 3.8-liter turbocharged engine, the wires that power the injectors must be attached in the firing order. These wires are color-coded.

Fuel Line O-Rings. Threaded connections between the fuel pump and the fuel rail use O-rings to reduce the possibility of leaks. When replaced, they should only be replaced with O-rings designed to tolerate exposure to gasoline.

Fuel Flex Hose. The fuel flex hose used on PFI is internally reinforced with steel mesh. No attempt should be made to repair it. It should be replaced if it becomes unusable or questionable.

WARNING: For personal safety, the factory recommended procedures should be carefully adhered to when servicing fuel lines or fuel line connections. The fuel is under high pressure, and failure to comply with factory-recommended procedures can result in a fuel leak and fire.

SUMMARY

This chapter has focused on port fuel injection systems that use one injector per cylinder and are triggered by the computer either in the firing order of the engine, or simultaneously. We've seen the advantages port injection provides in a more accurately metered fuel delivery, and the elimination of problems associated with distribution of an air/fuel mixture through the intake manifold. We have seen the method General Motors uses employing a cold-start injector to provide an initially richer mixture to aid in starting under low temperature conditions.

In this chapter, we considered the turbocharger boost controls for vehicles equipped with this accessory. These controls include a mechanism to modify boost using the wastegate as well as a more sensitive spark advance system using a knock sensor.

We have seen the introduction of a stepper motor controlled engine idle speed, using an air bypass to meter the amount of intake air around the closed throttle. We have also covered the EGR system on this system and how the computer monitors and controls this exhaust recirculation.

▲ DIAGNOSTIC EXERCISE

A sequentially port fuel injected vehicle has been towed into the shop with a customer complaint of "cranks, but won't start." As the technician is verifying the symptom, he notices that he does hear the fuel pump operate for 2 seconds when he first turns the ignition switch on. In other tests, he finds that, while power is supplied to the fuel injectors with the ignition turned on, there is no PCM pulse to any of the fuel injectors. At this point, what faults could be responsible for the no-start condition? What test(s) would you recommend be performed next in order to narrow down the list of possible faults?

REVIEW QUESTIONS

1. On PFI systems, only air is moving through the intake manifold when the engine is running, thus eliminating the concern for condensation of fuel vapors in the manifold runners. As a result, what systems that were used with TBI applications are now eliminated?
   A. Heated intake air (Thermac) system
   B. Early fuel evaporation (EFE) system
   C. Canister purge system
   D. Both A and B
2. What do most PFI systems use to prevent throttle blade icing?
   A. Heated intake air (Thermac) system
   B. Early fuel evaporation (EFE) system
   C. Canister purge system
   D. Coolant passage in the throttle body
3. In order to properly control the air/fuel ratio during cranking mode, the PCM relies primarily on input information relating to throttle position and which of the following?
   A. Engine coolant temperature
   B. Intake air temperature
   C. Engine load
   D. Exhaust oxygen content
4. During cranking mode, input from the TPS is important to the PCM in order to determine which of the following?
   A. How much to advance the spark timing
   B. How far to open the EGR valve
   C. Whether to initiate clear flood mode
   D. Whether to energize the fuel pump relay
5. What controls the fuel pressure regulator that causes it to vary fuel pressure between idle and WOT?
   A. The PCM
   B. A venturi vacuum signal
   C. A ported vacuum signal
   D. A manifold vacuum signal
6. Why is the fuel pressure increased as the throttle moves from idle to WOT?
   A. To provide more fuel flow through the injectors anytime they are open under heavy load conditions
   B. To keep the pressure differential across the injectors constant under all engine operating conditions
   C. To keep the flow rate through the injectors constant any time they are open
   D. Both B and C
7. When a PFI engine is shut down, fuel vapors from the intake ports condense and form fuel deposits on the throttle blades. This condition can create all *except* which of the following symptoms?
   A. Hard starting
   B. Engine overheating
   C. Erratic idle
   D. Hesitation and surging
8. What type of signal does an AC MAF sensor send to the PCM?
   A. An AC voltage that ranges from 30 to 150 volts
   B. A DC voltage that ranges from 30 to 150 volts
   C. A digital frequency that ranges from 30 to 150 Hertz
   D. A DC voltage that ranges from 0.5 to 5.0 volts
9. When diagnosing a "cranks, but won't start" symptom on a PFI engine, a technician finds

that there is no PCM pulse to the injectors. Which of the following problems could cause this?

A. TPS has lost its ground and is delivering a 5-volt signal to the PCM.
B. The REF signal from the distributor's pickup coil has been lost.
C. The ECT sensor circuit is open resulting in a 5-volt signal to the PCM.
D. The fuel pump relay is defective.

10. What types of sensors are used by a C3I system to produce the crank and cam signals?
    A. Permanent magnet
    B. Optical
    C. Hall effect
    D. Piezoresistive
11. During engine cranking, a Fast Start type III C3I ignition system enables the PCM to identify which ignition coil to fire first within which of the following?
    A. 120 degrees of camshaft rotation
    B. 360 degrees of camshaft rotation
    C. 120 degrees of crankshaft rotation
    D. 360 degrees of crankshaft rotation
12. A cold-start injector is controlled by which of the following?
    A. The starter circuit
    B. A thermo-time switch
    C. The PCM
    D. Both A and B
13. Which of the following is the purpose of the ESC module?
    A. To advance spark timing according to engine load and rpm
    B. To retard spark timing according to input from a detonation sensor
    C. To control fuel injector pulse width according to engine load
    D. To control idle rpm
14. On a PFI system, when does the PCM reference the count of the IAC's pintle valve by momentarily moving the valve to the closed position?
    A. Each time the ignition is turned off
    B. Each time the engine is started
    C. When the PCM sees 30 mph from the VSS
    D. When the PCM sees 600 rpm from the tach reference signal
15. On a system with a dual-bed, three-way catalytic converter, during closed loop operation the AIR system should direct air flow to which of the following?
    A. Intake manifold
    B. Exhaust manifold
    C. Oxidizing bed of the catalytic converter
    D. Atmosphere
16. An EGR valve is designed to allow some exhaust gas into the intake manifold in order to reduce which of the following?
    A. HC emissions
    B. CO emissions
    C. CO_2 emissions
    D. NO_x emissions
17. *Technician A* says that turbochargers increase engine performance by increasing the engine's volumetric efficiency.
    Technician B says that a wastegate is used to limit a turbocharger's boost pressure to avoid preignition and engine damage.
    Who is correct?
    A. A only
    B. B only
    C. Both A and B
    D. Neither A nor B
18. The TCC is designed to do all *except* which of the following?
    A. Increase performance potential during WOT operation
    B. Increase fuel economy
    C. Eliminate hydraulic slippage in the torque converter during cruise conditions
    D. Eliminate heat production in the torque converter
19. What is indicated if long-term fuel trim (block learn) numbers are higher than normal?
    A. The PCM is operating in open loop most of the time.
    B. The PCM is operating at base timing most of the time.

C. The PCM has learned to compensate for a lean tendency and is adding fuel to the base fuel calculation.
D. The PCM has learned to compensate for a rich tendency and is subtracting fuel from the base fuel calculation.

20. *Technician A* says that when replacing a PROM or CALPAK chip in a PCM, you should be careful not to touch the pins in order to avoid damage from static electricity.
Technician B says that when replacing a PCM with a removable PROM chip, you should check the service bulletins to be sure you have the latest PROM update.
Who is correct?
A. A only
B. B only
C. Both A and B
D. Neither A nor B

Chapter 11

Cadillac's Digital Fuel Injection

OBJECTIVES

Upon completion and review of this chapter, you should be able to:

- ❑ Define the added abilities of the PCM.
- ❑ Describe the operating modes of the PCM.
- ❑ Define the inputs to the PCM.
- ❑ Define the outputs that the PCM controls.
- ❑ Understand the function of the BCM.
- ❑ Explain the diagnostic procedures associated with the PCM and the BCM on a Cadillac DFI-equipped vehicle.

KEY TERMS

Accumulator
Ampule
Anode
Cathode
Grid
Modulated Displacement
Propagation
Vacuum Fluorescent

Cadillac introduced Digital Fuel Injection (DFI) in the middle of model year 1980. By 1981, the system had replaced all other fuel injection systems on eight-cylinder Cadillac engines except for the diesels. Before DFI, Cadillac used a system called simply Electronic Fuel Injection (a Bendix-patented, single-function, analog computer-controlled fuel injection system). The Cadillac EFI system was used from 1975 through early 1980. In 1979 and 1980, some California 6.0-liter engines came with an early version of what would become the Computer Command Control system, then called a Computer-Controlled Catalytic Converter. Just to be precise, any further use of the abbreviation EFI in this book refers to the GM TBI system. Since 1981, all four- and six-cylinder Cadillacs have used either CCC or EFI systems.

In 1987, Cadillac introduced a two-seat sports car, the Allante. It featured the 4.1-liter engine introduced earlier on other Cadillacs, but with a sequential port fuel injection system. This system is similar to the nonturbo sequential PFI system on the Buick V6, except for the following:

- It uses an oxygen sensor in each exhaust manifold. Two oxygen sensors allow the system to control the air/fuel ratio more closely. It can adjust the pulse width on one side of the engine independently of the adjustments on the other side.
- It uses a distributor and the HEI ignition system as the DFI system did, as opposed to a Computer Command Control system, Figure 11–1.

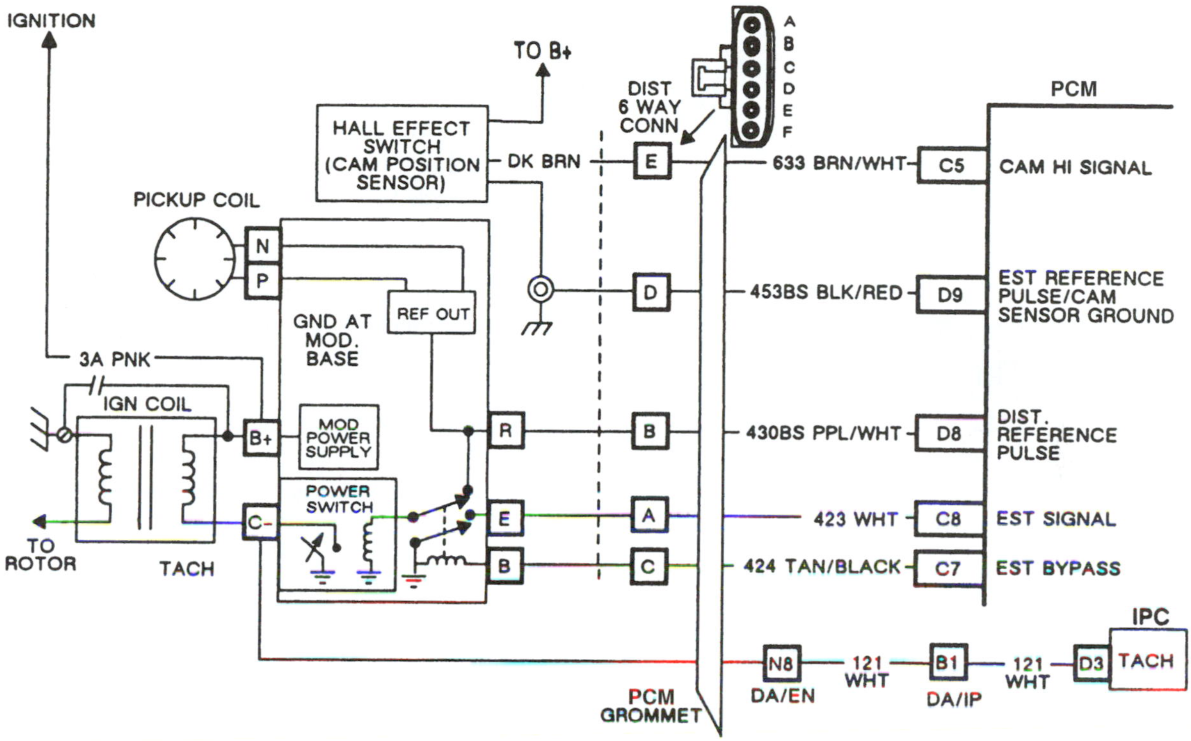

Figure 11–1 The 4.5-liter PFI/HEI ignition system. *(Courtesy of General Motors Corporation, Service Technology Group.)*

- A single, 180-degree vane, Hall effect switch is used in the distributor as a camshaft position sensor for referencing injector pulses.

The 4.1-liter engine was later enlarged to 4.5-liters, but it remained basically the same engine. The newer PFI system replaced the DFI system on increasing numbers of Cadillacs until 1990, by which time all 4.5-liter engines used the PFI system. Also beginning in 1990, Cadillac used the 5.7-liter throttle body injection (TBI) engine as used in Chevrolet Suburbans as an option for a vehicle with the towing package in full-sized, rear-wheel-drive cars.

The Cadillac DFI system works very similarly to the EFI system, which probably came directly from Cadillac DFI, minus a few features. DFI uses a dual throttle body injection (TBI) unit with two injectors. Do not, however, confuse a dual throttle body unit with the Crossfire EFI system. The Crossfire system uses two separate, one-barrel TBI units above an open plenum. The DFI's dual throttle body is configured like that of a conventional two-barrel carburetor, Figure 11–2.

An electric fuel pump in the tank provides fuel pressure and delivery. The pressure is then controlled by a pressure regulator in the TBI unit. Each injector pulses alternately with the other, with one pulse corresponding to a cylinder firing during normal driving.

The two unique features of the Cadillac DFI system are:

1. the type of functions it controls and monitors such as air conditioning and heating, fuel consumption information and cruise control, in addition to engine parameters.

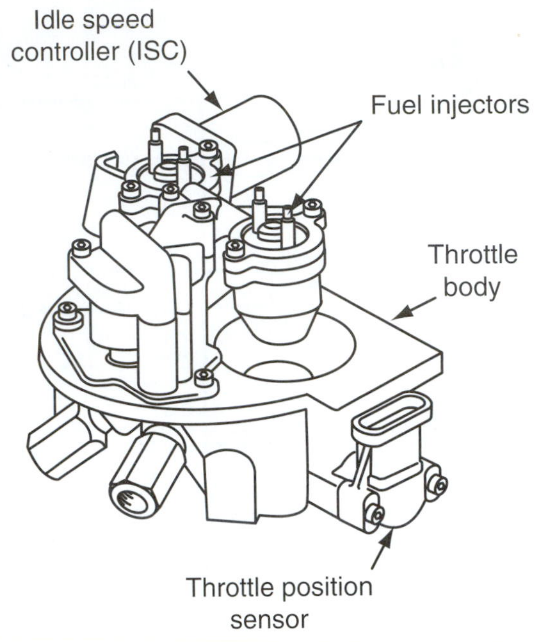

Figure 11–2 TBI unit.

Cadillac Allante

Cadillac introduced its two-seat sportscar, the Allante, in 1987, with the 4.1-liter engine and a sequential, port-injected fuel system. The system is similar to the nonturbocharged sequential PFI system on Buick V6, except that it uses an oxygen sensor in each exhaust manifold. This "stereo" oxygen sensor system allows the computer to control the air/fuel ratio more closely since it can control the pulse width on one side of the engine independently of what it does on the other side.

In 1985, Cadillac introduced a body computer (Body Control Module, BCM) on the new front-wheel-drive C-body car ("C" designates a particular size car body in GM's nomenclature) to work with the PCM. The body computer took over all A/C and related functions, data display and some diesel engine controls for vehicles with such engines. Although our focus here is on engine controls, we include a brief section on the BCM later in this chapter because of its interconnections with the engine control computer.

2. the DFI system has an elaborate self-diagnostic capacity regularly expanded since its introduction.

In model year 1981, the DFI system also controlled the **modulated displacement**, or 8-6-4 system, in which selected cylinders were disabled (by preventing the rocker arms from opening the valves) when power requirements were low. Under acceleration, all eight cylinders are used. During steady-state driving, when minimal power is required, only six cylinders are used; during very low power demand or deceleration, only four cylinders are used. The system was used only on the 6.0-liter engine, discontinued in 1982 for all but limousine and commercial chassis (mostly hearse, some ambulance) vehicles. At that time, the 4.1-liter engine was first introduced. The 8-6-4 system was often regarded as unreliable, troublesome to repair and less than successful at managing emissions and fuel economy objectives.

POWERTRAIN CONTROL MODULE (PCM)

The PCM for the DFI system has greater capacity than the computers used on other General Motors' systems of similar model years. It is located on the right side of the instrument panel and contains, like similar models, a removable PROM. All inputs fed into it and all functions it controls are shown in Figure 11–3.

PCM "Learning" Capacity

As driving conditions change and as a vehicle wears with age and use, sensor inputs change. Not all of them change in direct proportion to each other. For instance, if a car that is normally driven at or near sea level starts on a trip that takes it to a significantly higher altitude, the PCM will see:

- engine speed and road speed still match up as they normally do (provided the transmission

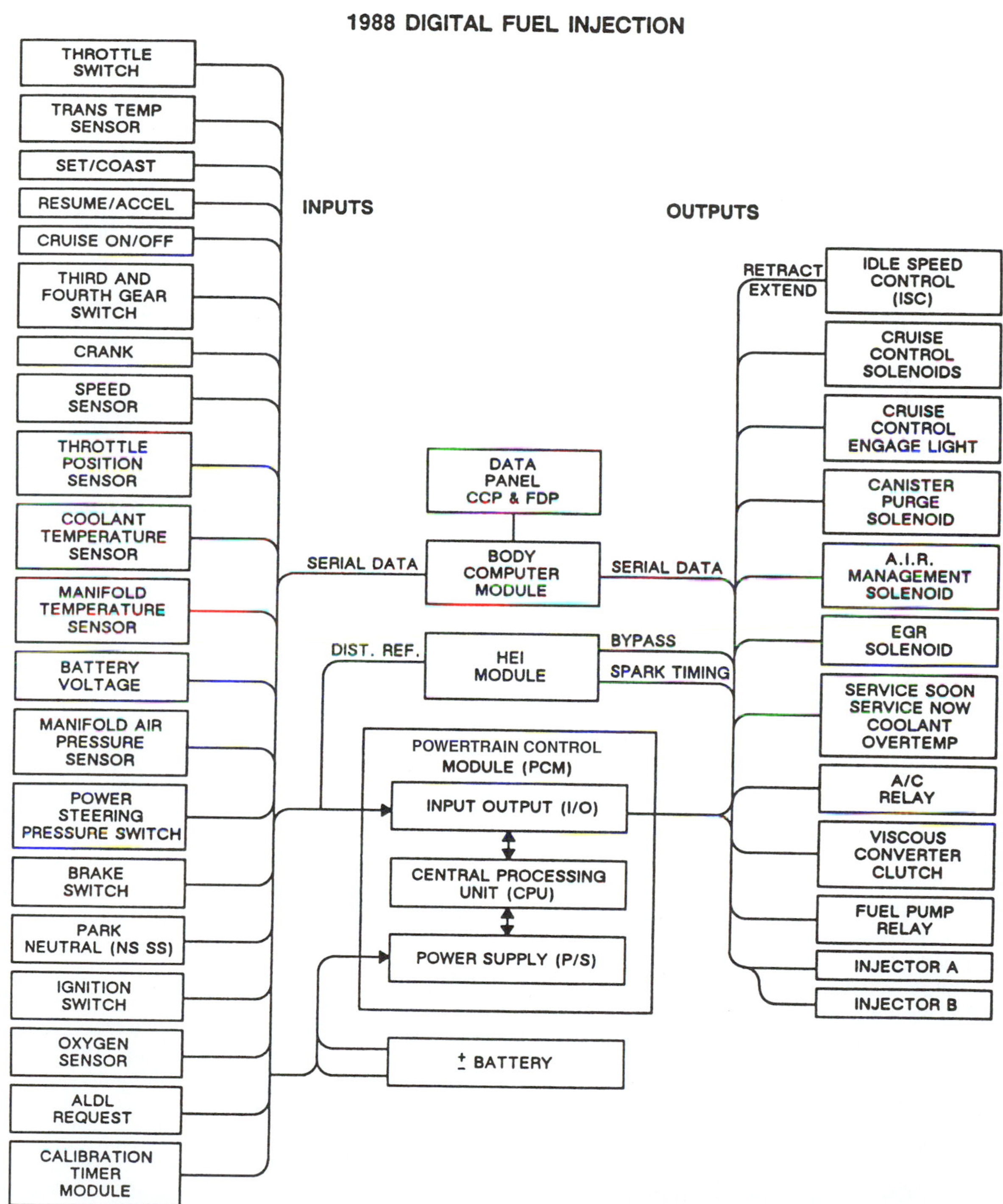

Figure 11–3 DFI overview. *(Courtesy of General Motors Corporation, Service Technology Group.)*

torque converter lockup clutch still applies), but the throttle position required to maintain a given speed is different.
- engine vacuum is lower than it normally is at the lower altitude throttle position.

If the trip is long enough that the new combination of values is consistent over time, the PCM will accept them as normal and will issue spark timing and other commands to provide optimum driveability, emissions performance and fuel economy under the changed conditions. If battery power is removed from the PCM, its volatile memory of what it had learned is erased, and performance will be noticeably different—and worse—until it can relearn the same data. If the battery has been so disconnected, drive the car at normal operating temperature, under part and moderate throttle conditions until normal performance is relearned.

Keep-Alive Memory (KAM)

The computer has two kinds of memory, read-only memory (ROM) and random-access memory (RAM). The ROM is encoded on the PROM and cannot be changed by the car computer in the course of its activity. The RAM is also called the keep-alive memory or KAM. The basic instructions for running the engine are encoded, hard-wired on the PROM, and this does not change. As the computer operates the engine, however, it builds a copy of this information in its KAM, and it can change this copy as it learns that other, slightly different settings will more successfully optimize the driveability, emissions and fuel economy performance. It stores these other settings in the keep-alive memory. If the KAM is erased or the data is somehow corrupted, the computer returns to the ROM data until it can relearn the appropriate KAM data.

OPERATING MODES

The PCM operates the DFI system in one of several different operating modes, depending on the prevailing driving conditions. While there are some similarities with other GM engine management systems, DFI is a special system.

Starting Mode

When the ignition is turned on, the PCM also turns on the fuel pump relay and checks engine coolant temperature and throttle position. Once the engine starts cranking, the computer pulses the injectors to produce an air/fuel ratio ranging from 1.5 to 1 at a temperature of –36°C (–33°F) to 14.7 to 1 at a temperature of 94°C (201°F).

Clear Flood Mode

If the engine floods, that is, if the spark plugs become so wet with fuel they short to ground without spark, pressing the throttle pedal to the floor (or to 80 percent or more of throttle range) and cranking the engine causes the injectors to pulse so briefly that an air/fuel ratio of 25.5 to 1 is delivered. This ratio is maintained until either the engine starts or the throttle is closed to below 80 percent.

Run Mode

Most driving is done in this mode, which includes open- and closed-loop conditions. When the engine is first started and exceeds 400 rpm, the system goes into open loop. The PCM calculates air/fuel requirements based on coolant temperature, manifold pressure, throttle position and engine rpm (in that order of importance). Oxygen sensor inputs are not used in open loop.

The PCM puts the system into closed loop once it sees:

- a varying voltage from the oxygen sensor crossing the 0.45 volt measure. This means the oxygen sensor is hot enough to work and is reflecting the mixture delivered to the intake manifold.
- coolant temperature above 70°C (158°F).
- at least 60 seconds have elapsed since engine startup, allowing the air and fuel delivery systems as well as all the sensors to stabilize.

The PCM uses oxygen sensor information to determine air/fuel mixture once the system is in closed loop to determine fuel injector pulse width. The injector opens for only a few milliseconds even during maximum fuel demand.

Acceleration Mode

Whenever the PCM sees the throttle position change rapidly open and this change is confirmed by a like rapid change in the manifold pressure, the PCM takes the system quickly out of closed loop and puts it into acceleration mode. The higher manifold pressure when the throttle opens slows the evaporation of fuel, so more fuel is needed than makes for a 14.7 to 1 air/fuel ratio. As long as the throttle position sensor indicates WOT (or more precisely 80 percent of WOT), the system will stay in acceleration enrichment mode until the throttle opening and engine load reduce, as reflected by the intake manifold vacuum.

Deceleration Mode

When the PCM sees the throttle close rapidly from the TPS return signal and when manifold pressure confirms the throttle has closed, the PCM reduces the amount of fuel injected. If the deceleration is considerable, the PCM will shut off fuel entirely. During the period of normal running that preceded the deceleration, a certain amount of fuel condensed on the intake manifold walls and floor. When the intake manifold pressure suddenly drops under deceleration, that condensed fuel suddenly evaporates, overly richening the delivered air/fuel mixture. The fuel shutoff is an attempt to correct for that as much as possible.

Battery Voltage Correction

If battery (charging system) voltage drops, the PCM will sense this and will make the following corrections to maintain the best possible operating efficiency:

- increase injector pulse width to maintain the proper air/fuel ratio
- increase idle rpm
- increase ignition primary circuit dwell time to maintain satisfactory spark performance

INPUTS

All the DFI inputs are listed in this section, but some of them are discussed more fully under the main General Motors chapters when they are shared. These include:

- engine coolant temperature sensor (ECT)
- throttle position sensor (TPS)
- oxygen sensor
- ignition switch
- crankshaft position signal

Others are discussed as required.

Manifold Absolute Pressure (MAP) Sensor

The earliest DFI systems used Bendix-built MAP and BARO sensors, located in the passenger compartment. Later versions adopted the standard General Motors units.

Distributor Reference Signal (REF)

The DFI system uses the HEI (and successor systems) pickup coil in the distributor as the source of the ignition base reference pulse.

Manifold Air Temperature (MAT) Sensor

The MAT sensor is similar to that used on other GM products. It measures the air/fuel mixture's temperature in the intake manifold.

Throttle Switch

This is the same as the ISC switch discussed in Chapter 8 where more can be learned about it. The DFI system uses an idle speed control motor like the one used on CCC systems instead of the idle air control motor used by EFI and PFI systems.

Battery Voltage

The PCM monitors battery voltage to determine when it needs to go into voltage correction mode, as described in this chapter, in the operating modes section.

Park/Neutral (P/N) Switch

The park/neutral switch indicates to the PCM when the transmission is in park or neutral as opposed to being in a forward or reverse gear. The computer uses this information to control idle speed, cruise control and the torque converter lockup clutch or viscous converter clutch system. The front-wheel-drive C-body cars, first introduced in 1985, use the 440-T4 transaxle, including a viscous lockup torque converter clutch. The transaxle and converter lockup clutch were both introduced in the 1985 model year (see Viscous Converter Clutch in the Outputs section of this chapter).

Third- and Fourth-Gear Switches

These switches screw into passages in the valve body; hydraulic pressure activates them. When the transmission shifts into third gear, the third-gear switch opens. When the transmission shifts into fourth gear, the fourth-gear switch opens. By monitoring these switches, the PCM knows whether the transmission is in third, fourth or one of the lower gears. The PCM uses this information to control the viscous converter clutch.

Viscous Torque Converter Lockup Clutch (VCC) Temperature Sensor

A thermistor-type temperature sensor is located in the transmission to monitor fluid temperature. This information affects the PCM's control of the viscous torque converter lockup clutch. In general, this information is used chiefly when the computer decides to leave the viscous torque converter locked up to prevent overheat damage under sustained high-load cruise conditions (like hill climbing or trailer towing) in which it would ordinarily disengage the clutch.

Generator Monitor

General Motors, for some reason, prefers the word *generator* to the word *alternator*, a preference found throughout the company's technical literature. The named device, however, is the same device everyone else calls an *alternator*. On most GM computer-controlled engine management systems, the computer monitors "generator" voltage through a wiretap into the fuel pump power circuit. Although this circuit is not directly wired to the generator (alternator), it reflects charging system voltage as most components on the system perceive it, a reliable reflection of the actual alternator output.

If the PCM sees less than 10.5 volts or more than 16 volts, it turns on the Malfunction Indicator Lamp (MIL) and sets a code 16 in its diagnostic memory. On vehicles with a BCM, all PCM codes are preceded by the letter "E" (reference to the PCM by its original name "Engine Control Module" or "ECM" prior to OBD II standardization) and all BCM codes are preceded by either the letter "F" or the letter "B," depending on the year and model.

Power Steering (P/S) Anticipate Switch

If power steering pressure reaches a specified pressure, it opens a hydraulically actuated switch. With the ignition on, battery voltage passes through the switch to the PCM. If the switch opens, causing the signal voltage at the PCM to drop to zero during idle, the PCM increases the throttle blade opening to support the additional engine load. If the switch opens above 40 mph, the PCM turns on the MIL and records code 40. This switch does not appear on DFI systems before model year 1985.

Cruise Control Enable (On/Off Switch)

When the cruise control switch is turned on, it supplies a 12-volt signal to the PCM. This makes the PCM operate the cruise control system. After the 1986 model year, Eldorado and Seville cruise control operation was directed by the BCM.

Cruise Set/Coast Switch

If the driver pushes the set/coast switch on the cruise control, a 12-volt signal is sent to the PCM. If the cruise control system is in a coast mode, the PCM will engage and maintain the current vehicle speed. If it is in an engaged mode when the switch is pressed, the PCM will allow vehicle speed to coast down until the button is released.

Cruise Resume/Accelerate Switch

When the resume/accelerate switch is moved toward resume, a 12-volt signal is sent to the PCM. The PCM resumes a controlled preset speed that was disengaged because the brakes were applied. If the switch is moved toward accelerate while the system is engaged at a particular speed, the PCM opens the throttle and accelerates the vehicle as long as the switch is held. Once the switch is released, the PCM maintains the speed at which it was released.

Cruise Control Brake Switch

Whenever the ignition switch is on, it supplies battery voltage to the brake switch. When the switch is closed (when the brake is applied), voltage feeds through the switch to the PCM. When the brake is applied, the switch opens. The voltage signal at the PCM disappears, and the PCM knows the brake is applied and modifies its outputs accordingly. Among these output commands is one to disengage the cruise control if it is engaged.

DLC Request

When someone grounds the test terminal of the DLC (Data Link Connector) under the dash, the PCM responds by putting the electronic spark timing at a fixed timing advance. This mode is the one used to check ignition timing.

Serial Data

Information to the PCM reflected on the display panels or from the BCM comes via a data link. It is most properly referred to as *serial data.* This information can concern a driver-selected heater or air-conditioning temperature. It can show fuel mileage, or it can come as information requested from the BCM.

OUTPUTS

Injectors

Each of the two injectors uses an electric solenoid, Figure 11–4. When this solenoid is pulsed by a signal from the PCM, the injector's fuel valve lifts off its seat by a few thousandths of an inch. Fuel sprays at a controlled pressure between 9 and 12 psi. The amount of fuel sprayed is a function of how long the valve stays open (in pulse width, or the time the valve is open. The width in question is the width along the horizontal time line of a graph).

Each of the injectors has its own power supply circuit and fuse. Voltage comes through the

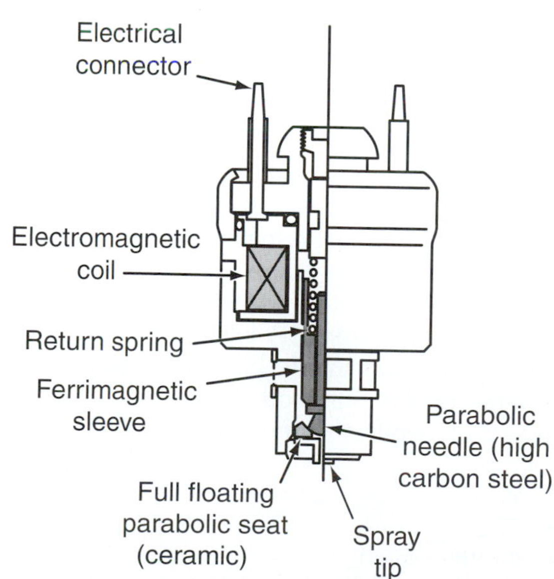

Figure 11–4 Injector comparison.

ignition switch, and the PCM grounds each injector individually for whatever period of time it calculates is appropriate.

Fuel Pressure Regulator. The pressure regulator provides an operating pressure ranging between 9 to 12 psi at the fuel injectors. It functions identically to the standard GM fuel pressure regulator, discussed Chapter 9.

Fuel Pump Operation

Fuel and fuel pressure come from a twin-turbine electric pump inside the fuel tank. The pump turns on and off following a fuel pump relay under the instrument panel. Its operation is similar to that used in the rest of the General Motors systems and described in Chapter 9. Unlike that system, however, the DFI fuel pump did not use an in-line oil pressure switch as a backup feature.

Electronic Spark Timing (EST)

The EST works the same way as on the EFI system discussed in the Outputs section of Chapter 9, except that DFI does not use a Hall effect switch.

Idle Speed Control (ISC)

The Cadillac DFI system uses the same kind of idle speed control motor that the regular General Motors CCC system uses. While the motor works and you adjust it the same way as those discussed in Chapter 8, the PCM uses a somewhat different strategy to control it. The PCM uses information from the following sensors and switches to control throttle blade position in any of the three operational modes (some of these are inputs that identify the operational mode):

- VSS
- TPS
- throttle switch
- ignition reference signal
- A/C clutch signal
- ECT signal
- MAP sensor signal
- P/N switch
- ignition switch

During Cranking and Initial Warmup. During startup cranking, the throttle is held in a predetermined position. Once the engine starts, the throttle is held in a predetermined fast idle position to allow intake manifold conditions to stabilize. If the engine is started hot, the fast idle period will be very short. If the engine is started cold, the fast idle period will be longer until the engine reaches operating temperature.

Curb Idle. When the engine is idling at normal operating temperature, curb idle is maintained at a predetermined speed. If engine load changes because of the A/C clutch cycling on and off, the transmission going into gear or neutral, or because of momentary high power steering pressure load, the PCM adjusts throttle opening to maintain the predetermined engine speed.

Deceleration. During deceleration, manifold pressure goes very low (or intake manifold vacuum goes very high, an equivalent description), and any liquid fuel in the manifold evaporates very quickly in the reduced pressure. This results in a very rich air/fuel mixture and high HC and CO exhaust emissions, unless enough air is let into the manifold to mix with the fuel vapors, as well as to reduce the vacuum. The PCM considers vehicle speed and manifold pressure to calculate the degree of deceleration and determine how much throttle opening is required to prevent the rapid fuel evaporation. During very hard deceleration, for example, if descending a hill in manual second, the PCM can also shut off fuel injection entirely.

AIR Management

The Air Injection Reaction (AIR) system on Cadillac's DFI is basically the same as for the rest of the General Motors vehicle systems. The DFI system uses a slightly different strategy to control it, however. The PCM de-energizes the divert solenoid and puts the system into divert under any of the following conditions:

- acceleration or WOT operation. (Some DFI AIR systems go into divert mode any time the vehicle speed exceeds 60 mph.)
- deceleration.
- extremely cold weather startup.
- if there are certain system failures that also turn on either of the service engine lights.
- high rpm, which can cause air pump pressure to exceed the calibration of the pressure relief valve built in to the control valve. Though this is not a controlled divert function, it has similar results.

Notice that any control valve solenoid or related electrical circuit failure also results in a fail-safe condition in which the system stays in divert mode. Such a failure should leave the valve in the safest operating mode, one that is unlikely to damage the catalytic converter. Of course, a mechanical failure in the valve, should it leave the air injection directed to the converter, will not have this fail-safe feature.

Torque Converter Clutch (TCC) or Viscous Converter Clutch (VCC)

The TCC is a hydraulically-actuated lockup clutch inside the transmission torque converter. When actuated, it eliminates the slippage in the torque converter to improve fuel economy and reduce the heat generated by the turbulence in the converter's automatic transmission fluid. When the lockup clutch is released, the torque converter works normally (though for some vehicles, engineers have used a torque converter with more converter slip than before to maximize torque multiplication and smoothness, knowing they could eliminate all the slip once the vehicle was underway at cruise speed). Application and release of the clutch are controlled primarily by the PCM. Read Chapter 8 for more information about converter lockup clutches. Some parameters by which the Cadillac DFI system controls the clutch are slightly different, and those are discussed here.

The VCC, introduced on the 1985 Cadillac front-wheel-drive C-body cars, works the same way as the TCC does except that it does not provide a 100 percent lockup between the engine and the transmission input shaft. It allows a smoother apply and release than the TCC. The VCC clutch is made up of a rotor sandwiched between two parts, called the cover and the body, Figure 11–5. The rotor and body surfaces facing each other have a series of fins that mesh together with a small space between each fin, Figure 11–6. The effective distance between them is very small, and the shared area is very large. This space is filled with a viscous silicone fluid that drives the rotor. When the VCC is applied, it allows about 40 rpm slippage at 60 mph, just enough to eliminate some of the torsional vibration between the engine and the transmission. The viscous clutch lockup is controlled the same way as a TCC clutch except that transmission fluid temperature plays a larger role in its control.

TCC/VCC Control Parameters. The PCM grounds the TCC/VCC apply solenoid circuit when:

- the criterion speed is reached according to the vehicle speed sensor (VSS). On some cars this can be as low as 24 to 36 mph depending on throttle position and transmission oil temperature.
- the engine has reached normal operating temperature. If the engine coolant temperature is 18°C (64°F) or more when the engine starts, the PCM will apply the converter clutch when the coolant reaches 60°C (140°F) for VCC vehicles. If the coolant is colder than 18°C at startup, the PCM uses a fixed-time delay before applying the clutch. The purpose of this delay is to improve low-temperature driveability until conditions allow lockup.
- the TPS signal is within a predetermined window that varies depending on what gear the transmission is in, among other factors. The TPS signal is compared to the vehicle speed signal, and the PCM selects different apply

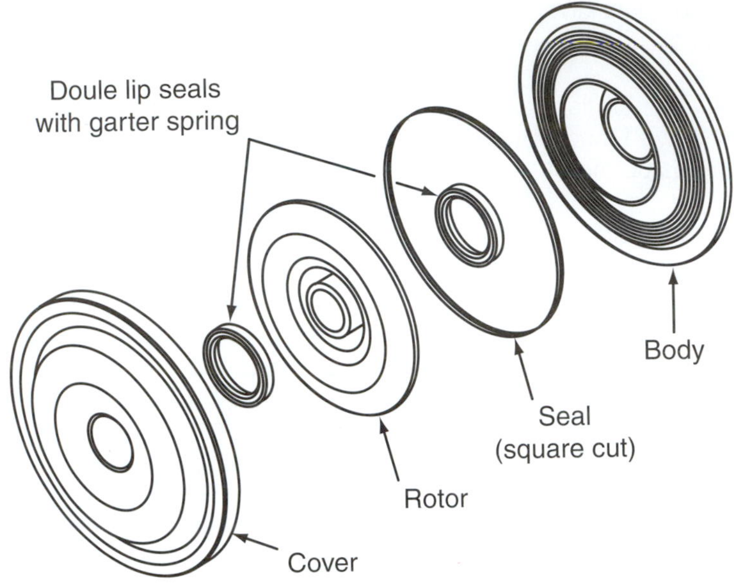

Figure 11–5 Viscous converter clutch (exploded view).

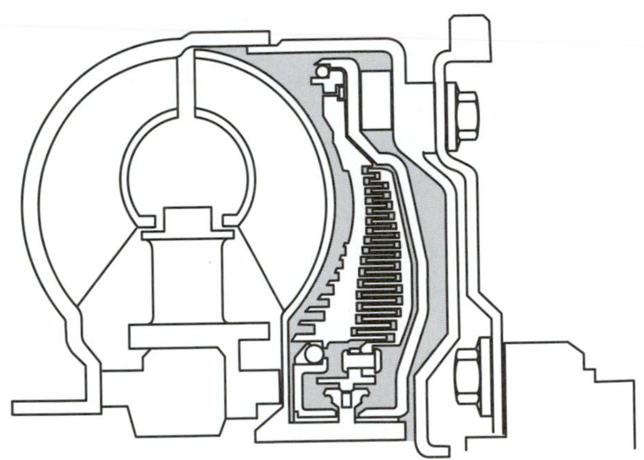

Figure 11–6 Viscous converter clutch.

speed thresholds for different combinations of TPS and VSS.

Two other signals can disengage the lockup clutch on VCC vehicles: the brake switch and the overtemperature protection switch, Figure 11–7.

Most TCC applications do not use an overtemperature switch. These switches are not controlled by the PCM but are in the apply solenoid circuit. Each must be closed before the circuit can close to the PCM. Both are normally closed during driving operations; but if the brake switch is opened when the brake is depressed or the overtemperature protection switch opens (when the transmission oil temperature exceeds 157°C or 315°F), the lockup clutch disengages.

On the 1987 and later Allante, the PCM also controls the transmission 2-3 and 3-4 shifts. The PCM does this through two solenoids it controls in the transmission's valve body. The solenoids block or open passages allowing oil to move under pressure to the 2-3 and 3-4 shift valves.

Exhaust Gas Recirculation (EGR)

When the EGR valve opens, it allows a measured amount of exhaust gas (mostly carbon dioxide and nitrogen, neither of them capable of sustaining flame) into the intake manifold. This inert gas displaces a certain amount of com-

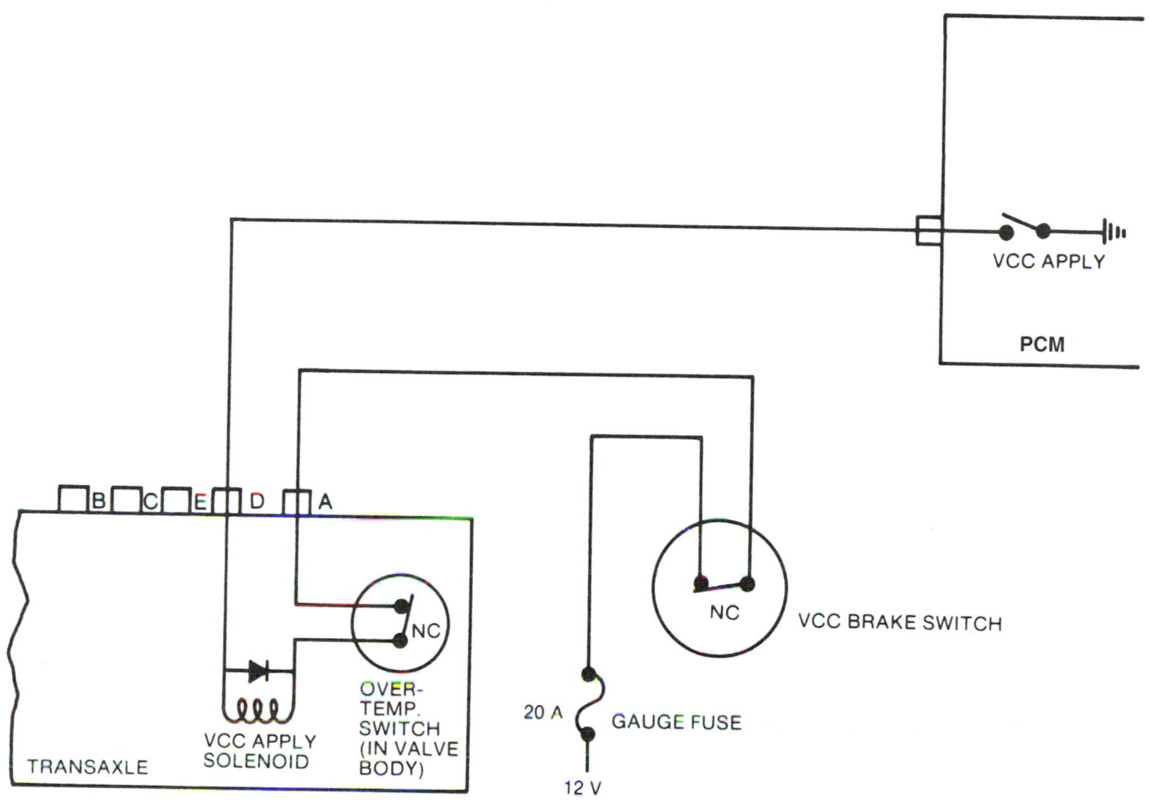

Figure 11–7 VCC electrical control circuit.

bustible intake mixture and has the effect of slowing the **propagation** of the flame through the combustion chamber. This slower flame burns at a lower temperature and thus reduces the production of oxides of nitrogen (NO_x)

The Cadillac DFI system uses a negative backpressure EGR valve, Figure 11–8. This type of valve has some self-modulation capacity. Exhaust pressure from the exhaust crossover (see early fuel evaporation in Chapter 8) passage in the intake manifold is sensed through a small hole in the lower portion of the valve itself. The hollow valve stem allows exhaust backpressure to get to a chamber beneath a second diaphragm below the main diaphragm. This pressure helps a small spring push the lower (second) diaphragm up and block a bleed hole in the main diaphragm. When the bleed hole is blocked, ambient air vented into the space between the two diaphragms no longer bleeds into the vacuum chamber above the main diaphragm. This allows the buildup of sufficient vacuum in the chamber (using manifold vacuum as the source) to pull the diaphragm up and open the EGR valve. When the valve opens, the same hole that admitted exhaust pressure also allows manifold vacuum in.

If the net result of the engine vacuum versus exhaust backpressure is a positive or near positive pressure, the lower diaphragm pushes up and opens the valve. If, however, the net result is a negative pressure, the lower diaphragm stays down, and the bleed valve opens. This keeps the EGR valve itself closed.

EGR Modulation. In addition to the EGR control provided mechanically by the negative backpressure EGR valve, the Cadillac DFI system

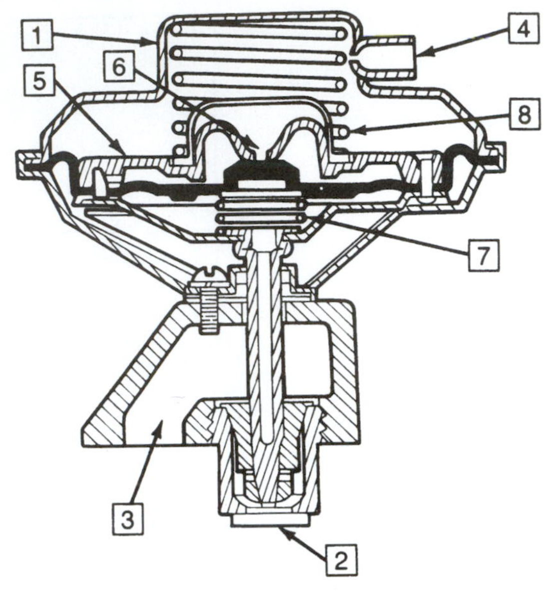

1	EGR VALVE	6	VACUUM BLEED HOLE
2	EXHAUST GAS	7	SMALL SPRING
3	INTAKE AIR	8	LARGE SPRING
4	VACUUM PORT		
5	DIAPHRAGM		

Figure 11–8 Negative backpressure EGR valve. *(Courtesy of General Motors Corporation, Service Technology Group.)*

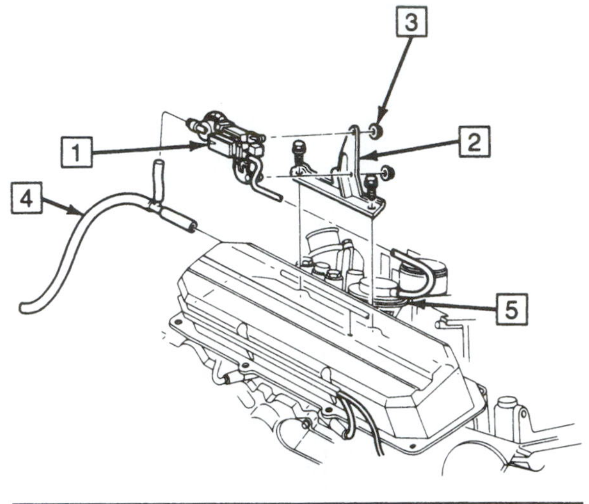

1	EGR SOLENOID
2	BRACKET
3	NUT
4	HOSE ASM.
5	EGR VALVE

Figure 11–9 EGR solenoid. *(Courtesy of General Motors Corporation, Service Technology Group.)*

on the 4.1-liter engine also controls vacuum to the EGR vacuum port. It does this with a duty-cycled solenoid that bleeds atmospheric pressure into the vacuum supply hose to the EGR valve when on, causing the valve to close, Figure 11–9. When the solenoid is turned off, it allows manifold vacuum to the EGR valve and causes it to open. By duty-cycling the solenoid, the PCM can control the amount of EGR opening and thus the EGR flow rate. The PCM varies the flow rate for different driving conditions to provide the best possible driveability characteristics while still maintaining control of the production of NO_x. A 10 percent duty cycle provides full vacuum to the EGR and a 90 percent duty cycle vents all vacuum to it, effectively closing the valve. The PCM uses information from the following sensors to determine the prevailing driving conditions and the amount of EGR needed:

- engine coolant temperature (ECT signal—no EGR when coolant temperature is low because there would be no NO_x production at those temperatures anyway)
- throttle position sensor (TPS signal)
- manifold pressure (MAP signal)
- idle control throttle switch (ISC signal—no EGR flow is needed at idle because combustion temperatures are not high enough to generate NO_x)
- ignition reference signal (REF)

Charcoal Canister Purge

Like most engine management systems, the Cadillac DFI system uses an evaporative emissions control system consisting primarily of a charcoal canister pneumatically plumbed to the vapor area of the fuel tank. The tank vents through the canister, and fuel vapors are stored in the activated charcoal in the canister. Once the engine is running in closed loop, the PCM can open the line to the intake manifold and allow vacuum to purge the fuel vapors in the canister, Figure 11–10. Because the PCM has no way to tell how much fuel, if any, is stored in the canister, it waits until the system is in closed loop to allow any canister purge to occur, so it can compensate for any extra fuel or air introduced by the canister.

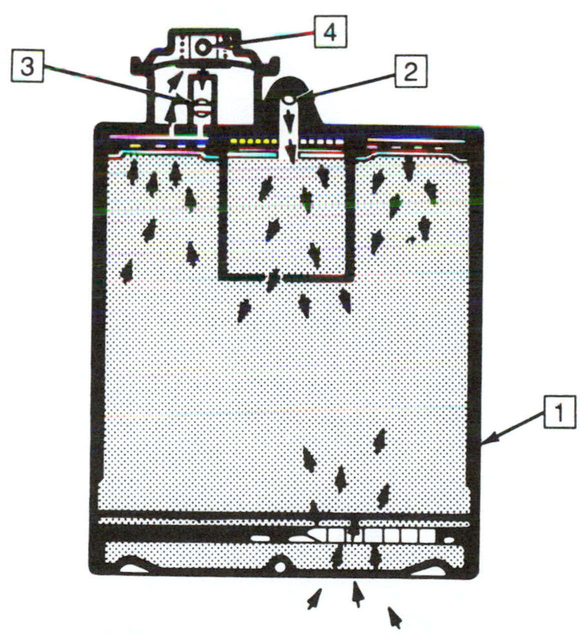

Figure 11–10 Charcoal canister. *(Courtesy of General Motors Corporation, Service Technology Group.)*

The canister purge is directly controlled by two valves, the control valve and a solenoid-operated valve in the vacuum line going to the control valve, Figure 11–11. The control valve contains a diaphragm and spring. The spring pushes the diaphragm down and closes the passage that would otherwise let air and fuel vapors to flow through the purge port. When manifold vacuum is applied to the top of the diaphragm, it lifts and opens the passage to the purge port. By turning the solenoid on or off, the PCM controls vacuum to the control valve. When the solenoid is energized, vacuum to the control valve is blocked. The PCM turns the solenoid off and allows purge when:

- the system is in closed loop and off idle.
- no faults are present triggering codes 13, 16, 44 or 45.

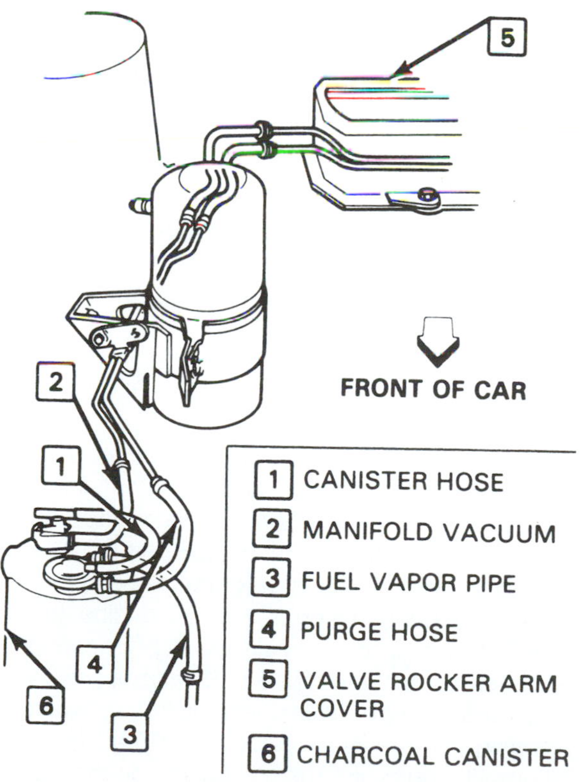

Figure 11–11 Charcoal canister hose connections. *(Courtesy of General Motors Corporation, Service Technology Group.)*

An electrical failure in the solenoid or its circuit allows purging to occur anytime there is enough intake manifold vacuum. This can cause:

- an overrich condition during warmup or idle.
- an overlean condition during warmup or idle (if the canister has no fuel vapors stored).
- dieseling after shutdown.

Early Fuel Evaporation (EFE)

The purpose of the EFE system is to apply heat to the area just below the throttle valves to prevent throttle blade icing and to increase fuel evaporation during the crucial engine warmup period. This helps to:

- reduce the richness requirement of the air/fuel ratio.
- improve driveability.
- reduce hydrocarbon and carbon monoxide emissions.

Rear-wheel-drive vehicles with Cadillac DFI have used exhaust heat for this purpose, not under the control of the PCM. There is a valve in the exhaust pipe on one side of the engine where it bolts to the exhaust manifold. A vacuum actuator containing a spring and a diaphragm opens and closes this valve. A thermal vacuum switch controls vacuum to the actuator diaphragm. When coolant temperature is below 49°C (120°F), the thermal vacuum switch routes vacuum to the EFE actuator. Vacuum applied to the actuator pulls the diaphragm against the spring and closes the EFE valve. This forces exhaust gasses from that side of the engine through a passage in the head and through the intake manifold EFE passage. The exhaust crosses beneath the plenum to the other head, where it travels through another passage to the exhaust pipe on that side. Once the coolant temperature goes above 49°C, a temperature-actuated vacuum control valve blocks vacuum to the actuator. The spring pushes the diaphragm back down and opens the EFE valve. Exhaust gas is no longer forced through the EFE passage.

The layout of front-wheel-drive cars makes this arrangement unsatisfactory, so they employ a ceramic-covered electric heating element under the TBI unit as a source of EFE heat, Figure 11–12. The heat element is powered through a relay (EFE relay) in turn controlled by the PCM.

Diagnostic & Service Tip

The pivots on any valve in an exhaust passage are obviously subject to considerable thermal stress. It is not uncommon on older vehicles to find the EFE valve has seized in either the open or the closed position (ordinarily closed). On a cold engine, the valve should move easily by hand, against the spring actuator. If an EFE valve is seized closed, it can cause excessive exhaust backpressure and hence, poor fuel economy. It can also cause oil caking on the adjacent areas of the intake manifold over the cam valley.

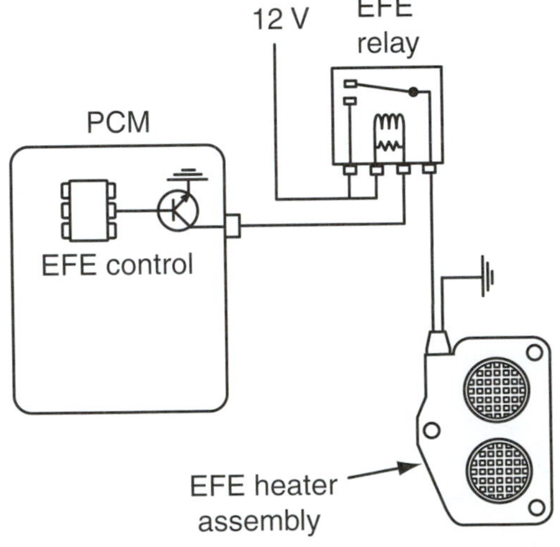

Figure 11–12 EFE grid.

The PCM activates the relay when all of the following conditions exist:

- intake air temperature (IAT signal) is lower than 75°C (167°F).
- engine coolant temperature (ECT signal) is below 106°C (223°F).
- battery voltage is above 10 volts (in conditions of undercharge, the heating element could easily rob more important components of electric power).

If any of these conditions is not met, the PCM turns the EFE relay off.

If the EFE is off during driving, the PCM will turn it on if all of these conditions are met:

- IAT is below 38°C (100°F).
- ECT indicates below 106°C (223°F).
- battery voltage is above 12 volts.

Once any of the above conditions is no longer met, the PCM turns the EFE relay off.

One other set of conditions causes the PCM to turn on EFE heat. If throttle position is open more than 30 degrees (zero is closed and 90 degrees is WOT) and IAT is less than 60°C (140°F), the PCM will turn on the EFE relay for at least 15 seconds to prevent throttle blade icing.

Cruise Control Vacuum Solenoid and Cruise Control Power Valve

When the ignition is turned on, the switch provides electric power to (among other things) the cruise control on/off switch and to the bleed switch, Figures 11–13 and 11–14. The cruise control on/off feeds power to:

- the resume/acceleration and set/coast switches.
- the cruise-on indicator light.

The brake switch feeds power to:

- the PCM (as a signal the brake is not applied).
- the cruise control vacuum solenoid.
- the cruise-engaged indicator light.

When the driver depresses the cruise control set/coast switch (at speeds above 25 mph), the selected speed is recorded in the PCM's memory, and the PCM grounds the cruise vacuum solenoid. The solenoid routes vacuum to the cruise control vacuum servo, which actually sets the throttle position through a vacuum servo diaphragm and spring, thus controlling engine speed. As vacuum is applied, the throttle opens. As vacuum is vented, the spring closes the throttle. The amount of vacuum applied to the servo is determined by the cruise control power valve, controlled in turn by the PCM. The PCM supplies a pulsed voltage signal to the power valve.

This is one of the very few instances in which the PCM supplies the power for a component instead of providing the ground. In this case, the reason for the difference is to ensure safety in case of a short circuit: if something goes wrong with the system, the cruise control will turn off rather than open the throttle any farther.

As the power valve turns on and off, the servo is alternately exposed to vacuum or atmospheric pressure. The PCM pulses the power valve at whatever rate is necessary to make the indicated speed from the VSS match the speed recorded in its memory. As the PCM begins to control vehicle speed, it also turns on the cruise control engaged light.

When the cruise control system engages, holding a set speed, it can disengage in any of three ways:

- The cruise control on/off switch can be turned off, in which case the selected speed is erased from the PCM's memory.
- The brake pedal can be depressed, causing the PCM to disengage the cruise control vacuum solenoid and vent the servo to atmospheric pressure. Moving the resume/accelerate switch to the resume position and releasing it after the brake is released causes the system to reengage.
- Many faults that cause the PCM to turn on an MIL also cause it to disable the cruise control system.

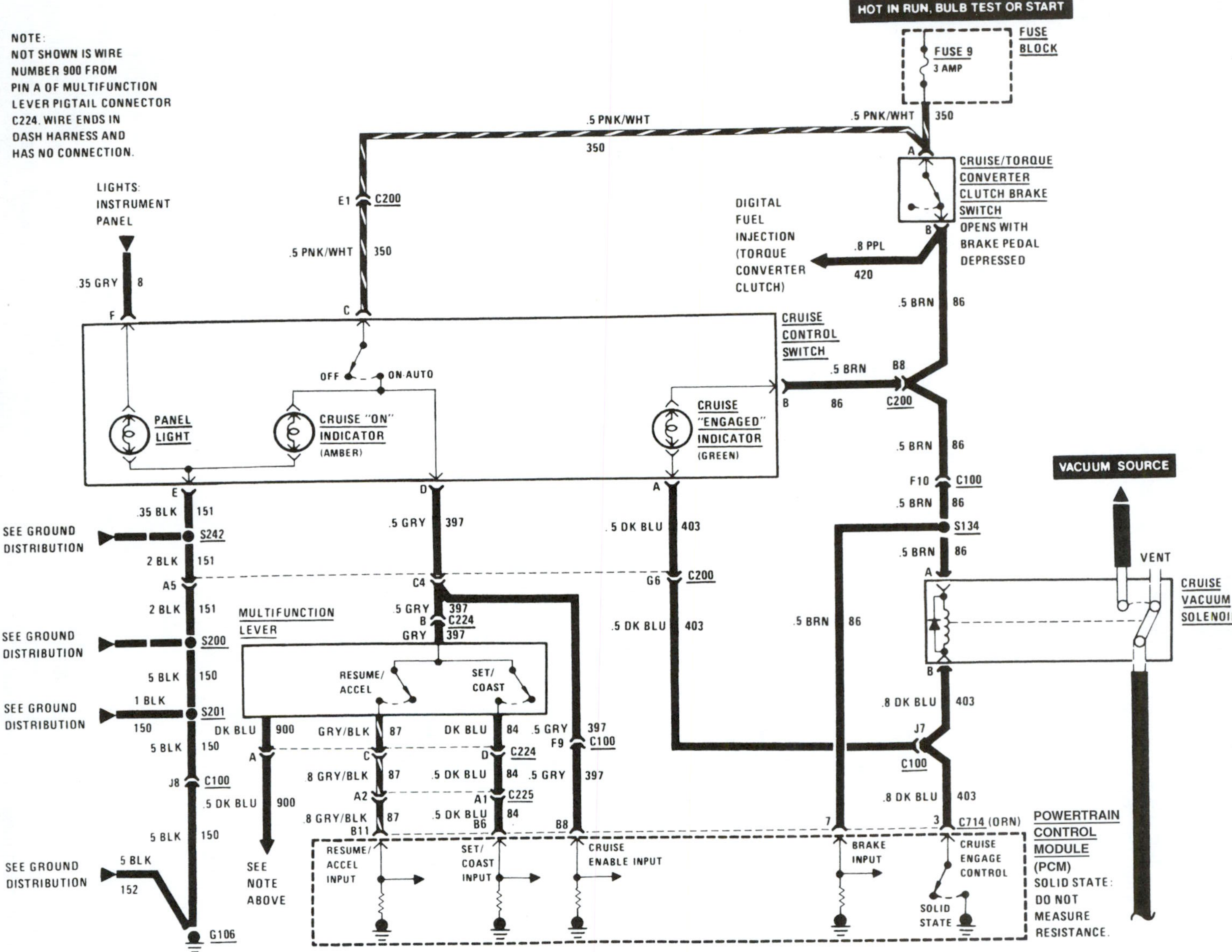

Figure 11–13 Cruise control circuit. *(Courtesy of General Motors Corporation, Service Technology Group.)*

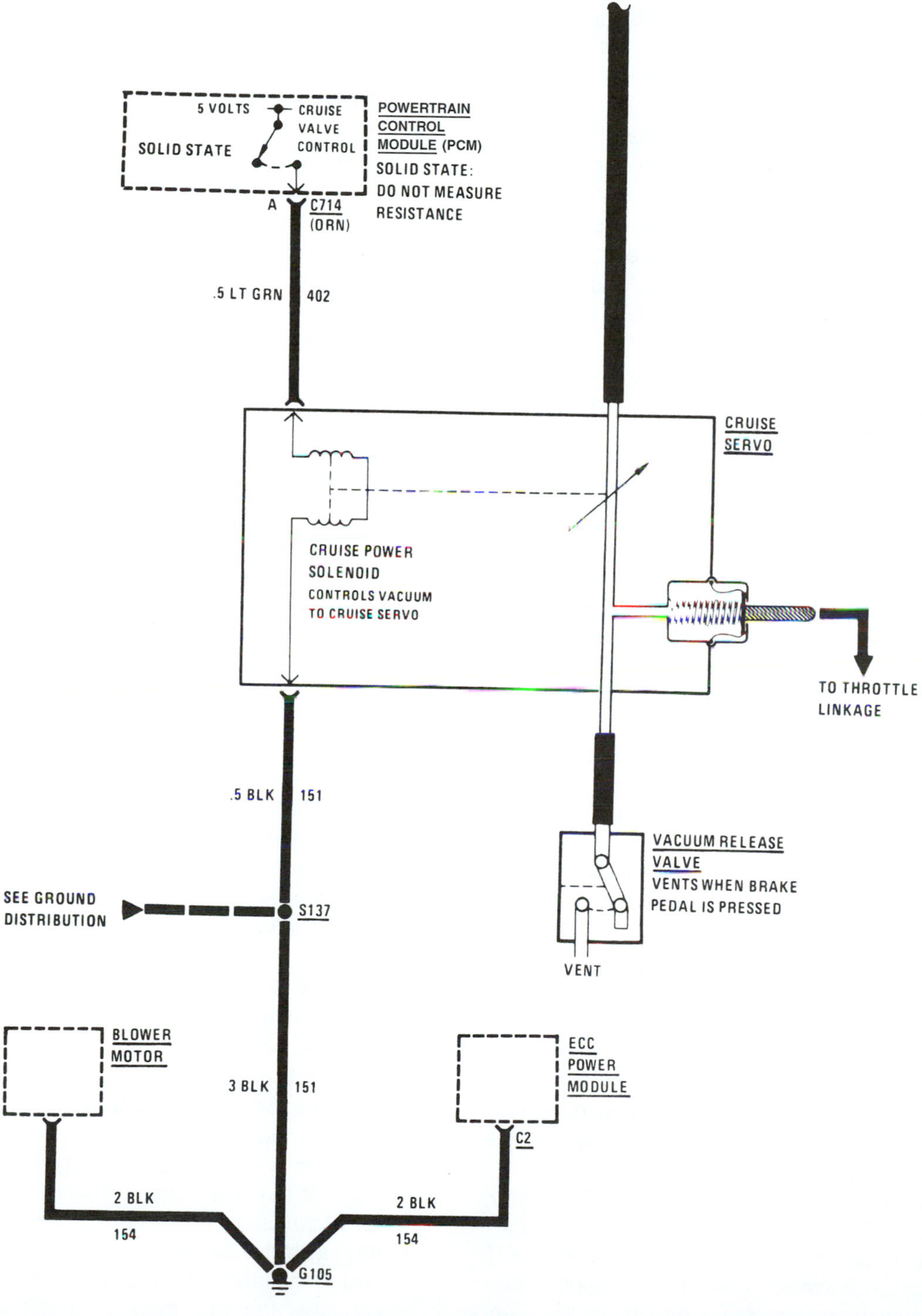

Figure 11–14 Cruise control circuit, continued. *(Courtesy of General Motors Corporation, Service Technology Group.)*

If the resume/accelerate switch is moved and held while the system is turned on, the cruise control system accelerates the vehicle until the switch is released. The system then engages whatever speed the vehicle was traveling when the switch was released. If the set/coast switch is depressed and held while the system is on, it disengages and the throttle closes. The system then reengages on whatever speed the vehicle is traveling when the set/coast switch is released.

Air-Conditioning (A/C) Relay and Cut-Out

Later model DFI systems use an A/C compressor control relay to turn the compressor clutch on and off. The relay control coil is itself grounded by the PCM, Figure 11–15. The PCM, however, gets instructions from the BCM as to when to ground or unground the relay. The climate control panel receives input from the driver. This information is used to determine the operating mode for the climate control system and is sent to the BCM by way of the serial data line. If the BCM determines that cooling is required, it uses its serial data line to the PCM to command that the compressor be turned on.

The BCM receives input from the A/C high side and low side temperature sensors plus the A/C low pressure switch. It determines whether to turn the compressor on or off according to temperature indications. It also commands the compressor to turn off if the low pressure switch opens, indicating the refrigerant quantity is low enough that continued operation might damage the compressor.

The PCM also has logic circuits that can control compressor engagement based on input from the TPS, the ECT, the VSS and the power steering pressure switch.

Earlier DFI systems did not use an A/C relay, and the only A/C-related activity of those EFI systems was to turn the compressor off during WOT operation. On those systems, the A/C compressor was energized and controlled directly by the Electronic Climate Control (ECC) module or the BCM, depending on the model and build year. When the PCM sees a WOT signal from the TPS, it commands the ECC module or the BCM to disengage the compressor. This provides more engine power for acceleration and reduces the heat load on the engine. Refer to Compressor Clutch Control and Programmer and Power Module in the Body Computer Module section of this chapter for more information on compressor clutch control.

Service and Coolant Overtemperature Lights

If either of the two modules, the PCM or the BCM, sees a fault in any of the circuits they monitor, they will turn on one of the diagnostic indicator lights, Figure 11–16. The light the computer illuminates depends on the nature of the problem detected.

BODY CONTROL MODULE (BCM)

The BCM is a microcomputer just like the PCM. It has the basic components familiar from the engine computer (ROM, RAM, PROM, input/output interface) and about the same calculating power as the PCM; however, it controls different functions in the vehicle. On the 1985 C-body Cadillac, where it was introduced, it performs the following functions:

- controls the electronic climate control (air-conditioning system)
- controls the electric cooling fans drawing air through the A/C condenser and radiator
- controls the power windows, trunk release, and electric sunroof
- provides and displays information for the driver
- controls the information display panel dimming for visual clarity in different driving conditions
- monitors the BCM system for faults, storing codes to identify the faults and in some cases provides fail-safe measures to compensate for a system failure

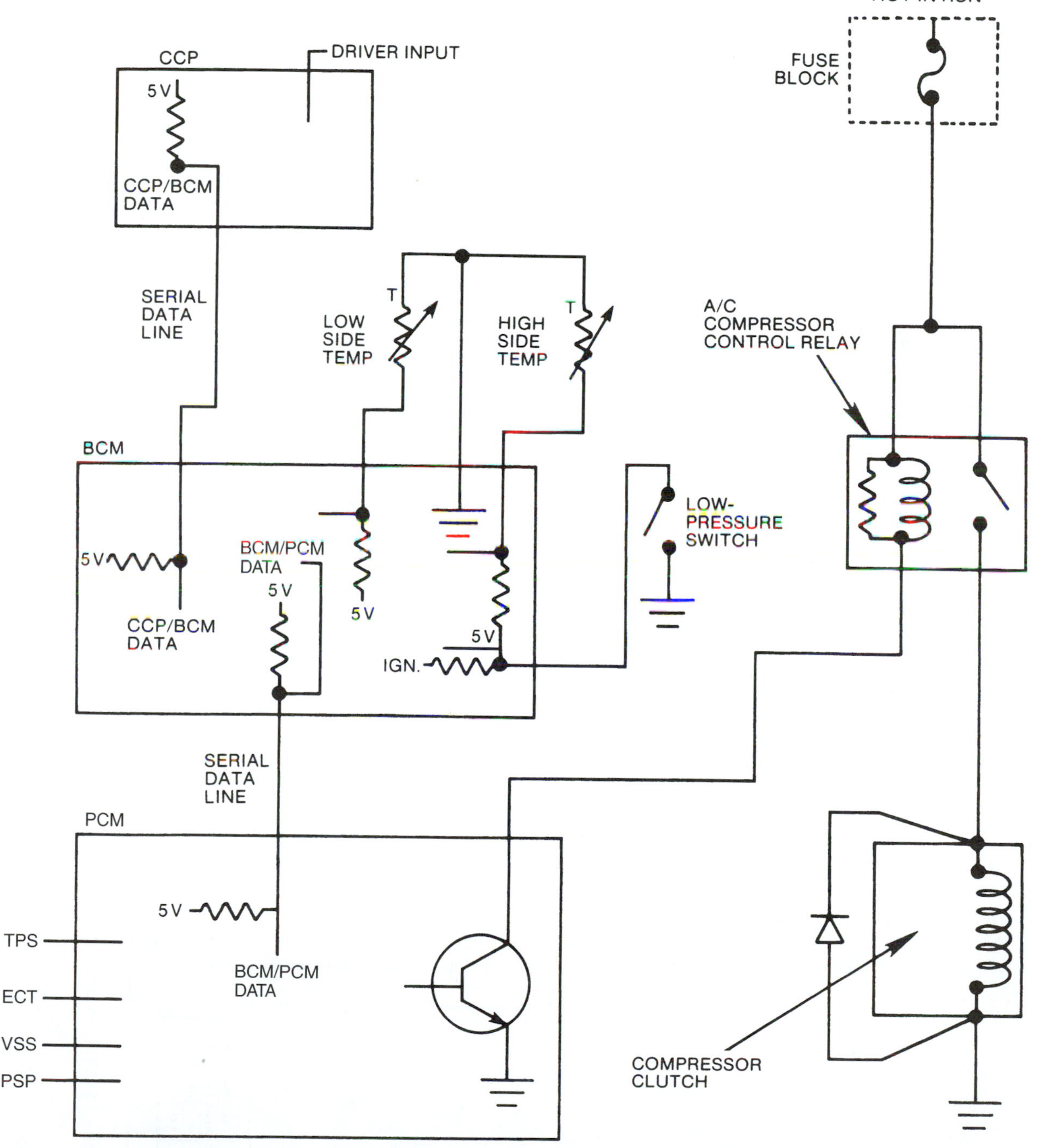

Figure 11–15 Compressor clutch control circuit.

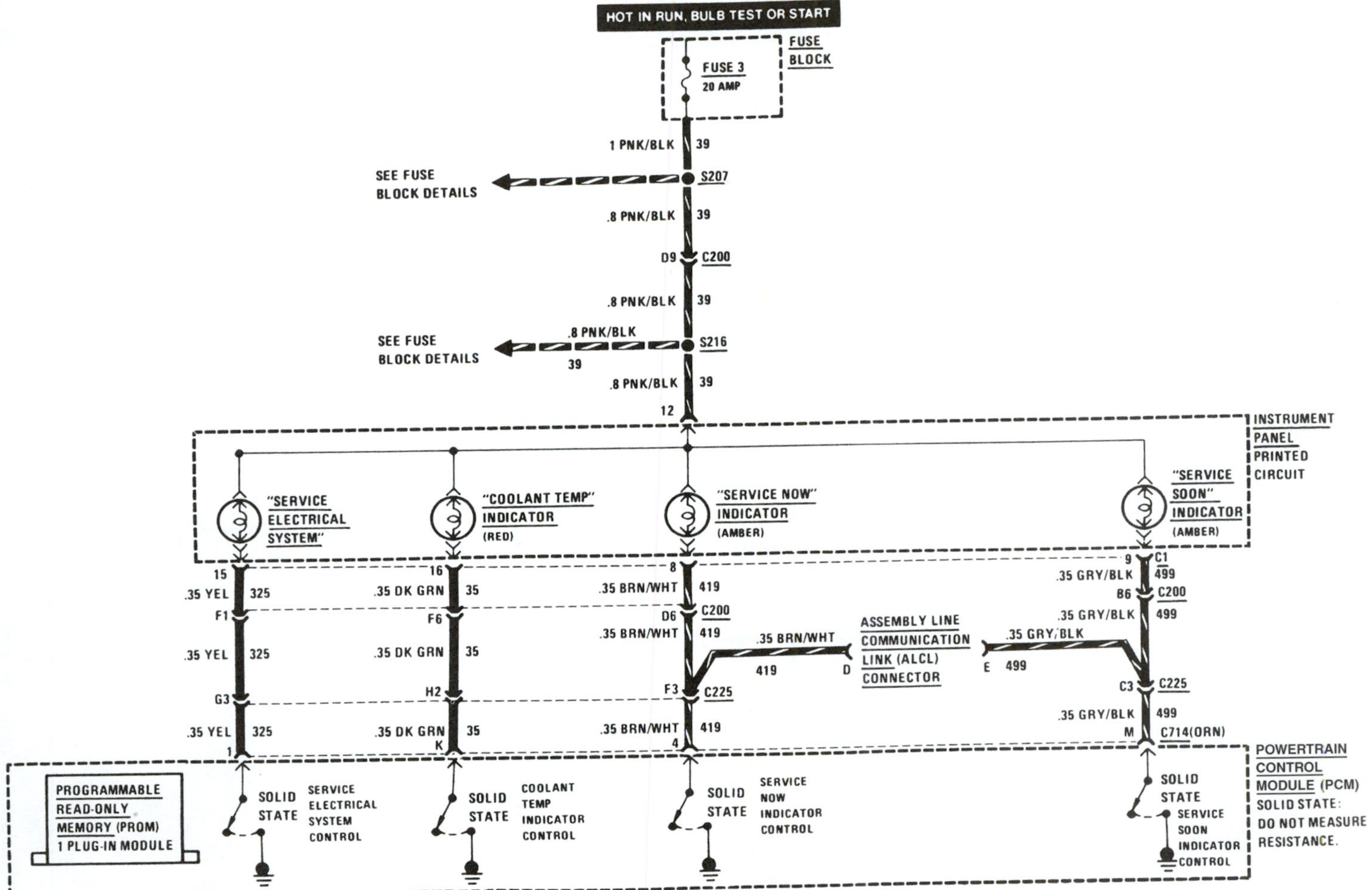

Figure 11–16 Indicator lights circuit. *(Courtesy of General Motors Corporation, Service Technology Group.)*

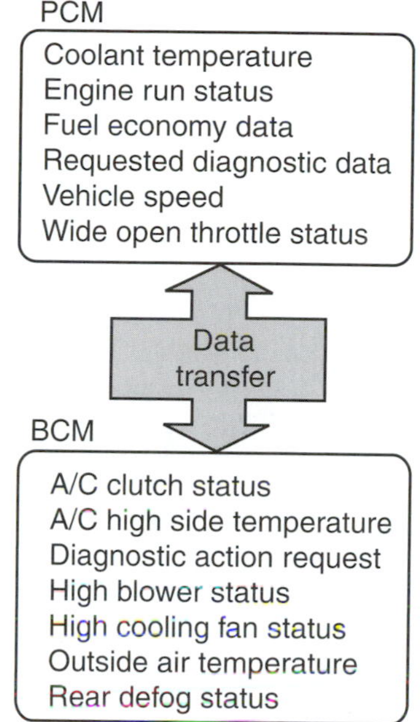

Figure 11–17 BCM/PCM interaction.

Besides its sensors and switches for input information and its actuators to activate, the BCM and the PCM interact with each other. In some cases, information in the PCM is sent by a data link to the BCM as input information, and it in turn can send outputs to the PCM as a PCM input. Figure 11–17 shows the kind of information the two computers exchange.

Electronic Climate Control (ECC)

To maintain the driver-selected temperature in the passenger compartment, the BCM selects the most appropriate air inlet and outlet modes and blower speed. It also monitors the status of the various ECC components.

ECC Inputs

Input information the BCM needs to control the ECC system comes through several sources, Figure 11–18.

Climate Control Panel (CCP). The driver selects the temperature desired and the operating mode (automatic, high fan, low fan, economy, front window defog, rear window defog) on the CCP. The CCP communicates this information in digital form through a data link to the BCM.

PCM. The PCM informs the BCM when the throttle is fully opened or when the coolant temperature goes above a specific value. In either case, the BCM commands the compressor clutch to disengage.

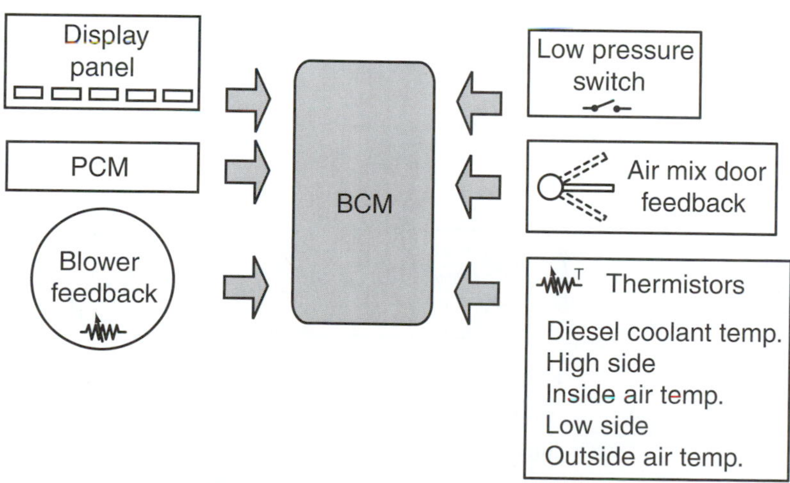

Figure 11–18 BCM information sources.

Temperature Sensors. Thermistors located in specific locations provide information about outside air temperature, inside air temperature and the temperature of the A/C refrigerant on both the low and high pressure sides of the A/C system. A switch located in the low pressure side of the system notifies the BCM if the A/C pressure goes too low.

Air Mix Door Position. A potentiometer attached to the air mix door in the air distribution system housing provides feedback information to the BCM and identifies where the door is. The air door mix blends heated and refrigerated air to produce the desired air temperature, Figure 11–19. By monitoring voltage at the blower motor, the BCM can determine the blower speed and increase or decrease it as needed, Figure 11–20.

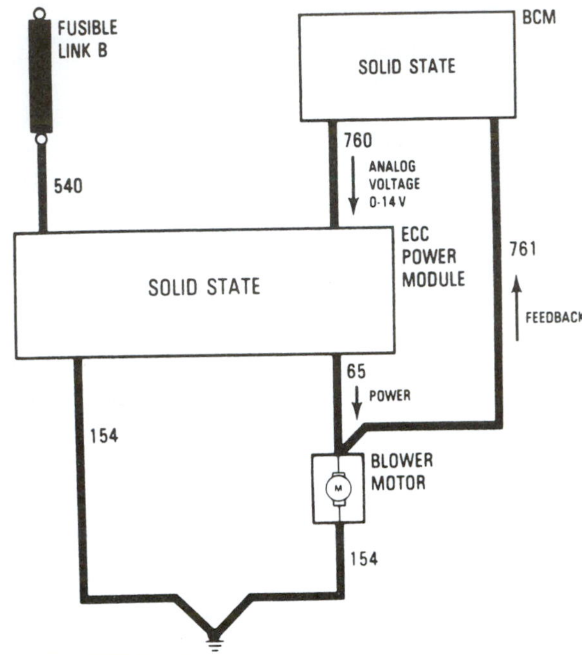

Figure 11–20 EEC blower motor control circuit. *(Courtesy of General Motors Corporation, Service Technology Group.)*

ECC Outputs

Program Number. When the engine first starts, the BCM compares inside air temperature to the driver-selected temperature. The computer then calculates what operational mode the ECC should use to achieve the selected air temperature. In its calculations, it factors in the effect of the outside air temperature. This calculation produces a program number between zero and 100. A low number means the inside air temperature is higher than the selected temperature and calls for cooling. A high number means just the opposite. The program number dictates blower speed and air distribution system door positions, Figure 11–19.

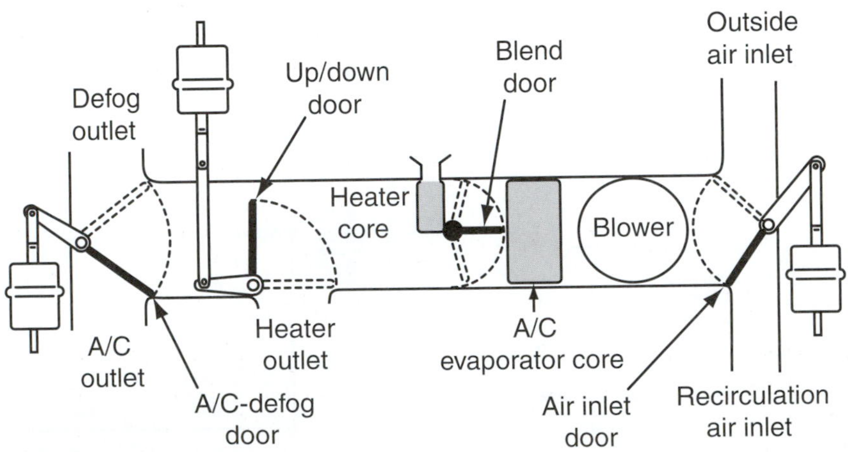

Figure 11–19 ECC mode doors.

As inside air temperature comes closer to the selected temperature, the BCM updates the program number, the blower speed decreases and the computer adjusts the position of the air distribution system doors. When the temperatures are the same, the program number reaches an equilibrium, and the BCM maintains the program number at this equilibrium point. The program number is displayed on the CCP only when the system is set to its diagnostic mode by a technician.

Air Mix Door. An electric motor moves the air mix door through the connecting linkage. All the other doors move by vacuum motors controlled by vacuum solenoids. To determine what the air mix door position should be, the BCM checks:

- the engine coolant temperature (information provided by the PCM) to determine what the heated air temperature will be.
- the air-conditioning refrigerant temperature on the low pressure side to determine what the cold air temperature will be. If the A/C is not operating, the BCM uses the outside air temperature instead.
- the program number already calculated.

The BCM uses the air mix door feedback signal to see whether the door moved to the commanded position. Because the linkage between the air mix door motor and the door itself is adjustable, it must be carefully adjusted to make sure its position coincides with the BCM-commanded position. The appropriate shop manual provides information for this adjustment.

Compressor Clutch Control. The BCM controls the compressor clutch. Before applying the clutch, it checks that:

- the outside air temperature is above 0°C (32°F).
- engine coolant and/or A/C system high pressure side refrigerant temperatures are not too high.
- refrigerant pressure (indicating the amount of refrigerant charge) is not too low.
- the throttle is not at the WOT position as signalled by the TPS.

When the compressor is engaged, the BCM constantly rechecks these factors and disengages the clutch if any of them fall above or below the threshold values.

To cycle the clutch, the BCM looks at outside air temperature and low pressure refrigerant temperatures. The compressor clutch engages when the low side temperature exceeds 10°C (50°F). The clutch disengages again when the low side refrigerant temperature drops to −1°C (30°F).

Programmer and Power Module. The BCM uses two interface devices for many ECC functions: the ECC programmer and the ECC power module, Figure 11–21. The BCM sends instructions for air door positions, heater water control, rear window defog and compressor clutch engagement information through a data link to the programmer. The rear window defogger is a heat element in the rear window glass powered through a relay. The programmer commands the actuators for each of these functions, Figure 11–22, except the compressor clutch. The compressor clutch on/off signal is sent from the programmer to the power module, and the power module actually engages or disengages the

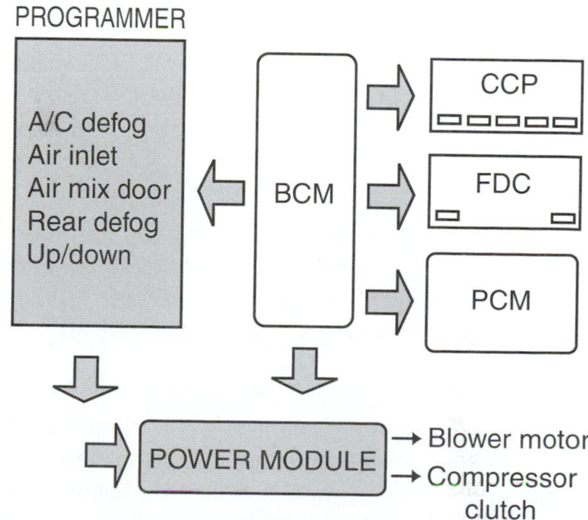

Figure 11–21 BCM interface devices.

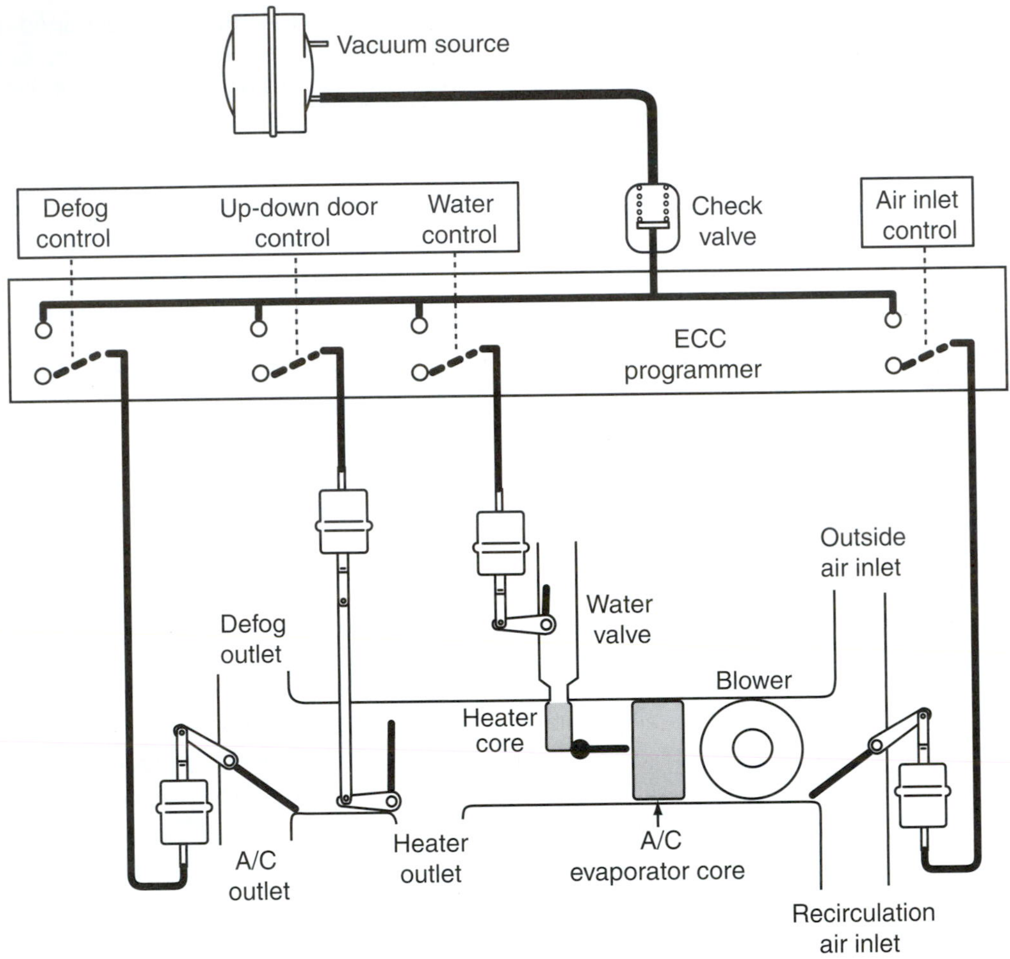

Figure 11–22 EEC programmer control of EEC mode.

clutch. The instructions to turn the compressor on or off, which the BCM sends to the programmer, are also sent to the PCM so it can anticipate the change in engine load and adjust engine speed, an adjustment that comes into play only at idle speed of course.

The BCM sends the blower motor instructions directly to the power module in the form of a variable voltage. The power module amplifies this signal and keeps it proportional to the voltage received from the BCM. It then issues that voltage as a command to the blower motor, Figure 11–20.

Electric Cooling Fans

Inputs. The BCM receives inputs for cooling fan operation from two sources: the A/C high pressure side thermistor, located in the condenser outlet, and from the engine coolant temperature information provided by the PCM.

Outputs. The speed of the two cooling fans is controlled by a pulse width modulated (PWM) signal produced by the BCM and sent to the fan control module, Figure 11–23. The module completes or opens the ground side of the fans' power

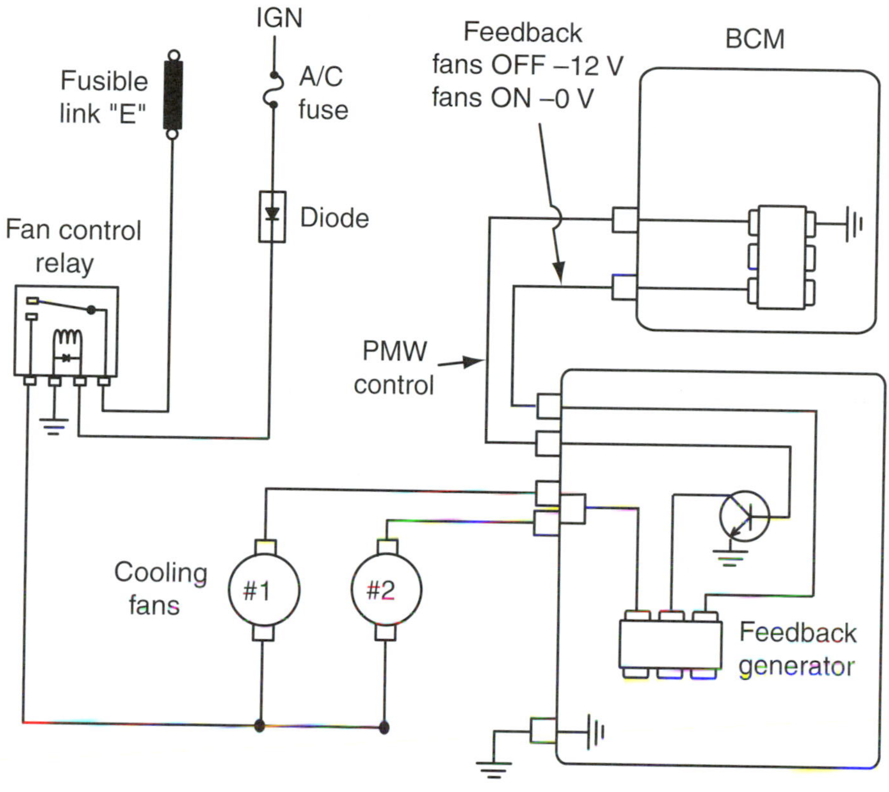

Figure 11–23 Cooling fans control.

circuit in response to the BCM signal. The PWM signal is cycled once every 32/1000 of a second. If the signal voltage is high during only a short portion of the on-period, the fan motors run at low speed, Figure 11–24. This is because their power is supplied in short pulses and they spend most of each revolution coasting, though the momentum of the blades keeps their speed near constant. The greater the proportion of each period during which the pulsed on-voltage is high, the faster the fan motors turn.

The fans run at 40 percent full speed when either the coolant temperature reaches 106°C (223°F) or when the A/C high-pressure refrigerant temperature reaches 61°C (142°F). The fans run at full speed when either the engine coolant tem-

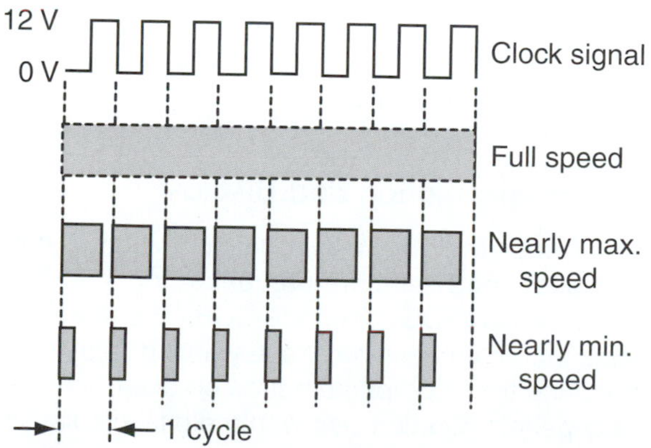

Figure 11–24 PWM command to cooling fan motors.

perature reaches 116°C (241°F) or the A/C high side refrigerant temperature reaches 72°C (162°F). They turn off completely when either the coolant temperature drops to 102°C (216°F) or the A/C high side refrigerant temperature drops to 57°C (135°F). The lower turn-off temperatures prevent the fans from dithering on and off if one of the temperatures hovers at the on-value.

The feedback signal sent to the BCM from the fan control module enables the BCM to monitor fan motor realization of the commands it was sent. This signal is nearly 12 volts when the fans are off and near zero volts when they are on. The BCM compares feedback voltage to the commands it sent previously; if they do not match, it sets fault code F41.

Retained Accessory Power

If the ignition is turned off (locked) or turned to the accessory position, the BCM provides power to the power windows, electric roof and the trunk release for ten minutes or until either a door opens or the courtesy lights turn off, Figure 11–25. Once the ignition is off, the BCM grounds the retained power accessory relay for 10 minutes. This relay then connects circuit 300 to the battery voltage from a constantly powered source. Circuit 300 powers the trunk release, power windows and electric sunroof. If a door switch, the courtesy light switch or the ignition switch turns on, the BCM ungrounds that relay, and the relay reconnects circuit 300 to the ignition run terminal. Circuit 300 will then be powered only when the ignition switch is turned on.

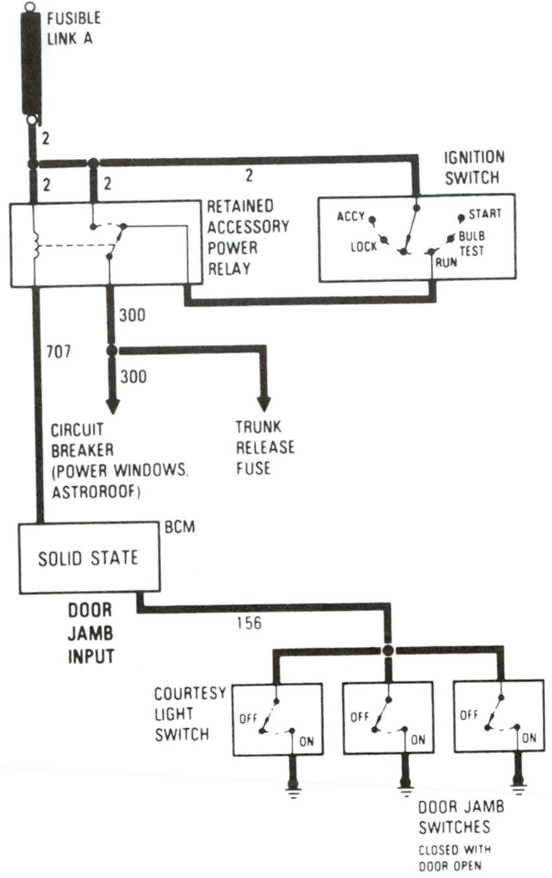

Figure 11–25 Retained accessory power circuit. *(Courtesy of General Motors Corporation, Service Technology Group.)*

Driver Information and Display

The BCM displays information about fuel consumption and temperature either on the Fuel Data Center (FDC) or on the CCP. The driver selects the information he or she wants displayed by pressing the appropriate button on one of the display panels. On vehicles with a digital display instrument cluster, the driver can choose to have the information displayed in either English or metric units by moving the English/metric switch on the digital cluster. Vehicles not equipped with the digital cluster display in either English or metric, depending on whether a wire (circuit 811) leading from terminal J2-13 of the BCM is connected to ground. If the wire is grounded, the information is displayed in metric units; if it is open, the information is in English units.

Fuel Data Inputs. The BCM gets information from the fuel gauge sender (a standard, variable potentiometer), the gauge fuse and the PCM. The potentiometer measures fuel level just as fuel gauge senders have done for many years.

An arm with a float on it moves the movable contact of the potentiometer as the fuel level changes. By comparing return voltage from the potentiometer to a voltage reading taken at the gauge fuse, the BCM can calculate the fuel level, Figure 11–26.

Vehicle speed and distance traveled (this information can be easily calculated from VSS information because 4004 VSS pulses equal 1 mile using the pulse generator-type VSS) plus injector pulse width information are sent from the PCM to the BCM through the data link.

Fuel Data Outputs. Normally the BCM displays fuel level through the FDC as explained previously. However, on signalled request it can also display:

- *Instantaneous fuel mileage.* If the INST button on the FDC panel is pressed, the BCM looks at fuel consumed (calculated from the injector pulse width data) and vehicle speed. From these two pieces of information, it can instantly calculate the current achieved fuel mileage at that moment.
- *Average fuel mileage.* The BCM constantly receives and updates its information regarding fuel used and distance traveled. This information is stored in memory. If the driver presses the reset button on the FDC panel, however, all the previous information is erased and the computer accumulates new data. If the driver pushes the AVG button, the BCM will calculate fuel mileage for the distance traveled since the last reset.
- *Fuel range.* When the RANGE button is pressed, the BCM checks remaining fuel from the fuel level information, calculates the average fuel use for the last 25 miles and calculates a distance that can be traveled with the remaining fuel. Obviously, if driving speed or other conditions significantly changes so fuel consumption goes up or down, this calculation will be incorrect.
- *Fuel used.* When the driver requests FUEL USED, the BCM calculates the fuel used since the last reset. This does *not* indicate fuel used since the last fill-up unless that coincides with the last reset.

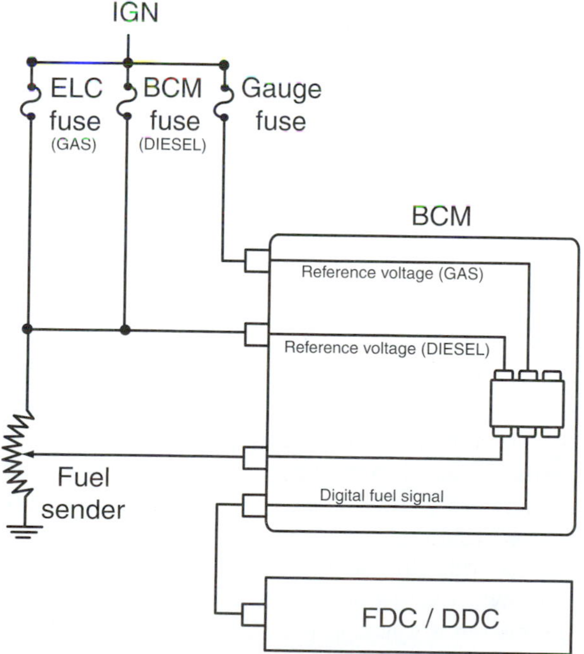

Figure 11–26 Fuel gauge circuit.

Temperature Display. Upon driver request, the BCM displays outside temperature as reported to it by the outside temperature thermistor. If outside temperature is not requested, the CCP displays inside, passenger compartment temperature.

Instrument Panel Display Dimming

The characters displayed by the FDC, the CCP, the radio and the optional digital cluster are displayed by a group of small **vacuum fluorescent (VF)** glass tubes. **Anodes** (conductors on which there is a positive potential voltage) are placed on one side of the tube so they form all of the vertical and horizontal bars of the alphanumeric characters. The anodes are coated with a fluorescent material. A series of thin, tungsten-coated wire strands are placed opposite the anodes on the other side of the tube. This side then serves as the **cathode** (a conductor on which there is a negative potential voltage). Between the anode and the cathode, a fine wire mesh called the **grid** is placed. The tube (**ampule**) is evacuated and then filled with argon or neon gas.

As current passes through the cathode, it becomes hot and causes the tungsten to give off a cloud of electrons. The BCM applies an amplified 17-volt positive potential to selected anodes. The electrons given off by the cathode are attracted to and bombard the energized anodes. As they strike an anode's fluorescent coating, it glows visibly and thus provides the digital display. The more electrons striking the fluorescent material, the more brightly it glows.

During daylight hours the BCM adjusts the VF displays to full brightness for maximum visibility. As the sunlight fades, the VF displays need less brightness to be seen in the darker surroundings. When the parking lights are turned on (either with the headlights or independently), the BCM gets that information as an input signal, Figure 11–27. When the BCM learns the headlights or parking lights have been turned on, it uses the headlight switch rheostat to determine how bright to make the display. The driver can set this rheostat throughout its range to select VF display brightness.

The BCM controls the display brightness by controlling a pulse-width-modulated voltage to the grid in the ampule. When the BCM applies a modulated voltage to the grid, it attracts or filters some of the electrons given off by the cathode and prevents them from striking the energized anodes.

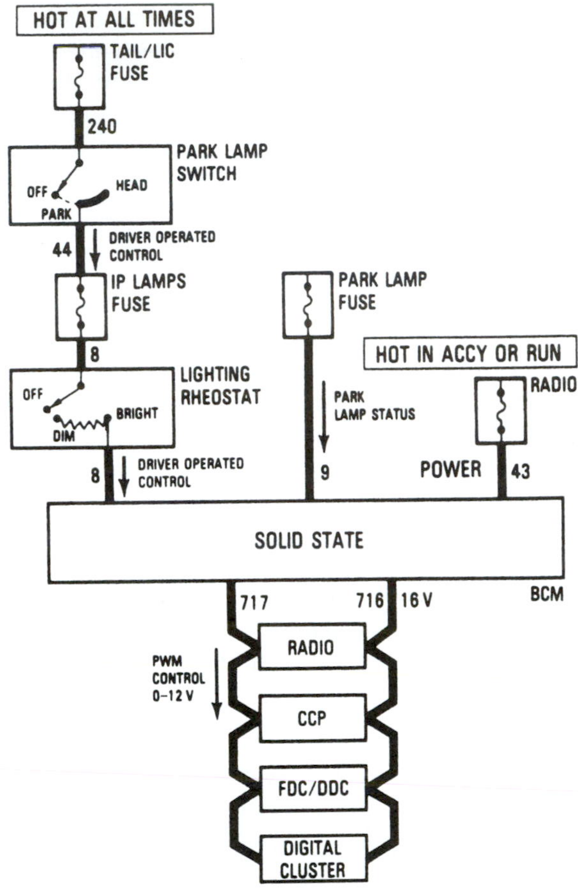

Figure 11–27 Vacuum fluorescent display dimming circuit. *(Courtesy of General Motors Corporation, Service Technology Group.)*

Diagnostic Testing and Fail-Safe Actions

The BCM monitors many of its sensor and output circuits for proper operation, such as the blower motor and the cooling fan motors. It also uses sensors specifically designed to report problems such as the low-pressure refrigerant switch in the A/C **accumulator**. If the BCM sees such a fault (voltage in a given circuit or from a sensor that is out of the expected range) in any of its related circuits, it records a code in its diagnostic memory. It may turn on the service air-conditioning light, or it may initiate a fail-safe action, depending on the circuit with the fault, Figure 11–28. For example, every time the BCM turns on the A/C compressor clutch, it checks the A/C low-pressure switch first. Should the switch be open, indicating low pressure, it will not engage the clutch. If the switch stays open for 30 seconds, it will set code F48 and the following fail-safe actions are initiated:

BCM DIAGNOSTIC CODES

Turns on Service A/C light	No telltale light	Code	Malfunction
	✔	F10	Outside temperature sensor circuit
	✔	F11	A/C high side temperature sensor circuit
	✔	F12	A/C low side temperature sensor circuit
	✔	F13	In-car temperature sensor circuit
	✔	F14	Diesel coolant sensor circuit
	✔	F30	CCP to BCM data circuit
	✔	F31	FDC/DDC to BCM data circuit
	✔	F32	PCM-BCM data circuit
	✔	F40	Air mix door problem
	✔	F41	Cooling fan problem
✔		F46	Low refrigerant warning
✔		F47	Low refrigerant condition
✔		F48	Low refrigerant pressure
	✔	F49	High temperature clutch disengage
	✔	F51	BCM PROM error

Comments:

F11 Turns on cooling fans when A/C clutch is engaged.
F12 Disengages A/C clutch.
F14 & F32 Turn on coolig fans.
F30 Turns on front defog at 75° F.
F41 Turns on Coolant Temp/Fans light when fans should be on.
F47 & F48 Switches from AUTO to ECON.

Figure 11–28 BCM diagnostic codes.

- The service air-conditioning light turns on for 30 seconds. The light will come on again every time the driver selects AUTO on the CCP and the next time the ignition switch is turned on.
- The BCM disables the compressor clutch until the ignition is turned off.
- The BCM switches the ECC system into ECON instead of AUTO if the AUTO mode is selected.

Keep in mind the purpose of fail-safe operation is to protect system components and that the codes initiating fail-safe measures cannot be set while the system is in diagnostics. So if a code is cleared while the system is in a diagnostic mode, the fail-safe measures will be defeated and it is possible to cause damage to a system component.

✔ SYSTEM DIAGNOSIS AND SERVICE

The self-diagnostic function of the DFI system works fundamentally the same way as other General Motors computer control systems, but it is more elaborate, monitoring and controlling more functions. When first introduced in mid-1980 (the same system remained throughout model year

1981), its self-diagnostic capacity was like that of a CCC system. Since then, however, the system's capacities have greatly expanded. As the PCM and/or BCM control the systems for which they are responsible, each continuously monitors operating conditions for faults. They compare the operating conditions against expected, preprogrammed operating condition standards. By doing this, certain circuit and component faults can be identified. When a fault is identified, a two-digit DTC, either with an E or an F prefix, is stored in the computer's memory. Codes with an F prefix (or, on some newer models, a "B") are stored by the BCM, Figure 11–28. Codes with an E prefix are stored by the PCM, Figure 11–29. When the system is in diagnostic mode, the stored code displays either on the CCP or the FDC, depending on the year and model of the car.

Types of Diagnostic Trouble Codes (DTCs)

Each computer can store a DTC either of two ways: as a current mode (sometimes called a *hard code* because it refers to a problem that currently exists) or as a history code (representing a fault that occurred at one time during engine operation but is not present at the time of diagnosis). When the system goes into diagnostic mode, all of the codes display once in numerical order from low to high; then they repeat. This time, however, only the current, hard codes display. From 1982 through 1984, DFI systems show all codes in the first two passes through the display and show only current hard codes in a third pass-through.

Service Lights and Fail-Safe

Faults the PCM recognizes (E-prefixed codes) and that require immediate attention turn on the service engine now light. Faults the PCM recognizes and those that need to be called to the driver's attention but are less urgent, bring on the service engine soon light. Certain F codes pertaining to the ECC bring on the service air-conditioning light.

If a fault that might cause damage to a circuit or component that can result in unacceptable performance during continued vehicle operation, the PCM or BCM will activate a fail-safe strategy. The fail-safe action is designed to either deactivate the faulty component or compensate for its malfunction or both.

Diagnostic Procedure

Before beginning a diagnostic procedure on a DFI-equipped vehicle, be sure that first you:

- have at least a general understanding of the operation of the PCM and/or the BCM and any of the subsystems in which the problem or problems seem to occur.
- are at least familiar with the self-diagnostic features of the system.
- have access to a service manual listing the diagnostic charts. Reading the codes without knowing what they mean will not help the technician.
- do a thorough visual inspection of the system in which the fault seems to occur and any related systems. Check for loose or broken vacuum lines, pinched or broken wires and loose connectors.

Diagnostic Guide. Having complied with the previous suggestions, continue your diagnostic procedure by answering the questions in the next four paragraphs:

- *Are the on-car self-diagnostic systems working?* If the service air conditioning, service engine now or service engine soon lights are not working properly, or if either of the data display panels fails to fully light up or respond properly, the self-diagnostic system check will refer to the appropriate chart for correcting the problem, Figure 11–30.
- *Are DTCs displayed?* If any DTCs are displayed, go to the code chart of the same number. If the fault is intermittent (the code was not repeated on the second pass, or third pass for earlier models), observe any notes

PCM/ECM DIAGNOSTIC CODES

Turns on SERVICE NOW light	Turns on SERVICE SOON light	No telltale light	Code	Malfunction
✔			E12	No distributor signal
	✔		E13	Oxygen sensor not ready
	✔		E14	Shorted coolant sensor circuit
	✔		E15	Open coolant sensor circuit
✔			E16	Generator voltage (out of range)
	✔		E18	Open crank signal circuit
	✔		E19	Shorted fuel pump circuit
✔			E20	Open fuel pump circuit
	✔		E21	Shorted TPS circuit
	✔		E22	Open TPS circuit
	✔		E23	EST/Bypass circuit problem (air)
	✔		E24	Speed sensor circuit problem
	✔		E26	Shorted throttle switch circuit
	✔		E27	Open throttle switch circuit
	✔		E28	Open 3rd or 4th gear circuit
	✔		E30	ISC circuit problem
✔			E31	Shorted MAP sensor circuit (air)
✔			E32	Open MAP sensor circuit (air)
✔			E34	MAP sensor signal too high (air)
	✔		E37	Shorted MAT sensor circuit
	✔		E38	Open MAT sensor circuit
	✔		E39	VCC engagement problem
	✔		E40	Open pwr. strg. press. circuit
✔			E44	Lean exhaust signal
✔			E45	Rich exhaust signal
	✔		E47	BCM - ECM data problem
✔			E51	ECM PROM error
		✔	E52	ECM memory reset indicator
		✔	E53	Distributor signal interrupt
		✔	E59	VCC temperature sensor circuit
		✔	E60	Transmission not in drive
		✔	E63	Car and set speed difference (high)
		✔	E64	Car acceleration too high
		✔	E65	Coolant temperature too high
		✔	E66	Engine rpm too high
		✔	E67	Cruise sw. shorted during enable

Comments: E16 & E24 disable VCC for entire ignition cycle.
E24 & E67 disable cruise for entire ignition cycle.
Cruise is disenabled with codes e16, E51 or E60–E67.

Figure 11–29 PCM (ECM) diagnostic codes.

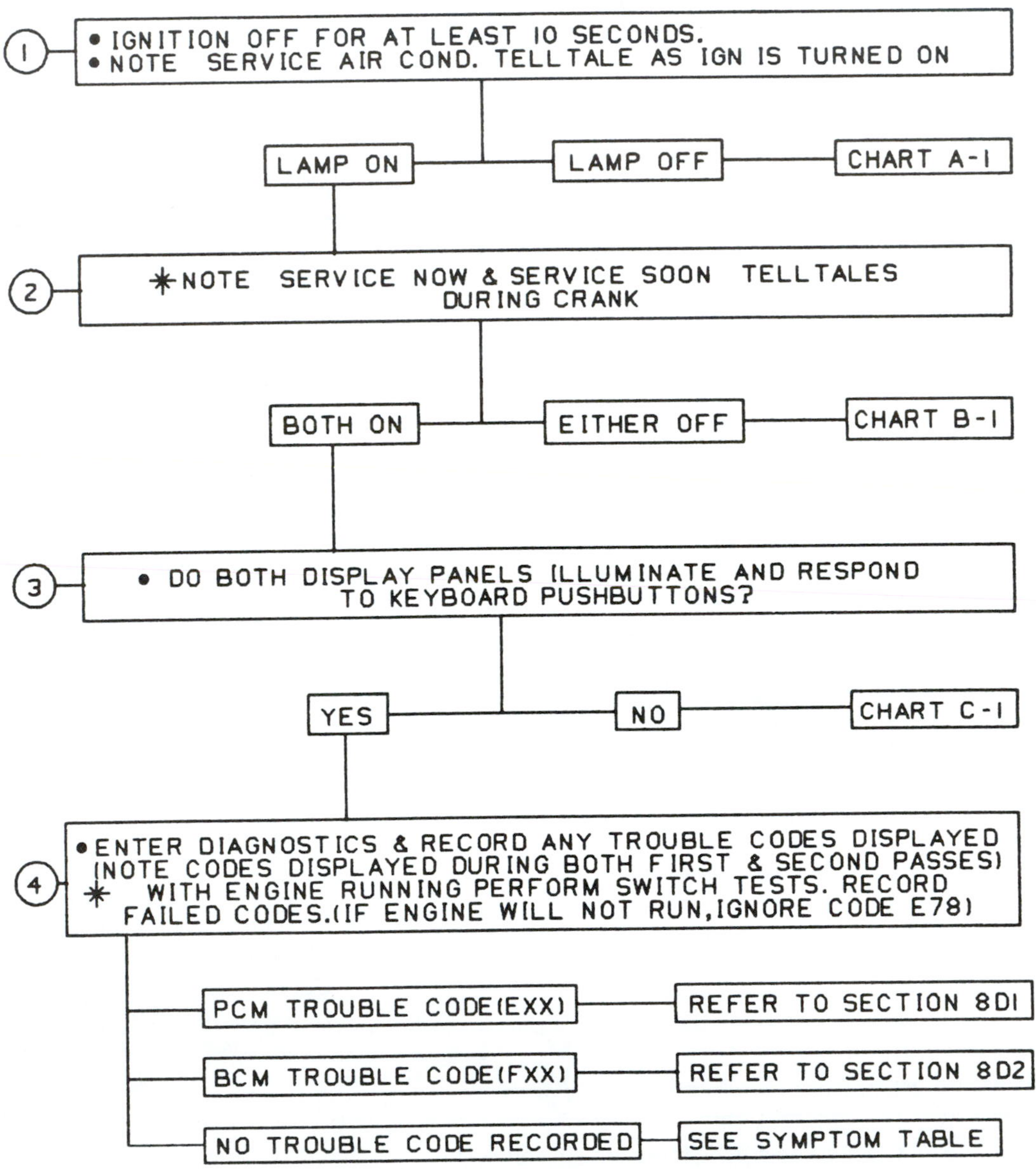

Figure 11–30 Self-diagnostic system check. *(Courtesy of General Motors Corporation, Service Technology Group.)*

concerning intermittent codes in the text that accompanies each code chart. The procedure for displaying the DTCs is explained later in this section. If no DTCs are stored, proceed to the next paragraph.

- *Do all switch tests pass?* After the DTCs have been displayed, perform the switch test as explained under Switch Test in this section. If any switch tests do not pass, go to the code chart identified by the corresponding number as the diagnostic display number (under Switch Tests in Figure 11–31) for the switch that did not pass. If all the switches do pass the test, go to the next paragraph.
- *Is the fuel system controlling air/fuel mixture correctly?* To answer this, use the field service mode, which is a part of the DFI system check, Figure 11–32. If the fuel system is not controlling air/fuel mixture properly, the DFI system check will refer you to the appropriate chart. If the fuel system is controlling mixture properly, the DFI system check will refer you to the symptom charts.

Using these simple steps at the beginning of the diagnostic procedure saves time and often prevents unnecessary parts replacement resulting from misdiagnosis.

Entering Diagnostic Mode. To enter diagnostics, turn the ignition on, then simultaneously depress the off and warmer buttons on the CCP, Figure 11–33. Hold the buttons until all display panel segments of both display panels light up. If any of the digit segments do not light up (they should all make squared-off computer 8's), the panel will have to be replaced. An 8 displayed by a panel with two segments inoperative can appear as a 3.

Display of Trouble Codes. After the display panel segment check ends, the FDC displays 8.8.8 followed by ..E. The ..E indicates that the first pass of the PCM codes will be displayed next and that all PCM codes will display. After the first pass, ..EE displays and indicates the second pass of the PCM codes is coming, in which only the hard codes will display. If all stored PCM codes are intermittent, the ..EE will not display and the second pass will not occur. If code E51 (a PROM fault) displays, it will continue to display until the diagnostic mode is exited or until the PROM fault is repaired and the code clears. While code E51 displays, the system will not advance to another diagnostic feature.

If no PCM codes are stored, the 8.8.8 will be followed by ..F, indicating that BCM codes come next. The same sequence occurs. The first pass displays all stored BCM codes followed by ..FF, and the second pass displays only hard codes. If no BCM hard codes are stored, the ..FF will not display and the second pass will not occur.

After all stored codes display, or if there are no stored codes, .7.0 will display. This indicates the system is ready for selection of the next self-diagnostic feature. The technician can select several different diagnostic tests at this point:

- PCM switch tests
- PCM data display
- PCM output cycling
- BCM data display
- ECC program override
- cooling fan override
- exit diagnosis or clear codes and exit diagnosis

PCM Switch Tests. The engine must be running to perform this test sequence. When the FDC displays .7.0, depress and release the brake pedal. The FDC display should advance to E.7.1. When it does, cycle the cruise control brake switch by tapping the brake pedal again within 10 seconds. As the switch cycles, the PCM monitors its operation to be sure it is functioning properly. If the switch does not cycle within 10 seconds, the PCM will record it as having failed. At the end of 10 seconds or after the switch has cycled properly, the display advances to the next switch test code. The same sequence repeats for each code and switch listed under Switch Test in Figure 11–31.

After the switch test sequence completes, the PCM displays on the FDC the code of any switch that did not pass. The failed switch code remains on display until the switch circuit is repaired and

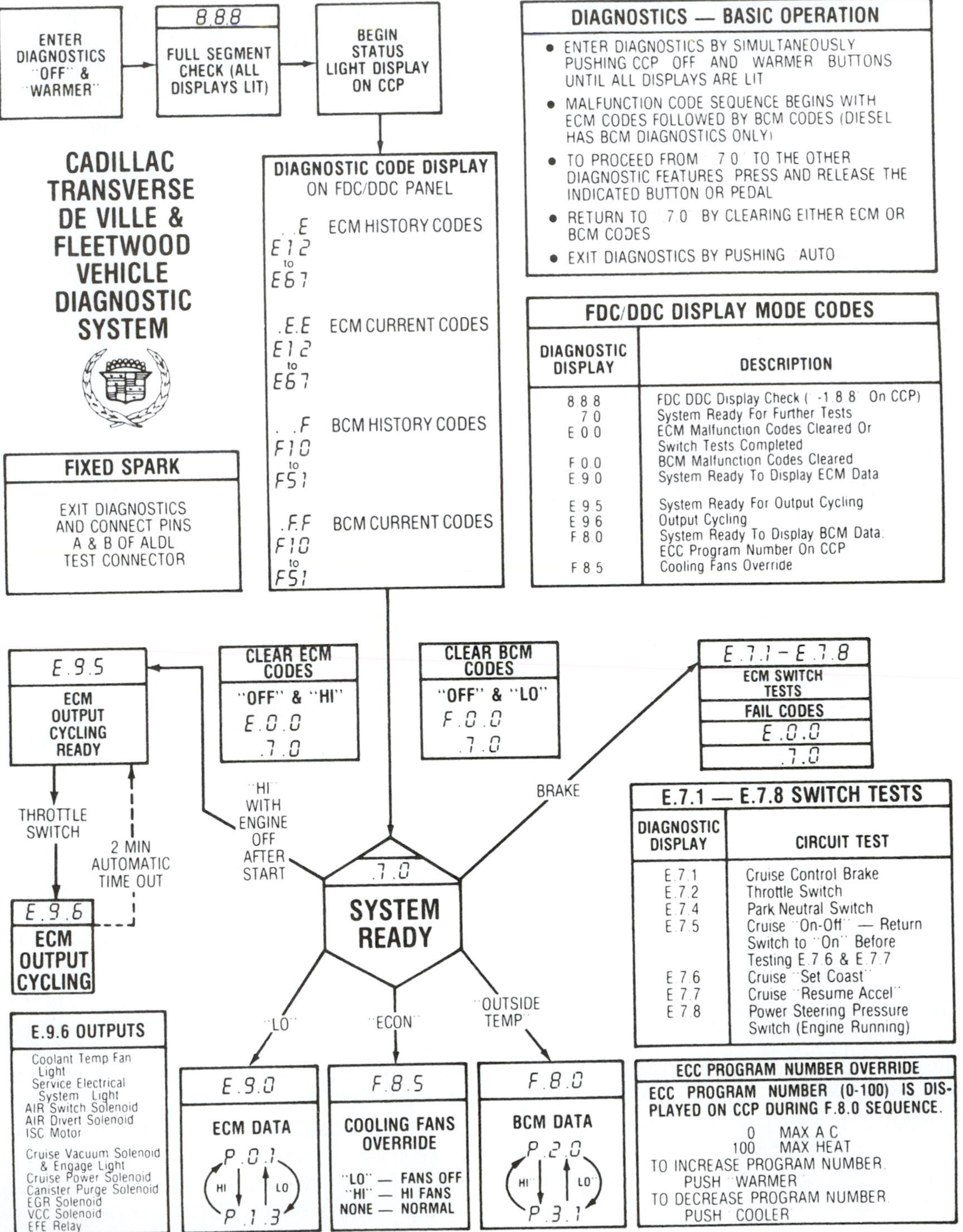

Figure 11–31 Diagnostic chart, part 1. *(Courtesy of General Motors Corporation, Service Technology Group.)*

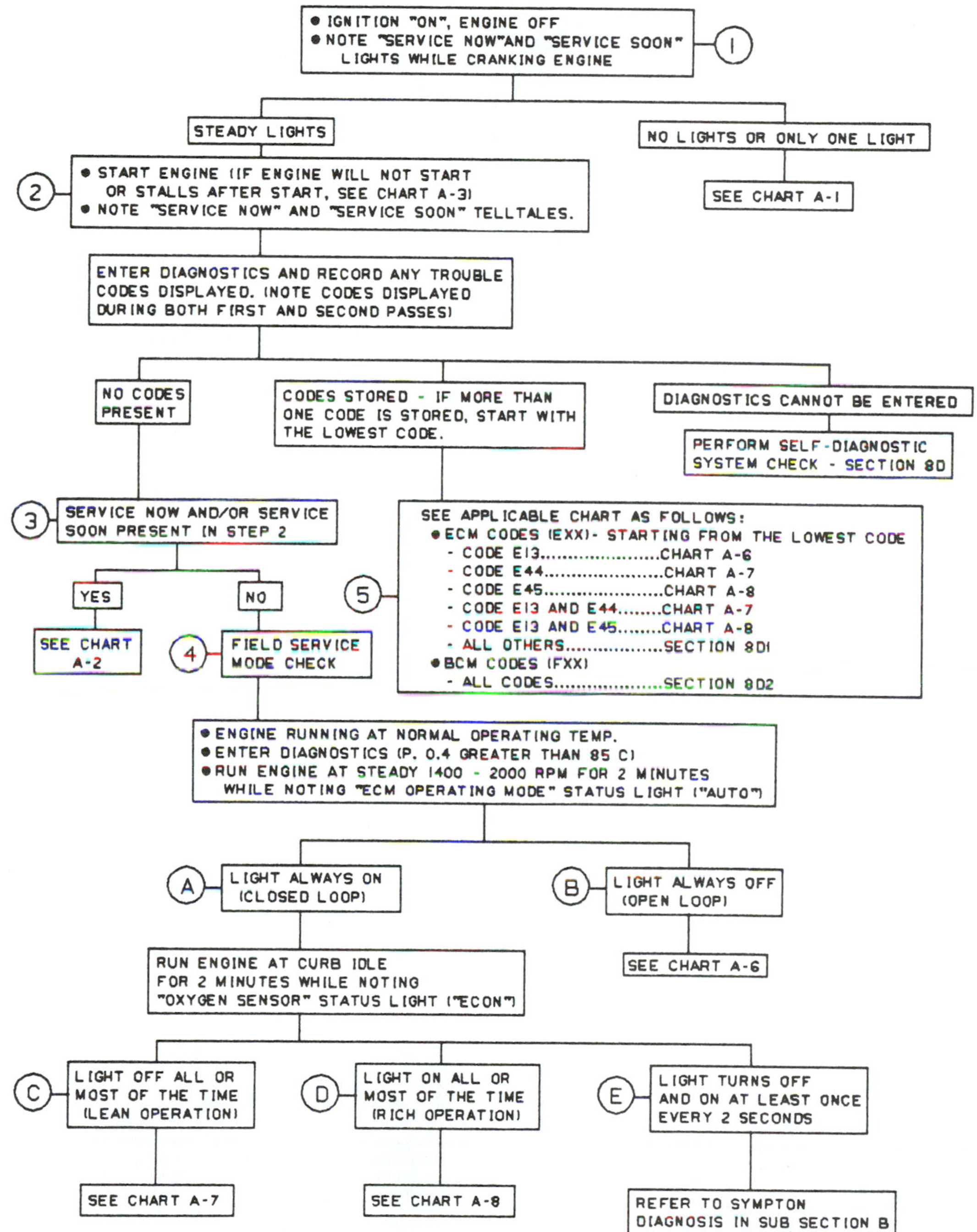

Figure 11–32 DFI system check. *(Courtesy of General Motors Corporation, Service Technology Group.)*

E.9.0 ENGINE DATA DISPLAY

PARAMETER NUMBER	PARAMETER	PARAMETER RANGE	DISPLAY UNITS
P.0.1	Throttle Position	-10 - 90	Degrees
P.0.2	MAP	14 - 109	kPa
P.0.3	Computed BARO	61 - 103	kPa
P.0.4	Coolant Temperature	-40 - 151	C
P.0.5	MAT	-40 - 151	C
P.0.6	Injector Pulse Width	0 - 99.9	ms
P.0.7	Oxygen Sensor Voltage	0 - 1.14	Volts
P.0.8	Spark Advance	0 - 52	Degrees
P.0.9	Ignition Cycle Counter	0 - 50	Key Cycles
P.1.0	Battery Voltage	0 - 25.5	Volts
P.1.1	Engine RPM	0 - 6370	RPM ÷ 10
P.1.2	Car Speed	0 - 255	MPH
P.1.3	ECM PROM I.D.	0 - 255	Code

F.8.0 BCM DATA DISPLAY

PARAMETER NUMBER	PARAMETER	PARAMETER RANGE	DISPLAY UNITS
P.2.0	Commanded Blower Voltage	-3.3 - 18.0	Volts
P.2.1	Coolant Temperature	-40 - 215	C
P.2.2	Commanded Air Mix Door Position	0 - 100	%
P.2.3	Actual Air Mix Door Position	0 - 100	%
P.2.4	Air Delivery Mode 0 Max A/C 4 Off 1 A/C 5 Normal Purge 2 Intermediate 6 Cold Purge 3 Heater 7 Front Defog	0 - 7	Code
P.2.5	In-Car Temperature	-40 - 102	°C
P.2.6	Actual Outside Temperature	-40 - 93	°C
P.2.7	High Side Temperature (Condenser Out)	-40 - 215	°C
P.2.8	Low Side Temperature (Evaporator In)	-40 - 93	°C
P.2.9	Actual Fuel Level	0 - 19.0	Gallons
P.3.0	Ignition Cycle Counter	0 - 99	Key Cycles
P.3.1	BCM PROM I.D.	0 - 255	Code

ECM PROM I.D.

ECM PROM I.D. is Parameter 1.3 of Engine Data and is displayed as a numerical code as follows: X X X

FINAL DRIVE RATIO
2 = 3.33:1
(2.97:1 Effective Ratio)

EMISSIONS SYSTEM
1 = Federal
2 = California
3 = Export
4 = Altitude

ECM PROM CALIBRATION
Number varies with individual calibration.

BCM PROM I.D.

BCM PROM I.D. is Parameter 3.1 of BCM Data and is displayed as a numerical code as follows: X X X

ENGINE SYSTEM
Blank = Gas
1 = Diesel

BCM PROM CALIBRATION
Numbers vary with individual calibration.

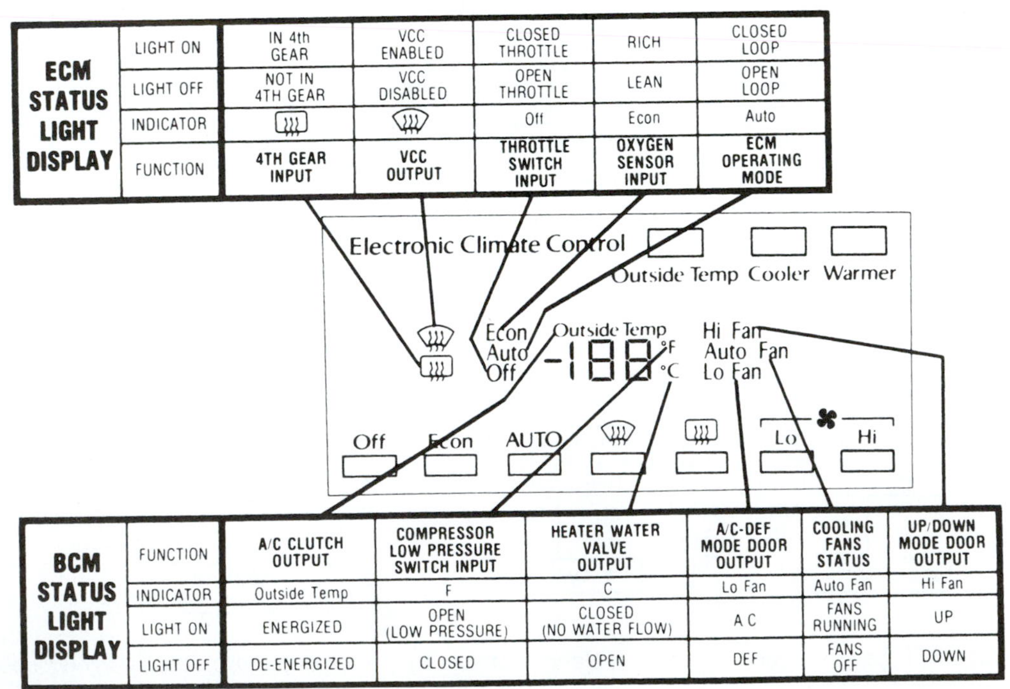

ECM STATUS LIGHT DISPLAY

LIGHT ON	IN 4th GEAR	VCC ENABLED	CLOSED THROTTLE	RICH	CLOSED LOOP
LIGHT OFF	NOT IN 4TH GEAR	VCC DISABLED	OPEN THROTTLE	LEAN	OPEN LOOP
INDICATOR	(rear defog symbol)	(front defog symbol)	Off	Econ	Auto
FUNCTION	4TH GEAR INPUT	VCC OUTPUT	THROTTLE SWITCH INPUT	OXYGEN SENSOR INPUT	ECM OPERATING MODE

BCM STATUS LIGHT DISPLAY

FUNCTION	A/C CLUTCH OUTPUT	COMPRESSOR LOW PRESSURE SWITCH INPUT	HEATER WATER VALVE OUTPUT	A/C-DEF MODE DOOR OUTPUT	COOLING FANS STATUS	UP/DOWN MODE DOOR OUTPUT
INDICATOR	Outside Temp	F	C	Lo Fan	Auto Fan	Hi Fan
LIGHT ON	ENERGIZED	OPEN (LOW PRESSURE)	CLOSED (NO WATER FLOW)	A/C	FANS RUNNING	UP
LIGHT OFF	DE-ENERGIZED	CLOSED	OPEN	DEF	FANS OFF	DOWN

Figure 11–33 Diagnostic chart, part 2. *(Courtesy of General Motors Corporation, Service Technology Group.)*

retested. After the test sequence has completed and all switches have passed, E.O.O will display, followed by .7.0 (meaning the system is ready for another diagnostic feature selection).

PCM Data Display. This mode is often called *engine parameters*. In this mode, the PCM displays the values it receives from selected sensors or switches. It can be entered by pressing the Lo button on the CCP while .7.0 displays on the FDC. When this mode is entered, E.9.0. appears on the FDC. The parameter codes, the parameters they represent, the value range and the units in which the parameters are expressed are shown in Figure 11–33 under Engine Data Display. To advance to the next higher parameter code number, press the Hi button on the CCP. Pushing the Lo button returns to the next lower parameter code number.

These parameter values can be used as follows:

- They can be compared to those of a vehicle known to be operating properly to see whether they are what they are supposed to be.
- They can be watched to see that they change as operating conditions change (for instance, the TPS parameter should change as the throttle moves).
- Some of them are called for in the diagnostic charts, where their values are interpreted as acceptable or faulty.

This diagnostic mode can be exited at any time and the display returned to the selection point, .7.0. by clearing the PCM or BCM codes (press Off and Hi for PCM codes or Off and Lo for BCM codes).

PCM Output Cycling. In this diagnostic mode, the PCM cycles all of its output actuators every 3 seconds with the exception of the cruise control power valve, which cycles continuously. Cycling continues for 2 minutes; then the system automatically reverts to PCM Output Cycling Ready (E.9.5 displays on the FDC). While the actuators cycle, you can observe that each component operates properly. To enter this output cycling mode, .7.0 must be displayed and the engine must be running.

- Turn the cruise on/off switch to the on position. This allows the cruise outputs to cycle.
- Turn the engine off. Within 2 seconds turn the ignition on.
- Press Hi on the CCP (E.9.5 displays on the FDC).
- Open and close the ISC throttle switch by depressing and releasing the throttle. The PCM output cycling mode is activated (E.9.6 displays).

This diagnostic mode can be exited at any time and the display returned to the selection point .7.0 by clearing the PCM or BCM codes (press Off and Hi for PCM codes, or Off and Lo for BCM codes).

BCM Data Display. As shown in Figure 11–33 under BCM Data Display, this mode displays the values sent to the BCM from most of its information sources as well as the value of some of its outputs. Figure 11–33 also shows the maximum range of each parameter and the units of measurement.

To go into the BCM data display mode, .7.0 must be displayed on the FDC. Depress and release the outside temp button on the CCP. The FDC displays F.8.0 briefly and then goes to parameter code P.2.0 or P.3.1 (the first and last parameters in the sequence). The parameter code displays for 1 second. Then the parameter value displays for 9 seconds. This sequence repeats with the same parameter until you select another. To advance to another parameter, push the Hi button on the CCP; to move back to a lower parameter code number, push Lo on the CCP.

These parameter values can be used as follows:

- They can be compared to those of a vehicle known to be operating properly to see whether they are what they are supposed to be.
- They can be watched to see that they change as operating conditions change (for instance, the commanded air mix door position should change as the selected temperature changes).
- Some of them are called for in the diagnostic charts, where their values are interpreted as good or at fault.

This diagnostic mode can be exited at any time and the display returns to the selection point, .7.0, by clearing the PCM or BCM codes (press Off and Hi for PCM codes, or Off and Lo for BCM codes).

ECC Program Override. During the BCM data display mode, the CCP displays a two-digit number representing the ECC program number. This number indicates the level of heating or cooling "effort" by the ECC system as explained in the BCM section of this chapter. The program number automatically changes according to changes ordered by the BCM following changes of inside temperature and/or the selected temperature changes.

The automatic calculation of the program number can, however, be overridden by pushing the warmer or cooler buttons, Figure 11–33. While in the BCM data display mode, holding the warmer button pushed in manually forces the program number up to a maximum of 100; 100 commands maximum heating. Pressing and holding the cooler button drives the program number down to a minimum of 0, which commands maximum cooling.

This override of the automatically calculated program number allows you to set it anywhere between zero and 100 to observe the reaction of the ECC system and the BCM data parameters. Once the program number has been overridden, it continues overridden until the BCM data display mode is cancelled.

Cooling Fan Override. This feature allows you to override the automatic control of the cooling fans' speed and manually command either high fans or fans off. To activate this feature, the FDC must be displaying .7.0. Depress and release the econ button on the CCP. The FDC then displays F.8.5. To command high fans, press and hold the Hi button on the CCP until the fans achieve full speed. Releasing the Hi button returns the fans to BCM automatic control. To command fans off, press and hold the Lo button on the CCP until the fans stop. Releasing the Lo button returns the fans to BCM automatic control. This feature can be exited at any time and the display returned to the selection point, .7.0, by clearing the PCM or BCM codes (press Off and Hi for PCM codes or Off and Lo for BCM codes).

Exit Diagnosis. Exit diagnosis by pushing AUTO on the CCP.

PCM/BCM Status Light Display. While in diagnostics, the ECC mode indicators on the CCP indicate that the mode in which the selected systems, operated either by the PCM or BCM, are operating (bottom portion of Figure 11–33). The operational mode is indicated by whether the designated light is on.

SUMMARY

This chapter has covered the development of the Cadillac DFI system, which parallels in many ways the other General Motors systems. Sensors and actuators follow the patterns in the regular CCC and successor systems, but the Cadillac system includes a variety of driveability enhancement measures specific to that brand. Perhaps most notable is the interconnection with the BCM, in charge of climate control and other features.

We covered the extended self-diagnostics of both computers, including the capacity to test-activate various outputs. Thus the technician can use the switch test, can access the PCM and BCM data, can override the cooling fans' circuits, and other tests.

▲ DIAGNOSTIC EXERCISE

A car is brought in with a complaint of running hot under virtually all driving conditions and poor power and fuel mileage. What is the most time-economical sequence of diagnostic steps to review how the engine is running and identify the problem? How is this changed if the computer self-diagnostics indicate no stored faults, current or memory?

REVIEW QUESTIONS

1. On a DFI system, the PCM monitors battery voltage and will turn on the MIL and set a DTC in memory if battery voltage does not stay within a range of which of the following?
   A. 12 to 14 volts
   B. 12 to 15 volts
   C. 10.5 to 15 volts
   D. 10.5 to 16 volts
2. Which of the following controls the cruise control function on a DFI system?
   A. A separate, dedicated control module that controls the cruise control function only
   B. The PCM only
   C. The BCM only
   D. Either the PCM or the BCM depending on the year and model
3. On a DFI system, the fuel pressure regulator provides an operating pressure in which of the following ranges?
   A. 6 and 9 psi
   B. 9 and 12 psi
   C. 12 and 15 psi
   D. 18 and 21 psi
4. *Technician A* says that, unlike many other GM vehicles, a DFI system does not use an oil pressure switch as a backup in the fuel pump circuit in case the fuel pump relay fails. *Technician B* says that in a DFI system, the fuel pump is mounted outside of the fuel tank, but in a location near the fuel tank.
   Who is correct?
   A. A only
   B. B only
   C. Both A and B
   D. Neither A nor B
5. General Motors began using a VCC on front-wheel-drive Cadillacs beginning in 1985. *Technician A* says that the VCC provides 100 percent lockup between the engine and the transmission input shaft when it is applied. *Technician B* says that the VCC allows about 40 rpm slippage at 60 mph when it is applied.
   Who is correct?
   A. A only
   B. B only
   C. Both A and B
   D. Neither A nor B
6. When the EGR valve is open, which of the following gasses makes up the majority of the gasses that it allows back into the intake manifold?
   A. Carbon dioxide and nitrogen
   B. Carbon dioxide and oxygen
   C. Carbon monoxide and oxygen
   D. Carbon monoxide and hydrocarbons
7. Concerning the EFE system on a DFI-equipped rear-wheel-drive Cadillac, all *except* which of the following statements are true?
   A. It uses an electric heat grid installed just below the throttle body.
   B. It helps prevent throttle blade icing.
   C. It increases fuel evaporation during engine warmup, resulting in a reduced fuel requirement and better driveability.
   D. It reduces hydrocarbon and carbon monoxide emissions.
8. At what vehicle speed does the DFI cruise control system respond when the driver depresses the "Set/Coast" switch?
   A. At speeds above 10 mph
   B. At speeds above 25 mph
   C. At speeds above 30 mph
   D. At speeds above 35 mph
9. After a DFI cruise control system has been turned on and engaged, all *except* which of the following things may disengage or turn off the cruise control system?
   A. The cruise control on/off switch is turned to OFF.
   B. The brake pedal is depressed.
   C. The throttle is momentarily moved to wide open throttle.
   D. The PCM senses certain faults with the cruise control system.
10. The BCM controls all *except* which of the following systems?

A. Electronic Climate Control (ECC)
B. Electric cooling fans that draw air through the A/C condenser and radiator
C. Power windows, trunk release and electric sunroof
D. VCC in the torque converter

11. A DFI-equipped vehicle that has problems with the Electronic Climate Control (ECC) system comes into the shop.
Technician A says that the Climate Control Panel (CCP) communicates information about driver-selected temperature and operating mode to the BCM digitally over a data link (data bus).
Technician B says that the PCM communicates information about throttle position and coolant temperature to the BCM digitally over a data link (data bus).
Who is correct?
A. A only
B. B only
C. Both A and B
D. Neither A nor B

12. In the ECC system, the power module is used by the BCM to control which of the following?
A. Air door positions
B. Heater water control
C. Rear window defog
D. Blower motor

13. Which of the following inputs are used by the BCM to properly control the cooling fans on a DFI system?
A. Engine coolant temperature and A/C low-pressure side temperature
B. Engine coolant temperature and A/C high-pressure side temperature
C. Engine coolant temperature and A/C evaporator temperature
D. Engine coolant temperature and A/C mode selector switch position

14. *Technician A* says that in a DFI system the BCM controls the speed of the two cooling fans through a pulse width modulated (PWM) signal that it communicates to the fan control module.
Technician B says that in a DFI system the BCM monitors the control of the cooling fans with a feedback signal from the fan control module.
Who is correct?
A. A only
B. B only
C. Both A and B
D. Neither A nor B

15. Information that may be displayed by the BCM on the Fuel Data Center (FDC) includes all *except* which of the following?
A. Instantaneous and average fuel mileage
B. Fuel level, fuel range and fuel used
C. Fuel required to arrival at current fuel mileage
D. Outside temperature

16. *Technician A* says that a liquid crystal display (LCD) is used to form the display characters on the FDC, the CCP and the radio.
Technician B says that the BCM controls the display brightness on the FDC, the CCP and the radio through a pulse width modulated voltage.
Who is correct?
A. A only
B. B only
C. Both A and B
D. Neither A nor B

17. To enter diagnostics through the CCP on a DFI system, the technician should first turn on the ignition switch. What two buttons should then be pushed simultaneously on the CCP?
A. OFF and COOLER
B. OFF and WARMER
C. OUTSIDE TEMP and WARMER
D. AUTO and OUTSIDE TEMP

18. During diagnostics on a DFI-equipped vehicle that has been brought into the shop, the CCP displays .7.0 after all of the stored DTCs have displayed. This indicates which of the following?
A. The ignition should be cycled off, and the code pulling should be repeated.

B. The PCM has failed a self-test.
C. The BCM has failed a self-test.
D. The system is ready for selection of the next self-diagnostic feature.

19. A DFI-equipped vehicle is brought into the shop.
Technician A says that during the ECC Program Override mode, the number "100" is used to command maximum cooling.
Technician B says that the Cooling Fan Override mode allows the technician to command "high fans" or "fans off."
Who is correct?
A. A only
B. B only
C. Both A and B
D. Neither A nor B

20. Which button on the CCP should be pushed to exit diagnostics on a DFI system?
A. AUTO
B. OFF
C. OUTSIDE TEMP
D. WARMER

Chapter 12

Advanced General Motors' Engine Controls

OBJECTIVES

Upon completion and review of this chapter, you should be able to:

- ❑ Understand the primary features of the Aurora/Northstar engine.
- ❑ Define the inputs and outputs that are used with the Aurora/Northstar engine.
- ❑ Describe the difference between the functions of a Powertrain Control Module (PCM) and a Vehicle Control Module (VCM).
- ❑ Explain how Central Multiport Fuel Injection (CMFI) or Central Sequential Fuel Injection (CSFI) is different from standard port fuel injection.
- ❑ Understand the primary features of the Generation III V8 engine.
- ❑ Understand the primary features of the Premium V6 engine.
- ❑ Explain how GM's Variable Valve Timing system functions.
- ❑ Describe some of the features of the Saturn 1.9L engine application.

KEY TERMS

Aurora/Northstar
Background Noise
Central Multiport Fuel Injection (CMFI)
Central Sequential Fuel Injection (CSFI)
EEPROM
Engine Metal Overtemp Mode
Hot Wire Air Flow Sensor
Misfire Detection
Primary/Secondary Oxygen Sensor
Quad Driver/Output Driver
Torque Management
Variable Valve Timing (VVT)
Vehicle Control Module (VCM)
Voltage Drop Test
Waste Spark

As with all manufacturers, General Motors continues to develop engines and computerized engine control systems that utilize more advanced levels of technology. In this chapter, we present some of their more recent computerized engine control systems. It is advantageous for the reader to be familiar with the information contained in the previous chapters.

AURORA AND NORTHSTAR ENGINE

While General Motors builds a large number of different engines of various sizes and configurations, in this section we focus on the engine used in the Oldsmobile **Aurora**, Figure 12–1. The same engine, in a slightly larger displacement, is also used under the name **Northstar** in certain Cadillac models, because it employs a computer control system that includes most of the systems used on other engines.

The Northstar V8 engine, first introduced in 1993, continues to be utilized on all Cadillac models except the Catera. Displacing 4.6 liters, the dual overhead camshaft, four valves-per-cylinder Northstar has two power ratings. The 275 horsepower version is standard in the DeVille, Eldorado and Seville, with a 300 horsepower version for the DeVille Concours, Eldorado Touring

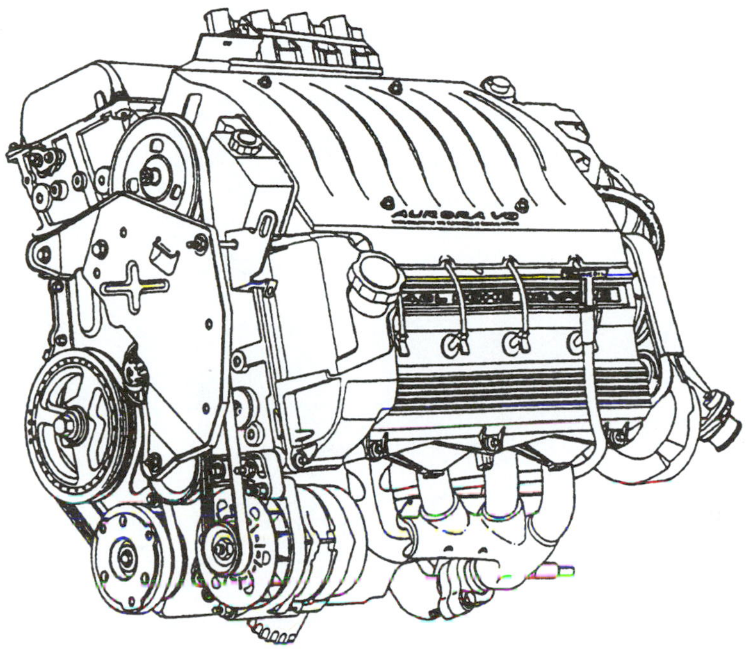

Figure 12–1 Oldsmobile Aurora/Cadillac Northstar engine. *(Courtesy of General Motors Corporation, Service Technology Group.)*

Coupe (ETC) and Seville Touring Sedan (STS). The Northstar motor has direct mount accessories (including the water-cooled generator) to reduce noise. The transmission features Performance Algorithm Shifting (PAS) that alters shift timing in response to lateral acceleration. PAS also causes the transmission to downshift into the proper gear (and holds it there) as the vehicle enters and executes a turn.

The 1998 Oldsmobile Aurora retains the L47 4.0 liter, dual overhead camshaft, 32 valve V8 and 4T80E transaxle. The aluminum engine develops 250 horsepower and features chain-driven camshafts, magnesium cam covers, a cooling system thermostat placed at the engine inlet and a PCM with dual microprocessors. The 4T80E transaxle is fitted with a viscous coupling torque converter clutch.

General System Features

The heart of this General Motors engine control computer systems lies in the Powertrain Control Module (PCM), Figure 12–2. It is a much faster and more powerful computer than those used previously to monitor and control engines for emissions quality, driveability and fuel economy. In addition, the PCM is responsible for all the diagnostic functions required for compliance with OBD II regulations.

One precaution worth remembering: with the considerably increased number of Diagnostic Trouble Codes (DTC) in the OBD II system, technicians are often startled to find a huge number of codes stored. Check to see what circuits are involved: often all the codes recorded are for a single circuit or a set of closely related circuits.

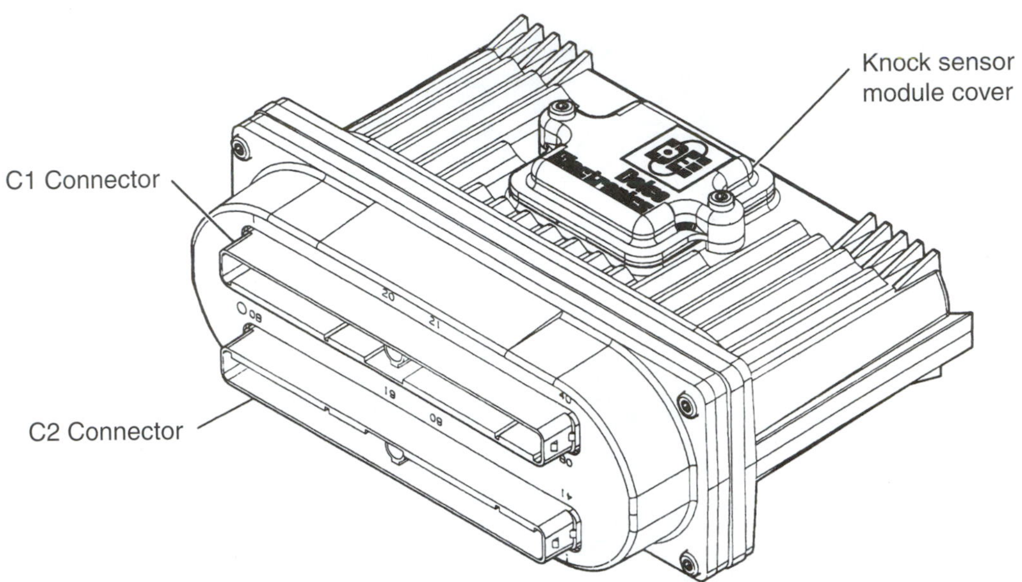

Figure 12–2 Powertrain control module. *(Courtesy of General Motors Corporation, Service Technology Group.)*

Figure 12–3 shows typical inputs and outputs for the system. Among the engine and vehicle functions the PCM controls are:

- fuel mixture control.
- ignition control.
- knock sensor system (detonation control).
- automatic transaxle shift functions.
- enabling the cruise control.
- regulating the output of the generator (alternator).
- evaporative emission (EVAP) purge (canister purge).
- exhaust gas recirculation (EGR).
- air injection reaction (AIR).
- A/C clutch engagement control.
- radiator cooling fan control.

Sensor Reference Voltage. The PCM also provides a reference voltage to the various sensors and switches it uses to collect information on how the engine is running. To protect both the PCM and the sensor, this voltage is "buffered" in the PCM before it is sent to the switches and sensors. Many voltmeters, particularly analog voltmeters, may not give an accurate reading of this reference voltage because their impedance is too low. To be sure of accurate measurements of reference voltage (and for that matter, of voltage throughout this system), use only digital voltmeters with a minimum of 10 megohm input impedance, specification J 39200.

CAUTION: To continue a precaution hopefully familiar from previous computer systems, when you work on these vehicles, you must be very careful to prevent damage to components by static electricity through the fingers. While it takes about 4,000 volts of static discharge for a person to feel even a slight zap, less than 100 volts static discharge—1/40th of that—can be enough to damage a computer chip or memory unit.

Quad Drivers/Output Drivers. Internally the PCM has many devices like analog-to-digital converters, signal buffers, counters, timers and special drivers. It controls most components, as

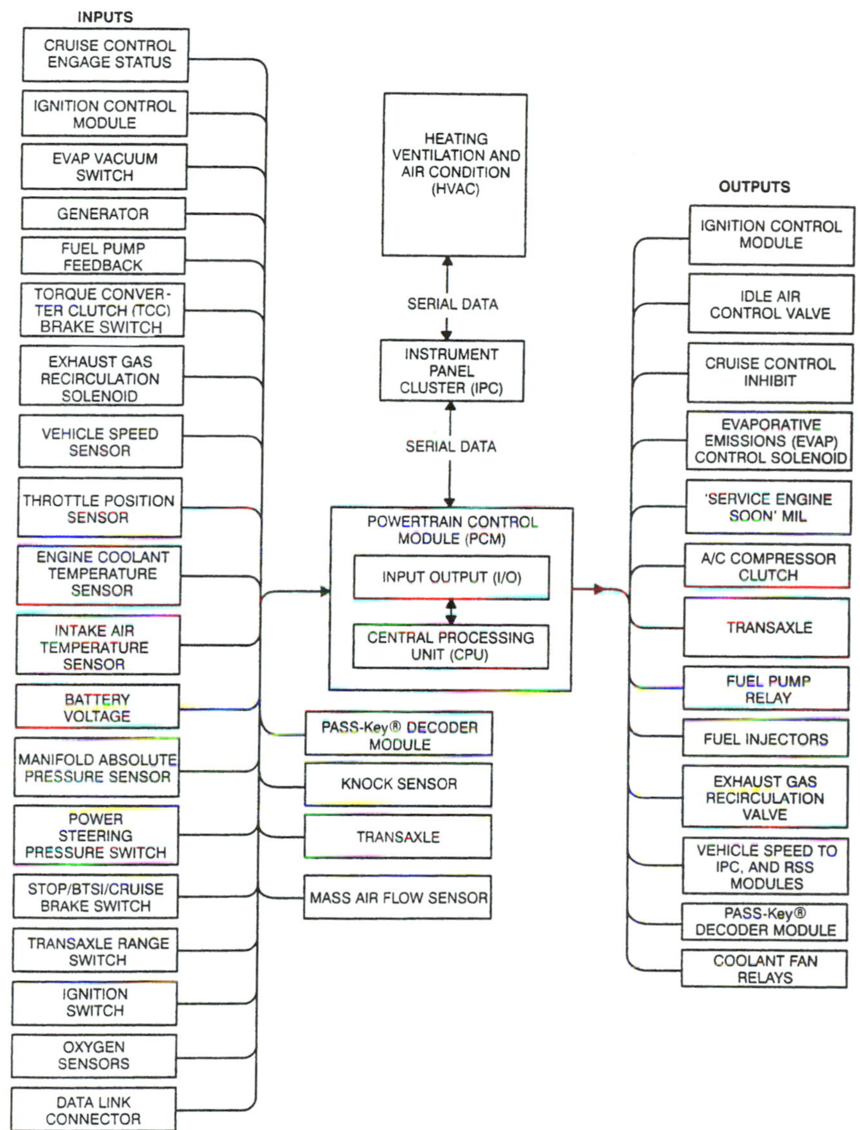

Figure 12–3 Powertrain control system inputs/outputs. *(Courtesy of General Motors Corporation, Service Technology Group.)*

car computers have for many years, by completing the circuit to ground (the components' power source usually comes from the ignition switch or some other source). These power transistors are effectively solid-state on/off switches, often called **quad drivers** or **output drivers**. If the switches are surface-mounted in a group of 4, they are quad driver modules; if the switches are surface-mounted in a group of up to 7, they are output driver modules. Certain switches may go unused on a particular vehicle, depending on its model year, accessory list and so on.

PCM Learning

The PCM has a much more advanced learning capacity than previous engine control computers, but resetting the system is relatively similar. Yet, if erroneous data has somehow become recorded in memory, the technician can no longer merely disconnect the battery for a few minutes while the computer's volatile memory (keep-alive memory) deletes. Changes to the memory data may require the use of a dedicated scan tool to make the deletions. Finer discrimination learning comes as the system performs various tests and experiments, while the vehicle is driven at part throttle, with moderate acceleration. Some vehicles systems also allow reprogramming of the control module without removal from the vehicle, a process called *flashing*.

Torque Management

Besides the functions normally provided by an engine control computer, this PCM also works to control, that is, reduce, engine torque under certain circumstances. Torque reduction is performed for three reasons:

1. to prevent overstress of powertrain components.
2. to limit engine power when the brakes are applied.
3. to prevent damage to the vehicle or powertrain from certain abusive driving maneuvers.

To calculate whether to employ its torque-limiting functions, the PCM analyzes manifold pressure, intake air temperature, spark advance, engine speed, engine coolant temperature, the engagement or disengagement of the A/C clutch through its control circuit, Figure 12–4, and the EGR status. It also considers whether the torque converter is locked up, what gear the transmission is in and whether the brakes are applied. When the numbers add up to torque reduction, the PCM's first strategy is to reduce spark advance to lower engine torque output. In some more extreme cases, the PCM can also shut off fuel to one or more cylinders to further reduce engine power.

There are several occasions when **torque management** is likely to come into play:

- during transaxle upshifts and downshifts
- when there is hard acceleration from a standstill (traction control)
- if the brakes are applied simultaneously with moderate to heavy throttle
- if the driver initiates abusive or stress-inducing actions, like shifting gears at high open throttle angles.

In the first two cases, the operation of the torque management will probably not be noticeable.

Traction Control. The PCM is also responsible for the vehicle's traction control system. If the computer learns from the wheel speed sensors that the front wheels are slipping during acceleration, it can individually apply the front brakes to prevent spin. In extreme cases, the torque management system can reduce engine torque also, disabling as many as seven of the fuel injectors to do so, Figure 12–5. This strategy will not be adopted, however, if any of the following conditions occur:

- engine coolant temperature is below –40°
- the engine coolant level is low (there is a sensor for that purpose)
- the engine is at a speed below 600 rpm

If the traction control system is disabled for one of the above reasons or a system failure, the TRACTION OFF light will come on in the driver's information panel.

The torque management system can sometimes have some surprising consequences. If a car is running under cruise control and encounters slippery pavement, perhaps slick ice, the wheel speed sensors will report extremely rapid acceleration. The traction control, among its other measures, will shut off the cruise control. Not only will it shut the cruise control off, but it will disable

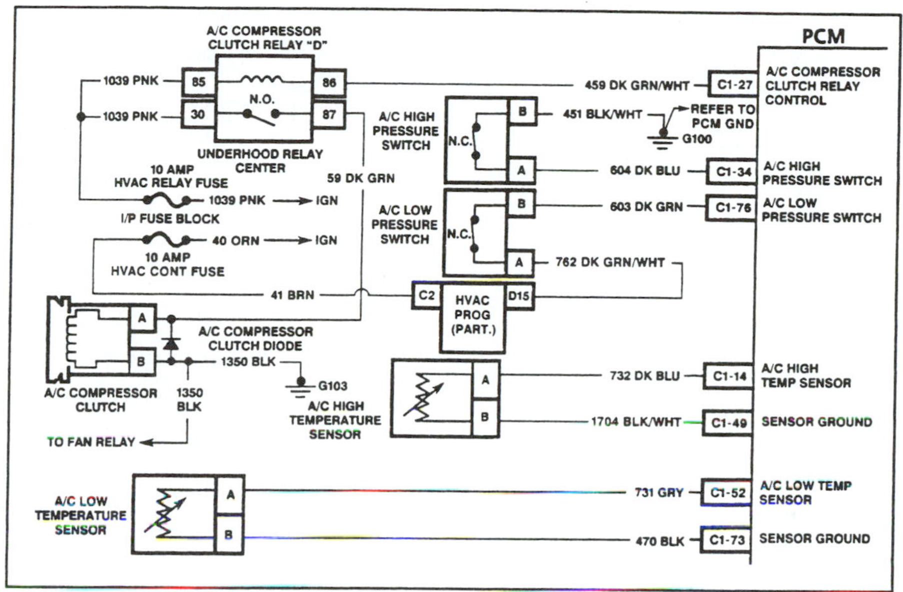

Circuit Description

This system consists of a HVAC control assembly, separate A/C refrigerant high and low pressure switches, separate A/C high and low temperature sensors, a control relay, the compressor clutch and the PCM.

When the HVAC control assembly is placed in the A/C mode, a request signal is sent to the PCM. The PCM will then energize the A/C clutch relay, unless abnormally high or low A/C pressure is detected by the A/C refrigerant pressure switches. The PCM also monitors the A/C low temperature sensors to cycle the A/C clutch "ON" and "OFF." If the PCM detects an A/C evaporator temperature less than 2°C (36°F), the A/C clutch will be prohibited from being enabled. The PCM will also energize the cooling fans "ON" when A/C is requested. Refer to "Electric Cooling Fan(s)," Section "6E3-C12". The A/C control relay is controlled by the PCM so that the PCM can increase idle speed before turning "ON" the clutch or disable the clutch when high engine coolant temperature is detected or high engine RPM is detected.

Diagnostic Aids

- If DTC P1653 is stored as history, it is most likely that one of the A/C pressure switches opened.
- If any PCM DTC(s) are set, perform those diagnostics before using this chart. The A/C clutch is disabled for many DTC(s).

Test Description

Number(s) below refer to the step number(s) on the Diagnostic Table.

3. The outside air temperature must be great enough for the A/C compressor clutch to engage.
5. This step uses the Scan Tool to determine if there is an open or short.
6. This step checks the power feed circuit.
7. This step bypasses the relay.

Figure 12–4 A/C compressor clutch control circuit. *(Courtesy of General Motors Corporation, Service Technology Group.)*

it, set a code and leave the subsystem disabled until the code is deleted by a technician. Various other problems, such as a slipping transaxle, can have the same effect.

Information Functions. The PCM also provides the driver with a variety of messages, about engine and transmission oil life or remaining coolant level, for example, some of which are engine-related. As these are not directly related to controlling the combustion in the engine, we omit them for space reasons.

Ignition and Fuel Startup. The engine uses a four-coil **waste spark** ignition system, as described in Chapter 10. When the engine is cranked, the ignition control module monitors crankshaft position sensor signals to determine which coil to fire first. Once the module determines the coil timing, it sends a fuel reference signal to

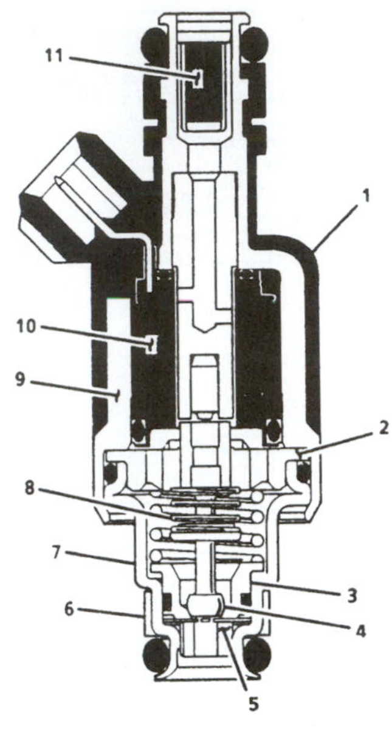

1 SOLENOID ASSEMBLY
2 SPACER & GUIDE ASM
3 CORE SEAT
4 VALVE - BALL
5 PLATE - DIRECTOR
6 BACKUP - O-RING
7 HOUSING - SPRAY
8 SPRING - CORE
9 HOUSING - SOLENOID
10 SOLENOID
11 FILTER - FUEL INLET

Figure 12–5 Fuel injector. *(Courtesy of General Motors Corporation, Service Technology Group.)*

the PCM. At reception of that pulse, the PCM pulses all eight injectors once simultaneously. It then leaves all the injectors off for two full engine crankshaft revolutions. This provides each cylinder with enough fuel for one power stroke. During those two revolutions, a camshaft position signal should have reached the PCM, and it can begin to trigger the injectors in the proper sequence, Figure 12–6. Should there be no camshaft position signal, the PCM will start fuel delivery in a random pattern, though the chances are 8 to 1 against its being correct. While it will not necessarily be correct, most of the 7 possible wrong sequences can support an engine running at somewhat reduced performance levels.

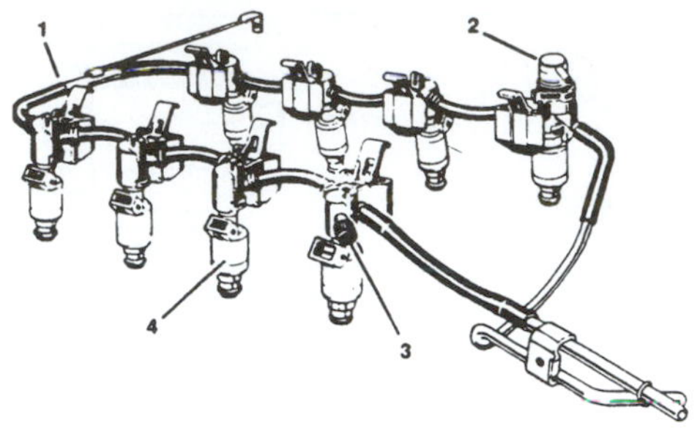

1 RAIL ASM - MFI FUEL
2 REGULATOR ASM - FUEL PRESSURE
3 CONNECTION ASM - FUEL PRESSURE
4 INJECTOR ASM - MFI FUEL

Figure 12–6 Fuel rail. *(Courtesy of General Motors Corporation, Service Technology Group.)*

INPUTS/OUTPUTS

Mass Airflow (MAF) Sensor

The mass airflow sensor uses a **hot wire airflow sensor** type device to measure the airflow rate, Figure 12–7. It works by maintaining the hot wire at a specified temperature above ambient temperature. A second wire senses the ambient temperature. The hot wire is placed in the airstream and cooled by the intake air. The current required to keep the wire at the specified

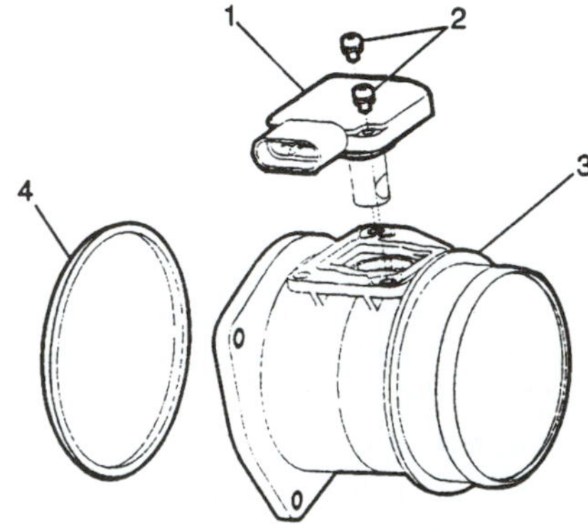

Figure 12–7 Mass airflow (MAF) sensor. *(Courtesy of General Motors Corporation, Service Technology Group.)*

temperature corresponds closely to the amount of air entering the intake manifold by mass. Internal circuits in the MAF sensor convert the current required to keep the hot wire at the specified temperature into a frequency signal sent to the PCM. This airflow information is used in the calculation of fuel injector pulse width.

Manifold Absolute Pressure (MAP) Sensor

The PCM gets engine load and barometric pressure information from the MAP sensor, Figure 12–8. This is the type commonly used. It receives a 5-volt reference signal from the PCM, reduces it in accordance with the sensed pressure, and returns the reduced voltage as an indication of manifold pressure. The MAP signal works inversely to what you might read on a vacuum gauge, that is, the higher the vacuum goes, the lower the output signal voltage will be. The higher-capacity PCM monitors MAP sensor signals against throttle position reports to determine if the MAP sensor is sticking. It will notice if throttle position changes as little as 3.2 degrees in 0.5 second. If the MAP signal does not change by a voltage corresponding to at least 4 kiloPascals during that period (assuming a base MAP reading of at least 17.3 kiloPascals), it will set a trouble code for the MAP sensor.

The computer uses MAP sensor information to calculate spark advance, fuel mixture (including acceleration enrichment) and sometimes to

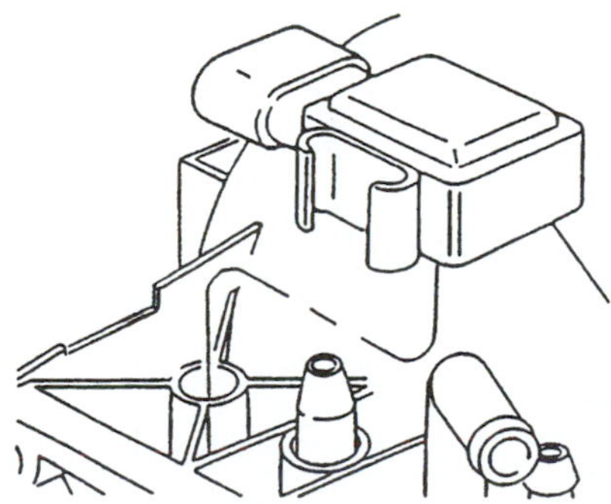

Figure 12–8 MAP sensor. *(Courtesy of General Motors Corporation, Service Technology Group.)*

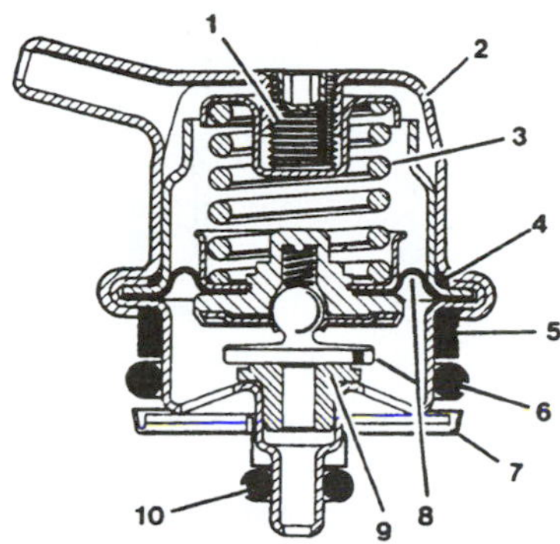

Figure 12–9 Fuel pressure regulator. *(Courtesy of General Motors Corporation, Service Technology Group.)*

determine whether to employ torque limitation measures. The fuel mixture is also affected by the pressure differential across fuel injectors, held steady by the fuel pressure regulator, Figure 12–9.

Temperature Sensors

Intake Air Temperature (IAT) Sensor. The intake air temperature sensor is a thermistor, a variable resistor, that varies its resistance depending on its temperature. Receiving a 5-volt signal from the PCM, it reduces that depending on its temperature, modifying the signal corresponding to the intake air temperature. As sensor temperature goes up, resistance goes down, so when the temperature is high, the signal will go to a low voltage.

Engine Coolant Temperature (ECT) Sensor. The system uses a dual-range coolant temperature sensor, Figure 12–10. The PCM has within it a 3.65 Kilo-ohm resistor and a 348-ohm

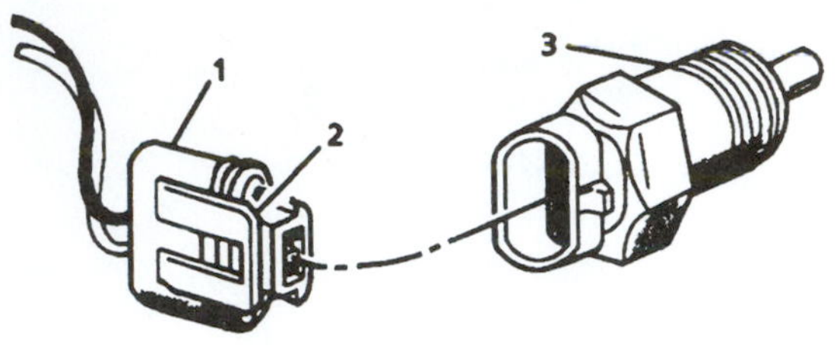

Figure 12–10 ECT sensor. *(Courtesy of General Motors Corporation, Service Technology Group.)*

resistor in a series circuit, Figure 12–11. On a cold engine, 5 volts (reference voltage) is passed through both of these resistors before being passed out to and through the ECT sensor itself. It is then connected to ground through the PCM. As with a standard ECT circuit, the PCM's voltage-sensing circuit monitors the voltage drop across the ECT sensor. As the engine coolant warms up, the resistance of the ECT sensor's thermistor decreases, thereby also decreasing the voltage drop across the ECT sensor as measured by the PCM. However, this also means that the voltage drop across the two resistors in series within the PCM is increasing. That is, because all of the applied reference voltage will be used up by all of the resistance in the circuit (if the circuit is complete), as the voltage drop across the ECT sensor decreases, it is shifted to being dropped across the two resistors in series within the PCM. At about 50°C (122°F) when the voltage drop across the ECT sensor has been reduced to about 0.97 volts (and the voltage drop across the two resistors in series is therefore about 4.03 volts), the PCM suddenly completes a circuit that bypasses the 3.65 Kilo-ohm resistor. This causes the reference voltage to be applied through only the remaining 348-ohm resistor on the way to the ECT sensor. This has the effect of suddenly shifting much of the voltage drop to across the ECT sensor, thereby suddenly increasing the PCM's measured voltage to a value closer to reference voltage again, Figure 12–12. As the engine continues to warm up, the PCM will see the voltage drop across the ECT sensor continue to decrease again. Ultimately, this has the effect of allowing

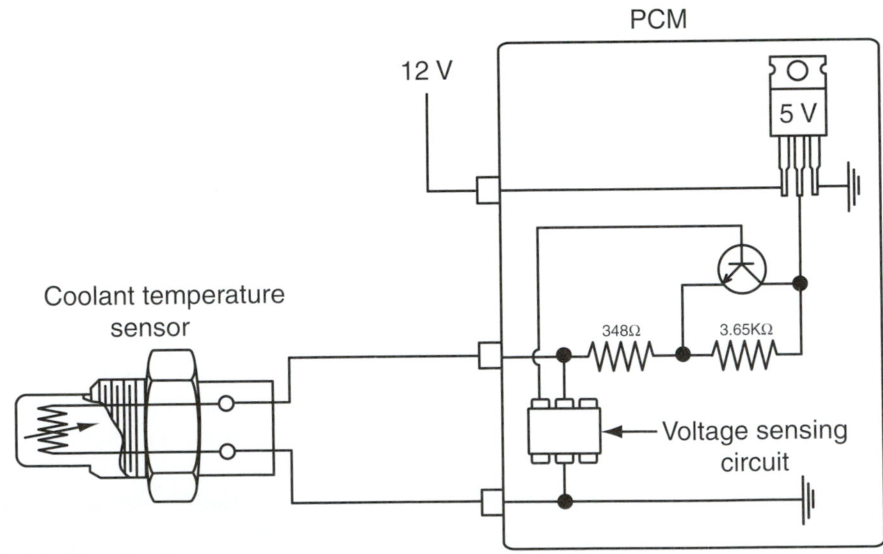

Figure 12–11 General Motors' dual-range temperature sensor circuit.

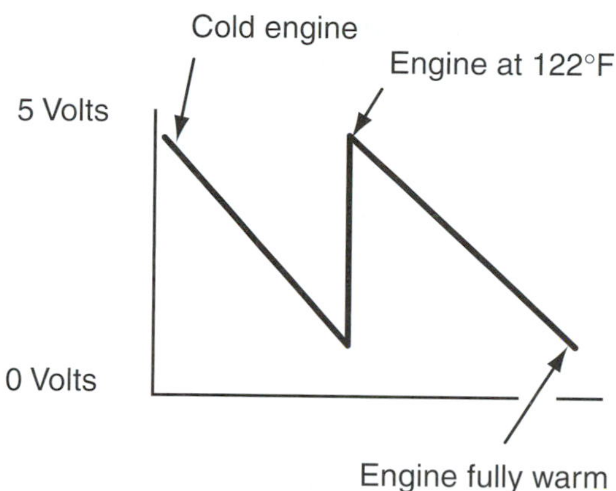

Figure 12–12 Dual-range ECT sensor waveform.

the PCM a more precise measurement of engine coolant temperature because the voltage drop change across the ECT sensor as the engine warms up exceeds the 5-volt reference voltage applied to that circuit.

The information from this sensor, of course, also determines the operation of the radiator cooling fan.

One of the internal audit tests the computer automatically performs is to check to see whether the engine is warming quickly enough. After the engine has run 4.25 minutes, it checks to see that the coolant temperature is 5°C (41°F) or greater. If it is not, or if the coolant temperature falls below that temperature for more than 3 seconds, a code is set. The timer ratchets backward during torque management or traction control activity. Ordinarily, the only reason that an engine does not reach this temperature within the specified time is because a thermostat is blocked open. This assumes, of course, that the engine has been run under moderate load during the elapsed time. If someone merely started an engine at –40 below and let it idle for 4.25 minutes, it might not gain the 80+ degrees, and a code would be set.

Throttle Position (TP) Sensor

The computer uses TPS information for calculating idle speed, fuel mixture, spark advance, deceleration enleanment and acceleration or WOT enrichment. This sensor returns a proportionate signal: the lower the return voltage, the lower the throttle angle. The computer also compares (1) changes in throttle position with its internal clock to determine how quickly the throttle has been changed, and (2) TPS with MAP information: any significant discrepancy indicates one or both circuits are faulty.

The throttle position sensor on this system is said to be self-adjusting, Figure 12–13. What that actually means is that it is not adjustable, but the computer learns what to expect from it and

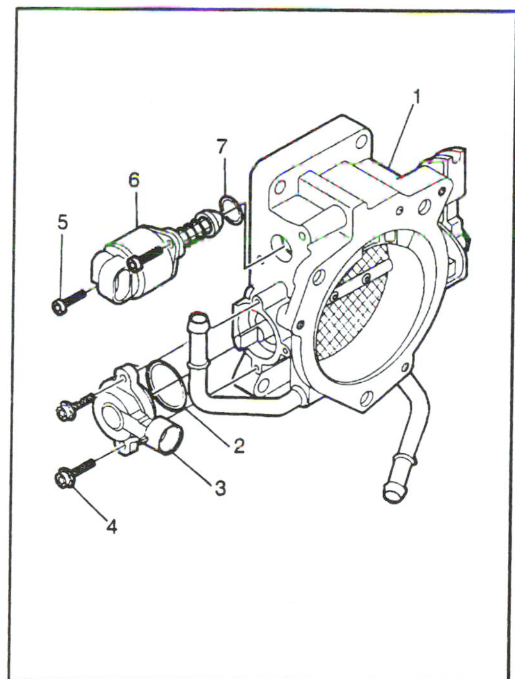

Legend
1. Body Assembly - Throttle
2. O-Ring - Throttle Position (TP)
3. Sensor - Throttle Position (TP)
4. Screw - TP Sensor Attaching
5. Screw - IAC Valve Attaching
6. Valve Assembly - Idle Air Control (IAC)
7. O-Ring - IAC Valve Assembly

Figure 12–13 Throttle position sensor. *(Courtesy of General Motors Corporation, Service Technology Group.)*

adjusts its calculation tables accordingly. For the self-adjustment, the computer assumes that the throttle is closed when the key is turned off with the throttle position signal at the same value twice in succession.

Heated Oxygen Sensor

The heart of any closed-loop fuel metering system is the oxygen sensor, and this system uses a special, heated sensor (to bring it up to operating temperature as quickly as possible), Figure 12–14. In addition, the computer sends it a reference voltage of 0.45 volts. As long as this is the return voltage, the PCM stays in open loop. Once the oxygen sensor gets warm enough (and newer sensors start to work at about 200°C or 392°F), the oxygen sensor signal will swing from rich to lean about once every 2 seconds once the PCM is controlling closed loop. A 2-second oxygen sensor cycle is actually fairly long for a functioning sensor: this is the point at which the new Aurora/Northstar system sets a code for a slow oxygen sensor. New sensors typically cycle faster than once per second.

With OBD II comes a second oxygen sensor, functionally equivalent to the first, but downstream of the catalytic converter. Its function is to check on the effectiveness of the converter. If all is well with the catalytic reduction and oxidation processes, things will be fairly uneventful at the downstream oxygen sensor. There should be little or no change of output voltage, provided the converter has removed the hydrocarbons and neutralized (reduced) the NO_x. When the second oxygen sensor starts reflecting the first, the catalytic converter is not doing anything; and the computer, recognizing this, will set a fault.

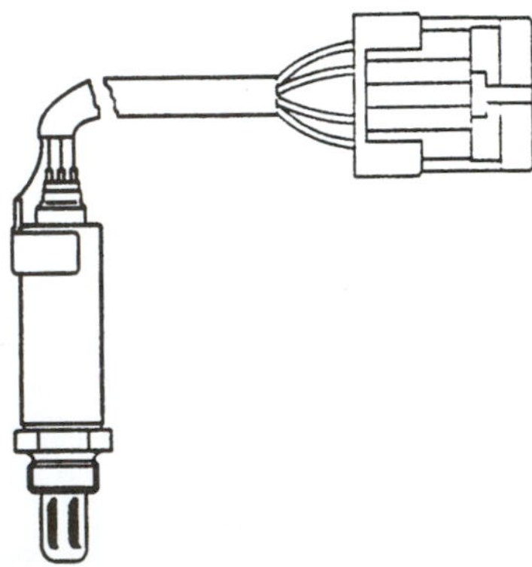

Figure 12–14 Heated oxygen sensor (HO_2S). *(Courtesy of General Motors Corporation, Service Technology Group.)*

On the Aurora engine as well as the Cadillac Northstar (and a number of other modern engines) there are two **primary oxygen sensors**, one on each cylinder bank. The individual oxygen sensors, obviously, report on the air/fuel mixture and combustion in the cylinders on that side, but there is nothing in principle different from a single sensor system. Keep in mind, however, that the system uses *three* oxygen sensors. The computer's trouble code recording capacity includes monitoring of each sensor and each sensor's heater circuit separately.

The PCM also monitors the oxygen sensor's heater circuit and reports opens or shorts in it as different trouble codes. Likewise, it compares closed-loop voltage cycling rates as a test of oxygen sensor function. If the cycle falls too long, this is stored as a fault.

Fuel Pump Feedback Circuit. The PCM gets a signal through the fuel pump relay confirming operation of the fuel pump. As on previous systems, the fuel pump is actuated for 2 seconds after the key is turned to ON to prime the injectors for starting. After that 2 seconds, the pump shuts off until there is a crankshaft or camshaft position signal. If the PCM detects a voltage below 2 volts on the fuel pump power circuit, it records that as a fault. The engine, of course, will probably not start because the pump is not turning, or not turning fast enough to pressurize the fuel.

Misfire Detection

Among the most important capacities mandated by OBD II are measures for the **misfire detection** and correction of misfire. A misfire means a lack of combustion in at least one cylinder for at least one "combustion event." A misfire pumps unburned fuel through the exhaust system, and while the catalytic converter can deal with an occasional puff of fuel vapor, a steady stream of it from a dead cylinder would soon overheat it and spill into the atmosphere. The PCM detects misfire by very closely monitoring minute changes in crankshaft speed. At higher speeds, these changes are measured over two engine revolutions for a comparison. If the system detects a dead cylinder, it will shut off the fuel injector to that cylinder, preventing catalytic converter overheating. Also, on some applications, the PCM receives input from the antilock brake system's wheel speed sensors in order to determine when perceived cylinder misfire is simply the result of drivetrain influences due to a rough road surface.

Keep in mind that shutting off an injector does have an effect on closed-loop feedback systems. The camshaft and valves still work to move air from the intake manifold to the exhaust manifold. The air pumped through that cylinder contains a higher percentage of oxygen than the exhaust of the air/fuel mixture burned in the other cylinders. Since the oxygen sensor does not know where oxygen is coming from, it will report an overlean condition to the computer, which will try to richen the mixture. On most engines, that much enrichment is beyond the capacity of the system, because no amount of fuel added will burn the pass-through oxygen through the dead cylinder. The computer will set a trouble code and turn on the malfunction indicator light. If the misfire is steady, the computer will consider that an event with a high probability of damage to the catalytic converter. It will flash the service engine soon light until the condition is no longer present or the engine is shut off. Such serious events are stored in the freeze frame and failure records for later retrieval and diagnosis.

When cylinder misfire is present, check to see whether two cylinders are involved. If there are, look for something shared by both cylinders, like the common coil or exposure to the same vacuum leak.

Knock Sensor (KS) Circuit

An engine can withstand a small amount of spark knock for a short time, and this computer factors that into its spark advance calculations. The knock sensor must report a detonation event lasting longer than 99.99 milliseconds for three continuous cycles. The response then is to retard spark until the knock disappears. The basic ignition (spark advance) map for the engine assumes a fuel with an octane equivalence of 87, but the system can manipulate spark (more or less advanced) to accommodate a higher or lower detonation resistance with different fuels. In general, the system will drive the spark advance right to the edge of knock to maximize the fuel economy and reduce emissions problems.

The computer's knock sensor circuit is also used to detect other knock sensor faults. Because the sensor is fundamentally a kind of microphone, and the computer sorts out what kinds of noise are knock and what is just ordinary engine noise, the **background noise** can be used to detect some circuit faults. If the background noise does not rise in an expected way with engine speed and load (as reflected by TPS and rpm through crank and cam sensors for a given period of time), it will record a knock sensor fault, even if there is no knock.

Crankshaft Position Sensors

The system uses two crankshaft sensors, Figure 12–15, on the front of the engine block between cylinders 4 and 6. Crankshaft position sensor A is in the upper crankcase, and crankshaft position sensor B is in the lower crankcase. Both sensors extend into the block and are sealed with O-rings. There are no adjustments.

The pickups work like distributor pickups. As the reluctor on the crankshaft moves over the sensor, the coil generates an alternating, polarity-reversing signal that appears (after it goes through the buffer circuit) to the computer as an on/off/on/off signal (almost like that of old-fashioned contact points!). The reluctor ring under

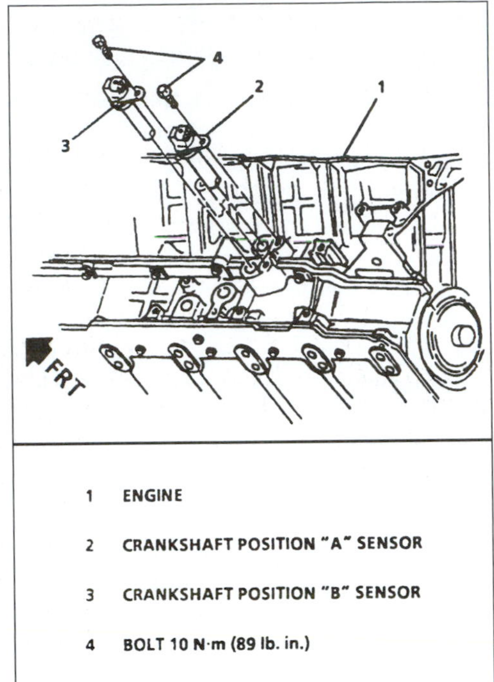

Figure 12–15 Crankshaft position sensors A and B. *(Courtesy of General Motors Corporation, Service Technology Group.)*

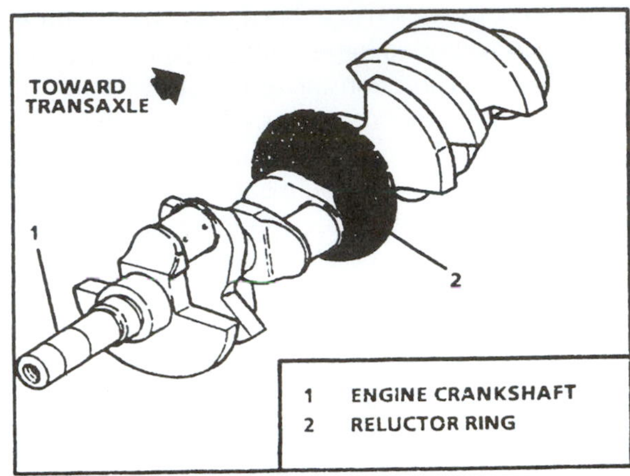

Figure 12–16 Crankshaft reluctor ring. *(Courtesy of General Motors Corporation, Service Technology Group.)*

the sensors has 24 evenly spaced notches and an additional 8 unevenly spaced notches for a total of 32, Figure 12–16. As the crankshaft rotates once, each sensor reports 32 on/off signals. But since the sensors are positioned 27 degrees of crankshaft rotation apart, this creates a pattern of on/off signals that enables the ignition control module to identify the crankshaft position.

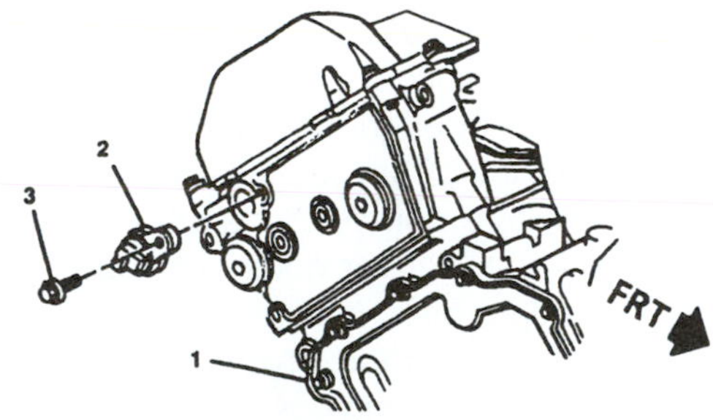

Figure 12–17 Camshaft position sensor. *(Courtesy of General Motors Corporation, Service Technology Group.)*

Camshaft Position Sensor (CAM) Signal

The camshaft position sensor is on the rear cylinder bank at the front of the exhaust camshaft, Figure 12–17. It extends into the cylinder head and is sealed with an O-ring. There are no adjustments.

As the camshaft turns, a steel pin on its drive sprocket passes the magnetic pickup of the sensor, Figure 12–18. This generates an alternating, polarity-reversing signal similar to the crankshaft position sensor signal. This on/off signal, in time with the camshaft, occurs once every other crankshaft revolution.

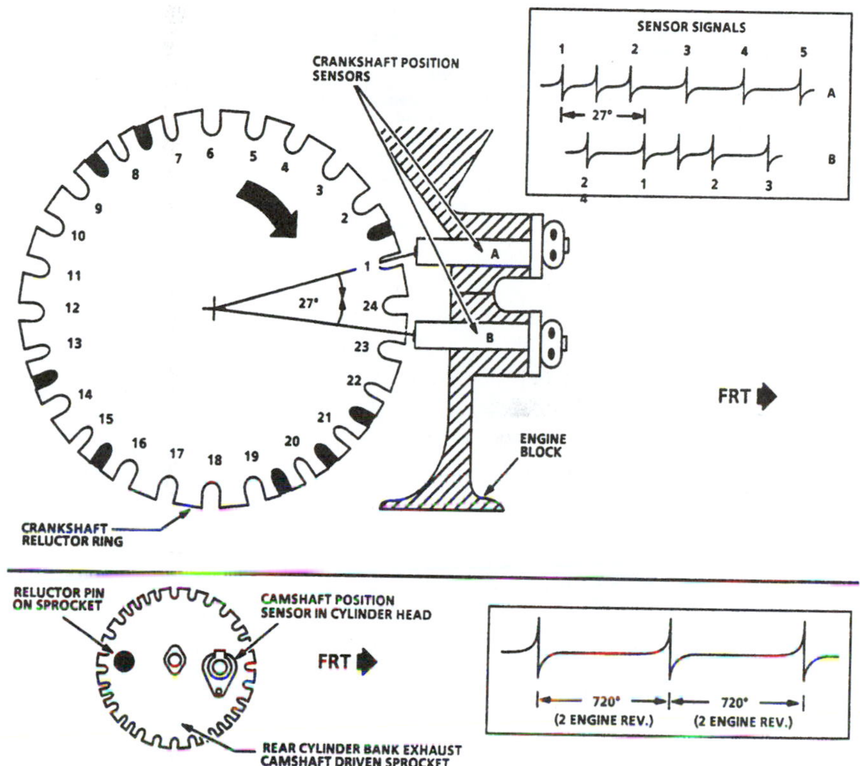

Figure 12–18 Crankshaft and camshaft position sensor configurations. *(Courtesy of General Motors Corporation, Service Technology Group.)*

Ignition Control Module

The ignition control module (IC) is on top of the rear camshaft cover, Figure 12–19. It performs these functions:

- monitoring the on/off pulses from the two crankshaft position sensors and the camshaft position sensor
- generating a 4x and 24x reference signal sent to the PCM for ignition control
- generating a camshaft reference signal sent to the PCM for fuel injection control
- providing an ignition ground reference for the PCM
- providing for the PCM's control of spark advance
- providing for "module mode," a limited means of controlling spark advance without PCM input, fundamentally a limp-home running mode.

The ignition module is not repairable; replacement is the only option if it develops a problem.

Ignition Coils

The system employs four coils to fire eight spark plugs in a waste spark ignition system, as explained in Chapter 10, Figure 12–20. On this

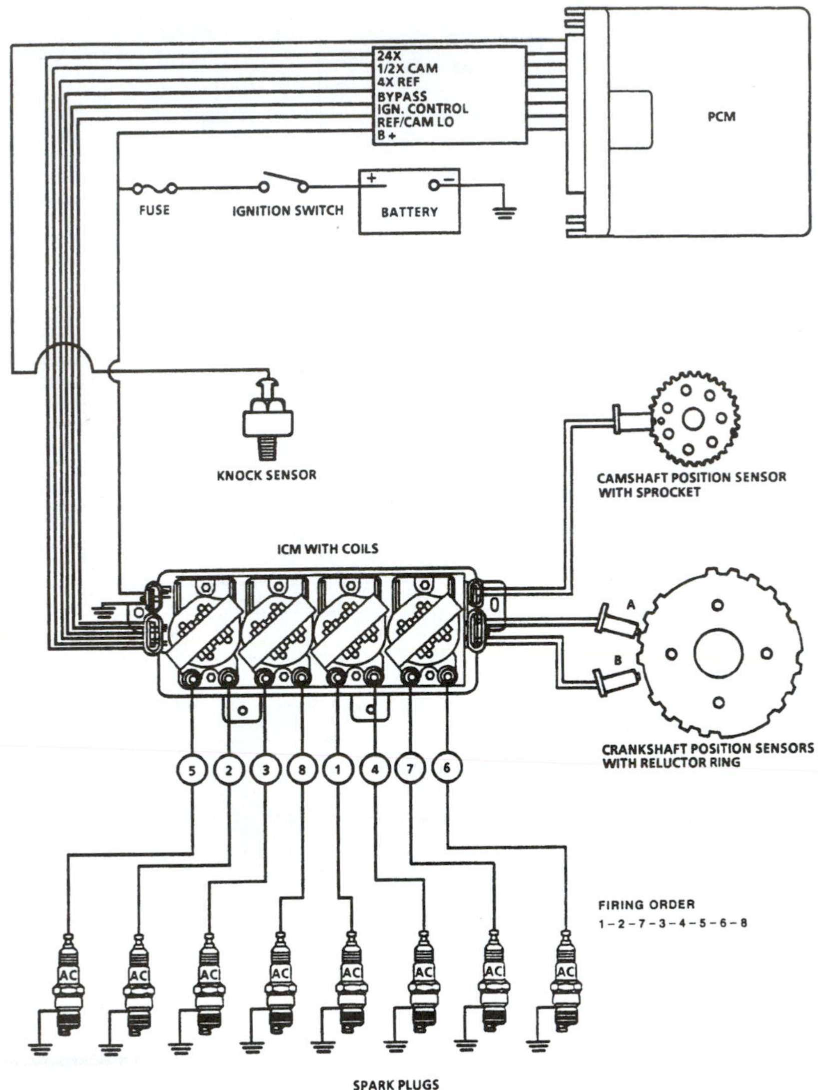

Figure 12–19 Ignition control system. *(Courtesy of General Motors Corporation, Service Technology Group.)*

system the coils are replaceable individually rather than as a complete coil pack.

Exhaust Gas Recirculation (EGR)

The linear EGR valve used in this system can be tested by the PCM during specific driving operations. When the proper test conditions are met, the PCM cycles the EGR from off to on and back, meanwhile monitoring the MAP signal. Obviously, opening another source of intake into the manifold should raise the intake manifold pressure somewhat (or similarly lower vacuum). The PCM repeats the test a number of times, comparing results to what is hardwired in its program. If the results do not agree, it sets a code.

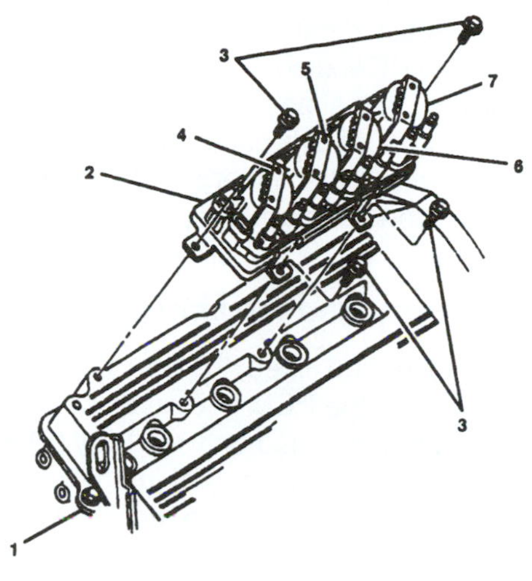

1 ENGINE
2 IGNITION CONTROL MODULE
3 BOLT 12 N·m (106 lb. in.)
4 5-2 IGNITION COIL
5 3-8 IGNITION COIL
6 1-4 IGNITION COIL
7 7-6 IGNITION COIL

Figure 12–20 Ignition control module with coils. *(Courtesy of General Motors Corporation, Service Technology Group.)*

Catalytic Converter Efficiency Test

As explained in the section on oxygen sensors in this chapter, the PCM compares the signal from the two primary oxygen sensors with the single oxygen sensor downstream of the catalytic converter. If the converter is working properly, there should be very little variation in the signal produced by the **secondary**, downstream **oxygen sensor**. Should the secondary sensor produce a signal similar to those generated by the primary oxygen sensors, the PCM sets a code indicating a failed catalytic converter test.

Evaporative Emission Test

The PCM monitors the function of the evaporative emissions system by monitoring the vacuum switch in the line between the canister and the solenoid. The charcoal canister is almost identical to those used in many vehicles for many years. When the engine draws 12 inches of vacuum, the switch is supposed to open and allow venting of the stored fuel vapors. The PCM monitors this by following the amount of time the vacuum switch is continuously open or closed during purging. When purging is supposed to occur, the PCM starts a timer. If the switch stays closed for at least 9 seconds, the test is failed; if the switch opens for at least 2 seconds continuously, the test is passed.

Idle Air Control System

The idle air control system uses a stepper motor that works the same as those on earlier systems, Figure 12–21. In this system, the PCM matches the position it sets the IAC motor to, with the achieved idle speed. If the idle speed is higher or lower than the desired speed by 80 rpm, the test is failed and a code is set.

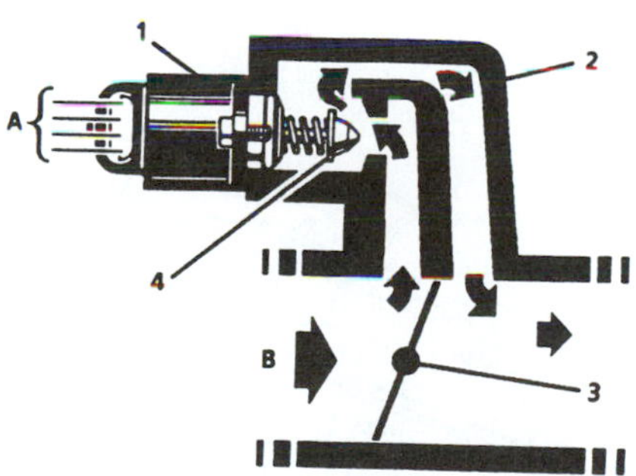

1 VALVE ASSEMBLY - IDLE AIR CONTROL (IAC)
2 BODY ASSEMBLY - THROTTLE
3 VALVE - THROTTLE
4 PINTLE - IAC VALVE
A ELECTRICAL INPUT SIGNAL
B AIR INLET

Figure 12–21 Idle air control (IAC) valve. *(Courtesy of General Motors Corporation, Service Technology Group.)*

Power Steering Pressure Switch

Normally the power steering switch is closed, but when hydraulic boost pressure in the P/S system rises from 450 to 600 psi, it opens. The purpose of the switch is to inform the PCM when to increase idle speed slightly to compensate for the additional load during, for example, parking maneuvers. The PCM monitors the switch information against its other inputs and sets a code if the switch is ever open at a vehicle speed above 45 mph as sensed by the vehicle speed sensor, Figure 12–22. This is a speed at which little or no power steering boost should be required.

PCM Selftest and Memory Test

Like computers in vehicles previously, this computer checks its internal circuits continuously for integrity. Likewise, it checks its nonremovable **EEPROM** for the accuracy of its data. In computer terms, it tests the checksum of its files against what they are supposed to be and sets a code if they are different. Besides the hardwired memory, the PCM also monitors its volatile keep-alive memory. If that has been improperly changed or deleted, this sets a code. Note: Such a code will be set when the battery is disconnected for more than a brief instant.

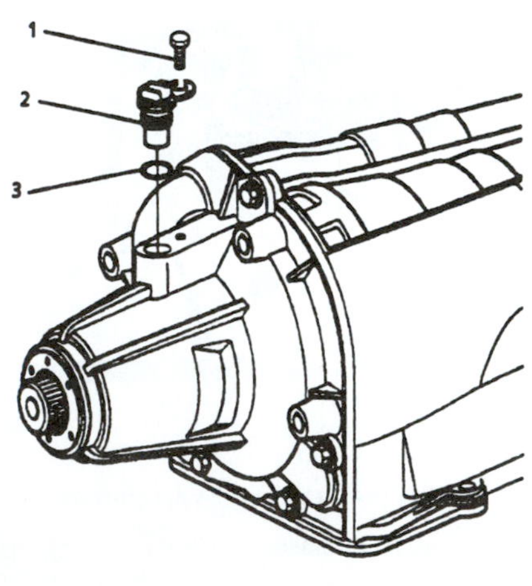

1 VEHICLE SPEED SENSOR SCREW
2 VEHICLE SPEED SENSOR
3 SEAL VEHICLE SPEED SENSOR

Figure 12–22 Vehicle speed sensor. *(Courtesy of General Motors Corporation, Service Technology Group.)*

PCM Memory and the EEPROM

Instead of the fixed, unalterable PROM of previous years, the PCM has an electrically erasable programmable read-only memory (EEPROM) it uses to store a large body of information. While the EEPROM is soldered into the PCM and is not service-removable, it can store data, like the PROM, without a continuing source of electrical power. The PCM uses it as a storage bank for its throttle position/learned idle control tables, its transaxle adapt values, its transmission oil life information, its cruise control learning, and more. The EEPROM includes several areas available to store this data, and the PCM moves the data to a good location if there is damage to an earlier place. It will store a code if internal damage prevents the EEPROM from storing the information. The same data is stored in the keep-alive memory, so even if there is an EEPROM failure, the driver will probably not notice any difference. If the EEPROM defect code is set, the PCM must be replaced since the EEPROM is not removable.

Pass-Key Decoder Module

As part of an antitheft system, the PCM gets information from the pass key decoder module to identify the key used for the ignition switch and steering lock. If the wrong key is used or if the system fails, the PCM disables the fuel injection system to prevent theft of the vehicle.

VEHICLE CONTROL MODULE (VCM)

The familiar PCM on a General Motors' application provides control over the powertrain, that is, performance of both the engine and transmission.

It may also control the speed control system. GM has now produced a new version of this computer known as a **Vehicle Control Module** or **VCM**, which, in addition to controlling all of the above systems, also provides control over the vehicle's antilock braking system (ABS). In the past, a separate control module was used to control the ABS system. Therefore, GM vehicles now may utilize either a PCM or a VCM, depending upon whether the vehicle has antilock brakes under the control of the engine computer. However, either term may now be used to signify the engine computer.

CENTRAL PORT FUEL INJECTION SYSTEMS

Central Multiport Fuel Injection (CMFI)

In the mid 1990s, General Motors introduced a new port fuel injection system known as **central multiport fuel injection** or **CMFI**. This system uses an injector assembly that consists of a fuel metering body, a fuel pressure regulator and a single TBI-style fuel injector, which is connected to individual poppet nozzles (one per cylinder) with nylon fuel tubes, Figure 12–23. This assembly is housed in the lower portion of a dual intake manifold assembly. The poppet nozzles, as in Figure 12–24, are simple, nonelectric, spring-loaded valves (normally closed) that are opened by applied fuel pressure from the single fuel injector solenoid. When opened, each poppet nozzle sprays fuel to the back side of its intake valve until fuel pressure is reduced enough to allow it to close. The PCM controls fuel delivery through all poppet nozzles by controlling the on-time of the single injector solenoid.

Another interesting feature is that the fuel pressure regulator is able to maintain a steady pressure differential across the injector and poppet nozzles without the existence of the normal vacuum hose used to index most fuel pressure regulators to manifold pressure/vacuum. This is because it is

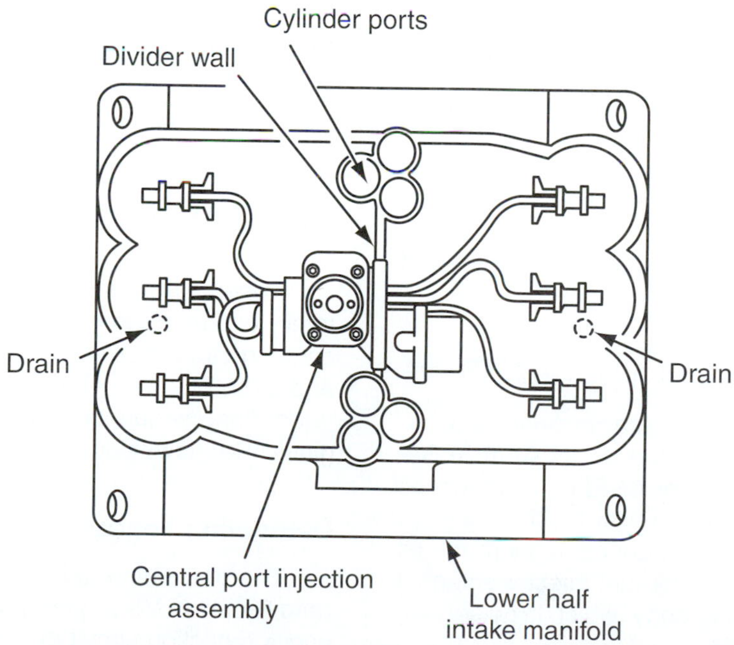

Figure 12–23 Central multiport fuel injection components in the lower half of the intake manifold.

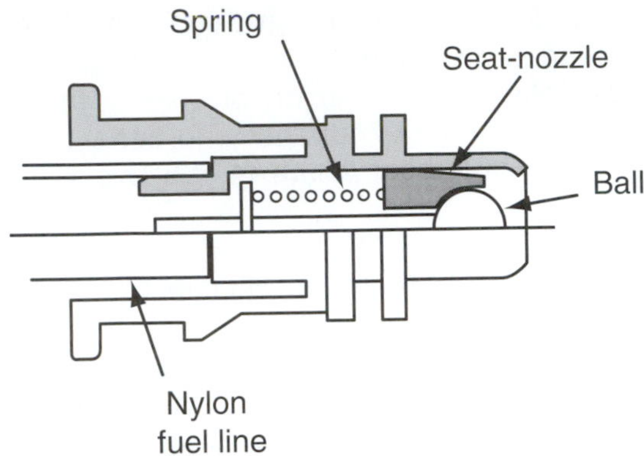

Figure 12–24 Poppet nozzle internal design.

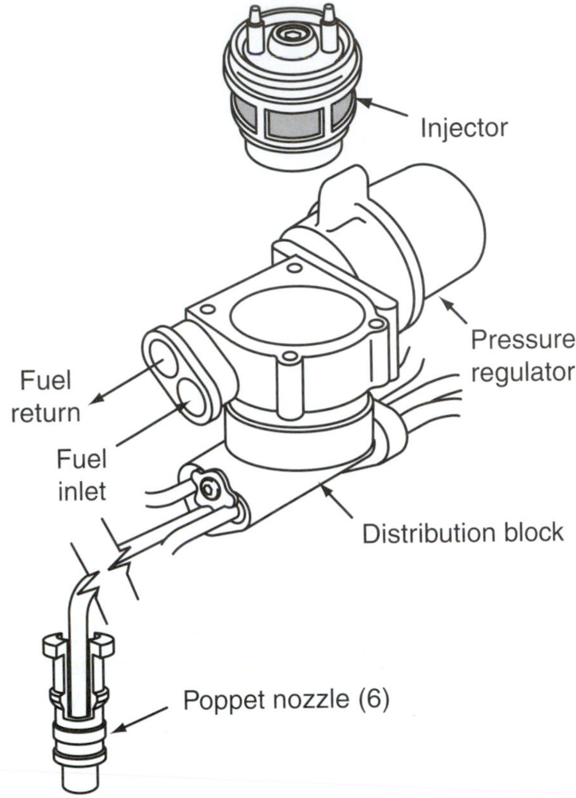

Figure 12–25 Central port injector operation.

located, as part of the fuel metering body, down in the lower intake manifold with its diaphragm already open to manifold pressure or vacuum, Figure 12–25.

The fuel injector solenoid in this assembly is operated as a peak and hold injector, as are most TBI-type fuel injectors. The injector's winding has low resistance (about $1\frac{1}{2}$ ohms), which allows more current to flow and thus faster response when it is energized by the PCM. Once opened, the PCM then reduces the injector's current flow to prevent overheating of the solenoid winding, while maintaining enough current to keep the injector open for the remainder of its intended pulse.

Central Sequential Fuel Injection (CSFI)

General Motors has also introduced a sequential version of CMFI known as **Central Sequential Fuel Injection** or **CSFI** (also known as CSI). This design is very similar to CMFI, except that each poppet nozzle is connected through its nylon tube to an individual fuel injector solenoid within the fuel metering body, which is pulsed sequentially by the PCM.

GENERATION III V8s

General Motors has released an entire new family of Generation III V8 engines. The engines come in various sizes, including 4.8L, 5.3L, 5.7L and 6.0L. The new Generation III V8s use a different firing sequence than previous GM V8 engines. The new firing order is 1-8-7-2-6-5-4-3.

Detonation Sensor

The new General Motors Generation III V8s (and Premium V6 engines) have a modified spark knock detection circuit in the PCM. Previously, a

5-volt DC source was used to detect circuit integrity for shorts and grounds. Measured at the PCM connector with a DC voltmeter, this voltage was approximately 2.5 volts. An AC voltage was produced by the piezoelectric crystal inside the spark knock sensor whenever spark detonation occurred. This AC voltage was placed on top of the DC voltage. The amplitude of the AC signal, which rode on top of the DC carrier, determined both when the ignition spark was retarded and the length of time it was retarded.

The new circuit looks the same as the old circuit but acts differently. The new circuit has eliminated the DC carrier voltage. The PCM monitors the circuit only for AC activity. The AC voltmeter reads 30 to 50 millivolts normally. The voltage produced by the spark knock sensor rises rapidly when spark knock is detected. It remains high until the timing is retarded and the spark knock subsides.

The Generation III V8 uses two spark knock sensors located in the engine block valley underneath the intake manifold. Previous GM dual knock sensor engines used a series/parallel circuit. The new engine uses two totally separate knock sensor circuits. Each knock sensor has its own separate power supply. The two knock sensors share no common computer sensor wiring.

The Generation III engine uses a new 32U PCM, which does not require the knock sensor module to be transferred from the previous unit when a new PCM is installed. The knock sensor module is soldered onto the printed circuit board of the new PCM.

Ignition System

The Generation III V8 ignition system is a coil-near-plug system. Eight coil/driver assemblies are mounted on the valve covers. A short high-voltage secondary wire completes the electrical circuit to the spark plugs. The system uses either Denso or Melco coil/driver assemblies on cars. The Delphi coil/driver assembly is used on trucks. The coil/driver assemblies must not be intermixed because of differences in internal current limiters and mounting.

Fuel System

The Generation III engines use a combination of all technologies for the fuel supply system. Trucks use a traditional vacuum biased pressure regulator located on the engine's fuel rail. Cars use a semi-returnless system. In this system, the pressure regulator is located inside the fuel filter container. The fuel filter container is located outside the fuel tank but in close proximity to it. On some model cars the fuel pressure regulator may be located inside the fuel tank itself.

The *engine coolant overtemp fuel disable* is a new Generation III feature that previously had been used only on the Northstar engine. If the PCM detects a temperature in excess of 270°F, the engine runs only on four cylinders. This allows every other cylinder in the firing order to act as a heat transfer pump. Then the system switches to the next four cylinders. It continues to switch until the engine temperature problem is resolved.

Idle Learn Procedure

Whenever the battery has been disconnected or a new PCM has been installed, the Generation III PCM requires the technician to perform an idle learn. Although the procedures vary for standard and automatic transmissions, the typical idle learn procedure looks like this:

1. Start the engine and let it warm up to a minimum of 176°F.
2. Shift the transmission into drive.
3. Allow the engine to idle for 5 minutes.
4. Turn on the air conditioning.
5. Allow the engine to idle for 5 minutes.
6. Shift the transmission into Park.
7. Allow the engine to idle for 5 minutes.
8. Turn off the air conditioning.
9. Allow the engine to idle for 5 minutes.
10. Turn engine off for 30 seconds. Restart.

Crankshaft Learn Procedure

In order to increase the accuracy of spark timing on its new Generation III engines, General Motors technicians must complete a procedure called crankshaft learn. Crankshaft learn must be performed whenever the PCM is replaced or reprogrammed. It also must be performed when a component such as the crankshaft, the crankshaft sensor or the engine block is replaced. Crankshaft learn demands the use of a hand-held scan tool called the Tech 2, or its equivalent. The miscellaneous functions menu of the Tech 2 has a selection called crankshaft relearn, which leads the technician through a step-by-step process. The procedure looks like this:

1. Wait until the engine has warmed up. The Tech 2 will tell you when you can proceed.
2. Rapidly accelerate the engine to the fuel cutoff point. Tech 2 will identify that speed for you. Release the accelerator pedal when the engine rpm drops.
3. During the coast back to idle rpm, the PCM will learn any variations in the notches cut into the crankshaft reluctor and will store them in the computer's memory.
4. A screen message will appear on the Tech 2 that tells the technician that the relearn procedure was successful.

Passlock Learn Procedure

Many new GM systems are equipped with a theft deterrent feature called passlock. The passlock system has no obviously apparent resistor like the pellet that was located in the key for several years. It uses a calibrated resistor, which is part of the passlock sensor and ignition switch lock cylinder. Rotation of the lock cylinder creates a voltage drop circuit. The body control module (BCM) then sends a password to the PCM. If the passlock sensor does not see the normal rotation of the lock cylinder, it either will not crank the engine at all or it will start the engine but immediately stall it. When the passlock cylinder, the BCM or the PCM is replaced, the system must relearn the passlock voltage. This can be done in two ways. The first method uses a hand-held scan tool called the Tech 2, or a similar tool. Follow the instructions for the BCM programming of the body controller. The second way a technician can perform passlock relearn is called auto-learn. Auto-learn is performed without the use of a Tech 2. The following five steps show you how to perform auto-learn:

1. Hook a charger to the battery and charge at a slow rate.
2. Turn the key on. Leave the key on until the security indicator turns off. This takes about 10 minutes.
3. Turn the key off for 5 seconds.
4. Repeat steps 2 and 3 two more times for a total of 3 cycles.
5. The vehicle will now start and run.

Throttle Actuator Control

The Generation III V8 engine was introduced on some GM vehicles in 1999. Generation III vehicles use an electronic throttle actuator control (TAC) system. The TAC system is used in both Camaro and Firebird (F body) and Corvette (Y body) vehicles as well as Chevrolet and GMC (C/K body) pickup trucks.

The components of the throttle actuator control system, shown in Figure 12–26, are a three-channel accelerator pedal position (APP) sensor, a dual-channel throttle position sensor (TPS), a throttle actuator control module and a DC bidirectional pulse width modulated (24 Hertz) motor.

The PCM commands the TAC module to open or to close the throttle blade in the intake manifold system. The TAC module reverses the 12-volt DC pulse width modulated signal and powers the bidirectional motor. The forward signal opens the throttle for normal acceleration; the reverse signal closes the throttle. This action can be used to create an anti-slip/traction operation that reduces wheel slip during heavy acceleration. This system needs no cruise control components other than on-off, set and resume switches.

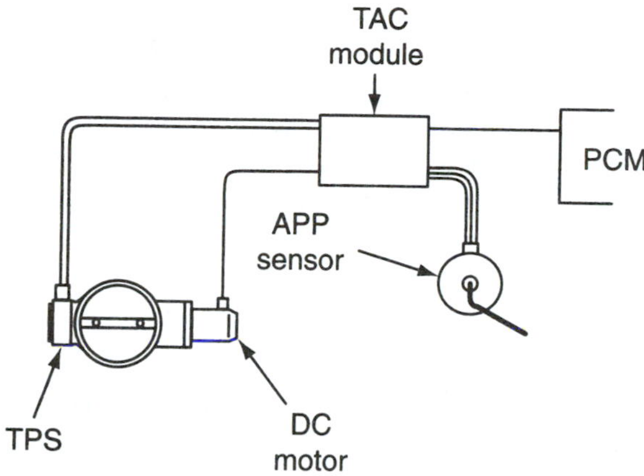

Figure 12–26 Throttle actuator control (TAC) components.

PREMIUM V6 ENGINE

Ignition System

The new 3.5L Premium V6 engine uses an ignition cassette for each side of the engine, Figure 12–27. The cassette is located on the overhead valve covers. This location eliminates the need for high-voltage secondary plug wires. The cassette contains three molded ignition coils, three rubber boots and a serviceable ignition module. The high-tension secondary windings of the coils made contact with the spark plugs through a spring inside the rubber boot.

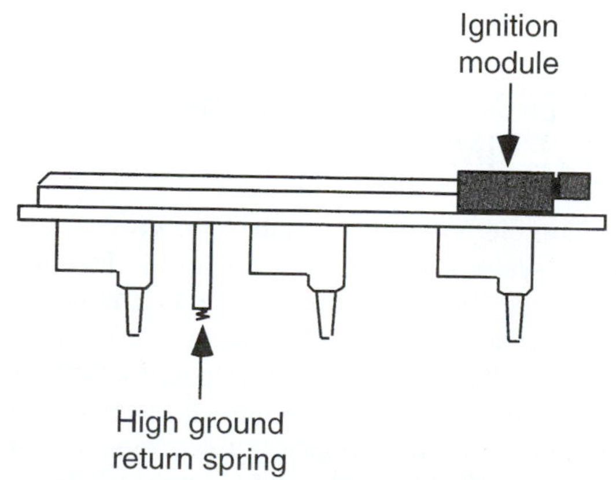

Figure 12–27 Ignition cassette.

This new ignition system uses a high ground return spring rather than the common ground of the engine block and wiring harness. If this high ground return spring is not installed, excessive radio frequency interference static on AM radio stations results. However, the engine will operate without a driveability problem.

Crankshaft Sensor

The crankshaft sensor for the 3.5L Premium V6 engine is two separate magneto resistive devices, Figure 12–28. The two sensors allow for greater spark timing accuracy, but each magneto resistive device continues to run if one of the two sensors fails to operate. The sensor contains six electrical terminals for the two sensors.

The two sensors use the same power and ground supplied by the PCM but have separate square wave output signals that are returned to the PCM, Figure 12–29.

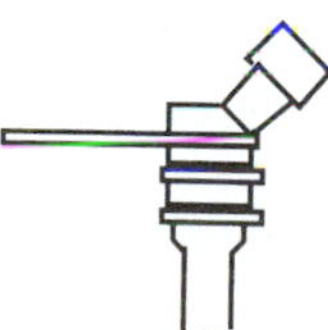

Figure 12–28 Crankshaft sensor.

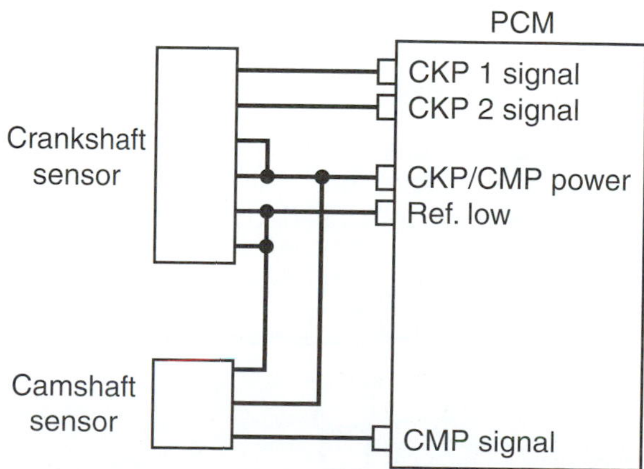

Figure 12–29 Crankshaft/camshaft sensor circuit.

Cam Sensor

The new cam sensor for the 3.5L Premium engine is a single magneto resistive device, Figure 12–30. The cam sensor is used to synchronize the ignition and fuel injection systems. The output of the sensor is a digital 5-volt signal similar to the crankshaft sensor.

The cam sensor determines the position of the cam gear when the key is turned on but before the engine runs. If the cam sensor fails, the crankshaft sensor determines that a pair of cylinders are at top dead center. However, the PCM will not know which cylinder is on the compression cycle and which cylinder is on the exhaust cycle. The PCM then fires a cylinder. If the engine starts, the PCM has chosen the right coil and cylinder. If the engine fails to start, the PCM reverses the sequence. Only the crank time was increased.

Fuel Tank Anti-Spitback Valve

A spring-loaded anti-spitback valve is located at the bottom of the fuel filler pipe. The technician must not attempt to siphon fuel from the fuel tank in the normal fashion or the valve will catch the siphon tube. An access panel in the floor behind the rear seat was developed to allow the technician to drain the fuel tank. This panel allows access to the fuel level sender, the fuel tank pressure sensor and the in-tank fuel pump. Be careful not to spill fuel in the interior of the vehicle.

Oxygen Sensors

The 3.5L Premium engine uses two heated oxygen sensors that reduce the time it takes the engine to achieve closed-loop operation; the heated oxygen sensors also prevent the system from falling out of closed-loop operation under certain conditions. Although both pre- and postsensors look alike, they are different. They must not be interchanged. The presensor is built by Delso, which is a joint venture of Delphi and Denso, and is rated at 12 watts. The postsensor is built by Delphi and its heater is rated at 18 watts.

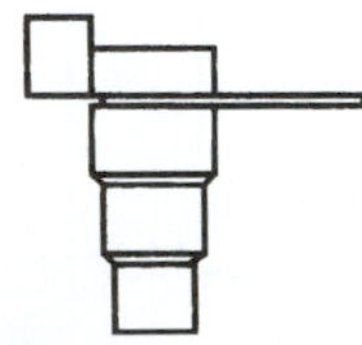

Figure 12–30 Cam sensor.

Injectors

The 3.5L Premium engine uses dual spray pattern injectors aimed at the dual intake valves of each cylinder. The electrical connectors must be rotated perpendicular to the fuel rail for proper alignment. Failure to align the injector spray may cause driveability or emission problems.

Powertrain Control Module

The new 32U PCM requires constant air movement over the finned aluminum housing. Its location inside the air cleaner housing helps keep the PCM cool.

Coolant Overtemp Protection

The 3.5L Premium engine uses a coolant overtemp protection strategy similar to the Generation III engines. Three cylinders are dropped alternately, bank to bank, every 60 engine cycles. There are no codes associated with this operation.

GENERAL MOTORS' VARIABLE VALVE TIMING (VVT)

The General Motors' **Variable Valve Timing (VVT)** system, as used on the 2002 Trailblazer (and other GM counterparts such as Envoy), uses a VVT mechanism on the exhaust camshaft to change exhaust valve timing. The camshaft is designed with a helical spline that allows the camshaft to change position in its relationship to the camshaft drive sprocket, Figure 12–31. The PCM controls a hydraulic solenoid, which can

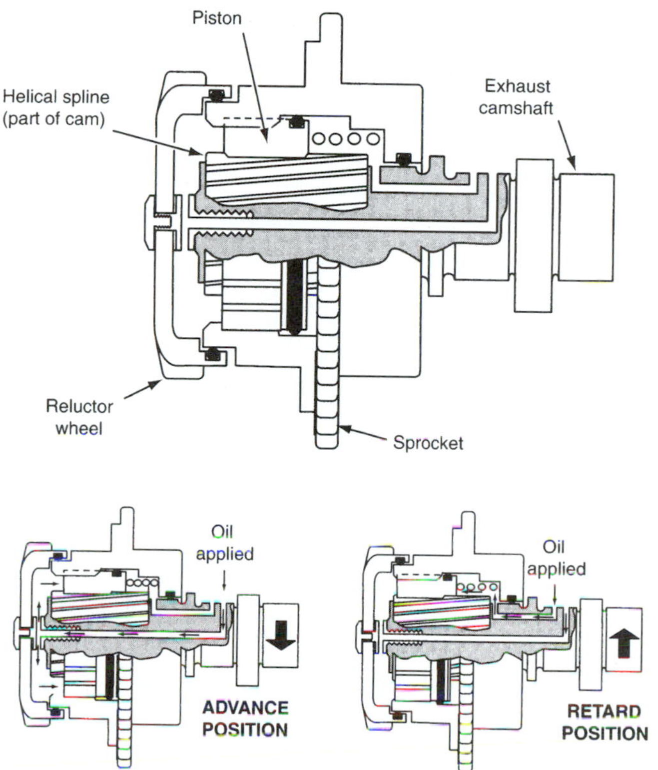

Figure 12–31 General Motors' variable valve timing system.

apply oil pressure to either side of a piston, thus moving it in either direction. In one direction, piston movement advances valve timing. In the other direction, piston movement retards valve timing.

The PCM, shown in Figure 12–32, monitors a Hall effect camshaft position sensor, which, in turn, monitors a reluctor wheel that has six notches that are spaced evenly at 60 degree intervals. A seventh notch is used as a synch notch. The Hall effect cam sensor is located at the front of the cylinder head, Figure 12–33. The PCM compares input from the camshaft position sensor to its input from the crankshaft sensor in order to detect how far advanced or retarded the camshaft is in its relationship to the sprocket that drives it, Figure 12–34.

At idle, the PCM positions the camshaft to provide smooth engine operation. At higher rpms, the PCM positions the camshaft to provide maximum performance. The camshaft can be adjusted up to 25 degrees.

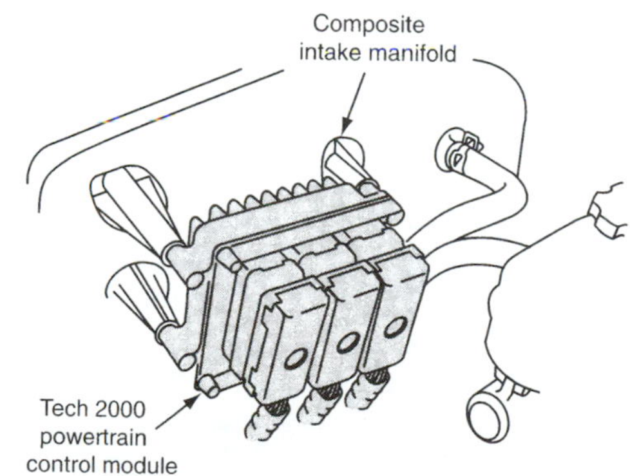

Figure 12–32 General Motors' Tech 2000 PCM.

The Trailblazer also uses a throttle actuator control (or drive-by-wire) system, Figure 12–35.

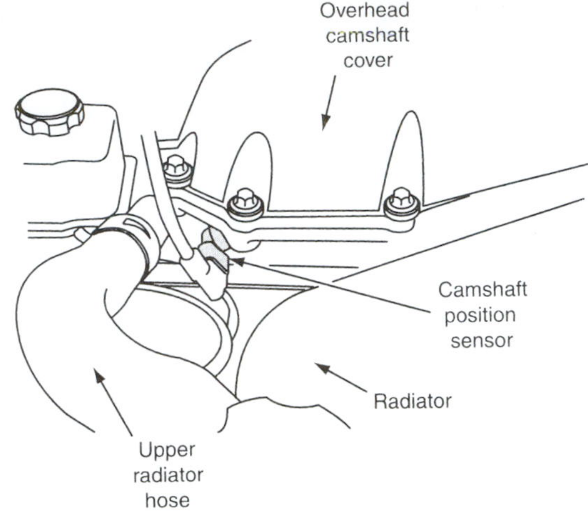

Figure 12–33 Hall effect camshaft position sensor.

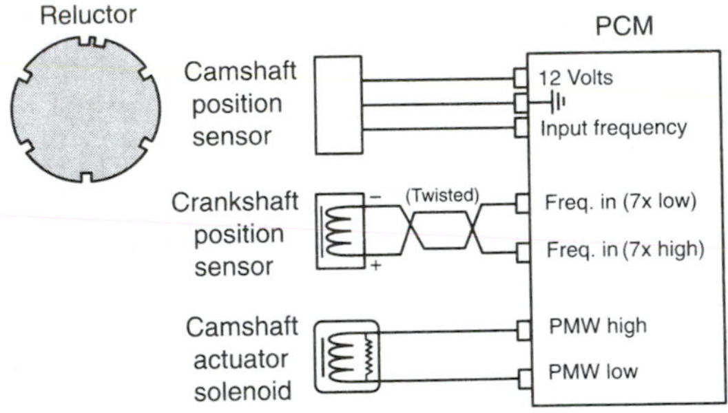

Figure 12–34 General Motors' variable valve timing sensors and solenoid.

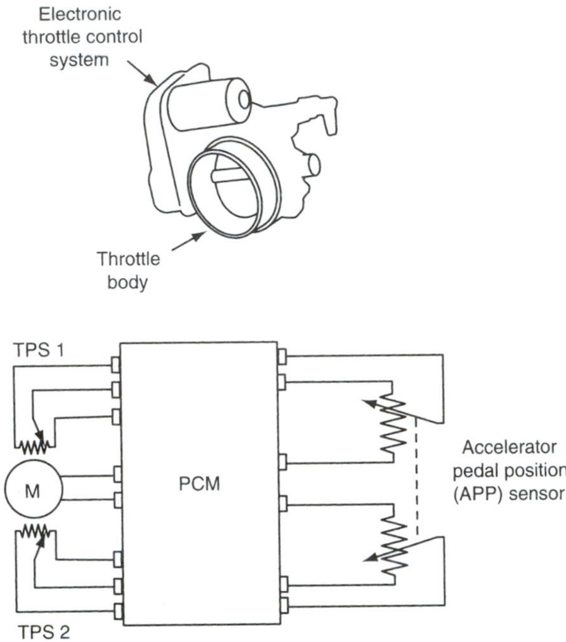

Figure 12–35 General Motors' two-wire APP drive-by-wire system.

This system is discussed earlier in this chapter. The Trailblazer uses dual accelerator pedal position (APP) sensors to monitor driver demand as well as dual throttle position sensors (TPS) to monitor the PCM-controlled position of the throttle plate.

SATURN

Spark Knock Sensor

The 2000 Saturn 1.9L engine uses a new flat response knock sensor that operates over a much wider frequency range. In addition to the knock sensor, the Saturn switched to a new PCM. This PCM is supplied by Motorola rather than by Delco Electronics. In the 1.9L engine, the new Motorola PCM is designed to filter out any signal that is not considered to be knock. Knock frequency detection for the 1.9L DOHC engine is 6 KHz to 8 KHz.

Heated Oxygen Sensor

The heated oxygen sensor that Saturn uses is now supplied by Denso rather than by Delphi. Technicians should be careful not to install an older heated oxygen sensor because of the wattage heat produced. There is a difference in the internal heater. An oxygen sensor with less wattage takes longer to reach the desired operating temperature. The longer warmup time may delay sensor activity and set a failure code.

Instrument Panel Cluster

The instrument cluster now receives data on the class two network line from the BCM, PCM, sensing diagnostic module (SDM) and the electronic brake traction control module (EBTCM). Instrument cluster warning lights, odometer and gauges are all controlled via the class two network line. The only hardwired components are the right and left turn signal indicator bulbs.

Theft Deterrent

The passlock system does not use a resistor in the ignition key the way previous theft deterrent systems did. The passlock lock cylinder contains two Hall effect switches that determine if the ignition switch has been activated correctly. The BCM monitors the passlock lock cylinder and searches for a specific voltage. This specific voltage sends a password on the class two network line to the PCM. The correct specific voltage enables the fuel injectors to continue operation. Failure to use the correct password results in a start and stall condition.

Body Control Module (BCM)

The 2000 "S" series vehicle uses a new BCM located in the center dashboard area. The new unit includes the chime/dome lamp module as well as the remote keyless entry (RKE) module. One feature new in 2000 is an audible warning chime produced by the BCM that comes on when the park brake is left on and the vehicle is driven above 3 mph. The transmission must be shifted out of Park to activate the daylight running lamps (DRL). Additionally, the battery saver feature is now available in all models. This feature works when the dome light is left on for 20 minutes with a door ajar. The light will be shut off but the computer will not be able to go to sleep, causing a higher than normal parasitic battery draw. Technicians must consider this condition when diagnosing a dead battery.

Diagnostic & Service Tip

While the PCM on this system has become much more complex than earlier, simpler engine control computers, it has also become much more reliable. Despite the widespread belief that almost any driveability problem is caused by "the computer," in fact the PCM itself is very rarely the cause of driveability or emissions problems. Even when the trouble codes indicate something in the computer, it is important to check the ground circuits for the computer and the engine itself. Because of the very low current/low voltage signals used for these components, it is a much more revealing test to check **voltage drop** rather than resistance when trying to eliminate problems. All ohmmeters are of questionable accuracy as they approach zero resistance, but voltmeters remain quite accurate even down to millivolts. Learn how to substitute voltage drop tests for resistance tests if you plan to do successful work on engine control computers.

✔ SYSTEM DIAGNOSIS AND SERVICE

WARNING: These vehicles are equipped with safety air bags. Any technician working on or around the circuits involved in the air bag system should review the sections in the service manual before performing any service on the vehicle. Failure to do so can result in personal injury from unexpected air bag deployment, as well as expensive damage to the vehicle.

The first step in any problem diagnosis is a careful visual inspection. As mentioned before, one of the most important points of this inspection is to check for grounds for the engine and the computer. Familiarize yourself with those locations (they vary with the vehicle) and learn to use the voltage drop test to check for resistance. While checking for grounds is often described, as here, as a visual inspection, it is, of course, nothing of the sort. Use voltage drop measurements to make these checks.

Very many intermittent problems are the result of poor ground connections. Check for leaking vacuum lines, loose connectors and each of the ordinary problems that might disable any car, computer-controlled or not. Check for damaged or collapsed air ducts, air leaks around the throttle body or sensor mating surfaces. Check the spark plug cables for cracks, hardness, proper routing and continuity. Most problems should be discovered in the course of a visual inspection.

Check that the PCM and MIL (Malfunction Indicator Lamp) lights are working properly. They should come on with the ignition switch as a bulb check. The next step is to check for stored codes. Connect a scan tool to the DLC (which is supposed to be visible from a squatting position at the driver's door). Following the instructions on the tool, read out any stored codes and record them.

Make sure the engine is at a comparatively low temperature. The Aurora/Northstar engines include a feature whereby at temperatures above 130°C (265°F), what is called ***engine metal overtemp mode***, the PCM will selectively disable four of the cylinders at a time to keep the engine temperature from reaching destructive levels. This injector disabling will be felt by the driver as a loss of power and roughness, of course. Verify the customer complaint and compare the stored codes to those found in the appropriate section of the manual.

The PCM includes a snapshot feature that records all the relevant sensor and actuator states at the time of any recorded fault. Often this stored information will point to the cause of the problem. The scan tool can also call up the freeze frame or failure records data for intermittents. You can check to see that the PCM is properly storing malfunctions by unplugging the MAP sensor at engine idle until the MIL comes on. It should store a code that remains in memory after the ignition switch is cycled off; otherwise the PCM itself is faulty.

SUMMARY

In this chapter we covered the newly developed extended sensors and actuator ranges for the new PCM used on the Aurora/Northstar engine. We learned about the electronic mechanism of the output drivers, successors to the earlier quad drivers that work similarly to energize the actuators by providing ground circuits.

We looked at the torque management system, which serves not only to control traction slip but also to cushion transmission shifts. The four-coil waste spark system was explained in some detail, indicating how the direct ignition system works firing opposite pairs of spark plugs simultaneously, one just before the power stroke and the other at the end of its exhaust stroke. We learned the sensor techniques the computer uses to identify which cylinders are the appropriate ones to fire at a given crankshaft position.

The chapter included the beginnings of the OBD II standardized selfdiagnostic system. This requires the addition of the downstream, secondary oxygen sensor, which monitors the effectiveness of the catalytic converter. We learned how a very precise measurement of crankshaft speed enabled the computer to detect misfire, and what sorts of countermeasures the computer uses to control emissions quality and protect the catalytic converter should there be such a misfire. And we learned about the EEPROM and how it is different from, and a successor of, the older, hardwired PROM. We began to understand the diagnostic procedures for this system and OBD II systems generally.

Additionally, we also covered General Motors' central port fuel injection systems known as central multiport fuel injection (CMFI) and central sequential fuel injection (CSFI) and the features that make these systems different from most port fuel injection systems. We also covered some of the features associated with the newer GM Generation III engines, Premium V-6 engines, and the Saturn 1.9L engine. And we discussed the basic concepts of GMs' variable valve timing system.

▲ DIAGNOSTIC EXERCISE

A vehicle with the Aurora/Northstar engine is brought into the shop with much reduced power shortly after startup. All three oxygen sensors report a lean mixture, though the spark plugs are all black with carbon. Even before you pull the OBD II codes, what kinds of problems could cause these symptoms?

REVIEW QUESTIONS

1. *Technician A* says that a person can feel a static discharge of about 4,000 volts or more. *Technician B* says that an electronic component such, as a computer's memory chip, can be damaged by a static discharge of as little as 100 volts.
   Who is correct?
   A. A only
   B. B only
   C. Both A and B
   D. Neither A nor B
2. *Technician A* says that modern PCMs have a more advanced learning capacity than previous PCMs did.
   Technician B says that disconnecting the battery will always erase all of the learned data from a modern PCM's memory.
   Who is correct?
   A. A only
   B. B only
   C. Both A and B
   D. Neither A nor B
3. On a Northstar/Aurora application, if the PCM senses that a front drive wheel has lost traction, it may individually apply the brakes at one wheel in order to regain traction. In extreme cases, it may also reduce engine torque by disabling up to seven of the fuel injectors *except* under which of the following conditions?
   A. When the engine coolant temperature is below –40°F
   B. When the engine coolant level is low
   C. When the engine speed is below 600 rpm
   D. All of the above
4. A Northstar engine in a vehicle is started cold and is then allowed to warm up. The ECT sensor voltage is being monitored. While the engine is warming, the voltage is decreasing. Then the voltage suddenly increases.
   Technician A says that this indicates that the ECT sensor is faulty and should be replaced.
   Technician B says that this engine likely has a dual-curve ECT circuit.
   Who is correct?
   A. A only
   B. B only
   C. Both A and B
   D. Neither A nor B
5. If the PCM detects cylinder misfire on an OBD II application, it will do which of the following?
   A. Shut off the fuel injector for the misfiring cylinder
   B. Keep the injector energized for the misfiring cylinder
   C. Turn off the spark for the misfiring cylinder
   D. Retard spark timing for the misfiring cylinder
6. On a Northstar application, the PCM will set a code in memory if the power steering pressure switch is open and vehicle speed is in excess of which of the following speeds?
   A. 10 mph
   B. 25 mph
   C. 30 mph
   D. 45 mph
7. On a Northstar application, if the PCM sets an EEPROM defect code in memory, what must the technician do?
   A. Replace the EEPROM
   B. Replace the PCM
   C. Reset the EEPROM by disconnecting the vehicle battery for 10 minutes
   D. Reset the EEPROM by removing it from the PCM, then reinstalling it

8. In addition to controlling the systems that are controlled by a PCM, which of the following does a VCM also control?
   A. The instrument gauge cluster
   B. The antilock brake system
   C. The electronic climate control system
   D. The memory seat function
9. *Technician A* says that a CMFI system uses a single fuel injector solenoid to deliver fuel to all cylinders through nylon tubes and individual poppet nozzles.
   Technician B says that a CSFI system uses multiple fuel injector solenoids that are pulsed sequentially by the PCM and deliver fuel to each cylinder through a nylon tube and poppet valve.
   Who is correct?
   A. A only
   B. B only
   C. Both A and B
   D. Neither A nor B
10. How is the Generation III V8 ignition system designed?
    A. It uses a distributor to distribute spark from a single ignition coil to each of the spark plugs.
    B. It is a waste spark ignition system that uses four ignition coils.
    C. It uses eight ignition coils placed on top of the spark plugs to eliminate the secondary ignition wires.
    D. It uses eight ignition coils placed near the spark plugs and uses secondary ignition wires.
11. The engine coolant overtemp fuel disable feature is found on which of the following?
    A. Northstar engines
    B. Generation III V8 engines
    C. Premium V6 engines
    D. All of the above
12. On a Generation III application, if the passlock cylinder, the BCM or the PCM is replaced, the computers must relearn the passlock voltage.
    Technician A says that this can be done using a scan tool.
    Technician B says that this can be done without the use of a scan tool.
    Who is correct?
    A. A only
    B. B only
    C. Both A and B
    D. Neither A nor B
13. Which of the following statements is *false* concerning the Throttle Actuator Control (TAC) system on a Generation III application?
    A. It allows the throttle pedal to directly control the position of the throttle blade through a cable.
    B. It allows the PCM to control the position of the throttle blade.
    C. It can be used to provide the antislip/traction control function.
    D. It can be used to provide the cruise control function.
14. *Technician A* says that the precat and postcat oxygen sensors on a Premium V6 engine are identical and may be interchanged.
    Technician B says that the Premium V6 engine uses dual spray fuel injectors that must be properly aligned with the electrical connectors perpendicular to the failure rail. Failure to do so may result in driveability or emission problems.
    Who is correct?
    A. A only
    B. B only
    C. Both A and B
    D. Neither A nor B
15. *Technician A* says that the PCM on a Premium V6 application requires constant airflow over its finned aluminum housing.
    Technician B says that the PCM on a Premium V6 application is located inside the air cleaner housing.
    Who is correct?
    A. A only
    B. B only
    C. Both A and B
    D. Neither A nor B
16. What component does the VVT mechanism change in relation to its drive sprocket on a 2002 Trailblazer?

A. The intake camshaft
B. The exhaust camshaft
C. Both camshafts
D. The crankshaft

17. On a 2002 Trailblazer, how many degrees of adjustment can be made with the VVT mechanism?
A. 10 degrees
B. 25 degrees
C. 40 degrees
D. 55 degrees

18. The battery saver feature on a Saturn shuts off the power to the interior dome light if a door has been left ajar for how long?
A. 5 minutes
B. 10 minutes
C. 20 minutes
D. 1 hour

19. Which of the following is the most reliable method of testing the PCM's power and ground circuits?
A. Use a test light to test circuit continuity.
B. Use a logic probe to test circuit continuity.
C. Use an ohmmeter to measure each circuit's resistance.
D. Use a DVOM to test voltage drop across each circuit.

20. *Technician A* says that when working on or around the circuits that control safety air bags, a technician should first review the air bag sections in the service manual before performing any service work on the vehicle. *Technician B* says that an unexpected air bag deployment can result in personal injury. Who is correct?
A. A only
B. B only
C. Both A and B
D. Neither A nor B

Chapter 13

Ford's EEC I, EEC II and EEC III

OBJECTIVES

Upon completion and review of this chapter, you should be able to:

- ❑ Define the engine applications of the EEC I, EEC II and EEC III systems.
- ❑ Describe the functions of the EEC I system.
- ❑ Describe the functions of the EEC II system.
- ❑ Describe the functions of the EEC III system.
- ❑ Understand how to perform the selfdiagnostic tests that are programmed into an EEC III PCM.

KEY TERMS

Aneroid
Barometric Pressure Sensor
Duraspark II

Ford introduced the EEC I system in 1978, using it on the 1978 and 1979 5.0-liter Lincoln Versailles. EEC II was introduced in 1979, limited to full-size Fords sold in California and Mercurys sold in all fifty states, all with 5.8-liter engines. Ford expanded the system as EEC III in 1980 with somewhat more extensive control applications, but its use was still limited to the V8 engines. In 1981, Central Fuel Injection (CFI) replaced feedback carburetors as part of the EEC III system on some engines. (Ford's CFI system is the equivalent of the TBI systems used by other manufacturers.) In 1982 EEC III was introduced for some light trucks.

Many of the component parts described in this chapter are also found on the later EEC IV system. This chapter covers only those controls, sensors and actuators found on the early systems. Because the earlier systems are simpler, it is easier to learn the basics of the Ford systems from them. Elements that are carried over into the later systems will only be repeated when there is a material difference in their construction or function.

ELECTRONIC ENGINE CONTROL (EEC) I

The most important factor to remember about the earliest Ford system, the EEC I, is that it does not control the air/fuel mixture, but only spark advance. The engines controlled use the variable venturi (VV) carburetor, which has no computer-actuated mixture control on these systems. There is, therefore, no oxygen sensor and no closed-loop operational mode. This system was the last, best effort of the mechanical carburetor system to maintain driveability while optimizing emissions.

Powertrain Control Module (PCM)

The Powertrain Control Module, or PCM, formerly known as the Electronic Control Assembly, or ECA, prior to OBD II standardization, is like later combustion control computers except that it:

- performs fewer functions.
- has no selfdiagnostic capacity.
- uses an externally mounted calibration unit to fit the unit to specific vehicles, accounting for curb weight, axle ratio, high altitude use and so forth, Figure 13–1. This is similar to a General Motors' PROM, except that it is much simpler.
- includes a fuel octane adjustment switch to retard ignition timing either 3 or 6 degrees if spark knock occurs.
- provides a 9-volt reference voltage to its sensors.

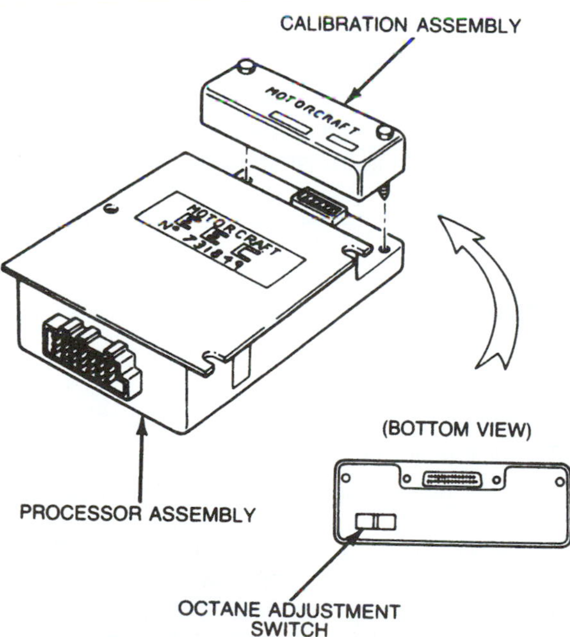

Figure 13–1 EEC I PCM and calibration assembly. *(Courtesy of Ford Motor Company.)*

Default Mode. If the PCM fails, it stops sending commands to the three actuators it controls. Ignition stays fixed at base timing; the Thermactor dumps air into the atmosphere as long as the engine is running, and the EGR valve stays closed. Obviously, neither driveability nor emissions quality will be at the desired levels, but the vehicle can be driven at reduced power and greater fuel consumption until it is diagnosed and repaired.

The PCM includes a power relay to protect it from reversed polarity, a protection retained on later Ford systems. Both the relay and the PCM itself are on the driver's side of the instrument panel, near the brake pedal.

Inputs

The PCM gets input from these sensors:

- engine coolant temperature (ECT) sensor
- manifold absolute pressure (MAP) sensor
- barometric pressure (BP) sensor
- throttle angle position (TAP) sensor
- crankshaft position (CP) sensor
- intake air temperature (IAT) sensor
- EGR valve position (EVP) sensor

The ECT, TAP and EVP sensors are very similar to the corresponding units on the later Ford systems. The IAT sensor is in the air cleaner and works similarly to the corresponding sensor or an EEC IV system.

MAP Sensor. The MAP sensor uses an **aneroid**, a sealed accordion-like capsule with a measured amount of gas that expands and contracts with changes in the surrounding pressure. As the pressure outside the aneroid increases, it compresses; as the pressure outside decreases, the aneroid expands. The aneroid is situated in a housing directly connected to intake manifold vacuum, Figure 13–2. With changes of intake manifold vacuum, the expansion and contraction of the aneroid moves the wiper of a potentiometer to signal the PCM with a return voltage proportional to manifold pressure.

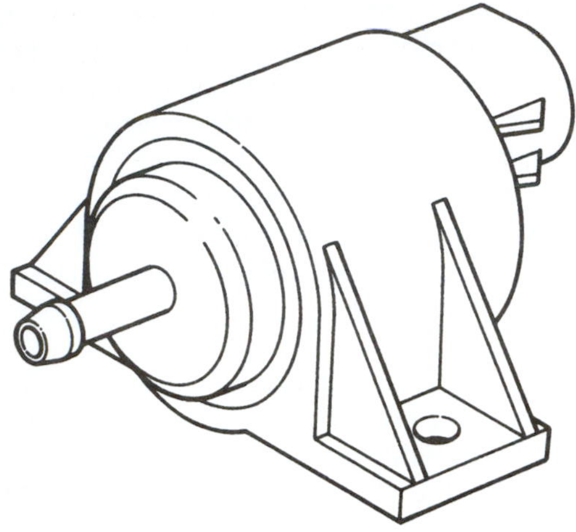

Figure 13–2 Aneroid manifold absolute pressure sensor. *(Courtesy of Ford Motor Company.)*

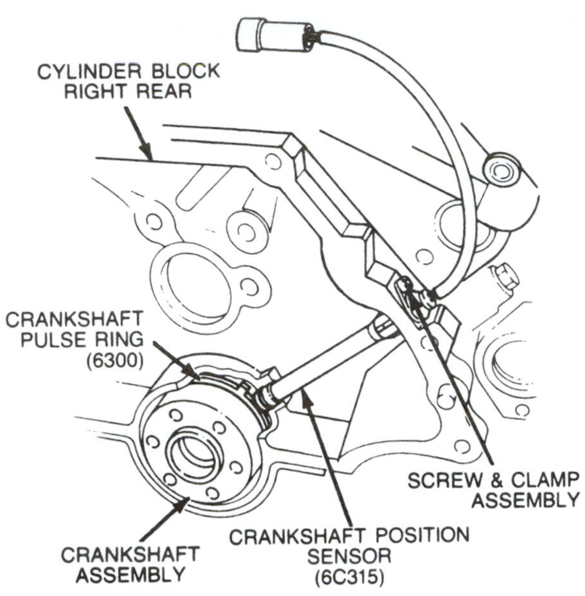

Figure 13–3 Crankshaft position sensor. *(Courtesy of Ford Motor Company.)*

BP Sensor. The **barometric pressure sensor** works just like the MAP sensor, except it is vented to the atmosphere rather than plumbed to the intake manifold. Its signal depends exclusively on the barometric pressure and the altitude of the vehicle.

CP Sensor. The crankshaft position sensor consists of a four-lobe pulse ring on the output end of the crankshaft, and a probe with a permanent magnet and a wire coil at its tip. This sensor is essentially a pickup coil removed from the distributor and relocated at the rear of the crankshaft, Figure 13–3. Of course, a pickup clocking crankshaft position monitors a shaft turning twice as fast as a conventional distributor, but the computer does not need to "know" what part of the engine cycle a particular cylinder is in, compression or exhaust. Since this system uses a (relatively unconventional) distributor to determine which cylinder gets the next spark, the computer need only calculate the appropriate spark advance.

The pickup probe extends into the back of the engine block leaving a small air gap between its tip and the pulse ring on the crankshaft. As the crankshaft turns and each lobe of the ring passes under the magnet in the probe's tip, the air gap is suddenly much smaller. This causes a sudden increase in the strength of the magnetic field at the magnet's pole. This sudden change in the magnetic flux field generates a voltage signal in the coil around the magnet. As the crankshaft continues to turn, this steady stream of voltage polarity-reversal signals is transmitted to the PCM, each pulse indicating a piston is at 10 degrees before top dead center, the ignition reference pulse.

Outputs

EEC I controls three output circuits:

1. ignition timing
2. Thermactor air control
3. EGR flow

Ignition Timing. The PCM monitors coolant temperature, intake air temperature, engine speed, throttle position and engine load (as reflected from the intake manifold absolute pressure) to calculate the proper ignition timing. Low intake air temperature sensor signals, for example, allow increased spark advance; and higher temperatures call for less spark advance. If intake air temperatures rise above 90°F, the spark advance is also retarded to avoid spark knock.

If the ECT sensor indicates overheating during idle, however, the spark advance is increased to raise the idle speed, turn the water pump faster and obtain more coolant flow (EEC I does not directly control idle speed through airflow, so advancing the timing is the only strategy to increase engine rpm, and thus, water pump speed and coolant circulation). If the throttle opening increases rapidly, spark advance is retarded. Spark advance is also retarded if the manifold pressure increases (intake manifold vacuum declines) even if there is no change in throttle position. This might occur if a vehicle started up a hill and the engine slowed. The PCM's spark timing command is sent to the standard **Duraspark II** ignition module, which interrupts primary coil current to fire the appropriate spark plug.

Revised Distributor Design

When the automotive world changed over to computerized control of spark timing, both of the traditional mechanical advances (vacuum and centrifugal) were eliminated. Referring back to the centrifugal advance mechanism, its job was not only to advance the timing of the spark as engine rpm increased, but it simultaneously performed the task of physically advancing the rotor's position in its relationship to the distributor cap to compensate for the increased spark advance. This mechanical movement of the rotor allowed the rotor to always be in the ideal relationship to the intended receiving electrode in the cap at the time that the spark was produced in the secondary winding, regardless of the amount of the advance that the centrifugal advance mechanism might add in.

When the centrifugal advance was eliminated, the ability of the rotor to compensate for increased spark timing was also eliminated. Therefore, as the PCM advances spark timing on a distributor-type application, the rotor is positioned farther and farther away from the intended receiving electrode in the cap at the time that the spark is delivered across it. If spark timing is increased so that the spark is delivered across the rotor while the rotor is still closer to the previous electrode than the upcoming intended electrode, the spark will jump the rotor-to-cap gap of the least resistance, therefore firing to the wrong electrode and resulting in a cylinder misfire. This is generally known as *cross-firing* within the cap and rotor assembly. Thus, when spark timing is under the control of a computer on a distributor-type ignition system, the PCM is normally limited to an increase in spark timing that does not allow the rotor to be caught at a position that equals or exceeds one-half of the distance between cap electrodes at the instant that the spark is delivered across it. This distance would equate to 22½ degrees on an eight-cylinder cap. (The distance between cap electrodes is 45 degrees.) This would double to 45 degrees of rotation at the crankshaft.

Ford engineers designed the PCM in the EEC I, EEC II and EEC III systems to have a spark advance program that was more aggressive than this. In order to avoid the rotor cross-firing to the wrong cap electrode during periods of heavy spark advance, these systems incorporate a bilevel cap and rotor design. The distributor cap was designed with eight electrodes, which attached to each cylinder's plug wire, but, on the underside of the cap, four of these electrodes are short and alternate with four electrodes that are long, Figure 13–4. The coil wire's electrode is designed on the underside of the cap as a plus sign or cross. The rotor has two electrodes that take turns receiving the spark from the coil wire's electrode as it turns. One of the rotor's electrodes is designed to deliver the spark to the cap's closest short (or shallow) electrode and the other delivers spark to the cap's closest long (or deep) electrode, Figure 13–5. As a result, the rotor effectively delivers successive sparks to electrodes spaced 225 degrees apart as it turns, Figure 13–6. That is, in Figure 13–6, as the rotor spins counterclockwise, it delivers successive sparks to every fifth cap electrode in order to accommodate the engine's firing order.

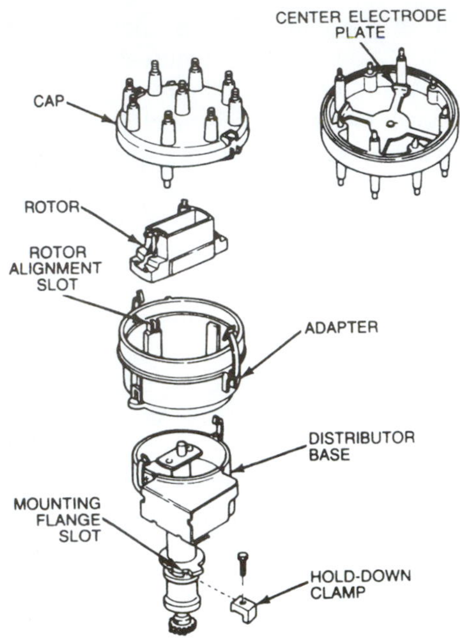

Figure 13–4 EEC I distributor assembly. *(Courtesy of Ford Motor Company.)*

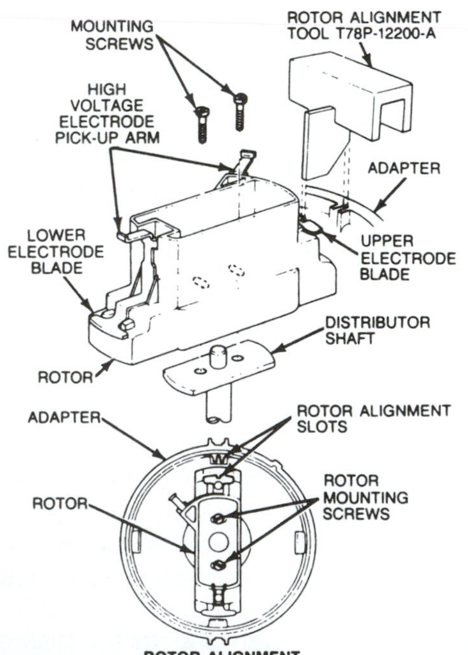

Figure 13–5 EEC I distributor rotor. *(Courtesy of Ford Motor Company.)*

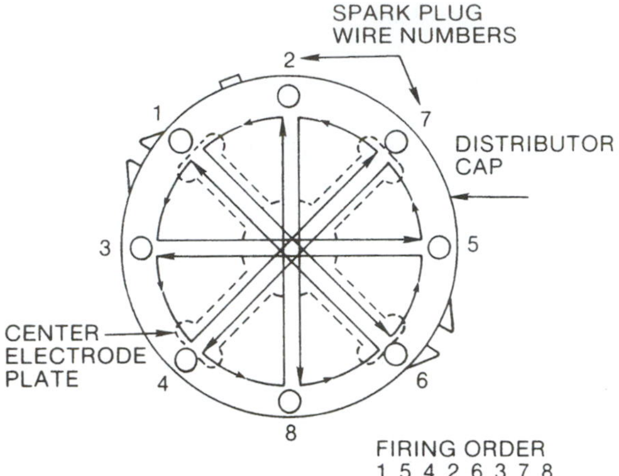

Figure 13–6 EEC I distributor firing order.

This distributor also has a notch (or slot) for the hold-down bolt so that a technician cannot rotate the distributor. Base ignition timing is not adjustable on EEC I, EEC II and EEC III applications. In fact, there is a decal on the distributor cap's adaptor that states that base timing is not adjustable. This statement is also reiterated on the emission decal. Because the crankshaft position (CP) sensor produces the tach reference signal, the distributor is, in fact, empty of primary ignition components. It is a distributor in the truest sense of the word and contains only a cap and rotor to distribute spark from a single coil to multiple cylinders. As a result, any attempt to turn the distributor housing does not actually change spark timing, but, instead, changes the relationship between the rotor and the cap at the time that the ignition coil produces the spark. This will result in cross fire and, therefore, cylinder misfire.

Because of the slight rotation of the distributor housing allowed by the tolerances designed into the hold-down notch, after a technician has tightened down the distributor housing, he should also perform a rotor alignment on these applications. This is done by positioning the crankshaft at a certain number of degrees ATDC (as per the service manual), then using the rotor alignment tool (Figure 13–5) to check the rotor's alignment with the cap. Two screws may be loosened in the upper portion of the distributor shaft to allow it to rotate slightly in proportion to the lower distributor shaft. Be sure to retighten them after this adjustment is complete. Misalignment may cause crossfiring within the cap and rotor assembly and may also shorten spark duration caused by an overly large rotor-to-cap gap. This adjustment does not affect base timing. That is, adjusting the rotor's relationship to the distributor cap cannot change when, in time, the ignition module turns off primary current flow, inducing the spark in the secondary winding. It only changes the rotor-to-cap relationship in effect at the instant the spark is produced, thereby only affecting spark distribution.

Because the tach reference signal is generated at the crankshaft (as with most distributorless systems), timing chain stretch does not affect ignition timing. As a result, there is no designed method to change base ignition timing. However, the calibration assembly found on the PCM contains a switch that can be used to instruct the PCM to back off of its original spark advance program either three or six degrees, Figure 13–1. This adjustment is referred to as an *octane adjustment*.

Finally, the original EEC I distributor included an enclosure for mounting a future miniaturized ignition module on the distributor. This feature was not carried over to EEC III and was never actually used. And when EEC IV finally mounted an ignition module on the distributor, it was of another design.

Thermactor Air Control. The EEC I system employs only a single-bed catalytic converter, so the Thermactor system does not supply air to the converter during warm engine operation. Air is supplied to the exhaust manifolds only during low-temperature conditions and is dumped into the atmosphere once normal operating temperature is reached. The PCM controls these operational modes with a solenoid-operated vacuum control valve, the Thermactor control solenoid. This normally closed valve controls manifold vacuum to the Thermactor bypass valve, Figure 13–7. The Thermactor pump delivers air to the bypass valve directly.

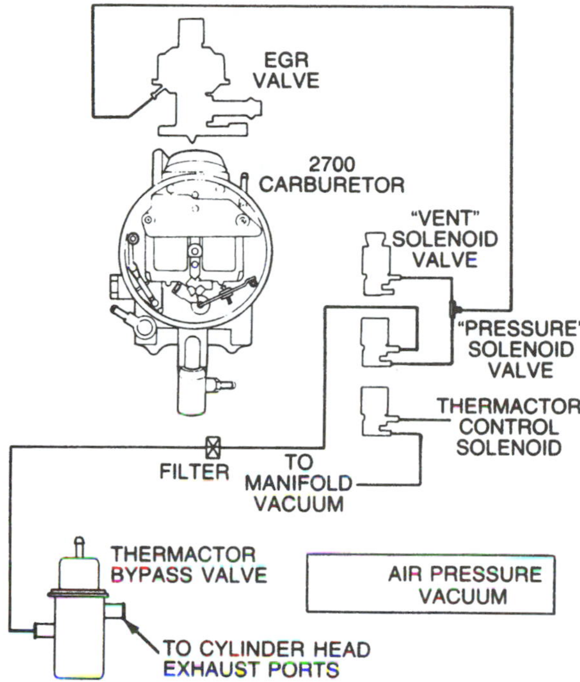

Figure 13–7 EGR and Thermactor system control. *(Courtesy of Ford Motor Company.)*

When the sensor in the air cleaner indicates to the computer that underhood temperature is low, the PCM grounds the Thermactor control solenoid circuit. The solenoid opens the vacuum valve, and vacuum is routed to the bypass valve. The bypass valve then routes Thermactor air into the exhaust manifolds to help burn off residual hydrocarbons. Once the engine warms to operating temperature, the PCM shuts off the solenoid, and blocks vacuum from the bypass valve. Thermactor air is then dumped to the atmosphere. The PCM also dumps Thermactor air during WOT conditions.

EGR Flow. The EEC I exhaust gas recirculation (EGR) valve opens by air pressure instead of by vacuum, Figure 13–8. An external tube supplies exhaust gas to the EGR valve connected to an exhaust manifold. To improve driveability and to extend EGR valve life, the temperature of the EGR gas is reduced by running the supply tube through a heat exchanger, the EGR cooler, Figure 13–9.

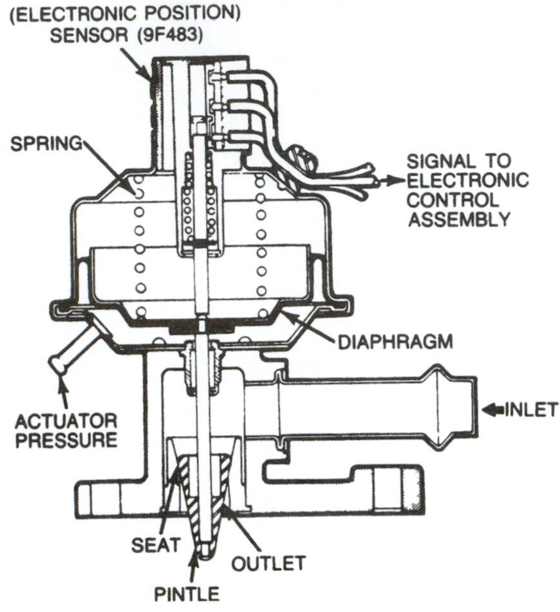

Figure 13–8 Pressure-operated EGR valve. *(Courtesy of Ford Motor Company.)*

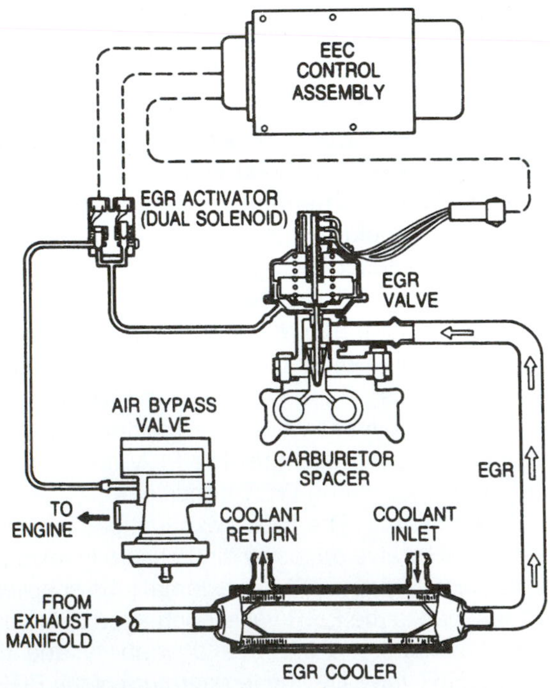

Figure 13–9 EGR cooler assembly. *(Courtesy of Ford Motor Company.)*

Coolant flows from a heater hose through the EGR cooler.

The PCM controls EGR flow through two solenoid valves, the normally closed pressure valve and the normally open vent valve, Figure 13–10. The PCM monitors information from its sensors and calculates how much EGR flow is needed. For example, suppose the inputs indicate increased EGR flow is needed: the PCM then energizes both solenoids. The normally closed pressure valve opens to apply pressure from the Thermactor system to the lower side of the EGR valve. The normally open vent valve closes to prevent the pressure escaping to the atmosphere. Now the

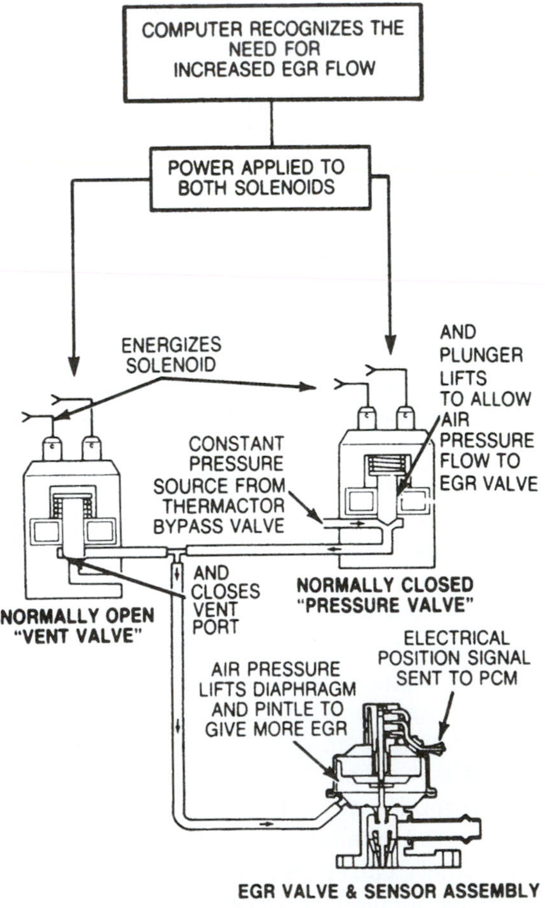

Figure 13–10 EGR control system. *(Courtesy of Ford Motor Company.)*

EGR valve opens wider until the EVP sensor signals the computer that the EGR valve is at the desired position. Then the PCM de-energizes the pressure solenoid. It closes and traps the existing pressure under the EGR valve diaphragm, keeping the EGR valve in a fixed position. This condition persists until operating conditions change as signaled by the various sensors (throttle position and intake manifold pressure changes) or until some of the trapped pressure leaks out and the EVP sensor signals the PCM the change in the EGR valve's position. If the PCM determines to decrease EGR flow, it de-energizes both solenoids, closing the pressure source and venting the existing pressure. If barometric pressure decreases, either because of weather changes or changes of the altitude of the vehicle (by driving up a hill), EGR flow reduces. The computer cuts off all EGR flow if the coolant is below 21°C (70°F) or above 110°C (230°F). Normally—except in the temperature extremes just described—the two solenoids are each cycled rapidly to maintain the calculated EGR valve delivery.

✔ SYSTEM DIAGNOSIS AND SERVICE

Early automotive engine control computers like the PCM in the Ford EEC I system had no self-diagnostic capacity. Special test equipment and procedures were developed and are listed in the appropriate Ford service manual for identifying and correcting driveability or other engine-related problems an owner may report. The special tests employ special tester Rotunda No. T78L-50-EEC-I (or its equivalent) in series with the PCM harness. There is a disconnect in the harness for just this purpose. Once the tester is installed, all signals to or from the PCM pass through the tester, not unlike a breakout box. A special DVOM (Rotunda No. T78-50) connects to the first special tool and provides the readings for each test in the sequence.

The first of the two special test sequences is performed after the engine warms up and with the ignition off. Eight test readings check the condition of each of the sensors, the power relay, the Thermactor control solenoid and part of the PCM's internal circuitry. In the second test sequence, the technician checks spark advance, EGR valve position and Thermactor air mode with the engine running at 1,600 rpm in park, with vacuum from an external source applied to the MAP sensor. This second test sequence checks the remaining PCM circuitry.

Rotor Alignment

Technicians who regularly service Ford vehicles have discovered correct rotor alignment is very critical to good engine performance. This dimension is one of the first things experienced technicians check when responding to complaints of poor engine performance on EEC I, EEC II or EEC III system vehicles.

The results of each reading and check are recorded and compared to test specifications provided with the tester. If all the results are within the specification limits, the system is functioning properly; if not, diagnostic procedures in the service manual guide diagnosis and repair of the fault in the identified circuit.

With these detailed tests, the problems of diagnosing EEC I failures are not too difficult, even in the absence of any selfdiagnostic aids. While individual tests of sensors and actuators are required, these are fewer in number and simpler in operation—hence easier and quicker to test—than the components employed on later Ford systems.

EEC II

The EEC II system is similar to EEC I with some minor differences in sensors and with significantly more control responsibility. The PCM and power relays are similar except the PCM includes more internal circuitry to perform the additional functions. It also provides a 9-volt reference signal to each of its sensors. New with the EEC II system are an exhaust oxygen sensor and an electronically controlled air/fuel mixture device. Hence with EEC II we find the first Ford system that can operate in closed-loop mode. Closed-loop means, on Ford systems as on others, that the computer controls the fuel/air mixture in response to feedback signals from the oxygen sensor, which in turn responds to the delivered mixture—hence *closing* the loop.

The "default" mode is now called *limited operational strategy* (LOS), as we will see in EEC IV. Other systems call this state the limp-home mode or the fail-safe mode or something similar. This mode is triggered by an electrical failure in some critical component or circuit, and in it all actuators are left in their deactivated mode. Driveability, fuel economy and exhaust emissions quality are all compromised in that state.

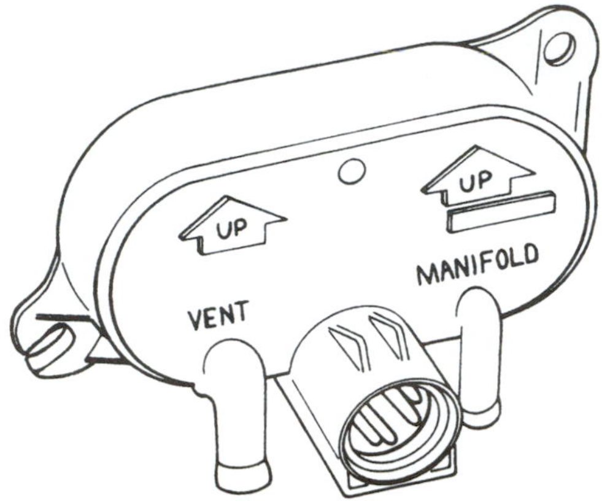

Figure 13–11 BARO and MAP sensor. *(Courtesy of Ford Motor Company.)*

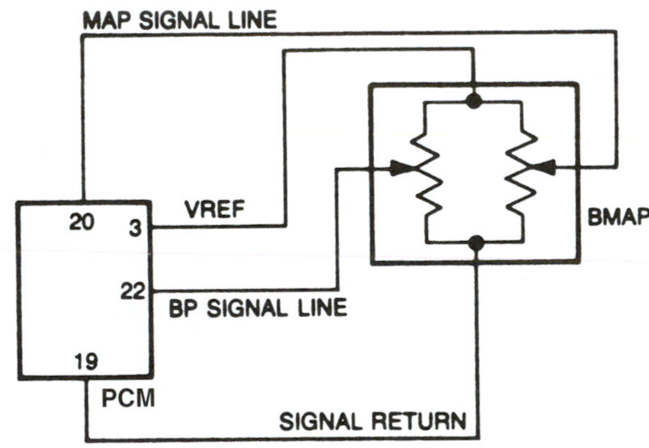

Figure 13–12 BARO and MAP sensor electric circuit. *(Courtesy of Ford Motor Company.)*

Inputs

EEC II sensors include:

- engine coolant temperature (ECT) sensor
- manifold absolute pressure (MAP) sensor
- barometric pressure (BP) sensor
- throttle position (TP) sensor
- crankshaft position (CP) sensor
- exhaust gas oxygen (EGO) sensor
- EGR valve position (EVP) sensor

ECT. All temperature-related information comes from the ECT sensor, which is identical to the previous system's sensor. EEC II uses no intake air temperature (IAT) sensor as the computer is able to optimize air/fuel mixture adequately without that information.

MAP and BP Sensors. The MAP and BP sensors work like those on the earlier EEC I system, except they are both in one housing, called a B/MAP sensor, Figures 13–11 and 13–12.

TP Sensor. The TP sensor works the same as on the earlier system, but the word *angle* is dropped from its name. Like its predecessor, the EEC II TP sensor is a potentiometer returning a signal proportional to the throttle opening.

CP Sensor. The CP sensor works the same way as its corresponding sensor on EEC I, except it is now at the more accessible front of the engine and is electrically shielded to protect its signal from electrical interference, Figure 13–13.

EGO Sensor. EEC II is the first of the Ford systems to employ an oxygen sensor, which is carried over to the later systems, also. As on vehicles built by other manufacturers, the oxygen sensor on Ford EEC II vehicles generates a low voltage signal proportional to the difference in residual oxygen in the exhaust gases, compared to the ambient air.

Figure 13–13 EEC II crankshaft position sensor. *(Courtesy of Ford Motor Company.)*

EVP Sensor. The EEC II EVP sensor is the same as that for the previous system.

Outputs

The PCM controls the following functions:

- air/fuel ratio
- ignition timing
- idle speed control
- EGR flow
- Thermactor air control
- canister purge

Each of these functions is either new to EEC II or somewhat different from the earlier versions on EEC I.

Air/Fuel Ratio. This function is new to EEC II. The fuel bowl of a carburetor is vented either through a balanced vent or an external vent to the atmosphere. The atmospheric pressure provides the force to drive the fuel through the jets and emulsion passages into the lower pressure region of the carburetor venturi in the throat. EEC II vehicles use the Model 7200 VV (variable venturi) carburetor, and use this pressure at the top of the fuel in the bowl to control the air/fuel mixture: raising the pressure slightly increases the flow of fuel slightly; reducing the pressure slightly correspondingly reduces the fuel flow. The specific means to do this consists of a restricted carburetor vent plus a vacuum passage allowing carburetor vacuum (called *control vacuum* in this application) to draw air from the bowl area and thus reduce the applied pressure, Figure 13–14.

The small amount of pressure drawn from the fuel bowl is controlled through a pintle valve controlled by a stepper motor. As the stepper motor extends the pintle valve, the vacuum passage is restricted. This permits more pressure to develop in the fuel bowl through the other vent, and the air/fuel mixture gets richer. If the pintle valve retracts, pressure in the fuel bowl is reduced and the air/fuel mixture gets leaner.

The PCM can selectively energize four separate coils in the stepper motor to run it forward or backward. The coils do not get a steady voltage applied, but rather a series of short pulses. Each pulse rotates the armature a specific number of degrees (hence it is a *stepper* motor). As the armature turns, the pintle valve either extends or retracts depending on which way the armature turns. Thus the PCM can transmit pulses to the

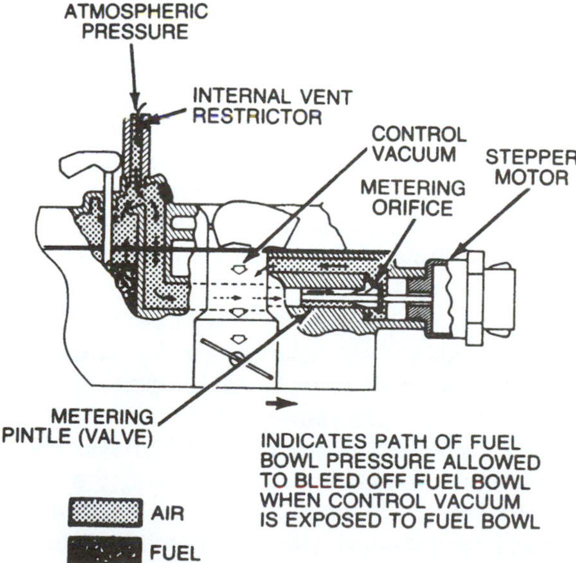

Figure 13–14 Variable venturi 7200 feedback carburetor mixture control. *(Courtesy of Ford Motor Company.)*

Diagnostic & Service Tip

A frequent problem on Ford variable venturi carburetors, including the feedback versions used with EEC II, is mechanical warping of the top of the carburetor, the air horn. If someone tightens the air cleaner hold-down bolt excessively, this can bend the metal very slightly. On many carburetors, this would make little difference, but on the variable venturi, any bending of the air horn can cause the fuel metering rod controls to stick. Unfortunately, in many cases the air horn does not return to its original position once the strain is relieved and replacement is the only option.

stepper motor to achieve whatever air/fuel ratio is required, within its range. The pintle has an extension range of 0.4-inch, and 120 pulses are required to move it from one extreme to the other.

Ignition Timing. The EEC II ignition system features the same coil and distributor as EEC I and a new Duraspark III ignition module, designed specifically to work with the EEC II PCM. The same precautions about rotor alignment apply.

Idle Speed Control. The idle kicker solenoid and idle kicker actuator first appear on EEC II and continue in EEC III and some applications of EEC IV, as we will see in Chapter 14. See this chapter for a full discussion.

EGR Flow. The EGR control solenoids, EGR vent (EGRV), and EGR control (EGRC) function like those in EEC I. However the EGRP (EGR pressure) solenoid used in EEC I was renamed. The word *pressure* was replaced by the word *control*. This name change occurred because the EGRC solenoid on EEC II systems is connected to manifold vacuum and the EGR valve opens by vacuum rather than by Thermactor system pressure. This arrangement continues in EEC III and some EEC IV applications. Experience showed that a vacuum-activated unit was more durable and reliable in service. Operation of the EGRV and EGRC solenoids is discussed further in Chapter 14.

Thermactor Air Control. The EEC II system uses a dual-bed catalytic converter and therefore requires an additional Thermactor system mode: the capacity to direct air to the converter in addition to the capacity to direct it to the exhaust manifold or the atmosphere. For this purpose, there are two control solenoids and a combination air management valve: a thermactor air bypass (TAB) solenoid, a thermactor air divert (TAD) solenoid, and a combination TAB/TAD air control valve, Figure 13–15. This system carries over into both EEC III and EEC IV and is discussed in greater depth in Chapter 14.

Canister Purge Solenoid. This function is new for EEC II, in response to new emissions regulations. The charcoal canister stores fuel vapors from the fuel tank to release them during normal operating conditions into the engine for consumption during combustion. The solenoid is a normally closed vacuum valve. The valve is in the purge hose connecting the canister to the intake manifold. When the PCM grounds the solenoid circuit, the solenoid opens and allows manifold vacuum to purge the solenoid's stored fuel va-

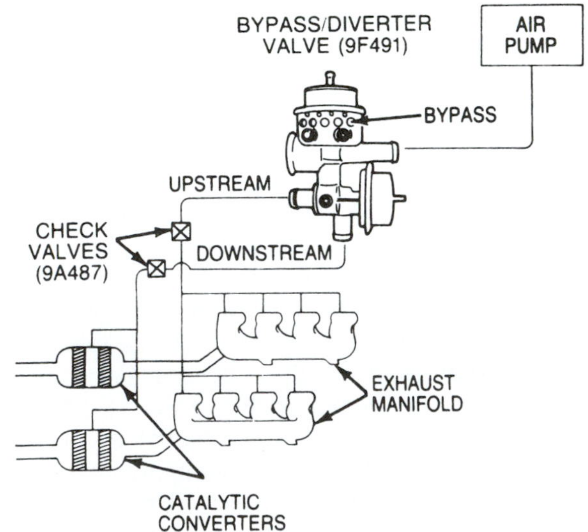

Figure 13–15 EEC II Thermactor control system. *(Courtesy of Ford Motor Company.)*

Diagnostic & Service Tip

If a Ford with EEC II constantly tries to lean the mixture, but always shows a high oxygen sensor output, check whether the charcoal canister purge is either stuck open or whether the canister itself is flooded with liquid fuel. (This can happen when a motorist tries to top off the tank beyond the automatic shutoff.)

pors. This canister purge system continues with EEC III and some forms of EEC IV.

✔ SYSTEM DIAGNOSIS AND SERVICE

The EEC II system, like its predecessor EEC I, has no selfdiagnostic capacity. Diagnosing malfunctions is performed similarly to diagnosing malfunctions on an EEC I system. Special test equipment or its equivalent, listed in the service manual, is required for this diagnosis. Essentially, a technician tries to limit the possible causes to one or a few control circuits and then tests each of them individually.

EEC III

Carbureted EEC III vehicles have the same components as on EEC II. The air/fuel mixture control device on the carburetor and the PCM itself have been modified, however. The PCM became more complicated, and there are some early electronic fuel injection systems with the EEC III system. The fuel injection system is the throttle body type that Ford technical literature most often called Central Fuel Injection. In the Ford EEC III literature, the terms *electronic fuel injection* and *central fuel injection* are used interchangeably to mean the same system. Once the Ford Motor Company went to the EEC IV system, electronic fuel injection usually refers specifically to multipoint fuel injection.

PCM

For EEC III the PCM operates in one of three operating strategies: the base engine strategy, the modulator strategy and the limited operational strategy.

Base engine strategy covers normal operating conditions, divided into four modes:

1. cranking
2. closed throttle operation
3. part throttle operation
4. wide-open throttle operation

The PCM determines the current operating conditions from the sensor input signals and selects the appropriate actuator commands according to its specific vehicle calibration program.

The modulator strategy compensates for conditions requiring more extreme calibration commands. These conditions are:

- cold engine
- overheated engine
- high altitude

The limited operational strategy is like that of the EEC II system on carbureted vehicles. For CFI-equipped vehicles the LOS keeps the injectors operating, but at a fixed pulse rate to produce a full, rich mixture. LOS is, as we have seen before, similar to a limp-home mode, enabling the vehicle to move under its own power, but with unsatisfactory performance, economy and emissions quality. Extended driving under these conditions could damage the oxygen sensor and the catalytic converter, as well as produce combustion chamber carbon deposits.

Inputs

Each EEC III sensor receives a 9-volt reference signal from the PCM, which the sensor modifies according to the state of whatever it monitors and returns a corresponding reduced voltage signal to the computer. The only difference between the EEC III sensors and those for EEC II is that the EEC III system with CFI includes an Intake Air Temperature (IAT) sensor bolted into the number seven intake manifold runner. The IAT sensor carries over to some EEC IV applications and is described in Chapter 14.

Outputs

With the following exceptions, all the EEC III actuators are exactly like those used for EEC II:

EEC III 7200 VV Air/Fuel Mixture Control. The fuel mixture control works somewhat differently on EEC III than on EEC II. A stepper motor and pintle valve work the same way, though the stepper motor is slightly different. The pintle valve, however, does not control fuel bowl pressure. Instead the pintle valve controls the amount of air bled into the main metering system, a variable main metering air bleed, Figure 13–16. Air enters a passage at the top of the carburetor from inside the air cleaner. This passage leads to an orifice, which the stepper motor and pintle valve control. When the pintle extends, less air is allowed through the orifice; when the pintle retracts, more air passes through. After passing through the orifice, the air channels around the throats of the carburetor to the other side of the carburetor, to the main metering system. The more air that bleeds into the main metering system, the less fuel flows into the carburetor throat.

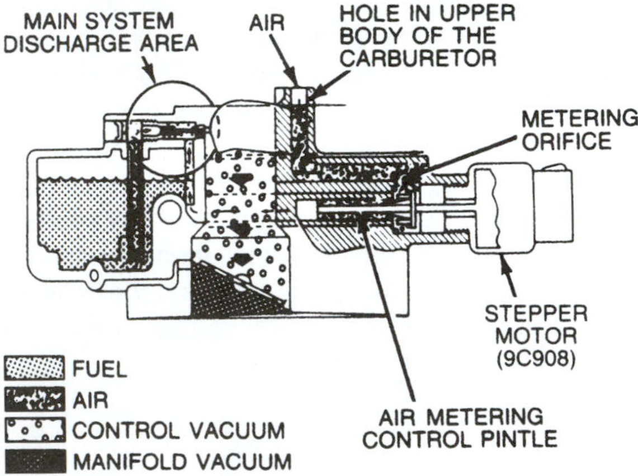

Figure 13–16 Variable venturi 7200 feedback carburetor, emulsion air control system. *(Courtesy of Ford Motor Company.)*

This difference makes use of the fact that to control air/fuel mixture you can vary either the air or the fuel to get the ratio desired. By introducing more air into the emulsion passages, the fuel delivered to the carburetor venturi is itself filled with a greater proportion of small bubbles of air. This effectively leans or richens the mixture as the computer controls the amount of air in those passages.

CFI Injectors. The fuel delivery system and the operation of the CFI injector solenoids for EEC III carry over to some of the EEC IV applications and are described in Chapter 14.

Second-Generation Distributor. Later EEC III systems have a second-generation distributor, Figure 13–17. It has the upper- and lower-level electrode blades seen in the first EEC distributor rotor, but the rotor shape is changed to conical rather than the original rectangular. The rotor's pickup arms and the cap's center terminal are redesigned, but they still work the same way to pick up and deliver spark alternately every 225 degrees of distributor rotation.

CFI Canister Purge. Four different canister purge systems are used on EEC III CFI systems. Consult the service manual for the specific vehicle for information on how each specific purge system works. The general operation and purpose of each canister purge system is the same.

✔ SYSTEM DIAGNOSIS AND SERVICE

The EEC III PCM was Ford's first to incorporate a selfdiagnostic program. Although it is not capable of storing diagnostic trouble codes (DTCs) in memory (and therefore cannot help the technician solve an intermittent problem), it is capable of running a Key On, Engine Running

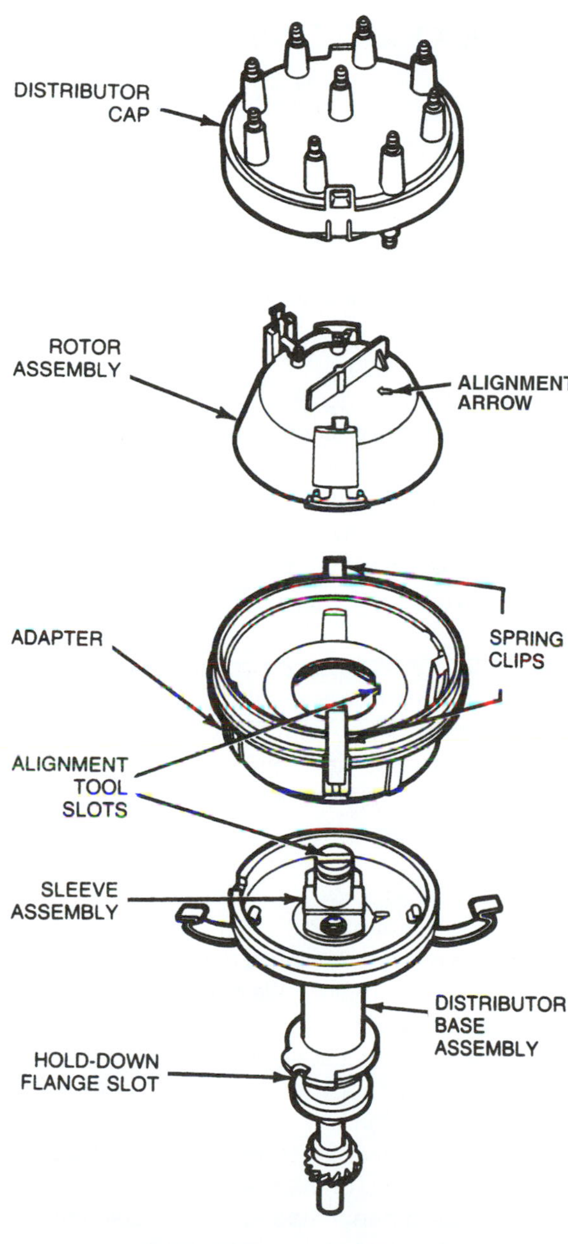

Figure 13–17 EEC III second-generation distributor design. *(Courtesy of Ford Motor Company.)*

(KOER) self-test and displaying the resulting DTCs. These DTCs, of course, represent hard faults identified during the self-test procedure. It is also capable of performing one functional test—a computed timing test.

To perform the KOER self-test, first warm the engine until the upper radiator hose is hot and pressurized. Then run the engine at 2,000 rpm for 2 minutes in order to be sure that the O_2 sensor is hot enough to operate.

At this point, you are ready to trigger the self-test. To do this, with the engine idling, connect a vacuum pump to the BARO port of the B/MAP sensor. Then pump the pressure level down to 20 inches of mercury of vacuum, hold this vacuum level for 8 to 10 seconds, and then release the vacuum. The PCM perceives this sudden change in its barometric pressure value as an instruction to enter into a KOER self-test and proceeds to perform both an electrical test and a functional test of many of the system's components.

At the end of the self-test, the PCM counts out DTCs to the technician by pulsing (energizing, then de-energizing) the thermactor air bypass (TAB) and thermactor air diverter (TAD) solenoids on the right-hand valve cover. However, because earlier pulses are simply for the purpose of the PCM to test these solenoids, it is important to know when the pulses intended to represent DTCs begin. To do this, tee a vacuum gauge into the throttle kicker (TK) used by the PCM to increase idle rpm. As long as you see vacuum applied to the TK, the pulses are not intended to represent DTCs. As soon as the vacuum level at the TK drops to 0 inches of mercury, pulses of the TAB and TAD solenoids are now intended by the PCM to represent code output.

To count the pulses of the TAB and TAD solenoids, connect an analog voltmeter between the negative side of either solenoid and battery negative (in parallel to the PCM so that the voltmeter will not load the PCM's circuits). When the PCM energizes the solenoid, the voltmeter will read 0 volts; then when the PCM releases the ground, the voltmeter will read applied voltage. By counting the pulses to 0 volts and back, DTC output can be read. Like the EEC IV system described in Chapter 14, code 11 indicates a "system pass."

The computed timing test is a functional test that allows the technician to test the PCM's ability to control spark timing. Even though base timing is not adjustable on this application, there are timing marks on the front crankshaft pulley. During the KOER running self-test, but prior to the loss of vacuum at the TK, computed timing should read between 27 degrees and 33 degrees BTDC as measured with a timing light. This is the equivalent of a fixed base timing of 10 degrees BTDC with the PCM adding in 20 degrees. A tolerance of + or – 3 degrees is allowed. This test is similar to the computed timing test described in Chapter 14 for the EEC IV system.

SUMMARY

In this chapter, we have covered the early Ford computerized engine control systems. With EEC I, we saw the use of the computer to govern relatively simple elements, like spark advance, EGR flow and Thermactor air mode. With EEC II we saw the introduction of Ford's first oxygen sensor and the beginnings of closed-loop operation. EEC III saw the beginnings of selfdiagnostics, at least in the sense of the capacity to report problems currently observed. For each of these systems, we saw Ford's unique new-design distributor, and we looked at the different methods used to govern fuel mixture with both the variable venturi carburetor and the early fuel injection systems.

▲ DIAGNOSTIC EXERCISE

A car with the Ford EEC III system is brought in, running, but poorly. How would you set about determining whether it was in LOS? What elements not controlled by the computer would you inspect?

REVIEW QUESTIONS

1. In what year was Ford's first EEC system introduced?
   A. 1975
   B. 1976
   C. 1978
   D. 1980
2. Ford's EEC I system controls all *except* which of the following?
   A. Air/fuel ratio
   B. Ignition timing
   C. Thermactor air control
   D. EGR flow
3. On Ford EEC I, EEC II and EEC III systems, what is the reference voltage the PCM sends to the sensors that require it?
   A. 5 volts
   B. 9 volts
   C. 12.6 volts
   D. 14.2 volts
4. On an EEC I system, which of the following sensors uses an aneroid assembly that, in turn, moves a wiper within a potentiometer?
   A. MAP
   B. BP
   C. CP
   D. Both A and B
5. Which of the following is used to generate the ignition reference pulse to the PCM on an EEC I engine?
   A. A Hall effect sensor within the distributor
   B. A permanent magnet pickup coil within the distributor
   C. A permanent magnet pickup coil mounted at the rear of the crankshaft
   D. A permanent magnet pickup coil mounted at the front of the crankshaft
6. What was the first Ford system to employ an oxygen sensor?
   A. EEC I
   B. EEC II
   C. EEC III
   D. EEC IV

7. How does a variable venturi 7200 feedback carburetor on an EEC II system control the air/fuel mixture?
   A. It uses a stepper motor to meter a control vacuum signal to the fuel bowl, thereby reducing atmospheric pressure in the fuel bowl.
   B. It uses a stepper motor to meter bleed air into the main metering system.
   C. It uses a duty-cycle solenoid to meter bleed air into the main metering system.
   D. It uses a duty-cycle solenoid to meter fuel through the main jets into the main fuel well.
8. Central Fuel Injection was first used with which Ford system?
   A. EEC I
   B. EEC II
   C. EEC III
   D. EEC IV
9. Which of the following is used to generate the ignition reference pulse to the PCM on an EEC II (or EEC III) engine?
   A. A Hall effect sensor within the distributor
   B. A permanent magnet pickup coil within the distributor
   C. A permanent magnet pickup coil mounted at the rear of the crankshaft
   D. A permanent magnet pickup coil mounted at the front of the crankshaft
10. What selfdiagnostic tests are programmed into the PCM on an EEC II system?
    A. The PCM can store fault codes in memory.
    B. The PCM can perform a Key On, Engine Running (KOER) self-test.
    C. Both A and B
    D. Neither A nor B
11. *Technician A* says that some EEC III systems use a 7200 variable venturi feedback carburetor to control the air/fuel ratio.
    Technician B says that some EEC III systems use a Central Fuel Injection system to control the air/fuel ratio.
    Who is correct?
    A. A only
    B. B only
    C. Both A and B
    D. Neither A nor B
12. How does a variable venturi 7200 feedback carburetor on an EEC III system control the air/fuel mixture?
    A. It uses a stepper motor to meter a control vacuum signal to the fuel bowl, thereby reducing atmospheric pressure in the fuel bowl.
    B. It uses a stepper motor to meter bleed air into the main metering system.
    C. It uses a duty-cycle solenoid to meter bleed air into the main metering system.
    D. It uses a duty-cycle solenoid to meter fuel through the main jets into the main fuel well.
13. An EEC III-equipped vehicle is brought into the shop with a complaint of poor engine performance.
    Technician A says that someone may have previously misadjusted the base ignition timing.
    Technician B says that the rotor-to-distributor cap alignment adjustment should be checked by verifying base ignition timing with a timing light.
    Who is correct?
    A. A only
    B. B only
    C. Both A and B
    D. Neither A nor B
14. What selfdiagnostic tests are programmed into the PCM on an EEC III system?
    A. The PCM can store fault codes in memory.
    B. The PCM can perform a Key On, Engine Running (KOER) self-test.
    C. Both A and B
    D. Neither A nor B

Chapter 14

Ford's Electronic Engine Control IV (EEC IV)

OBJECTIVES

Upon completion and review of this chapter, you should be able to:

- ❏ Describe the major aspects of the EEC IV PCM.
- ❏ Understand the operating modes that pertain to the EEC IV PCM.
- ❏ Define the inputs associated with an EEC IV system.
- ❏ Define the outputs under the control of the EEC IV PCM.
- ❏ Describe the basic operation of an Integrated Vehicle Speed Control (IVSC) system.
- ❏ Understand the proper approach to system diagnosis and service on an EEC IV-equipped vehicle.

KEY TERMS

A4LD (Automatic Four-Speed Light-Duty) Transmission
Adaptive Strategy
AXOD (Automatic Transaxle Overdrive)
Capacitor
Diagnostic Routines
Engine Calibration Assembly
PIP Signal
Quick Test
SPOUT
Star Tester
Thick Film Integrated (TFI)
Torque Converter Lockup Clutch

The Electronic Engine Control IV (EEC IV) system is the fifth generation of Ford's computerized engine control systems. It was first introduced in late 1982. By 1985 the EEC IV system was the only computerized engine control system Ford was using except for a few 5.8-liter police cars and some Canadian models. By 1985, the EEC IV system was also used on most Bronco, Ranger, E-series and F-series light trucks.

The EEC IV system's primary function is to control the air/fuel ratio. This is achieved through either a feedback carburetor, a single-point injection system (which Ford calls Central Fuel Injection) or multipoint injection (which Ford calls Electronic Fuel Injection [EFI]). The throttle body unit for CFI is shown in Figure 14–1 and the EFI unit is shown in Figure 14–2. To control these and in some cases other functions, the computer collects information about the engine and vehicle operating parameters from a variety of sensors, Figure 14–3.

POWERTRAIN CONTROL MODULE (PCM)

The EEC IV system computer, known under OBD II standards as the Powertrain Control Module (PCM), was known prior to OBD II standardization as the Electronic Control Assembly (ECA). It has considerably more capacity than previous Ford engine control computers, Figure 14–4. It is the first to include KAM, a keep-alive memory (an expansion of its hardwired random-access mem-

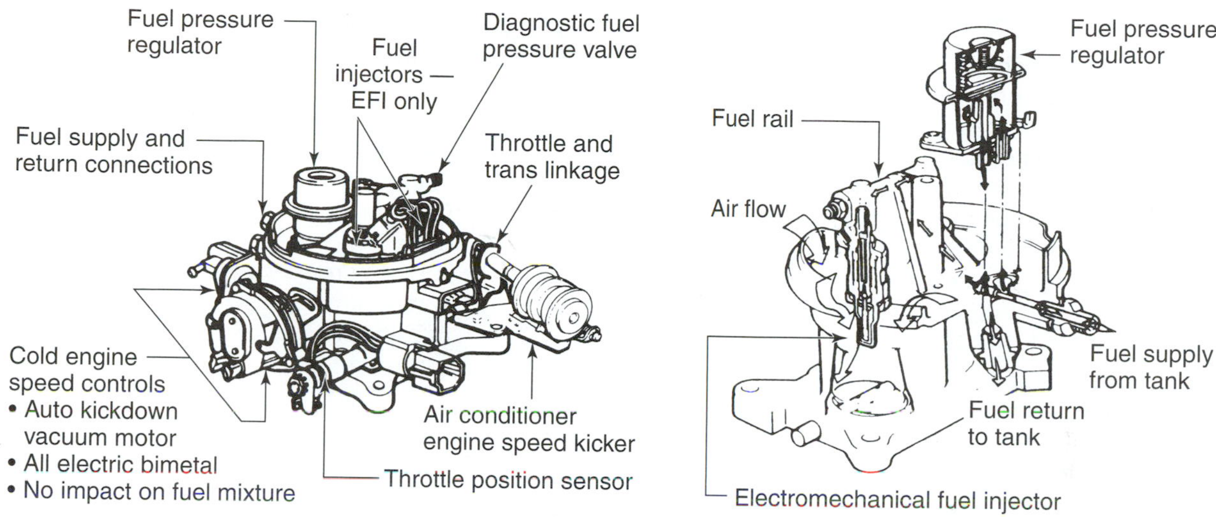

Figure 14–1 Central fuel injection unit. *(Courtesy of Ford Motor Company.)*

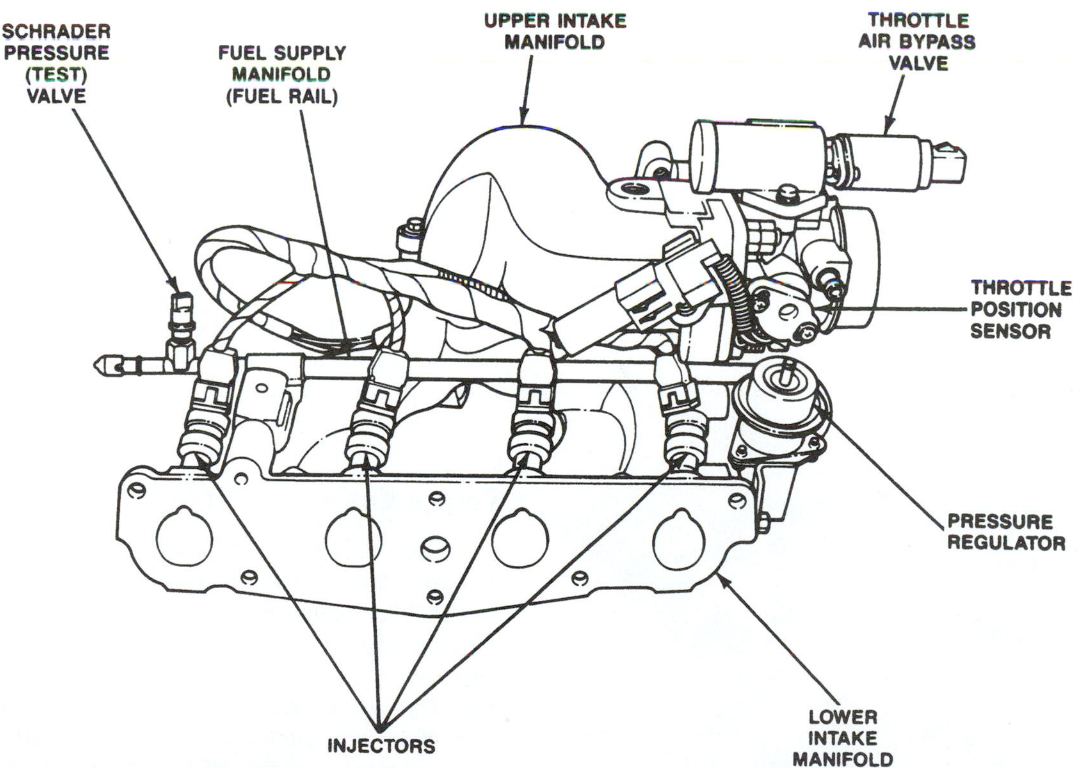

Figure 14–2 Multipoint fuel injection manifold (1.6-liter). *(Courtesy of Ford Motor Company.)*

Inputs

Engine coolant temperature
Manifold absolute pressure and/or barometric pressure
Throttle position
Engine speed and crankshaft position
Exhaust gas oxygen
EGR valve position

The following sensors are unique to specific engines:

Air charge temperature
Vane air temperature
Vane airflow
Idle tracking switch
Transmission position
Inferred mileage sensor
Knock sensor
Ignition diagnostic monitor
Clutch engaged switch
Brake on/off switch
Power steering pressure switch
Air-conditioning clutch
Ignition switch

Powertrain Control Module (PCM)

Power Relay

Outputs

Air/fuel mixture control device, carburetor solenoid or injector
Ignition timing control
Idle speed control
Thermactor airflow control
Canister purge control
EGR flow control
Torque converter clutch control
Turbocharger boost control
A/C and cooling fan controller module
Wide-open throttle A/C cut-off
Inlet air temperature control
Variable voltage choke
Temperature-compensated accelerator pump
Shift indicator light
Fuel pump relay

All components are not used on any one system.

Figure 14–3 EEC IV system overview.

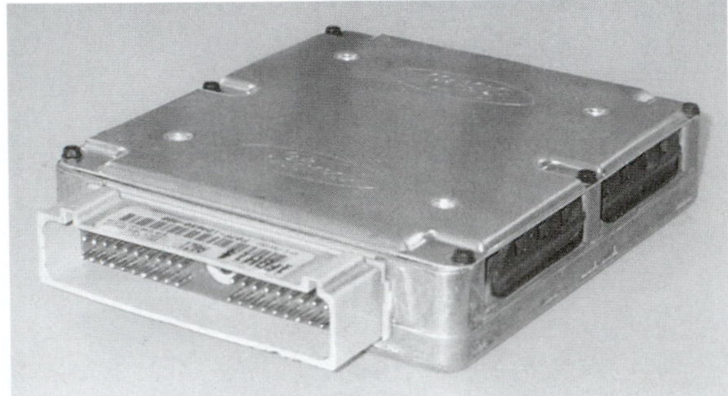

Figure 14–4 EEC IV Powertrain Control Module (PCM).

ory) enabling it to store codes related to faults that it has previously observed but that are no longer present. This capacity was not true for the first application of EEC IV on the 1.6-liter engine. The number of codes it can recognize is greatly increased compared to the computers in earlier Ford systems.

Engine Calibration Assembly

The PCM's **engine calibration assembly** contains the necessary programming to fine-tune the PCM's commands to the specific needs of that vehicle and engine's weight, axle ratio and transmission. The calibration assembly is similar to a GM PROM, except that unlike those systems' unit, the Ford calibration assembly is an integral part soldered into the PCM, and it cannot be removed or serviced separately.

OPERATING MODES

The PCM controls the engine in one of several operational modes or strategies:

Base Engine Strategy. The PCM uses this mode to control warm engine calibration through the wide range of operating conditions occurring in normal driving. Prior to 1986, the calibration for warm engine operation was a coolant temperature of 88°C (190°F) or higher. To reduce the time spent running in open-loop conditions, that threshold was lowered in 1986 to about 54°C (130°F) or higher. This warmup mode is divided into four submodes:

1. cranking
2. closed-throttle operation
3. part-throttle operation (closed loop)
4. wide-open throttle (WOT)

Input information enables the PCM to recognize these operational modes. Once it recognizes the state of the engine, the operational mode, the PCM issues calibration commands to the appropriate actuators, commands designed to produce the best results in terms of emissions, fuel economy and driveability.

If the engine threatens to stall while in the base engine strategy mode, the PCM will employ an underspeed response feature.

MPG Lean Cruise. When predetermined criteria are met during cruise conditions on some engine applications (beginning in 1988), the PCM takes the system out of closed loop to an even leaner "dead-reckoning" (no feedback) open-loop air/fuel ratio. The purpose of this modification is to achieve better fuel economy.

Modulator Strategy. Operating conditions requiring significant compensation to maintain good driveability cause the PCM to modify the base engine strategy. Conditions causing such modifications are:

- cold engine
- overheated engine
- high altitude

Limited Operational Strategy (LOS). When a component failure prevents the system from functioning in a normal strategy, the PCM enters an alternative strategy designed to protect other system components, such as the catalytic converter, yet still provide enough driveability to keep the vehicle running until it can be repaired. If the PCM's central processing unit (CPU) fails, the PCM will operate in a fixed mode: there will be no spark advance, no EGR and air from the Thermactor (Ford's air injection system) will be dumped to the atmosphere.

Adaptive Strategy

This feature, first introduced in 1985, enables the PCM to constantly adjust some of its original calibrations, those programmed into it by engineers based on ideal conditions. Whenever conditions are less than ideal due to variations in manufacturing tolerances, wear or deterioration of sensors or other components that affect the PCM's control of the most critical functions, the **adaptive strategy** puts an adjustment or adaptive factor into the calculation process. Adaptive strategy could also be called a learning capacity or a self-correction ability. The PCM learns from past experience so it can better control the present conditions. Learning starts when the engine is warm enough to go into closed loop and is in a stabilized mode.

Adaptive Fuel Control. This adaptive strategy concerns air/fuel ratio. If the injectors become fouled with deposits or if wear in the pressure regulator causes a slight loss of fuel pressure, the result is less fuel injected into the cylinder than should be during the pulse width the computer has calculated for those conditions. When the system is in closed loop, the oxygen sensor reports a lean condition to the PCM. The PCM responds by commanding a wider injector pulse width to maintain the proper 14.7 to 1 air/fuel ratio.

The adaptive fuel control recognizes that a wider pulse width than what the original calibration designated is required. If this condition continues for several minutes, the adaptive fuel control modifies the original calibration to provide a wider pulse width so the calibration is more on target and less correction is needed to bring the mixture to stoichiometry. Notice that all forms of adaptive strategy, or learning, are in response to signals from the oxygen sensor.

How the Calibration Modification Is Achieved. The ROM (read-only memory, the hardwired original set of computer instructions) contains an engine load (MAP sensor signal) versus the engine rpm table used as the starting point for calculating pulse width, Figure 14–5. The table is divided into separate cells. Each cell is assigned a base value of 1. Cross-referencing the current MAP value with the current rpm value, the microprocessor identifies the cell that best represents current engine load and speed conditions. The value in that cell is used as a multiplier in the microprocessor's calculations to determine the correct pulse width command for the injectors under the specific conditions then present.

Since this table is stored in the ROM, it cannot be changed or erased. There is, however, another copy of this table in the keep-alive memory (KAM). The microprocessor uses that copy to work from, and the base values in each of the cells of the KAM copy of the table can be modified according to what the computer learns through its adaptive strategy, or learning.

If the pulse width calculation, using the base value in a given cell, does not consistently produce the desired 14.7 to 1 air to fuel ratio as reflected in the oxygen sensor's return signal, the adaptive fuel control adds to or subtracts from the base value in the KAM copy of the table, Figure 14–6. If the air/fuel ratio tends to be too lean, the adaptive fuel control adds to the base value. If the air/fuel ratio tends to be too rich, the adaptive fuel control subtracts from the base value. This is Ford's version of fuel trim and is similar to GM's Block Learn and Integrator discussed in Chapter 10.

At idle, adaptive fuel control works the same way it does at other engine speeds. The microprocessor uses the same table for pulse width calculations at idle, but it uses cells that are used only during idle conditions. Depending on engine application, there are either four or six cells on the MAP versus rpm table for idle pulse width calculations. The table could look something like the one shown in Figure 14–7. The idle cells, however, are not selected by MAP and rpm; they are selected by the transmission being in drive or neutral, by the A/C being on or off and in some cases by the A/C control switch being on or off.

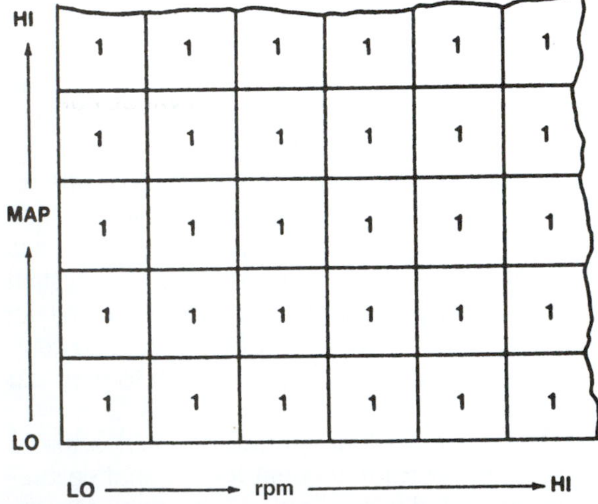

1	1	1	1	1	1
1	1	1	1	1	1
1	1	1	1	1	1
1	1	1	1	1	1
1	1	1	1	1	1

Figure 14–5 Adaptive fuel control table base values.

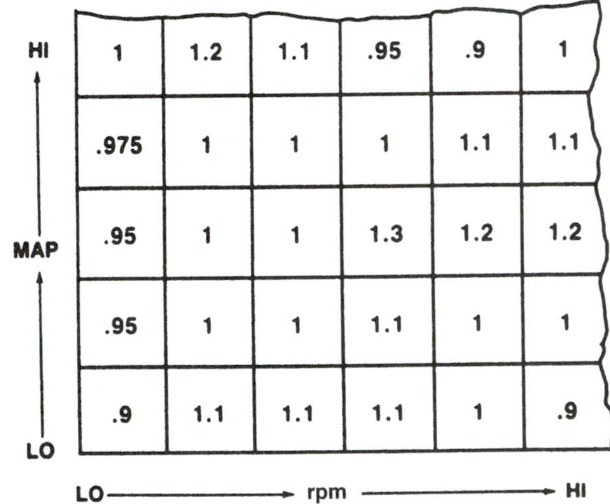

1	1.2	1.1	.95	.9	1
.975	1	1	1	1.1	1.1
.95	1	1	1.3	1.2	1.2
.95	1	1	1.1	1	1
.9	1.1	1.1	1.1	1	.9

Figure 14–6 Adaptive fuel control table with adapted cell values.

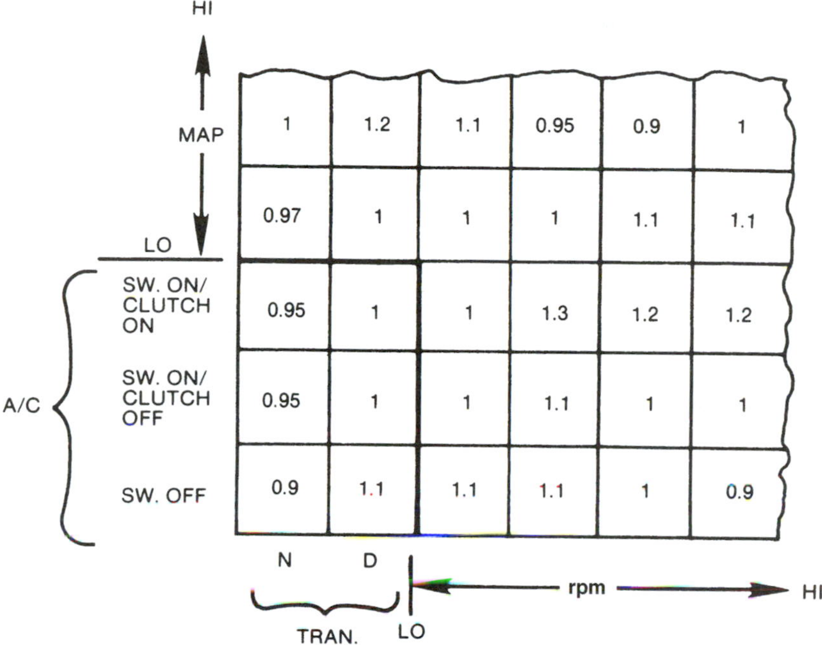

Figure 14–7 Idle fuel control cells on MAP versus rpm table.

Adaptive Idle Speed Strategy. Adaptive idle speed strategy is used only with CFI and EFI systems, not with carbureted engines. The idle speed strategy calculates the commands sent to the idle speed control actuator. This device controls the amount of air admitted to the induction system bypassing the throttle valve. If the commands sent to the actuator fail to produce the calibrated idle speed, the PCM substitutes a new value in place of the one in the originally calibrated values used for the calculation. The value replaced is obtained from an idle table in the KAM similar to the MAP versus rpm table used for fuel/air mixture and discussed previously, Figure 14–8. When idle speed is corrected, the adaptive idle speed strategy monitors the revised value needed to make the correction and records it in place of the originally recorded value in the KAM idle table.

Adaptive idle speed strategy learning occurs only under the following conditions:

- idle speed is stabilized during base idle conditions (warm engine, automatic transmission in drive, manual transmission in neutral and A/C off).
- system is in closed loop.
- a specific time has elapsed since idle mode began. (This time varies for different engine applications.)

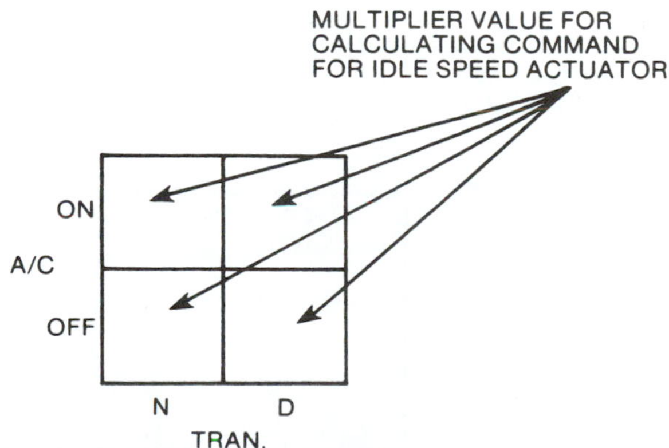

Figure 14–8 Idle speed table.

Because the information learned by the adaptive strategy is stored in the KAM, it can be used or revised as needed. The learning process works constantly during stabilized, closed-loop conditions.

Other adaptive strategies incorporated into the EEC IV systems since 1985 include adaptive spark control (the computer learns to modify the range and shape of the spark advance map) and adaptive oxygen sensor aging (the computer learns how to compensate for the gradually declining signal frequency of the oxygen sensor). Their operation is similar to those just described.

Adaptive strategy also tests regularly for the possibility that corrupt data (false values) could have gotten into the KAM. If any data corruption is found, those values are erased and replaced with the original corresponding values from the ROM. Learning can then modify the replacement values as needed.

Power Relay

A power relay is used to supply operational power to the PCM, Figure 14–9. The control coil

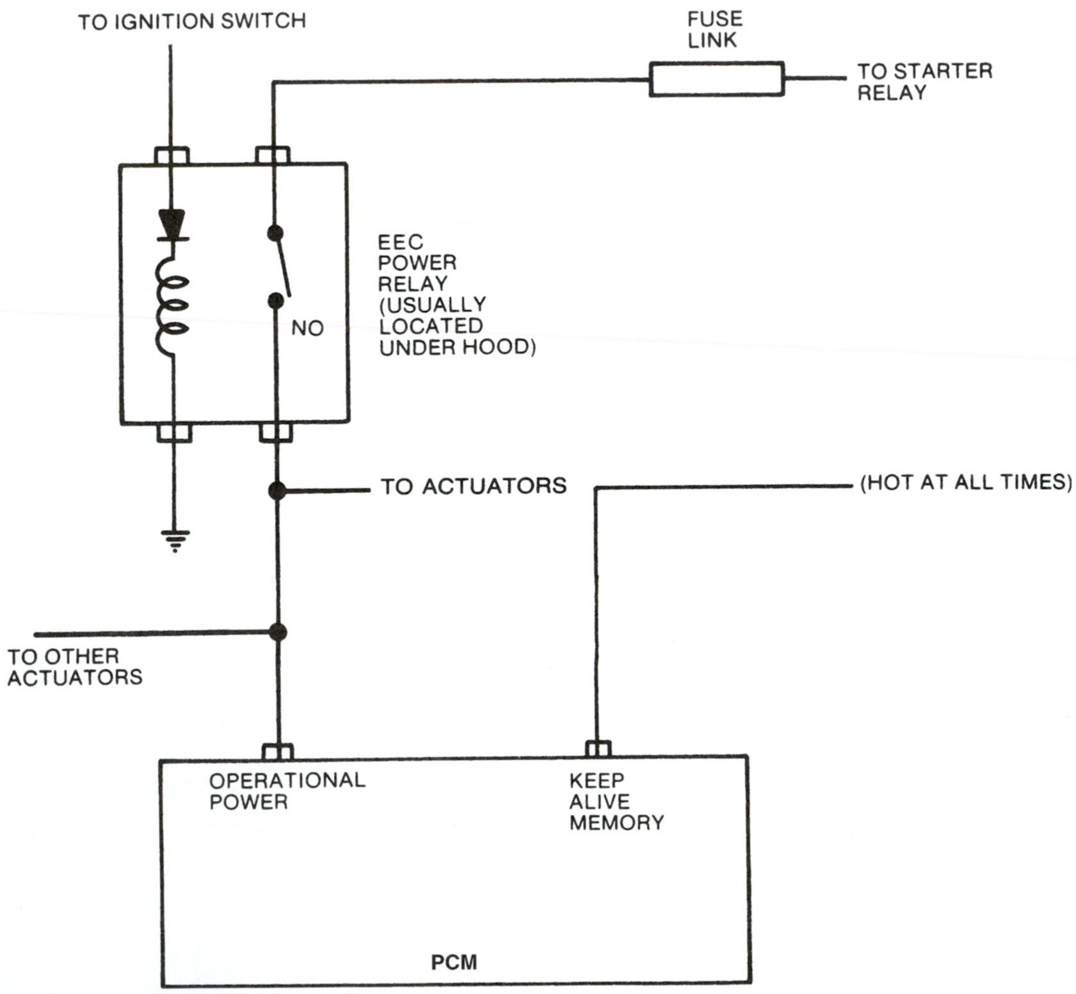

Figure 14–9 Power relay circuit.

of this relay is powered directly through the ignition switch. Once the ignition switch is turned on, the contact points of the relay, which are normally open, close and supply power to the PCM. The control coil of the relay is in series with a diode. If the battery's power is reversed, the diode blocks current through the control coil and prevents the contact points from closing. This protects the power circuits of the PCM from damage caused by reverse polarity.

Beginning in 1986 on some applications and becoming more widespread in 1987, the EEC power relay was integrated into a unit called the Integrated Relay Control Module (IRCM), Figure 14–10. This unit typically contains the EEC power relay, the fuel pump relay, the low-speed Electro-Drive fan (EDF) relay (cooling fan relay), the high-speed Electro-Drive fan (HEDF) relay and the solid state A/C compressor clutch relay. Excepting the EEC power relay, these relays are under the control of the PCM. The IRCM is normally located just above the radiator and cooling fan and just below the emission decal with its plastic support.

The power to the KAM does not come through the power relay, but directly from a separate battery source that is always hot. If this power supply gets disconnected, all learned adaptive strategy will be erased. Keeping this circuit powered requires very little current (only a few milliamps) and does not discharge the battery even if the vehicle is not operated for several days. In fact, internal current in the battery will discharge it much more quickly than the minuscule KAM current draw.

But because the KAM is not protected by the power relay, its circuitry is designed to be less susceptible to reverse polarity damage. Care should always be taken to prevent such mistakes, however.

INPUTS

Engine Coolant Temperature (ECT) Sensor

The ECT sensor is a thermistor-type sensor, Figure 14–11. Its voltage signal to the PCM normally ranges from 4.5 volts when very cold to 0.3 volt when the engine is hot. Its resistance values are:

- –40°F = 269 kilohms
- 32°F = 96 kilohms
- 77°F = 29 kilohms
- 248°F = 1.2 kilohms

The information it provides influences the PCM's calibrations commands controlling:

- air/fuel mixture ratio.
- idle speed.
- EGR.
- Thermactor air.
- canister purge.

Figure 14–10 Integrated Relay Control Modules (IRCM).

Figure 14–11 Engine coolant temperature (ECT) sensor.

- choke voltage.
- temperature-compensated accelerator pump.
- upshift light.

Pressure Sensors

Ford's pressure sensors operate differently than those discussed in earlier chapters. They use a pressure-sensitive variable **capacitor**, Figure 14–12. The capacitor is formed by two conductive plates separated by a thin air space and a thin layer of insulating material. One plate, the positive, is a thin copper film on a round, rigid ceramic bed with a thin film of insulating material sprayed over the copper. The other plate, the negative, is also a thin copper plate on the underside of a round ceramic disk, but this disk is slightly flexible. The flexible ceramic disk is mounted on top of the rigid ceramic bed with a thin adhesive strip around the entire circumference. The adhesive strip makes an airtight seal and spaces the two copper plates a few thousandths of an inch apart.

This assembly makes up the ceramic body of the sensor and is mounted in a plastic housing. The area below the ceramic body is sealed from the area above it. A small hole through the ceramic bed vents the area between the copper plates to the sealed, reference pressure below the ceramic bed. The space above the ceramic body is exposed to the monitored pressure.

Electronic circuitry and components are attached to the bottom of the rigid ceramic bed.

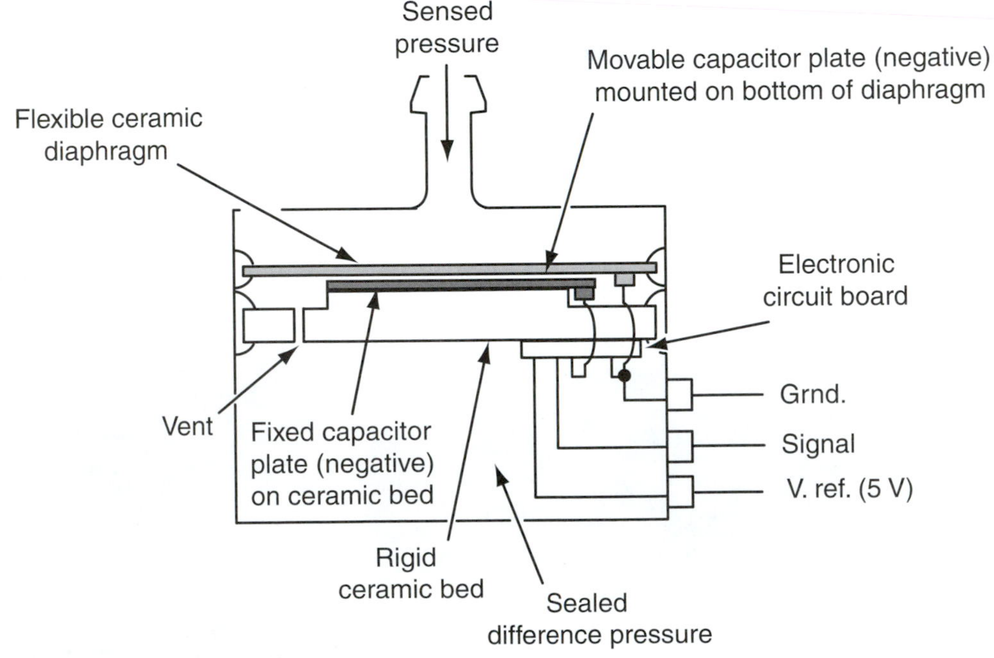

Figure 14–12 Typical variable capacitor pressure sensor.

Each capacitor plate has a wire that connects it to the electronic circuitry. Additional wires connect the circuitry and the negative plate to the PCM.

If the pressure sensed is lower than the reference pressure between the plates, the diaphragm flexes outward, moving the plates farther apart. This *lowers* the capacitance of the capacitor. As the pressure sensed increases, the diaphragm is forced inward, moving the plates closer together. This *increases* the capacitance of the sensor. The electronic circuit, powered by the reference voltage (VREF) from the PCM, measures the capacitance of the variable capacitor and sends a corresponding electrical signal back to the PCM.

Manifold Absolute Pressure (MAP) Sensor. In addition to being a variable capacitor, the MAP sensor used on EEC IV systems is a frequency-generating device, Figure 14–13. An additional chip in the sensor's electronic circuit produces a frequency signal corresponding to the capacitance. The capacitance, as we have seen, is a function of the relative distance of the plates and thus of manifold pressure. This frequency signal is what is sent to the PCM.

Figure 14–13 Manifold absolute pressure sensor.

The resulting frequency signal this sensor delivers to the PCM is actually a square wave that oscillates between 0 and 5 volts at the defined frequency. Therefore, this MAP sensor is sometimes referred to as a digital MAP sensor. The signal is turned on the same amount of time that it is turned off, regardless of its frequency, resulting in a 50 percent on-time. That is, whether it is cycling at a frequency of 100 hertz. (typical no-load idle) or whether it is cycling at 159 hertz (typical barometric pressure or WOT operation at sea level), the voltage will be at 5 volts for one-half second of every second and at 0 volts for one-half second of every second. The measured frequency indicates how many full on and off cycles of the voltage occur in the time frame of one second.

Ultimately, much like an analog MAP sensor delivers a high voltage to the PCM when the sensed pressure is high and a low voltage when the sensed pressure is low, this digital MAP sensor delivers a high frequency signal to the PCM when the sensed pressure is high and a low frequency signal when the sensed pressure is low.

If you apply a voltmeter directly to the sensor's output circuit, you will see a steady voltage signal of about 2.5 volts. This will change very little regardless of manifold pressure because it is not the voltage of the output that conveys the information. Actually the signal does change (more or fewer pulses per second) in response to changes in manifold pressure, but the frequency is much more rapid than even the fastest voltmeter can display reading the signal. The meter effectively just gives an average voltage reading, which is not what the computer is looking at.

During the lowest manifold pressure conditions (highest vacuum, lowest load), the sensor's signal frequency is about 92 cycles per second (92 hertz). At WOT, the manifold pressure is nearly equal to atmospheric pressure. At sea level the frequency is about 159 hertz. The PCM reads the frequency as a manifold pressure value. The MAP sensor's input affects:

- air/fuel ratio.
- EGR flow.
- ignition timing.

Those EEC IV engine applications using a manifold pressure sensor do not use a separate barometric pressure sensor. Rather, the PCM uses a different strategy, using the MAP sensor as a barometric pressure sensor during two specific conditions. During the brief period of time between when the ignition is turned on and the starter starts to crank, the PCM takes a reading from the MAP sensor. The pressure in the intake manifold is atmospheric pressure during this period. The PCM stores this MAP reading in its RAM memory as a barometric pressure reading. In addition, during WOT operation, manifold pressure is almost the same as atmospheric pressure. While there is a slight difference because of intake system aerodynamic friction, this difference is predictable for a specific engine. Thus, during WOT, the PCM will adjust the MAP reading for the difference between manifold pressure and atmospheric pressure. This reading will then be used to update the barometric pressure reading stored in RAM. Obviously this will be very close to the start-up reading unless the vehicle has changed altitude without using WOT, or driven so long the weather has changed substantially.

Also worth mentioning is the fact that the MAP sensor was mounted in the fender well area or fairly low on the firewall through 1986. Because fuel vapors could react with the protective gel within the sensor itself causing it to loosen, the gel would eventually make its way into the vacuum hose and plug it, resulting in overly rich air/fuel mixtures. Beginning in the 1987 model year, the location was changed to the top of the firewall area as part of the effort to keep fuel out of the sensor.

Barometric Pressure (BP) Sensor

Again, as stated earlier, engine applications that utilize a MAP sensor do not use a full-time BP sensor, but, instead, use the MAP sensor to update the BP values during engine start or WOT operation. However, engines with a vane air flow (VAF) meter must utilize a full-time BP sensor, because, while the VAF meter measures engine load by measuring the volume of intake air, it cannot measure the density of the air as affected by barometric pressure. Additionally, beginning with the 1989 model year when Ford introduced the mass air flow (MAF) sensor and through the 1991 model year, a full-time BP sensor was used with the MAF sensor on some applications, even though a MAF sensor can compensate for barometric pressure values. For example, full-time BP sensors were used with MAF sensors on automatic transmission versions of the 4.0L V6 through 1991, but the manual transmission version used the MAF sensor alone. By the 1992 model year, the BP sensors were eliminated on MAF sensor applications.

In those applications that used a full-time BP sensor, the sensor itself was essentially a MAP sensor in its design, except that a loose fitting collar was placed around the atmospheric vent to keep someone from fitting a vacuum hose to the vent, Figure 14–14. That is, because it looks exactly like the MAP sensor used on other applications and because the PCM recognizes the resulting signal as representing full-time barometric pressure values, the collar was present in order to keep someone from inadvertently mis-

Diagnostic & Service Tip

Many of the connectors used on Ford systems have locking tabs that are inadvertently broken by technicians who try to pull outward on the locking tabs in order to release them. These locking tabs are stamped with the word PUSH and should be pushed toward the body of the component, resulting in a quick, easy release of the connector. The components that typically have this style of connector include the MAP and BP sensors, TFI ignition modules, EEC power relays and fuel pump relays.

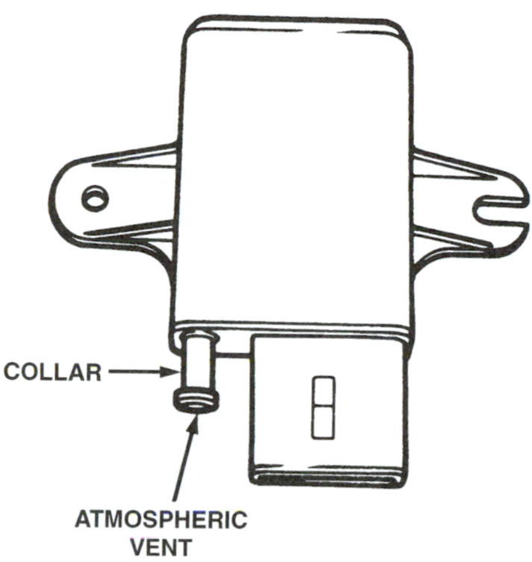

Figure 14–14 Barometric pressure sensor. *(Courtesy of Ford Motor Company.)*

taking this full-time BP sensor as a MAP sensor. The BP sensor's input affects:

- air/fuel ratio.
- EGR flow.

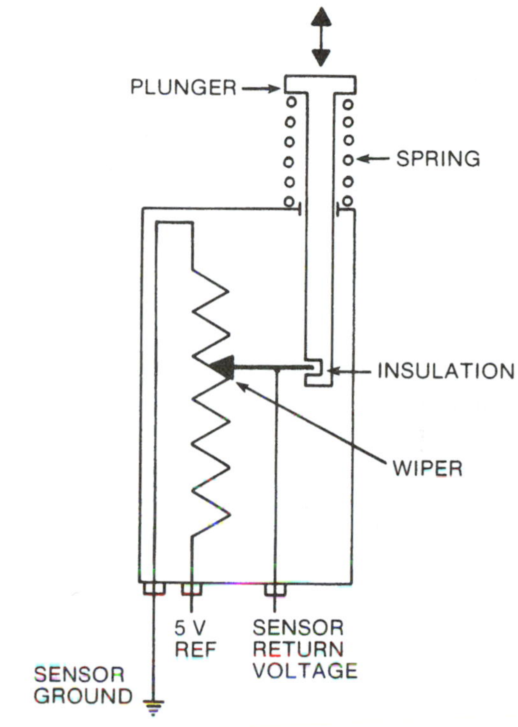

Figure 14–15 Linear throttle position sensor diagram.

Throttle Position Sensor (TPS)

The TPS is a variable potentiometer monitoring the position and movement of the throttle butterfly plate. It provides a return reduced voltage signal to the PCM indicating the throttle position and tracing its recent movement.

Linear TPS. A few carbureted EEC IV engine systems use a linear TPS, Figure 14–15. A cam on the throttle shaft contacts a plunger on the TPS and pushes it in as the throttle shaft rotates, Figure 14–16. The linear TPS is located inside of a housing attached externally to the carburetor and has an adjustment screw on the bottom end of the housing for correcting the voltage adjustment at idle. This design is typically found on 1984 Ford Tempo and Mercury Topaz models and is similar to the linear TPS found inside of General Motors' Rochester Quadrajet and Dualjet feedback carburetors, except for the electrical connector.

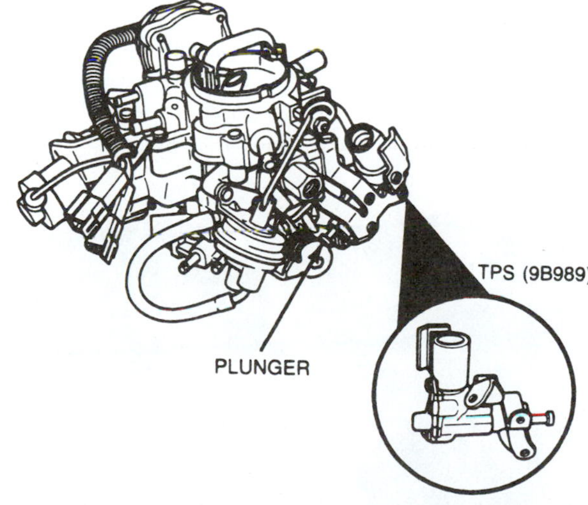

Figure 14–16 Linear throttle position sensor. *(Courtesy of Ford Motor Company.)*

Diagnostic & Service Tip

The most frequent problem by far with throttle position sensors is a bad spot on the variable resistor, causing a high or low voltage spike in the return signal. Although this spike can be observed on a sensitive voltmeter, a lab scope is the most secure way to identify this problem because it will draw a trace of the signal, and any voltage spike will be immediately visible.

The next most frequent problem with throttle position sensors is loss of resistance-free connection through the ground circuit. This will not let the information signal get reduced to the proper voltage, either leaving it at full reference voltage (and causing starting problems because the computer assumes the throttle is wide open and initiates clear flood mode), or simply raises the output signal corresponding to the voltage drop through the high resistance connection. The latter situation throws off the signal consistently on the lower range and makes it completely inaccurate in the upper range.

Rotary TPS. The rotary TPS is functionally similar to the linear type, but the resistor forms an arc around the axis of the throttle shaft instead of a straight line nearby. The wiper pivots from the center of the shaft and more directly parallels the movement of the throttle butterfly plate since there is no tangential "cam" effect. The rotary TPS is mounted on the side of the carburetor or throttle body, so the throttle shaft directly engages and drives the potentiometer wiper, Figure 14–17. Beginning in 1985, all EEC IV systems use the rotary type TPS. Some are adjustable; some are not. For either type, the PCM sends a 5-volt reference input to the TPS. Its return signal range is from 0.5 to 4.5 volts. Most rotary TPS sensors in normal operation transmit a signal between 1.0 volt at closed-throttle idle and 4.0 volts at WOT. The PCM uses the TPS primarily to identify the driving mode (idle, cruise or WOT), but it also can tell how quickly the throttle has been moved in some applications.

Profile Ignition Pickup (PIP) Sensor

A Hall effect switch generates the **PIP signal** (see Hall effect switch in the introductory chapter). This switch is called the PIP sensor and provides the PCM information about engine speed and

Figure 14–17 Rotary throttle position sensors.

crankshaft position. On engines with a distributor, the PIP sensor is in the distributor, Figure 14–18.

In 1986, Ford introduced sequential electronic fuel injection (SEFI) on the 5.0-liter engine in passenger cars. In a sequential fuel injection system, the computer pulses each injector individually in the firing order. To do so, the computer must know where the engine is in its cycle, so it can pulse, for example, the number three injector during the intake stroke for cylinder number three. So the PCM can identify the engine's place in the cycle, Ford engineers made one of the PIP sensor trigger vanes narrower than the others. As the narrow vane passes between the magnet and the crystal, a shorter on-time occurs in that part of the signal, Figure 14–19. This shorter pulse is called the signature PIP. The PCM recognizes the signature PIP as the second cylinder in the firing order. Being able to recognize any given cylinder in the firing order, combined with a hardwired memory of the firing sequence and having a PIP signal counting program, enables the PCM to pulse each fuel injector in the proper sequence.

In 1989, Ford introduced a distributorless ignition system (DIS) on applications such as the 3.8-liter supercharged engine, the 3.0-liter special high-output (SHO) engine and the 2.3-liter engine used in the Ranger pickup truck, Figure 14–20. On these applications, the PIP sensor is located on the front of the engine, just behind the crankshaft pulley and hub assembly, Figure 14–21. A vaned cup on the back of the pulley triggers the Hall effect PIP sensor switch, Figure 14–22. Since the switch is mounted on the crankshaft where it monitors two crankshaft revolutions per engine cycle, only one vane is needed for each pair of companion cylinders. As the leading edge of the vane moves between the Hall effect

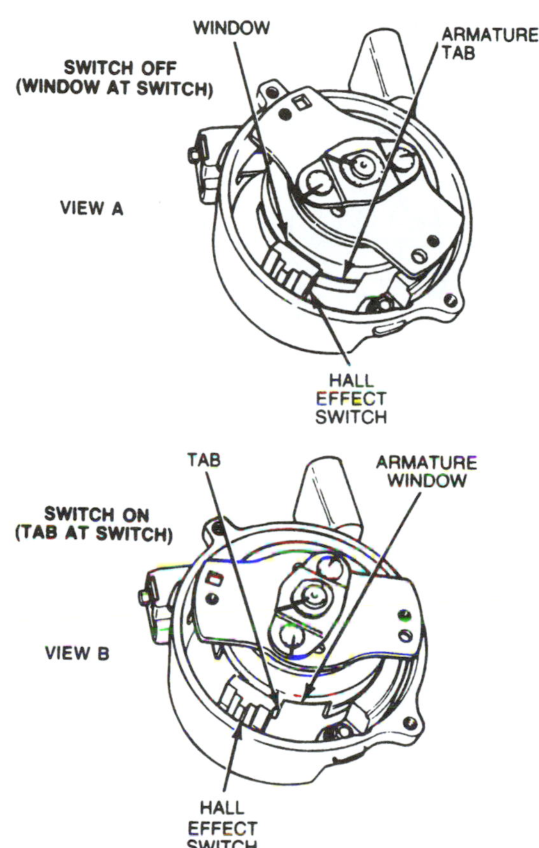

Figure 14–18 Hall effect switch in distributor. *(Courtesy of Ford Motor Company.)*

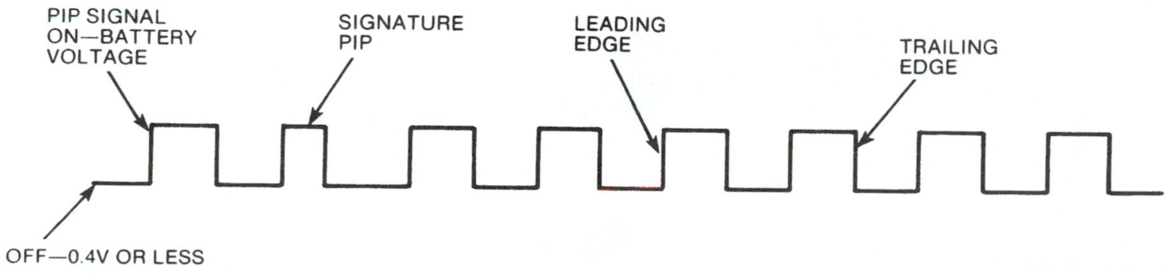

Figure 14–19 PIP signal for 5.0-liter sequential injection engine.

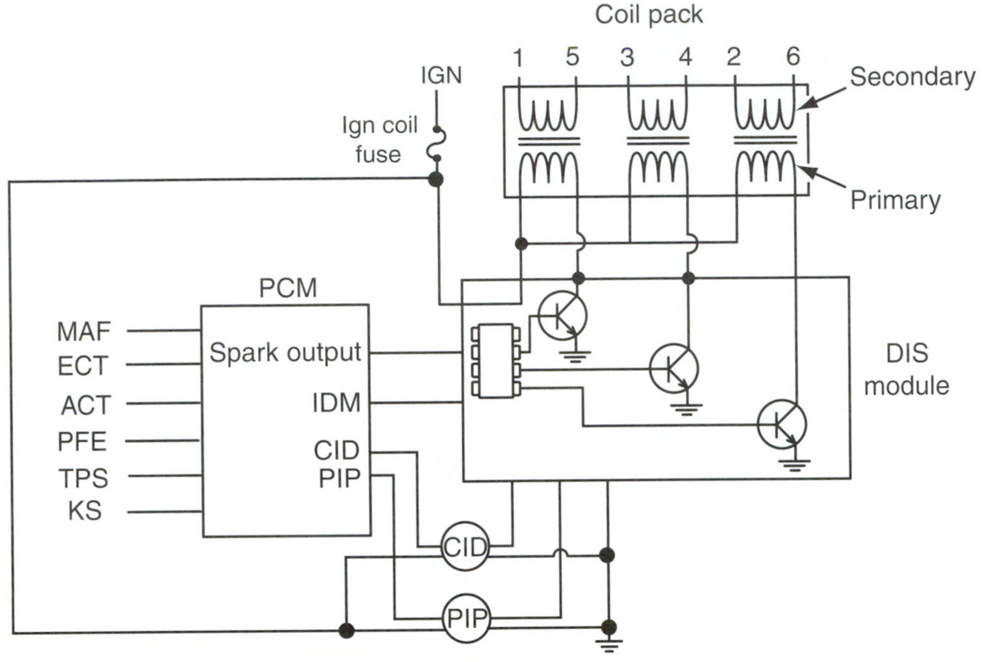

Figure 14–20 Distributorless (direct) ignition, supercharged 3.8-liter engine.

Figure 14–21 DIS components, 3.0-liter SHO engine. *(Courtesy of Ford Motor Company.)*

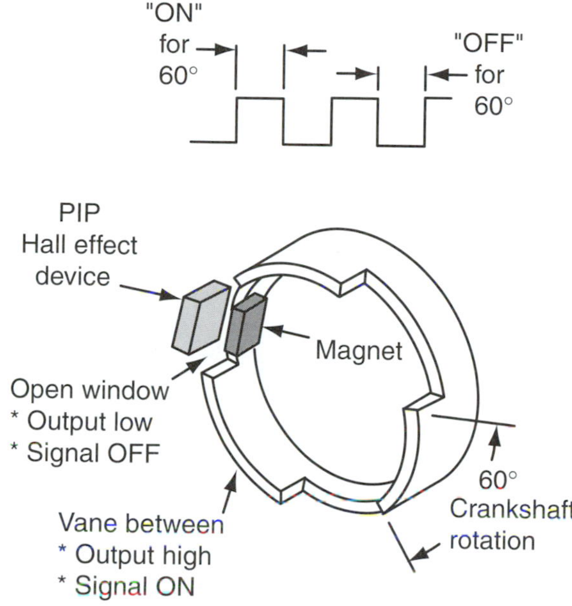

Figure 14–22 PIP sensor, supercharged 3.8-liter and SHO engines.

switch magnet and the crystal, the PIP signal rises quickly to battery voltage, Figure 14–23. The vanes on most engines are configured so a PIP signal goes to battery voltage each time a piston reaches 10 degrees BTDC. As the trailing edge of the vane moves from between the magnet and the crystal, the PIP signal drops to 0.4 volt or less.

Cylinder Identification (CID) Sensor

Distributorless ignition systems (or an equivalent term, *direct ignition systems*) use a separate coil for each pair of companion cylinders. Companion cylinders are cylinders 360 degrees apart in the 720 degrees of the engine cycle. For each pair of companion cylinders, one approaches the top of its compression stroke while the other approaches the top of its exhaust stroke; the positions reverse next time. This ignition system (one of the waste spark family of ignition systems) fires the spark plugs in each companion cylinder simultaneously, one for its combustion and power stroke, the other (the "waste" spark) at the end of its exhaust stroke. The PCM calculates the proper spark advance and instructs the DIS module when to fire the coil. But the DIS module must determine which coil to fire. To know that, it must have information about which cylinder is ready for combustion ignition.

The CID sensor is also a Hall effect switch. On the 3.0-liter SHO engine, the CID sensor is mounted on the right end of the rear head; one of the overhead camshafts drives it, Figure 14–21. On the 3.8-liter engine the CID sensor mounts where the distributor used to be on a more conventional engine, driven directly by the camshaft. On both these engines, the camshaft drives a single vane cup, Figure 14–24. As the leading edge of the vane moves between the Hall effect switch magnet and crystal, the CID signal rises to battery voltage. This tells the DIS module and the PCM that the number one cylinder is at 26 degrees ATDC on its power stroke. As the trailing edge of the vane moves out from between the Hall effect switch magnet and crystal, the CID signal drops to 0.4 volt or less. This in turn tells the DIS module and the PCM that the number one cylinder is at 26 degrees ATDC on its intake stroke. Either transition, voltage either rising or falling, identifies the position of the number one cylinder. The DIS module uses this information during cranking to determine which coil to fire first.

Because both of these engine types are sequentially fuel injected (the 3.8-liter only after model year 1988), the PCM uses the CID signal to determine the injector pulse sequence, just as it uses the signature PIP on sequentially fuel injected engines using a distributor.

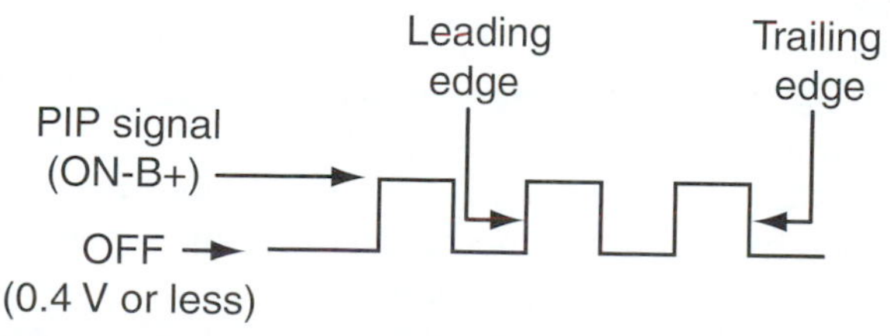

Figure 14–23 PIP signal.

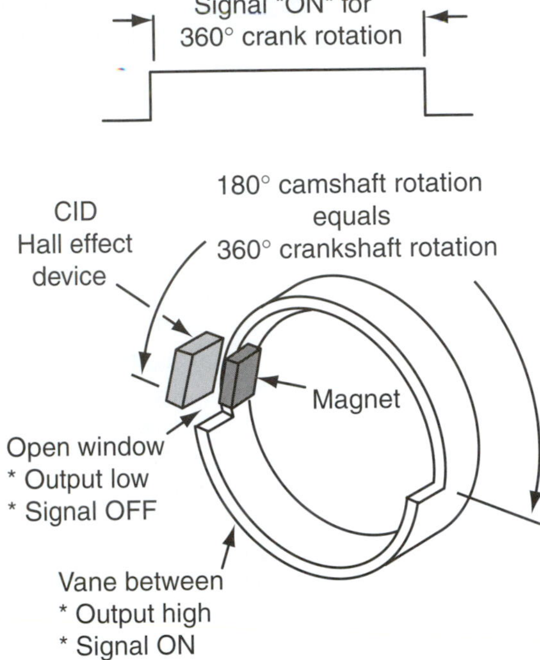

Figure 14–24 CID sensor, supercharged 3.8-liter and 3.0-liter SHO engine.

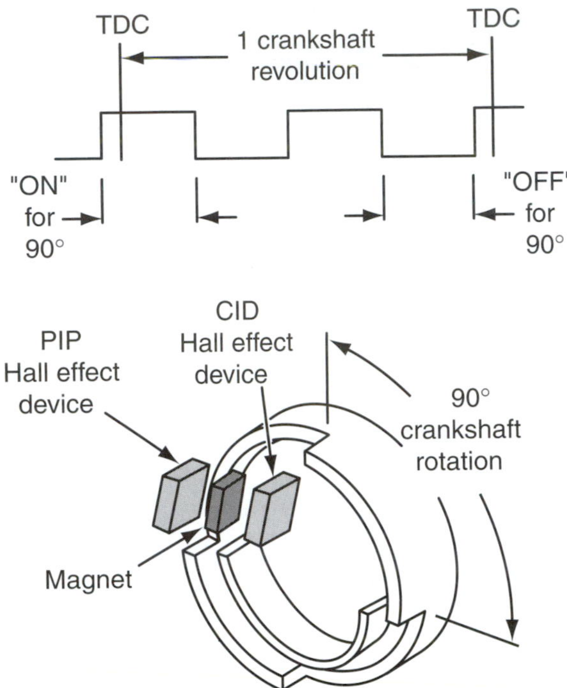

Figure 14–25 PIP sensor, 2.3-liter, dual plug engine.

On the DIS 2.3-liter engine, the CID sensor is combined with the PIP sensor behind the crankshaft pulley and hub assembly, Figure 14–25. The 2.3-liter engine's CID sensor works like those on the 3.0- and 3.8-liter engines with one important exception. Because the single vane triggering the CID sensors is on the crankshaft pulley, it produces two high CID and two low CID signals per engine cycle, Figure 14–26. Therefore, none of the transitions in the CID signal uniquely identifies the number one cylinder. When the signal goes to the high voltage, either the number one or the number four cylinder is at 10 degrees ATDC on its power stroke. The DIS module knows to fire coil 2 in response to the next spark timing command from the PCM. When the signal goes low, either cylinder number three or cylinder number two is at 10 degrees ATDC, and the DIS module knows to fire coil 1 in response to the next spark timing command from the PCM. The CID sensor is sometimes called a *cam sensor*, though obviously on the four-cylinder engine, this is a misnomer.

Ignition Diagnostic Monitor (IDM)

The IDM, not used on all EEC IV systems, is a wire feeding the tach signal from the negative side of the ignition coil primary windings to the PCM. On a TFI IV ignition system, this wire with a 22 kilohm resistor runs from the coil directly to the PCM, Figure 14–27. On DIS engines, the tach signal is picked up from the DIS module and fed through a 20 kilohm resistor to the PCM, Figure 14–20. The resistor protects the PCM from the coil's high voltage spike when the primary current is cut off to trigger the spark.

The IDM is like an echo of the PCM's spark

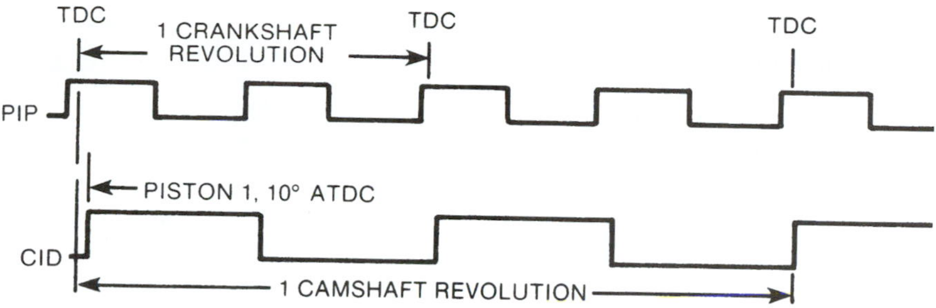

Figure 14–26 CID and PIP sensor signals.

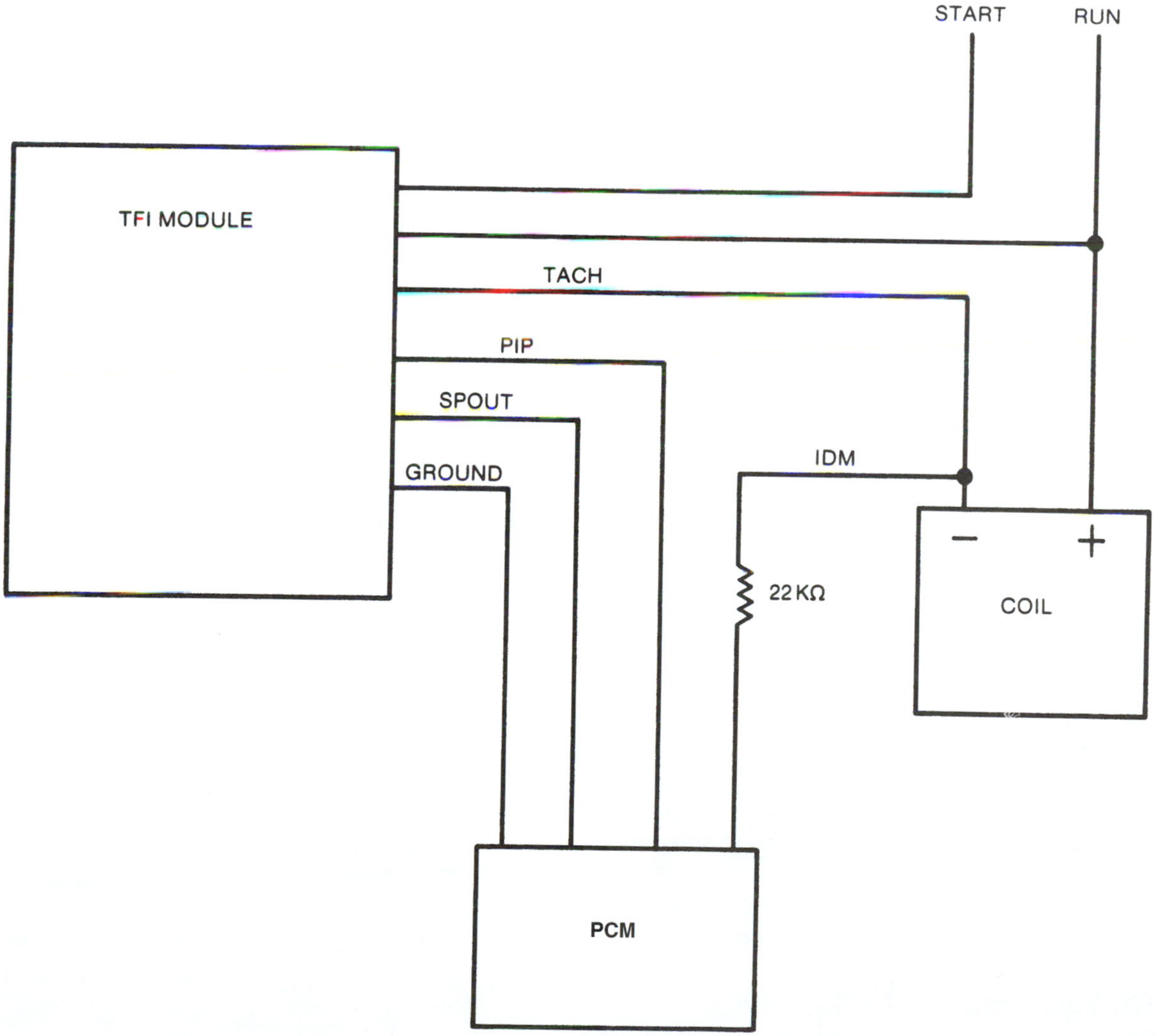

Figure 14–27 Ignition feedback schematic to PCM.

output (**SPOUT**) command. The PCM compares the IDM signal to the PIP signal to check that:

- the SPOUT command was carried out.
- the calculated timing advance circuit is working properly.

If the PCM sees a fault, it stores a service code in its continuous monitor memory.

Exhaust Gas Oxygen (EGO) Sensor

The EGO looks similar to the oxygen sensing unit used on other computer-controlled engine management systems and works exactly the same way. It generates a low-voltage signal in response to the difference between exhaust residual oxygen and the oxygen in the ambient air. As the fuel mixture controls modify the delivered mixture in response to the oxygen sensor's signal, the oxygen signal in turn completes the closed loop and responds to the new mixture with a corrected feedback signal.

Intake Air Temperature (IAT) Sensor

The IAT sensor (prior to OBD II standardization, known as the air charge temperature or ACT sensor) measures the temperature of the air in the intake manifold that mixes with the fuel. The sensor uses a thermistor as a temperature-sensing element just as the ECT (engine coolant temperature) sensor does. On the IAT sensor, however, the sensing end of the housing has openings to allow intake air to come into direct contact with the thermistor coils, Figure 14–28. The IAT sensor is screwed into the intake manifold (or into the air cleaner on 2.8-liter engines). The PCM uses IAT sensor information to calculate the air/fuel mixture and spark timing advance. Voltage and resistance values for the IAT sensor are the same as those for the ECT. The IAT sensor is used on most Ford engine applications, but not on all.

EGR Valve Position (EVP) Sensor

The EVP sensor is a linear potentiometer mounted on top of the EGR valve. Its plunger sits on the EGR valve diaphragm and pintle assembly, and it informs the PCM as to the pintle's position—closed, open, or somewhere in between, Figure 14–29. The EVP sensor receives a 5-volt reference voltage from the PCM and returns a voltage value that is low when the valve is closed to a value of about 4.5 volts when the valve is wide open. Two versions of this sensor are in use—a black one that carried over from EEC III and was used through 1987, and a white (light gray) one that was introduced in 1985. With the valve fully closed, the black one returns a voltage of 1 volt, and the acceptable range is between .8 volt and 1.2 volts. (On an EEC III application, the same EVP sensor returns a voltage of about 1.8 volts with the valve closed due to the 9-volt reference.) The white EVP sensor returns a voltage of

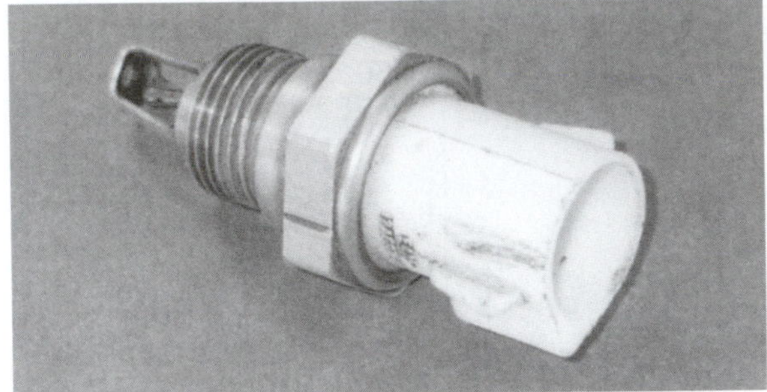

Figure 14–28 Intake air temperature (IAT) sensor.

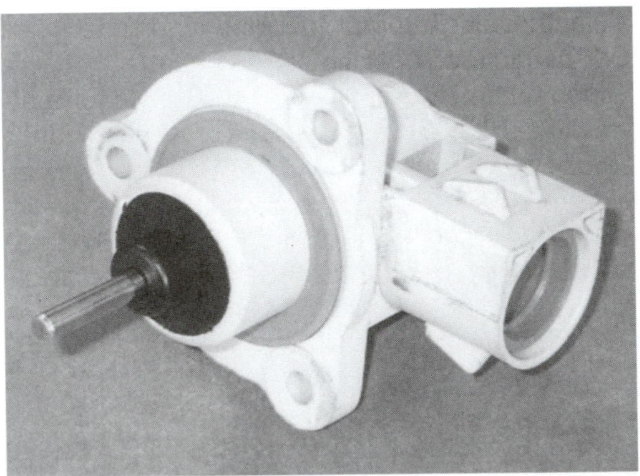

Figure 14–29 EGR valve position (EVP) sensor.

about .4 volt when the valve and sensor are new with an acceptable range of between .3 volt and .7 volt. Because these sensors are interchangeable physically, care must be taken to use the correct part number on a given application.

As carbon builds up within the EGR valve's pintle area keeping the valve from fully seating, the EVP sensor returns a higher voltage to the PCM. Also, sometimes the EGR valve's metallic diaphragm will develop an indentation at the EVP sensor's contact area and return a lower voltage to the PCM. Either of these conditions, when severe enough to push the voltage signal out of the proper range, can set a fault code during system diagnosis.

The PCM uses the EVP sensor's voltage signal to calculate the probable flow rate for any set of driving conditions. Unfortunately, any carbon buildup in the EGR passages can result in some inaccuracy due to the fact that this sensor measures only the EGR valve's position, not flow rate.

The EVP sensor's input is used to:

- calculate air/fuel ratio.
- calculate spark advance timing.
- adjust or correct EGR flow.
- set a service code for an EGR valve that fails to open or close when so actuated.

Pressure Feedback EGR (PFE) Sensor. The PFE sensor, Figure 14–30, is a variable capacitor working as described in the preceding section, Pressure Sensors. It senses exhaust pressure in a chamber just under the EGR valve pintle, Figure 14–31. When the EGR valve closes, the pressure in the sensing chamber is equal to exhaust backpressure. When the EGR valve opens, pressure in this chamber drops because it is then exposed to intake manifold pressure, which is always lower, while the restriction at the other end of the chamber limits the rate at which pressure from the exhaust system can enter the chamber. The more the EGR valve opens, the more the pressure in the chamber drops.

The PFE sensor, through its internal electronic circuitry, provides an analog voltage signal to the PCM that the PCM can compare to values on its look-up table to determine EGR flow.

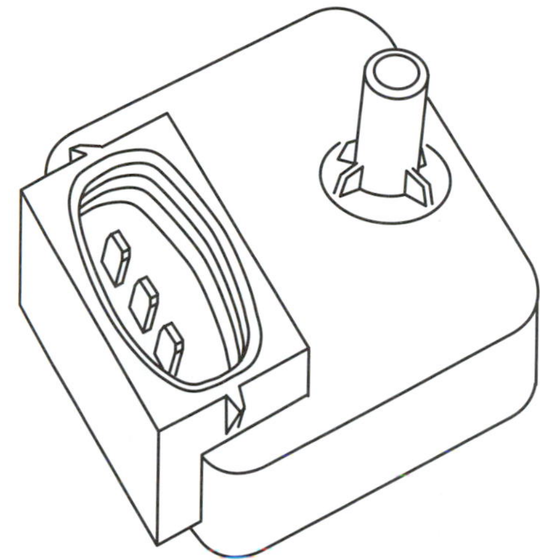

Figure 14–30 Pressure feedback EGR (PFE) sensor.

The PFE differs from the EVP in that it monitors actual EGR flow, while the EVP monitors only EGR opening. The EVP leaves the PCM to assume there is no carbon buildup or other restrictions hampering EGR flow. While this may be true for new engines, it is rarely the case for those in service for any length of time.

The PCM uses PFE information to:

- fine tune its control of EGR valve opening.
- more accurately control air/fuel ratio.
- modify ignition timing.

Vane Meter

The vane meter is used on some multipoint fuel injection applications, such as the 1.6-liter turbocharged and nonturbocharged engines, the 2.3-liter turbocharged engines and the 2.2-liter engine in the Ford Probe. It fits between the air cleaner and the throttle body, and it measures the velocity of air flowing into the engine's induction system and the temperature of that air. Both of these measurements are made by sensors in the vane meter.

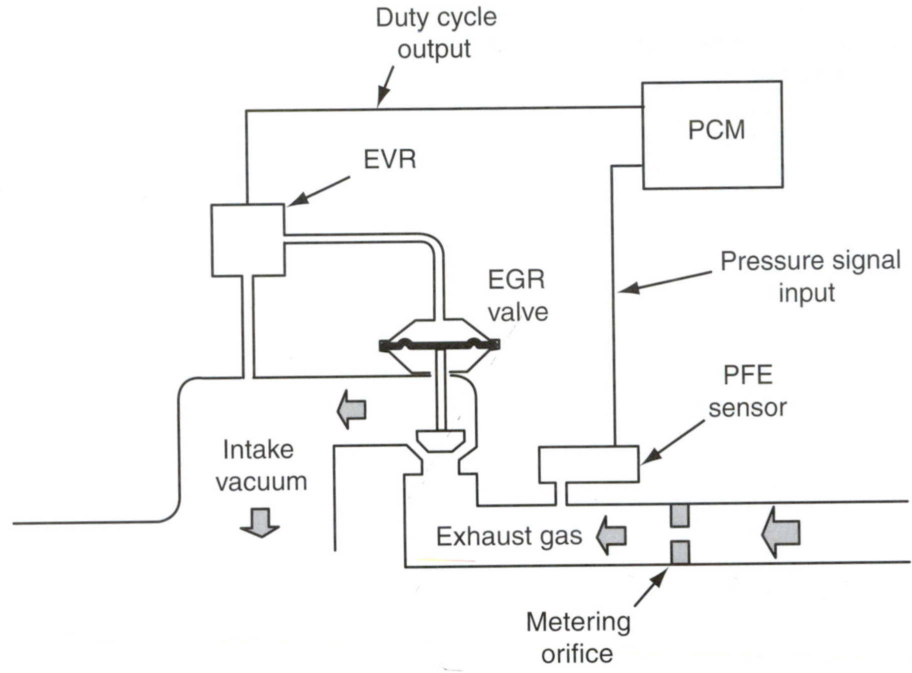

Figure 14–31 Pressure feedback EGR sensor and EGR flow.

Vane Airflow (VAF) Sensor. The air inlet opening of the vane meter is closed by a spring-loaded door, a vane, Figure 14–32. As the throttle valve opens with the engine running, air moving into the induction system forces the vane open; the more airflow, the wider the vane opens. The vane hinges on a pivot pin. As the vane opens, it moves the wiper of a variable potentiometer, which functions as an airflow sensor for the PCM. The greater the airflow, the higher the VAF sensor's signal voltage.

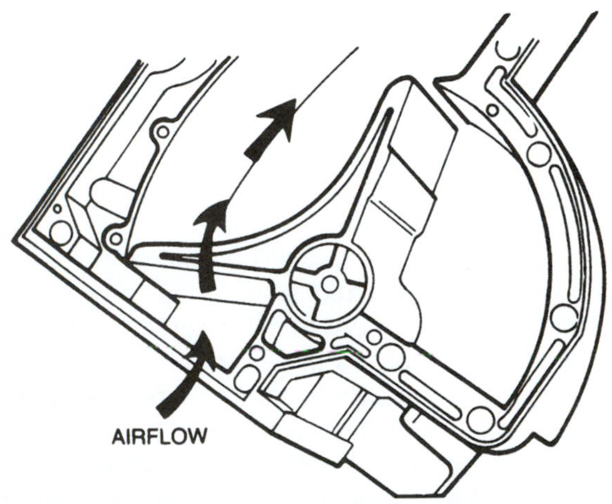

Figure 14–32 Vane airflow sensor. *(Courtesy of Ford Motor Company.)*

As the vane opens, its compensator flap pushes into a specially designed cavity below the airflow passage. There is a lull or pulse between intake strokes, particularly on four-cylinder engines, and the vane has a tendency to dip closed slightly during those periods. The sealed space behind the compensator flap acts as a damper to reduce vane flutter.

Vane Air Temperature (VAT) Sensor. This sensor is effectively the same as the IAT sensor used on some other engine applications, except it is located in the vane meter's air inlet opening, Figure 14–33.

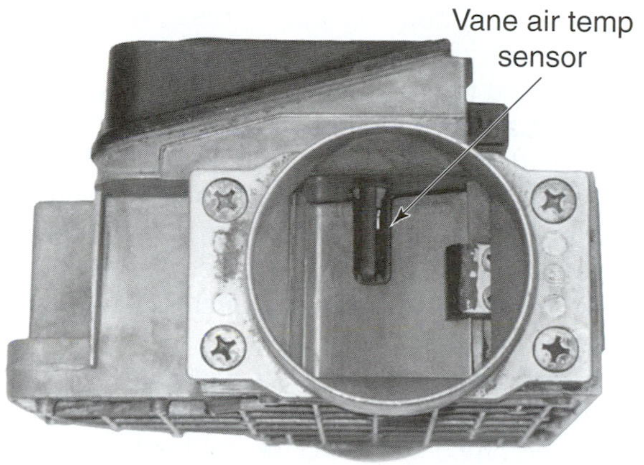

Figure 14–33 Vane airflow (VAF) meter with vane air temperature (VAT) sensor.

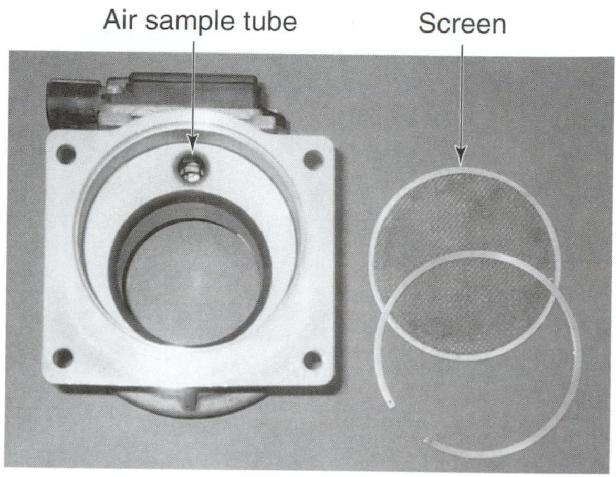

Figure 14–34 Mass airflow (MAF) sensor with air sample tube.

Speed Density Formula. The PCM compares inputs from the VAF, VAT and BP sensors (or the MAP sensor, depending on the engine application and the kind of sensors employed), along with those from the TPS and EVP sensors to its hardwired look-up charts. In this way it can calculate the engine's air intake volume and flow rate. This information, along with estimates of the engine's volumetric efficiency at different engine speeds and loads, is what the PCM uses to calculate exactly how much fuel should be injected. This is the Ford application of the speed density formula (see Speed Density Formula in the introductory chapter).

Mass Airflow (MAF) Sensor. Beginning in 1988, certain engine systems employed a MAF sensor instead of a MAP sensor (see Mass Airflow Sensor in Chapter 2). The Ford unit was built by Hitachi of Japan. It is different from the Hitachi unit used by General Motors. The Ford MAF sensor is in the air tube between the air cleaner and the throttle body. A small air sample tube in the top of the MAF sensor air passage directs air across a hot wire and a cold wire, Figure 14–34. The cold wire includes a thermistor used to measure the temperature of the incoming air. The hot wire wraps around a ceramic element and is coated with a super-thin layer of glass. A module mounted on the MAF sensor body keeps it 200°C above the ambient air temperature.

Current flow through the sensor varies from 0.5 to 1.5 amperes, depending on the air's mass airflow rate. Voltage varies from 0.5 volt at 0.225 kilograms (0.5 pound) of air per minute to 4.75 volts at 14.16 kilograms (31.23 pounds) of air per minute. The MAF sensor's input to the PCM affects calculations for:

- air/fuel mixture.
- ignition spark timing.

Idle Tracking Switch (ITS)

In the body of the idle speed control motor assembly (which is used on some carbureted and some CFI engines) is a normally closed switch, Figure 14–35. Whenever the throttle closes, the throttle lever presses against a plunger extending from the nose of the assembly. The pressure forces the plunger to move slightly back into the assembly and to open the switch. When the throttle opens, the switch closes again. The PCM monitors the switch and can determine when the throttle is open or closed. It uses this information to control the ISC motor, both to determine when to employ it and what to set it at.

Figure 14–35 Idle speed control (ISC) motor.

Transmission Switches

The PCM must know whether the vehicle is in gear, in park or in neutral. This information affects:

- idle fuel and air control strategy for automatic transmission vehicles. (The torque converter loads the engine, even if the vehicle is not moving.)
- response to rapid closing of the throttle. (If the vehicle is in gear, the PCM will issue appropriate commands for deceleration.)

For most automatic transmissions, the neutral drive switch, the same switch used to open the starter circuit when the transmission is in gear, serves as an input to the PCM, Figure 14–36. When the transmission is in neutral or park, the switch closes. A 5-volt reference signal is fed through a resistor in the PCM, output pin 30 to ground through the N/D switch. When the switch closes, voltage on the switch side of the resistor is low. By monitoring this voltage, the PCM knows whether the transmission is in gear. A diode in the switch prevents battery voltage from being applied to the PCM circuit when the starter solenoid is energized.

The **AXOD (automatic transaxle overdrive)** uses three different switches to identify what gear the transmission is in. They are:

- Neutral Pressure Switch (NPS).
- Transmission Hydraulic Switch (THS) 3-2.
- Transmission Hydraulic Switch (THS) 4-3.

These are normally open pressure switches screwed into the transmission valve body so they are exposed to the hydraulic pressure applied to specific bands or clutches in the transmission. When hydraulic pressure is applied, the switch closes and grounds the circuit. The PCM supplies voltage to each switch through a resistor, Figure 14–37A. By monitoring the voltage in each switch circuit, the PCM can identify each transmission gear position with the exception that first cannot be distinguished from second, Figure 14–37B.

A transmission temperature switch (TTS) is employed on some AXOD transaxles to alert the PCM if the transmission fluid temperature goes too high, Figure 14–37A. The TTS is normally closed, but it opens at 135°C (275°F). It also bolts directly into the valve body. If the PCM sees the TTS open during an operating condition in which it normally keeps the **torque converter lockup clutch** disengaged, such as a long uphill climb with a large throttle opening, it will apply the clutch to reduce the heat produced by high-torque turbulence in the torque converter.

Manual transmission applications can be taken out of gear either by the clutch or by the gearshift. Thus, the system uses two switches, Figure 14–36. A Neutral Gear Switch (NGS) attaches to the shift linkage. It remains closed in neutral. A Clutch Engaged Switch (CES) attaches to the clutch pedal linkage and closes when the clutch pedal is depressed (when the clutch is disengaged). Because the NGS and CES are in parallel, they must both be open for the PCM to determine that the vehicle is in gear. If either switch closes, voltage at pin 30 goes low, and the PCM knows the geartrain is disengaged.

Inferred Mileage Sensor (IMS)

Beginning in 1985 some light truck EEC IV vehicles came with an electronic module (IMS) containing an E-cell. The IMS is energized by the power relay. After a specified amount of ignition-on time, the E-cell opens the IMS's feedback circuit to the PCM. The PCM responds by changing programmed calibrations to direct Thermactor air

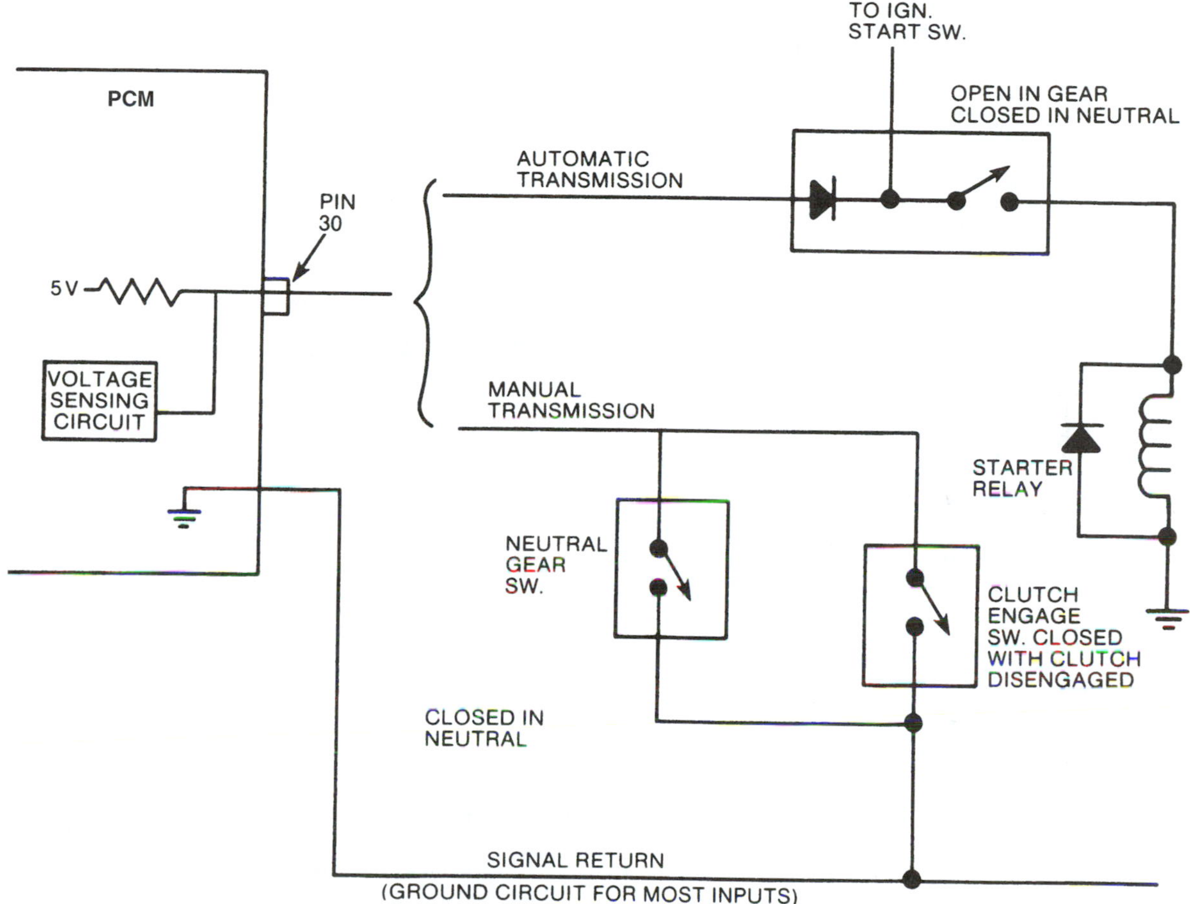

Figure 14–36 Transmission load switches.

to the exhaust manifold for longer periods of time. It also changes the EGR flow rate, both measures to compensate for the "inferred" engine wear assumed to have occurred.

The IMS should not be confused with the emission maintenance warranty/extended useful life (EMW/EUL) module. There are two types of these latter units, each used on Ford trucks. The early type used an E-cell depleted after the ignition was on for a total time equivalent to about 60,000 miles. When the E-cell depleted, it triggered an indicator light on the instrument panel. This was supposed to alert the driver to replace the EGR valve and oxygen sensor. On later models, the E-cell in the module was replaced by a small microprocessor that could be reset. Replacing the EGR valve or oxygen sensor at set intervals is no longer required on other EEC IV systems applications. Replacement is required, of course, when those components fail or degrade beyond useful function.

Knock Sensor (KS)

The EEC IV engine applications using a knock sensor use a standard, piezoelectric type, Figure 14–38. When excited by vibrations in the frequency characteristic of detonation (knock), it

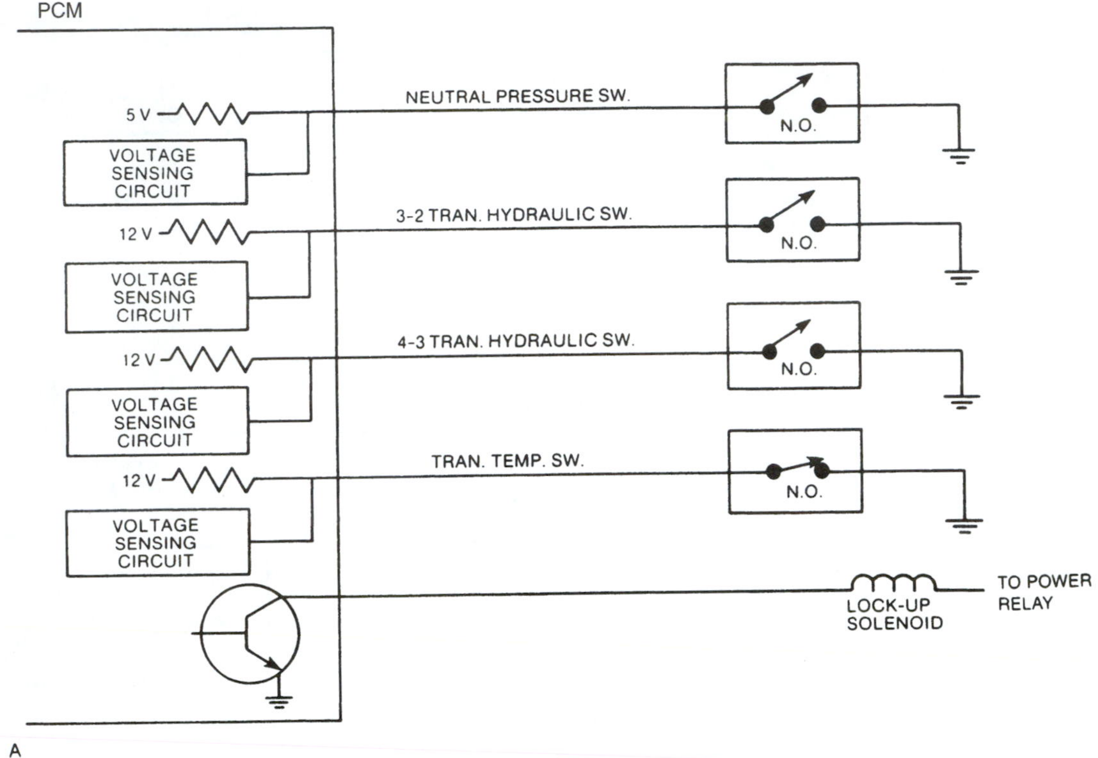

Gear Position	Neutral Pres. Sw.	3-2 Tran. Hyd. Sw.	4-3 Tran. Hyd. Sw.
Park/Neutral	Open (5 V)	Open (12 V)	Open (12 V)
First	Closed (0-1 V)	Open (12 V)	Closed (0-1 V)
Second	Closed (0-1 V)	Open (12 V)	Closed (0-1 V)
Third	Closed (0-1 V)	Closed (0-1 V)	Closed (0-1 V)
Fourth	Open (5 V)	Closed (0-1 V)	Open (12 V)
Reverse	Closed (0-1 V)	Open (12 V)	Open (12 V)

B

Figure 14–37 AXOD transmission switches.

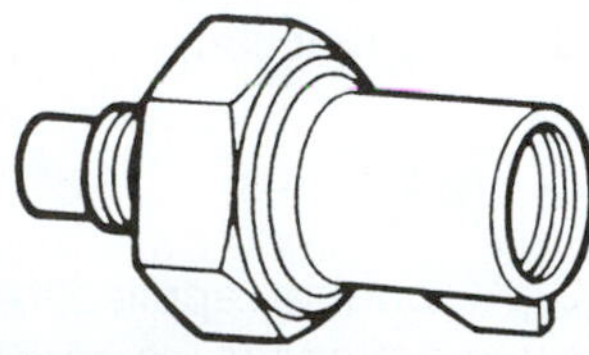

Figure 14–38 Knock sensor. *(Courtesy of Ford Motor Company.)*

sends a signal to the PCM, which can then retard spark timing advance to correct for the condition.

Vehicle Speed Sensor (VSS)

The vehicle speed sensor, a magnetic pulse generator driven by the transmission speedometer output gear, produces 8 cycles for each revolution, or 16 AC signals per revolution, or 128,000 signals per mile to the PCM. The PCM will modify this signal, convert it to a vehicle speed value and store it in memory. In addition to vehicle speed control and depending on vehicle application, this information may be used for the control of the transmission torque converter lockup clutch, coolant fan control, and to identify deceleration conditions.

Two versions of a magnetic VSS were used. One version was used when the vehicle had onboard electronics, such as the PCM, but still retained a mechanical speedometer assembly. This VSS allowed a speedometer cable to still be attached to it, Figure 14–39. The other version had no provision for a speedometer cable to be attached to it, Figure 14–40, and was used on those vehicles that had an electronic speedometer and odometer assembly with either an analog or a digital display.

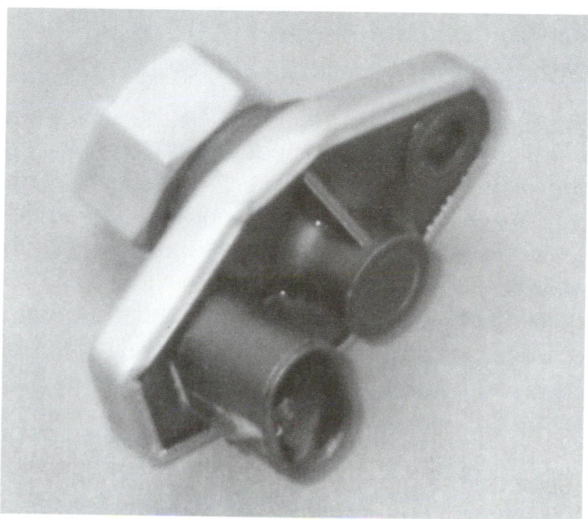

Figure 14–40 Vehicle speed sensor used on applications having an electronic speedometer/odometer assembly.

Brake On/Off (BOO) Switch

Vehicles with the **A4LD (automatic four-speed light-duty) transmission**, such as Rangers and Broncos, use the BOO switch (located in the stop lamp switch assembly) to signal the PCM when the brakes are applied. The A4LD transmission employs an PCM-controlled torque converter lockup clutch. When the switch closes, the PCM releases the lockup clutch. If the brakes are applied during idle and the transmission is in gear, the PCM may also raise idle speed to compensate for the increased engine load of the torque converter and brake booster. During a prolonged idle with the brakes applied, the PCM may also disengage the A/C compressor clutch. Use of the BOO switch has become more widespread on newer model vehicles to the point that it is used on most models today.

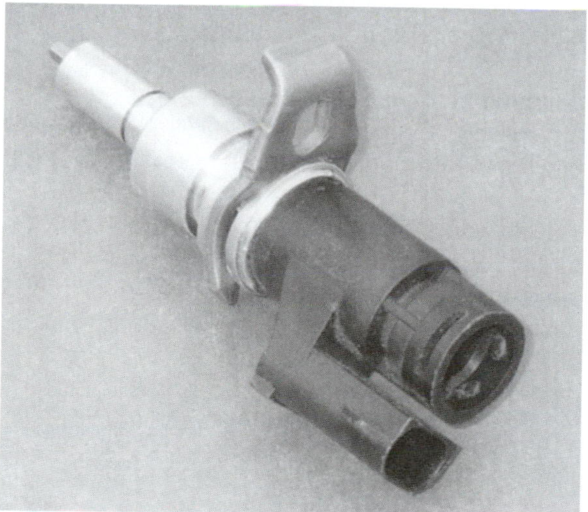

Figure 14–39 Vehicle speed sensor used on applications that retain a speedometer cable to operate a mechanical speedometer.

Power Steering Pressure Switch (PSPS)

The power steering pressure switch (PSPS) is located in the high-pressure side of the power steering system and is normally closed. At a pressure between 400 and 600 psi, the switch opens. When the PCM senses that the PSPS circuit is open, it raises idle speed to compensate for the additional load caused by the hydraulic steering boost.

Air-Conditioning Demand (ACD) and Air-Conditioning Clutch Cycling Switch (ACCS)

On many applications, when the driver selects an A/C operating mode, the ACD switch (in the climate control mode switch assembly) informs the PCM that such a mode has been selected. Then the PCM watches the ACCS to be able to respond with an increase in idle speed to compensate for A/C compressor load.

Ignition Switch

When the ignition switch turns on, the power relay activates. The power relay then supplies current to the PCM and most of the circuits that the PCM controls. The relay's power to the PCM acts as an input and signals the PCM to begin its programmed functions.

OUTPUTS

Air/Fuel Mixture Control

The EEC IV system employs three different types of fuel metering devices: carburetors (referred to as feedback carburetors), throttle body injectors (referred to as Central Fuel Injection) and port fuel injectors (referred to as Electronic Fuel Injection).

Electronic Feedback Carburetors. Three different carburetors have been used with EEC IV engine applications. They each use a duty-cycled solenoid to control the air/fuel mixture within the range available to the control unit. Each feedback carburetor (FBC), however, uses a somewhat different feedback control solenoid. Each feedback control solenoid gets power from the power relay, and its function is controlled by the PCM, by grounding the circuit to energize it, Figure 14–41.

The Motorcraft 2150A-2V, Figure 14–42, uses a solenoid that introduces or blocks air from the air cleaner into the fuel vacuum passages, as the solenoid cycles alternately on and off. These passages provide bleed air to the idle and main metering systems and determine the richness of the fuel introduced into the carburetor's venturis. The air introduced by the solenoid is in addition to the air introduced by the fixed air bleeds found in prefeedback carburetors and retained, in modified form, on these. If the solenoid were turned off completely (a duty cycle of zero), additional air would be blocked, and the air/fuel

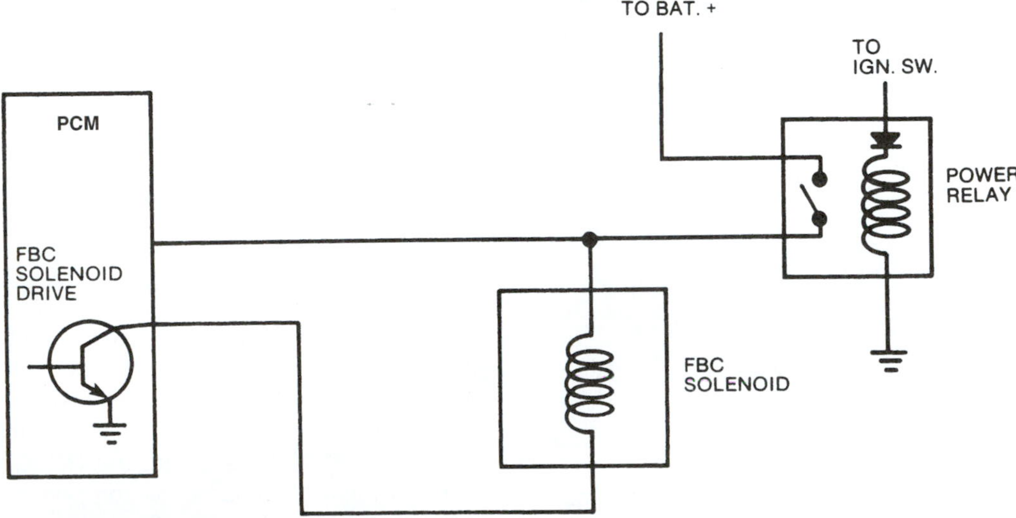

Figure 14–41 FBC solenoid circuit.

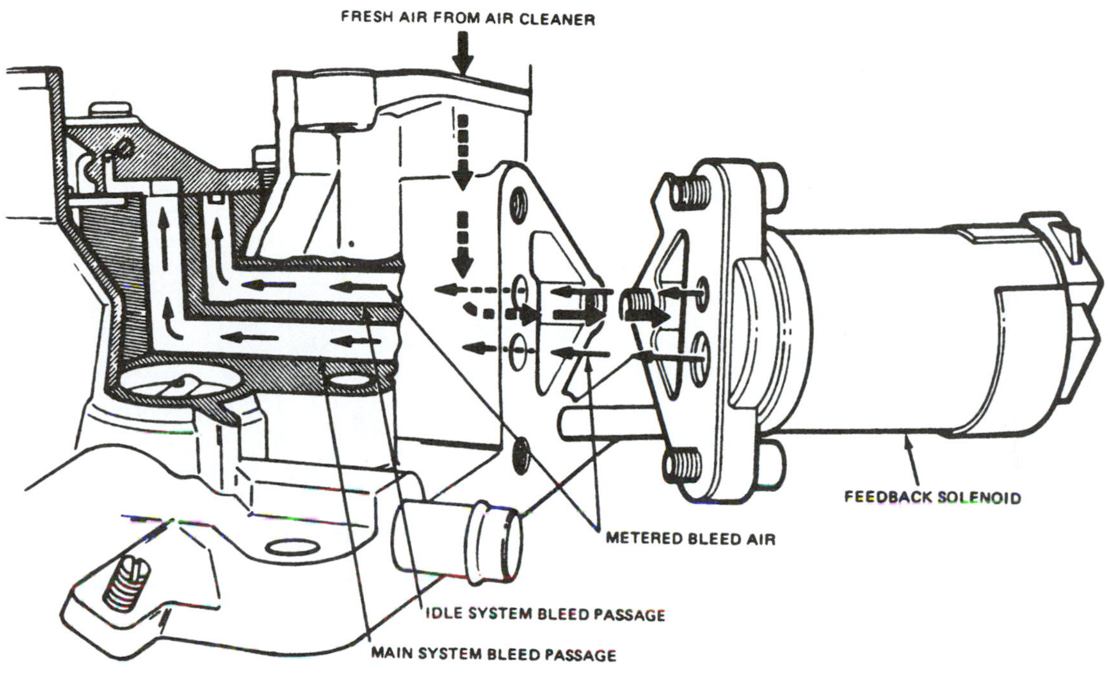

Figure 14–42 Feedback carburetor (FBC) solenoid (2150-2V carburetor). *(Courtesy of Ford Motor Company.)*

ratio would go full rich. If the solenoid were on constantly (a duty cycle of 100 percent), the mixture would go full lean. The PCM controls the duty-cycle to optimize the delivered fuel/air ratio. In open loop the duty cycle is determined by a hardwired constant value, depending on coolant temperature, engine load, engine speed and throttle position. In closed loop it varies constantly, depending on the feedback signal from the oxygen sensor.

The Carter YFA-1V carburetor has a feedback control solenoid that looks different from the one used on the Motorcraft 2150A, but its function is identical.

The Holley 6149-1V carburetor is controlled by a remotely mounted feedback control solenoid, Figure 14–43. Its control of the idle circuit air/fuel ratio is much like that of the solenoids for the two carburetors above. Its duty cycle controls the amount of air introduced into the idle speed circuit, Figure 14–44. The main metering mixture control, however, works differently.

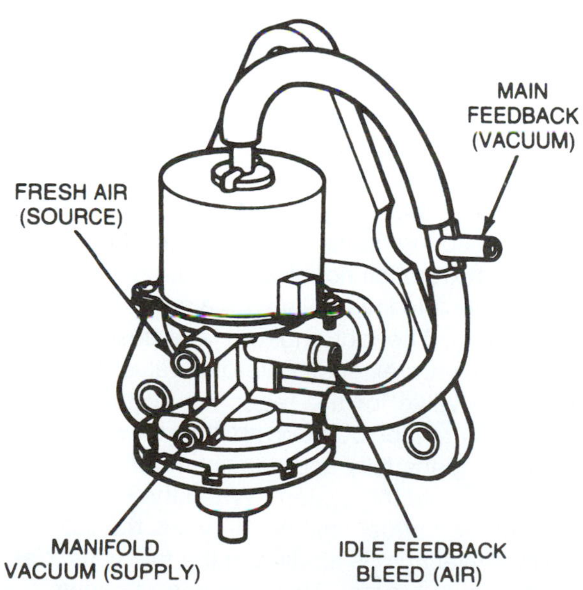

Figure 14–43 Remote duty cycle solenoid. *(Courtesy of Ford Motor Company.)*

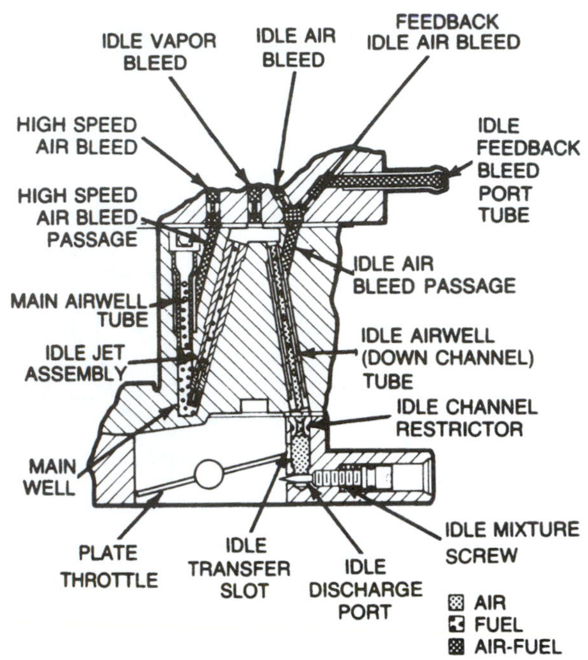

Figure 14–44 Idle feedback metering. *(Courtesy of Ford Motor Company.)*

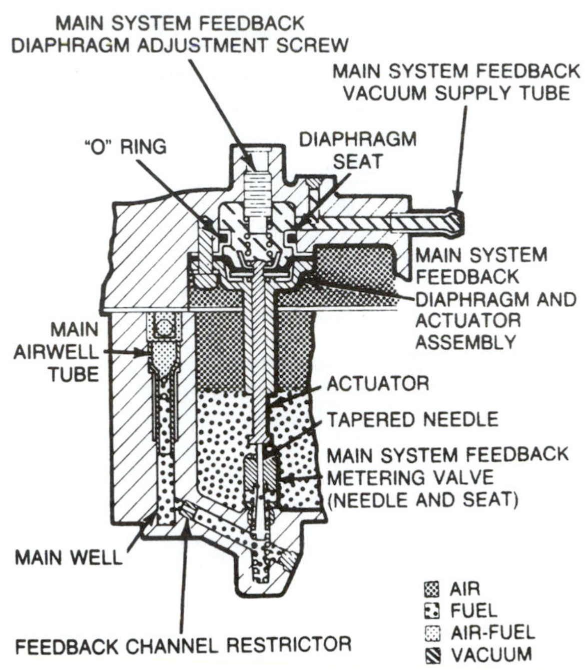

Figure 14–45 Main feedback metering. *(Courtesy of Ford Motor Company.)*

The Holley 6149-1V main metering circuit has a solenoid controlling vacuum to a diaphragm in the air horn section of the carburetor, Figure 14–45. The diaphragm is part of a fuel control valve assembly. It can easily be mistaken for a power valve assembly, but its function is quite different. A spring holds the diaphragm down, making the mixture full rich. Extending down from the diaphragm into the fuel bowl is a rod with a foot at the bottom. The foot engages and pushes down on the upward extending tip of a valve (the main metering valve for the feedback system) in the floor of the fuel bowl. With this valve pushed down, it opens fully and additional fuel flows into the main metering system. When the PCM wants to lean the air/fuel mixture, it cycles the solenoid so more vacuum is applied to the fuel control assembly diaphragm. This lifts it and allows the valve in the fuel bowl floor to close partially, thus leaning the delivered mixture.

Central Fuel Injection (CFI). Two kinds of CFI systems are used with EEC IV applications: a high-pressure system and a low-pressure system. Each uses a throttle body unit mounted on the manifold in the place where a carburetor would be.

The solenoid operated injectors are opened by electrical command pulses from the PCM. The injector's control circuit is similar to that of the feedback carburetor solenoid. Electrical power comes directly from the power relay, and the PCM grounds the circuit, activating the injector, for whatever pulse width (on-time) it calculates is needed for the current driving conditions. Fuel sprays from a single injector (low-pressure system, Figure 14–46) or from two injectors (high-pressure system, Figure 14–1) directly over the throttle valves. During cranking, the high-pressure CFI system pulses both injectors simultaneously in response to each PIP signal. After the engine is

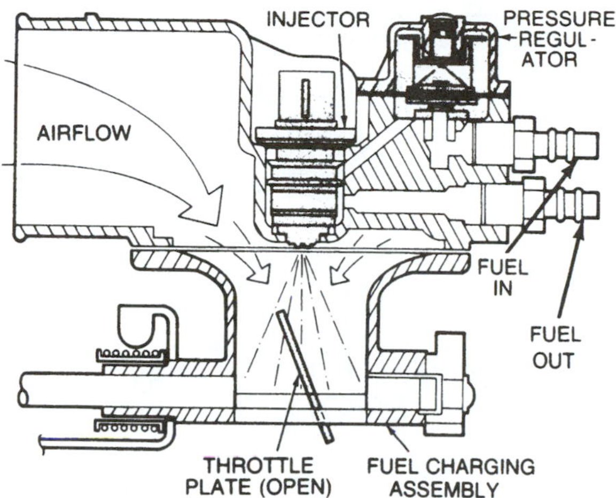

Figure 14–46 Low-pressure central fuel injection unit. *(Courtesy of Ford Motor Company.)*

started, they pulse alternately at a fixed frequency. Because the fuel pressure is constant and the injector always opens the same amount (about 0.01 inch), the air/fuel mixture is controlled entirely by the injector pulse width.

During the mid-1980s, the 3.8-liter V6 used two oxygen sensors, one on each exhaust manifold. The input from both oxygen sensors was averaged by the PCM and then used to fine-tune the injector's pulse width.

Electronic Fuel Injection (EFI). Ford's nonsequential form of multipoint fuel injection was known as EFI and used group injection, also known as bank-to-bank fuel injection. That is, the PCM pulsed the injectors in two groups. For example, on a V8 engine, one group was comprised of the front and rear injectors at one cylinder head and the two centrally located injectors at the other cylinder head. Conversely, the other bank of injectors was comprised of the other four injectors. Each group was pulsed every 720 degrees of crankshaft rotation, alternating with the other group which was pulsed on the opposite crankshaft rotation. Each pulse delivered 100 percent of the fuel requirement for that engine cycle.

Sequential Electronic Fuel Injection (SEFI). Ford introduced sequential port fuel injection on the 5.0-liter V8 engine in 1986 model year vehicles and on the 3.8-liter V6 engine in 1988 model year vehicles. By the late 1980s, most of their passenger cars had sequentially fuel-injected engines, although the light truck line was not sequentially injected until the 1990s. In the SEFI system, as with other makes, the injectors are pulsed one at a time in the engine's firing order.

Fuel Supply System

Fuel Pump. The low-pressure CFI system uses a fuel pump mounted in the fuel tank. The fuel pump relay controls it, and the pump delivers fuel to the fuel delivery assembly—the throttle body unit.

The CFI high-pressure and EFI systems use one of two fuel delivery systems. Most use a single high-pressure, in-tank electric pump that feeds fuel at about 39 psi to the pressure regulator. Vehicles with a greater distance between the fuel tank and the engine, such as pickup trucks or vans, can use two electric fuel pumps to ensure adequate fuel pressure to the injectors without the risk of vapor lock. A low-pressure "lift pump" is mounted in the tank. It pumps fuel to an in-line high-pressure pump, which increases the pressure to about 39 psi and sends the fuel to the pressure regulator.

Either of the high-pressure pump systems can generate an unregulated pressure of more than 100 psi if the fuel line is blocked. If that pressure is reached, a pressure relief switch shuts off the pump.

Carbureted EEC IV vehicles use the conventional mechanical fuel pump on the side of the engine block.

Fuel Pressure Regulator. A pressure regulator similar to those described in previous chapters controls fuel pressure to the injectors, Figure 14–47. Manifold vacuum (MAP) is routed to the MPI system pressure regulator so that it varies fuel pressure at the injectors in response to throttle position, or more properly in response to load as reflected in the intake manifold vacuum. This delivered pressure can range from a low of about 30 psi

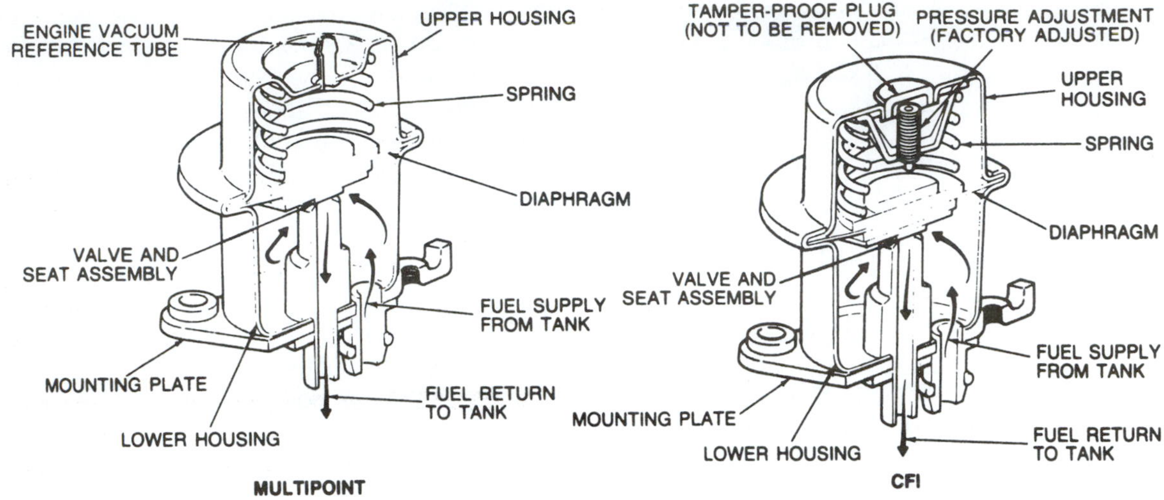

Figure 14–47 Fuel pressure regulators. *(Courtesy of Ford Motor Company.)*

at high vacuum (low manifold pressure) conditions to as high as 39 psi at WOT (high manifold pressure). On turbocharged engines, the fuel pressure at the injectors during boost conditions can allow a delivered fuel pressure as high as 50 psi.

The CFI systems do not employ a manifold pressure-controlled regulator, and their fuel pressure stays fairly constant. The high-pressure CFI system maintains between 38 and 40 psi. The low-pressure system (on the 2.3 HSC engine) maintains about 14.5 psi at the injector tip. There is no need to modulate pressure according to load with these systems because they spray fuel ahead of the throttle, into air at ambient barometric pressure.

Fuel Pump Relay. The fuel pump relay controls the fuel pump and is in turn controlled by the PCM, Figure 14–48. When the ignition switch is turned on, the PCM grounds the fuel pump relay coil. If the PCM does not receive a PIP signal indicating engine cranking, it will turn the fuel pump relay off after about 1 or 2 seconds. Once the engine is running, the PCM keeps the fuel pump relay energized until the ignition is turned off or until the PCM no longer receives a PIP signal.

In series with the fuel pump relay is an inertia switch, Figure 14–49. Its purpose is to disable the fuel pump if the vehicle is involved in an accident to reduce the chance of a fuel fire. The inertia switch consists of a steel ball sitting in a metal

Diagnostic & Service Tips

Anytime a fuel-injected Ford has suddenly refused to start, the technician should first check the inertia switch. Sometimes these switches have even been triggered off by another driver's parking by Braille, running into the Ford with enough force to trigger the switch, but not enough to cause damage. Less frequently, badly worn shock absorbers on the rear axle can allow enough of a bump to turn the inertia switch off.

This switch, sometimes located in the trunk, can also be damaged by moisture, so it is good practice to check to see if the contacts are still clean where the harness connects. Never run a jumper wire around an inertia switch except for an in-shop test. Under no circumstances should a vehicle be released to a motorist with the inertia switch disabled.

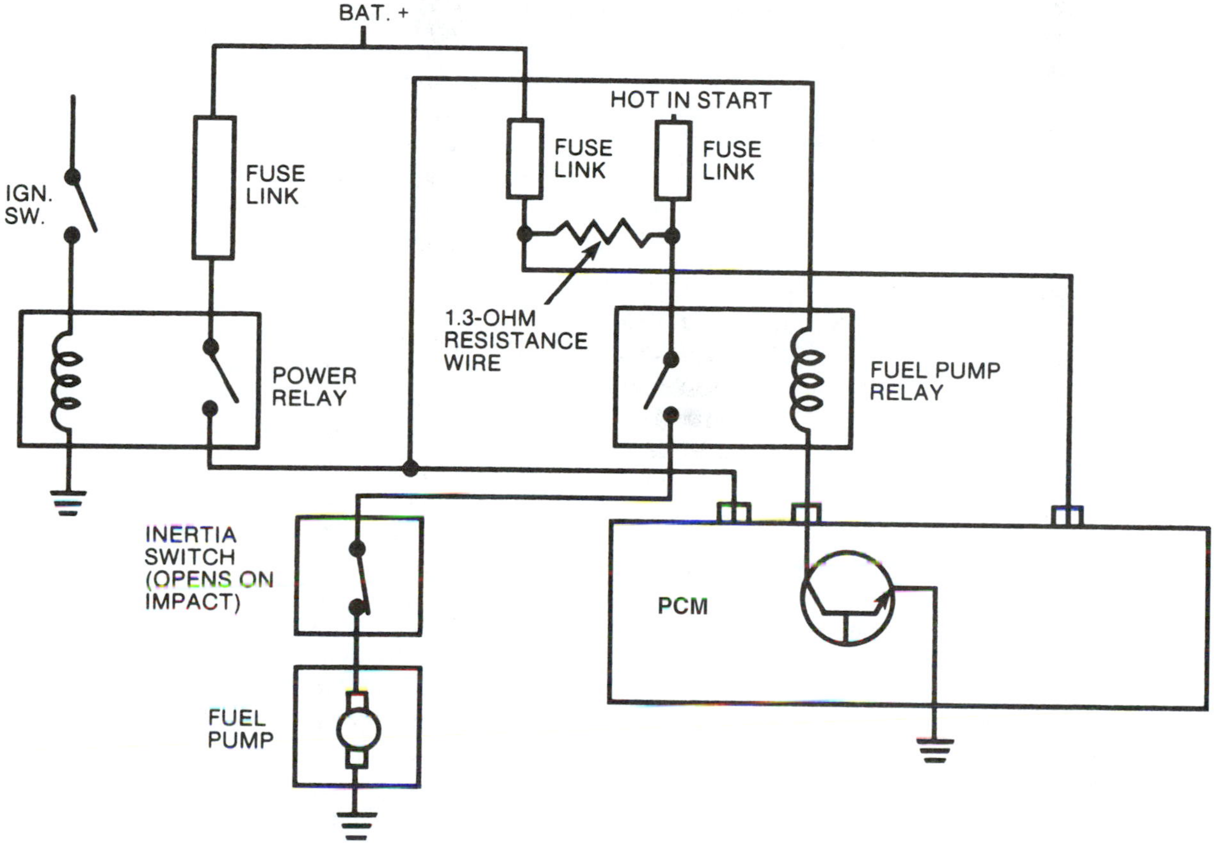

BAT. +
HOT IN START
IGN. SW.
FUSE LINK
FUSE LINK
FUSE LINK
1.3-OHM RESISTANCE WIRE
POWER RELAY
FUEL PUMP RELAY
INERTIA SWITCH (OPENS ON IMPACT)
PCM
FUEL PUMP

Figure 14–48 Fuel pump control circuit.

TARGET PLATE
BALL
ELECTRICAL CONTACTS
MAGNET
SWITCH TERMINALS
RESET LEVER POSITION FOR OPEN SWITCH
CLOSED SWITCH

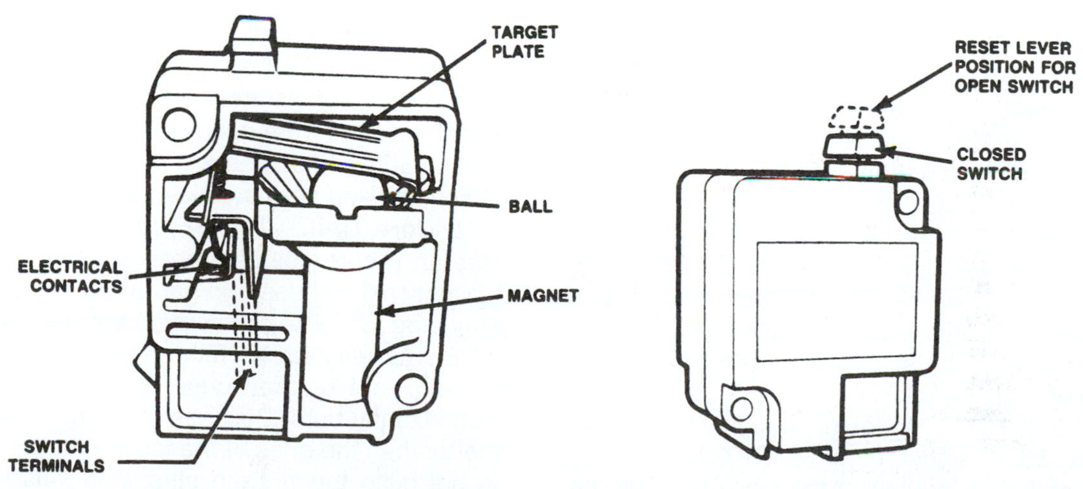

Figure 14–49 Inertia switch. *(Courtesy of Ford Motor Company.)*

bowl. A magnet holds the ball at the bottom of the bowl. In the event of a jolt severe enough to dislodge the ball, the ball strikes a lever above it. This lever triggers an overcentering spring opening the contacts of the inertia switch and holding them open. This opens the circuit to the fuel pump and turns it off. It also pops up a reset button on the top of the inertia switch. Pushing the reset button back down reverses the overcenter switch and closes the contacts again, enabling the fuel pump to run if the PCM energizes the fuel pump relay and sends it current.

Thick Film Integrated (TFI-IV) Ignition System

The **thick film integrated (TFI-IV)** ignition system takes its name from the TFI module bolted to the distributor housing. *Thick film* refers to the type of chip on which the module's circuit is printed. Its function is to turn the ignition coil primary circuit on and off and to control primary circuit dwell time. The spark output command from the PCM is the module's signal to open the primary circuit and fire the spark. If the spark output circuit is open, the TFI module will open the primary circuit in response to the PIP signal. This, of course, results in base timing only, with much less spark advance than is called for in practically all driving conditions.

Spark Output (SPOUT). The PCM receives the PIP signal from the Hall effect switch in the distributor through the TFI module, Figure 14–50. From this signal the PCM determines engine rpm and crankshaft position. The PCM then modifies this signal (changes its spark advance timing) to achieve the best results considering the engine's speed, temperature, load, the atmospheric pressure, the EGR flow and the air temperature. The time-modified signal is sent to the TFI module as the spark output (SPOUT) signal or command.

The SPOUT signal is a digital (on-off or square wave) signal, Figure 14–19. The leading edge of the on portion of the signal causes the TFI-IV module to open the primary ignition circuit and allows the coil's collapsing magnetic field to generate the high voltage in the secondary windings to fire the spark plug.

If the SPOUT circuit is open, the TFI module will select the base or reference signal from the PIP signal for its spark firing timing.

Computer Controlled Dwell (CCD). In early TFI-IV ignition systems, dwell (the duration of the time the coil primary circuit is energized) is controlled by the TFI-IV module. In other words, the TFI-IV module decides when to turn the coil on, and as we saw in SPOUT, the PCM tells it when to turn the coil off and fire the plug. The TFI-IV module includes circuitry that monitors engine rpm and increases dwell as engine speed increases. Beginning in 1989 for a few TFI-IV applications, the PCM instead controlled the ignition dwell. The trailing edge of the SPOUT signal tells the TFI-IV module to close the primary circuit, and the leading edge tells it to open the primary circuit and fire the plug, Figure 14–19.

The reason why the system varies the dwell with engine rpm is that all that is desired from the coil is full spark for the next plug firing. It takes a certain amount of time for the coil to fully build the magnetic field, the collapse of which generates the high voltage secondary current. To keep the coil energized longer than needed is simply to use electric power as resistance heat to make the coil hot. Dwell control began much earlier than computer controls, beginning at least with the 1976 versions of electronic ignition. Even earlier, ballast resistors were used to reduce electric current through ignition primary circuits for the same reason: to keep from overheating the coil.

E-Core Coil. The TFI-IV ignition system uses an E-core coil differently shaped from the conventional coil used on prior Ford Motor Company products. The coil core resembles two capital E's turned face to face. The coil windings, molded in epoxy, are wound around the E's center horizontal bars. Because the center bars are shorter than the ones at the top and bottom, they do not quite touch each other and thus form a

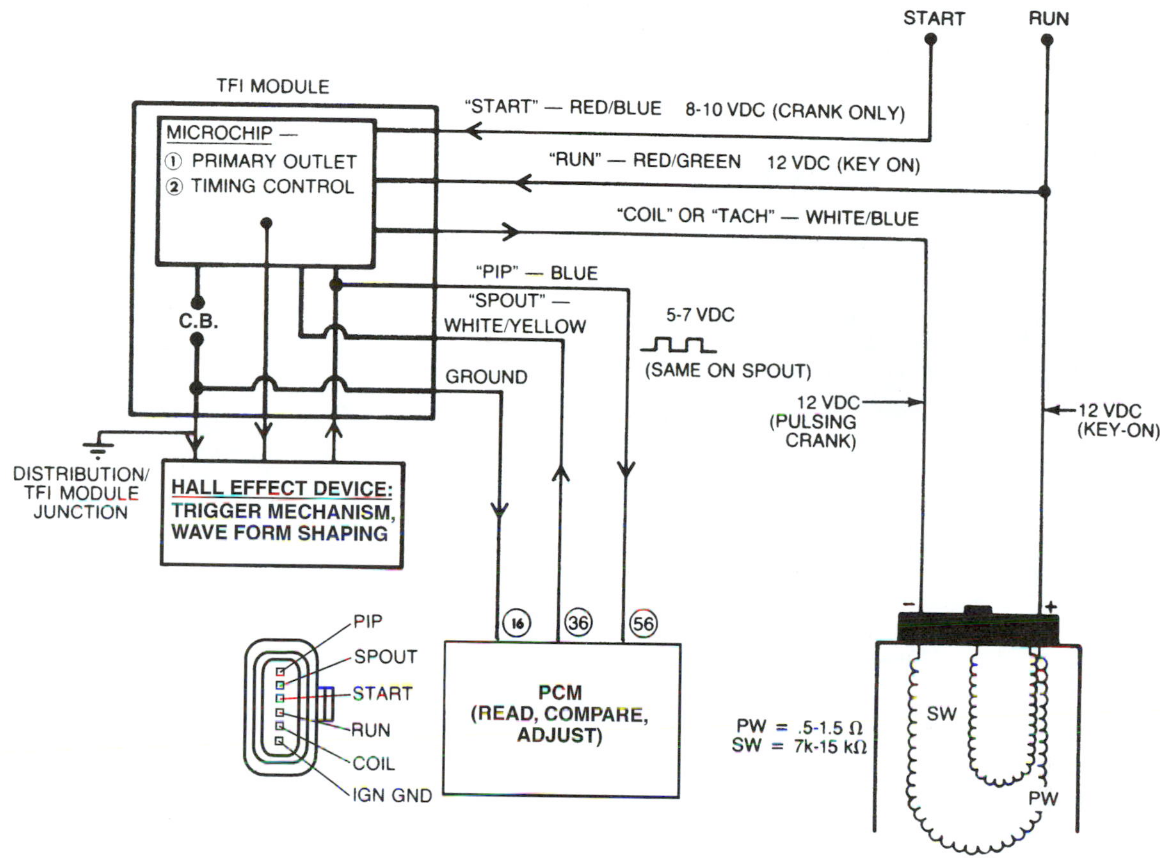

Figure 14–50 TFI ignition circuit. *(Courtesy of Ford Motor Company.)*

small air gap between them. The smaller air gap (compared to the distance between the ends of a conventional longitudinal coil core) helps the coil develop a much stronger magnetic field quickly. With no primary ballast resistor and the small core air gap, the E-coil core draws high primary circuit amperage (5.5 to 6.5 amps) and can put out as much as 50,000 volts. Even though secondary circuit amperage is very small, technicians should be very careful never to allow ignition secondary voltage to shock them. On modern high energy ignition systems, such a shock can be more than just painful.

Octane Rod Adjustment

The TFI-IV distributor has an octane rod, which is inserted from the outside of the distributor and is used to position the Hall effect sensor, Figure 14–51. Of course, the position of the Hall effect sensor is one variable that affects base timing. Ford Motor Company has issued Technical Service Bulletins (TSBs) on certain applications to change the octane rod with one of a different length, which, in return, retards base timing a selected 2, 3 or 6 degrees, depending on the octane rod used. The octane rod is stamped with a number representing the amount of spark timing retard the rod will create.

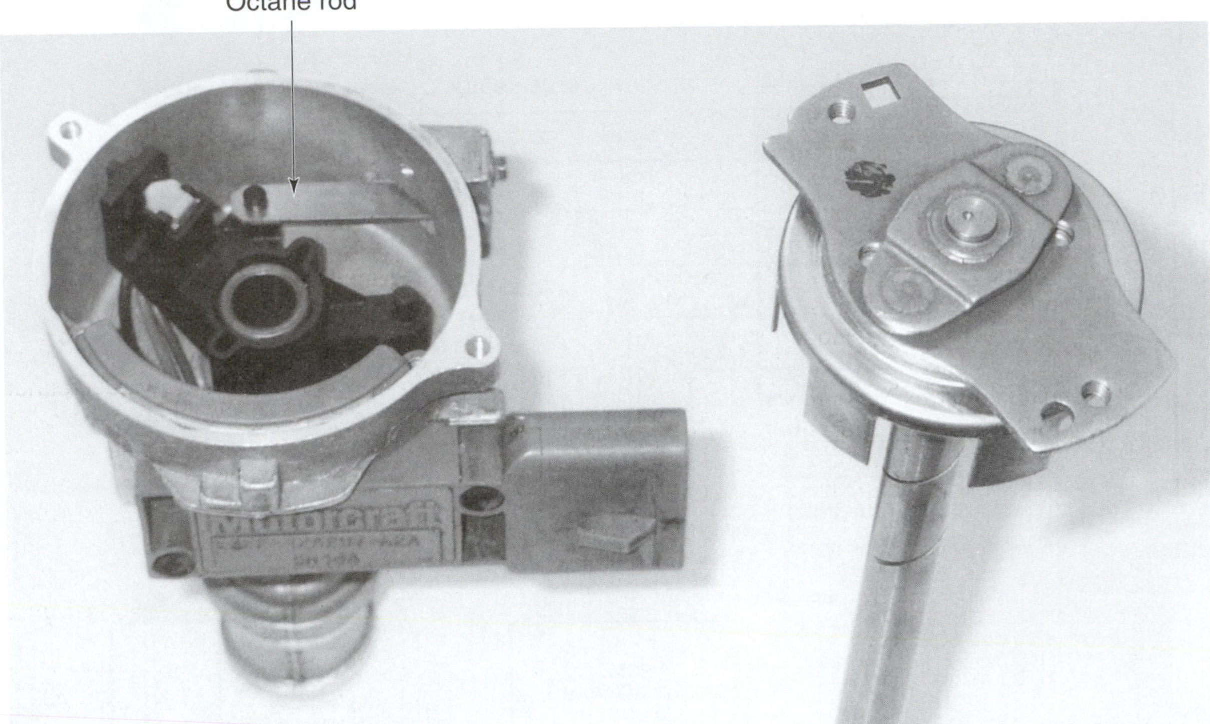

Figure 14–51 A TFI-IV distributor.

Diagnostic & Service Tip

When a TSB is issued to instruct the technician to replace the octane rod with a retard rod, some technicians have taken a short cut and simply retarded base timing by turning the distributor housing and ignored the instruction to replace the octane rod with the retard rod provided. During normal base timing readjustments, when the distributor housing is turned in order to retard base timing in the primary circuit, the rotor-to-distributor cap relationship is automatically changed to compensate for the change in spark timing. When the engineers issue a TSB to replace the octane rod as a method of changing base timing without rotating the distributor housing (that is, change the primary circuit element without changing the secondary circuit element to compensate), this has the effect of correcting the rotor-to-distributor cap relationship. Ultimately, a technician should evaluate underlying concepts very carefully before deciding to short cut any instructions provided by the manufacturer.

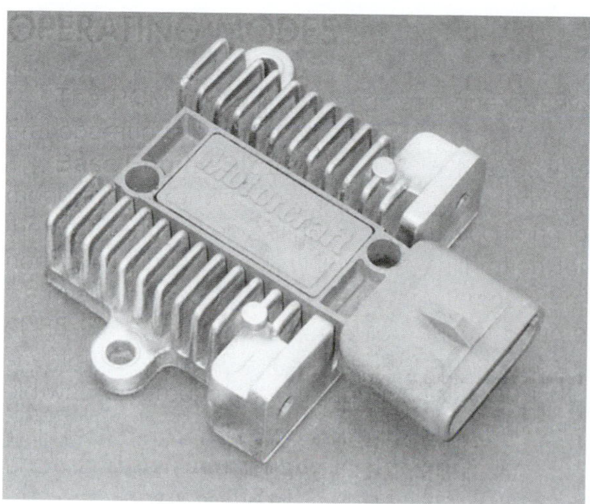

Figure 14–52 A TFI-IV ignition module mounted remotely in a heat sink, rather than on the distributor housing.

Thick Film Integrated (TFI-IV) Ignition System—Closed Bowl. In the 1988 model year, Ford engineers introduced another version of the TFI-IV system that relocated the TFI ignition module to a remote location in a large heat sink to better allow the module to dissipate heat, Figure 14–52. The TFI module's "hole" in the side of the distributor bowl was then closed up, thus the name *Closed Bowl TFI-IV* system.

Electrically, in the closed bowl version of the TFI-IV system, the Hall effect sensor creates two PIP signals: PIP-A is sent directly to the PCM and PIP-B is sent directly to the TFI module. The TFI module still uses the SPOUT signal from the PCM to fire the ignition coil as per the TFI-IV Universal Bowl system.

Distributorless Ignition System (DIS)

3.0-Liter SHO and Supercharged 3.8-Liter Engines. In these systems, the PCM controls both dwell and timing. When the starter turns the engine, PIP and CID signals are fed to the DIS module and to the PCM. Once the DIS module receives a CID signal, it turns on coil 2 by providing a ground for the coil's primary circuit, Figure 14–20, in response to the next SPOUT trailing edge that it sees, Figure 14–53. Coil 2, Figure 14–54, builds a magnetic field while the DIS module waits for the next SPOUT leading edge. When the next SPOUT leading edge occurs, the DIS module opens coil 2's primary circuit and fires spark plugs 3 and 4.

The CID signal transition the DIS module waited for before it turned on coil 2 could have been a leading edge or a trailing edge, because either transition indicates that cylinder number 1 is at 26 degrees ATDC on either its power stroke or its intake stroke. Following the CID transition, the DIS module selects coil 2 because coil 2 fires spark plugs 3 and 4 simultaneously, and cylinders 3 and 4 are the next cylinders at TDC, one on its compression stroke and the other on its exhaust stroke.

Having opened the primary circuit and fired coil 2, the DIS module now waits for the next trailing edge of the SPOUT signal. When that comes, the module grounds coil 3. When the leading edge of the SPOUT signal occurs next, the DIS opens the primary circuit of coil 3, firing its two plugs. Next it turns coil 1 on and off in response to the next SPOUT signal from the PCM. Obviously, the PCM controls when the cylinders fire by the time it raises the SPOUT signal voltage, and it controls ignition coil dwell by controlling how long the SPOUT signal voltage is low.

Once the DIS module has identified the number 1 cylinder, a counter circuit in the module keeps track of the coil firing sequence. The DIS module continues to monitor the CID signal to verify its cylinder counting circuit. The PCM continues to use the CID signal while the engine runs to control injector pulsing.

2.3-Liter, Dual Plug DIS. The DIS on the 2.3-liter engine works similarly to the 3.0-liter SHO and the supercharged 3.8-liter engines, but it has some unique features of its own,

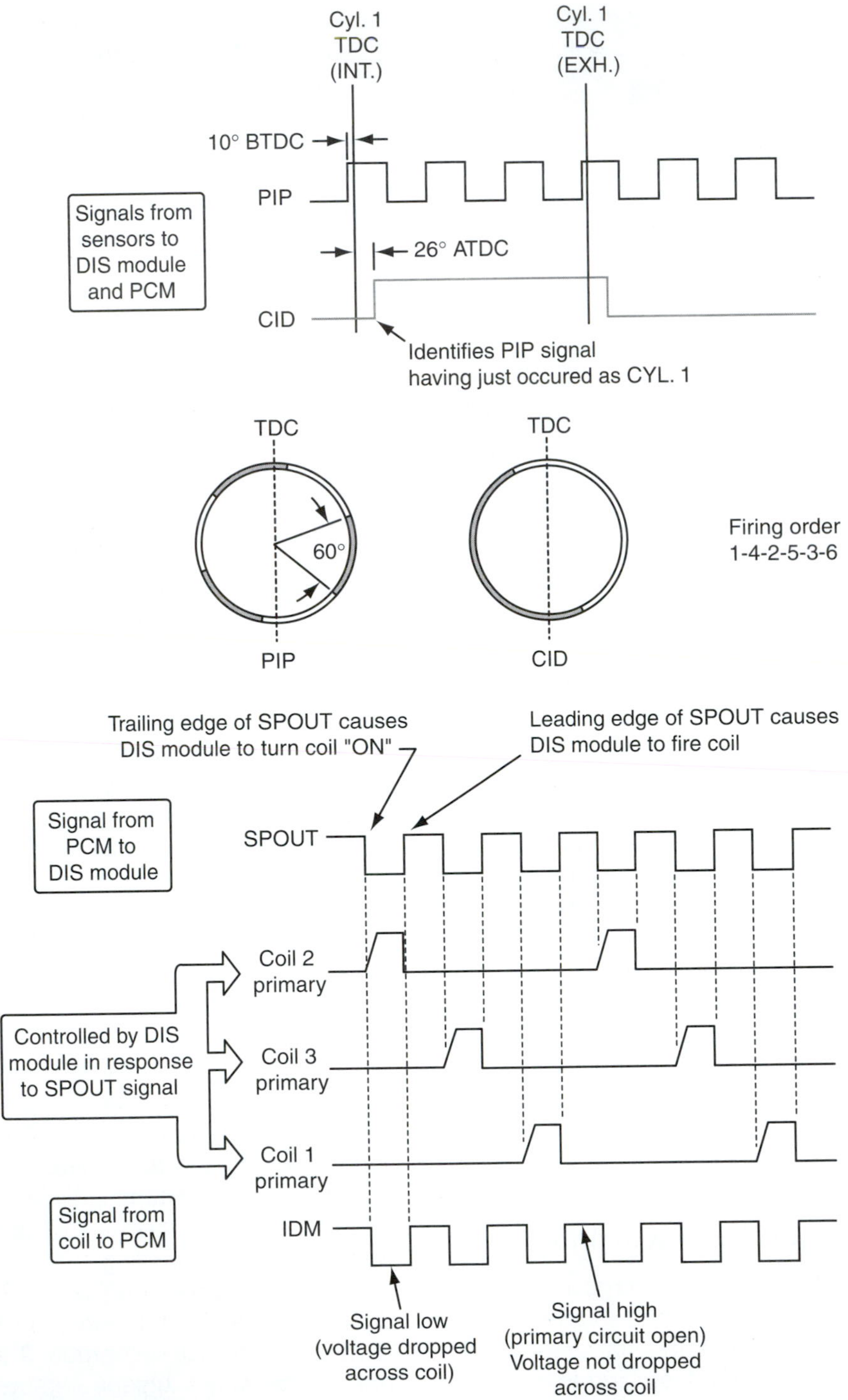

Figure 14–53 DIS coil firing 3.0-liter SHO and supercharged 3.8-liter engines.

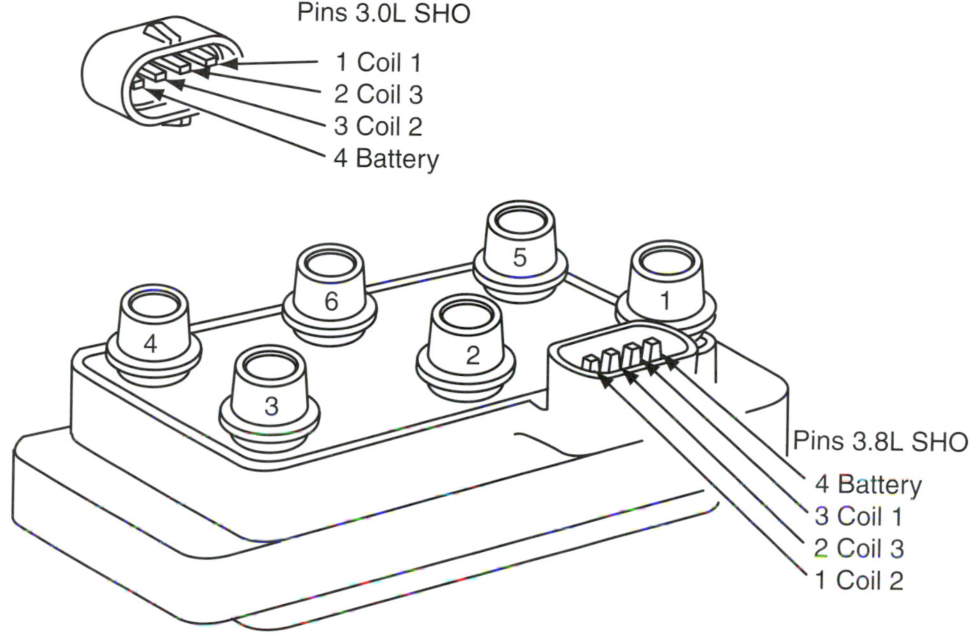

Figure 14–54 DIS coil pack, supercharged 3.8-liter and 3.0-liter SHO engines.

Figure 14–55. Because the 2.3-liter engine with DIS uses two spark plugs per cylinder (to achieve cleaner burning of the intake charge during combustion), the system has two separate coil packs: one on the right side of the engine and one on the left, Figure 14–56. The right coil pack contains coils 1 and 2. Coil 1 fires the right side spark plugs in cylinders 1 and 4, Figure 14–57. Coil 2 fires the right side spark plugs in cylinders 2 and 3. Coil 3 fires the left side spark plugs in cylinders 1 and 4, while coil 4 fires the left side spark plugs in cylinders 2 and 3.

After the engine starts, both plugs fire in each cylinder together: coils 1 and 3 fire together with coil 1 firing right plugs 1 and 4, and coil 3 firing left plugs 1 and 4. During startup cranking, only the right coil pack fires. To prevent the left coil pack from working, the PCM sends a 12-volt signal to pin 6 of the DIS module, Figure 14–57. This is called the Dual Plug Inhibit Signal. With 12 volts at pin 6, the DIS module does not ground coils 3 and 4. After the PCM can tell the engine is running (from the rpm signal), it removes the 12-volt DPI signal, and the DIS module fires both coil packs. There is no spark advance during cranking.

As shown in Figures 14–53 and 14–55, when the SPOUT signal goes low, the primary coil circuit turns on; there is a large volt drop across the primary coil winding, so the voltage at the coil negative terminal, where the IDM originates, is near zero. Therefore, the IDM signal is low. When the SPOUT signal goes high, the coil primary circuit is open, and there is no volt drop across the winding; voltage at the coil negative terminal is battery voltage, and the IDM signal is high. While the 2.3-liter, dual plug engine is cranked and the system is in dual plug interrupt mode, the IDM signal is inverted, Figure 14–55. By monitoring the IDM signal, the PCM can tell whether the DPI command is being responded to.

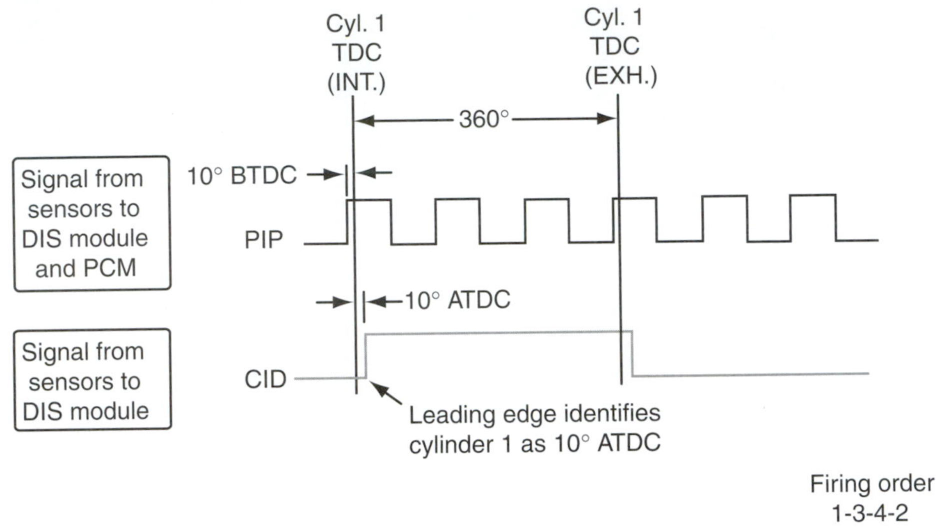

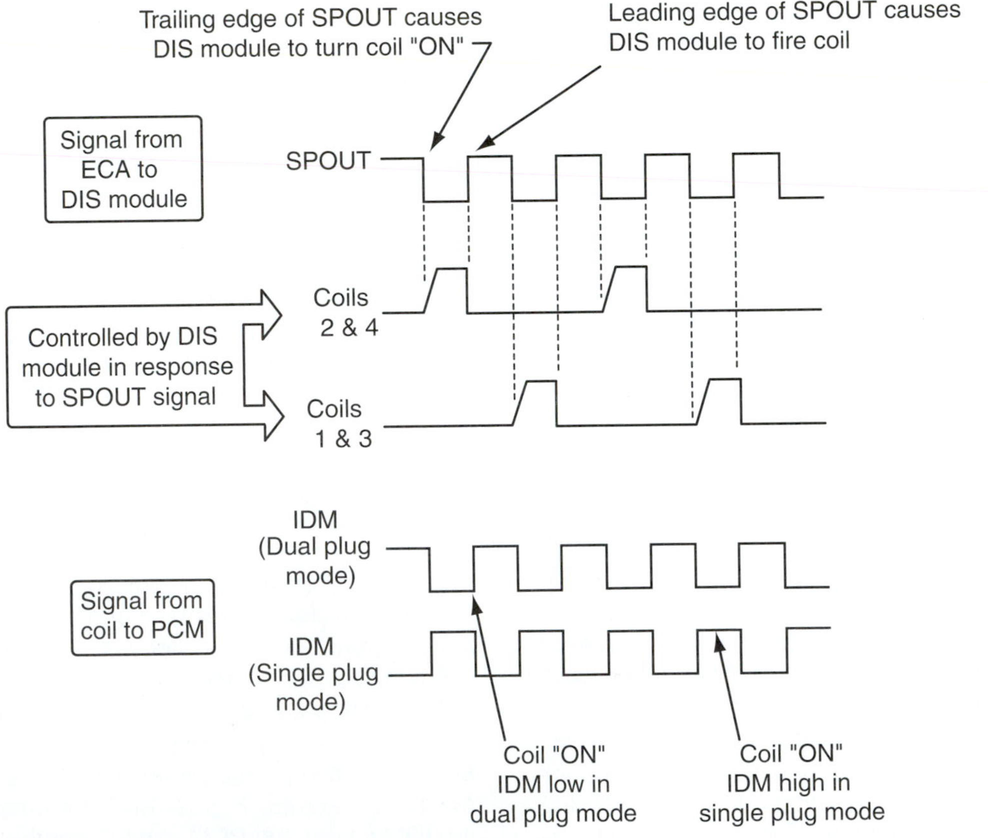

Figure 14–55 DIS coil firing, 2.3-liter, dual spark plug engine.

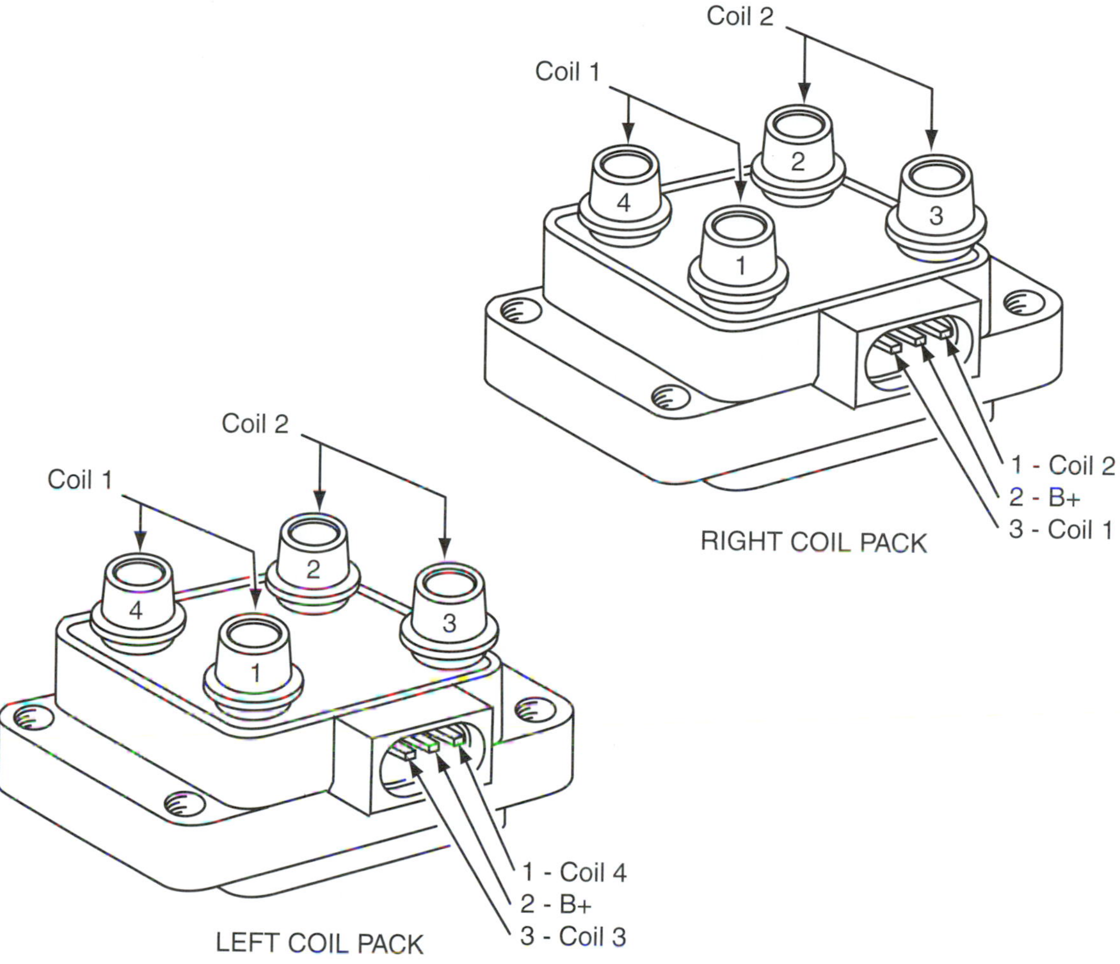

Figure 14–56 DIS coil packs, 2.3-liter dual plug engine.

Because the 2.3-liter, dual plug engine does not use sequential fuel injection, the PCM does not need to monitor the crankshaft position in the engine cycle. So on this system, the CID signal is not sent to the PCM.

On either system, if when the engine starts cranking, the CID signal does not arrive at the DIS module, the module randomly selects the first coil to fire in response to a SPOUT signal. Once the first coil is selected, the other coils follow in their normal sequence. If the random choice is wrong—if the computer guessed wrong when selecting a coil—the engine will not start. Each time the starter motor stops and restarts, the DIS module makes another random selection. When the random selection turns out to be the right coil, the engine will start. Since there are only two possibilities, it is unlikely that a great number of trials will be necessary on one of these systems.

If the SPOUT signal fails, the DIS module fires the coils in response to the PIP signal. Like the TFI-IV module, the DIS module has a current limiter circuit. Primary current is limited to 5.5 amps +/– 0.5 amp.

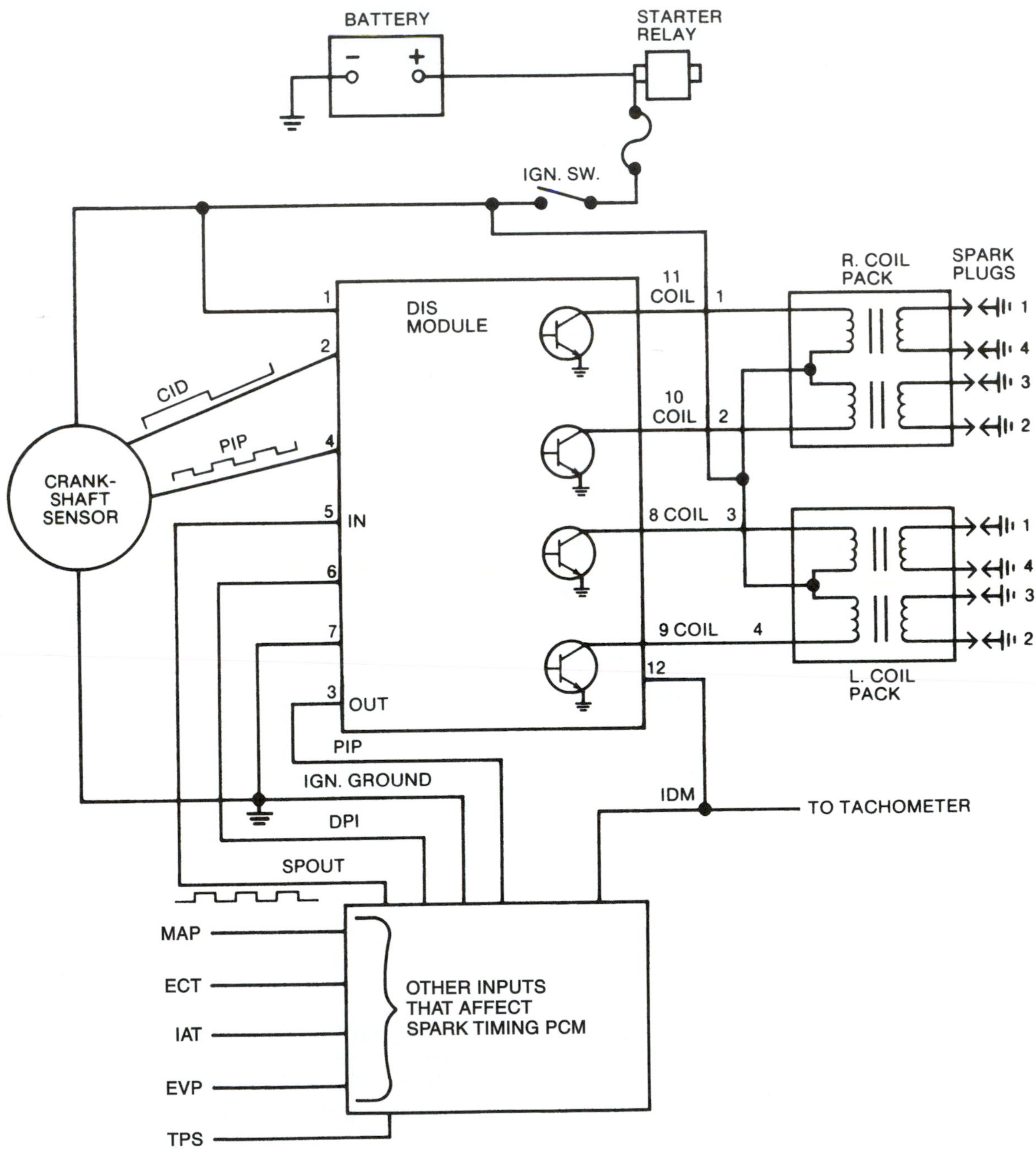

Figure 14–57 2.3-liter, dual plug DIS schematic.

Electronic Distributorless Ignition System (EDIS)

In 1991, Ford introduced a second-generation distributorless system known as Electronic Distributorless Ignition System or EDIS on the 1.8-liter and 4.6-liter passenger car engines and also on the 4.0-liter light truck engine. This system is still a waste spark ignition system, similar to Ford's DIS system, but the crankshaft and camshaft sensors are variable reluctance sensors instead of Hall effect. These sensors therefore produce an AC voltage pulse for each tooth that they monitor.

The crankshaft sensor monitors 35 teeth on the harmonic balancer, one tooth spaced every 10 degrees of crankshaft rotation for a total of 36 teeth, minus one tooth to create a 20-degree space or blank spot. The blank spot informs the PCM of the position of the crankshaft, and then the PCM is able to count off the remainder of the crankshaft rotation in 10-degree increments. This allows the PCM to calculate when each pair of pistons is approaching TDC. This information is needed to allow the EDIS ignition module to fire the correct ignition coil. The signal from the crankshaft sensor was originally called the Variable Reluctance Sensor (VRS) signal. Under OBD II standards, it is now referred to as the Crankshaft Position (CKP) signal.

The camshaft signal is used on the two sequentially fuel injected engines, the 1.8-liter and the 4.6-liter passenger car engines. This signal allows the PCM to determine which stroke a given cylinder is on for purposes of pulsing the correct fuel injector. This sensor monitors one tooth per camshaft rotation and produces the Camshaft Position (CMP) signal.

The command signal from the PCM to the EDIS ignition module (which carries instructions as to when to fire each coil) is called *Spark Angle Word* or SAW. This signal equates to the SPOUT signal in the TFI-IV and DIS ignition systems.

Base Timing Check

Both the DIS and EDIS ignition systems have fixed base timing, which is not adjustable. However, because a technician is expected to *check* base timing before performing the computer timing check (described in the SYSTEM DIAGNOSIS AND SERVICE section later in this chapter), Ford still provided a means to take the system to base timing. In the DIS ignition system, the technician simply pulls the shorting bar in the SPOUT circuit. Similarly, in the EDIS ignition system, the technician simply pulls the shorting bar in the SAW circuit. This shorting bar, when removed from its connector, simply opens the electrical circuit that it normally completes. When the shorting bar is removed, the engine runs at base timing. This shorting bar should always be reinstalled after testing is complete. For more information on the SPOUT/SAW connector, see the SYSTEM DIAGNOSIS AND SERVICE later in this chapter.

Octane Adjustment

Because both the DIS and EDIS ignition systems have nonadjustable base timing, both systems also employ a circuit known as the *Octane Adjustment* circuit. This circuit contains a shorting bar that looks like the shorting bar in the SPOUT/SAW circuit. When the shorting bar is removed, the PCM calculates computed timing at 3 degrees to 4 degrees less than the original spark advance program calls for under a given engine operating condition. This allows the technician to correct for slight detonation problems. If the engine still detonates, then the customer should be advised to use a fuel that is rated at a higher octane. When detonation occurs, this shorting bar may be left out of its circuit and put in the glove box of the vehicle. In order to differentiate between the shorting bar in the octane adjustment circuit and the shorting bar in the SPOUT/SAW circuit, wire colors should be checked and an electrical schematic should be used.

Knock Sensor Response. On systems using a knock sensor, the strength of the knock sensor's signal is proportional to the severity of the spark knock. When a knock signal is received by the PCM, it retards timing one-half degree per engine revolution until the knock disappears.

Idle Speed Control

Three different systems are used for idle speed control among various EEC IV applications:

Throttle Kicker. Throttle kicker is a vacuum-actuated device containing a diaphragm, a spring and a plunger. It is mounted on the carburetor or CFI unit. On applications where the throttle kicker is used, normal idle speed is controlled by a conventional idle stop screw. However, when vacuum is applied to the diaphragm, the plunger extends and opens the throttle blade slightly to increase idle speed. It is sometimes referred to as a Vacuum-Operated Throttle Modulator (VOTM).

The PCM controls a solenoid that in turn controls vacuum to the throttle kicker. Part of the Throttle Kicker solenoid (TK solenoid) assembly is a normally closed vacuum valve. When the solenoid energizes, the vacuum valve opens and allows vacuum to the throttle kicker. The throttle kicker is activated during the following operating conditions:

- when engine temperature is below a specific temperature
- when engine temperature is above a specific temperature
- when the A/C compressor clutch is engaged
- when the vehicle is above a specific altitude

DC Motor Idle Speed Control (ISC). The ISC motor is a small, reversible electric motor. It is part of the assembly including the motor, a gear drive and a plunger, Figure 14–58. When the motor turns in one direction, the gear drive extends the plunger; when the motor turns in the opposite direction, the gear drive retracts the plunger. The ISC motor mounts so the plunger can contact the throttle lever. The PCM controls the ISC motor and changes the polarity applied to the motor's armature to control the direction it turns, Figure 14–59. When the idle tracking switch opens (indicating throttle closed), the PCM commands the ISC motor to control idle speed. The ISC provides the correct throttle opening for cold or warm engine idle.

When the throttle opens and closes the ITS, the PCM commands the ISC motor to fully extend the plunger to function as a dashpot. In other words, when the throttle closes again during deceleration, the plunger prevents the throttle from closing completely. The pressure on the plunger opens the ITS, and the motor in turn retracts the plunger to the normal idle position. This delayed throttle closing is to allow additional air into the manifold and completely evaporate fuel condensed

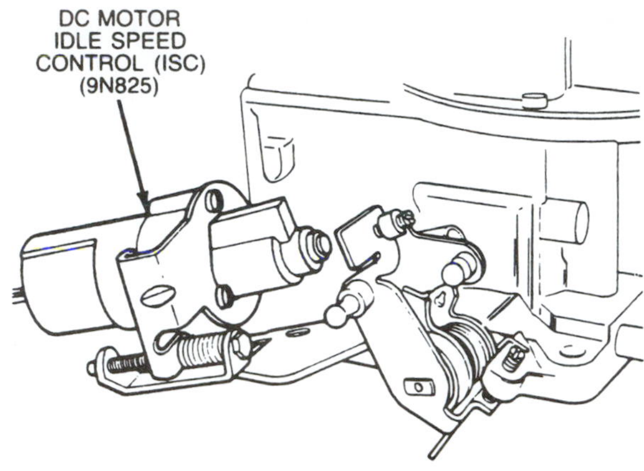

Figure 14–58 ISC motor assembly. *(Courtesy of Ford Motor Company.)*

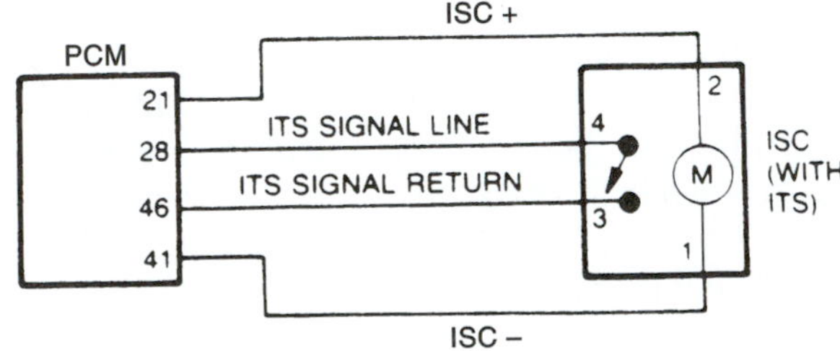

Figure 14–59 ISC motor circuit. *(Courtesy of Ford Motor Company.)*

on the manifold walls from the sudden high vacuum and temperature drop. The effect is to prevent a sudden very rich mixture on deceleration.

With the ignition off, the PCM commands the ISC motor to retract the plunger fully, closing the throttle completely and preventing dieseling. Once the engine has completely stopped, the PCM commands the plunger to the fully extended position in preparation for the next engine start.

Idle Air Bypass Valve Solenoid. This idle speed control device is used on multipoint fuel injection systems. It includes an PCM-controlled solenoid operating a pintle valve, Figure 14–60. This bypass valve is so attached that when it is open, air passes from in front of the throttle butterfly blade through the pintle valve to the intake manifold side of the throttle. The PCM controls this valve by duty-cycling the solenoid. As the solenoid pulses, the valve opens. The PCM cycles the solenoid at whatever duty cycle the idle speed requires.

During a cold start, the PCM holds the idle air bypass solenoid valve at a 100 percent duty cycle (valve held wide open). During cranking, hot or cold, the bypass valve provides sufficient air so it is not necessary for the driver to touch the throttle. It also provides a throttle dashpot function as described previously.

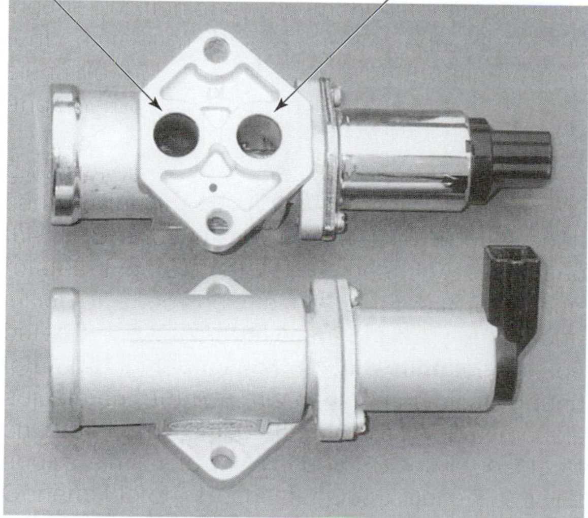

Figure 14–60 Throttle air bypass solenoid valves.

Thermactor Air Management

The Thermactor system is the injection system to deliver air to the exhaust manifold or the catalytic converter to aid in the reduction of HC and CO. During operations requiring a rich air/fuel mixture such as cold startup or WOT, Thermactor air is dumped back into the atmosphere to avoid overheating the catalytic converter. One such condition for EEC IV and most other Ford electronic engine control systems is during engine idle. For this period, Ford tends to program the system for a rich idle mixture to maintain a good idle quality and prevent stalls.

EEC IV systems use one of two Thermactor systems. One is PCM-controlled and is discussed here. The other is a pulse injection system using the negative pressure between exhaust pulses in the exhaust manifold (most effective and most common on four-cylinder engines) to draw air from the air cleaner past check valves into the exhaust manifold. This system is sometimes called *Thermactor II* and is not PCM controlled.

Thermactor Air Bypass (TAB) and Thermactor Air Divert (TAD) Valves. The air pump runs constantly, driven by a belt. This pump supplies air to the bypass valve, which either directs it to the divert valve or dumps it to the atmosphere, Figure 14–61. The divert valve can either direct the Thermactor air upstream to the exhaust manifold or downstream to the catalytic converter. The valves are activated by vacuum diaphragms controlled by PCM-controlled solenoids.

TAB and TAD Solenoids. The TAB and TAD solenoids control whether vacuum is applied to the TAB and TAD valves. If the engine starts with a coolant temperature below 10°C (50°F), the PCM will leave the TAB solenoid circuit ungrounded. Vacuum to the bypass diaphragm is blocked, and the diaphragm spring holds the valve in the down position, bypassing pump air to the atmosphere. Once engine temperature reaches 10°C, the PCM grounds the TAB solenoid circuit,

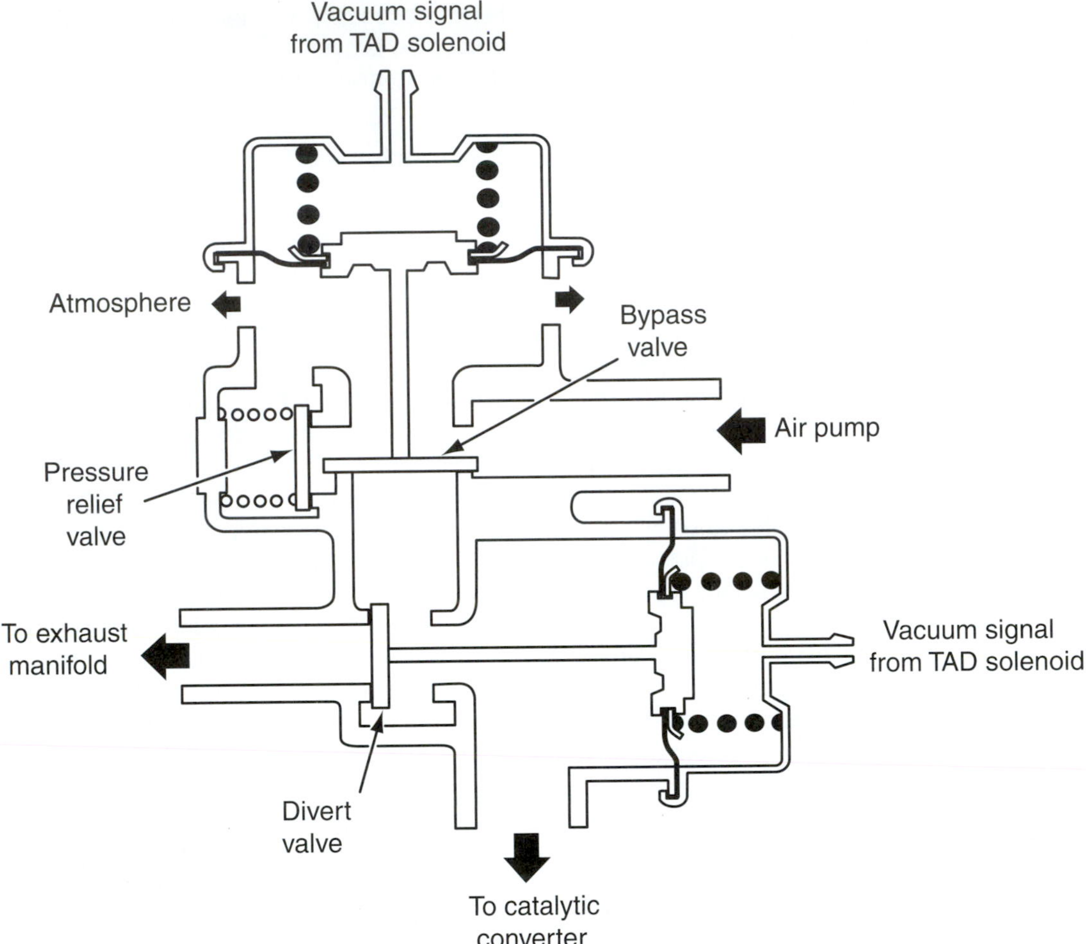

Figure 14–61 Combination TAB/TAD valves.

Figure 14–62. Now vacuum is routed to the diaphragm, lifting the bypass valve. Injection air is now directed to the divert valve. At the same time, the PCM grounds the TAD solenoid, and it in turn directs vacuum to the divert valve diaphragm. The divert valve pulls back to close off the passage to the catalytic converter and direct the air to the exhaust manifold. Once the coolant reaches approximately 88°C (190°F), the system goes into closed loop, the PCM deactivates the TAD solenoid and air is directed to the catalytic converter, behind the reducing bed and upstream of the oxidizing bed.

Certain operating conditions put the system back into bypass mode to protect the catalytic converter from overheating. These are:

- idle
- failure of the oxygen sensor signal to switch between rich and lean frequently enough (indicating that air/fuel mixture is not properly controlled or that the oxygen sensor is cooling, or has become unresponsive with age and exhaust deposits)
- acceleration or WOT operating conditions

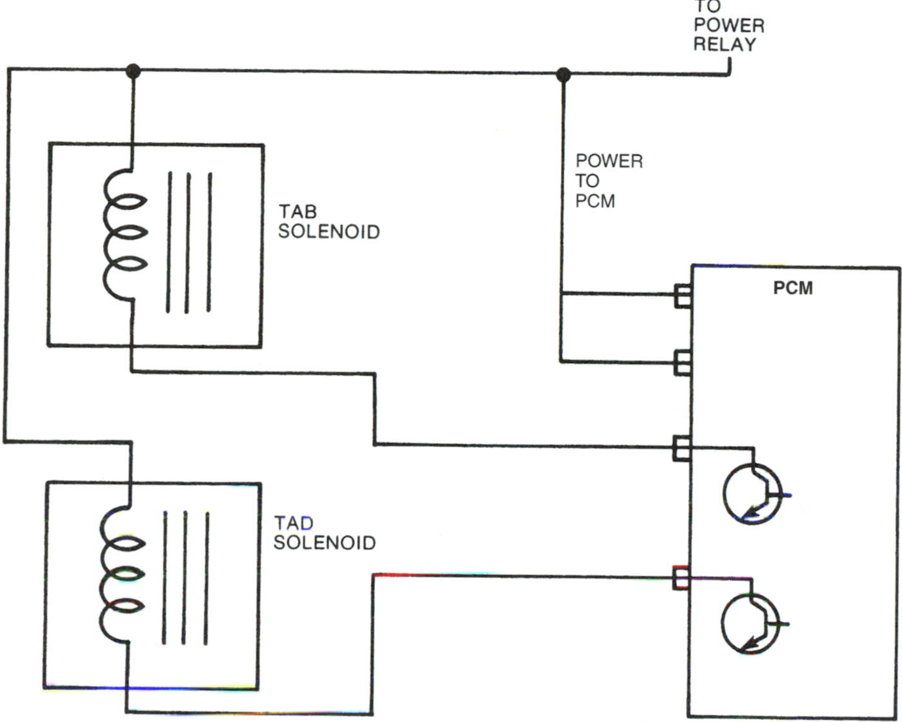

Figure 14–62 TAB/TAD solenoid control circuit.

Diagnostic & Service Tip: Closed Loop Identification.

Often a technician wants to know whether a computer-controlled system has gone into closed loop. While it is possible to check the oxygen sensor for its characteristic fluctuating low voltage signal, on most vehicles including EEC IV Fords, the fastest way to check for closed loop is to see where the air injection system is routing the air. If the air is bypassed to the atmosphere or routed to the exhaust manifold, the system is in open loop. Only if the air is injected into the catalytic converter is the system in closed loop. This quick trick can save considerable time, but keep in mind that if there is anything wrong with the TAD solenoid circuit, or its slave diaphragm and vacuum circuit, the test may be inaccurate.

Canister Purge

Three types of canister systems are used on EEC IV vehicles. Some fuel-injected engines (both CFI and multipoint) use a constant purge system with no PCM control. Some engines use an in-line, PCM-controlled, solenoid-operated vacuum valve. The third uses an PCM-controlled solenoid controlling vacuum to a ported vacuum switch (a temperature-controlled vacuum valve), which in turn controls the canister purge valve and an exhaust heat control valve.

In-line Canister Purge (CANP) Solenoid. This is a simple, normally closed, solenoid-controlled vacuum valve in the purge line connecting the charcoal canister to the intake manifold, Figure 14–63. Once the engine reaches normal operating temperature, the PCM duty cycles the solenoid. The duty cycle depends on the vehicle's operating conditions. This allows the canister purge to work at a controlled rate.

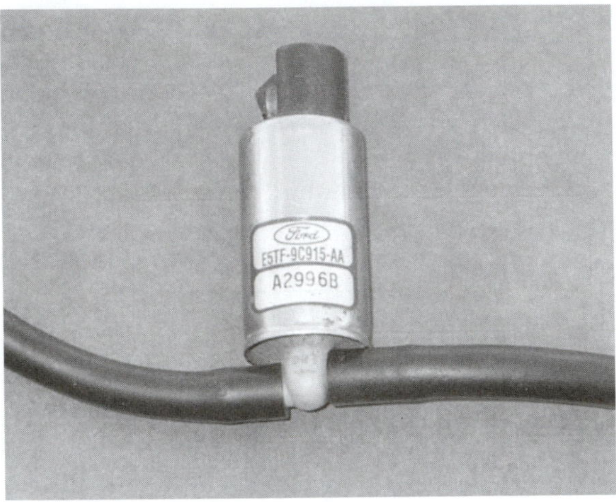

Figure 14–63 Canister purge solenoid.

Canister Purge/Heat Control System. This system (very limited application) gets two-for-one from an PCM function. The PCM controls a normally closed, solenoid-operated vacuum valve. When the solenoid activates, vacuum is routed to the center port of a thermal vacuum switch, Figure 14–64. If the engine is cold, the thermal vacuum switch directs the vacuum to the top port, which connects to the heat control valve actuator. The actuator closes the heat control valve and forces exhaust gases through the crossover passage and warms the intake manifold plenum.

Once the engine warms, the wax pellet in the lower portion of the thermal vacuum switch expands and raises the valve up. It closes the top port and opens the lower port to the vacuum signal. The heat control valve opens; then the purge control valve opens and allows purging of the charcoal canister.

EGR Control

The PCM controls four slightly different EGR systems, depending on engine application and vehicle.

EGR Control (EGRC) and EGR Vent (EGRV) Solenoids. This system is a carryover from EEC III and is designed to control the amount of EGR flow (the EGR valve can be open, closed or anywhere in between). A position sensor (EVP) above the EGR valve tells the PCM where the EGR valve is. Two solenoids control vacuum to the EGR valve, Figure 14–65. The normally closed control solenoid allows manifold vacuum to the EGR valve when energized. The normally open vent solenoid allows atmospheric pressure into the vacuum line when not energized.

To *apply* additional vacuum to the EGR valve, the PCM energizes both solenoids, thereby increasing the level of vacuum being used to pull the EGR valve open and causing it to open further than it currently is. With both solenoids energized, the EGRC solenoid opens and allows vacuum to be applied to the EGR valve and the EGRV closes and therefore allows the system to hold vacuum. If this mode of operation is held long enough, the EGR valve will fully open.

To *maintain* the current vacuum level at the EGR valve and thereby maintain the current position of the EGR valve, the PCM energizes only the EGRV solenoid. With the EGRC solenoid de energized, additional vacuum is not applied to the EGR valve, and with the EGRV solenoid energized, EGR vacuum is not vented to atmosphere, but instead is held within the EGR valve. That is, both solenoid-operated valves are closed and the current level of vacuum is trapped at the EGR valve. The PCM may choose to hold the system in this mode for extended periods of time as engine performance conditions permit.

To *release* some of the currently trapped vacuum from the EGR valve and thereby allow the EGR valve to close up further than it currently is, the PCM de-energizes both solenoids. The EGRC solenoid does not allow vacuum to be applied to the EGR valve, but the EGRV solenoid does allow vacuum to be released. If this mode of operation is held long enough, the EGR valve will fully close. The *release* mode is also the default mode if the system goes into its failure mode known as Failure Mode Effects Management (or FMEM) as described later in this chapter, or if the system becomes otherwise incapable of controlling the EGR solenoids.

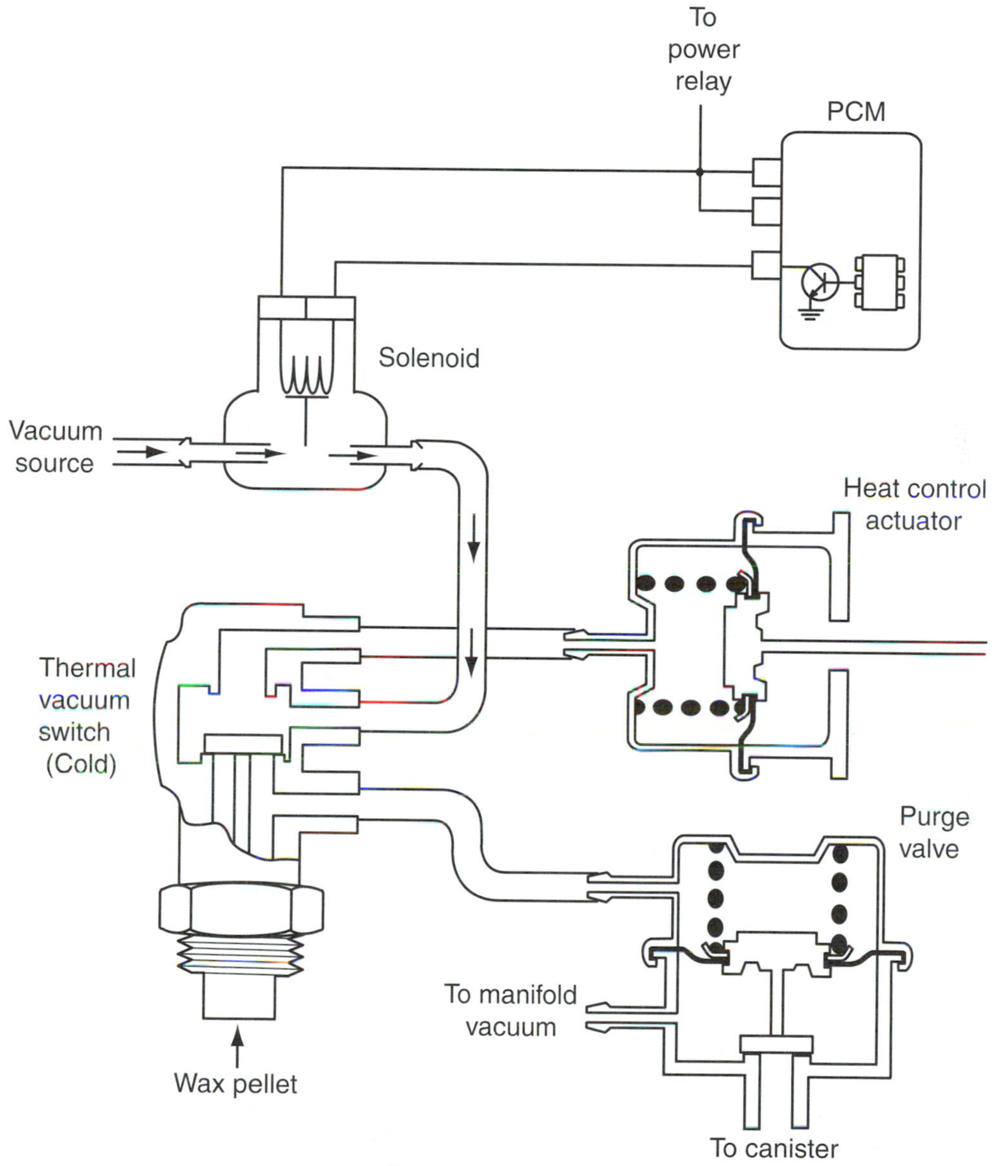

Figure 14–64 Canister purge/heat control system.

It is important to understand that the PCM does not operate these solenoids on either a duty cycle or on a pulse width. Rather, the PCM modulates the solenoids on an *as needed* basis, similar to most vacuum-controlled cruise control systems. Ford engineers refer to this method of control as "dithering the solenoids." Both solenoids cycle in order to achieve and maintain the desired EGR valve position. Throughout all of this, the EVP sensor provides a feedback signal to the PCM as to the current position of the EGR valve. Therefore, this is sometimes referred to as

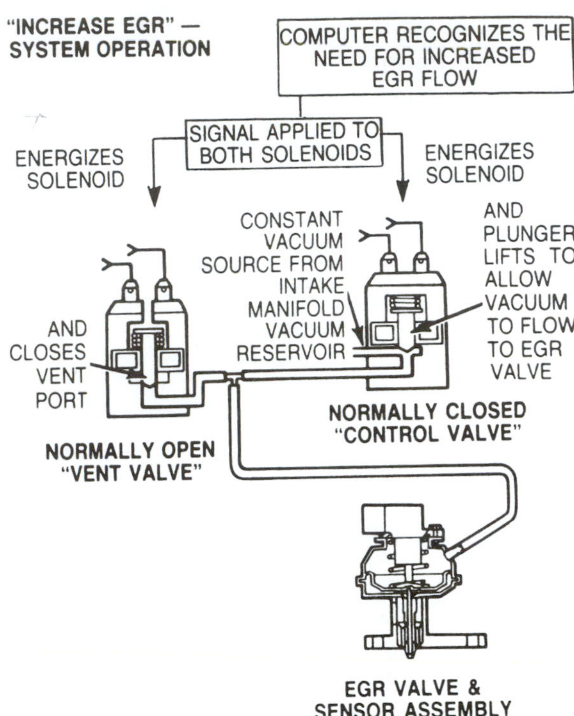

Figure 14–65 EGR control/EGR vent solenoids. *(Courtesy of Ford Motor Company.)*

a closed-loop EGR system. This system allows the PCM more accurate control of the air/fuel mixture than a system that only assumes the EGR valve's position because of the role that these exhaust gasses play in affecting how much air gets into the combustion chamber. Additionally, this system allows the PCM to set fault codes during diagnostics if it has been tampered with, or if, for some other reason, is not working properly.

EGR Shutoff Solenoid. This system uses one vacuum control solenoid operating like the vacuum control just described except it uses ported vacuum instead. When the solenoid is energized, vacuum moves the EGR valve; when the solenoid is not energized, vacuum is blocked. With this system the EGR valve may or may not have an EVP sensor. Those with the sensor work like the EGRV and EGRC solenoids combined into one, while those without the sensor are not modulated. They are either open or closed.

EGR Shutoff Solenoid with Backpressure Transducer. This system uses one normally closed vacuum control solenoid like the system described above. The strength of the ported vacuum signal to the EGR valve is controlled, however by an exhaust backpressure variable transducer. The transducer is tied into the EGR valve's vacuum supply line and controls an air bleed into the line, Figure 14–66. Pressure signals are connected to the transducer from the exhaust system. The backpressure signal comes from a tap in the tube routing exhaust gas to the EGR valve. The control pressure signal comes from a point just downstream from the metering orifice allowing exhaust into the EGR supply pipe. These two pressures act on a valve in the transducer to position it to control the air bleed.

When exhaust pressure is low and vacuum is high, the air bleed opens and causes the vacuum signal to the EGR valve to weaken. As exhaust pressure increases and manifold vacuum decreases, the air bleed becomes more closed and strengthens the vacuum signal to the EGR valve. This way, the transducer controls how far the EGR valve opens and the consequent EGR flow. This system does not employ an EVP sensor.

EGR Vacuum Regulator (EVR)

This EGR control device, introduced in 1985, is simpler than the previous EGRC/EGRV control system. It works in a similar way, however. Like the EGRC/EGRV solenoid tandem, it allows the PCM to open the EGR valve to any degree through modulation of the signal. Since 1985, the EVR has become the standard EGR device used on most Ford engines.

The EVR is a single unit with two vacuum ports, Figure 14–67. The lower port connects to either manifold or ported vacuum depending on which engine is in the vehicle. The upper port connects to the EGR valve. The vented cover allows air to flow through the filter and down the hollow, electromagnetic core. This allows atmospheric pressure to be present above the metal disk at the bottom of the core at all times. The disk

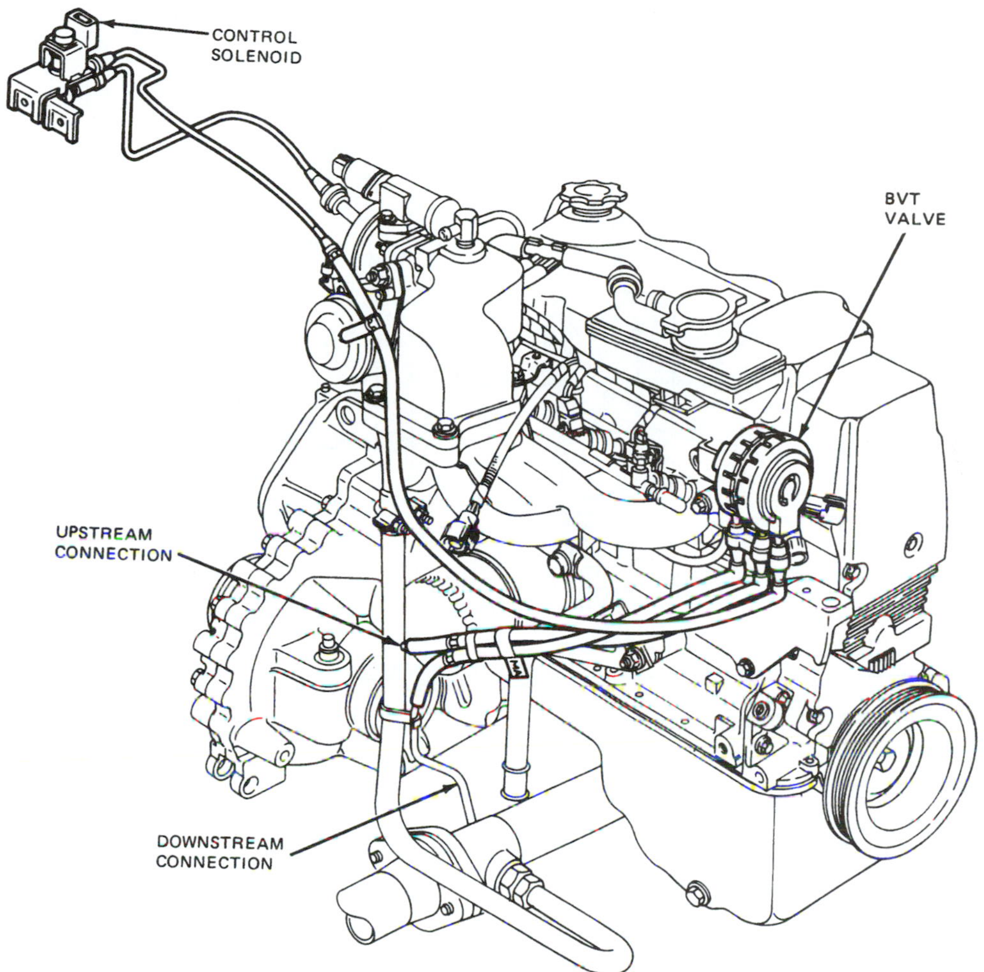

Figure 14–66 Backpressure variable transducer. *(Courtesy of Ford Motor Company.)*

is held against the bottom of the core by a weak spring. When the engine runs and the EVR is not energized, engine vacuum creates a low pressure area below the disk. With this pressure differential, the spring cannot hold the disk against the bottom of the core tightly enough to keep atmospheric pressure from filling the vacuum chamber. With atmospheric pressure in the vacuum chamber, of course, the EGR valve closes.

When the PCM wants to open the EGR valve, it grounds the EVR. The magnetic field produced in the coil pulls the metal disk tightly against the bottom of the core, sealing off the chamber below from atmospheric pressure. Engine vacuum now applies the EGR valve. By rapidly cycling the EVR on and off, the PCM can control the amount of vacuum applied to the EGR and thus how far the EGR valve opens and the amount of gas recirculated.

The EVR is often referred to as a solenoid. But because its electromagnetic core does not move, it is really not a solenoid but rather a simple electromagnet.

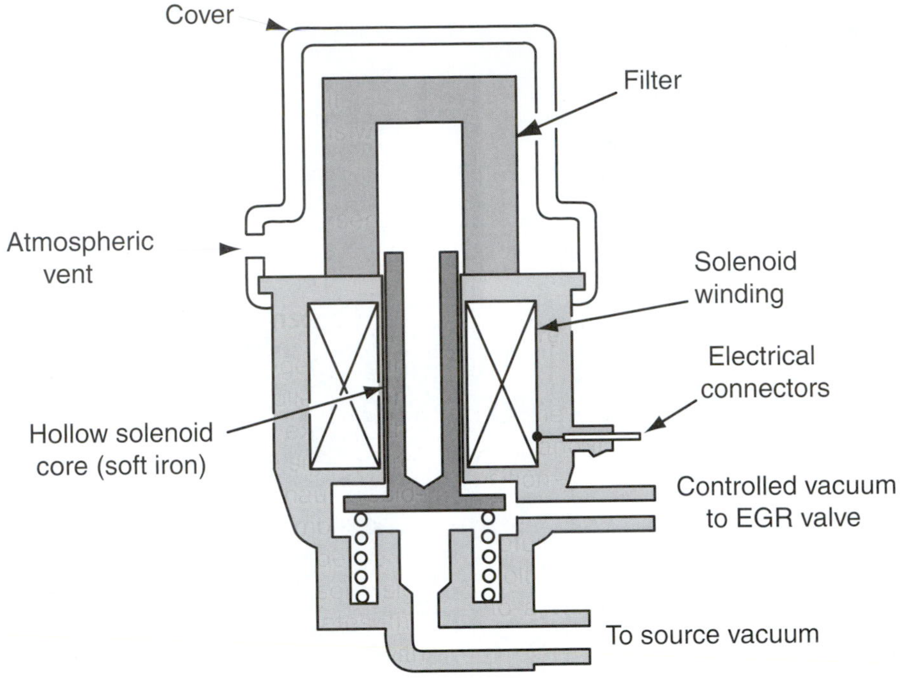

Figure 14–67 EGR vacuum regulator (EVR) solenoid.

Torque Converter Clutch Control

Some Ford automatic transmissions such as the A4LD, AXOD, 4EAT and E40D employ a PCM-controlled torque converter lockup clutch. In each case, the PCM controls the converter clutch with a solenoid in the transmission, Figures 14–67 and 14–68. This solenoid moves a valve in the transmission valve body, a valve routing transmission fluid under pressure into the torque converter and applying the lockup clutch.

The hydraulic circuitry applying the torque converter lockup clutch varies slightly between the different transmissions, but they work similarly. Figure 14–69 shows the hydraulic clutch apply circuit for the E40D transmission in the clutch-released mode. Line pressure to the converter clutch regulator valve feeds oil under pressure to the converter clutch control valve. With the converter clutch control valve in the right position as shown, line pressure oil is fed into the

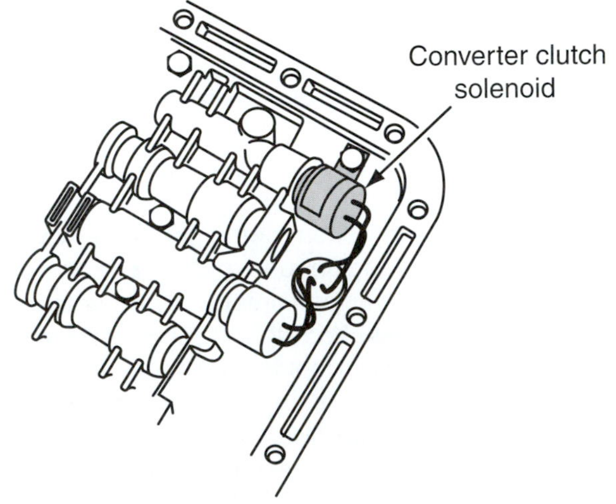

Figure 14–68 Torque converter lockup clutch solenoid. *(Courtesy of Ford Motor Company.)*

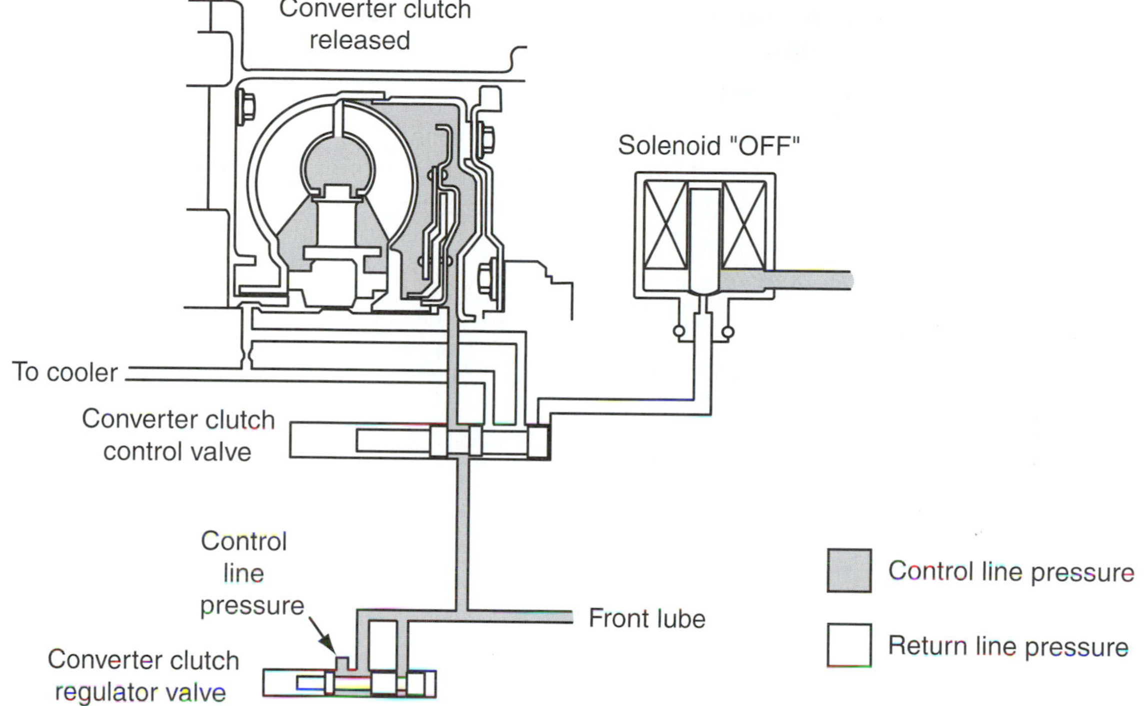

Figure 14–69 Torque converter lockup clutch released.

converter through the release passage between the transmission input shaft and the stator support. This introduces the oil into the converter on the engine side of the converter clutch plate. The pressure forces the clutch plate away from the engine side of the converter housing. The oil then flows over the clutch plate at its circumference and exits the converter through the apply passage between the stator support, and then flows to the transmission oil cooler.

When the PCM energizes the converter clutch solenoid, Figure 14–70, line pressure is applied to the right end of the converter clutch control valve. It then moves left. This reroutes the oil under pressure coming from the converter clutch regulator valve. Line pressure oil now feeds into the converter through the apply passage. With the transmission oil introduced into the converter on the transmission side of the clutch plate, the pressure must flow over the circumference of the clutch plate to get to the exhaust passage and exit the converter. This moves the clutch plate toward the engine side of the converter housing. As the friction material at the outer circumference of the clutch plate contacts the machined surface of the converter housing, it seals off the flow of oil to the exhaust passage. With no pressure on the engine side of the clutch plate and line pressure on the transmission side, the converter clutch plate is forced against the converter housing with enough force to lock it up and drive the transmission input shaft at engine speed.

There are slight variations in the torque converter clutch control strategy depending on the engine and transmission, but in general the following conditions must exist for the PCM to apply the torque converter clutch:

- engine warmed up to normal operating temperature

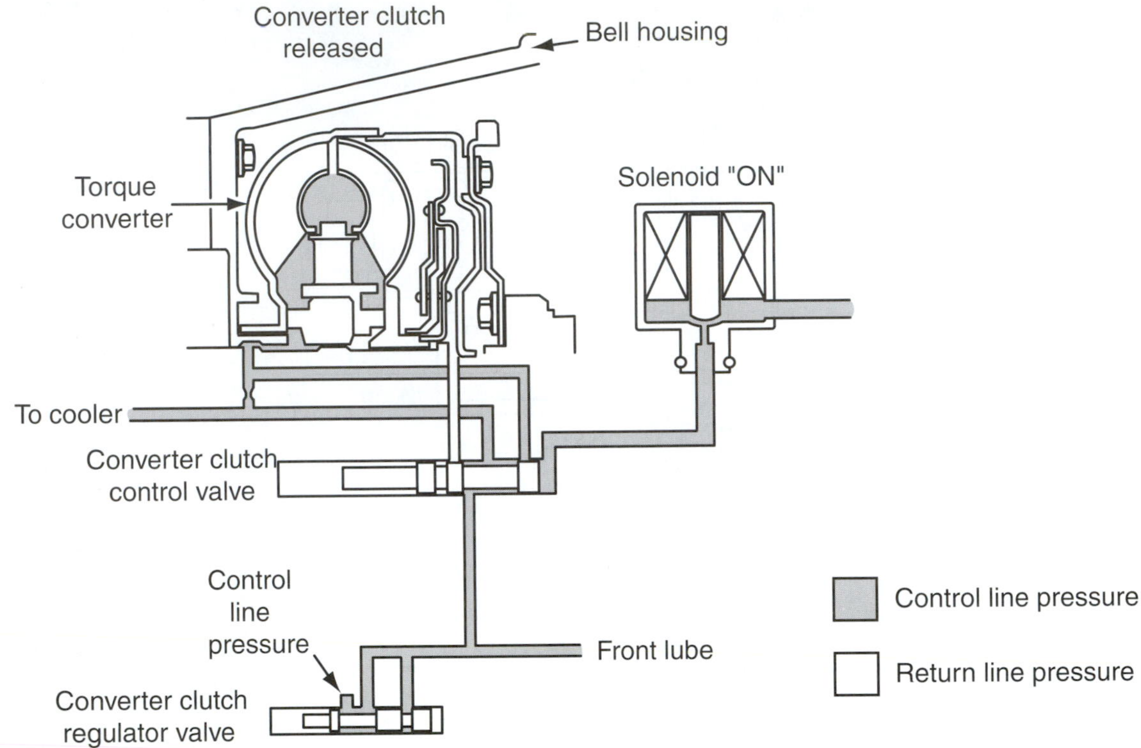

Figure 14–70 Torque converter lockup clutch applied.

- engine not under heavy load
- part throttle cruise
- engine within a predetermined rpm range
- vehicle above a predetermined speed

The solenoid will deactivate when:

- the BOO switch indicates the brakes are applied.
- the throttle position sensor indicates WOT.
- there is severe deceleration.

Turbocharger Boost Control Solenoid

Turbocharged 2.3-liter engines with EEC IV control employ an PCM-controlled boost control solenoid that allows boost pressure to rise from the normal 10 psi to as much as 15 psi. Boost pressure is controlled by a wastegate that opens at a specified pressure and allows some exhaust to bypass the exhaust turbine wheel, thus controlling the amount of boost pressure. The wastegate is controlled by a spring and diaphragm assembly (an actuator). Intake manifold pressure is routed to the actuator diaphragm. At the predetermined pressure, the diaphragm compresses the spring and moves a rod opening the wastegate.

If the PCM is satisfied with the current driving conditions, it activates the normally closed solenoid and cycles it at 40 hertz. When the solenoid activates, it bleeds off pressure from the wastegate actuator and allows a higher intake manifold pressure before the wastegate opens.

A/C and Cooling Fan Control

Some engine applications use a separate electronic control module to control the A/C compressor clutch and the engine cooling fan (or fans). Others, typically rear-wheel-drive vehicles with traditionally "north-south" mounted engines and belt-driven cooling fans, use a relay to disable the A/C clutch during WOT operation.

A/C and Cooling Fan Electronic Module. The module, Figure 14–71, is usually located in the passenger compartment. It gets input information from:

- the PCM (WOT signal).
- a coolant temperature switch (which closes at 105°C [221°F].
- the brake switch (brakes applied).
- the A/C switch (A/C on).

The module has a ground terminal and gets power from the battery through the ignition switch.

When the A/C switch is on (assuming the ignition is also on), the module provides power to the compressor clutch. During WOT operation, the PCM signals the module, which then disables the A/C clutch by deactivating it. If the WOT condition continues, the clutch stays disabled for about 30 seconds before normal A/C operation resumes. If the coolant temperature is below

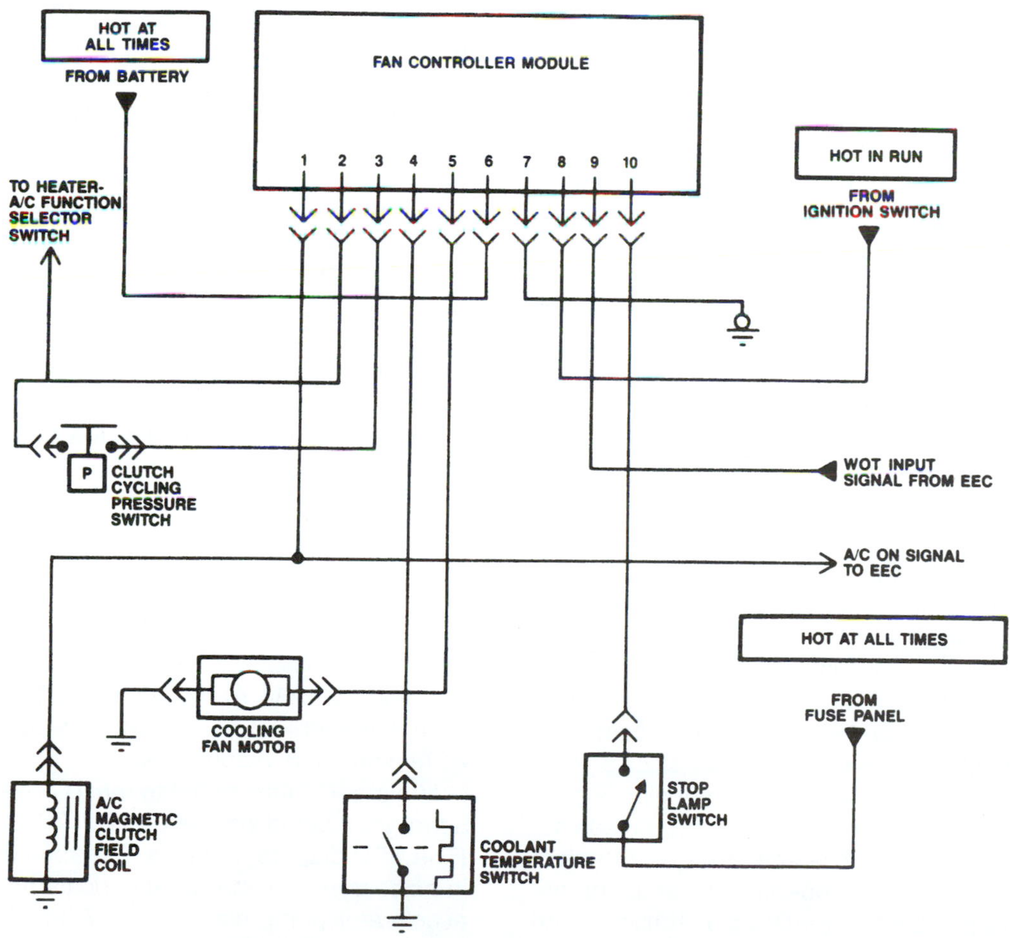

Figure 14–71 A/C and radiator fan control module. *(Courtesy of Ford Motor Company.)*

105°C, the module also disables the engine cooling fan in response to the WOT signal. If the coolant switch is closed, the module overrides the WOT signal from the PCM by refusing to disable the cooling fan.

If the vehicle has automatic transmission and power brakes, depressing the brake pedal signals the module to disable both the A/C clutch and the cooling fan for 5 seconds. This is to prevent an engine stall from the load combined with the suddenly closed throttle.

Wide-Open Throttle A/C Cutout Relay. The normally closed WOT cutout relay is in series with the A/C switch and the clutch coil. During WOT operation, the PCM grounds the WOT relay coil. The relay opens and thus disables the A/C clutch.

Inlet Air Solenoid (IAS)

The 2.8L V6 applications with a feedback 2150-2V carburetor (1984–1985 Ranger/Bronco II and early 1986 Aerostar) use an inlet air solenoid, or IAS. This consists of a PCM-controlled solenoid to control the vacuum signal to the Thermostatic Air Cleaner (TAC). That is, the PCM controls the vacuum signal to the vacuum actuator (vacuum motor) located in the snorkel of the air cleaner. It should be noted that the PCM controls the IAS in response to the Intake Air Temperature, or IAT, sensor (originally known by Ford engineers as the Air Charge Temperature, or ACT, sensor), which, on this application, is located within the air cleaner assembly. When removing the air cleaner on these vehicles, it is best to remove the IAT sensor from the air cleaner housing so that it may be left intact electrically to avoid setting any false DTCs in the PCM's memory.

The IAS function was traditionally controlled by a bimetallic vacuum valve in the air cleaner or a wax pellet thermostat. When the engine is cold, the PCM energizes the solenoid and routes vacuum to the actuator. The actuator moves a door to close the snorkel opening to ambient air. Heated air is drawn instead from a shroud around the exhaust manifold, through a duct to an opening in the bottom of the snorkel. This heated air helps evaporate fuel as it is introduced to the intake manifold.

As the engine and the air around it warm, the heated air mode is not needed. The PCM then deactivates the solenoid, vacuum to the actuator is blocked and the door moves to open the snorkel's ambient air inlet and close the duct to the hot air.

Variable Voltage Choke (VVC)

The 2.8L V6 applications with a feedback 2150-2V carburetor (1984–1985 Ranger/Bronco II and early 1986 Aerostar) use a Variable Voltage Choke, or VVC. This consists of a PCM-controlled choke relay used to precisely control the electric choke thermostat. The choke cover holds an electric heating element to apply heat to the choke's bimetallic spring. When the engine starts at low temperature, the PCM duty cycles the relay every 2.5 seconds. When the relay contacts close, battery voltage applies to the choke heater element. When the engine reaches 27°C (80°F), the PCM holds the relay on continuously (100 percent duty cycle). This allows the choke to stay on while the engine is at low temperature, but turns it off as quickly as the engine can warm up and run properly without it. This avoids the emissions problems caused by the superrich mixture when the choke is on.

Temperature-Compensated Accelerator Pump (TCP) Solenoid

The 2.8L V6 applications with a feedback 2150-2V carburetor (1984–1985 Ranger/Bronco II and early 1986 Aerostar) use a Temperature Compensated Pump, or TCP. This consists of a PCM-controlled vacuum solenoid, which, in turn, controls a vacuum signal to the TCP vacuum diaphragm located on the front of the carburetor. (The TCP diaphragm has a Y-shaped cover held in place by three screws and located next to the accelerator pump diaphragm.) When the engine is below 35°C (95°F), the PCM does not ground the solenoid, and the accelerator pump functions normally. Once the engine coolant temperature reaches the specified temperature, the engine needs less fuel as the throttle is opened. The PCM energizes the TCP solenoid and routes vacuum to the TCP diaphragm. With vacuum applied, the pump cannot deliver as much fuel to the accelerator pump discharge nozzle.

Engine Fuel Injector Cooling Tube

In 1987, the 4.9-liter, six-cylinder in-line engine was introduced with multipoint fuel injection instead of a feedback carburetor. Because the 4.9-liter is not a cross-flow engine (the flow of gasses does not enter as intake on one side and leave as exhaust on the other), the injectors are located just above the exhaust manifold. This exposes the injector and its connecting fuel line to much more heat than on most systems, particularly during the heat soak following a hot engine shutdown. Exposing the injector to this much heat can result in fuel vaporization in the injector and adjacent fuel line (vapor lock) or even seizing of the injectors. Either condition could result in hard starting, rough running after startup, or no start at all. To prevent this, an injector cooling system was put on the 4.9-liter engine.

A tube runs almost the entire length of the engine, just above the injectors, Figure 14–72. Jets from the tube point to each of the first five injectors (injector six is not as exposed to exhaust manifold heat). A small blower blows air into the tube. The far end of the tube is closed so the cooling air is directed at the injectors. The fan is powered by an injector blower relay controlled by a module. This module is connected to a temperature-sensitive switch attached to the fuel rail. If the engine is shut down with a rail temperature of 77°C (170°F), the switch triggers the module, which turns on the fan. The module allows the fan to run for a maximum of 15 minutes. It turns the fan off if the temperature switch opens or if the ignition is turned on before the 15 minutes have elapsed. This system is not part of, nor controlled by, the EEC IV system. It does, as we have seen, affect startup and driveability after heat soak.

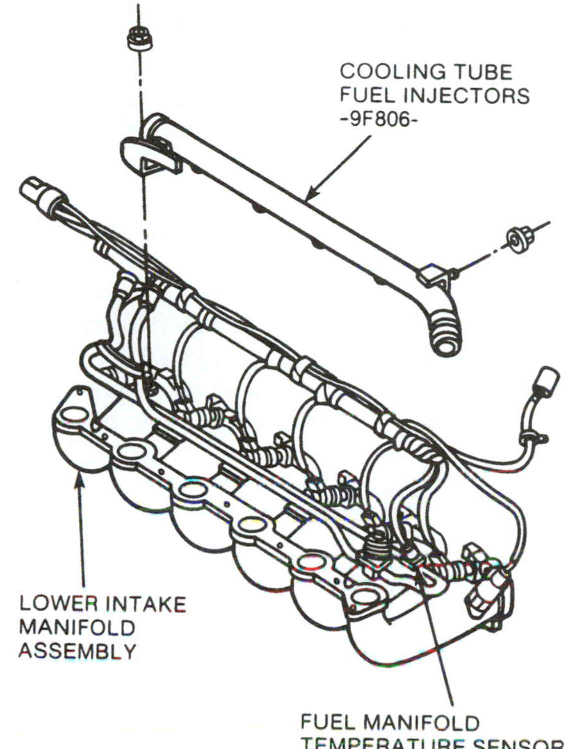

Figure 14–72 4.9-liter EFI engine fuel injector cooling tube. *(Courtesy of Ford Motor Company.)*

Shift Light Indicator

Some EEC IV-controlled vehicles with manual transmissions feature a shift indicator light on the instrument panel to indicate to the operator the best time to shift to the next higher gear to optimize emissions quality. To a lesser extent, this also indicates the best shift point for fuel economy. The PCM uses information concerning engine speed, manifold vacuum and engine temperature to control the light. When the PCM sees an engine speed greater than 1,800 rpm with a light engine load, it grounds the indicator light circuit. The PCM will not turn on the light while the coolant temperature is low. When the transmission goes into top gear, a switch in series with the PCM and the indicator light opens to prevent the light from coming on again.

Integrated Vehicle Speed Control (IVSC)

Beginning with 1986 Ford Taurus and Mercury Sable vehicles, the speed control function (cruise control) became integrated into the functions under the control of the PCM. This feature, added to the Crown Victoria in 1988 and to the Thunderbird in 1989, was known as Integrated Vehicle Speed Control or IVSC. (In the 1991 model year, Ford engineers discontinued the use of IVSC as other systems increased the demand for PCM terminals.) If the IVSC function is turned on, the PCM will control how much vacuum is applied to the servo through two solenoids. The speed control vacuum control (SCVAC) solenoid is normally closed and is energized by the PCM to apply vacuum to the servo, which in turn pulls the throttle open. The speed control vent (SCVNT) solenoid is normally open and, when de-energized, allows servo vacuum to be vented, thus allowing the throttle to close up. Modulation of these two solenoids is identical to the EGRC and EGRV solenoids described in greater depth earlier in this chapter. Concerning the speed control function, the driver communicates with the PCM through the speed control command switches (SCCS) located on the steering wheel.

Top Speed Limiting

Some vehicles with the 5.0-liter engine, beginning in 1988, include a top speed limit. The PCM is programmed to control maximum vehicle speed to a value within the vehicle's original equipment tires. The PCM achieves this by reducing the fuel injectors' pulse width once the specified vehicle speed is reached.

Programmed Ride Control (PRC). Beginning in 1987, certain Fords were equipped with a PRC system. A switch on the instrument panel allows the operator to select either a plush or firm ride. The shock absorbers have adjustable dampening valves, which are adjusted by rotating a small shaft extending up to the top of each shock absorber. A small electric motor mounted on top of each shock absorber turns the shaft. The motor is controlled by a PRC module. If, while operating in the plush mode, certain operating conditions occur, the module will automatically adjust the shock absorbers to the firm position. Some of the information that identifies those conditions is supplied to the PRC module by the EEC IV's PCM.

Information supplied directly to the PRC module from sensors includes:

- hard braking—supplied by a pressure switch on the brake system
- hard cornering—supplied by a sensor on the steering shaft

Information supplied by the EEC IV PCM includes:

- more than 90 percent of WOT.
- speeds above 83 miles per hour.

✔ SYSTEM DIAGNOSIS AND SERVICE

One of the outstanding features of the EEC IV system is its self-diagnostic capacity. The diagnostic features that are programmed into the EEC IV PCM, combined with the diagnostic procedures in the service manual, are very effective if the procedures are performed by a knowledgeable technician. These procedures may seem confusing at first because of their extent and complexity, but studying and practicing the procedures make them a powerful and effective tool in diagnosing EEC IV system complaints.

Malfunction Indicator Light (MIL)

Beginning in 1987, a limited number of Ford vehicles were produced with a malfunction indicator light (MIL). The MIL is located in the instrument panel and is used by the PCM to alert the driver to a fault within the EEC IV system. During self-testing of the EEC IV system, the MIL will also blink out the diagnostic trouble codes (DTCs).

Diagnostic Routines

The Ford engine performance diagnostic service manual, known as *Shop Manual Volume H*, or for short as the *H Manual*, contains a section called ***Diagnostic Routines***. As with any system, the first step of diagnosis is to verify the driveability complaint or symptom. Then the technician should always begin by looking up the symptom in the charts in the "Diagnostic Routines" portion of the service manual. The charts in this portion reflect the likely order of probable problem areas for each symptom encountered according to Ford engineers. For example, for some symptoms, EEC IV system diagnostic procedures are high priority. For other symptoms, EEC IV system diagnosis is much farther down the list, and other items are recommended as priority checks such as cooling system checks or other basic engine checks.

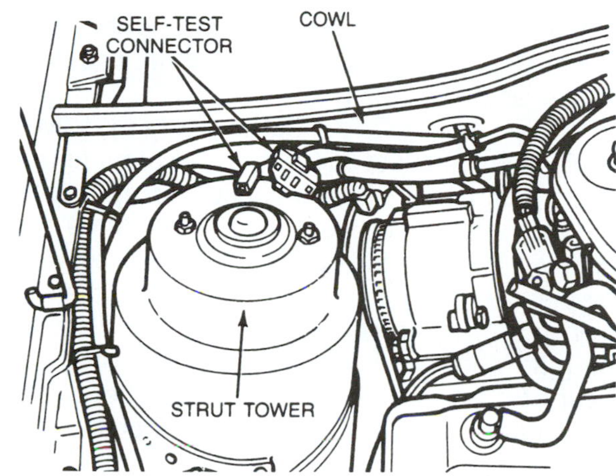

Figure 14–73 Self-test connector. *(Courtesy of Ford Motor Company.)*

EEC IV Equipment Hookup

One of the advantages in EEC IV system diagnosis is that, while a scan tool may be used to perform all diagnostic procedures (and is recommended), Ford engineers also provided a means for the technician to perform all self-tests and functional tests using manual methods. The only feature, of course, that cannot be accessed through manual methods is the data stream.

Both the manual methods and the scan tool utilize the self-test connector found in the engine compartment, Figure 14–73. This connector is known under OBD II standards as the Diagnostic Link Connector (DLC). While the DLC was designed for Ford's **Star Tester** and Super Star II Tester (Ford's early scan tools), it will also work with Ford's New Generation Star tester (NGS) and Worldwide Diagnostic System (WDS) as well as aftermarket scan tools. In fact, some aftermarket scan tools allow the technician to select a Super Star II Tester mode in addition to the engine test mode shown in the scanner's menu. This allows the technician to connect to and test electronic systems other than what may appear in the scanner's menu, for example, Ford's electronic suspension system. Whereas most modern scan tools prompt the technician when interaction is required, the Super Star II Tester mode does not prompt the technician when technician interaction is required during the self-test procedures, similar to using Ford's Super Star II Tester or manual methods. The technician must be familiar with Ford diagnostic concepts to use the Super Star II Tester mode.

The following procedures describe the manual methods when applicable. Understanding of the manual methods allows the technician to comprehend what a scan tool does in order to interact with this system

Slow Codes and Fast Codes

DTCs may be read by the pulsing of an analog meter's swing needle, the pulsing of an LED on a scan tool, or the pulsing of the MIL. Ford refers to these pulses as "slow codes." For example, a pulse, pause, pulse would equate to a DTC 11. And two pulses, a pause, and three more pulses would equate to a DTC 23. All Ford EEC IV PCMs are capable of producing slow code output. Both the Ford Star Tester and the Ford Super Star II Tester are capable of reading slow codes, as well as the NGS.

When the EEC IV PCM is ready to deliver DTC output, the first codes received are referred to as "fast codes." These codes exactly imitate the slow codes, but are pulsed at 100 times the speed of the slow code pulses. They are followed by the slow code output. The fast codes are designed to be interpreted by a scan tool and cannot be read when using manual methods. They appear as a flicker of the scan tool's LED or of the MIL. When using an analog voltmeter to read code output, the fast codes are seen as a very short sweep of the needle. All EEC IV PCMs, except a few of the very earliest ones, produce fast code output. Ford's Super Star II Tester and NGS are capable of interpreting fast codes as are most aftermarket scan tools.

EEC IV Self-Test Connector

The DLC (self-test connector) in the engine compartment is used to initiate (or trigger) the various tests. It may also be used with an analog voltmeter, if desired, to read code output. Or the MIL may be used for this purpose. The primary DLC contains six female spade terminals. Similar connectors associated with other systems, including the antilock brake system and the electronic suspension system, may be found elsewhere on the vehicle. But only the EEC IV system DLC has an extra single terminal protruding from the same harness as the primary DLC. This extra terminal is called self-test input or STI. It is through grounding this terminal that the self-tests are triggered. A scan tool grounds this terminal internally when a self-test is selected from the menu.

If the primary 6-terminal DLC is held facing you and its overall shape, (which is shaped almost like an arrow) is pointing upward (with the two terminals on top and the row of four terminals on the bottom), Figure 14–74, its terminals can be numbered in the same manner that you would read a book: left-to-right and top-to-bottom. DLC terminal 2 (known as Signal Return or SIG RTN) is connected to the ground-side of the sensors and connects to PCM terminal 46, and then it is grounded through the PCM. A short jumper wire can be easily made with a male spade terminal on each end, and then used to connect Signal Return to STI in order trigger any of the self-tests.

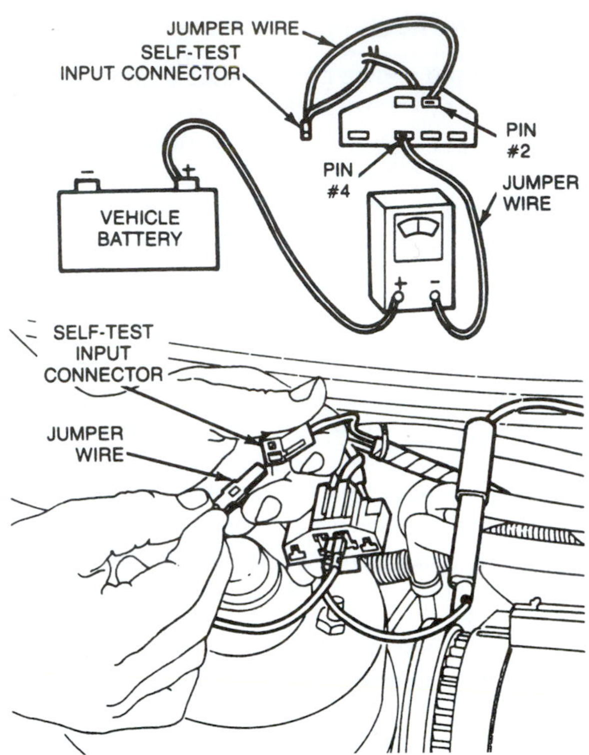

Figure 14–74 Voltmeter connection to self-test connector. *(Courtesy of Ford Motor Company.)*

To count code output from the PCM, simply count flashes of the MIL. If the vehicle is an earlier model year that does not have an MIL (or if you prefer not to use the MIL), you may connect an analog voltmeter between the battery and terminal 4 of the DLC, known as "Self-Test Output" or STO. Make up a short second jumper wire with a male spade terminal to connect to STO, and then use it to connect the voltmeter's negative lead to STO. The voltmeter's positive lead should be connected to the battery positive terminal. When the PCM is ready to count code output on terminal 4, it will complete the ground path to energize the voltmeter in pulses, as if the voltmeter were an output actuator. Count the meter pulses in order to interpret the code. It is acceptable to connect a DVOM (on a DC voltage scale) or a logic probe to STO for the purpose of counting code output, although code output pulses are difficult to interpret with the DVOM unless the DVOM has a bar graph display.

CAUTION: *Do not connect a standard test light to STO!* It will draw more current than the PCM's output driver in this circuit is designed to handle. As a result, the PCM will be damaged and will have to be replaced before code output can be obtained with any of these methods, including with the use of a scan tool or through the MIL. Also, when using either an analog meter or a DVOM, verify that the meter is not set to an amperage scale and verify that the leads are not connected to an amps jack. Ammeters operate effectively as a jumper wire and also destroy the PCM's circuitry!

EEC IV Engine System Performance Code Pulling Procedural Concepts

Four procedures are associated with code pulling on an EEC IV application.

Key On, Engine Off Self-Test. The first one is called a Key On, Engine Off (or KOEO) self-test. The resulting DTCs are hard faults and are known as *KOEO on-demand* faults. These are the first DTCs that should be tested through pinpoint testing as per the appropriate flow chart.

Key On, Engine Running Self-Test. The second code pulling procedure is to perform a Key On, Engine Running (or KOER) self-test. The resulting DTCs are also hard faults and are known as *KOER on-demand* faults. These DTCs should be tested next through pinpoint testing as per the appropriate flow chart.

Continuous Memory Codes. The third code pulling procedure is to pull memory codes from the PCM's memory that set in recent history, known as *continuous memory* fault codes. With the EEC V system to follow (as explained in Chapter 15), pulling continuous memory fault codes appears as a separate menu item on a scan tool's display and can be selected individually. But with EEC IV, continuous memory fault codes appear automatically following the KOEO self-test. That is, these DTCs are the last codes received following any KOEO self-test, and pulling continuous memory codes cannot be done on an EEC IV system without performing a KOEO self-test first. When the KOEO self-test is performed, the continuous memory DTCs received should simply be recorded at that point in time. Continuous memory DTCs should be tested lastly through pinpoint testing as per the appropriate flow chart, following testing of both KOEO and KOER hard faults.

Erasing Continuous Memory Codes. The fourth procedure associated with fault codes is for the technician to erase any continuous memory codes from the PCM's memory. (KOEO and KOER on-demand faults do not need to be erased, as they are never stored in the PCM's memory. The method to delete these codes is to properly perform the repair, then rerun the self-test to verify that the faults are no longer present.)

EEC IV Engine System Performance Code Pulling Procedures

Performing a KOEO Self-Test and Pulling Continuous Memory Codes. To initiate a KOEO self-test, the engine should be properly warmed up (if possible), and then the ignition should be turned off for at least 10 seconds. Ford's reasoning behind this is that if any form of self-test has been initiated since the last time the ignition was turned on and the PCM was powered up, another attempt to trigger a self-test will not do so, but instead causes the PCM to enter a functional test known as Continuous Monitor Mode. Because applications that use a motor to control idle speed (feedback carburetors and Central Fuel Injection) must keep the PCM alive after the ignition is turned off for about 8 seconds, this 10-second shutdown rule should be followed. On EFI (port injected) applications, because a duty cycle solenoid is used to control idle speed, the PCM powers down immediately when the ignition is turned off.

Once the engine has been warmed up and the PCM has been powered down, turn the ignition to the run position. Then jumper from SIG RTN (DLC terminal 2) to STI to trigger the KOEO self-test. The PCM will then perform a self-test, energizing and de-energizing each of the outputs that it controls while checking voltage drop in each circuit. Technician interaction is not required during the KOEO self-test, except on manual transmission 2.5L and 4.9L engines where the clutch pedal must be held fully depressed during the self-test or false codes will be generated. Also, all accessories should be turned off. If the A/C is turned on during the KOEO self-test, a false code is generated. DTCs will follow. The first codes received are the fast codes (which are unreadable unless a scan tool is being used), followed by the slow codes. The pulsing of the slow codes begins with the KOEO on-demand DTCs and ends with the continuous memory DTCs. In order for the technician to distinguish between the KOEO on-demand DTCs and the continuous memory DTCs, a code 10 is used as a separator code, Figure 14–75. While this appears as a DTC 10 on a scan tool, if slow codes are being counted, it is represented by a single pulse due to the fact that the zero will not have any pulses representing it. As a result, Ford writes any code numbers that contain zeroes with the zero in parenthesis in order to show that the zero will not display during slow code output. The fault codes are displayed as a two-digit code on earlier models, with codes 11 through 98 being available. Ford introduced three-digit codes in 1991, with codes 111 through 998 being available. Unlike many other manufacturers, Ford has a DTC that means "system pass." (Other manufacturers, such as GM or Chrysler, use the "absence of DTCs" between anticipated codes to mean "system pass.") This system pass code is DTC 11 in the two-digit code system and DTC 111 in the three-digit code system. Code 11 or 111 will either be received, or one or more failure codes will be received. Code 11 or 111 will never be received with a failure code within any one of the three diagnostic areas. That is, while you could have a system pass code in the KOEO self-test area and simultaneously get failure codes in both the KOER self-test and continuous memory areas, you could not get both a system pass code and a failure code in the KOEO self-test area at the same time. This also holds true for the KOER self-test and the continuous memory as well.

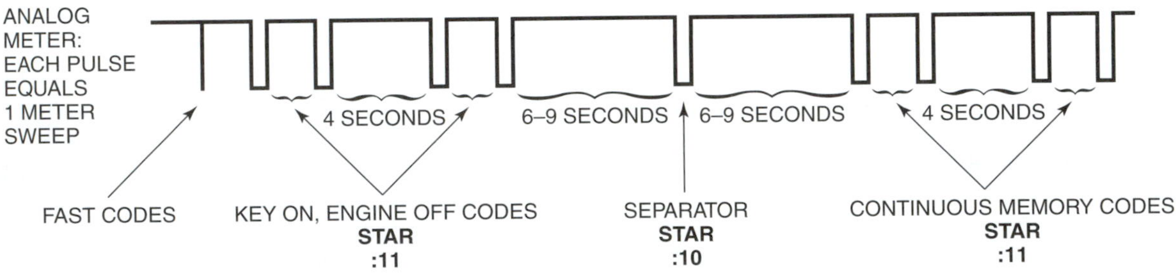

Figure 14–75 Key on, engine off and continuous memory code format. *(Courtesy of Ford Motor Company.)*

Following a KOEO self-test, the first codes displayed are the KOEO on-demand faults, either a system pass code or one or more fault codes. Once complete, each code repeats one more time. Then the separator code 1(0) displays (a single pulse). It does not repeat. Then the continuous memory codes display, again, either a system pass code or one or more fault codes. They, too, repeat one more time. At that point code output ends. At this time, any KOEO on-demand DTCs should be pinpoint tested and repaired, and any continuous memory DTCs should be written down for future reference. The KOEO on-demand codes must display a system pass (code 11 or 111) before a KOER self-test can be performed successfully.

Performing a KOER Self-Test. To perform the KOER self-test, the engine must be fully warmed up and running with the upper radiator hose hot and pressurized. Then, before initiating the self-test, run the engine at 2,000 rpm for 2 minutes to ensure that the oxygen sensor(s) are hot and ready for testing. Then let the engine return to idle, turn off the ignition for 10 seconds, restart the engine and trigger the KOER self-test by placing a jumper wire between SIG RTN and STI. The PCM will now perform functional tests of many of its sensors, actuators and controlled systems. For example, the PCM will test the oxygen sensor's output and then use the oxygen sensor(s) to help it perform a functional test of the Thermactor system, directing airflow to the atmosphere and then to the exhaust manifold using the TAB valve (discussed earlier in this chapter), and then directing air downstream to the catalytic converter and then back to the exhaust manifold using the TAD valve. And the PCM will test its ability to increase engine rpm through the idle speed control device.

The KOER self-test requires technician interaction. As soon as the technician triggers the KOER self-test, a cylinder ID code is generated, Figure 14–76. This code serves three purposes: it allows the technician to know that the engine performance KOER self-test has been correctly initiated, it tells the technician how many cylinders the PCM is designed for and it is also the technician's cue to interact with the PCM by helping it test up to four different circuits that contain switches.

The cylinder ID code 2(0) indicates a 4-cylinder PCM, code 3(0) indicates a 6-cylinder

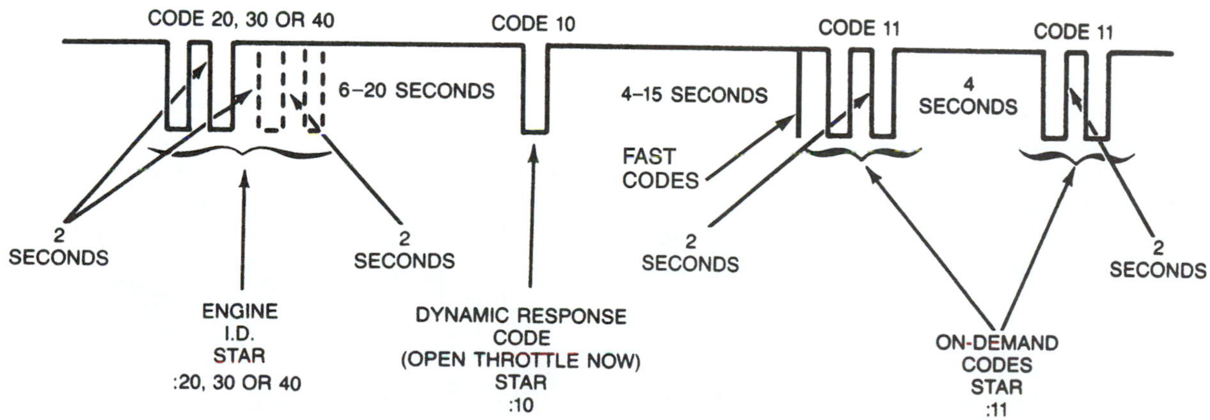

Figure 14–76 Typical service code display during engine running segment of self-test. *(Courtesy of Ford Motor Company.)*

PCM and code 4(0) indicates an 8-cylinder PCM. As soon as the cylinder ID code displays, the technician needs to push the brake pedal in order to allow the PCM to test the BOO switch, then quickly turn the steering wheel one-half turn and back to center steer in order to allow the PCM to test the PSPS. If they exist on the particular vehicle, the technician must also turn on, then turn off the Overdrive Cancel Switch (OCS), and then operate the 4-wheel drive switch to 4X4 High and back to 2-wheel drive. These circuits may be tested in any order. Then the PCM continues to test the system while the technician relaxes for a few moments. About the same time that the engine rpm is reduced back to idle, some PCMs will display a code 1(0). This is the technician's cue to interact with the PCM one more time by quickly snapping the throttle to WOT and back. This allows the PCM to test the response of the knock sensor and the MAP sensor. Ford refers to this as a "Dynamic Response" test. (Thus, a code 1(0) during a KOER self-test has a very different meaning from the separator code 1(0) displayed during a KOEO self-test.) Finally, whether or not the PCM requests a Dynamic Response, fault code output will follow, first displaying the fast codes (unreadable except by a scan tool), then displaying the slow codes. These codes consist of either a system pass code 11 or 111, or one or more fault codes. Any displayed codes repeat one more time.

Any DTCs displayed within any one of the three areas are displayed by the PCM in the order that they are to be repaired.

Erasing Continuous Memory Codes. Finally, to erase memory codes from the PCM's memory, a KOEO self-test should be initiated. Then as soon as the fast codes (or first slow code) display, the test should be prematurely aborted by disconnecting the jumper wire between SIG RTN and STI prior to the code output completing. This action instructs the PCM to erase all continuous memory DTCs. If erasing of continuous memory codes is selected on a scan tool's menu, the scan tool will do this internally.

EEC IV Engine System Performance Functional Tests

There are four functional tests, or diagnostic aids, programmed into an EEC IV PCM. With most (though not all) functional tests, the PCM does not determine pass/fail status of a component/circuit, nor does it generate DTCs (the exception to this is Ford's power balance test). Rather, it is up to the technician to determine pass/fail status of the circuit/component and simply use the functional test as an aid in making that determination. The technician may be directed to a particular functional test by a flow chart, or may choose to use a functional test without the service manual's direction.

Computed Timing Check. This functional test checks the ability of the PCM to properly control computed ignition timing. But first, base timing must be verified to be correct and readjusted if necessary. To do so, the SPOUT circuit must be opened electrically by disconnecting the male/female connector near the distributor, Figure 14–77 (1983 to 1985 models), or pulling the shorting bar in the SPOUT circuit, Figure 14–78 (1986 to 1995 models).

Once the base timing is verified, reinstall the SPOUT connector. Then perform a KOER self-

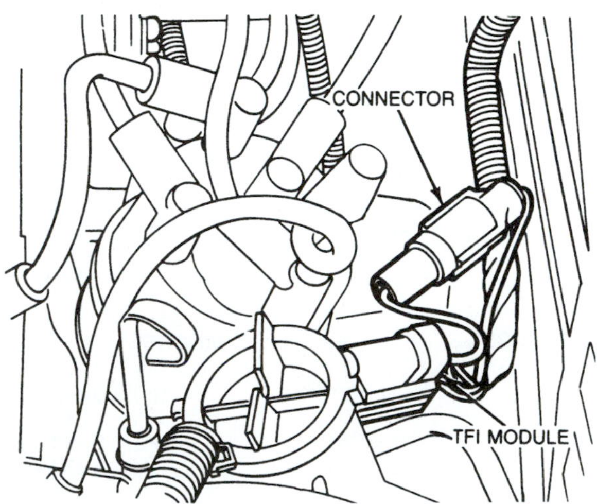

Figure 14–77 Early SPOUT connector. *(Courtesy of Ford Motor Company.)*

Figure 14–78 SPOUT connector with shorting bar, 1986 and newer.

test. Following the last code output, wait 10 seconds to verify that the last code has been received. The PCM will now add 20 degrees of spark advance to base timing for 2 minutes following the completion of the KOER self-test or until the self-test is deactivated, whichever occurs first. The specification is base timing plus 20 degrees with a tolerance of + or – 3 degrees. That is, if base timing was verified at 10 degrees BTDC, then timing advance during the Computed Timing Check should be between 27 degrees and 33 degrees BTDC. (This is the only time that the technician can place a specification on computed timing.)

Continuous Monitor Mode (Wiggle Test). This functional test aids the technician in identifying intermittent faults within the EEC IV system, particularly with the sensor circuits. For this test to work, no hard faults must be present within these circuits. To initiate this test, run either a KOEO or KOER self-test (depending upon which state you need the engine in during this functional test), and then after the last code is output, wait 10 seconds. At this point, deactivate and reactivate the self-test by disconnecting and reconnecting the jumper wire between SIG RTN and STI. (The scan tool will do this internally when this functional test is selected.) The test is now active and will alert the technician to any sensed fault. Tap on the component in question, wiggle its connector and wiggle the harness all the way from the component to the PCM while watching for the analog voltmeter's swing needle to go to battery voltage, or for the scan tool's LED to illuminate.

Output State Check. This functional test allows the technician to command the PCM to energize and de-energize several actuators, particularly those associated with emission systems. This allows the technician to test the ability of the PCM to energize these actuators and also allows the technician to perform other tests (such as vacuum tests associated with vacuum solenoids) while this mode is active. To initiate this test, perform a KOEO self-test, wait for the final code output and then wait 10 seconds. (Do not deactivate the self-test.) Now move the throttle to WOT and back to idle to energize these actuators. Move it to WOT and back to idle again to de-energize these actuators. This action may be repeated as needed.

SEFI Cylinder Balance Test. This functional test provides an easier method for the technician to perform a power balance test than using an ignition scope, but applies only to sequentially fuel-injected EEC IV systems from 1985 through 1995. The PCM uses each fuel injector, in turn, to disable fuel delivery to each of the cylinders as it measures rpm drop. It also automatically stabilizes ignition timing, engine rpm and electric cooling fan operation prior to beginning the test. At the end of the test, the PCM compares cylinder readings and reports the results to the technician through code output.

To initiate this test, perform a KOER self-test. (Of course, the engine should be fully warm.) After the last code is output, wait 10 seconds, then slightly tip in the throttle and release. The slight change in TPS voltage instructs the PCM to enter the cylinder balance test. After raising and stabilizing engine rpm, the PCM begins dropping the cylinders, one at a time, beginning with the highest numbered cylinder. Then it reduces engine rpm and report the results. DTC 9(0) indicates that all cylinders passed. DTCs 1(0) through 8(0) indicate failures with cylinders 1 through 8 respectively. The 1986 and newer models can run this test to each of three different failure specification levels, Figure 14–79.

If, after the first test, a DTC 9(0) is received indicating that all cylinders passed, there is no need to initiate further levels of the test. But if a failure code is received, DTC 5(0) for example, indicating that cylinder 5 failed the test, then simply, once again, slightly tip in and release the throttle to initiate the next level of the test. If now a DTC 9(0) is received, cylinder 5 is slightly weak and simply failed to drop at least 65 percent of the rpm drop recorded on the strongest cylinder. But, if at the end of the second test, DTC 5(0) is still received, then slightly tip in the throttle and release it one more time to initiate the final test. If now DTC 9(0) is received, cylinder 5 did not drop at least 43 percent of the rpm drop recorded on the strongest cylinder and is perceived as moderately weak. If, however, at the end of the final test, DTC 5(0) is still received, cylinder 5 did not drop even 20 percent of the rpm drop recorded on the strongest cylinder and is perceived as a dead or nearly dead cylinder. Any further tests initiated beyond this point will use the 20 percent specification level. A DTC 77 indicates that the test should be restarted. This test quickly isolates a weak cylinder and indicates the level of failure, but it is still up to the technician to determine the cause of the failure, whether it is mechanical, ignition or fuel related.

EEC IV Engine System Performance Quick Test

When directed to by the *H Manual*'s Diagnostic Routines section, the technician should perform a **Quick Test** as defined in the *H Manual*. The Quick Test is defined as a seven-step procedure, as follows:

Step 1: Visual check, vehicle preparation
Step 2: Equipment hookup (either scan tool or manual method)
Step 3: KOEO self-test
Step 4: Computed timing check
Step 5: KOER self-test
Step 6: Continuous memory code diagnosis (retrieved during the KOEO self-test)
Step 7: Diagnosis by symptom

The Quick Test therefore includes these seven items that Ford engineers consider of equal importance. Ultimately, steps 3 through 7 can direct the technician to an appropriate flow chart for diagnosis. (Ford's defined Quick Test does not simply equate to a self-test, but, rather, includes both self-tests along with other diagnostic aids.)

Test Level			***DTC 9(0) = "System Pass"***
One (spec = 65%)	**Two (spec = 43%)**	**Three (spec = 20%)**	
DTC	***DTC***	***DTC***	***Indication***
9(0)	(not needed)	(not needed)	All cylinders passed
5(0)	9(0)	(not needed)	Cyl. #5 is slightly weak
5(0)	5(0)	9(0)	Cyl. #5 is moderately weak
5(0)	5(0)	5(0)	Cyl. #5 is dead/nearly dead

Figure 14–79 SEFI cylinder balance test.

EEC IV IVSC (Speed Control) System Performance Code Pulling Procedures

On those EEC IV vehicles that have IVSC, the PCM does not store speed control-related DTCs in memory as it does with engine performance DTCs. However, it will perform both KOEO and KOER self-tests relating to the speed control function. DTC output from the PCM may still be read in the methods previously described or a scan tool may also be used.

Performing an IVSC KOEO Self-Test. To initiate an IVSC KOEO self-test, simply turn on the ignition, and then trigger the self-test by depressing the speed control on switch. (The jumper wire between SIG RTN and STI is not used to trigger this test as it was for both of the engine performance tests. Rather, depressing the On switch is recognized by the PCM as the trigger as long as the engine is not running.) Immediately a code 1(0) should be received, indicating that the IVSC self-test has been entered. Then, within 15 seconds, depress Off, Resume, Set/Accel and Coast. Also depress both the brake and clutch pedals. (Do not touch the throttle.) Code output will follow, first the fast codes and then the slow codes. Again, DTC 11 indicates a system pass. (This system was discontinued the year that Ford introduced the three-digit code system.)

Performing an IVSC KOER Self-Test. To initiate an IVSC KOER self-test, the engine must be fully warm and running, then a double trigger is used to trigger the self-test. First, depress the speed control on switch. Then, within 15 seconds, use a jumper wire to connect SIG RTN to STI. If successful, the PCM will immediately respond with a DTC 1(0), indicating that the IVSC KOER self-test has been entered. No technician interaction is required during this self-test. The PCM will test its ability to increase engine rpm using the speed control servo this time (rather than the idle speed control device as with the engine performance KOER self-test). After the engine slows down, code output will follow, first the fast codes and then the slow codes. Again, DTC 11 indicates a system pass.

EEC IV Data Stream

In 1989, Ford added a data stream capability to its California models. In 1991, this was added to its federal models. This simply added the ability of a scan tool to read data stream values, including sensor values and actuator commands, and is similar to GM's data stream.

EEC IV Breakout Box

An EEC IV breakout box may be installed to aid in pinpoint testing the various circuits/components. Many flow charts call for the technician to connect a breakout box. If resistance checks are to be performed, or if actuators are to be energized manually from the breakout box, the PCM is to be left disconnected. That is, the breakout box should be connected to the EEC IV electrical harness only. The PCM should be connected to the breakout box if the system needs to be operational and if voltage, frequency, duty cycle or pulse width checks are to be made.

Pinpoint Testing

Ultimately the goal of any diagnostic approach is to correctly identify and repair the problem. To this end, the *H Manual* contains pinpoint test charts or flow charts that the technician can be directed to by any of the engine performance or IVSC self-tests, by a Continuous Memory DTC, by a failure of the Computed Timing Test or by a symptom. Many aftermarket service manuals utilize separate flow charts, using the letter O to refer to DTCs generated by a KOEO self-test, a letter R to refer to DTCs generated by a KOER self-test and a letter C to refer to Continuous Memory DTCs.

Failure Mode Effects Management (FMEM). FMEM is a strategy programmed into the PCM, beginning in mid-1986, to maintain a good engine performance if one of the most critical sensors fails. Prior to 1987, such a sensor failure would either stop the engine or put the system into Limited Operational Strategy (in which the engine runs with no spark advance and a fixed air/fuel mixture). With FMEM Strategy, the PCM will, in most cases, be able to replace the failed sensor's input with a plausible substitute value from another sensor. It will also set a code 98 in its Continuous Self-Test memory. Code 98, in 1987, indicates that a fault exists and that FMEM is functional.

The operator may not know a fault exists except for the following:

- idle speed may be fixed
- the cooling fan may run all the time
- newer vehicles have an MIL on the instrument panel that will come on when the fault occurs

All 1987 and later EEC IV vehicles employ FMEM Strategy. Figure 14–80 shows the sensors included in the FMEM Strategy and the action taken in response to each one's failure.

Failed Sensor	Action Taken
MAP	Substitute TPS value. If TPS is faulty, use a fixed constant value. Disable Adaptive Learning. Disable EGR. Fixed idle speed.
TPS	Substitute fixed values during cranking, MAP value after engine starts.
ECT	Substitute IAT value during warm-up, fixed value after warm-up. Disable EGR. Disable Adaptive Learning. Fixed idle speed.
IAT	Substitute ECT value during warm-up, fixed value after warm-up. Disable EGR. Disable Adaptive Learning. Fixed idle speed.
EVP	No substitute. Disable EGR.

Figure 14–80 FMEM strategy.

Self-Test Improvements. Several improvements have been made in the Self-Test over the years. Some new codes have been added, and some codes have been made more reliable. The Self-Test and the continuous monitor have the capacity to check more functions. Consequently, code numbers may change meaning from year to year. The technician must have current published information for the vehicle being worked on.

Diagnostic & Service Tips

Here are a few final suggestions when working on Ford vehicles with an EEC IV system:

- Pinpoint procedures should not be used unless the Quick-Test directs you to do so. Each Pinpoint procedure assumes a fault has been detected in the circuit considered. To use them without being so directed by the Quick-Test can result in unnecessary and thus ineffective repairs and the replacement of functional parts. Do not replace parts unless specific test procedures indicate they are defective.
- Do not connect a test light at the PCM harness connector unless specifically directed to do so by a particular test procedure. Otherwise, damage to the PCM or to sensors or actuators could result.
- Isolate both ends of a circuit and turn the ignition off before testing for shorts or continuity unless otherwise directed by a specific test procedure. This same rule applies to solenoids and switches. All circuit resistance tests must be made with the circuit electrically separated from the rest of the system, unless you are using voltage drop techniques.
- An open circuit is any resistance greater than 10,000 ohms and a good connection is any path with less than 5 ohms, unless specifically detailed otherwise.

Adaptive Strategy and Driveability. If for some reason the continuous battery power to the PCM is removed, the adaptive strategy learning is erased. The calibration values stored in the KAM revert to the original look-up table values in the read-only memory. Because the learned values compensated for some engine wear or component inefficiency, the vehicle will possibly have driveability problems such as surging, hesitation or poor idle until the PCM can relearn.

A parallel problem—one not always understood by inexperienced technicians—is that if a component replaced was one the adaptive strategy had compensated for, the vehicle might have driveability problems with the new, fully functional component until the PCM has an opportunity to learn the characteristics of the new or repaired part. To avoid this problem after replacement or repair of significant system components, interrupt battery power to the KAM for at least 3 minutes during or after such a repair. This is particularly important for major engine component replacements.

MAP/BP Relocation. Beginning in 1987, MAP and/or BP sensors were repositioned so they sit in a horizontal position with the vacuum line sloping downward to its manifold connection. This prevents fuel migration into the sensor. Fuel getting into these sensors has been responsible for many pressure sensor failures.

SUMMARY

In this chapter, we have covered the three operational strategies of the EEC IV system, the oldest built of the Ford engine control systems (since 1982). We have reviewed the major sensors, for engine coolant temperature, engine speed, throttle position, manifold pressure and residual oxygen in the exhaust. We have seen the ignition system unique to Ford, with the SPOUT and PIP signals. We saw how they work and what their input information signals to the PCM look like.

We have reviewed all the actuators and the three different kinds of fuel metering devices on EEC IV systems and the mixture solenoids providing air to the emulsion circuits. We saw which components most directly affected emissions and which were more concerned with driveability or fuel economy. Finally, we looked at the major aspects of the diagnostics and repair of the EEC IV system.

▲ DIAGNOSTIC EXERCISE

An EEC IV vehicle with a SEFI system (sequential port fuel injected) comes into the shop because it failed an emission test for high CO (rich air/fuel mixture), but it seems to perform okay. The technician discovers that the fuel pressure regulator will not hold vacuum due to a small tear in the diaphragm and is apparently leaking unmetered fuel into the intake manifold through the vacuum hose. After getting the owner's approval, a new pressure regulator is installed. After the engine is started, it appears to have developed a lean surge and actually runs worse than before the pressure regulator was replaced. What is the most likely problem, and what should be done to correct it?

REVIEW QUESTIONS

1. *Technician* A says that an EEC IV PCM uses its keep-alive memory (KAM) to store fault codes related to faults that occurred but are no longer present.
   Technician B says that in an EEC IV system the engine calibration assembly is an integral part of the PCM that is soldered into it, and it cannot be removed or serviced separately.
   Who is correct?
   A. A only
   B. B only
   C. Both A and B
   D. Neither A nor B
2. What is the ability of the EEC IV PCM to learn to compensate for variations in manufacturing tolerances, wear or deterioration of sensors and other components called?

A. Base Engine Strategy
B. Modulator Strategy
C. Limited Operational Strategy
D. Adaptive Strategy

3. Which of the following sensors in an EEC IV system require a frequency meter in order to interpret its signal to the PCM?
   A. ECT
   B. MAP
   C. TPS
   D. EVP
4. What is the name of the signal generated by the Hall effect sensor in an EEC IV distributor?
   A. PIP
   B. SPOUT
   C. IDM
   D. CID
5. What is the name of the signal used in an EEC IV system to allow the PCM to verify that the spark timing command was carried out and to verify whether the calculated spark timing advance is working properly?
   A. PIP
   B. SPOUT
   C. IDM
   D. CID
6. What is the name of the signal used in an EEC IV system to allow the DIS ignition module to know which ignition coil to fire?
   A. PIP
   B. SPOUT
   C. IDM
   D. CID
7. Which of the following sensors monitors EGR flow rate?
   A. IAT
   B. EVP
   C. PFE
   D. VAF
8. All except which of the following inputs are switches?
   A. ITS
   B. VAT
   C. BOO
   D. PSPS
9. *Technician A* says that the VSS generates DC voltage signals.
   Technician B says that the VSS generates 64,000 pulses per mile at 60 mph and will generate fewer pulses per mile at slower speeds.
   Who is correct?
   A. A only
   B. B only
   C. Both A and B
   D. Neither A nor B
10. *Technician A* says that some Ford pickup trucks and vans use multiple fuel pumps, a low-pressure "lift pump" mounted in the fuel tank and a high-pressure in-line pump.
    Technician B says that Ford low-pressure CFI systems are regulated to 14.5 psi and use a single fuel injector.
    Who is correct?
    A. A only
    B. B only
    C. Both A and B
    D. Neither A nor B
11. A fuel-injected Ford vehicle will crank, but not start. The technician verifies that the fuel pump does not run when the ignition is turned on. What is the next thing that the technician should do?
    A. Replace the PCM.
    B. Follow the flow chart for diagnosing the fuel pump relay.
    C. Remove the fuel pump from the fuel tank for testing.
    D. Check the inertia switch.
12. *Technician A* says that the SPOUT signal is an analog voltage signal.
    Technician B says that the SPOUT signal is used by the PCM to command the TFI-IV ignition module to fire the ignition coil.
    Who is correct?
    A. A only
    B. B only
    C. Both A and B
    D. Neither A nor B
13. *Technician A* says that on the Thermactor system, the TAB valve is under the control of

the PCM-controlled TAB solenoid and is used to direct airflow to the TAD valve or to atmosphere.

Technician B says that on the Thermactor system, the TAD valve is under the control of the PCM-controlled TAD solenoid and is used to direct airflow upstream to the exhaust manifold or downstream to the catalytic converter. Who is correct?

A. A only
B. B only
C. Both A and B
D. Neither A nor B

14. On an EEC IV system, Continuous Memory codes occur automatically after performing which of the following?
    A. KOEO self-test
    B. KOER self-test
    C. Computed timing check
    D. SEFI cylinder balance test
15. What is the purpose of connecting an analog voltmeter between battery positive and STO during diagnostics?
    A. To initiate the KOEO and KOER self-tests
    B. To read code output
    C. To perform a voltage drop test on the positive side of the PCM
    D. To read base ignition timing
16. *Technician A* says that when diagnosing an EEC IV engine performance symptom, the KOEO self-test must indicate a system pass before a KOER self-test can be performed successfully.
    Technician B says that the engine performance KOER self-test requires technician interaction in order to be performed properly.
    Who is correct?
    A. A only
    B. B only
    C. Both A and B
    D. Neither A nor B
17. EEC IV DTCs consist of which of the following?
    A. Single-digit codes
    B. Two-digit codes
    C. Three-digit codes
    D. Either two-digit or three-digit codes depending on the year and model
18. Initiating a KOEO self-test and aborting it prematurely (after code output begins but before it completes) is a method of instructing an EEC IV PCM to do which of the following?
    A. Erase KOEO on-demand fault codes
    B. Erase KOER on-demand fault codes
    C. Erase continuous memory fault codes
    D. All of the above
19. On a SEFI cylinder balance test, an EEC IV PCM does which of the following?
    A. Stabilizes engine rpm, ignition timing and cooling fan operation
    B. Turns off one fuel injector at a time and measures rpm drop
    C. Compares the cylinders against the strongest cylinder
    D. All of the above
20. *Technician A* says that FMEM is a strategy programmed into an EEC IV PCM in order to maintain good engine performance in case one of the most critical sensors were to fail.
    Technician B says that FMEM is an EEC IV strategy that will use another sensor to substitute a good value for a failed sensor.
    Who is correct?
    A. A only
    B. B only
    C. Both A and B
    D. Neither A nor B

Chapter 15

Ford's Electronic Engine Control V (EEC V)

OBJECTIVES

Upon completion and review of this chapter, you should be able to:

- ❑ Describe the major aspects of the EEC V PCM.
- ❑ Define the inputs used with the EEC V system.
- ❑ Understand how the PCM controls the fuel injection system and the differences associated with both types of returnless fuel injection systems.
- ❑ Describe the various types of ignition systems used with the EEC V system including both types of distributorless systems.
- ❑ Define the emission systems used with the EEC V system.
- ❑ Describe some of the advanced features that have been introduced in EEC V applications, including variable camshaft timing (VCT) and split port induction (SPI).
- ❑ Describe some of the advanced features that Ford has introduced on the Lincoln LS including the new idle air bypass valve that is used in conjunction with air-assisted fuel injectors, the PCM-controlled hydraulic cooling fan and the fail-safe cooling system.

KEY TERMS

Alternate Fuel Compatibility
Freeze Frame/Snapshot
Inertia Switch

Ford Motor Company took a unique approach to meeting OBD II standards. Rather than modifying an existing system to meet OBD II standards (as most other manufacturers did), they designed a new engine computer system, known as Electronic Engine Control V (EEC V), with OBD II specifically in mind. The EEC V system carries over from EEC IV many of the features with which you are already familiar (see Chapter 14), and adds several other features under OBD II standards in order to improve a technician's diagnostic effectiveness (see Chapter 6).

ENGINE CONTROLS

Powertrain Control Module (PCM)

The EEC V PCM is a 104-pin processor, Figure 15–1 (compared to an EEC IV processor which contained 60 pins). By making each of the pins (terminals) smaller in size and arranging them closer together in four rows, Ford engineers actually were able to design the 104-pin connector to be physically no larger than the EEC IV connector had been. The PCM controls fuel delivery

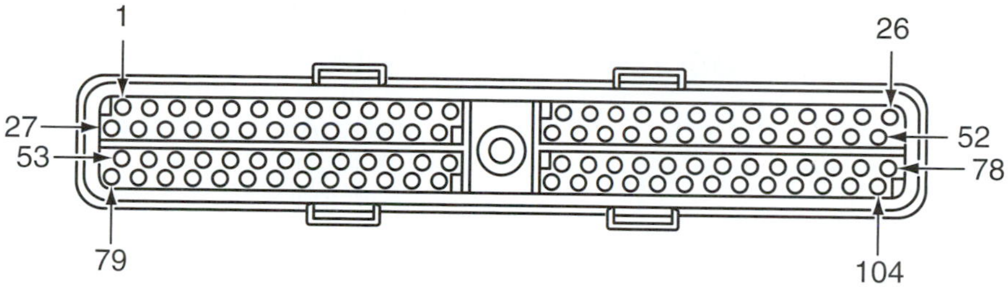

Figure 15–1 EEC V PCM 104-pin harness connector.

and ignition timing throughout the range of the engine's operational capacity. It monitors the various sensors and switches whose information is relevant to calculating the proper values for fuel injector pulse widths and firing sequence, as well as triggering actuation of the various components that perform its combustion commands.

The EEC V PCM determines the gradual wear and age of the vehicle over time and any changes in altitude or barometric pressure; it can then make compensatory adjustments in its own programs to correct for whatever needs to be altered because of the changed input information.

The principal inputs to the PCM are the throttle position sensor, the mass air flow sensor, the intake air temperature sensor, the engine coolant temperature sensor, the oxygen sensor, the camshaft position sensor, the knock sensor, the power steering pressure switch, the vehicle speed sensor and the idle air control valve.

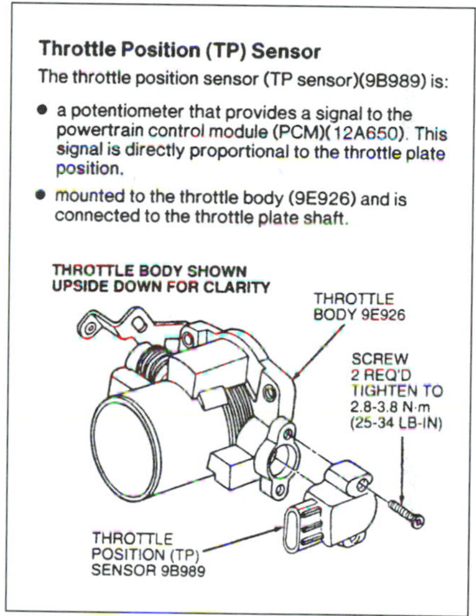

Figure 15–2 Throttle position (TP) sensor. *(Courtesy of Ford Motor Company.)*

INPUTS

Throttle Position Sensor. On the side of the throttle body is a variable potentiometer that provides a return signal to the PCM directly proportional to the position of the throttle plate, the open angle, Figure 15–2. This works in the same way TP sensors have since 1981, and the tests are still quite similar—you check for continuous variation in the resistance and return voltage throughout the throttle travel.

Mass Air Flow (MAF) Sensor. The Ford EEC V system uses a hot-wire type mass air flow sensor, Figure 15–3. The hot wire in the airstream is kept at a constant temperature above the ambient air, and a hot wire sensing element measures exactly how much current is required to maintain that heat. The sensor then sends an analog voltage signal to the PCM, which corresponds to the mass of the intake air. Of course, as with all such air mass sensors, it is very important to make sure there are no downstream air leaks as any intake air that does not go through the sensor will result in a false signal to the computer. The MAF is a single service unit: it cannot be repaired as a set of parts.

Mass Air Flow (MAF) Sensor

The mass air flow sensor (MAF sensor)(12B579):

- is located between the engine air cleaner (ACL)(9600) and the throttle body (9E926).
- uses a hot wire sensing element to measure the amount of air entering the engine. Air passing over the hot wire causes it to cool.
- sends an analog voltage signal to the powertrain control module (PCM)(12A650) to determine the intake air mass. The powertrain control module will then calculate the required fuel injector pulse width in order to provide the desired air / fuel ratio.

The mass air flow sensor hot wire sensing element and housing are calibrated as a unit and must be serviced as a complete assembly.

MASS AIR FLOW (MAF) SENSOR 12B579

Figure 15–3 Mass air flow (MAF) sensor. *(Courtesy of Ford Motor Company.)*

Idle Air Control Valve. The idle air control valve, Figure 15–4, is a normally closed solenoid (similar to the one used with the EEC IV system) and is operated by the PCM on a duty cycle in order to control idle speed. The duty cycle on-time determines the amount of bypass air that flows around the throttle plates.

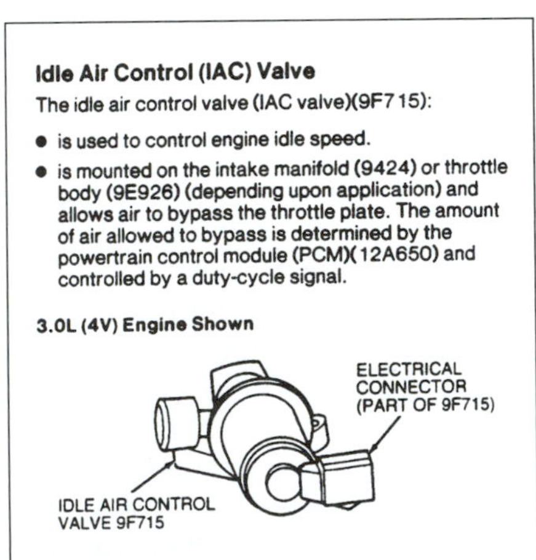

Idle Air Control (IAC) Valve
The idle air control valve (IAC valve)(9F715):
• is used to control engine idle speed.
• is mounted on the intake manifold (9424) or throttle body (9E926) (depending upon application) and allows air to bypass the throttle plate. The amount of air allowed to bypass is determined by the powertrain control module (PCM)(12A650) and controlled by a duty-cycle signal.
3.0L (4V) Engine Shown
ELECTRICAL CONNECTOR (PART OF 9F715)
IDLE AIR CONTROL VALVE 9F715

Figure 15–4 Idle air control (IAC) valve. *(Courtesy of Ford Motor Company.)*

Intake Air Temperature (IAT) Sensor. This sensor, Figure 15–5, is a variable resistance thermistor changing in response to the temperature of the intake air charge.

Engine Coolant Temperature (ECT) Sensor. Like the intake air temperature sensor, the ECT, Figure 15–6, is also a variable resistance thermistor. It changes its resistance in response to engine coolant temperature. Both signal the PCM with a voltage corresponding to the value of the temperature measured.

Heated Oxygen Sensors (HO_2S). The oxygen sensor system is different from earlier computer control systems chiefly in that there are four of them in tandem. Two are before the catalytic converters—used as previously for control of closed-loop fuel trim—and two are *after* the catalytic converters to monitor the effectiveness of the catalytic converters, Figure 15–7. All four are electrically heated to reach operating temperature as quickly as possible. The downstream,

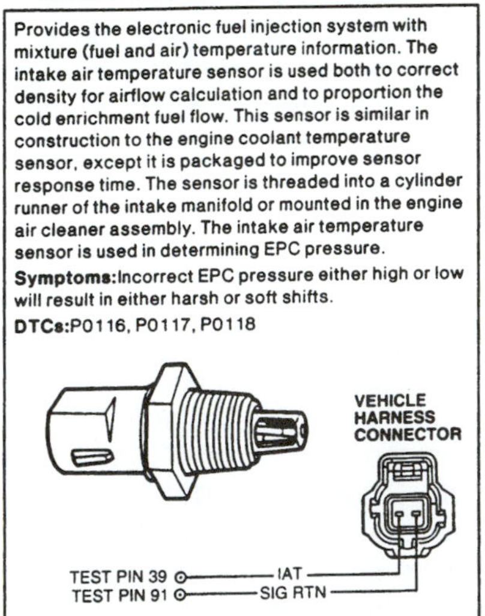

Intake Air Temperature (IAT) Sensor 12A697:
Provides the electronic fuel injection system with mixture (fuel and air) temperature information. The intake air temperature sensor is used both to correct density for airflow calculation and to proportion the cold enrichment fuel flow. This sensor is similar in construction to the engine coolant temperature sensor, except it is packaged to improve sensor response time. The sensor is threaded into a cylinder runner of the intake manifold or mounted in the engine air cleaner assembly. The intake air temperature sensor is used in determining EPC pressure.
Symptoms: Incorrect EPC pressure either high or low will result in either harsh or soft shifts.
DTCs: P0116, P0117, P0118
VEHICLE HARNESS CONNECTOR
TEST PIN 39
IAT
TEST PIN 91
SIG RTN

Figure 15–5 Intake air temperature (IAT) sensor. *(Courtesy of Ford Motor Company.)*

Engine Coolant Temperature (ECT) Sensor 12A648:

Detects the temperature of engine coolant and supplies the information to the powertrain control module. The ECT sensor is threaded into the heater outlet fitting or cooling passage on the engine. For engine control applications, the ECT signal is used to modify ignition timing, EGR flow, and air-to-fuel ratio as a function of engine coolant temperature. On electronic instrument cluster applications, the ECT output is used to control a coolant temperature indicator. The ECT is used to control torque converter clutch operation.

Symptoms: Torque converter clutch will always be off, resulting in reduced fuel economy.

DTCs: P0116, P0117, P0118

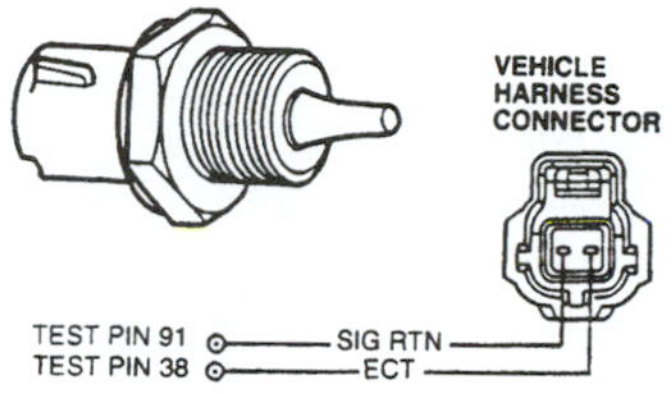

VEHICLE HARNESS CONNECTOR

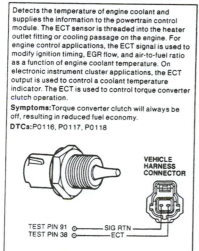

TEST PIN 91 — SIG RTN

TEST PIN 38 — ECT

Figure 15–6 Engine coolant temperature (ECT) sensor. *(Courtesy of Ford Motor Company.)*

secondary oxygen sensors should, if all is well, have a relatively flat signal. It does not matter much whether the voltage is high or low (typically they will slowly wander). What is important is that they do not reflect the kind of back and forth voltage fluctuation every second or two that the primary oxygen sensors do. If the secondary signal is a reflection of the primary sensors' signal, that means the catalytic converter is not doing anything to the exhaust gases.

Vehicle Speed Sensor (VSS). The Ford vehicle speed sensor, Figure 15–8, uses a magnetic pick-up to send a rapid on/off pulse, the frequency of which corresponds to the vehicle's speed. The PCM uses this information, among other things, to calculate fuel mixture and ignition timing. Since the new transaxles are also, in large measure, electronically controlled, the VSS signal plays a role in that function.

REMOVAL AND INSTALLATION (Continued)

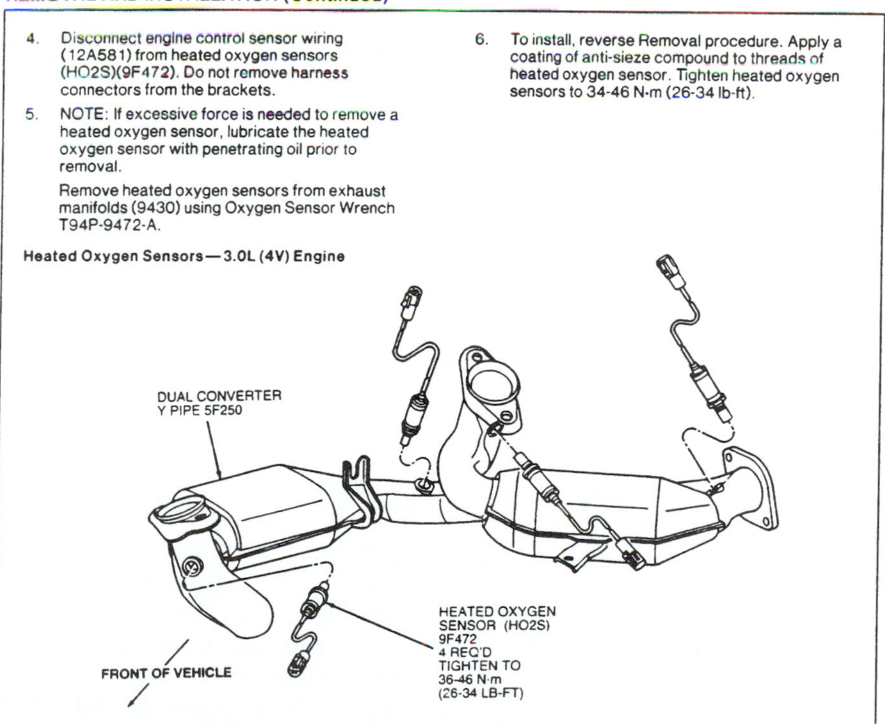

4. Disconnect engine control sensor wiring (12A581) from heated oxygen sensors (HO2S)(9F472). Do not remove harness connectors from the brackets.
5. NOTE: If excessive force is needed to remove a heated oxygen sensor, lubricate the heated oxygen sensor with penetrating oil prior to removal.

   Remove heated oxygen sensors from exhaust manifolds (9430) using Oxygen Sensor Wrench T94P-9472-A.
6. To install, reverse Removal procedure. Apply a coating of anti-sieze compound to threads of heated oxygen sensor. Tighten heated oxygen sensors to 34-46 N·m (26-34 lb-ft).

Heated Oxygen Sensors—3.0L (4V) Engine

Figure 15–7 Heated oxygen sensors. *(Courtesy of Ford Motor Company.)*

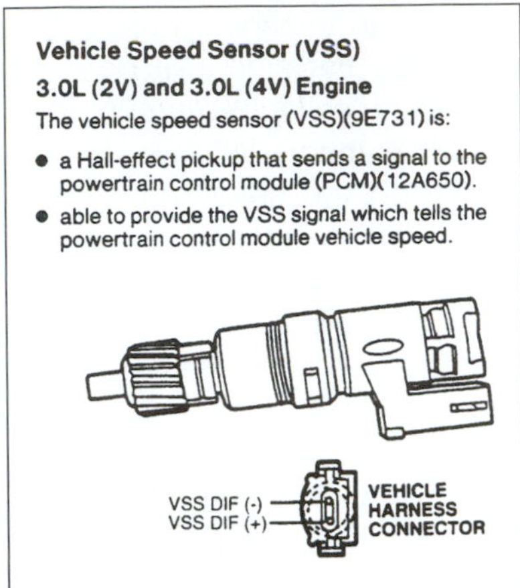

Vehicle Speed Sensor (VSS)
3.0L (2V) and 3.0L (4V) Engine
The vehicle speed sensor (VSS)(9E731) is:
• a Hall-effect pickup that sends a signal to the powertrain control module (PCM)(12A650).
• able to provide the VSS signal which tells the powertrain control module vehicle speed.
VSS DIF (-)
VSS DIF (+)
VEHICLE HARNESS CONNECTOR

Figure 15–8 Vehicle speed sensor (VSS). *(Courtesy of Ford Motor Company.)*

Power Steering Pressure (PSP) Switch 3N824:

Is used on certain applications to signal the powertrain control module when the power steering pressure exceeds a specific limit. Then the powertrain control module will adjust idle speed to compensate for this added load on the engine. The PSP signal is used as an input to the powertrain control module to assist in determining proper EPC pressure during increased engine loads.

Symptoms:

- Failed ON—EPC slightly high, firm engagements, firm shifts, harsh coastdown shifts.
- Failed OFF—EPC pressure slightly low during increased loading of the vehicle power steering.

DTCs:P1650, P1651

VEHICLE HARNESS CONNECTOR

TEST PIN 91 — SIG RTN

TEST PIN 31 — PSP

Figure 15–10 Power steering pressure (PSP) switch. *(Courtesy of Ford Motor Company.)*

Camshaft Position (CMP) Sensor. On the two-valve-per-cylinder engine, the camshaft position sensor is a single Hall effect magnetic switch activated by a vane driven by the camshaft. On the four-valve-per-cylinder engine, the camshaft position sensor is a variable reluctance sensor triggered by the high point on the left exhaust camshaft sprocket. In both cases, the sensor provides the PCM with information used to synchronize the fuel injection sequence with the cylinder firing order.

Knock Sensor (KS). The knock sensor, Figure 15–9, is the usual piezoelectric crystal type, which sends a voltage to the PCM if there is detonation in the engine. This information is used to retard spark advance and eliminate the knock.

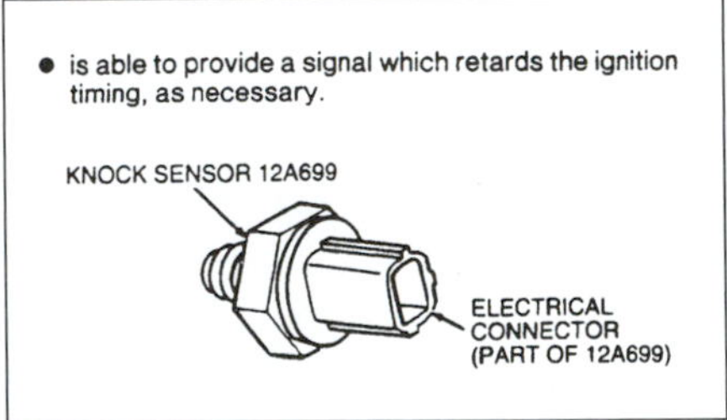

Figure 15–9 Knock sensor. *(Courtesy of Ford Motor Company.)*

Power Steering Pressure (PSP) Switch. The power steering pressure switch, Figure 15–10, is installed in the boost side of the power steering pump. Normally it is electrically closed, but when hydraulic boost builds to the point where it could be a load on the engine, the switch opens. This provides the computer with information relevant to setting the idle speed.

Crankshaft Position (CKP) Sensor. The crankshaft position (CKP) sensor, Figure 15–11, produces the tach reference signal for the ignition system. It consists of a variable reluctance sensor (permanent magnet) and produces an AC voltage signal. It monitors a toothed gear with one tooth spaced every 10 degrees of crankshaft rotation, except for one blank space where a tooth was left out. This is commonly referred to as a "36-minus-1-tooth" arrangement. The tip of the sensor and the toothed gear may be located either inside the timing gear cover, as on the 3.0-liter engine, or the sensor may monitor teeth that are located on the harmonic balancer and are then, of course, external of the timing gear cover, as on the 4.0-liter engine.

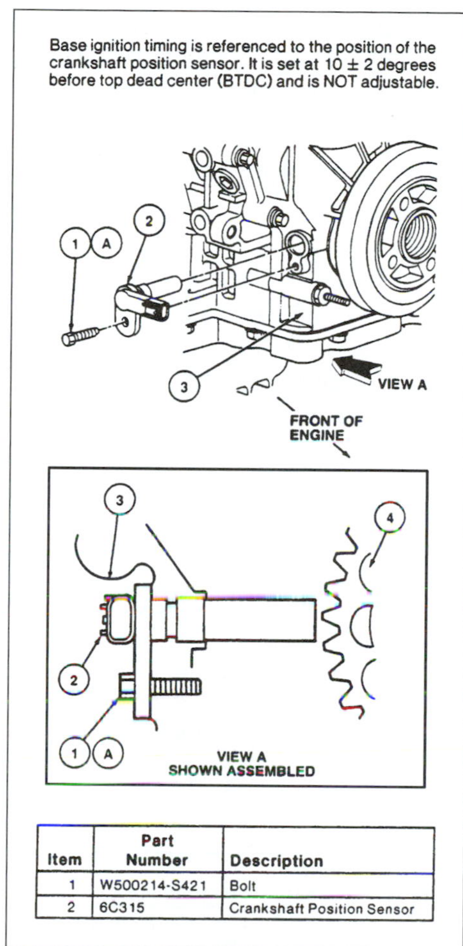

Item	Part Number	Description
1	W500214-S421	Bolt
2	6C315	Crankshaft Position Sensor

Figure 15–11 Crankshaft position sensor. *(Courtesy of Ford Motor Company.)*

The missing tooth identifies the position of the crankshaft for the PCM. Then, through the other teeth, the PCM is able to count off crankshaft rotation in 10-degree increments. This arrangement is considered to be a high-resolution signal and provides enough pulses per cylinder firing to allow the PCM to determine whether a cylinder is misfiring by watching the frequency of the pulses, as per OBD II standards. Even on an 8-cylinder engine, the PCM is able to count off nine teeth per cylinder firing (excepting, of course, the missing tooth). This signal is used by the electronic distributorless ignition system (EDIS) to identify each pair of pistons as they approach TDC. In the waste spark application, this information allows the PCM to fire the correct ignition coil.

OUTPUTS

Fuel Supply System

Fuel Pump Relay. The fuel pump relay is part of the constant control relay module, Figure 15–12. When the ignition is first turned on, the

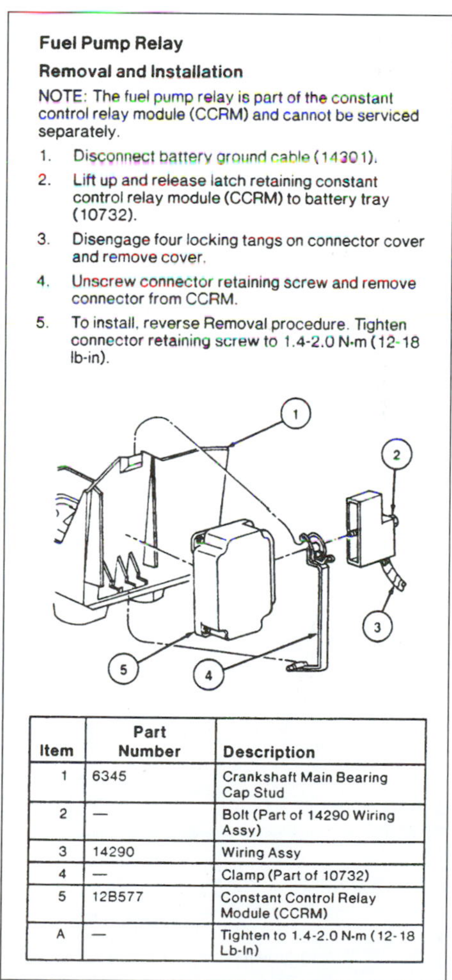

Fuel Pump Relay

Removal and Installation

NOTE: The fuel pump relay is part of the constant control relay module (CCRM) and cannot be serviced separately.

1. Disconnect battery ground cable (14301).
2. Lift up and release latch retaining constant control relay module (CCRM) to battery tray (10732).
3. Disengage four locking tangs on connector cover and remove cover.
4. Unscrew connector retaining screw and remove connector from CCRM.
5. To install, reverse Removal procedure. Tighten connector retaining screw to 1.4-2.0 N-m (12-18 lb-in).

Item	Part Number	Description
1	6345	Crankshaft Main Bearing Cap Stud
2	—	Bolt (Part of 14290 Wiring Assy)
3	14290	Wiring Assy
4	—	Clamp (Part of 10732)
5	12B577	Constant Control Relay Module (CCRM)
A	—	Tighten to 1.4-2.0 N-m (12-18 Lb-In)

Figure 15–12 Fuel pump relay (constant control relay module). *(Courtesy of Ford Motor Company.)*

PCM instructs the relay module to turn the pump on briefly to prime the injectors. If there is an ignition signal, the pump will stay on to maintain fuel pressure. If not, the pump is automatically shut off in a second or two. Besides the fuel pump relay, the system also employs an **inertia switch**, Figure 15–13. The purpose of this switch is to disable the fuel pump in case of an accident to prevent the possibility of a gasoline fire from the system under pressure.

Ford's inertia switches have improved in recent years, but it is not uncommon to find a car that will not start because the switch has been triggered by something less than an accident, perhaps because someone was forcibly bumped while parked. The switch contains a weight in a cone. The weight is held in place by a magnet. If a shock to the car dislodges the weight, it flips a switch that disconnects the fuel pump circuit. On the top of the switch is a reset button for just such occurrences. Since the switch is located in the trunk, it is also not uncommon to find some corrosion damage to the connector. Most experienced Ford technicians look at the inertia switch first when they find a no-start.

REMOVAL AND INSTALLATION (Continued)

6. To install, reverse Removal procedure. Tighten screws to 1.6-2.2 N·m (14-19 lb-in).

Figure 15–13 Inertia switch. *(Courtesy of Ford Motor Company.)*

Fuel Injectors. The injectors used in this system, Figure 15–14, are methanol and ethanol compatible. The O-rings that seal them in the

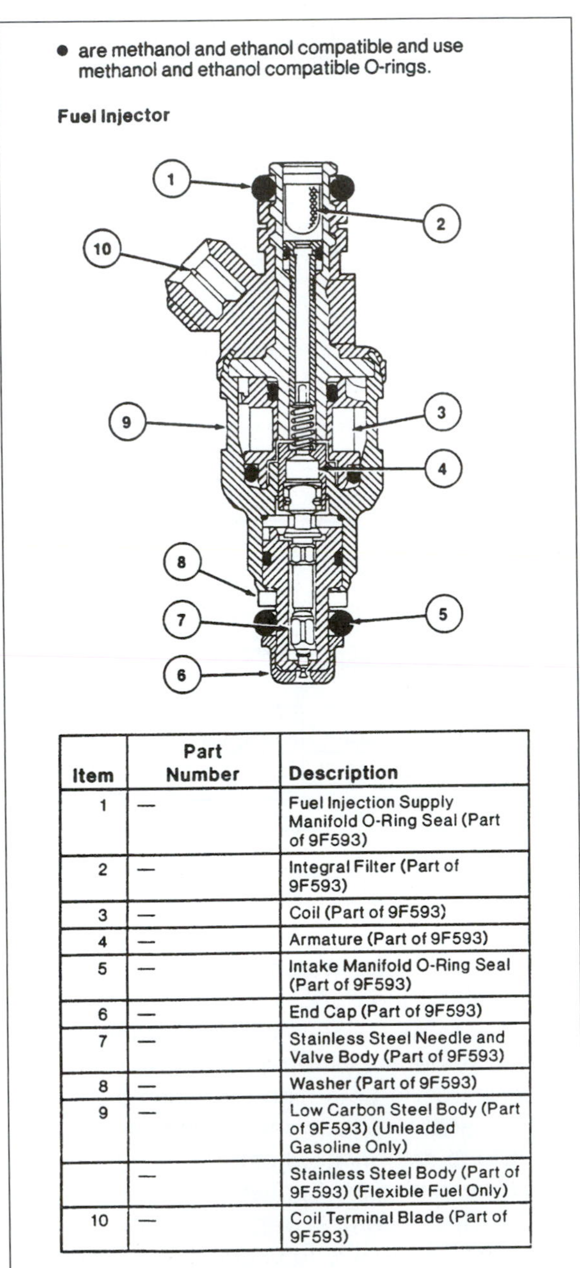

Item	Part Number	Description
1	—	Fuel Injection Supply Manifold O-Ring Seal (Part of 9F593)
2	—	Integral Filter (Part of 9F593)
3	—	Coil (Part of 9F593)
4	—	Armature (Part of 9F593)
5	—	Intake Manifold O-Ring Seal (Part of 9F593)
6	—	End Cap (Part of 9F593)
7	—	Stainless Steel Needle and Valve Body (Part of 9F593)
8	—	Washer (Part of 9F593)
9	—	Low Carbon Steel Body (Part of 9F593) (Unleaded Gasoline Only)
	—	Stainless Steel Body (Part of 9F593) (Flexible Fuel Only)
10	—	Coil Terminal Blade (Part of 9F593)

Figure 15–14 Fuel injector cutaway. *(Courtesy of Ford Motor Company.)*

manifold are likewise methanol and ethanol compatible, as should be any replacements. These injectors use a pintle designed to resist deposits from fuel additives as well. The PCM grounds each injector's circuit in turn, completing its circuit and opening the nozzle. The injection occurs during the intake stroke in the firing order of the engine.

Fuel Pressure Regulator. Also designed for **alternate fuel compatibility**, the fuel pressure regulator connects to the fuel rail (or fuel injection supply manifold, as Ford literature prefers to describe it) downstream of the injectors, Figure 15–15. It regulates the fuel pressure at the injectors through the interaction of its diaphragm-operated relief valve. One side of the diaphragm senses fuel pressure, and the other is subject to manifold pressure (vacuum). The initial fuel pressure is determined by the specially calibrated spring preload. The purpose of the arrangement is to maintain the fuel pressure drop through the nozzles to a constant pressure differential, regardless of the changes in the intake manifold pressure, and to make it simpler to calculate the fuel injection pulse width. As long as the pressure differential is constant, a fixed amount of fuel will flow for a fixed pulse width. If the pressure differential were allowed to vary, the PCM would have to compensate for different fuel quantity. Pressure is reduced from the pump's maximum outlet pressure by venting enough back to the tank to keep the rail charged at the design pressure.

- nominal fuel pressure is established by a spring preload applied to the diaphragm.
- balances one side of the diaphragm with manifold pressure to maintain a constant fuel pressure drop across the fuel injectors.
- bypasses and returns excess fuel to the fuel tank (9002).

Fuel Pressure Regulator—Cross Section

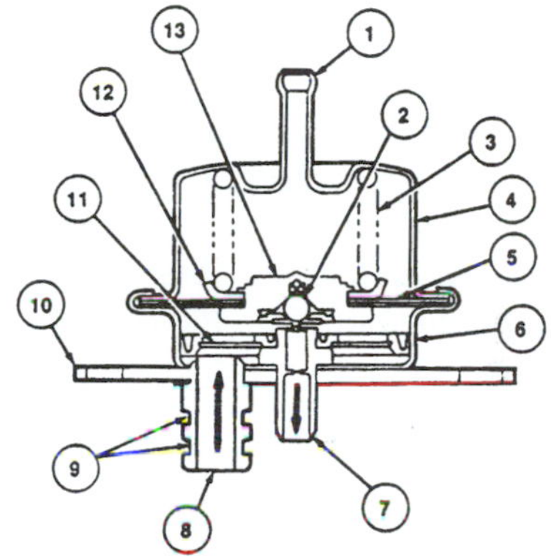

Item	Part Number	Description
1	—	Engine Vacuum Reference Tube (Part of 9C968)
2	—	Ball Seat (Part of 9C968)
3	—	Spring (Part of 9C968)
4	—	Upper Housing (Part of 9C968)
5	—	Diaphragm (Part of 9C968)
6	—	Lower Housing (Part of 9C968)
7	—	Fuel Outlet (Return Tube) (Part of 9C968)
8	—	Fuel Inlet (Supply Tube) (Part of 9C968)
9	—	O-Ring Grooves (Part of 9C968)
10	—	Mounting Plate (Part of 9C968)
11	—	Fuel Filter Screen (Part of 9C968)
12	—	Spring Seat (Part of 9C968)
13	—	Valve Assembly (Part of 9C968)

1996 Taurus, Sable June 1995

Figure 15–15 Fuel pressure regulator—cross section. *(Courtesy of Ford Motor Company.)*

Returnless Fuel Injection System

Similar to the DaimlerChrysler and General Motors' systems, the Ford fuel pressure regulator has been traditionally located on the fuel rail. Fuel in excess of that needed for engine requirements is returned to the fuel tank through a return line. This excess fuel returns heat absorbed from the engine compartment back to the fuel tank. In an effort to reduce evaporative emissions from the fuel tank, beginning in 1998 some Ford applications use a returnless fuel system. The new fuel pressure regulator is located on top of the fuel pump module, which, in turn, is located down into the fuel tank. Fuel that is in excess of engine requirements and therefore bypasses the regulator is immediately returned to the fuel tank. Any fuel that actually leaves this assembly, being pumped into the supply line, is not returned to the fuel tank.

Because this fuel pressure regulator is not indexed to manifold pressures, the fuel rail pressure remains constant under all driving conditions, which means that the fuel pressure differential across the fuel injectors varies as driving conditions vary, thus varying the flow rate of the injectors. (The pressure differential is the difference between the fuel rail pressure and the manifold pressure at the outlet tips of the injectors.) For example, when manifold pressure is high, as at WOT, the pressure differential across the injectors is decreased, therefore the flow rate is decreased. But when manifold pressure is low, as at idle or during deceleration, the pressure differential across the injectors is increased, therefore the flow rate is increased. Therefore, with a returnless fuel injection system, the PCM must first calculate what the pressure differential is across the injector (and the resulting flow rate) under each driving condition before it can determine what the injector's pulse width should be.

Electronic Returnless Fuel Injection System

Beginning in the 2000 model year, some applications such as the 2.0-liter split port induction (SPI) and the 2.0-liter ZETEC engine (both discussed later in this chapter) use an electronic returnless fuel injection system. A fuel pressure sensor is mounted at the fuel rail in the engine compartment, Figure 15–16, and is used by the PCM to monitor fuel rail pressure. A PCM-controlled fuel pump driver, located under the rear seat, sends a pulse width modulated signal to the in-tank fuel pump. This varies the fuel pump speed to maintain the proper fuel pressure without the use of a fuel pressure regulator. An added advantage of this design is that fuel rail pressure may be read through the data stream with use of a scan tool.

Ignition Systems

While some TFI-IV distributor ignition systems carried over from EEC IV into some EEC V applications, by 1999 the only application that retained a distributor was the 3.3-liter Villager. With the exception of the 2.0-liter Probe, the distributor applications continued to use a separate, dedicated ignition module to control the ignition coil. The 2.0-liter Probe allowed the PCM to directly control the ignition coil.

The electronic distributorless ignition system, or EDIS, carried from EEC IV over into EEC V (as

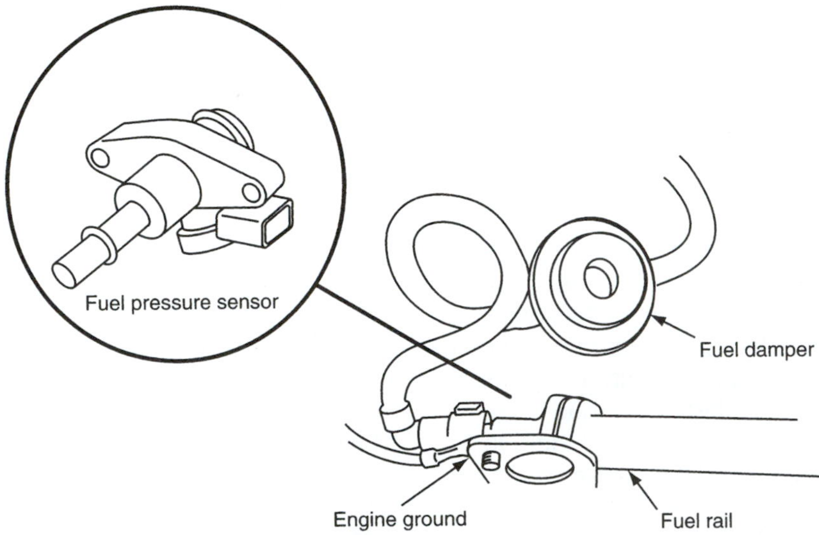

Figure 15–16 This pressure sensor monitors fuel system pressure as an input to the PCM.

described in Chapter 14). The EDIS system is a waste spark system, using one ignition coil per two cylinders. A given coil fires two spark plugs simultaneously, one on a cylinder near the end of its compression stroke and the other (the "waste spark") on a cylinder near the end of its exhaust stroke. All coils are formed into one component called a *coil pack*, Figure 15–17. Most 1996 and newer EEC V EDIS applications (except for the 1996–1997 Windstar) allowed the PCM to directly control the ignition coil primary circuit, without the use of an external separate ignition module. By 1998, all EDIS applications allowed the PCM to directly control the ignition coil.

EEC V has also incorporated another format of a distributorless ignition system, commonly known by Ford as *coil-per-plug* or CPP. This system may also be referred to as *coil-on-plug* or COP. This version uses one ignition coil per each cylinder. Each coil sits directly on top of a spark plug so that the secondary ignition wires are eliminated. The advantage of this design is that the potential for induced voltages in nearby circuits from secondary wiring has been reduced. This design was introduced on 1998 vehicles with such engines as the 3.0-liter V6 and the 3.9-liter V8, as well as the popular Triton modular 5.4-liter engine used on light truck applications. By 1999, Ford used the CPP system on many of its engines. All of the CPP systems allow the PCM to directly control each ignition coil without the use of an external separate ignition module.

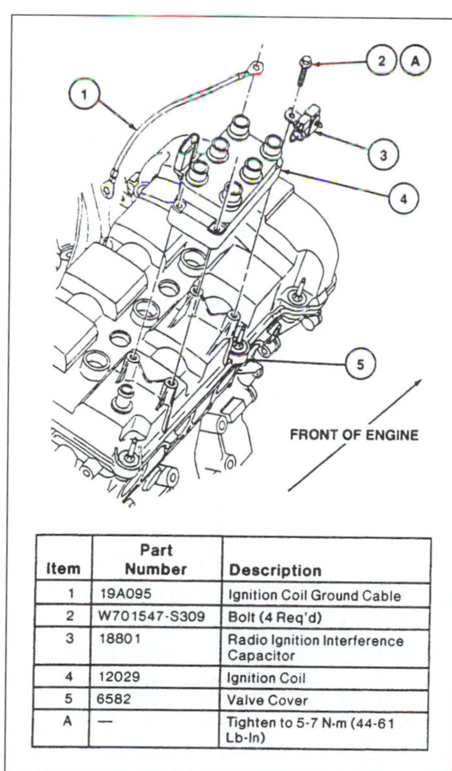

Item	Part Number	Description
1	19A095	Ignition Coil Ground Cable
2	W701547-S309	Bolt (4 Req'd)
3	18801	Radio Ignition Interference Capacitor
4	12029	Ignition Coil
5	6582	Valve Cover
A	—	Tighten to 5-7 N·m (44-61 Lb-In)

Figure 15–17 Ignition coil pack. *(Courtesy of Ford Motor Company.)*

EMISSION CONTROLS

Certain system components do not make the car go or stop or get good mileage, but they are necessary to prevent illegal toxic exhaust so that the car is allowed on the road. Some emissions components, like the PCV, work independently and have no more connection to the PCM than the glove box latch, while others are controlled as precisely as the spark advance or the fuel injection pulse width. We look at them next.

Secondary Air Injection. While the engine is warming up to a temperature at which the system can go into closed loop and regulate the air/fuel mixture according to the responses of the oxygen sensor, the air injection pump injects air into the exhaust manifold to burn as much residual fuel (hydrocarbons) as possible to minimize undesirable emissions, Figure 15–18. Once the engine and oxygen sensor are warm enough to go into closed loop and the sensor is generating its characteristic oscillating low voltage wave, the air injection switches to the catalytic converter. The switch from the exhaust manifold is the easiest way to check to see whether the system can achieve closed-loop operation.

Like other manufacturers' systems, the Ford air injection system shifts into divert mode under circumstances when the injected air could cause backfire or damage.

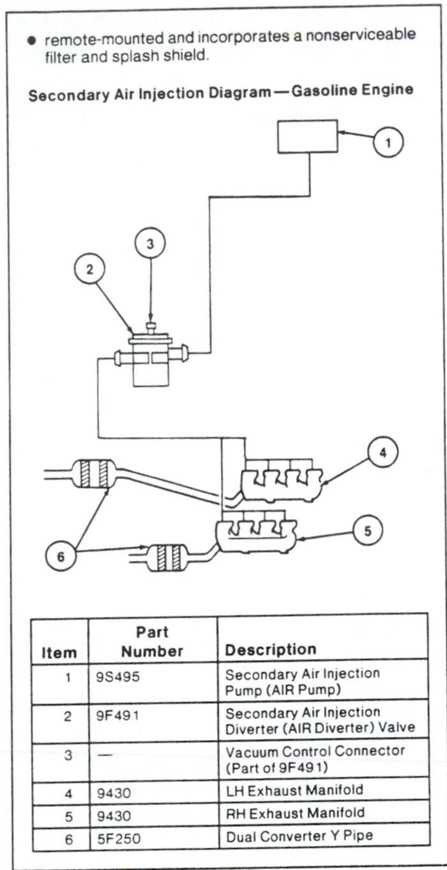

- remote-mounted and incorporates a nonserviceable filter and splash shield.

Secondary Air Injection Diagram—Gasoline Engine

Item	Part Number	Description
1	9S495	Secondary Air Injection Pump (AIR Pump)
2	9F491	Secondary Air Injection Diverter (AIR Diverter) Valve
3	—	Vacuum Control Connector (Part of 9F491)
4	9430	LH Exhaust Manifold
5	9430	RH Exhaust Manifold
6	5F250	Dual Converter Y Pipe

Figure 15–18 Secondary air injection diagram. *(Courtesy of Ford Motor Company.)*

Exhaust Gas Recirculation (EGR) System. To prevent the formation of oxides of nitrogen, NO_x, which can only form at high combustion temperatures, the system meters a certain amount of now-inert exhaust gas back through the intake manifold. The effect slows the burning of the fuel and keeps combustion temperatures below the temperature at which nitrous oxides form.

The Ford EEC V system uses a differential pressure feedback EGR (DPFE) sensor, Figure 15–19, (also known as a Delta Pressure Feedback EGR sensor) to continuously monitor the EGR flow rate. In the EEC IV system (described in Chapter 14), the pressure feedback EGR (PFE) sensor replaced the EGR valve position (EVP) sensor because the EVP sensor could not measure actual EGR flow rate, which is, of course, affected by carbon buildup in the EGR passages. While the PFE sensor is able to measure EGR flow rate, it still could be affected if an unexpected change in exhaust backpressure occurred.

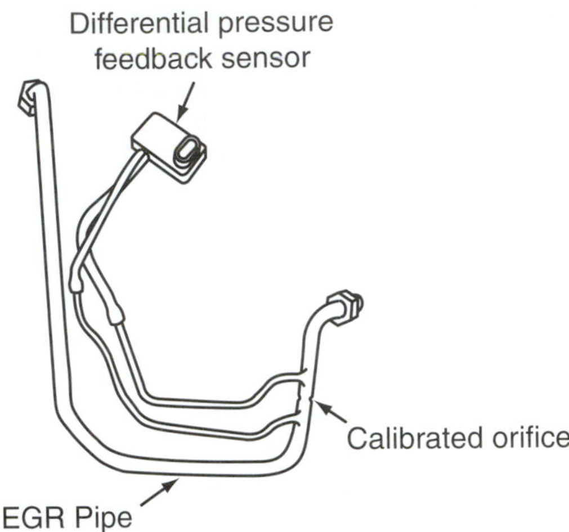

Figure 15–19 This EGR pipe assembly shows the calibrated orifice and pressure feedback sensor.

The DPFE sensor used in the EEC V system continues to measure the pressure in the EGR passage just before the EGR valve and after a restrictor placed in the passage, just as the PFE sensor did. If the EGR valve were fully closed, this would likely be a positive pressure, above atmospheric pressure. But if the EGR valve were wide open, this would likely be a negative pressure, below atmospheric pressure. In addition to measuring this pressure, the DPFE sensor also indexes this pressure value to raw exhaust pressure just before the restrictor, Figure 15–20. It actually measures the *difference* between raw exhaust pressure and the pressure between the restrictor and the EGR valve. In this way, it is able to measure accurately the EGR flow rate, even if the exhaust pressure were to change. This allows the PCM to control the air/fuel ratio more precisely, due to the fact that EGR gasses displace incoming air into the engine. In controlling the EGR valve, the PCM operates an EGR valve regulator (EVR) solenoid on a duty cycle, Figure 15–21. The EVR

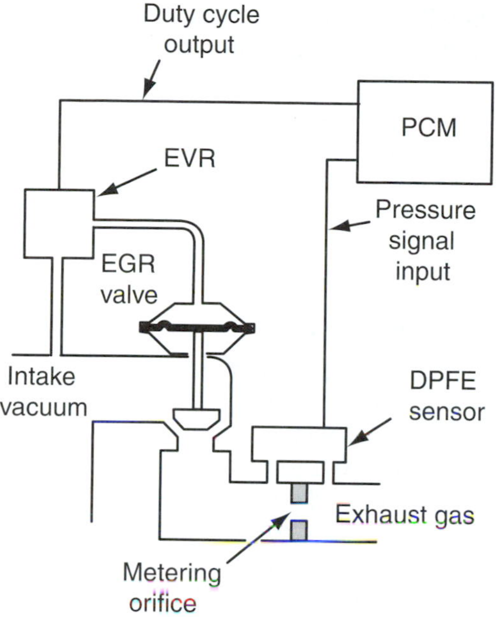

Figure 15–20 Differential pressure feedback EGR (DPFE) flow diagram.

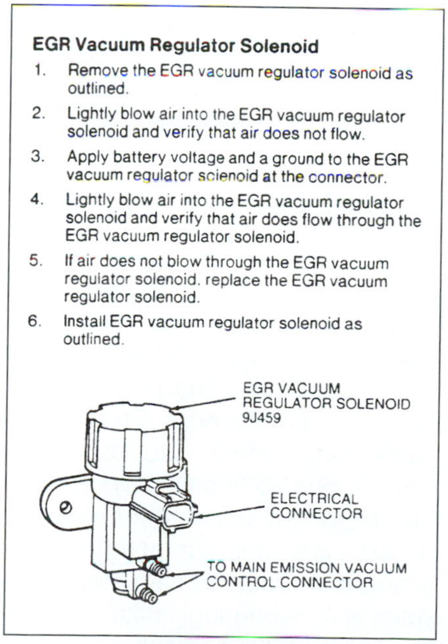

EGR Vacuum Regulator Solenoid

1. Remove the EGR vacuum regulator solenoid as outlined.
2. Lightly blow air into the EGR vacuum regulator solenoid and verify that air does not flow.
3. Apply battery voltage and a ground to the EGR vacuum regulator solenoid at the connector.
4. Lightly blow air into the EGR vacuum regulator solenoid and verify that air does flow through the EGR vacuum regulator solenoid.
5. If air does not blow through the EGR vacuum regulator solenoid, replace the EGR vacuum regulator solenoid.
6. Install EGR vacuum regulator solenoid as outlined.

EGR VACUUM REGULATOR SOLENOID 9J459

ELECTRICAL CONNECTOR

TO MAIN EMISSION VACUUM CONTROL CONNECTOR

Figure 15–21 EGR vacuum regulator solenoid. *(Courtesy of Ford Motor Company.)*

carries over from EEC IV and is described in Chapter 14.

ADVANCED EEC V FEATURES AND APPLICATIONS

ZX2 with Variable Camshaft Timing (VCT)

Ford introduced a new Escort model for 1998, the ZX2 sport coupe. The standard ZX2 engine is the 2.0-liter ZETEC-VCT (variable camshaft timing) in-line four-cylinder, an engine first seen (and still installed) in Ford Contour and Mercury Mystique vehicles.

For 1998 the ZETEC cylinder block has a new lower reinforcing member (called a *ladder frame*) to reduce perceived engine noise by half. The VCT system is also new for 1998—the first application of variable camshaft/ valve timing by Ford Motor Company. The VCT system allows intake camshaft timing to be changed (advanced and retarded) while the engine is running, which provides numerous benefits including smoother idle, better low-speed fuel economy and lower emissions. An additional benefit for the ZETEC engine is that all external EGR system hardware could be eliminated–camshaft timing can be adjusted to increase "valve overlap" which provides EGR within the engine cylinders.

The ZETEC-VCT features a cast-iron cylinder block, an aluminum, 16-valve cylinder head and a cast aluminum "structural" oil pan. Other notable features include a belt-drive system for the camshafts, intake valves with hollow stems and a distributorless ignition system. As installed in the Escort ZX2, the engine is rated at 130 horsepower. Contour and Mystique versions are rated at 125 horsepower.

2.0L with Split Port Induction (SPI)

The base 2.0-liter engine installed in Ford Escort and Mercury Tracer vehicles is fitted with a knock sensor for 1998. The single overhead engine has a cast-iron block and cylinder head. The camshaft is belt-driven and has roller followers.

Although the engine has only two valves per cylinder, its performance approaches that of a multivalve engine due to a split port induction (SPI) system. The SPI system cylinder head has two separate runners to each intake valve—one of the runners being significantly larger than the other. At engine speeds below 3,000 rpm intake air is directed through the smaller diameter passage only, increasing air velocity. This improves low-speed performance (and fuel economy) and reduces emissions. When engine speed exceeds 3,000 rpm, the larger passage is also used, allowing the maximum amount of air to enter the cylinders. This improves engine high-speed performance. The 2.0-liter SPI engine is fitted with a distributorless ignition system and is rated at 110 horsepower.

3.0L Duratec Engine

AX4N transaxle modifications (a higher stall torque converter and revised axle ratio) improve throttle response on the 3.0L Duratec V6, available in the 1998 Ford Taurus and the Mercury Sable. The Duratec features an aluminum alloy cylinder block, cylinder heads and intake manifold, as well as chain-driven, dual overhead camshafts and four valves per cylinder. A PCM-controlled short/long runner induction system optimizes both low- and high-speed power output. Also notable are a roller camshaft followers, a distributorless ignition system and the "Securilock" passive antitheft system (PATS) which disables the ignition system if the correct ignition key is not used. The 3.0L Duratec develops 200 horsepower.

2.5L Duratec Engine

The Duratec 2.5L V6 is available in the Ford Contour and the Mercury Mystique. A compact 60 degree V6, the Duratec features an aluminum alloy cylinder block and intake manifold. Cylinder heads are also made of aluminum and have dual overhead camshafts and four valves per cylinder. Camshafts are chain driven. A PCM-controlled short/long runner induction system optimizes both low- and high-speed power output. Also notable are a distributorless ignition system and a Securilock passive antitheft system (PATS) which disables the ignition system if the correct ignition key is not used. The Duratec 2.5L develops 170 horsepower.

3.0L Vulcan Engine

The 3.0L Vulcan camshaft-in-block V6 installed in some Ford Taurus, Mercury Sable and Ford Windstar vehicles is a carryover powerplant this year. The 60-degree "v" block is made of cast iron, as are the cylinder heads. The intake manifold is cast aluminum and the camshaft is manufactured using "assembled" camshaft technology. The Vulcan V6 is rated at 145 horsepower.

Two flex-fuel versions of the 3.0L Vulcan V6 are also available on Ford Taurus vehicles—one intended for use with methanol and/or gasoline, the other for ethanol and/or gasoline. These engines have a slightly higher compression ratio than the gasoline-only Vulcan and also have different air intake and fuel delivery systems. Flex-fuel engine power ratings are also slightly higher (5–10 horsepower, depending on emission package) than the gasoline-only engine.

3.4L Super High Output (SHO) Engine

The 3.4L V8 engine is available only in the Taurus SHO. The 3.4L cylinder block is made of cast aluminum, with a 60-degree "v" between banks. Cylinder heads are also made of aluminum and feature chain-driven, dual overhead camshafts and four valves per cylinder. The engine has a short runner/long runner induction system with a separate secondary throttle valve for each cylinder. Other important features of the engine are a vibration-reducing counterrotating balance shaft, direct-acting nonhydraulic cam followers, and a distributorless coil-on-plug ignition system. The SHO V8 is rated at 235 horsepower.

4.6L SOHC/DOHC Engines

The 4.6L SOHC V8 is available in the Ford Mustang GT, Ford Crown Victoria, Mercury Grand Marquis and Lincoln Town Car. The engine has a deep-skirted, cast-iron block and aluminum cylinder heads with two valves per cylinder. The single overhead camshafts use "assembled" technology and are chain driven. The connecting rods are made of powder metal and utilize "cracked" cap technology. Other significant features include an aluminum timing cover and a distributorless ignition system.

The Mustang GT version of the 4.6L SOHC engine is rated at 225 horsepower. Mustangs fitted with the 4.6-liter V8 engine and an automatic transmission qualify as transitional low emission vehicles (TLEV).

Crown Victoria and Grand Marquis versions, with slightly different manifolding, are rated at 200 horsepower with single exhaust, 215 with dual exhaust. On the Town Car, the engine is rated at 205 horsepower with single exhaust, 220 with dual exhaust.

Also available is the 4.6L DOHC InTech V8. The high performance engine features an aluminum cylinder block, cylinder heads, front timing cover and intake manifold (short/long runner type). Crankshaft main bearing caps are retained using six bolts per cap (two are cross bolts); connecting rods feature powder metal and "cracked" cap technology. Dual overhead camshafts are fitted, operating four valves per cylinder. Each exhaust camshaft is driven by its own chain off crankshaft sprockets. Two hydraulic tensioners are used to take up chain play. Intake camshafts are also chain driven—each by a secondary chain connected to the exhaust camshaft on that cylinder bank. Each of the secondary chains also has a hydraulic tensioner. Other significant features include roller rocker arms, torque-to-yield (TTY) crankshaft main bearing cap, connecting rod cap, and cylinder head bolts and a distributorless ignition.

The InTech V8, when installed in the Lincoln Mark VIII, is rated at 280 horsepower; in the LSC it is rated at 290 horsepower. The transverse-mounted version in the Lincoln Continental is rated at 260 horsepower. An even higher performance version of the InTech, fitted with special cylinder heads, camshafts and intake and exhaust systems is fitted to the Mustang Cobra and is rated at 305 horsepower.

EEC V System Truck Engines

The four cylinder engine available in Ford Ranger pickups was new for 1998. The engine displaces 2.5 liters and is based on the 2.3-liter powerplant. Technical features include a cast-iron cylinder block, an aluminum two-piece intake manifold and a distributorless ignition system. The new 2.5-liter's power output is 146 horsepower.

The Mercury Mountaineer SUV received a new standard engine for 1998, the 4.0-liter SOHC V6 introduced in 1997 on the Ford Explorer. While based on the 4.0-liter pushrod engine, virtually no components are shared between the two powerplants.

The 4.0L SOHC engine has a cast-iron cylinder block containing some unique features. An "assembled" intermediate shaft is located where the camshaft would be located on a camshaft-in-block engine. This intermediate shaft is driven by a chain-and-sprocket setup off the crankshaft nose. The intermediate shaft is then used to drive the overhead camshafts using two more chain-and-sprocket sets. The left-hand camshaft is driven from the front end of the intermediate shaft, while the right-hand camshaft is driven from the rear end.

The cylinder heads are made of aluminum and have pressed-in sintered iron valve guides and seats. The camshafts are "assembled" from steel tubes and sintered iron lobes and operate two valves per cylinder through roller followers. An oil spray tube mounted to each head is used to direct pressurized oil to valvetrain components.

The oil pan consists of two separate pieces, an upper "ladder frame" and a lower cover plate.

The ladder frame is an aluminum casting that bolts to both the cylinder block and the crankshaft main bearing caps to provide structural strength. The lower cover plate is made of stamped steel. Engines built for installation in four-wheel-drive vehicles have a chain-driven balance shaft assembly mounted to the bottom of the cylinder block, inside the oil pan.

Other notable features of the 4.0L SOHC engine are a two-piece thermo-plastic intake manifold with a low/high speed tuning valve, cast aluminum cam covers, hydraulic chain tensioners, forged steel connecting rods with integral oil spray passages and a distributorless ignition system. The 4.0L SOHC engine is rated at 205 horsepower.

Also new for 1998 was a bifuel version (natural gas/propane) of the 5.4-liter Triton V8 for the F-Series pickup and Econoline van. The engine shares the same technical features as the gasoline-fueled 5.4-liter Triton described later in this section, but has some unique components for bifuel. Some of these are the intake manifold, fuel injectors and throttle body.

The 3.0-liter camshaft-in-block V6 available in Ford Ranger pickup trucks received a new upper intake manifold for 1998, resulting in a 14 percent increase in engine torque output. The 3.0-liter engine features a cast-iron block and cylinder heads. Roller lifters are used for reduced valvetrain friction. The 3.0-liter V6 is rated at 150 horsepower.

The 4.0-liter camshaft-in-block V6 is available in the Explorer SUV and the Ranger pickup. The 4.0-liter engine has a cast-iron cylinder block and heads. The lower intake manifold and structural oil pan are made of cast aluminum. The upper intake manifold is made of a thermo-plastic. Other notable features are a cast steel camshaft, roller lifters, forged steel connecting rods with oil spray passages, a stamped steel windage tray and distributorless ignition system. The 4.0-liter pushrod engine is rated at 160 horsepower.

The 4.2-liter V6 engine is available in Ford F-Series pickups and Econoline vans. The 4.2 is a stroked version of the 3.8-liter V6 used in the Windstar minivan, and has the same features (including the split port induction system). The 4.2-liter V6 is rated at 210 horsepower.

The 4.6-liter Triton V8 is available in the Ford F-Series pickup, Expedition SUV and Econoline van. The first of Ford's "modular" truck engines, the 4.6-liter Triton features a cast-iron, deep-skirted block and aluminum cylinder heads and two valves per cylinder. The single overhead camshafts use "assembled" technology and are chain driven. The connecting rods are made of powder metal and utilize "cracked" cap technology. Other notable features are an aluminum timing cover and a distributorless ignition system.

The F-Series pickup version of the engine is rated at 220 horsepower, the Expedition version at 215.

The 5.4-liter Triton modular V8 is available in Ford F-Series pickups, the Ford Expedition and Lincoln Navigator. It shares the same technical features as the 4.6-liter Triton engine on which it is based, but also adds an intake manifold tuning system and a distributorless coil-on-plug ignition system. The 5.4-liter engines installed in F-Series pickups are rated at 235 horsepower. The Expedition and Navigator version is rated at 230 horsepower.

The 5.8-liter camshaft-in-block V8 is available only in the Ford Heavy-Duty F-Series pickups. The engine has a cast-iron block and cylinder heads. Roller lifters are used to reduce valvetrain friction. The 5.8-liter engine develops 210 horsepower.

Intake Manifold Tuning Valve

The Ford 3.0L/4V engine uses a computer-controlled 12-volt motor to operate a door between the two intake manifold plenums. The intake manifold tuning valve (IMTV) has two positions. Normally the intake plenum is divided into two separate areas designed to deliver a smooth idle. When activated, the motor connects the two intake manifold plenums. By joining the two plenums, air waves from one plenum bounce into the other plenum and deliver a pulsation enrichment charge, which increases engine performance at higher speeds.

LINCOLN LS

Powertrain Control Module (PCM)

The 2000 Lincoln LS has a new 150-pin Powertrain Control Module (PCM). This PCM, shown in Figure 15–22, has three separate harness connectors. These harness connectors are the engine pin connector, the transmission connector and the cowl connector. The 60-pin engine connector is used for inputs and outputs to provide engine control. The 32-pin transmission connector is used for inputs and outputs to provide transmission control. The 58-pin cowl connector is used for inputs and outputs to manage support services, such as the electronic speed control.

Unlike previous control modules, this control module no longer needs a breakout box to aid in diagnosis. All diagnostic information is available on the vehicle communication network.

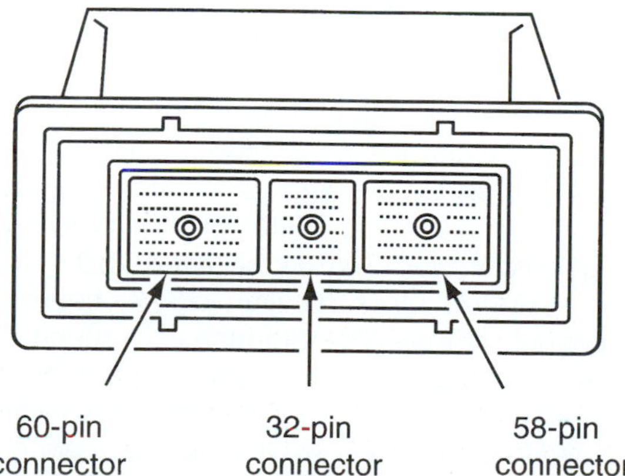

Figure 15–22 The new 150-pin PCM manages engine, transmission and support services.

Idle Air Bypass Valve/Air-Assisted Fuel Injectors

The Lincoln LS 3.9L engine has a modified idle air bypass valve, as shown in Figure 15–23. During cold engine operation, part of the idle bypass air travels to passages in the intake manifold that lead to the fuel injectors. Idle bypass air is added to the air/fuel mixture at the tip of the fuel injector. This additional air increases atomization and, as a result, lowers emission levels. Air injection hose connections can be released by first depressing the tab on the connector and then by pulling straight back until the fitting is released.

Computer-Controlled Cooling Fan

The Lincoln LS uses a gerotor hydraulic cooling fan. The engine drives a hydraulic pump, which supplies the fluid flow to operate the cooling fan. The pump is mounted on the right side of the engine. The fluid reservoir is located on the

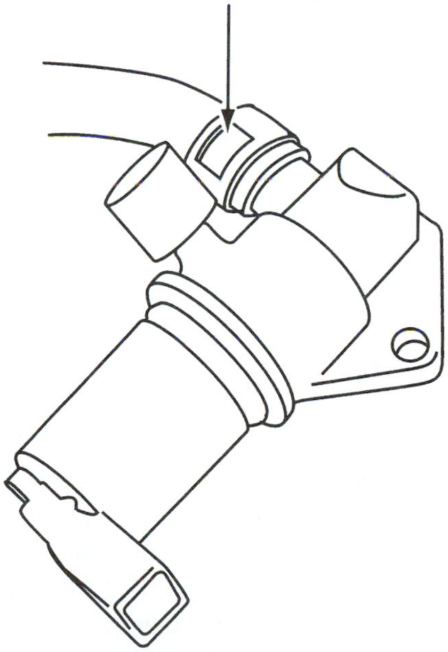

Squeeze tabs and separate

Figure 15–23 Air injection tubes are released from the idle air bypass valve by depressing the clamp tabs.

right fender well. The hydraulic fluid is routed through a PCM-controlled pressure variable orifice (PVO) solenoid on its way to the fan. The cooling fan runs continuously as long as the engine drives the pump. The PCM uses pulse width modulation technology to control the fluid flow through the PVO. This fluid flow determines the speed of the fan. The system uses a special fluid, part number WSA-M2C195-A. It should not be mixed with any other type of hydraulic fluid.

Auxiliary Coolant Pump

The latest Lincoln LS model is equipped with an electric auxiliary coolant pump in addition to the gerotor hydraulic cooling fan. The auxiliary coolant pump is used whenever the dual automatic temperature control module (DATCM) requests heat. The electric pump increases coolant flow into the dual-zoned heater core during idle. The electric auxiliary coolant pump has a single-speed DC motor. The DATCM controls the water flow with two pulse width modulated water control valves. This eliminates the need for temperature blend doors and electric modulators. If coolant is lost and the auxiliary coolant pump runs dry, it cannot be serviced; it must be replaced.

Combination Cooler

The Lincoln LS uses a three-in-one combination cooler assembly to provide separate cooling for the hydraulic gerotor cooling fan fluid, the transmission fluid and the power steering fluid. The combination cooler is located in front of the radiator.

Fail-Safe Cooling

The Lincoln LS fail-safe cooling strategy is similar to that in other Ford vehicles. Whenever engine temperature rises above 260°F, the PCM disables two injectors at a time. The disabled cylinders are charged only with air. The air-only charge reduces engine temperature. If the temperature rises above 330°F, the PCM shuts off all injectors until the engine temperature drops below 310°F. The PCM stores diagnostic trouble code 1299.

Fuel Damper

The 3.9L Lincoln LS engine is equipped with a dual fuel damper. The fuel line damper reduces fuel line pressure fluctuations caused by the fuel injector opening and closing.

Inertia Fuel Cutoff Switch

The inertia fuel cutoff switch has been used by Ford for many years. This switch turns off the power to the fuel pump in case of an accident. It is located in the front right foot well behind the trim panel.

✔ SYSTEM DIAGNOSIS AND SERVICE

Many of the diagnostic concepts that Ford utilized in the EEC IV system (and described in Chapter 14) carried over into the EEC V system. EEC V also integrated the necessary OBD II standards. As a result, the EEC V system surpasses past systems in its ability to help the technician diagnose the system.

Manual methods of code pulling are eliminated with this OBD II system. That is, connector terminals at the DLC (located under the left end of the instrument panel as per OBD II standards) should never be shorted with a jumper wire, but, instead, a scan tool should always be used.

Diagnosis should begin by connecting Ford's New Generation Star Tester (NGS), Worldwide Diagnostic System (WDS), or other OBD II scan tool to the DLC. Once the scan tool has been programmed with the VIN information, a menu displays all of the diagnostic features available for the particular application. These features include a Key On, Engine Off (KOEO) on-demand self-test, a Key On, Engine Running (KOER) on demand self-test, the ability to pull continuous memory DTCs from the PCM's memory, and the ability to use the scan tool to erase continuous memory DTCs. No longer must a KOEO self-test be performed in order to pull continuous memory DTCs (as with the EEC IV system described in Chapter 14). Rather, continuous memory DTCs now appear in the scan tool's menu as a separate menu item. Continuous memory DTCs should always be erased with a scan tool, never by disconnecting the battery.

Ford's Quick Test is defined with the EEC V system as consisting of the following items:

- visual check
- vehicle preparation and equipment hookup
- KOEO on-demand self-test
- KOER on-demand self-test
- pulling continuous memory DTCs.

About the only functional test that carried over from the EEC IV system is an active command mode (somewhat similar to the EEC IV output state check) that allows a technician to command the PCM to energize/de-energize various actuators, although a scan tool must now be used. Several actuators have been added to what the EEC IV system allowed the technician to command. (Other functional tests associated with the EEC IV system, such as the SEFI cylinder balance test, have been dropped due to the addition of certain OBD II-required monitors such as the misfire monitor that allows the PCM to be acutely aware of cylinder misfire problems.)

A parameter identification mode has been added to allow the technician to access certain values, analog and digital inputs/outputs, calculated values and system status information. Diagnostic features also added include an On-Board System Readiness (OSR) test that allows the technician to view the status of all OBD II monitors (see Chapter 6 for more information on OBD II) and whether they are complete or incomplete. Incidentally, Ford has a unique DTC, P1000, that indicates that the OBD II monitors have not been completed since the last battery disconnect. Also, per OBD II standards, **Freeze Frame/Snapshot** data is also available with any stored DTCs by accessing the generic OBD II feature on the scan tool's menu. This data allows the technician to know what many of the sensors were reporting when a fault occurred, as well as other pertinent information such as fuel trim values.

SUMMARY

In this chapter we covered the features of the new Ford PCM, the OBD II-compliant computer and the engine controls it manages. We looked at the new inputs and outputs of the system, and we covered the sensors and actuators carried over or modified from the previous Ford engine management systems: the throttle position sensor, the hot-wire mass air flow sensor, the intake air temperature sensor, the engine coolant temperature sensor, the oxygen sensor, the camshaft position sensor, the crankshaft position sensor, the knock sensor, the power steering pressure switch, the vehicle speed sensor and the idle air control valve.

We reviewed the provisions built into the system for alternative fuel capacity, using various blends of fuel that are less than 100 percent gasoline. We have seen the application of the standardized OBD II self-diagnostics program, including the freeze frame/snapshot memory functions.

We've seen how the Ford system's emissions controls have been revised and updated, and we've grown more familiar with the standardized OBD II diagnostic systems.

▲ DIAGNOSTIC EXERCISE

An EEC V vehicle is towed into the shop. The complaint of "engine cranks, but won't start" is verified. One technician prepares to test the ignition system for spark using a spark tester. Another technician prepares to connect a fuel pressure gauge to the fuel rail in order to test fuel system pressure. What easier check are these technicians forgetting that should be performed first?

REVIEW QUESTIONS

1. *Technician A* says that the EEC V PCM is able to determine the gradual wear and aging of the vehicle that occurs over time.
   Technician B says that the EEC V PCM is able to compensate for gradual wear and aging of the vehicle by making adjustments in its programs.
   Who is correct?
   A. A only
   B. B only
   C. Both A and B
   D. Neither A nor B
2. What type of sensor does the EEC V system use to measure engine load?
   A. Vane air flow (VAF)
   B. Manifold absolute pressure (MAP)
   C. Mass air flow (MAF)
   D. Pressure feedback EGR (PFE)
3. *Technician A* says that the purpose for the oxygen sensor located after the catalytic converter is to act as a backup sensor for air/fuel mixture control in case the pre-cat sensor fails.
   Technician B says that the purpose for the oxygen sensor located after the catalytic converter is to monitor the catalytic converter's effectiveness.
   Who is correct?
   A. A only
   B. B only
   C. Both A and B
   D. Neither A nor B
4. The power steering pressure (PSP) switch input primarily affects the PCM's control of which of the following?
   A. Idle speed
   B. Injector pulse width
   C. Spark timing
   D. The fuel pump relay
5. *Technician A* says that the crankshaft position (CKP) sensor allows the PCM to identify which pair of pistons is approaching TDC, enabling it to fire the correct ignition coil in the waste spark ignition system.
   Technician B says that the camshaft position (CMP) sensor provides the PCM with the information needed for fuel injection sequencing with the engine's firing order.
   Who is correct?
   A. A only
   B. B only
   C. Both A and B
   D. Neither A nor B
6. What is the purpose of the inertia switch used in the EEC V system?
   A. To disable the starter circuit in case of impact
   B. To remove power from all fuel injectors in case of impact
   C. To shut down the ignition system in case of impact
   D. To shut off the fuel pump in case of impact
7. An EEC V vehicle is brought into the shop with a fuel injection system that vents fuel past a pressure regulator through a return line back to the fuel tank.
   Technician A says that the fuel pressure regulator is indexed to manifold pressure (vacuum) in order to increase the flow rate of the injectors while under heavy load.
   Technician B says that the fuel pressure regulator is indexed to manifold pressure (vacuum) in order to maintain a constant pressure differential across the injectors so as to maintain a constant flow rate anytime that the injectors are energized.
   Who is correct?
   A. A only
   B. B only
   C. Both A and B
   D. Neither A nor B
8. What is the primary purpose for designing a vehicle with a returnless fuel injection system?
   A. To reduce vapor lock potential
   B. To reduce how much engine compartment heat is transferred to the fuel tank
   C. To increase fuel economy
   D. To increase engine performance

9. In a returnless fuel injection system, because the fuel pressure regulator is not indexed to intake manifold pressure, which of the following will result?
   A. The fuel pressure differential across the fuel injectors will be constant under all driving conditions.
   B. The fuel rail pressure will be constant under all driving conditions.
   C. The PCM will have to calculate the pressure differential across the injectors (and the injector's flow rate) before it can calculate the required pulse width.
   D. Both B and C
10. Which of the following ignition systems may be used with an EEC V computer system?
    A. TFI-IV distributor ignition system
    B. EDIS waste spark distributorless ignition system
    C. CPP (COP) distributorless ignition system
    D. All of the above
11. What is the primary advantage of a CPP (COP) distributorless ignition system?
    A. The PCM has more accurate control of ignition timing than with a waste spark ignition system.
    B. The potential for secondary voltage to induce a voltage in a nearby circuit has been reduced.
    C. Mounting ignition coils over the spark plugs make the plugs easier to change at the recommended service intervals.
    D. CPP (COP) systems can continue to operate if the crankshaft position (CKP) sensor were to fail.
12. When does the secondary air injection system direct air flow to the exhaust manifold(s)?
    A. During engine warmup
    B. During closed-loop operation
    C. During heavy load conditions when the air/fuel mixture is rich
    D. All of the above
13. *Technician A* says that a DPFE sensor is used on EEC V systems to measure the physical position of the EGR valve and cannot compensate for reduced EGR flow rate caused by carbon buildup.
    Technician B says that a DPFE sensor is used on EEC V systems to measure EGR flow rate, and is even able to compensate for changes in exhaust backpressure.
    Who is correct?
    A. A only
    B. B only
    C. Both A and B
    D. Neither A nor B
14. The Variable Camshaft Timing (VCT) system used on 1998 and newer 2.0-liter ZETEC engines is able to do which of the following?
    A. Vary the timing of the intake camshaft
    B. Vary the timing of the exhaust camshaft
    C. Vary the timing of the both the intake and exhaust camshafts
    D. Vary the timing of the intake camshaft and simultaneously vary when the distributor produces the tach reference pulse
15. Which emission system is physically eliminated by the use of the Variable Camshaft Timing (VCT) system due to its ability to control valve overlap?
    A. PCV
    B. Secondary air injection
    C. EGR
    D. Evaporative and canister purge system
16. *Technician A* says that on the 1998 and newer 2.0-liter base engine, the Split Port Induction (SPI) system allows the air intake velocity to be increased at lower engine speeds, thereby increasing low-speed engine performance and fuel economy.
    Technician B says that on the 1998 and newer 2.0-liter base engine, the Split Port Induction (SPI) system allows more air to enter the cylinders at higher engine speeds, thereby increasing high-speed engine performance.
    Who is correct?
    A. A only
    B. B only
    C. Both A and B
    D. Neither A nor B

17. The Lincoln LS 3.9L engine has a modified idle air bypass valve. What is the purpose of this modification?
    A. To deliver some idle bypass air through an air manifold to each of the fuel injectors
    B. To add idle bypass air to the air/fuel mixture at each of the fuel injectors
    C. To increase fuel atomization and reduce emissions
    D. All of the above
18. If engine temperature rises above 260°F on a Lincoln LS with the fail-safe cooling strategy, the PCM will do which of the following?
    A. Turn off one fuel injector at a time until engine temperature drops
    B. Turn off two fuel injectors at a time until engine temperature drops
    C. Turn off three fuel injectors at a time until engine temperature drops
    D. Turn off all fuel injectors until engine temperature drops
19. *Technician A* says that when diagnosing an EEC V system, DTCs may be pulled manually by connecting a jumper wire between two terminals at the Data Link Connector (DLC). *Technician B* says that when diagnosing an EEC V system, in order to perform either self-test or pull continuous memory DTCs, an OBD II scan tool must be used.
    Who is correct?
    A. A only
    B. B only
    C. Both A and B
    D. Neither A nor B
20. On the EEC V system, Ford's DTC P1000 indicates which of the following?
    A. The PCM has failed and should be replaced.
    B. The PCM has detected a fault on the input side that should be diagnosed and repaired.
    C. The PCM has detected a fault on the output side that should be diagnosed and repaired.
    D. The PCM has not seen the OBD II monitors completed since the last battery disconnect.

Chapter 16

Chrysler's Oxygen Feedback System

OBJECTIVES

Upon completion and review of this chapter, you should be able to:

- Describe the Chrysler oxygen feedback system operating modes.
- Define the inputs that are used with the Chrysler oxygen feedback system.
- Define the outputs that are controlled by the Chrysler oxygen feedback system.
- Understand how the Chrysler oxygen feedback system is designed to be put into self-diagnosis as well the various diagnostic features.

KEY TERMS

Inductance
Transducer

In 1972, Chrysler was the first carmaker to switch from a mechanical points-and-condenser ignition system to electronic ignition. So the company was ready with a computer-controlled engine system in 1979 when they began systematic installation of an oxygen sensor feedback system on many vehicles. The first system was on the California models with the slant-six engine. It combined the previous Chrysler spark advance control with an oxygen sensor that controlled a mixture control solenoid. The next year the system was extended to cover four- and eight-cylinder engines as well. By 1981 it applied to federal-specification vehicles too. By that date, it also had idle speed control functions along with EGR and air injection actuation. As the system developed, the shift indicator light, the radiator cooling fan and the vacuum solenoid controlling the carburetor secondary barrel on cars built for high altitudes all followed. Figure 16–1 shows a system overview.

COMBUSTION CONTROL COMPUTER

In the earliest systems, the combustion control computer was in the air cleaner, similar to the previous lean burn system's control unit. Like this unit, the early combustion control computers had problems with vibration and the variability of temperature at the air cleaner. Later versions located the computer in various more secure places in the engine compartment. Eventually, of course, Chrysler engineers would move the computer into the passenger compartment for some systems.

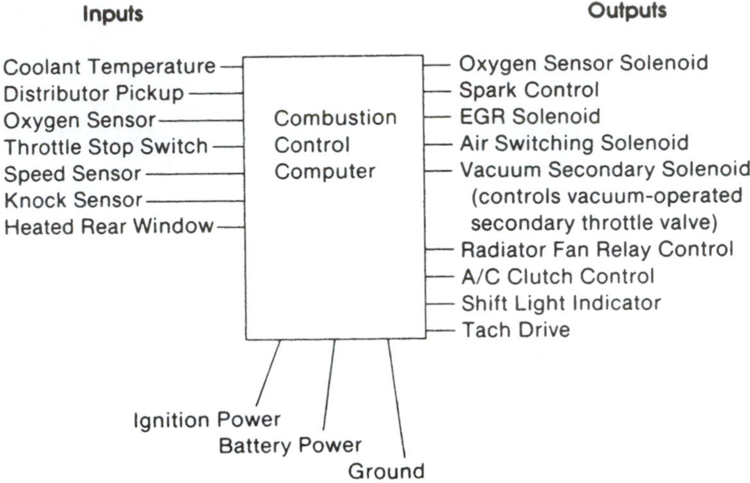

Figure 16–1 Oxygen sensor feedback system overview.

Chrysler literature sometimes refers to the earliest controller as the "spark control computer." From 1985 on, self-diagnostics facilitate system diagnosis.

Open Loop

In open loop, the computer does not use the oxygen sensor's signal, but calculates the fuel/air mixture from other sensors' inputs. The system is in open loop if any of the following conditions exist:

- the coolant is below the required operating temperature (120°F) (49°C).
- the oxygen sensor is below the required operating temperature (600°F) (315°C).
- the engine is at idle speed.
- the engine is accelerating the car (low manifold vacuum, relatively high load).
- the engine is decelerating the car (high manifold vacuum, relatively low load).
- the engine is restarted when hot.
- the oxygen sensor or its circuit (as well as some other sensors' and actuators' circuits affecting mixture control) have failed.

Closed Loop

In closed loop, the computer determines fuel/air mixture from the oxygen sensor signal as well as from the other sensor signals. The system goes into closed loop only after *all* of these conditions are true:

- the coolant reaches a specified temperature (on most vehicles about 120°F (49°C)).
- the oxygen sensor produces a recognizable voltage (beginning about 600°F (315°C)). This voltage ranges from 0.3 to 0.9 volt. The higher the voltage, the richer the mixture; the lower the voltage, the leaner the mixture. A new oxygen sensor cycles back and forth more often than once per second; a long-used sensor begins to slow its cycles.
- a predetermined time has elapsed since the engine started.

INPUTS

Coolant Temperature

Chrysler has used three different kinds of coolant temperature sensors on different oxygen sensor feedback systems: a thermistor, a dual thermistor and an on/off switch. See Figure 16–2.

Thermistors. A single element thermistor was first used on six- and eight-cylinder engines and later on four cylinders. About 1985, some transverse engines came with dual element thermistors, Figure 16–3. A thermistor is a semiconductor that changes electrical resistance as its temperature changes. The Chrysler system uses a negative temperature coefficient (NTC) thermistor, which decreases resistance as its temperature increases. The engine computer reads the voltage drop across the NTC thermistor. As its temperature increases, the thermistor's resistance and the voltage signal to the computer both decrease.

The dual element thermistor sends one signal for the fuel mixture calculation function (in general, the warmer the engine, the leaner the mixture can be with acceptable driveability, up to reasonable limits), along with spark, EGR and canister purge. The second signal controls the radiator fan relay. Some oxygen sensor feedback systems use a separate temperature sensor in the radiator inlet tank to control the radiator fan relay.

Temperature Sensing Switch. Most four-cylinder engines up to 1983 use a temperature sensitive switch to signal the computer once the closed-loop criterion coolant temperature is reached.

Charge Temperature

Some six- and eight-cylinder engines have a temperature sensor in the intake manifold to report the intake air temperature. Because the heated air inlet system brings the intake air temperature up to normal operating temperature long before the coolant reaches operating temperature, some computers employ both a charge mixture temperature sensor and a coolant temperature sensor. This allows the computer to set an air/fuel mixture leaner than would be indicated by the coolant temperature alone.

Distributor Pickup

On the four-cylinder engines, engine speed and crankshaft position information comes from

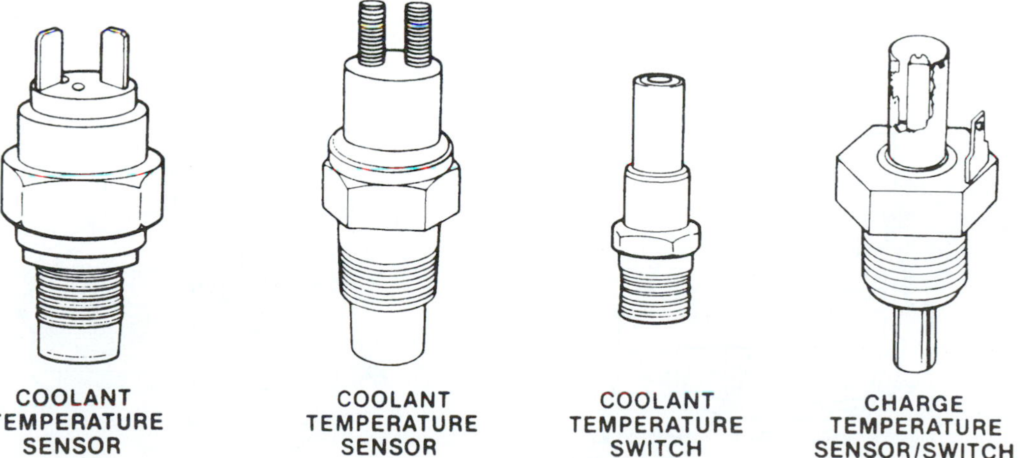

Figure 16–2 Temperature sensors/switches. *(Courtesy of Chrysler Corp.)*

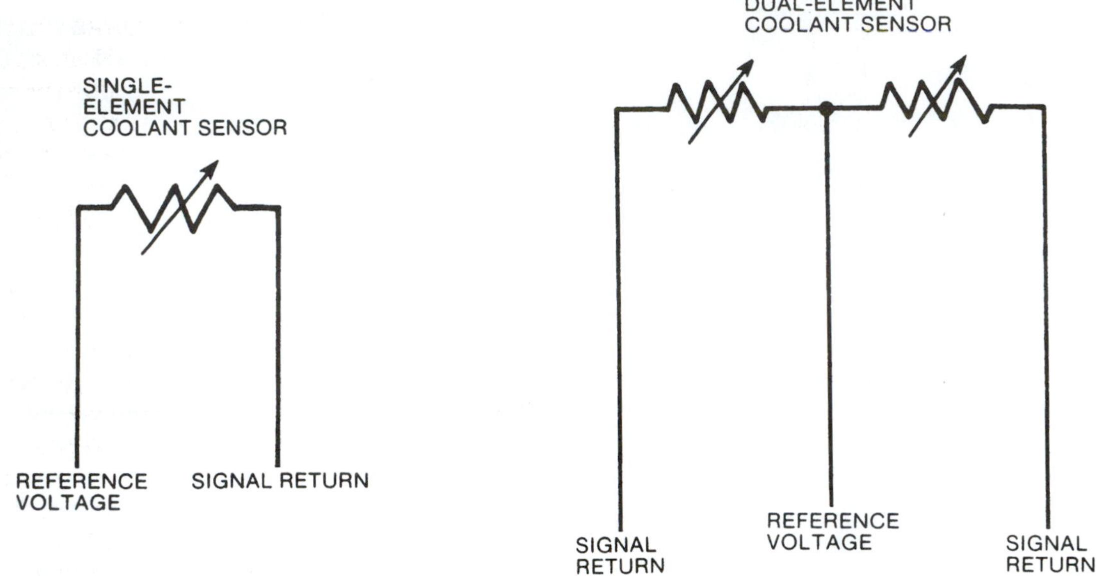

Figure 16–3 Coolant sensor schematics.

the Hall effect switch in the distributor, Figure 16–4. This is an on-off, direct current signal. On the six- and eight-cylinder engines, this information comes from the dual permanent magnet pickup coils in the distributor, a voltage-polarity-reversing alternating current. Note that the Hall effect distributor, unlike most other computer inputs, receives an eight-volt reference signal from the computer, not the standard five volt. Very late model Hall effect distributors (and some other sensors on Chrysler products) get a 9+-volt reference signal to accommodate a different kind of

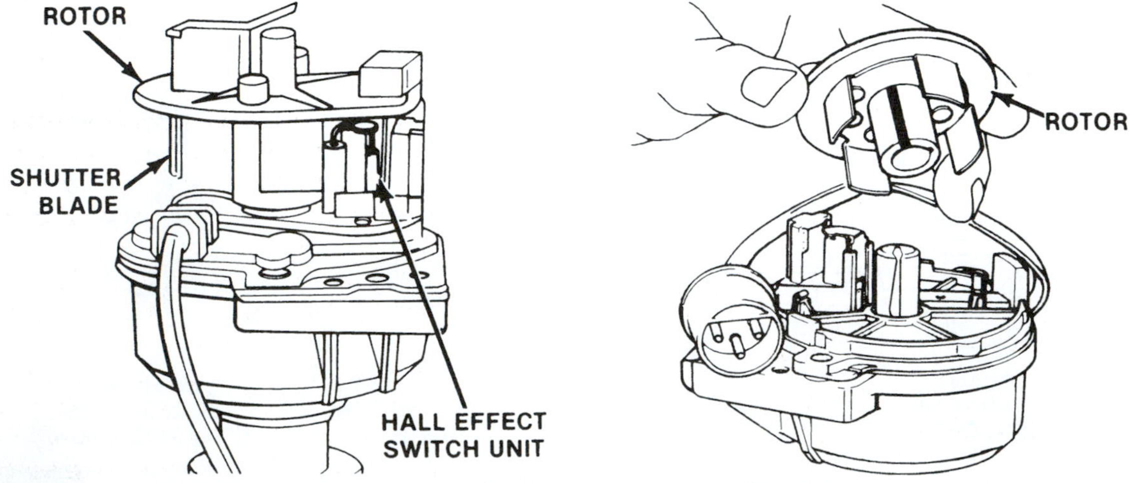

Figure 16–4 Hall effect distributor switch. *(Courtesy of Chrysler Corp.)*

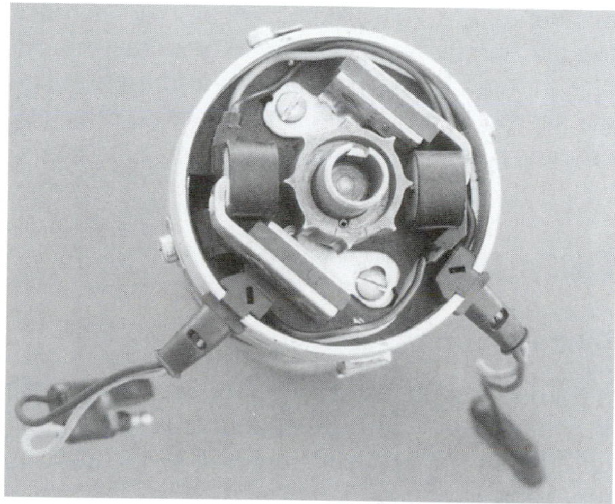

Figure 16–5 A distributor with dual permanent magnet pickups.

transistor. The inductive pickups in the six- and eight-cylinder pickups generate their own alternating current signal, Figure 16–5.

Vacuum Transducer

The computer gets intake manifold vacuum information, which corresponds inversely to engine load, from the vacuum transducer, Figure 16–6. In response to differences in intake manifold vacuum, the vacuum **transducer** changes the electrical properties of a coil through which the computer sends a reference alternating current. The intake manifold vacuum works against a spring to slide a movable metallic core in the electrical coil. As the force of the manifold vacuum increases, the metallic core is withdrawn; as the vacuum decreases, the spring drives the core back into the coil.

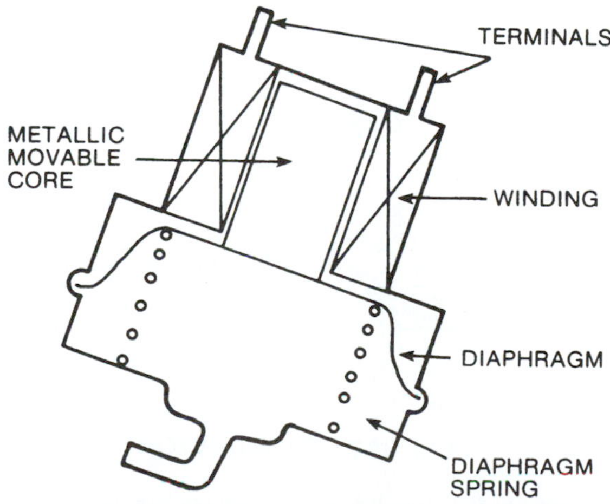

Figure 16–6 Vacuum transducer.

The computer passes an alternating current through the coil, and the alternating magnetic field produced by that current produces an **inductance**, an electromotive force produced by the varying current in the coil. Inductance is a kind of dynamic resistance or impedance specific to alternating current circuits through a coil. The alternating current itself builds up this resistance to further current flow as a result of its fluctuating magnetic field. This inductance changes as the metallic core changes position within the coil. The deeper the core is in the coil, the lower is the inductance of the coil. The computer measures voltage drop through the coil, and that corresponds to engine load.

Oxygen Sensor

The early Chrysler feedback systems use a single wire, unheated oxygen sensor. Like all oxygen sensors, it consists of a small ceramic thimble exposed to ambient air on the inside and exhaust on the outside. A superthin layer of zirconium on each side generates a low voltage, low amperage current reflecting the difference between the oxygen in the two gasses.

Throttle Stop Switch

The early Chrysler systems do not use a throttle position sensor, but instead a throttle position switch, Figure 16–7. A button switch on the throttle kicker (or dashpot) tells the computer when the driver has removed his or her foot from the accelerator pedal: the circuit completes, and the signal is returned. There was one exception: in 1979 Chrysler systems used a throttle position

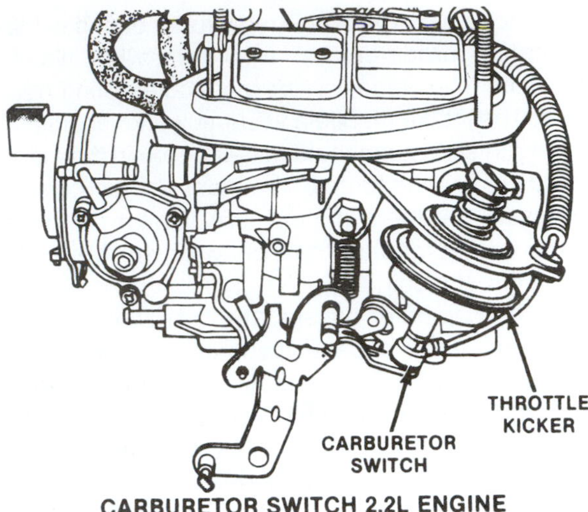

Figure 16–7 Throttle stop switch. *(Courtesy of Chrysler Corp.)*

transducer, which provides somewhat more detailed information about foot and throttle position.

Speed Sensor

An inductive pulse generator, driven by the transmission output shaft, signals the computer information about the vehicle speed. Like all inductive sensors, an increased speed of the signal is accompanied by increased voltage and frequency.

Detonation Sensor

On high-performance four-cylinder engines and on eight cylinders with four-barrel carburetors, a piezoelectric knock sensor signals the computer when frequencies characteristic of detonation occur. Again, this is an alternating current signal. This sensor is ordinarily called the *knock sensor*.

Heated Rear Window

Many Chrysler vehicles, particularly those with transverse four-cylinder engines, include an input from the rear window electric heater. When this heater is on, the computer can then increase the idle throttle setting to accommodate the additional load of the heavy current draw through the alternator, at just the time when the engine is producing its least torque.

OUTPUTS

Mixture Control

The O_2 feedback system has used three different mixture control solenoids on four different carburetors. Chrysler literature calls this the "O_2 feedback solenoid."

Holley Carburetor, Model 6145. This carburetor, the feedback version of the old simple one-barrel Holley, was used in 1979 and 1980 for the California models with the slant-six engine, see Figure 16–8. The system controls idle mixture through the idle air bleed volume and the main metering mixture by metering fuel, similar to the operation of the solenoid used on the Holley 6149. Air for the idle and low-speed system air bleed comes through a fixed orifice plus a passage controlled by a metering rod. A vacuum diaphragm controls the metering rod. A second diaphragm controls a spring-loaded rod and "foot"

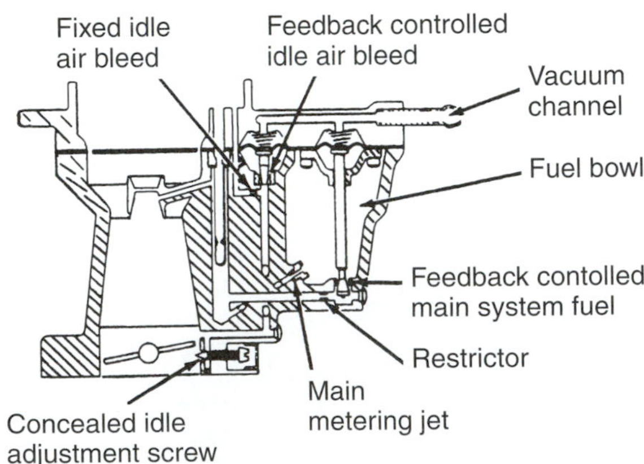

Figure 16–8 Holley 6145 mixture control solenoid circuit. *(Courtesy of Chrysler Corp.)*

to modify fuel in the main metering circuit. The spring drives the rod down. If the vacuum against the diaphragm allows the rod to come down, the foot opens a valve that meters additional fuel into the main circuit.

Vacuum to the two diaphragms is simultaneous, and as vacuum increases, the mixture leans. At wide-open throttle there is effectively no vacuum, and the mixture goes full rich. A remote, computer-controlled solenoid controls the vacuum, Figure 16–9. The lower portion of the unit is a simple vacuum regulator valve. It gets full manifold vacuum and reduces it to 5 inches of mercury (in. Hg). The 5 in. Hg vacuum is then applied to a valve seat higher in the assembly. A conical tip on the lower end of the solenoid armature closes this valve seat when the computer does not energize the solenoid. With the solenoid armature in this position, the regulator's output port connects to the vent port, making the output port vacuum zero. When the computer energizes the solenoid, the armature moves up, and a second conical tip on the top of the armature closes off the vent port. The lower passage opens, and the "reference" 5 in. Hg vacuum is now connected to the output port. The computer duty cycles the solenoid at whatever rate will maintain the proper fuel/air mixture, as called for by the oxygen sensor and reflected in the applied vacuum—between 0 and 5 in. Hg, at the two diaphragms in the carburetor. High vacuum leans the mixture in both idle and main metering circuits; low vacuum richens the mixture for both circuits.

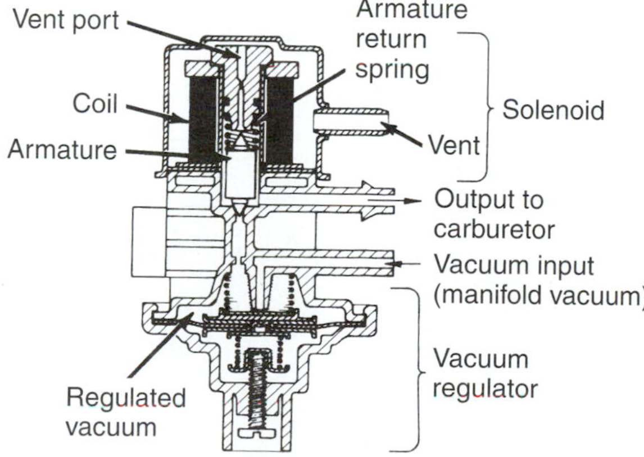

Figure 16–9 Solenoid-operated vacuum regulator for mixture control. *(Courtesy of Chrysler Corp.)*

Holley 6520 and Later Applications of the Holley 6145. In 1980, Chrysler applied an oxygen sensor and the two-barrel feedback Holley 6520 carburetor to four-cylinder cars. In 1981, they applied the same type of solenoid to six-cylinder engines with the Holley 6145 carburetor. This solenoid is essentially the same one used in the Holley carburetors used on some General Motors' CCC applications.

Carter BBD and Thermo-Quad. Carter carburetors on Chrysler vehicles got mixture control solenoids in 1980. Both the BBD two-barrel and the Thermo-Quad mixture solenoids work by a controlled air bleed for the idle and main metering circuits.

Air flows in the air horn above the carburetor venturi through a passage to the mixture solenoid, Figure 16–10. When the computer energizes the solenoid, air flows through the air horn passage and into the idle or main metering circuits (whichever is drawing air and fuel). Since the fuel emulsion includes more air, the mixture delivered to the cylinders goes lean. When the computer opens the mixture solenoid circuit and turns it off, the mixture goes lean. The computer's "choice," of course, follows the signal from the oxygen sensor, Figure 16–11.

Ignition Timing

The early Chrysler O_2 feedback system still follows the strategy of the still-earlier lean-burn system. The computer completes the primary ignition circuit by grounding the negative terminal. During cranking, the distributor pickup signal triggers the computer to open the primary circuit and fire the plugs at base (reference) timing.

Once the cold engine starts, the computer gradually adds spark advance, based on sensor inputs for coolant temperature, engine speed and intake manifold vacuum. The calculations are all from its look-up memory, and the feedback system

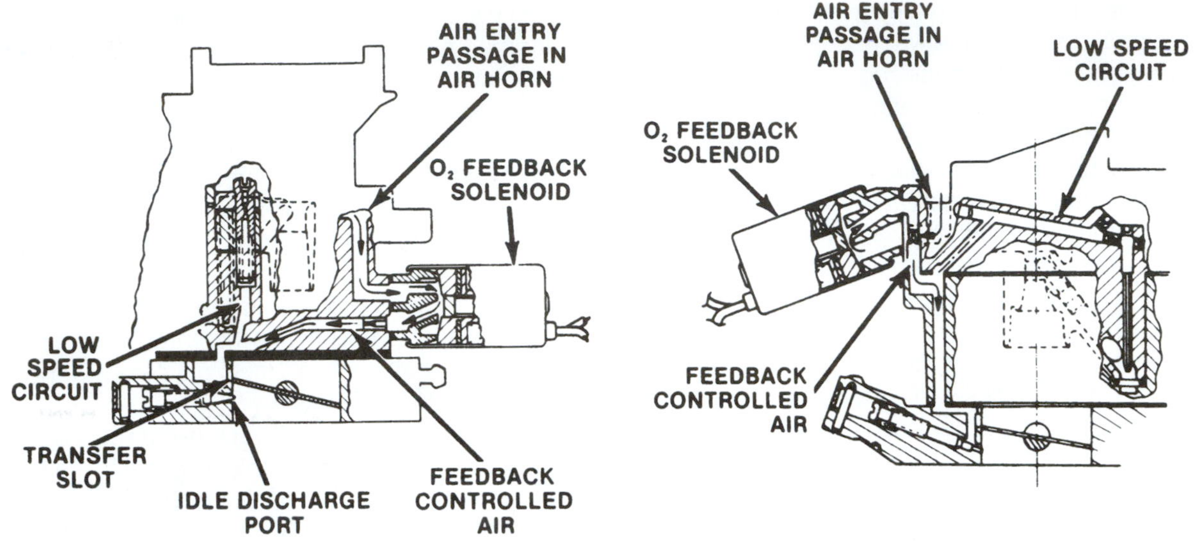

Figure 16–10 Carter carburetor mixture control, idle circuit. *(Courtesy of Chrysler Corp.)*

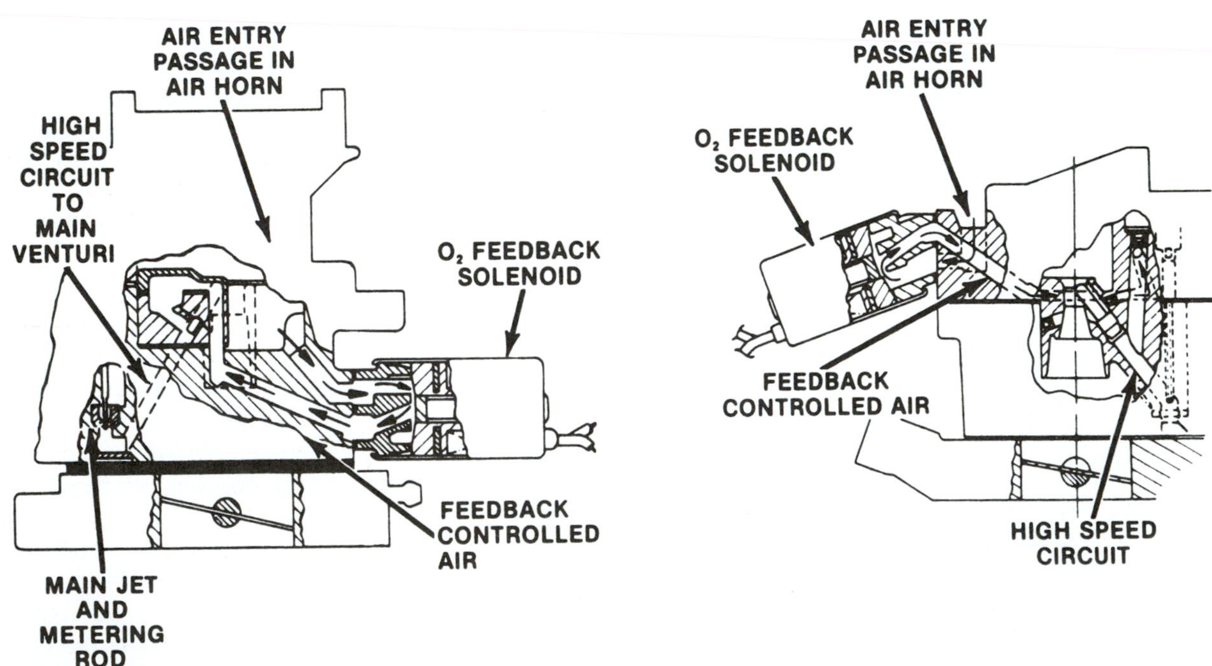

Figure 16–11 Carter carburetor mixture control, main metering circuit. *(Courtesy of Chrysler Corp.)*

plays no role in spark timing (naturally there is *no* oxygen sensor signal when the system is cold). Once the coolant reaches a specific criterion temperature, the computer determines timing advance from manifold vacuum and engine rpm. If the throttle switch circuit closes—indicating a closed throttle—spark advance drops to base timing.

EGR Solenoid

Only some of the early Chrysler O_2 feedback systems use the computer to control EGR. On these, the computer controls a solenoid that allows or shuts off vacuum to the EGR valve. Vacuum is shut off when:

- the engine is cold.
- the throttle is only slightly open.
- the engine has just restarted warm (and for the next minute or so).
- the throttle is wide open (approximately 80 percent or more). Of course, with WOT there is very little intake manifold vacuum, so even if the system malfunctioned, relatively little EGR gas would be drawn into the intake manifold.

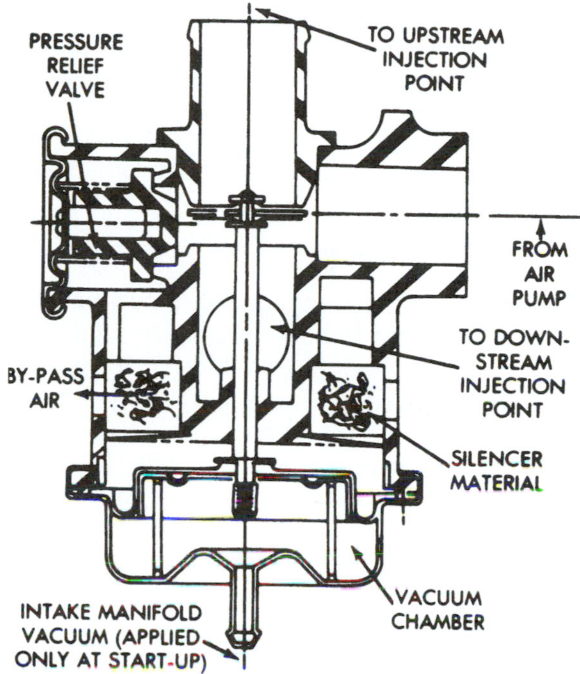

Figure 16–12 Air injection switch/relief valve. *(Courtesy of Chrysler Corp.)*

Air Switching Solenoid

On early Chrysler O_2 feedback systems controlling the air injection system, the computer controls a solenoid that applies or disconnects the vacuum to the switch relief valve, Figure 16–12. During cold engine operation, the computer routes vacuum to the valve diaphragm and directs air from the air pump into the exhaust manifold. Once the engine gets warm enough to run in closed loop, the computer stops vacuum to the valve diaphragm and directs injected air downstream to the catalytic converter. For a brief period after each engine start, warm or cold, the computer sends vacuum to the valve to inject air into the exhaust manifold. The system includes no provision to dump air into the atmosphere except as a pressure relief.

On systems where the computer does not control the air injection system, the same strategy results from the action of a coolant temperature-controlled vacuum valve in the place of the computer-controlled solenoid. On all air injection systems, the objectives are to route air to the exhaust manifold when the engine is cold, to the catalytic converter when the system is in closed loop, and to the atmosphere when there is no need for the air injection and/or there is risk of backfire.

Vacuum Secondary

The Holley 6520 uses a progressive secondary barrel, opened by vacuum. Some of the later applications of this carburetor include a solenoid-operated vent valve on the vacuum line.

When the computer opens this valve, this action locks out the carburetor secondary barrel until:

- the coolant reaches 140°F (60°C).
- vacuum drops below a specific criterion value.
- engine speed is above a specific rpm.

Radiator Fan Relay

On later model vehicles with transverse engines and electric fans, a radiator fan relay supplies power for both the cooling fan motor and for the air-conditioning compressor clutch. Normally the circuit is open, but when either the computer or the coolant temperature switch closes it, the fan turns on to draw air through the radiator until the coolant reaches the low limit of the switch/sensor.

Shift Light Indicator

On 1985 and 1986 vehicles with manual transmissions, the computer operates an instrument light that indicates to the driver the optimum time to make upshifts. "Optimum" in this case refers to optimum emissions and fuel economy, not optimum performance or engine life.

Tachometer Drive

Some models include an electric tachometer in the instrument panel. The computer generates the tach signal from the distributor side of the ignition coil. This is essentially an on/off signal, and frequency corresponds directly to engine speed.

Other engine control features often found on feedback systems, but not necessarily controlled by the computer, are:

- a vacuum-powered throttle kicker (dashpot).
- a choke heater.
- a carburetor bowl vent solenoid.

These devices usually get their power directly from the ignition switch and are grounded directly.

✔ SYSTEM DIAGNOSIS AND SERVICE

Diagnostic Procedure 1979–1984

Self-diagnostics did not appear on the Chrysler systems until 1985. Prior to that model year, diagnosis of a no-start or driveability problem followed driveability test procedures outlined in a separate booklet for each system application (by vehicle model and by build-year).

Driveability diagnosis falls into three parts:

1. verification of the complaint
2. visual inspection of the vehicle
3. performance of the designated test steps

The driveability test procedure outlines the visual inspection and the test steps. The test procedures also fall into three parts:

1. no start
2. cold driveability test
3. warm driveability test

If the engine will not start and the visual inspection is complete, begin with the first test in the no-start section. If the complaint applies only to cold driveability, begin with the first cold driveability test and likewise for a problem that occurs only when the engine is warm.

If you begin with the first test of the appropriate section and then go to the next test indicated by the results of the test just completed, the test procedure takes you through all the components and systems that could cause the driveability complaint, including systems not controlled by the computer. This, of course, is both the strength and the weakness of the early Chrysler system: granted, they thought of everything that can go wrong when they wrote the book, and going through the procedure will find everything. But it can be time consuming if the problem is at the end of the list. So later Chryslers include self-diagnostics.

Diagnostic Procedure, 1985 and Later

In the 1985 and later vehicles, Chrysler diagnostic tests include self-diagnostics, using fault codes and other features. Failures of the most vital sensors and actuators record specific fault codes.

Chrysler's Diagnostic Readout Box (DRB), tool C4805, or other aftermarket scan tool is recommended to perform these diagnostic procedures, although two of these procedures can also be performed manually without the use of a scan tool. The 1985 system has three diagnostic procedures:

1. Diagnostic test mode: The computer reads out any stored diagnostic trouble codes (memory DTCs) on the scan tool. A start of message code 88 appears first, followed by any fault codes. Once all fault codes have displayed, then an end of message code 55 appears. This procedure may also be initiated manually by turning on the ignition switch 3 times within 5 seconds, leaving it in the on position the third time. When pulling memory DTCs manually, the DTCs are retrieved by counting pulses of the malfunction indicator light (MIL), originally known by Chrysler prior to OBD II as the Power Loss Lamp (PLL). When counting codes on the MIL, code 88 is not displayed. Fault codes are received first, followed by code 55. If code 55 is displayed first, then no fault codes are stored in the computer's memory. These codes do not repeat. (Chrysler does not have a code that means "system pass," but rather, like General Motors, a system pass is indicated by the absence of fault codes.) Also, it should be noted, 1984 models allowed the technician to pull codes manually, but code pulses had to be retrieved by counting pulses of a red LED within the logic module under the instrument panel. An access hole for viewing is provided in the module's case once the cover is removed. The 1984 models did not provide an MIL on the instrument panel as 1985 and newer models did.
2. Switch test mode: This mode allows the technician to verify that the computer is able to read certain input switch signals, for example, the stoplight switch and the A/C select switch. This mode may be selected through a scan tool's menu, or it may also be performed manually by simply operating the respective switches after code pulling is complete (after code 55 has flashed) and then watching the MIL on the instrument panel as each switch is operated. The computer turns the MIL on and off as each switch is turned on and off, somewhat similar to a technician operating certain switches during Ford's self-tests (see Chapter 14) except that with Ford's system, the computer determines pass or fail. In Chrysler's switch test it is up to the technician, through watching the MIL, to determine pass or fail.
3. Actuator test mode (ATM): This mode allows the technician to choose an actuator circuit to be energized and de-energized by the computer. Once the ATM mode is activated, the computer energizes and de-energizes the selected actuator once every 2 seconds. Some outputs are turned on for 2 seconds, then turned off for 2 seconds. Others are simply quickly pulsed once every 2 seconds. This mode requires the use of a scan tool and cannot be performed manually without a scan tool. Also, once an actuator is selected from the scan tool's menu, if the computer is able to energize it, it may assist the technician in identifying the particular component, acting as a component locator. For example, selecting the "cooling fan relay" from the scan tool's menu would quickly allow the technician to determine which relay is the cooling fan relay, provided that the computer is able to energize it.

If the problem is not a no-start and if the visual inspection reveals nothing amiss, the diagnostic test mode is the usual first step in the self-diagnostic driveability test procedure.

The 1986 system added a sensor test mode to the other three modes. In this mode, the computer reads out the sensor input values through the readout box.

Reading Mixture Solenoid Calibration

A voltmeter connected across the carburetor mixture solenoid (sometimes called the O_2 solenoid) reflects the mixture commands from the computer. A voltage reading close to battery voltage indicates a lean command; a voltage reading closer to zero indicates a rich command, either because the oxygen sensor signal indicates lean or because operating conditions (such as high load or acceleration) require a rich mixture. Of course, since the actuator command is an on/off signal, the voltage reading is actually an average that reflects the duty cycle of the mixture solenoid. Hence an average reading of 7 volts would indicate approximately a 50 percent duty cycle. Notice also the system is set to go rich in the event the mixture control fails: this retains driveability until the vehicle can be repaired.

Diagnostic & Service Tip

Overcompensation. If a saturated charcoal canister or fuel-diluted crankcase oil introduces excess fuel vapor into the intake system, the system will overcompensate. It will drive the mixture too lean while in closed loop, leaner in fact than the actuator system can deliver. This problem frequently occurs when a motorist overfills a fuel tank, pumping long after the nozzle shutoff. That practice can allow liquid fuel to fill the vapor absorption lines and eventually the charcoal canister. Likewise, a vacuum leak, for example from a leaking PCV hose, can drive the system overrich.

SUMMARY

In this chapter, we have considered Chrysler's original computerized engine control systems, originally built in 1979. Beginning with the oxygen sensor feedback system, we learned the concepts of open and closed loop and how they govern the engine's intake air/fuel mixture, and the resulting exhaust emissions.

Among the major sensors the computer uses to calculate its outputs are the coolant and charge temperature sensor input signals. Fine-tuning of the intake air/fuel mixture follows the feedback signals from the oxygen sensor, and the spark advance is advanced as much as compatible with good driveability through the use of a knock sensor on some models.

▲ DIAGNOSTIC EXERCISE

An early Chrysler computer-controlled vehicle is brought into the shop with a stuck choke. It never opens beyond about halfway. What are the problems you might expect to find from such a condition? How would the computer attempt to compensate?

REVIEW QUESTIONS

1. What year did Chrysler begin using electronic ignition systems to replace mechanical points-and-condenser ignition systems?
   A. 1969
   B. 1972
   C. 1974
   D. 1981
2. *Technician A* says that in order for a Chrysler O_2 feedback system to enter closed loop, coolant temperature must be above about 120°F and the O_2 sensor must be heated to about 600°F.
   Technician B says that in order for a Chrysler O_2 feedback system to enter closed loop, a predetermined time interval must have elapsed since the engine was started.

Who is correct?
A. A only
B. B only
C. Both A and B
D. Neither A nor B

3. All *except* which of the following was used as a coolant temperature sensor on Chrysler O_2 feedback systems?
   A. Potentiometer
   B. Thermistor
   C. Dual thermistor
   D. On/off switch
4. On Chrysler O_2 feedback systems, what type of pickup is used in a four-cylinder distributor?
   A. Optical sensor
   B. Hall effect switch
   C. Single permanent magnet pickup
   D. Dual permanent magnet pickups
5. On Chrysler O_2 feedback systems, what type of pickup is used in a six- or eight-cylinder distributor?
   A. Optical sensor
   B. Hall effect switch
   C. Single permanent magnet pickup
   D. Dual permanent magnet pickups
6. On Chrysler O_2 feedback systems, what type of sensor is used to measure engine load?
   A. Throttle stop switch
   B. Rotary throttle position sensor (TPS)
   C. Vacuum transducer
   D. Manifold absolute pressure (MAP) sensor
7. On most Chrysler O_2 feedback systems, what type of sensor is used to measure throttle position?
   A. Throttle stop switch
   B. Rotary throttle position sensor (TPS)
   C. Vacuum transducer
   D. Manifold absolute pressure (MAP) sensor
8. *Technician A* says that Chrysler O_2 feedback systems use an oxygen sensor that contains a heating element and therefore has either three or four wires attached to it.
   Technician B says that Chrysler O_2 feedback systems use an zirconium oxygen sensor that generates a small voltage according to the difference of oxygen in the ambient air and oxygen in the exhaust gas.
   Who is correct?
   A. A only
   B. B only
   C. Both A and B
   D. Neither A nor B
9. *Technician A* says that on Chrysler O_2 feedback systems, the speed sensor produces an increased voltage at an increased frequency as vehicle speed increases.
   Technician B says that on Chrysler O_2 feedback systems, the detonation sensor produces direct current (DC) voltage pulses in response to engine knock.
   Who is correct?
   A. A only
   B. B only
   C. Both A and B
   D. Neither A nor B
10. Why does the computer receive an input signal from the heated rear window system on many Chrysler applications?
    A. So that it can increase throttle opening and idle speed when the heated rear window is turned on.
    B. So that it can compensate for the heavy load on the alternator when the heated rear window is turned on.
    C. So that it can reduce spark timing when the heated rear window is turned on.
    D. Both A and B.
11. *Technician A* says that the mixture solenoid on a Carter BBD feedback carburetor controls an air bleed to the idle and main metering circuits.
    Technician B says that the mixture solenoid on a Carter Thermo-Quad feedback carburetor controls an air bleed to the idle and main metering circuits.
    Who is correct?
    A. A only
    B. B only
    C. Both A and B
    D. Neither A nor B

12. *Technician A* says that, on a Chrysler O_2 feedback system, once the engine is warmed up, spark timing is determined primarily by manifold vacuum and engine rpm.
    Technician B says that, on a Chrysler O_2 feedback system, if the throttle is closed, the computer returns spark timing to base timing.
    Who is correct?
    A. A only
    B. B only
    C. Both A and B
    D. Neither A nor B
13. *Technician A* says that the computer allows vacuum to the EGR valve when the engine is cold in order to reduce NO_x emissions.
    Technician B says that the computer shuts off vacuum to the EGR valve when the engine is operating at wide open throttle.
    Who is correct?
    A. A only
    B. B only
    C. Both A and B
    D. Neither A nor B
14. *Technician A* says that the computer routes air flow from the air injection system downstream to the catalytic converter when the engine is cold.
    Technician B says that the computer routes air flow from the air injection system upstream to the exhaust manifold when the engine is warmed up and running in closed loop.
    Who is correct?
    A. A only
    B. B only
    C. Both A and B
    D. Neither A nor B
15. Why does the computer control a shift indicator light to indicate to the driver the optimum time to make upshifts?
    A. To reduce emissions
    B. To increase fuel economy
    C. To obtain optimum engine performance
    D. Both A and B
16. When using a voltmeter to read the computer's command to the carburetor's mixture solenoid, a reading of 7 volts indicates an approximate duty cycle of which of the following?
    A. 10 percent
    B. 25 percent
    C. 50 percent
    D. 90 percent

Chapter 17

DaimlerChrysler Single-Point and Multipoint Fuel Injection Systems

OBJECTIVES

Upon completion and review of this chapter, you should be able to:

- ❑ Describe the various DaimlerChrysler systems and PCMs used with fuel injection as well as their operating modes.
- ❑ Define the inputs used with a DaimlerChrysler fuel-injected system.
- ❑ Define the outputs controlled by a DaimlerChrysler fuel-injected system.
- ❑ Identify the two multiplexing systems used on DaimlerChrysler vehicles.
- ❑ Understand how a DaimlerChrysler fuel-injected system is designed to be put into self-diagnosis as well the various diagnostic features, both pre-OBD II and OBD II systems.

KEY TERMS

Adaptive Memory
Direct Ignition/Waste Spark Ignition
Inductive Pickup/Hall Effect Pickup
Limp-In Mode
Logic Module and Power Module
Self-Diagnostics/On-Board Diagnostics
Sequential Fuel Injection
Single-Point/Multipoint Fuel Injection

While Chrysler produced a few Imperials with an electronic fuel injection system in the early 1980s, the first current versions of Chrysler fuel injection appeared on the 1984 2.2-liter engines. These were the **single-point** system, which Chrysler calls Electronic Fuel Injection (EFI), and the **multipoint** system, which Chrysler calls Multi-Point Injection (MPI). The 2.2-liter MPI system is often called the *turbo* system because it uses a turbocharger. Since the original engine offerings, however, other power plants have come with either the EFI or MPI system. The two systems are quite similar, so this chapter covers them together. The text points out variations between the two.

Before 1986, the single-point system used high fuel pressure, about 36 psi. In 1986 a low-pressure (14.5 psi) injector with a ball-shaped valve replaced the previous unit with its pintle-style valve. This lower pressure injector is found on many other fuel injection systems besides Chrysler's.

POWERTRAIN CONTROL MODULE (PCM)

Logic Module/Power Module

From 1984 to 1988, two separate modules controlled the Chrysler electronic fuel injection systems. The **logic module** with the central processor was inside the passenger compartment, Figure 17–1. The **power module** controls

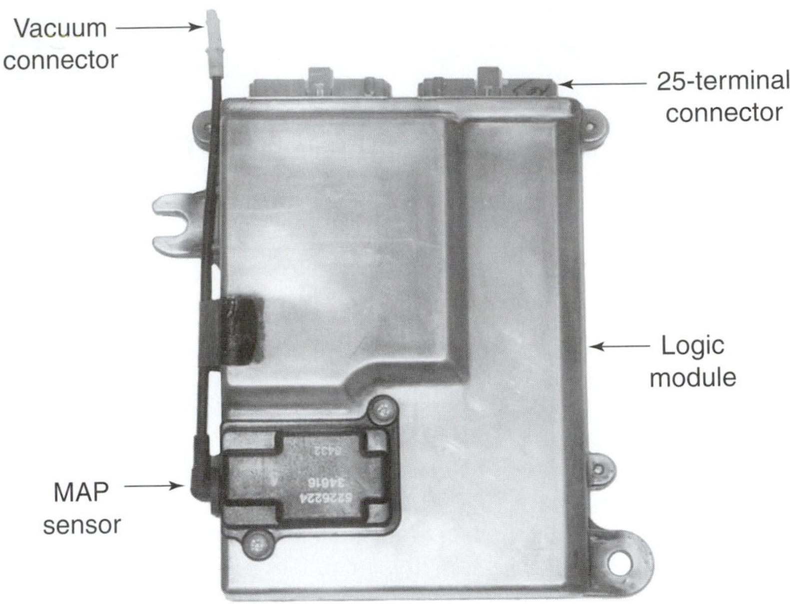

Figure 17–1 Logic module.

the actuators by switching the ground-side circuits for the ignition coil, fuel injector or injectors and the automatic shutdown relay, Figure 17–2.

The logic module gets most of its electric power from the power module, with a separate circuit through a fuse for memory retention. The logic module sends a 5-volt reference signal to its sensors and makes all the system's calculations, Figure 17–3.

Besides controlling the actuators, the power module also controls the charging system (as built in 1985). It does so by providing ground for

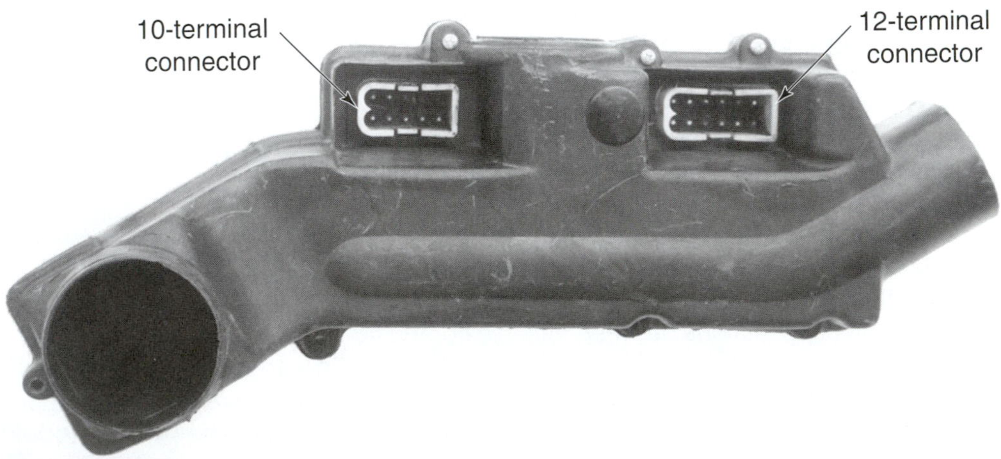

Figure 17–2 Power module.

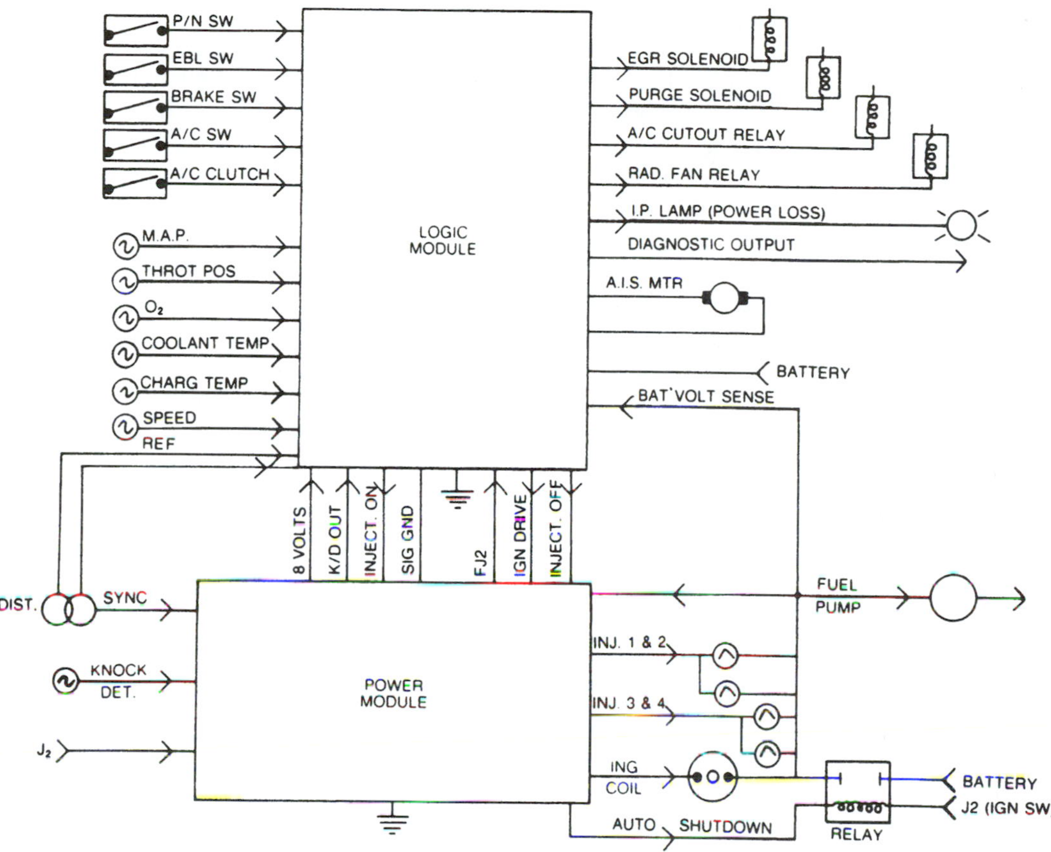

Figure 17–3 A 2.2-liter turbo EFI system schematic. *(Courtesy of DaimlerChrysler Corp.)*

the alternator's field coil. Because all of the actuator and alternator circuits are relatively high current compared to the information signal currents in the logic module, the power module is separate and is located in the engine air intake tube to carry off the heat these high current transistors convey.

The logic module signals the power module when to open and close the primary ignition circuit and when and how long to turn on the injector/injectors. The power module also provides a constant 8 volts to the logic module and to the distributor Hall effect pickup unit.

When the ignition switch is on (run), it powers the power module through a fused circuit. Chrysler literature refers to this circuit as J2. On the printed circuit board of the power module is a rail that acts as a fuse and protects the module against reversed polarity. Should a reversed polarity occur, the fuse blows. A reserve circuit with a diode continues to supply J2 power to the power module with a slight voltage drop because of the resistance of the diode.

Single-Module Engine Control (SMEC)

In mid-1987 in some 3.0-liter engine applications, the logic module and the power module were both placed as separate printed circuit boards in a single unit, the SMEC module. Physically, the power module is placed in a cavity created within the logic module and then the two modules are contained within one housing. Engine intake air flows through the modules' housing to remove heat generated by the modules. Jumper wires are simply run between the two modules' connectors, Figure 17–4. Together, of course, these two modules constitute the PCM. By 1988, most Chrysler vehicles used the SMEC system.

The only significant technical change is the increase to between 9.2 and 9.4 volts at the logic board component of the SMEC. It uses a different kind of transistor (metal oxide semiconductor field effect transistor) on the power board requiring the higher voltage. The battery temperature sensor—used by the system to calculate the proper charging voltage—has been moved to the logic board.

Single-Board Engine Controller (SBEC)

In 1989, Chrysler introduced a system that took the two separate printed circuit boards of the SMEC system and combined them by forming all components onto a single circuit board. This system was given the name Single-Board Engine Controller or SBEC. The SBEC PCM is also located in the engine compartment to allow engine intake air to carry heat away from it. The SBEC PCM is recognizable through its single electrical connector, Figure 17–5. By 1990, most Chrysler vehicles used the SBEC system.

In 1991, Chrysler introduced their second generation of SBEC, known as SBEC II. By 1992, most Chrysler vehicles used the SBEC II system, including passenger cars, minivans, light trucks and Jeeps. The SBEC II PCM still required intake air flowing through it for cooling.

In 1995, as OBD II standards were being set in place, Chrysler replaced the SBEC II system on their passenger cars and minivans with the SBEC III system. The SBEC III PCM is contained within a shielded case in order to prevent electromagnetic interference (EMI) and radio frequency interference (RFI). Additionally, the SBEC III case dissipates heat through external aluminum fins so that it no longer requires intake air flowing through it for cooling. In 1996, Chrysler also replaced the SBEC II system on Jeeps, Dodge trucks and the Dodge Viper with a new system known as the Jeep/Truck Engine Controller system, or JTEC system. In 1998, with enhanced OBD II standards requiring certain changes, Chrysler upgraded the SBEC III system on their passenger cars and minivans to an SBEC IIIA system. In each of these systems the PCM purposefully has different pin arrangements so that an incorrect replace-

Figure 17–4 Single module engine controller (SMEC) PCM.

Figure 17–5 Single-board engine controller (SBEC) PCM.

ment PCM cannot be inadvertently used. DaimlerChrysler continued to use both the SBEC IIIA system and the JTEC system on their respective vehicles through the 2001 model year.

DaimlerChrysler has now introduced a new, state-of-the-art engine and transmission control system on certain 2002 model year vehicles, known as the Next Generation Controller, or NGC system. This NGC PCM not only controls engine performance, but also replaces the transmission control module, or TCM, that has been used alongside the PCM with previous systems. The NGC system is expected to replace all SBEC IIIA and JTEC systems by the 2005 model year.

Features

Automatic Shutdown (ASD) Relay. The ignition switch energizes the ASD relay when it is in the run position. The power module energizes the ASD coil by grounding its return circuit, and the ASD relay powers the electric fuel pump, the ignition coil and the injector(s).

The fuel pump has its own independent ground and turns on immediately. The ignition coil and injectors eventually get power from battery positive, but no current flows through their circuits until the power module grounds their activation circuits. From 1985 through 1987 the ASD relay was internal to the power module. In 1984 and after 1987, it is a separate relay, outside the power module, Figure 17–6. Mounting it outside

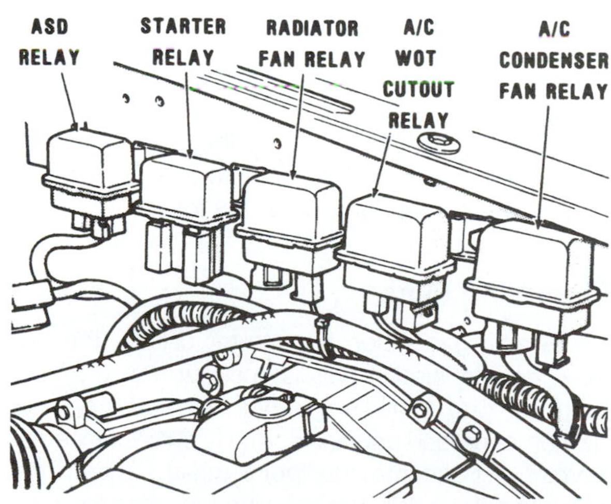

Figure 17–6 Typical location, ASD relay. *(Courtesy of DaimlerChrysler Corp.)*

saves module space and improves module reliability by the removal of the heat it generates.

Beginning in 1990, the ASD relay was also used to provide power for the heating element of the heated oxygen sensor(s) on the vehicle. Beginning in 1996, SBEC systems began using a dedicated fuel pump relay to provide power for the fuel pump. But the system retained the ASD relay that now provides power for the fuel injectors, the ignition coil primary winding, the heated oxygen sensor(s) and the alternator field windings. Both relays are de-energized if the tach reference signal is lost.

Self-Diagnostics or On-board Diagnostics. **Self-diagnostics** or **on-board diagnostics** mean essentially the same thing. The PCM monitors its most important input and output circuits for faults and improbable readings, and records fault codes in memory when an inappropriate value appears or an actuator fails.

Adaptive Memory. **Adaptive memory** is the ability of an engine control computer to assess the success of its actuator and sensor signals, and modify its internal calculations to correct them if the desired result is not achieved. The PCM can modify some of its programmed calibrations ("maps") for fuel metering to compensate for production tolerance variations and changes in barometric pressure or vehicle altitude.

There are limits to the range of adjustment available to the adaptive memory capacity. Sometimes mechanical problems—a collapsed exhaust system, a burnt valve—throw the engine performance so far off that no countermeasure available to the computer can correct for it.

Operating Modes

Starting. When the starter cranks the engine, double injector pulses occur to richen the intake mixture. The single-point injector on the nonturbocharged engines normally pulses twice per revolution, and the multipoint injectors normally pulse in pairs, each pair spraying fuel once for each revolution. The double-pulsing during crank occurs only for a programmed time interval to avoid flooding. At this time, coolant temperature alone determines the pulse width. Once the engine starts, the PCM provides fuel enrichment depending on coolant temperature and manifold pressure. The startup enrichment "decays" to base enrichment over a programmed time interval. Once the coolant reaches the closed-loop criterion temperature, of course, the feedback system controls mixture.

Primer Function. As soon as a driver turns the ignition on with the 3.0-liter engine, all six injectors pulse fuel into the intake ports. This improves starting and occurs regardless of coolant temperature, though if the coolant is warm, less fuel will spray. If the engine is cold, relatively more fuel sprays.

Open Loop. Open loop—when the oxygen sensor does not determine fuel injection pulse width—occurs when:

- the coolant is below a criterion temperature.
- the oxygen sensor is below operating temperature (approximately 600°F/315°C).
- the oxygen sensor does not switch back and forth across a switch point of about 0.45 volt.
- the vehicle is at wide-open throttle (WOT) or under similar high acceleration or load.
- when the turbocharger boost on the multipoint system reaches 1 psi or more.
- when the vehicle is decelerating or coasting with the throttle closed.
- when an engine with the single-point injector is idling.
- when there is a major sensor or actuator malfunction preventing feedback mixture control.

Closed Loop. Each system will go into closed loop only if:

- the coolant has reached the criterion temperature.
- the oxygen sensor produces a usable signal.
- a vehicle-specific time has passed since startup.

Limp-in Mode. The PCM's self-diagnostics circuits include the capacity to monitor incoming signals from the most important sensors. If one of them sends a signal out of the expected range or no signal at all, the self-diagnostics consider the sensor inoperative and shift the engine management system into **limp-in mode**. In this mode, the PCM substitutes a probable value for that of the failed sensor, sets a fault code and turns on the malfunction indicator light (MIL), formerly known as the power loss lamp, on the dashboard. Obviously, the PCM cannot substitute for the distributor's Hall effect sensor signal; the engine just stops. But it can do so for coolant temperature or MAP sensors.

In limp-in mode, the engine will not develop full power nor operate with full efficiency, but it can still run. The sensors whose failure can trigger limp-in are:

- Manifold absolute pressure (MAP) sensor—The computer can create a simulated value from the inputs from the throttle position sensor and the engine speed.
- Throttle position sensor (TPS)—The computer uses the MAP sensor signal to create a substitute value.
- Engine coolant temperature (ECT) sensor—The computer uses the charge temperature (intake manifold fuel/air mixture temperature) sensor signal as a substitute.
- Intake air temperature (IAT) sensor—The computer can usually manage engine performance without this information. While a fault code is set in the case of an IAT sensor failure, the MIL does not come on.

If the distributor reference signal stops, so does the engine. There is then no way to time ignition spark.

Diagnostic & Service Tip

Certain mechanical problems can throw a system out of closed loop repeatedly. Suppose a car develops a series of bad oxygen sensors in a relatively short time. An experienced computer control technician will check for causes external to the system, too. This particular problem often comes when a head gasket is starting to fail and seeps coolant into the combustion chamber or exhaust. The silicone in the coolant forms a superthin, impermeable layer on the exhaust surface of the sensor, "poisoning" it. Less frequently, coolant silicone from a burst hose or some other contaminant will get into the atmospheric side of the sensor. More rarely, silicone poisoning occurs when a technician uses older silicone-based gasket sealers.

INPUTS

Manifold Absolute Pressure (MAP) Sensor

The MAP sensor is a piezoresistive pressure sensor. A piezoresistor changes its electrical resistance in response to the slight bending that changes in pressure cause in it. The PCM uses its signal as a barometric pressure sensor during periods of key on/engine off or during certain other conditions on a few models as explained in the Outputs section of this chapter, Figure 17–7.

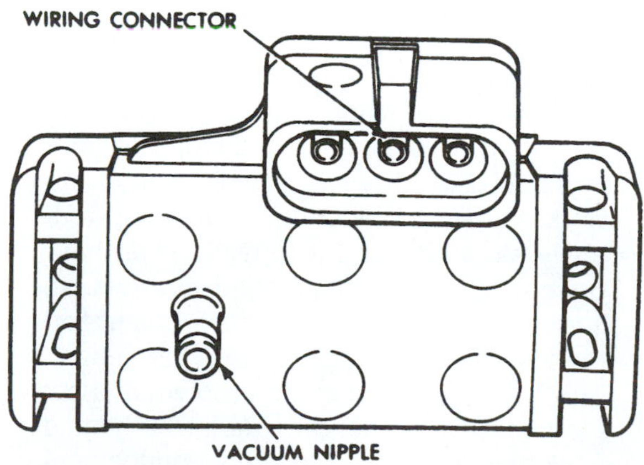

Figure 17–7 MAP sensor. *(Courtesy of DaimlerChrysler Corp.)*

The MAP sensor is usually located on the logic module in the two-module system.

The technician should be clear what manifold absolute pressure is. Formerly the term *intake manifold vacuum* was used. But that can be misleading (particularly after the widespread introduction of turbocharged cars). The manifold absolute pressure is the actual pressure of the air in the intake manifold. Often, of course, it is lower than ambient pressure, but it is still air pressure. If a turbocharger or supercharger increases the pressure above ambient pressure, it is still *manifold absolute pressure*. This dimension corresponds very closely to engine load.

Some earlier vehicles had a small bleed hole in the vacuum line to the MAP sensor to prevent condensation from collecting in the line. Those MAP sensors are calibrated to allow for the vacuum leak.

Diagnostic & Service Tip

The early system used the MAP sensor as a barometric sensor when the ignition was on but the engine off, and it retained that value in memory for the duration of that trip. However, if the trip involved a significant change of altitude—up or down a sizable mountain—the engine might run very poorly until it was shut off and restarted, with no codes set because there was no sensor or actuator circuit. A driver or roadside mechanic who believed that it was bad policy to shut off an engine that was still running, however poorly, would never get the engine right until the car returned to the beginning altitude, or stalled or ran out of gas and stopped on its own.

Baro Read Solenoid

In 1985, Chrysler added a baro read solenoid, which allowed the PCM to use the MAP sensor to update its barometric pressure values without having to wait for a restart of the engine. If the PCM energizes the baro read solenoid, then the MAP sensor is isolated from the manifold vacuum/pressure signal and is simultaneously vented to the atmosphere.

Diagnostic & Service Tip

The early MAP sensor arrangement had two problems frequently seen in repair bays. In colder climates in the winter, the system would set a code and drop into limp-in mode. But after the vehicle was in the work bay for a while, the problem mysteriously stopped without repairs. What happened was this: dirt plugged the vacuum line vent; the moisture froze and plugged the line, blocking vacuum from the sensor. The computer noticed the problem and set the code. Once the car was in the heated work bay, the ice plug melted, and the sensor worked normally.

Throttle Position Sensor (TPS)

The TPS is a rotary potentiometer—a variable resistor—at the end of the throttle shaft on the side of the throttle body. Depending on the position of the throttle, the sensor modifies the 5-volt reference signal to something less that corresponds with the throttle's current position, and returns that signal to the computer. Some engine management systems also keep track of how quickly the throttle moves, to more accurately provide acceleration enrichment when the pedal is suddenly floored, or to shut off fuel if the throttle is suddenly closed at high engine speed.

Oxygen (O_2) Sensor

The early Chrysler systems used a single-wire oxygen sensor. Newer systems use a heated sensor with additional wires for an internal heater element.

Engine Coolant Temperature (ECT) Sensor

The ECT sensor is a single-element, negative temperature coefficient (meaning the resistance goes down as the temperature goes up) thermistor screwed into the thermostat housing, Figure 17–8. As it warms up, its resistance goes down. At –4°F/–20°C its resistance is 11,000 ohms, and it ranges to 800 ohms at 195°F/90.5°C, full operating temperature on these systems. Prior to OBD II standardization, the ECT sensor was known by Chrysler as a coolant temperature sensor or CTS.

Intake Air Temperature (IAT) Sensor

The IAT sensor is a single-element thermistor (again negative temperature coefficient-type) screwed into a runner of the intake manifold. It has about the same resistance range as the coolant temperature sensor and looks similar. On early models its input contributes to cold engine enrichment calculations; otherwise it serves as a backup to the ECT sensor. On later multipoint systems, its input helps control the air/fuel mixture when cold and boost control at all times. Prior to OBD II standardization, the IAT sensor was known by Chrysler as a charge temperature sensor or charge temp sensor.

Throttle Body Temperature Sensor

The low-pressure, single-point system does not include the charge temperature sensor, but instead a throttle body temperature sensor to measure temperature at the throttle body, Figure 17–9. This temperature will ordinarily be very close to intake air temperature, and this signal helps calculate the mixture for hot restarts.

Range-Switching Temperature Sensor (Dual Range Temp Sensor)

To fine-tune the fuel/air mixture and spark timing even further, beginning in 1992 Chrysler changed to range-switching temperature sensors for coolant and air temperature, Figure 17–10.

The purpose of the change is to telescope the sensitivity of the sensor in the area of greatest importance: around operating temperature. It also accommodates the problem that the thermistor's reaction to changes in temperature is not linear, that is, there is more change in the cooler (lower) temperatures than at the upper end. Unfortunately, for engine management purposes, we're more interested in knowing about the higher temperatures in fine detail for air/fuel mixture control.

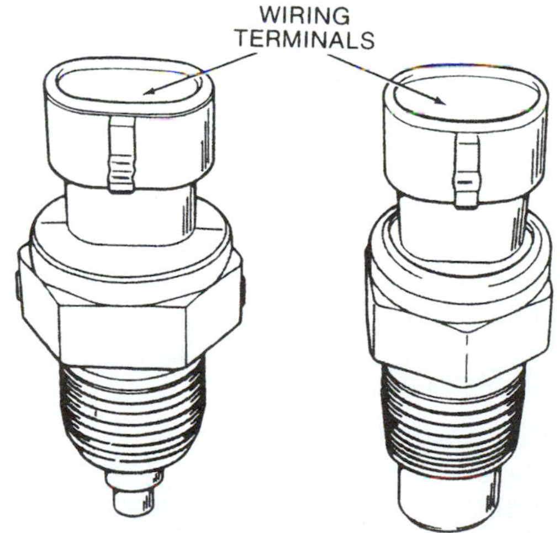

Figure 17–8 Charge and coolant temperature sensors. *(Courtesy of DaimlerChrysler Corp.)*

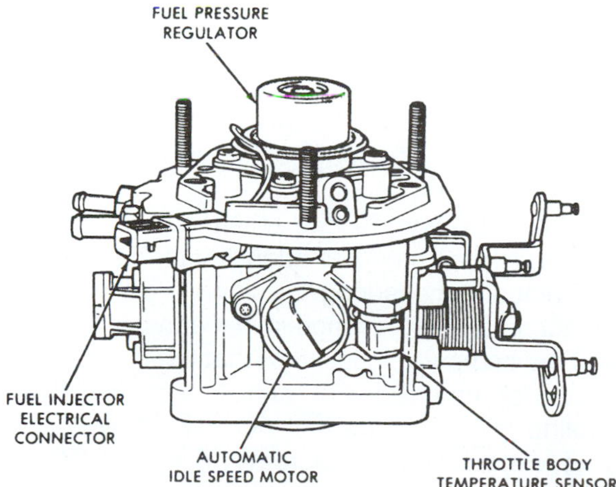

Figure 17–9 Low-pressure single-point throttle body, with throttle body temperature sensor. *(Courtesy of DaimlerChrysler Corp.)*

Temp. of	Sensor Resistance	Temp. Change	Resistance Change	OHM Change Per Degree	Volt Drop Across Sensor	
– 20	156,667 Ω				4.7 V	WITH 10,000 Ω FIXED RESISTANCE
		10 °	50,388 Ω	5038.8		
– 10	106,279 Ω				4.57 V	
40	25,714 Ω				3.6 V	
		10 °	6,302 Ω	630.2		
50	19,412 Ω				3.3 V	
110	4,577 Ω				1.57 V	
		10 °	1,244	124.4		
120	3,333□				1.25 V	
SENSOR CIRCUIT SHIFT (909 Ω FIXED RESISTANCE RATHER THAN 10,000 Ω)						
140	2,338 Ω				3.6 V	WITH 909 Ω FIXED RESISTANCE
		10 °	406 Ω	40.6		
150	1,932 Ω				3.4 V	
200	839 Ω				2.4 V	
		10 °	125 Ω	12.5		
210	714 Ω				2.2 V	
240	435 Ω				1.62 V	
		10 °	64.5 Ω	6.45		
250	371 Ω				1.45 V	

Figure 17–10 Temperature, resistance and voltage drop change with range-switching temperature sensor.

The sensor works like this: Below 125° Fahrenheit, there is a fixed 10,000-ohm resistor in series with the sensor's thermistor, Figure 17–11. At about 125° Fahrenheit, the PCM turns on a 1,000-ohm resistor in parallel with the 10,000-ohm fixed unit. This toggles the resistance to 909 ohms.

This spreads out the resistance more over the range the computer is looking for. Of course, it is actually measuring the voltage drop across the thermistor. This range shift means, of course, that the computer must know that the change that occurs around 125° Fahrenheit is not a sudden cooling of the engine but the resistance switch. This "expectation" is programmed into the PCM's memory.

When the second resistor is switched into parallel, the total fixed resistance is now considerably lower than that of the thermistor. There is a wider range of voltage drop available to be monitored and greater accuracy obtainable.

TMAP Sensor

Chrysler 2.0L, 3.2L and 3.5L engines use a new sensor called the TMAP, Figure 17–12. This sensor combines the functions of the intake air temperature and the manifold absolute pressure sensor into one component. The TMAP sensor is located at the intake manifold.

TMAP Strategy

Due to slow MAP sensor response when the driver suddenly moves the throttle, newer DaimlerChrysler PCMs use a strategy that uses the TPS signal to initially calculate MAP values as the throttle is moved, resulting in a quicker PCM response to throttle movement. This strategy is referred to

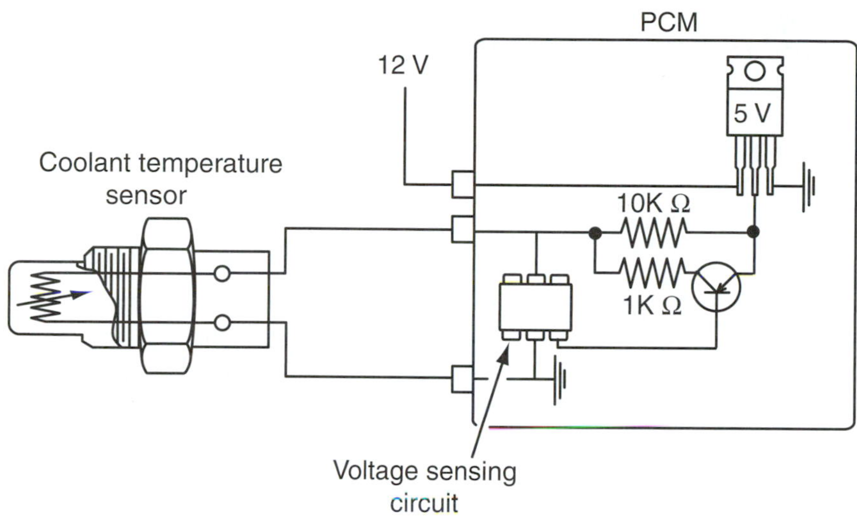

Figure 17–11 Range-switching temperature sensor electrical schematic.

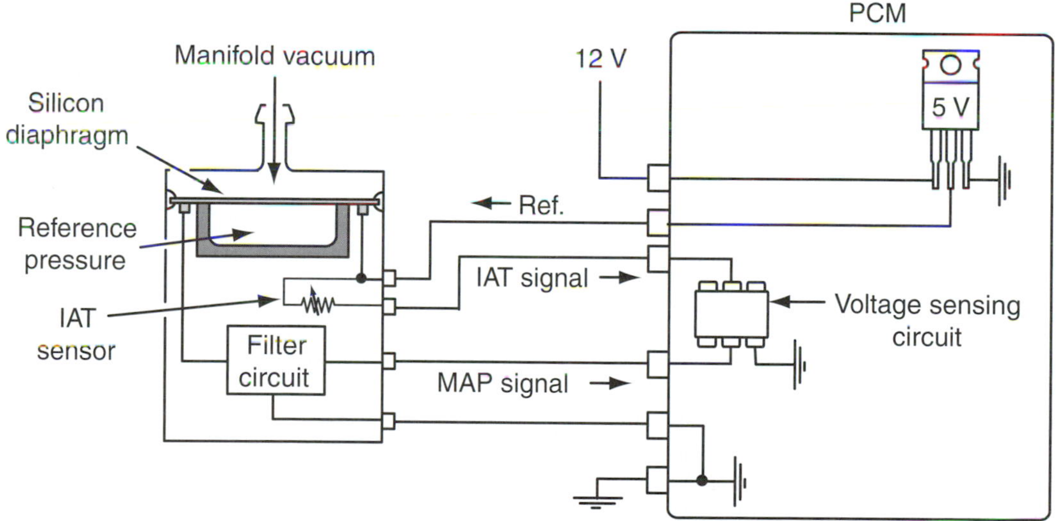

Figure 17–12 TMAP sensor.

as the TMAP strategy and should not be confused with the TMAP sensor.

Mass Air Flow (MAF) Sensor

DaimlerChrysler introduced a mass air flow (MAF) sensor on the 2001 Chrysler Sebring and the 2001 Dodge Stratus, similar to the sensors that GM and Ford began using in the 1980s. DaimlerChrysler's MAF sensor is made by Mitsubishi and is called the *Mitsubishi Ultimate Karman Airflow Sensor* or MUKAS. It replaces the speed density formula for *calculating* air intake and measures air intake with a hot-wire type sensor. It does this by measuring the cooling effect of the air's molecules on a heated wire element. This hot wire is kept heated to a certain temperature above the temperature of the ambient air entering the engine as measured by the IAT sensor. The IAT sensor allows the MAF sensor to be indexed to ambient air temperature so that only the *quantity* of air molecules, not their temperature, affects the hot wire.

The DaimlerChrysler MAF sensor produces a square-wave variable digital frequency by grounding out a 5-volt reference voltage from the PCM. The frequency at which it does this is measured by the PCM in order to know measured air intake. As a result, a Hertz meter should be used to measure the resulting signal, similar to GM MAF sensors and Ford EEC IV digital MAP sensors. If the ignition is turned on, but the engine is not running, the voltage is pulled to ground and does not oscillate.

Distributor Reference Pickup (REF Pickup)

A Hall effect switch in the distributor generates the interrupted-direct-current reference pickup signal from which the PCM determines engine speed and crankshaft position. On the dual module single-point system, this same signal goes to both modules. Without the signal, the power board removes ground from the ASD relay, shutting off the ignition coil, injectors and fuel pump. On the dual module multipoint systems, only the logic board gets the distributor reference signal, but the logic board uses it to instruct the power board when to turn on the injectors.

A Hall effect position sensor has specific differences from an inductive pickup, used by other manufacturers and for other purposes. The **Hall effect pickup** (sensor) generates a clean on/off signal and does not vary either in accuracy or output voltage with rpm. An **inductive pickup** (sensor), on the other hand, produces an alternating current, and the point of polarity reversal is the reference point. An inductive sensor may produce very low voltage signals at low speed, and very high voltage at higher speed. It can also change position as the speed increases, because of the time required for the magnetic field to reverse. A Hall effect sensor, however, does not generate its own current. It is dependent on the reference signal sent to it: the circuit is more complex and more things can go wrong.

Distributor Sync Pickup (SYNC Pickup)

Most multipoint systems use a second Hall effect switch/sensor in the distributor, below the reference pickup, Figure 17–13. This switch has a shutter wheel with only one vane instead of four. The one vane, however, covers 180 degrees, half of the shutter wheel. The leading and trailing

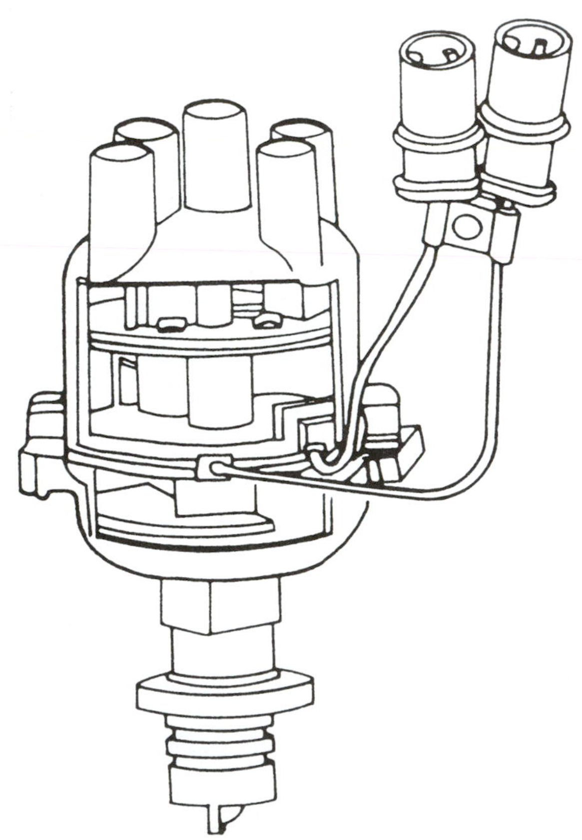

Figure 17–13 Distributor for turbocharged four-cylinder engine, with reference and sync Hall effect pickups. *(Courtesy of DaimlerChrysler Corp.)*

edges of the vane each provide a signal (on or off) to the PCM. The PCM uses the SYNC signal along with other information to calculate pulse width. The power board uses it to determine which pair of injectors (1 and 2 or 3 and 4) to turn on when the instruction from the PCM arrives.

Optical Distributor. A 3.0-liter V6 engine, built by Mitsubishi, appeared on some Chrysler vehicles in 1987. This was Chrysler's first naturally aspirated (nonturbocharged) engine with multipoint fuel injection. The computer for this engine uses a speed-density formula to calculate fuel quantity in a way similar to the way it is done on the 2.2 turbocharged engines. A speed-density system does not require a vane airflow meter or a mass airflow meter to measure air.

The engine management system with this engine does not use the Hall effect distributor, but an optical distributor with a pair of optical sensors, Figure 17–14. A disc with two sets of slits rotates with the distributor shaft. The inner set has six slits, called "low data rate" slits. The outer set has one slit for every two degrees of crankshaft rotation except for one small blank spot with no slits. This section alerts the PCM where the number one cylinder is. The high data rate from this sensor allows the computer to track minute changes in engine speed.

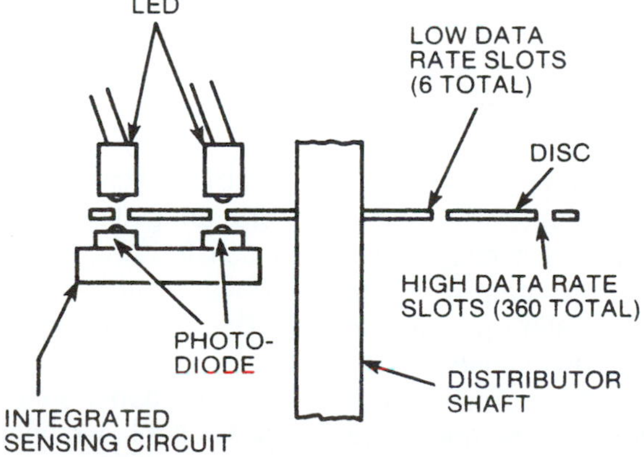

Figure 17–14 Optical distributor pickup.

Two light-emitting diodes (LEDs) are on one side of the disc, and the receptors are on the other, Figure 17–15. When one of the slits aligns with an LED, the light strikes a photodiode in the receptor, generating a small voltage applied to the base of a transistor.

We will follow the "low data rate" reference signal for the explanation, though the "high data rate" circuit works the same way: the transistor works as a ground switch for the 5-volt reference signal sent through a resistor by the PCM. When the transistor turns on from the photocell pulse to its base, the voltage drops and the voltage difference across the resistor increases. When the slit in the distributor disc moves and blocks the light, the photocell shuts off; the transistor opens the reference ground circuit, and the voltage drop across the resistor disappears. Each time the voltage drops toward zero, the PCM knows the next piston in the firing order has reached a specific position in its cycle. The "high data rate" sync circuit works the same way, each signal indicating the crankshaft has rotated two degrees.

Below 1,200 rpm, the PCM uses the sync signal to calculate spark timing and injector pulse. Above 1,200 rpm, it uses the reference signal. The reason for the switch (besides the easier task of monitoring the lower data rate at higher engine speeds) is that there are significant changes in engine speed at each power stroke, but slight differences in compression ratio, spark plug condition and fuel atomization mean different changes for each cylinder. The high data rate signal allows the PCM to monitor these changes and adjust timing and injection pulse accordingly. At higher engine speeds, there is much less variation in rpm between different cylinders' power strokes, so such detailed adjustments are not necessary.

Direct Ignition System (DIS). In 1990, Chrysler introduced its own 3.3-liter 60-degree V6 for some of its passenger cars and vans, Figure 17–16. This engine uses a distributorless, ***direct ignition system*** similar to General Motors and Ford distributorless systems. The idea of using one coil to fire a pair of plugs—one coil for every two cylinders—goes back to motorcycles, in which

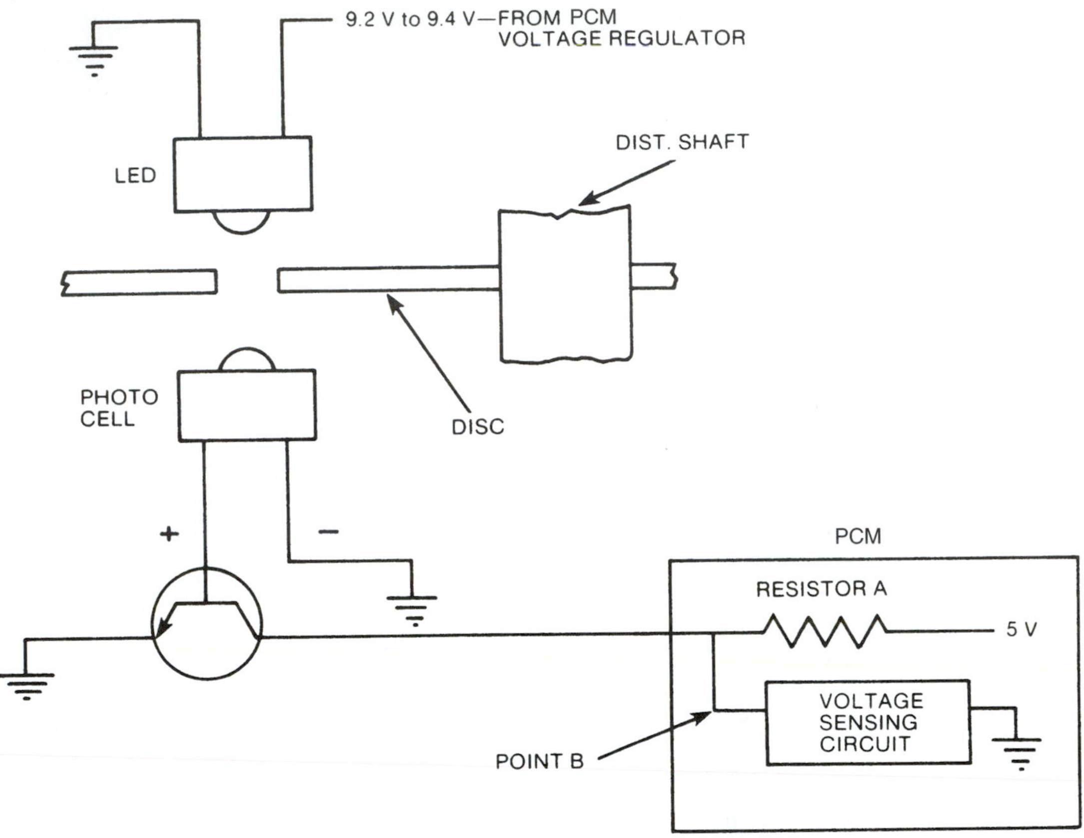

Figure 17–15 Optical distributor sensor schematic.

high rpm limited available dwell time to build up the coil's field and where space for the coils was even more limited than on cars. This system is often called a ***waste spark ignition***. Both plugs on the same coil fire simultaneously, one at the end of its compression stroke just before the power stroke; the other at the end of its exhaust stroke. Because of the low resistance in the exhaust stroke combustion chamber, the "waste" spark requires less voltage than the active one.

The electrical difference between a waste spark distributorless direct ignition system and a conventional setup is that the ignition secondary (high voltage) windings are not directly grounded. Instead, one plug gets a negative polarity spark and the other a positive polarity. Of course, from the point of view of the air/fuel mixture, the polarity of the spark does not matter: any hot spark of either polarity sets the mixture burning.

The Chrysler direct ignition system, like the GM and Ford systems, uses Hall effect crankshaft and camshaft position sensors, an ignition module and a coil pack with a coil for every 360-degree pair of cylinders (those with exactly alternating power strokes). The coil pack bolts over the ignition module.

The camshaft position sensor mounts on the timing cover and triggers off slots on the camshaft timing gear, Figure 17–17. The slots on the timing gear are coded so signals from the sensor identify

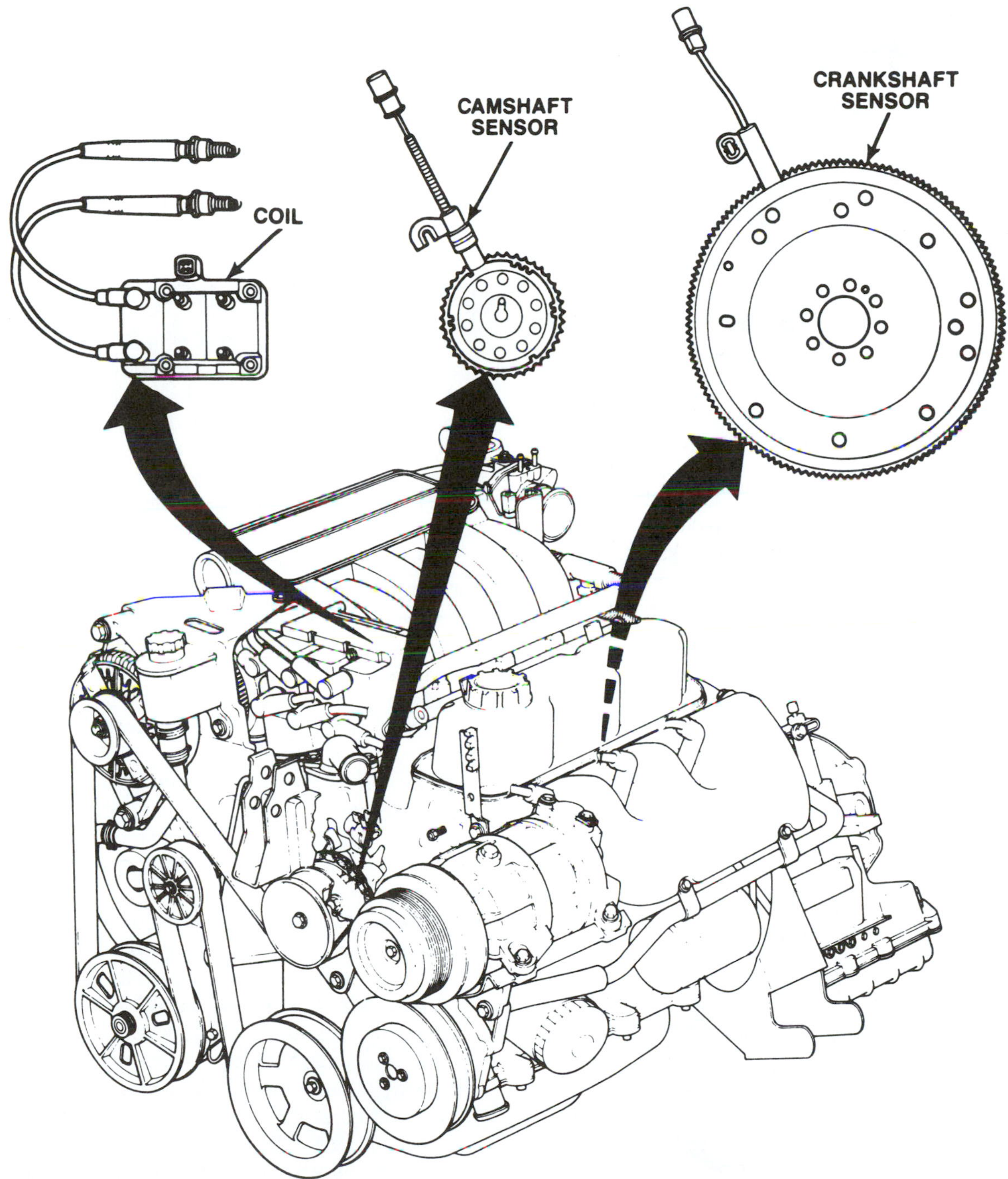

Figure 17–16 3.3-liter direct ignition system with camshaft and crankshaft position sensors. *(Courtesy of Daimler-Chrysler Corp.)*

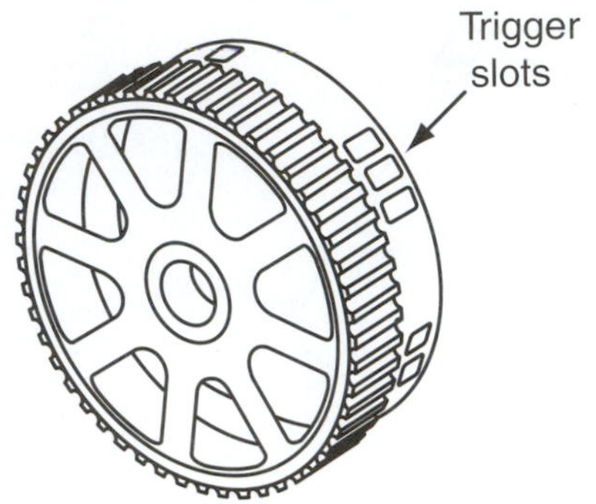

Figure 17–17 Six-cylinder camshaft position (CMP) trigger. *(Courtesy of DaimlerChrysler Corp.)*

which pair of companion cylinders the ignition module should fire next in response to the computer's spark timing command. This signal can also determine which injector the computer pulses next. Because the camshaft sensor can identify individual cylinders (unlike the crankshaft sensor), the computer can use its signal to begin firing spark plugs and pulsing fuel injectors within the first crankshaft rotation.

The crankshaft sensor mounts on the bell housing and triggers off slots on the torque converter drive plate, Figure 17–18. There are four slots per pair of companion cylinders. The computer uses these signals to determine crankshaft position and engine speed. It uses this information in determining ignition timing, injector timing and injector pulse width.

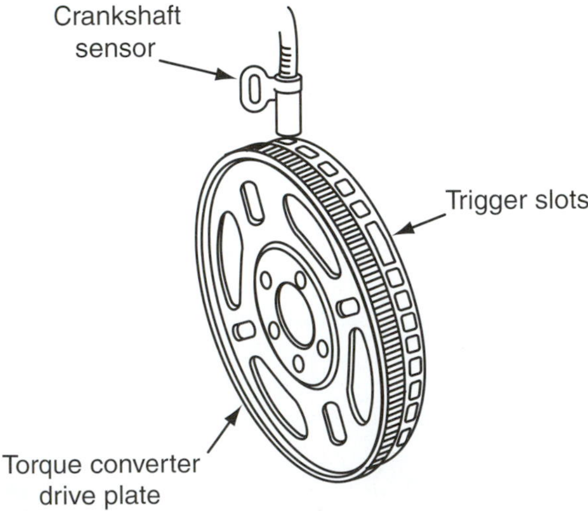

Figure 17–18 Crankshaft position (CKP) trigger. *(Courtesy of DaimlerChrysler Corp.)*

Figure 17–19 shows the waveforms generated by Chrysler's typical Hall effect camshaft position (CMP) and crankshaft position (CKP) sensors.

Vehicle Speed Sensor (VSS)

The speed sensor is a simple on/off switch that cycles eight times per speedometer cable rotation. The computer sends it a steady 5-volt reference signal which it interrupts (producing an interrupted-direct-current) at a frequency corresponding to transaxle output shaft speed. The vehicle speed sensor mounts to the transaxle at the base of the speedometer cable, Figure 17–20.

In 1992 Chrysler began using a Hall effect vehicle speed sensor (VSS), Figure 17–21. In order to power up, this sensor receives power and ground, then delivers a digital square wave to the PCM at a rate of 8,000 pulses per mile. The power supply is delivered to the VSS from the PCM over the same circuit that carries power to the CKP and CMP Hall effect sensors, 8 volts on early models and 5 volts on newer versions. Therefore, if the VSS were to short its power circuit to ground internally, the engine would not start due to the simultaneous loss of power at the CKP and CMP sensors.

Newer DaimlerChrysler vehicles also use permanent magnet vehicle speed sensors that monitor iron teeth and notches on the output shaft of the transaxle and produce AC voltage pulses that are sent to the PCM. As a result, depending on the application, newer DaimlerChrysler products may utilize either a Hall effect VSS or a permanent magnet VSS.

Also, prior to OBD II standardization, Chrysler referred to the VSS as a "Distance Sensor" due to the fact that the *number* of pulses produced only represents distance. The PCM must then combine the number of these pulses with time by

Figure 17–19 Typical waveforms generated by Chrysler's CMP and CKP Hall effect sensors. *(Courtesy of DaimlerChrysler Corp.)*

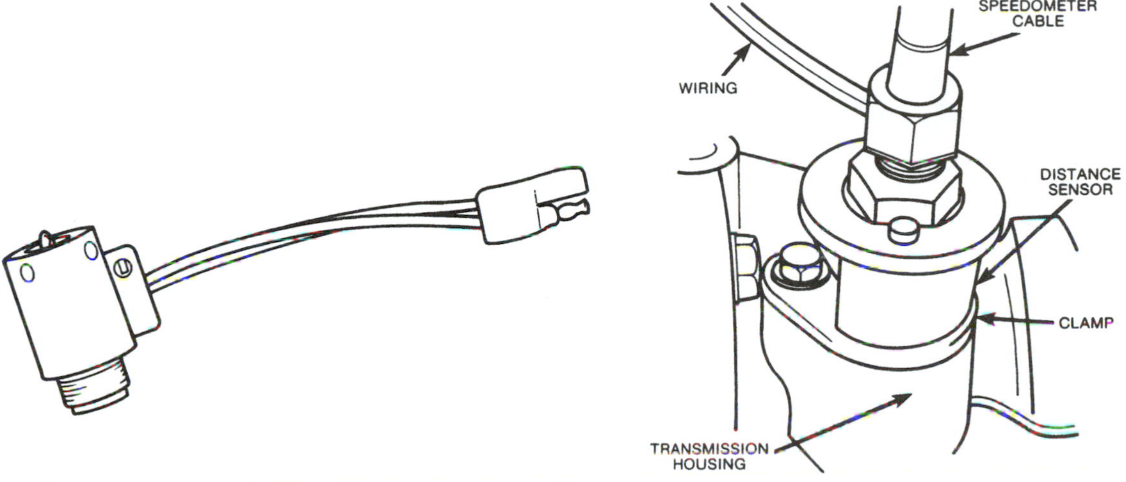

Figure 17–20 Vehicle speed sensor. *(Courtesy of DaimlerChrysler Corp.)*

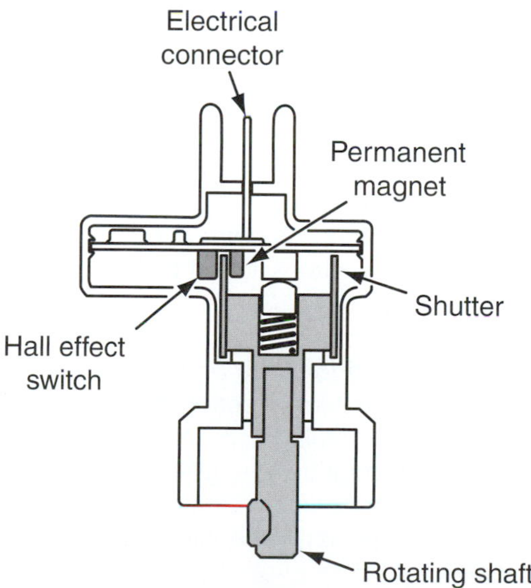

Figure 17–21 Hall effect vehicle speed sensor (VSS).

looking at the *frequency* of the pulses in order to know vehicle speed.

Detonation (Knock) Sensor

The Chrysler knock sensor is of the piezoelectric type, bolted into the intake manifold. When excited by vibrations typical of detonation, it sends a true alternating current signal to the computer. This signal then begins countermeasures, which reduce spark advance by individual cylinders and/or reduce turbocharger boost.

Chrysler 3.2L and 3.5L engines have changed from dual knock sensors to a single broadband knock sensor. This new sensor has a broader operating range of 5 kiloHertz to 20 kiloHertz. Unlike the previous single wire knock sensor, the new unit uses two wires that provide the single broadband knock sensor with its own high ground. This means that the sensor no longer shares the

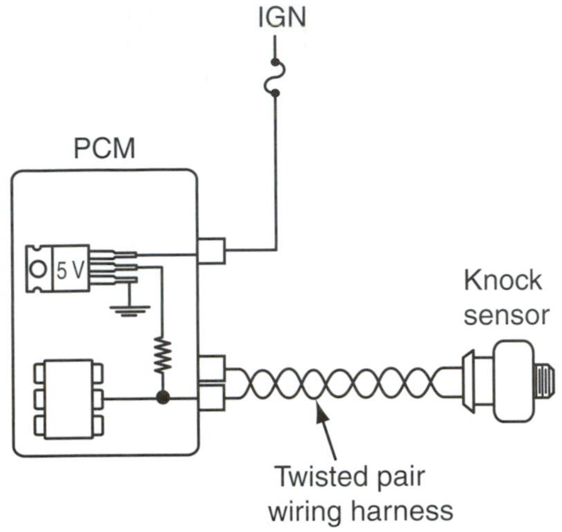

Figure 17–22 Knock sensor circuit.

common ground of all other electrical circuits in the vehicle. The two sensor wires are twisted to improve the shield from unwanted signals, Figure 17–22.

Battery Temperature Sensor

A thermistor in the PCM, adjacent to the battery, signals to the computer the battery's ambient temperature. The computer then uses this information to regulate alternator output and battery charging rate.

Charging Circuit Voltage

The PCM senses charging system voltage by monitoring the fuel pump power feed circuit and regulates alternator output accordingly. As we learned in the preceding paragraph, it also factors in the battery temperature to determine what the proper charging voltage should be.

Switch Inputs

Park/Neutral Switch (P/N). The P/N switch tells the PCM whether the transmission is in gear. Its input influences idle speed. Normal idle spark advance is canceled when the transmission is in neutral or park.

Electric Backlite (Heated Rear Window). Whenever the heated rear window switch is on, the PCM increases the idle speed to compensate for the additional load the heating current places on the alternator.

Brake Switch. Should the throttle position sensor/switch fail, the computer looks to the brake light switch as a signal that the throttle is closed.

Air Conditioning (A/C). Whenever the A/C is on, the A/C control circuit sends the PCM computer a signal. The PCM then increases idle speed to compensate for the additional compressor and alternator load.

Air-Conditioning Clutch. When the A/C system cycles the compressor clutch on and off at idle, the A/C clutch switch provides a signal for the computer to adjust the idle speed and compensate for variations in load. Later models combine both air-conditioning inputs into one signal.

Battery. One terminal of the PCM gets power directly from the battery to sustain power in the keep-alive memory (KAM) when the ignition is off.

OUTPUTS

Injector/Injectors

Figures 17–23 and 17–24 show the low and high pressure fuel injectors used in the single-point injection systems. Each type is a solenoid-operated device.

In normal operation, the single-point injector pulses fuel twice for each engine revolution. The early multipoint fuel injection systems on four-cylinder engines pulsed one pair of injectors during the first revolution of a full cycle and the other pair during the second revolution. As explained before, this injection schedule was doubled during cranking.

The 3.0 and 3.3-liter V6 engines pulse injectors in sequential pairs. In each case, the logic

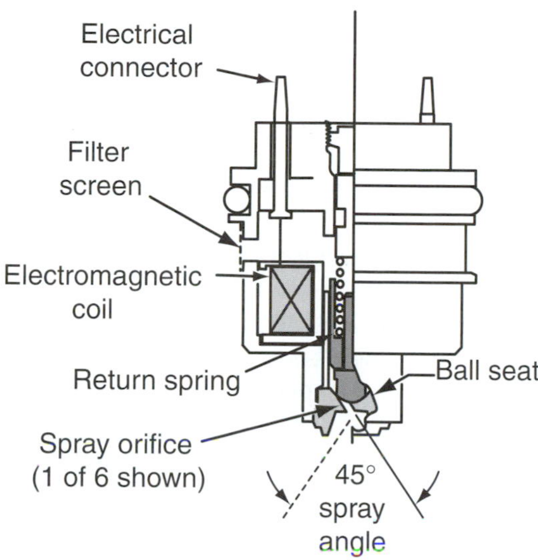

Figure 17–23 Low-pressure injector.

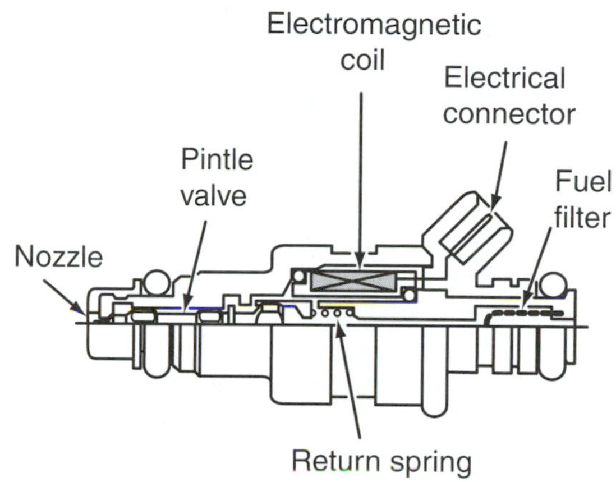

Figure 17–24 Typical high-pressure fuel injector.

module calculates the pulse width based on input signals from:

- ECT.
- MAP.
- RPM.
- TPS.
- oxygen sensor—in closed loop.
- IAT—during cold enrichment.
- VSS—during deceleration.
- fuel enrichment and enleanment factors programmed into the PCM.

In contrast to the more common wiring practice, the 1984 to 1987 Chrysler single-point systems have a fixed constant ground. The power module controls injection by switching power on and off (positive side switching). From 1987, the single-point system used ground-side switching, making it compatible with the multipoint systems.

Modern DaimlerChrysler fuel injection systems employ **sequential fuel injection**, a system in which each injector is pulsed individually in the engine's firing order. Not only does this design use an individual terminal of the PCM to operate each injector, but the PCM now needs to know TDC compression information for each cylinder. This information is taken from a CMP sensor, or from the pickup of a distributor geared into the camshaft.

Electric Fuel Pump

A permanent magnet-type direct current electric motor drives the vane-type fuel pump, both submerged in fuel at the bottom of the fuel tank. One check valve functions as a pressure relief valve; the other, in the outlet port, prevents fuel from running in either direction when the pump is off. This keeps the line full of pressurized fuel when the engine is off and reduces fuel vapor problems.

With the ignition switch on, the PCM grounds the fuel pump relay, powering the fuel pump. Unless there is a distributor reference signal within 2 seconds, the relay will lose its ground and shut off the fuel pump. Once shut off, the relay stays off until the distributor reference signal arrives at the computer, or the key is cycled off and on again.

A few vehicles, such as the Shelby GLH turbocharged Omnis, use two fuel pumps, one in the tank and another near the engine. As these vehicles sometimes use more fuel, they are more subject to vapor lock. The tandem electric pumps reduce that possibility.

Fuel Pressure Regulator. To make air/fuel mixture calculations possible, the fuel injection system uses a fuel pressure regulator to keep the difference between the fuel rail and the intake manifold constant, Figure 17–25. This is often described as a "pressure differential across the injector." Then the amount of fuel will correspond closely to the pulse width, with no allowances needed for pressure differences.

The different systems use different pressures and different (though similar) pressure regulators. The pre-1986 single-point system maintains a fuel pressure difference of 36 psi; the later low pressure system maintains a difference of 14.5 psi. The multipoint systems maintain a difference of 55 psi. Fuel rail pressure on a DaimlerChrysler multipoint system can vary from 42 to 62 psi.

On the 3.0L engine, the fuel pressure was raised to 46 psi to avoid vapor lock problems during hot restarts.

Returnless Fuel System

By 1996 Chrysler fuel injection systems had changed further in two specific ways: fuel pressure is now held constant at approximately 49 psi by a mechanical regulator independent of the PCM and of intake manifold vacuum. Second, the fuel injection plumbing is one way: there is no return line.

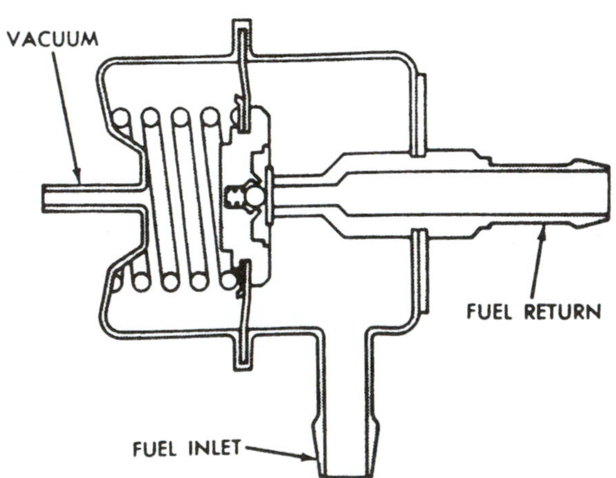

Figure 17–25 Fuel pressure regulator, multipoint-turbocharger system. *(Courtesy of DaimlerChrysler Corp.)*

Traditionally, the fuel pressure regulator has been located on the fuel rail or return line. Fuel in excess of engine requirements is returned to the fuel tank through a return line. Returned excess fuel may return the heat it absorbed while it was in the hot engine compartment. In an effort to reduce evaporative emissions from the fuel tank, all DaimlerChrysler engines will use a returnless fuel system. The fuel pressure regulator is located on top of the fuel pump module inside the fuel tank. Fuel that leaves the fuel tank will not be returned; excess fuel is simply dumped back into the tank, Figure 17–26.

Ignition Timing

The PCM calculates ignition timing from information reported by:

- the coolant temperature sensor.
- the distributor reference pulse (crankshaft/camshaft position sensors).
- the manifold absolute pressure sensor (MAP).
- the barometric pressure (startup MAP reading).

At warm idle, the PCM first uses spark advance manipulation to control normal idle speed fluctuations. This action is referred to by DaimlerChrysler as "active timing." If a specific amount of spark timing change does not put the idle speed at the design rpm, the PCM will use the automatic idle speed motor to change the amount of air entering the engine at idle.

Beginning in 1985, the multipoint system included a BARO-read (ambient barometric pressure) solenoid that the PCM used to briefly vent the vacuum line to the MAP sensor. During the time the line was vented, the MAP sensor read ambient barometric pressure. The computer can now update this reading as often as every 30 seconds if needed, though it is very unusual for barometric pressure or altitude to change much that quickly. This vent cycle occurs at least once each

Diagnostic & Service Tip

These changes have service caution relevance. Make sure you relieve the fuel pressure before opening any part of the fuel system. The higher pressure can spray more fuel farther and harder. Also, it may be necessary to remove the fuel filler cap when working on the fuel system. The pressure it retains can be enough to push fuel through the lines, with the attendant fire risk should that line be open. Keep in mind that should the system be in any actuation test mode that energizes the ASD relay, there will be full fuel delivery for as long as 7 minutes. That can just about empty a fuel tank.

time the throttle closes and engine speed is below a certain rpm. Vehicles with a turbocharger must monitor and control ignition timing very precisely, and barometric pressure significantly affects allowable spark advance.

Ignition Coils

DaimlerChrysler vehicles use one of three ignition coil designs: single coil, DIS coil packs and coil-on-plug.

Single Coil. The single coil is used with a distributor to distribute the spark to all cylinders. This is known as distributor ignition (DI). Chrysler DI systems may either use optical or Hall effect pickups within the distributor, as described in the "Inputs" portion of this chapter.

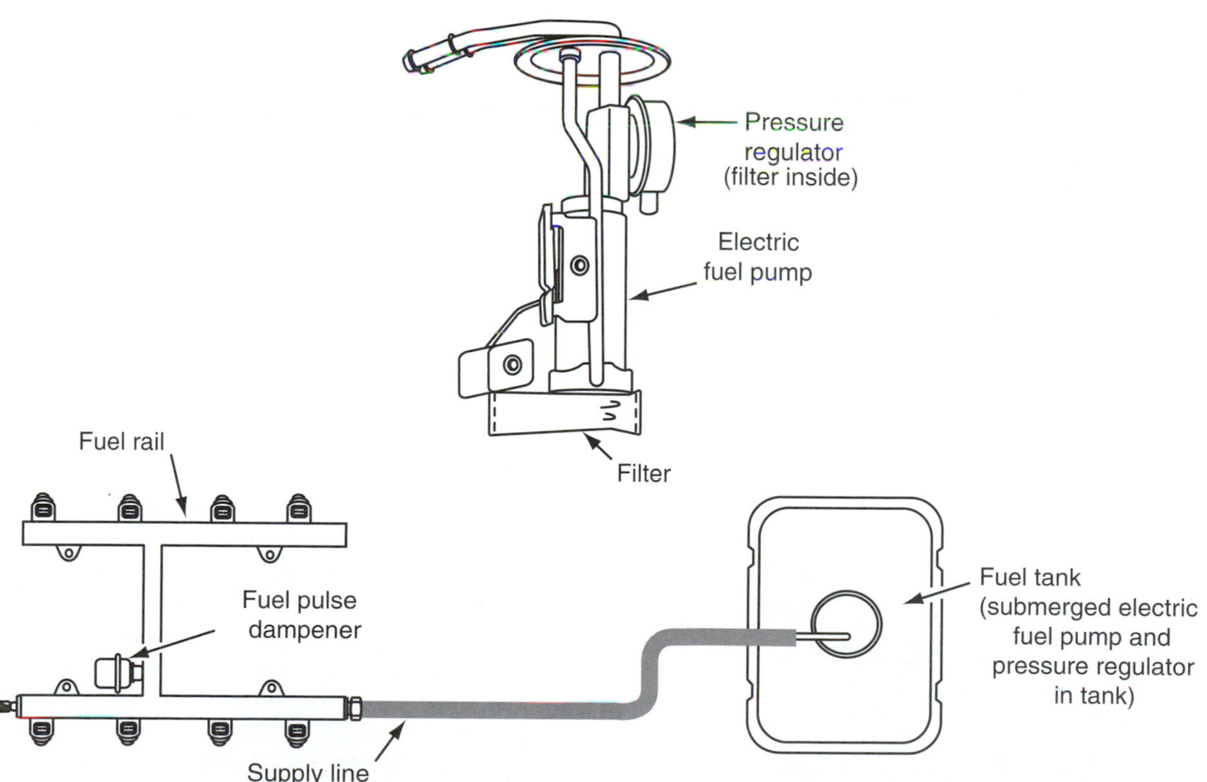

Figure 17–26 Returnless fuel system/tank unit.

DIS Coil Packs. Many modern DaimlerChrysler applications are using electronic ignition (EI) systems, that is, systems that do not use a distributor to distribute spark, but, instead, use multiple coils and distribute spark *electronically* by choosing which coil to fire. DaimlerChrysler uses two versions of EI systems: waste spark and coil-on-plug. The waste spark system uses multiple coils, all formed into one component called a *DIS coil pack*. Each coil within the coil pack is connected to (and fires) the spark plugs on two companion cylinders, one being near the end of its compression stroke and the other being near the end of its exhaust stroke. This system is known as the direct ignition system (DIS). The Hall effect CMP and CKP sensors used with this system are described in more detail in the Inputs portion of this chapter.

Coil-On-Plug. Chrysler 2.7L, 3.2L and 3.5L engines use a coil-on-plug (COP) ignition system, Figure 17–27. The COP EI system uses the same Hall effect CMP and CKP sensors as the DIS EI system does, described in further detail in the Inputs portion of this chapter. Using one ignition coil for each cylinder and installing it on top of the spark plug eliminates secondary ignition wiring, thereby reducing the potential for the ignition system to induce voltages into nearby circuits.

In this system, battery voltage is fed directly to each ignition coil. The PCM controls the ground side of each coil. The primary winding of the coil has low impedance with a resistance of 0.5 ohm.

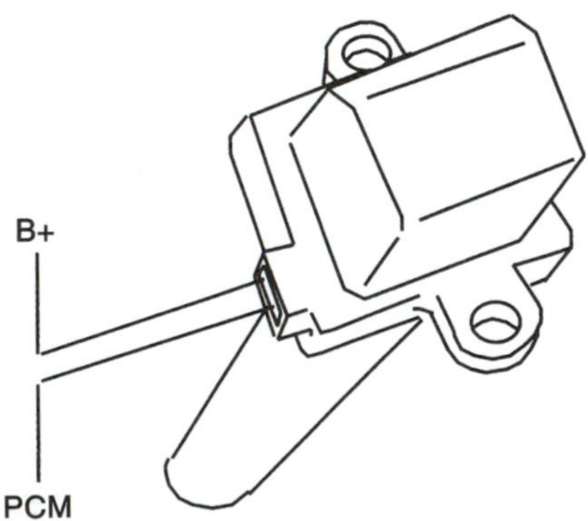

Figure 17–27 Coil-on-plug.

This low resistance provides quicker magnetic saturation of the coil. However, it also provides current much higher than that normally used in a computer-controlled circuit. In order to prevent damage to the solid state PCM circuitry, DaimlerChrysler incorporated a current regulation feature. Here is how it works. Whenever the engine temperature is below 176°F, as it is for a cold start, the primary ignition current is 7.7 to 9.0 amps. When the engine operates above that cold start temperature, the primary current is limited to 6.0 to 6.8 amps. When the technician monitors the dwell time, also called the on-time, of the primary circuit, there is a great difference in millisecond time between crank speed and actual run speed. Once the engine reaches actual run speed, the dwell time of all coils should be close.

Detonation Control. From the beginning of the turbocharged four cylinder with multipoint fuel injection, the Chrysler multipoint system has had the ability to retard the timing on just the cylinder with knock. If knock occurs, the PCM has already recorded which cylinder was just fired. It then retards the spark for just that cylinder slightly. If knock continues, the PCM will begin to reduce turbocharger boost.

Chrysler's Detonation Control

DaimlerChrysler put strong emphasis on performance in their turbocharged vehicles, is reflected in their detonation control strategy. You must prevent detonation, of course, or the engine will destroy itself. There are only two ways to stop detonation on a turbocharged engine under boost: reduce the boost or retard the timing. If you reduce the boost, you lose performance; if you retard timing, you raise exhaust temperature, which was already critical under boost. DaimlerChrysler's solution of retarding timing to one cylinder only raises exhaust temperature only slightly and allows the other three cylinders to keep producing at their full capacity.

Wastegate Control Solenoid

Before 1985, turbocharged engines used three types of boost control. The first was the wastegate, as described in the Outputs section of Chapter 10. The wastegate actuator, Figure 17–28, opens the wastegate at a boost pressure of 7.2 psi. The second method has the boost pressure sensed by the MAP sensor. If boost exceeds 7.2 psi, after snap acceleration for example, the PCM recognizes the overboost and skips fuel injection pulses until boost pressure drops to 7.2 psi. The third method uses the electronic engine speed governor: if rpm goes above 6,650, the PCM stops fuel injection until the rpm drops to 6,100.

In 1985, a fourth method was included. The PCM can operate a boost control solenoid that vents the line conveying manifold pressure to the wastegate actuator. If operating conditions are favorable, the PCM will pulse the boost control and bleed off some of the pressure. Under these circumstances the boost pressure can actually go to 10 psi before the actuator opens the wastegate. The PCM reviews these inputs for this calculation:

- barometric pressure.
- engine speed (rpm).
- engine coolant temperature (ECT).
- intake air temperature (IAT).
- detonation (knock sensor).

The PCM also keeps track of the engine's "detonation history," how inclined it is to knock and under what circumstances as well as how long it has been in boost.

Revised Boost Control

Beginning in 1988, the turbo boost actuator control on the Turbo 1 engine works directly from the turbocharger itself rather than from the intake manifold pressure. This keeps the wastegate open

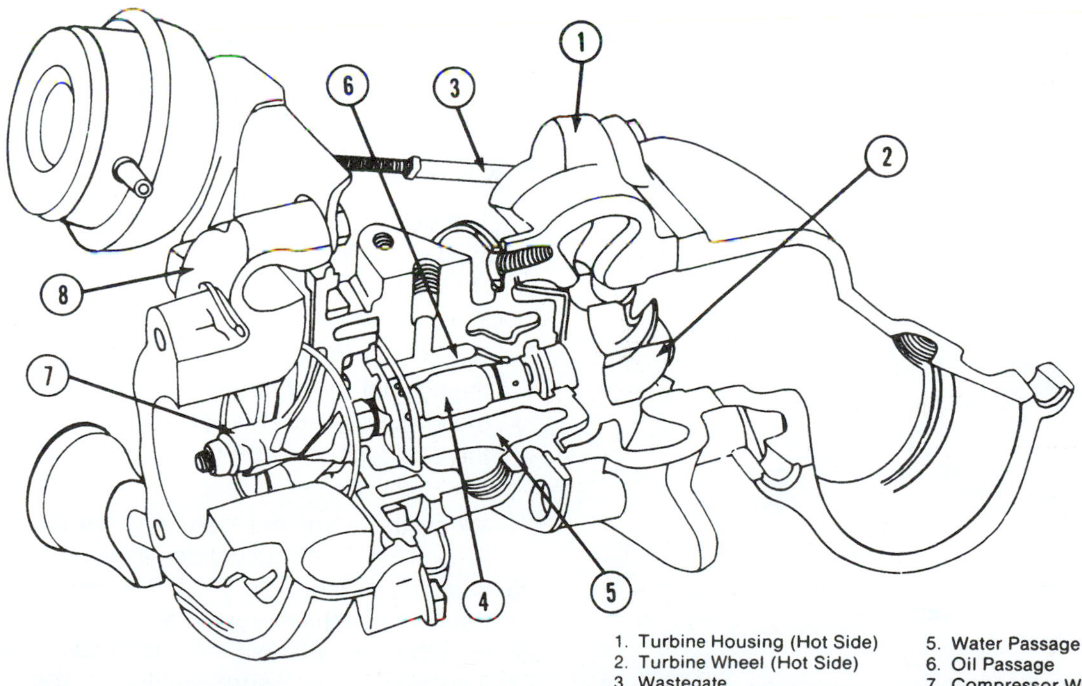

Figure 17–28 Turbocharger. *(Courtesy of DaimlerChrysler Corp.)*

during part-throttle operation and eliminates boost completely except at wide-open throttle. The advantages are:

- exhaust backpressure is lower.
- the intake air/fuel charge is cooler.
- the tendency to knock is reduced.
- part-throttle fuel economy improves.

Automatic Idle Speed (AIS) Motor

The AIS motor is a reversible electric motor, Figure 17–29. The PCM controls it based on information from the:

- TPS.
- ECT.
- speed sensor.
- P/N switch.
- brake switch.

The AIS motor moves its valve to control bypass air around the throttle plate in the throttle body. Even with the valve in its closed position, enough air gets past the throttle plate and through the throttle body to keep the engine idling at low speed with no load. The PCM directs the AIS motor to set the valve at different positions for different operating conditions. When the vehicle is decelerating, the valve opens to prevent engine stall and to prevent condensation of fuel on the intake port walls, as well as possible backfire.

In later years, DaimlerChrysler introduced a stepper motor version of the AIS actuator. This actuator moves a pintle valve in order to control throttle bypass air. The AIS stepper motor contains two windings that connect to the PCM with four wires. The PCM has the ability to apply current to the two windings in sequence and also has the ability to reverse polarity to each winding as necessary. The PCM can move the AIS stepper motor through a total of 255 steps, from fully closed to fully open, in order to control idle speed. The stepper motor rotates a precise amount for each pulse that the computer sends it, so by tracking the number of pulses, the PCM knows the exact position of the air bypass valve. This provides the PCM with increased precision in controlling idle speed.

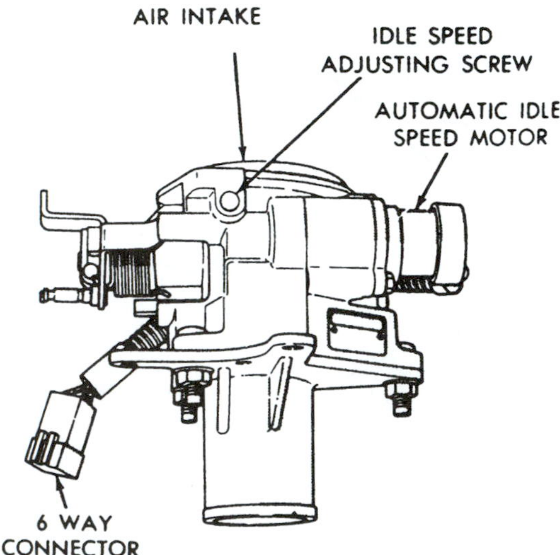

Figure 17–29 Multipoint throttle body showing AIS motor. *(Courtesy of DaimlerChrysler Corp.)*

EGR Control Solenoid

This solenoid controls ported vacuum to open the EGR valve. When the PCM energizes it, vacuum to the EGR valve is blocked. The computer blocks vacuum when:

- the coolant temperature is below 70°F/21°C.
- the engine speed is below 1,200 rpm.
- the engine is at wide-open throttle.

When the computer de-energizes the solenoid, allowing EGR actuation, a backpressure transducer also controls the EGR valve, Figure 17–30. Some vehicle systems have a vacuum bleed in the line between the transducer and the EGR to prevent pressure buildup in the EGR valve.

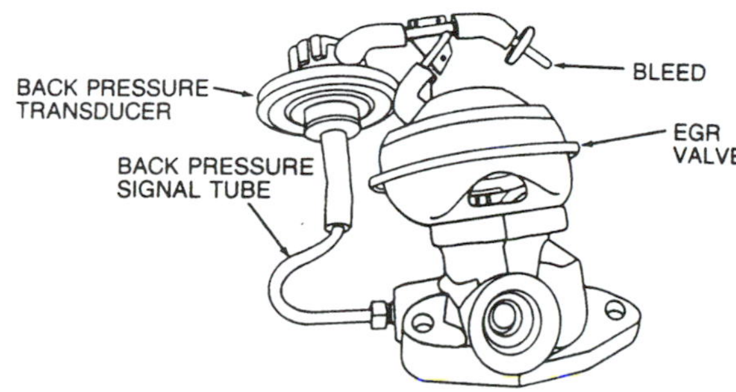

Figure 17–30 EGR valve and transducer. *(Courtesy of DaimlerChrysler Corp.)*

EGR Valve Temperature Sensor

Chrysler installed a thermistor temperature sensor on California versions of the 1.5-liter multipoint injection engine in the 1988 Dodge Colt. The purpose of this sensor is to detect an EGR valve that does not open when actuated. When an EGR valve opens, hot exhaust gas flows through it, and the temperature of the EGR mounting flange should go up sharply. If the PCM sends an actuation signal to the EGR solenoid and does not quickly get a return signal from the EGR temperature sensor, it will store a fault code and turn on the MIL. Because of changes in cam profile that have reduced valve overlap, Chrysler was able to eliminate the EGR valve on the turbocharged 2.2-liter engine in 1988.

Canister Purge Solenoid

The charcoal canister stores vapors from the fuel tank. So long as the coolant temperature is below 180°F/82°C, the PCM blocks the vent line to the throttle body by energizing a solenoid on the line. If the canister is not purged, it will gradually fill and will not be able to store any more vapors. If it purges too early or at other inappropriate times, it can drive the air/fuel mixture richer than the mixture control strategies can correct for.

Except for the 3.0L NS body style, all DaimlerChrysler engines now use a pulse width modulated purge solenoid valve to control the flow from the EVAP canister. The solenoid control circuit operates at a frequency of 200 MHz and is located in the PCM.

Radiator Fan Relay

The radiator fan relay supplies power for both the radiator fan and the A/C compressor clutch. The PCM grounds the relay coil whenever the coolant reaches a specific temperature or when the air conditioning is turned on. Both fan and compressor clutch are controlled by grounding their relay circuits.

Charging Circuit Control

Beginning in 1985, Chrysler vehicles have alternator output regulated by the PCM, thus replacing the voltage regulator. The PCM calculates the output based on the alternator's current output and on the battery temperature. The PCM controls the alternator output to between 12.9 and 15 volts. When the key is turned on, an electronic voltage regulator (EVR) circuit inside the PCM uses pulse width modulation to control the alternator field current. The PCM monitors the alternator output on a sensing wire from the ASD relay. If the charging rate cannot be monitored by the PCM, the field circuit duty cycle is limited to 25 percent so there is no high-voltage surge that might damage computer and sensor components, Figure 17–31.

A/C Cutout Relay

The A/C cutout relay is on the ground side of the A/C compressor clutch circuit, between the clutch and the A/C switch. When open, it stops the A/C clutch circuit. The PCM closes the compressor engagement circuit *except:*

- when the engine is operated at wide-open throttle.
- when engine speed is below 500 rpm.

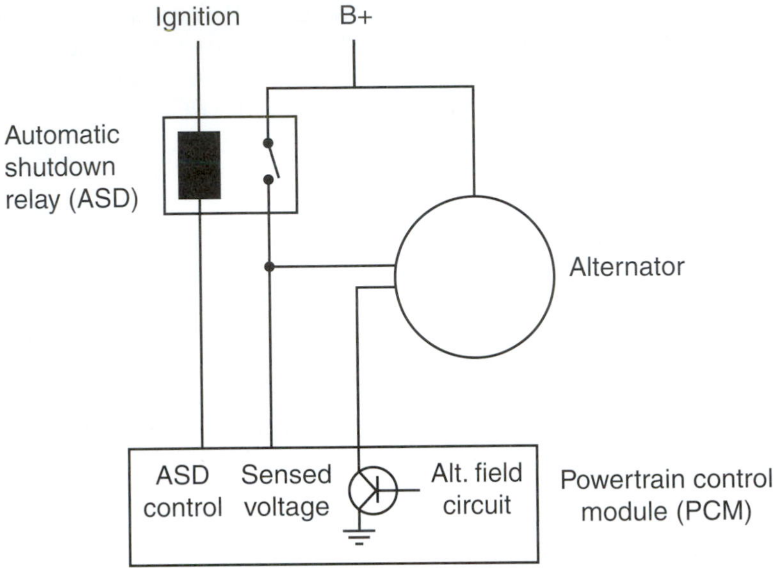

Figure 17–31 PCM alternator field control circuit.

- when the engine is cranked with the A/C switch on. The compressor cutout relay circuit remains open for 10 to 15 seconds after the engine starts.

Torque Converter Lockup Clutch

Beginning in 1988, the Torqueflite transmission came with a lockup torque converter for some vehicles. The PCM activates it under specific conditions:

- When the coolant temperature is above 150°F, *and*
- the park/neutral switch indicates the transmission is in gear, *and*
- the brake switch indicates the brakes are not applied, *and*
- the throttle angle, as reported by the TPS, is above a certain minimum.

The converter clutch works hydraulically, controlled by the computer. The actuation solenoid mounts on the valve body transfer plate and receives third clutch oil when the transmission goes into third gear. If the computer activates the solenoid by grounding its circuit while third clutch oil pressure is applied, the torque converter locks the crankshaft to the input shaft. If not, the open solenoid bleeds the oil pressure off, and the torque converter clutch releases. This works in a way very similar to the GM and Ford systems.

Malfunction Indicator Light (MIL)

The DaimlerChrysler malfunction indicator light (MIL) was formerly known as the Power Loss Lamp (prior to OBD II standardization) and serves the same purpose as the MIL on a GM or Ford vehicle. If the computer has encountered a fault in any of its sensors, actuators, circuits or internally in itself, it stores a code and (for most faults) turns on the MIL and sets the system in limp-in mode. The lamp also comes on when the key first turns the ignition on as a bulb check. If the fault disappears within one driving trip, the lamp stays on until the next startup. Then the lamp will be off and the system will come out of limp-in, but the code will stay in the computer's memory for the next 30 key off/on cycles without that problem. In the 1988 and later models, if a sensor fails, the computer generates its own substitute value for that sensor's readings and lights the lamp. If the sensor comes back into the proper range, however, the computer goes back into normal operation without the need for ignition cycling to exit limp-in mode. On Chrysler OBD II applications, the DTC will remain on for most recorded faults until the PCM has seen 40 warmup and cool-down cycles (as measured by the ECT sensor) without the fault reoccurring.

Vehicle Maximum Speed Governor

Also beginning in 1988, vehicles with 3.0-liter and 2.2-liter Turbo I engines have a vehicle maximum speed governor in the computer's program. Because of the speed rating of the tires installed on the vehicle, the computer shuts off all fuel if the vehicle exceeds 118 mph.

SMEC/SBEC-Controlled Cruise Control

Most of the DaimlerChrysler SMEC/SBEC-computer-controlled systems include the cruise control in the engine management system. The system is similar to the Ford system described in Chapter 14 of this book. One additional feature of the Chrysler variation is a third solenoid called a dump solenoid. The ignition switch powers the dump solenoid through a fuse, the cruise control on/off switch and the brake switch. Unlike most computer actuators, the dump solenoid has a fixed ground. Either turning the cruise control switch off or stepping on the brake will interrupt power to the dump solenoid, immediately bleeding off vacuum to the servo and letting the throttle return spring pull the linkage to idle.

DAIMLERCHRYSLER MULTIPLEXING SYSTEMS

Chrysler began networking on-board computers together over a data bus on their vehicles starting with the 1988 model year. Their first multiplexing system, known as Chrysler Collision Detection or C2D, was used from 1988 through 2000 and used a twisted pair of wires for serial data communication between the various computers. DaimlerChrysler then introduced their second-generation multiplexing system on 1998 models. This system is known as the Programmable Communication Interface or PCI system and was installed on most DaimlerChrysler vehicles by 2001. The PCI system uses an OBD II standardized protocol known as a J1850 Variable Pulse Width single-wire system. Additional information is provided in Chapter 7 of this textbook. Some additional product knowledge on these systems is contained within the following paragraphs.

Chrysler Collision Detection (C2D) System. Each module (node) can broadcast information at any time the network is free. To resolve the problem of two modules attempting to send simultaneously, Chrysler C2D systems prioritize each message. Each message starts with an ID code, a byte (an eight-digit binary number) that conveys the message's priority, content type and size in bytes, Figure 17–32. The message ID byte with the most zeroes to the left of the first 1 has the highest priority and gets broadcast first. In 1992, Chrysler multiplexing systems transferred information at a rate of 7812.5 baud (7812.5 digital bits per second).

Different vehicles may have different components wired to their "internal internet," but on many Chrysler products these include:

- body control computer.
- engine controller (SBEC).
- vehicle theft alarm (VTA).
- electronic vehicle information center (EVIC).
- engine node (a module in the front of the vehicle with sensors providing air temperature, and direction of travel).
- electronic transmission controller (EATX).
- mechanical instrument cluster (MIC).
- traveler.
- overhead console.

Magnetic Interference. The data bus itself consists of two wires twisted together to reduce

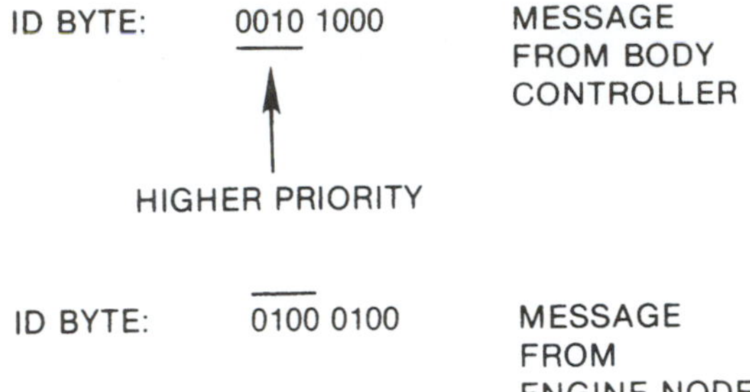

Figure 17–32 C2D chip priority code.

magnetic interference and the generation of false information by nearby electric components. Since the wires are twisted together, any magnetic induction affects both wires at the same time, canceling out most of the problem.

Programmable Communication Interface (PCI)

The PCI system appears on the 1998 LH, the 1999 WJ and the 2000 year models. The star hub for this data bus is located in the junction block above the BCM. Communication tests between all controllers on the bus can be done at this central point. The voltage of this bus is approximately 7.5 volts. Each module on the bus supplies its own bias and ground. The PCM is the main module and uses a one kilohm resistor for termination.

✔ SYSTEM DIAGNOSIS AND SERVICE

A thorough diagnostic procedure is published by DaimlerChrysler Corporation for each model year and application. This driveability test procedure should be in the service manual you use on any particular car.

On-Board Diagnostics

Besides directing the combustion activity of the engine, the PCM also monitors specific input and output circuits for faults (voltage values out of range or that do not change appropriately). If a fault is detected, the computer will store a two-digit code in its diagnostic memory. These codes and much other diagnostic information can be obtained by putting the system in one of its diagnostic modes using a Diagnostic Readout Box (tool C-4805 or other scan tool). On some systems the fault codes can be read out and the switches tested with the power loss lamp itself.

Test Modes

Diagnostic Test Mode. Connect the scan tool to the diagnostic test terminal. Early terminals are under the hood, later ones may be found in the passenger compartment. Turn the ignition on-off-on-off-on in 5 seconds or less. The readout box then displays any stored codes. When it is finished with everything stored in its memory, it displays code 55.

Switch Test Mode. Once the preceding test is complete and code 55 shows, make sure that all system input switches are turned off. Then turning on any of the input switches will cause the displayed code 55 to change as long as the PCM gets the incoming signal. Turning the switch off makes the code 55 reappear.

Circuit Actuator Test Mode (ATM). After the diagnostic test mode is completed and code 55 shows in the scan tool, press the ATM button on the scan tool until the desired ATM code number is displayed. When the button is released, the ATM will begin. The PCM cycles the designated actuator on and off at 2-second intervals for 5 minutes or until you turn the ignition off. This provides ample time to examine the actuator directly and be sure it is working properly. Each ATM code is a two-digit number representing each of the actuator circuits this mode can test. They are listed in the service manual with the fault codes.

Sensor Test Mode. Beginning in 1986, this test mode allows choosing an individual sensor, and the scan tool will display the return voltage signal that the sensor is sending the PCM at that moment.

Engine Running Test Mode. This test checks a number of sensors and actuators in use. Connect the scan tool to the diagnostic test connector and start the engine. Observe the scan tool display. Once the oxygen sensor is hot enough, the display will switch between 0 (lean) and 1 (rich). The scan tool can also command the idle speed motor to increase idle speed. On turbocharged models, striking the intake manifold near the knock sensor with a tool as explained in the service manual will make the number 8 appear alongside the switching 0-1. The number 8, of course, indicates that ignition timing has been retarded.

Test Mode 10. The random-access memory (RAM) of the SMEC/SBEC computer is less volatile than that of the logic module. Removing battery power from it, one of the procedures used to clear the memory for logic module-equipped vehicles, may not erase the stored fault codes for some time. So for these systems, there is another sensor test, mode 10. When the technician selects test mode 10, all fault codes are erased. Needless to say, fault codes should not be erased until after they have been written down for later diagnosis.

The SMEC/SBEC system also has a few additional fault codes and some changes from previous diagnostic procedures. For example, some of the harness connectors are hard to probe with test instruments, so as a result the revised diagnostic procedures may require disconnection of the harness and the performance of open circuit tests.

Testing Without the Scan Tool. Cycling the ignition switch on and off three times in five seconds triggers the logic module to flash out codes through the MIL if the scan tool is not connected. It flashes on and off to indicate the code number. For example, two flashes separated by a short pause of about one-half second, followed by a pause of about 2 seconds and then three flashes, separated again with a short pause indicates code 23. Once the codes are flashed out, the system goes into switch test mode. As long as it receives signals from the switches, the logic module will flash the MIL on and off in response to the input switches being correspondingly cycled. The other test modes, however, cannot be performed without the scan tool.

Driveability Test Procedure

The driveability test procedure is designed to use the test modes just described and to help diagnose engine performance complaints not caused by internal computer problems. It should be used whenever driveability problems are present. The first step, however, is to verify the problem. Ordinarily, the technician should drive the vehicle under the same conditions in which the owner discovers the problem.

The second step is a visual inspection. The driveability test in the DaimlerChrysler literature provides a guide for this, specific to the vehicle being diagnosed. If no problem is found during visual inspection, the most appropriate driveability test section should be used. If the engine does not start, obviously, use the no-start section. If the problem occurs only when the engine is cold, use the cold driveability section and so on. Once a driveability test section is begun, start with the beginning of the test and follow each step to the end.

WARNING: Because of the high fuel pressure in many of these systems, it is very important to relieve the pressure before opening the fuel system. Serious eye and fire hazards are present if a fitting is loosened while the system is still pressurized. In addition, use only the correct part numbers for fuel system hose and hose connections when making repairs.

OBD II Diagnostics

Chrysler OBD II systems require an OBD II scan tool for all diagnostic features, although, initially, Chrysler did retain manual methods of code pulling in parallel to OBD II scan tool methods on early OBD II systems. That is, on Chrysler's early OBD II systems, the diagnostic trouble codes (DTCs) may be pulled manually through cycling of the ignition switch and counting pulses of the MIL (as in pre-OBD II systems) or a scan tool may be used. However, cycling of the ignition switch produces only two-digit codes as per their pre-OBD II systems. In order to obtain the OBD II five-digit codes, an OBD II scan tool must be used. On later systems, the manual method has been eliminated and all diagnostic functions must be done through an OBD II scan tool.

Once connected to the Diagnostic Link Connector (DLC) under the driver's side of the instrument panel, an OBD II scan tool may be used to access several on-board computers, including the PCM, Transmission Control Module (TCM), Body Control Module (BCM), Mechanical Instrument Cluster (MIC) and the Antilock Brake Control Module (ABCM). Each of these modules provide the technician with the ability to access memory DTCs, perform functional tests such as the Actuator Test Mode (ATM) and perform certain other tests depending on the system that is selected. For example, if the PCM is selected, the technician may adjust engine rpm in 100-rpm increments (typically from about 800 rpm to about 2,000 rpm) through the scan tool's menu. And, in addition to DTCs, data stream information and functional tests, PCM freeze frame data is also available. Or if the ABCM is selected, the technician may use the scan tool to bleed the ABS modulator assembly (part of the brake bleeding procedure). The TCM also allows the technician to select the vehicle's tire size from a menu in order to correct VSS calibration when tire size has been changed. In order to know what tests are offered with a particular scan tool, simply bring up the test menu for any of the modules that appear on the display screen. As you might expect, DaimlerChrysler's third-generation diagnostic readout box (DRB III) allows the technician to perform many more functions than a generic (aftermarket) OBD II scan tool allows.

SUMMARY

In this chapter, we have covered single- and multipoint fuel injection systems used by DaimlerChrysler. We have seen the role of the PCM in controlling combustion in the engine under all driving conditions. Likewise, we reviewed the way the computer controls the idle speed under various conditions, how it controls ignition timing and how it prevents knock. The text explained how the DaimlerChrysler system controls turbocharger boost to optimize both power and engine life, first retarding spark to a specific cylinder, and only should that fail, to end knock by reducing turbocharger boost. The incorporation of control of the charging system output voltage was discussed. This accommodates battery condition when very cold, and can also smooth idle by reducing charge at that low-torque engine speed. We also reviewed the sensors such as the battery temperature sensor and the way the computer uses the voltage in the fuel pump circuit to calculate the alternator field strength. We've seen the role of the automatic shutdown relay, and we've discussed limp-in mode, self diagnostics and how to understand and repair any problems discovered.

▲ DIAGNOSTIC EXERCISE

Describe the PCM system problems that would be caused by a high resistance battery negative connection to the engine block and the kinds of tests you would use to determine if this was the problem.

REVIEW QUESTIONS

1. The first full-function engine control system for Chrysler was the dual module system. *Technician A* says that the logic module sends a 5-volts reference voltage to the sensors, makes calculations for the system based upon the input signals and provides output instructions for the power module.

Technician B says that the power module controls the output actuators according to commands from the logic module.
Who is correct?
A. A only
B. B only
C. Both A and B
D. Neither A nor B

2. Following the dual module system, the next engine control system that Chrysler used was which of the following?
   A. The JTEC system
   B. The SBEC system
   C. The NGC system
   D. The SMEC system
3. The engine control system used by Chrysler on Dodge trucks, the Dodge Viper and Jeeps beginning in 1996 was which of the following?
   A. The JTEC system
   B. The SBEC IIIA system
   C. The NGC system
   D. The SMEC system
4. The engine control system that DaimlerChrysler introduced in 2002 is which of the following?
   A. The JTEC system
   B. The SBEC IIIA system
   C. The NGC system
   D. The SMEC system
5. Technician A says that, depending upon the year and model, the fuel pump on a DaimlerChrysler vehicle may be controlled by the ASD relay.
   Technician B says that, depending upon the year and model, the fuel pump on a DaimlerChrysler vehicle may be controlled by a dedicated fuel pump relay.
   Who is correct?
   A. A only
   B. B only
   C. Both A and B
   D. Neither A nor B
6. The ASD relay may provide power for all of the following components *except* which of the following?
   A. The fuel injectors
   B. The Hall effect CKP and CMP sensors
   C. The ignition coil primary winding
   D. The O_2 sensor's heating element
7. *Technician A* says that the barometric pressure value in the PCM's memory is obtained from the MAP sensor.
   Technician B says that the Baro Read solenoid is used by the PCM to allow it to update the barometric pressure value in its memory at any time.
   Who is correct?
   A. A only
   B. B only
   C. Both A and B
   D. Neither A nor B
8. What is the primary purpose for incorporating a range-switching (dual range) temperature sensor on some DaimlerChrysler vehicles?
   A. Its circuit uses less current than a standard temperature sensor, therefore reducing the load on the alternator.
   B. It allows the PCM to use the same temperature sensor to measure both engine coolant temperature and intake air temperature.
   C. Its circuitry is simpler and therefore less confusing to a technician.
   D. It gives the PCM greater ability to fine tune the air/fuel ratio and spark timing due to its greater precision in measuring temperature.
9. What two sensors are sometimes combined into a single component on certain Chrysler vehicles?
   A. TPS and MAP
   B. IAT and MAP
   C. ECT and MAP
   D. TPS and ECT
10. In what model year did Chrysler begin using MAF sensors instead of MAP sensors as the primary engine load sensor on certain vehicles?
    A. 1985
    B. 1989
    C. 1994
    D. 2001

11. What type of tool should be used to measure and interpret the resulting signal from Chrysler's MAF sensor?
    A. A DC voltmeter
    B. An AC voltmeter
    C. A duty cycle meter
    D. A Hertz meter
12. An optical sensor within a distributor contains all *except* which of the following components?
    A. A permanent magnet
    B. An LED
    C. A disc with slits in it
    D. A photocell
13. What type of sensor is used on Chrysler distributorless ignition applications to monitor crankshaft position (CKP) and camshaft position (CMP)?
    A. Permanent magnet
    B. Hall effect
    C. Optical
    D. Piezoelectric
14. What type of VSS is used on Chrysler vehicles since 1992?
    A. Permanent magnet
    B. Hall effect
    C. Optical
    D. Piezoelectric
15. What type of detonation (knock) sensor is used on Chrysler applications?
    A. Permanent magnet
    B. Hall effect
    C. Optical
    D. Piezoelectric
16. Why are returnless fuel injection systems being used on Chrysler vehicles?
    A. To reduce the potential for vapor lock
    B. To reduce evaporative emissions from the fuel tank
    C. To reduce the load on the fuel pump
    D. To improve cold engine performance
17. An EI ignition system distributes spark electronically through the use of multiple ignition coils.
    Technician A says that a Chrysler EI system may use one ignition coil per two spark plugs.
    Technician B says that a Chrysler EI system may use one ignition coil per each spark plug.
    Who is correct?
    A. A only
    B. B only
    C. Both A and B
    D. Neither A nor B
18. *Technician A* says that, beginning in 1988, Chrysler turbocharger boost is controlled so as to eliminate boost completely except at wide-open throttle.
    Technician B says that, beginning in 1988, Chrysler turbocharger boost is controlled so as to provide boost during cruise and light acceleration in order to increase fuel economy.
    Who is correct?
    A. A only
    B. B only
    C. Both A and B
    D. Neither A nor B
19. In order to properly control alternator output, which of the following inputs is considered by the PCM?
    A. Monitored alternator output
    B. Battery temperature
    C. Throttle position
    D. Both A and B
20. *Technician A* says that, depending on year and model, a Chrysler vehicle may have either of two multiplexing systems, known as C2D or PCI.
    Technician B says that in order to pull the five-digit OBD II DTCs on a Chrysler vehicle, a scan tool is not needed. Simply turn on the ignition switch three times within 5 seconds and read the flashes of the MIL.
    Who is correct?
    A. A only
    B. B only
    C. Both A and B
    D. Neither A nor B

Chapter 18

European (Bosch) Engine Control Systems

OBJECTIVES

Upon completion and review of this chapter, you should be able to:

- ❑ Describe the two basic types of Bosch fuel injection systems.
- ❑ Explain the distinction between the Motronic system and the earlier Bosch fuel injection systems.
- ❑ Identify the various operating modes of the Motronic system and the operating conditions that they relate to.
- ❑ Define the inputs used with a Motronic fuel injection system.
- ❑ Define the outputs controlled by a Motronic fuel injection system.
- ❑ Understand the basic diagnostic steps of a Motronic fuel injection system.

KEY TERMS

Continuous Injection
Lambda
Motronic
Pulsed Injection

The Robert Bosch Corporation does not build cars or engines, but components and control systems. Carmakers buy these specially-engineered components and control systems from the Bosch Corporation to use on their own vehicles. An early pioneer in fuel injection systems, the Bosch Corporation makes many of the products found on most German and Scandinavian cars, as well as those on many American and Japanese vehicles. Bosch also licenses several of its systems for manufacture by other firms, so an understanding of the Bosch systems in turn provides an understanding of many systems built by other sources as well. A modern English car with a Lucas control system, for example, is actually controlled by Bosch designed components and controllers.

SYSTEM OVERVIEW

The Bosch company has been involved in fuel injection since the 1920s, when they began with diesel injectors. Their experience with gasoline injection goes back to military vehicles in the late 1930s. One of the earliest of their systems for passenger cars was the direct, diesel-style injection system used on a few of the gasoline-fueled Mercedes-Benz SLs. That system was a mechanical forced injection system very similar to diesel injection systems used on trucks.

All the Bosch engine management systems are computerized developments of earlier fuel injection systems, mostly mechanical. While our focus in this book is on computer-controlled systems, it would be difficult to understand what the Bosch engineers were doing without some explanation of the earlier mechanical systems on which their computers work.

While Bosch has built a version of throttle body injection for a limited number of European manufacturers, all of the systems used on vehicles imported to the United States have been multipoint injection systems. The advantages of multipoint are the same for Bosch as for any other injection system builder, and these advantages have been described in Chapter 10.

Bosch systems work in one of two basic ways: pulsed injection systems and continuous systems. D-Jetronic and L-Jetronic systems are pulsed, Figure 18–1; K-Jetronic systems are continuous (also referred to as Continuous Injection System, CIS), Figure 18–2. The D comes from the German word for (intake manifold) pressure, *Druck*, the major sensor in that system. The L comes from the German word for air, *Luft*, the airflow sensor or meter being that system's major sensor. K comes from the German word for continuous, *Kontinuerlich*, because the fuel flows continuously from that system's injectors as long as the engine is running. When the letter "E" is added, it indicates the system is electronic, that is, controlled by a computer. The term ***Motronic*** applies to the more fully

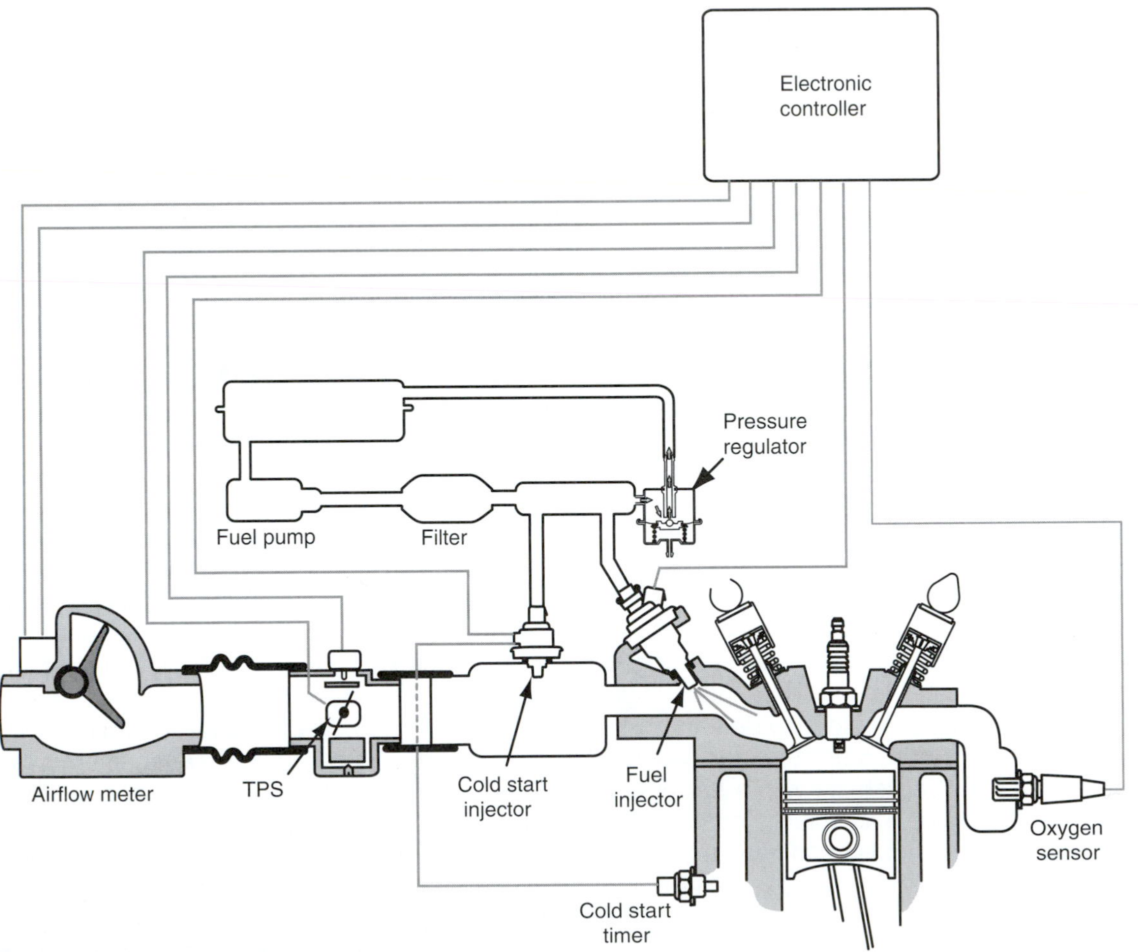

Figure 18–1 Pulsed injection system.

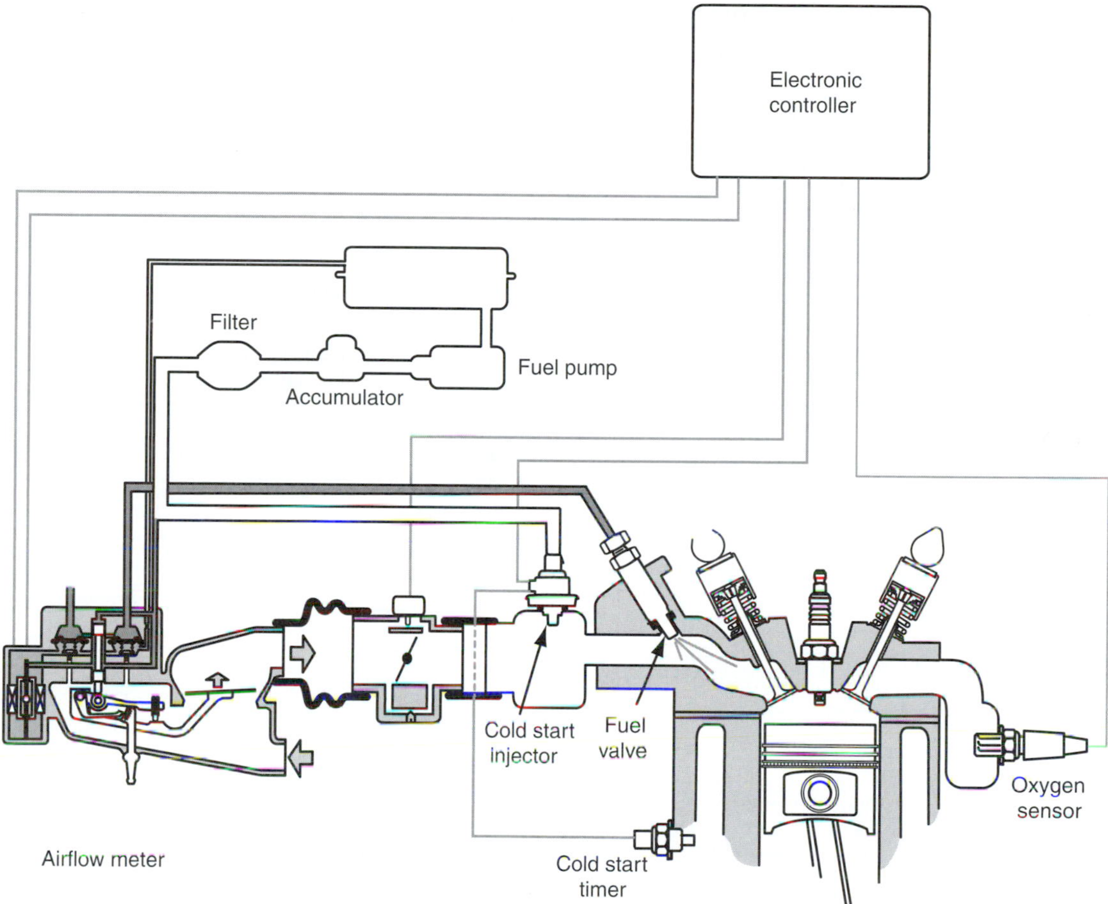

Figure 18–2 Continuous injection system.

developed engine management systems, covering spark timing, fuel injection and some other vehicle functions. In general, a Motronic system uses a single computer to manage all engine functions, while the earlier systems used independent controllers, typically one for ignition and another for fuel injection.

The **pulsed injection** systems work either sequentially or in clustered groups, varying fuel delivery by pulse width just as domestic systems do. The continuous systems spray fuel from port injectors whenever the engine is running, varying the amount sprayed to correspond with the amount of air entering the engine. Both types of systems have been elaborated into computer-controlled systems, with improved driveability, fuel economy and emissions quality. Once warmed up to operating temperature and driving normally, the system controls the air/fuel mixture by a closed-loop feedback process, Figure 18–3.

The **continuous injection** (K- and KE-Jetronic) systems are the most common of the Bosch systems, used on virtually all European makes at different times. Incoming air lifts or lowers (depending on the vehicle manufacturer) a plate suspended on a lever in a specially shaped cone, Figures 18–4 and 18–5. The movement of the plate (in the original mechanical systems)

Functional diagram taking as an example the L-Jetronic equipped with Lambda closed-loop control.

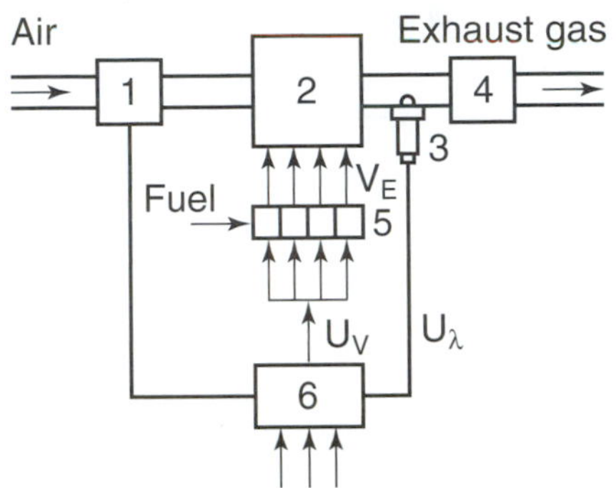

1 Airflow sensor
2 Engine
3 Lambda sensor
4 Catalytic converter
5 Injection valves
6 Control unit with closed-loop controller

U_λ Probe voltage
U_V Valve control voltage
V_E Quantity of fuel injected

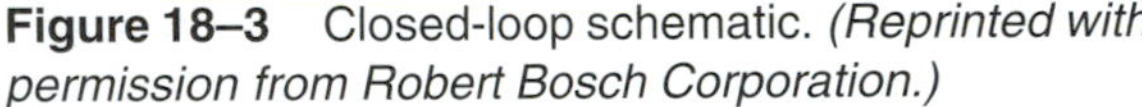

Figure 18–3 Closed-loop schematic. *(Reprinted with permission from Robert Bosch Corporation.)*

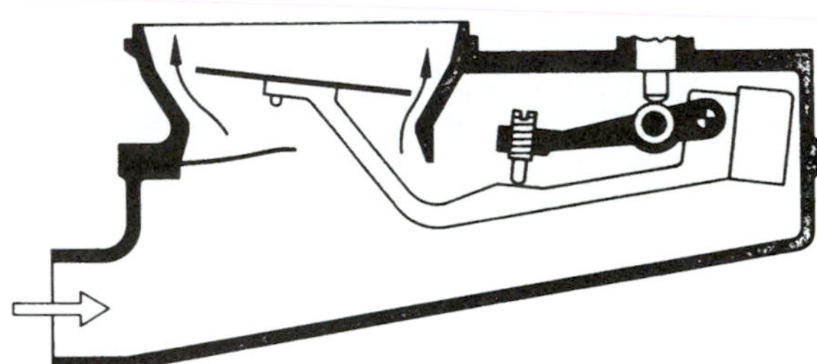

Figure 18–4 K-Jetronic air cone. *(Reprinted with permission from Robert Bosch Corporation.)*

completely determines the amount of fuel injected through the fuel distributor. In the later, computerized systems the control unit can vary the mixture by varying the pressure differential between the upper and lower parts of the fuel distributor in response to the signals from sensors, thus controlling the delivered air/fuel mixture, Figure 18–6. The control unit also employs programming for enrichment under cold engine, acceleration and some other driving conditions.

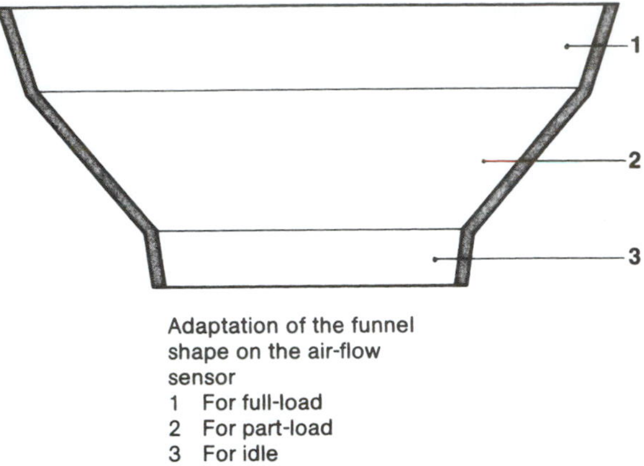

1
2
3
Adaptation of the funnel shape on the air-flow sensor
1 For full-load
2 For part-load
3 For idle

Figure 18–5 K-Jetronic air cone taper for different airflows. *(Reprinted with permission from Robert Bosch Corporation.)*

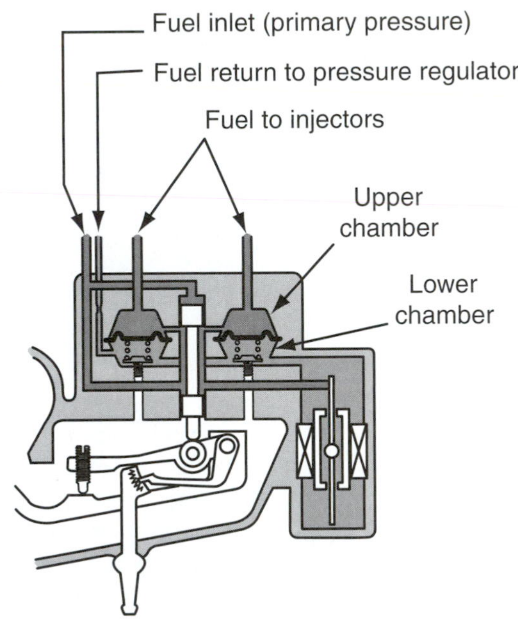

Figure 18–6 Air sensor plate and fuel plunger.

Mixture Control Unit

The major fuel and air component of the continuous injection systems is the mixture control unit, Figure 18–7. This combines the airflow sen-

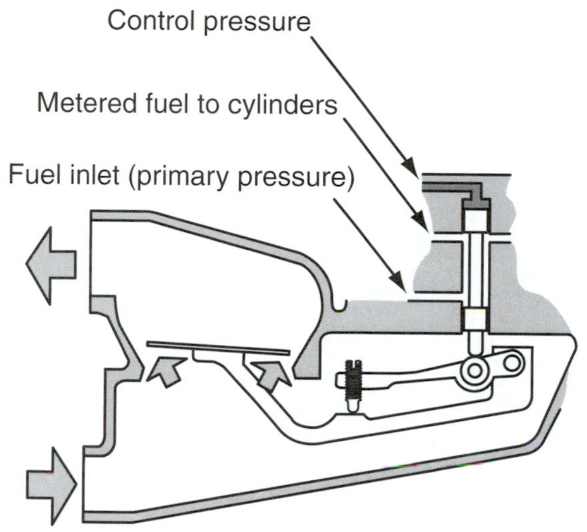

Figure 18–7 Mixture control unit.

sor plate and the fuel distributor. As the incoming air lifts (or on some models lowers) the plate, it moves a lever directly connected to the fuel distributor. The lever lifts a precisely machined plunger in a barrel with special slots to meter the fuel to each cylinder, Figure 18–8. The farther the plunger is raised, the more fuel passes the plunger, through the barrel and into the cylinders, corresponding to the increased amount of air moving past the air plate.

The original K-Jetronic systems worked mechanically to adjust the air/fuel mixture. Once the system became computer controlled to adjust the mixture in response to signals from sensors, the fuel distributor was modified to include a pressure actuator, a device to vary the pressure difference between the upper (delivery) and lower (control and return) halves of the fuel distributor, Figure 18–9.

The control unit does this by varying the current (amps) flowing through the pressure actuator, which then controls fuel bypass and the pressure difference and delivery rate. The current passes through an electromagnetic coil that raises a restrictor baffle in the fuel return line, determining the control pressure. Electromagnetism is directly proportional to the current, regardless of the voltage. The pressure difference between the two halves of the fuel distributor flexes a thin metal diaphragm which allows fuel to pass into the injector lines. The more the diaphragm flexes down, the more fuel is injected and vice versa, Figure 18–10. By returning more fuel to the tank,

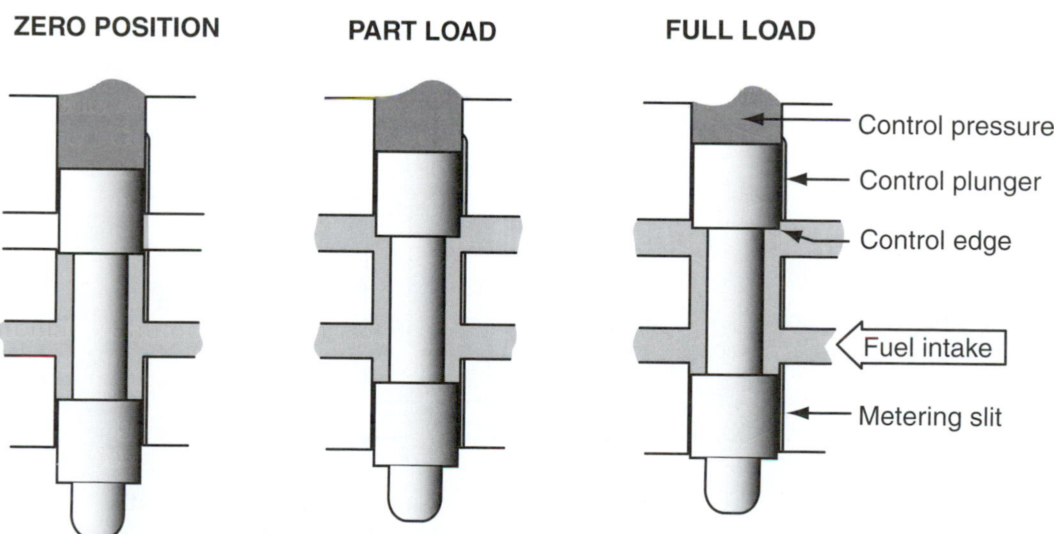

Figure 18–8 Fuel plunger.

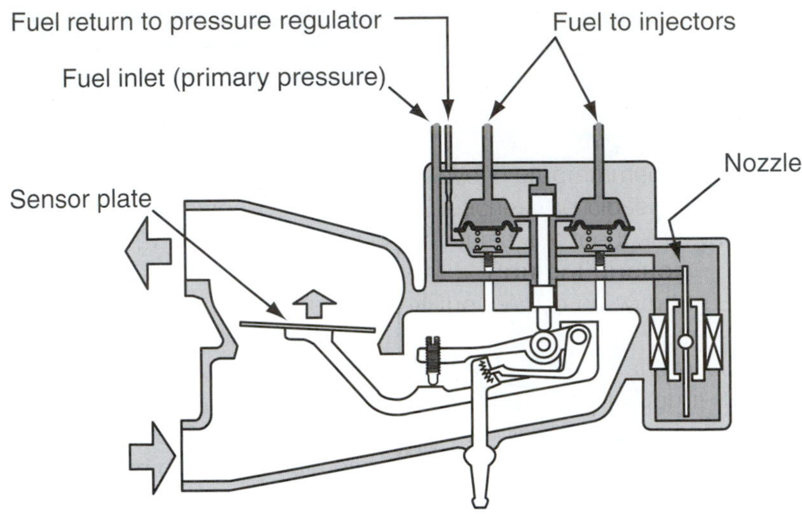

Figure 18–9 Pressure actuator.

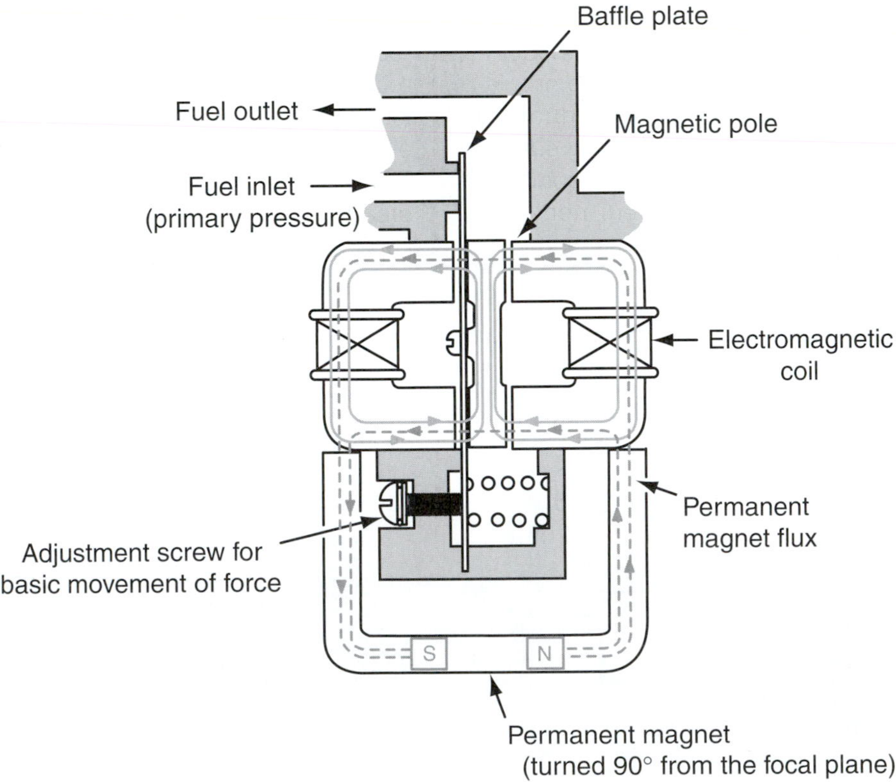

Figure 18–10 Operation of fuel distributor diaphragm.

the actuator reduces the pressure difference, flexes the diaphragm up and reduces fuel delivery. By restricting return fuel flow to the tank, the actuator increases the pressure difference, flexes the diaphragm down and increases fuel delivery.

Bosch continuous injection systems are often called mechanical fuel injection systems, but it should be noted that they are primarily hydraulic in the way they work.

Most continuous injection systems are installed separately from the engine, on a bracket in the intake air line. On certain V-form engines, notably Mercedes-Benz and Volvo, the control unit is mounted directly on the engine, much like a carburetor.

The injectors in a continuous system spray whenever the fuel pump is operating (powered as on domestic systems through a relay). On many systems, the injectors include a special vibrating needle in the tip which aids in atomizing the fuel. Often the action of these vibrating needles is audible as a hiss with the hood open and the engine at idle. This is normal and does not require any corrective action.

PULSED SYSTEMS

The pulsed-injection systems (D-Jetronic, L-Jetronic, and LH-Jetronic) vary the air/fuel mixture by varying pulse width, just as with domestic systems. As we will see, the sensors and actuators work very similarly.

The earliest of the modern Bosch systems was the D-Jetronic. Named for the German word for pressure, D-Jetronic uses intake manifold pressure as the principle input for the calculation of how much fuel to inject. While Bosch left this system years ago after tests and experience showed that air mass measurements were much more accurate for such a calculation, the intake manifold pressure information is nonetheless used on many systems as we have seen in earlier chapters.

The D-Jetronic systems are also unique in their use of analog computers. Check Chapters 2 of this book for more information about analog versus digital computers. A few of the early L-Jetronic systems also used an analog computer. These systems work more slowly and by variable voltages rather than by faster, digital systems. D-Jetronic systems only control fuel injection and have no control over ignition. Most of them do receive a signal from the distributor (from a second set of contact points in early models) to time the fuel spray.

Diagnostic & Service Tip

Because Bosch D-Jetronic systems calculate fuel delivery entirely by intake manifold pressure, a vacuum leak on these systems does not cause a bad idle from a lean mixture (unless there is a large vacuum leak to just one cylinder). However, a vacuum leak does cause a high idle speed in a way that can be puzzling to technicians unfamiliar with the system. Checking for vacuum leaks is done the same way as with other fuel induction systems.

The pulsed injection Bosch systems do not employ the airflow plate or fuel distributor from the K-Jetronic systems. Instead they measure the incoming air volume (or in later systems, mass), and the control unit calculates the proper pulse width under all engine conditions and opens the injectors to deliver the correct amount of fuel. The injectors for these systems are identical to those in domestic fuel injection/engine management systems. The most important sensor for this operation (after the engine is started and warmed up) is the airflow meter. While this is similar to the airflow meter used on some Ford systems, it is a Bosch-engineered component, so we describe it in some detail here.

Airflow Meter

The airflow meter is basically a lightly sprung swinging door in the airstream, sharing a pivot shaft with a potentiometer that modifies the return voltage signal to the computer, Figure 18–11.

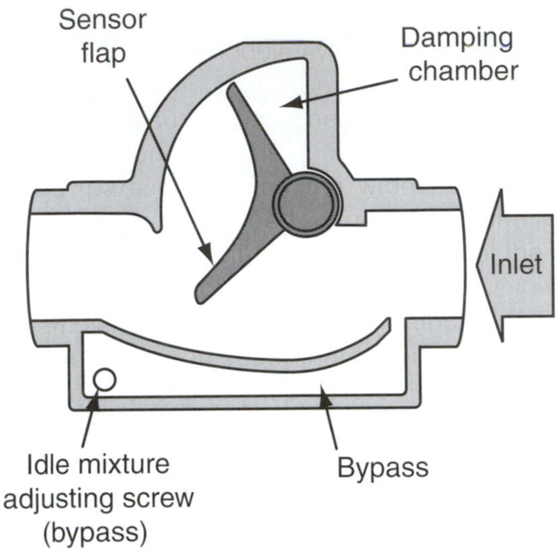

Figure 18–11 Airflow meter.

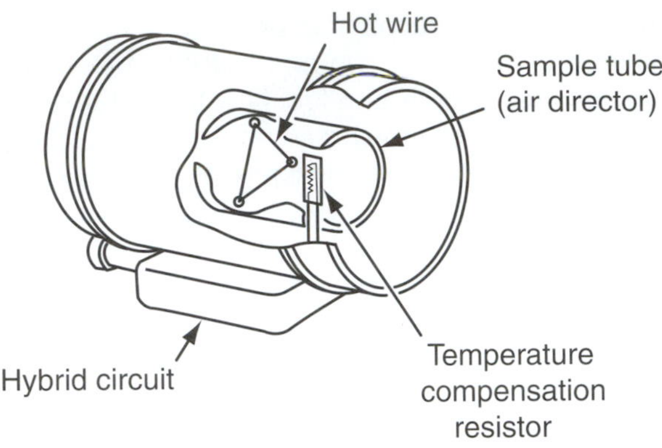

Figure 18–12 Air mass sensor.

Paired with the airflow meter door is a second, equal-sized door in a blind passage. The second door serves two functions: on four-cylinder engines, the individual cylinder pulses can be so distinct that they create pulses in the intake manifold, moving the sensor back and forth with each pulse. This is particularly likely at low engine speeds. The second door serves as a damper to prevent this sensor door pulsing. If the door were allowed to pulse, obviously, the computer would receive inaccurate signals about the amount of air taken into the engine. On many of the swinging-door airflow meters there is an additional door or pop-out plug in the center of the flap to allow a backfire to occur without destroying the expensive airflow meter when the air door would slam shut.

A later version of the L-Jetronic system, called the LH-Jetronic, uses a hot-wire air mass sensor, that works just like hot-wire air mass sensors used in domestic engine management systems, Figure 18–12.

MOTRONIC

The Motronic system is Bosch's first multifunction engine management system. In addition to controlling the air/fuel mixture, it controls ignition timing and dwell. The system may also be designed to control idle speed, EGR operation, evaporative emissions, automatic transmission and a turbocharger, depending on vehicle application.

Bosch and Imports

Of the imported European cars with a comprehensive computerized engine control system, most use a Bosch system. It first appeared on selected BMWs. Most of the systems used on Japanese cars imported into the United States are built using Bosch patents. Practically all the gasoline engine multipoint injection systems for domestic vehicles use Bosch components or patents.

CONTROL UNIT

The Bosch Motronic system is similar to the L-Jetronic system and like its predecessor controls fuel metering with pulse width modification. The Motronic's computer is a digital unit with the ability to control additional functions, such as spark timing. An overview of the Motronic system is provided in Figure 18–13.

The control unit's mounting brackets also serve to dissipate heat from the power transistors that drive the ignition and injectors, Figure 18–14. A 35-pin connector connects to the vehicle's system harness.

Main Relay

A main relay, similar in operation to the power relay used on Ford EEC systems (see Chapter 14), powers the control unit as soon as the ignition is turned on. This relay also includes a diode to protect the control unit against voltage surges and accidental polarity reversals.

On-Board Diagnostics

The first few model year applications of Motronic do not feature any form of self-diagnosis except if a fault is detected in the closed-loop operating circuit, a calculated pulse width aimed at maintaining an air/fuel mixture near stoichiometric is used. Later versions, particularly those imported after the implementation of OBD II regulations, include extensive self-diagnostics.

OPERATING MODES

The control unit is programmed with different operational strategies as driving conditions change. Most of the differences in strategy occur during open-loop mode.

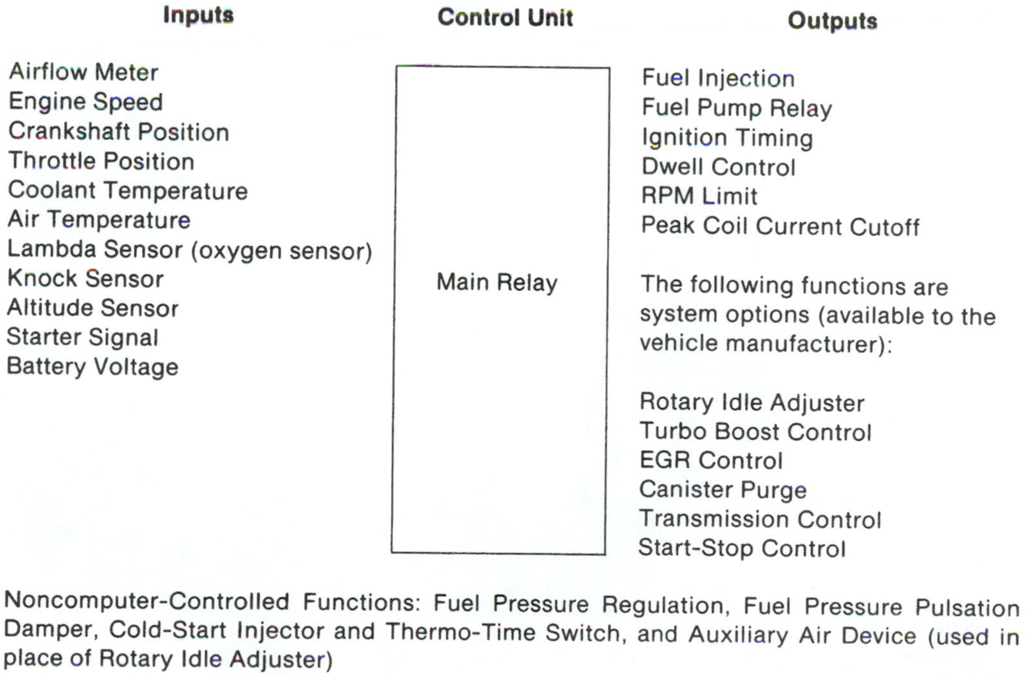

Inputs

Airflow Meter
Engine Speed
Crankshaft Position
Throttle Position
Coolant Temperature
Air Temperature
Lambda Sensor (oxygen sensor)
Knock Sensor
Altitude Sensor
Starter Signal
Battery Voltage

Control Unit

Main Relay

Outputs

Fuel Injection
Fuel Pump Relay
Ignition Timing
Dwell Control
RPM Limit
Peak Coil Current Cutoff

The following functions are system options (available to the vehicle manufacturer):

Rotary Idle Adjuster
Turbo Boost Control
EGR Control
Canister Purge
Transmission Control
Start-Stop Control

Noncomputer-Controlled Functions: Fuel Pressure Regulation, Fuel Pressure Pulsation Damper, Cold-Start Injector and Thermo-Time Switch, and Auxiliary Air Device (used in place of Rotary Idle Adjuster)

Figure 18–13 Overview of Motronic system.

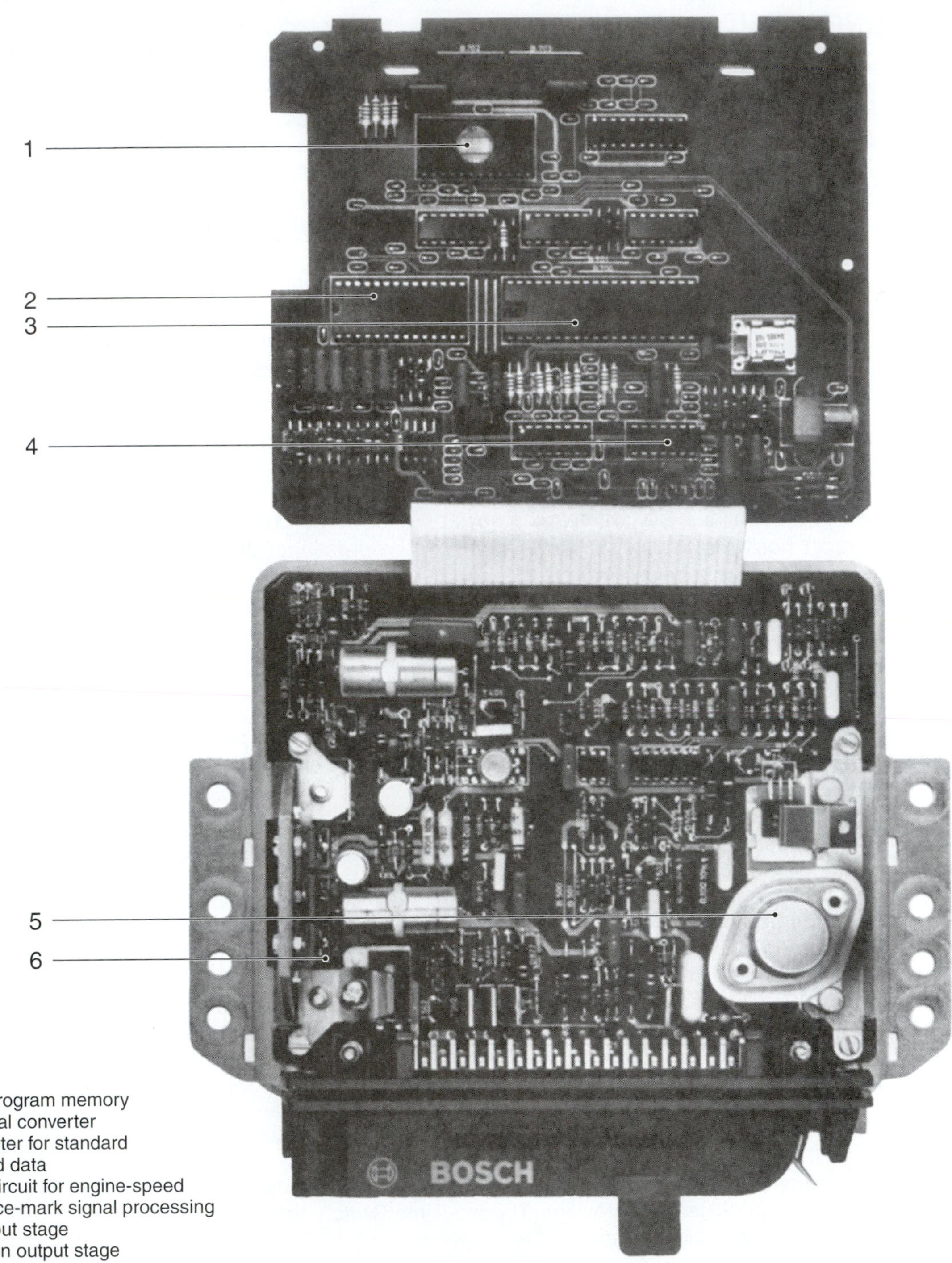

Figure 18–14 Control unit. *(Reprinted with permission from Robert Bosch Corporation.)*

Cranking

Two fuel-metering programs can be employed during engine cranking. One is based on cranking speed, the other on temperature. At lower crank speeds, the fuel quantity injected stays constant regardless of airflow fluctuations (input from the airflow sensor is not reliable in these conditions because of pulsing caused by individual cylinder intake strokes). At higher cranking speeds, air intake diminishes slightly as a result of lower volumetric efficiency, so fuel quantity injected is also reduced. During cold cranking, the control unit adds an enrichment program in addition to the speed-dependent program. Some versions of the Bosch systems include a cold-start injector in the intake manifold, Figure 18–15, while some merely add injection pulses at the port injectors. Ignition timing is also adjusted depending on cranking speed and coolant temperature. The cold-start injector system is very similar to that used on several American vehicles: it sprays a priming load of fuel into the cold intake manifold upstream of the main injectors. The extra distance the fuel must travel to reach the cylinders affords slightly more time for the gasoline to vaporize, making startup easier.

On applications without a cold-start injector, the control unit pulses the injectors several times per crankshaft revolution during cranking instead of once per engine cycle, the way it normally works. This should both enhance fuel evaporation and avoid flooding the spark plugs with wet fuel. The enrichment decays to zero over a specific and small number of engine revolutions. The number depends on the engine coolant temperature when cranking begins. If the rpm reaches a preset value before the specified number of revolutions has occurred, the cold-cranking enrichment program stops anyway. The preset rpm value also depends on the engine coolant temperature when cranking begins.

Diagnostic & Service Tip

Sometimes when a vehicle has been used in a warm climate for a season, a certain amount of moisture can collect in the fuel line to the cold-start injector. This moisture can then allow rust to form since the injector does not have the occasion to turn on and flush out the contamination. When the cold weather returns in the fall, the injector turns on but is plugged with rust, and the car won't start. At this point, you must service the cold start injector by cleaning or replacing it.

After Startup

As an engine starts cold, the intake port surfaces and combustion chamber walls, ceiling and piston top are also cold. Fuel can condense on these surfaces, leading to poor combustion. During the brief time it takes for these areas to warm, the control unit initiates a poststart enrichment program to maintain a good idle quality and improve throttle response. Once this begins, the enrichment also decays within a brief time to zero. This period of time depends on coolant temperature at the time of startup crank. The control unit also advances the ignition timing relatively more with the engine cold than it does in other circumstances. All of this is to provide the most efficient engine power with the least possible amount of fuel and exhaust emissions. The advanced spark also helps warm the combustion chamber temperature as quickly as possible.

Warmup

As the engine warms up, and the poststart period has expired, additional enrichment is based on a combination of engine coolant temperature and load. An idle speed increase is applied to prevent stalling, improve driveability and hasten warmup. The idle speed increase occurs during the poststart mode and continues into the warmup mode. As an added means of bringing the oxygen sensor and catalytic converter to

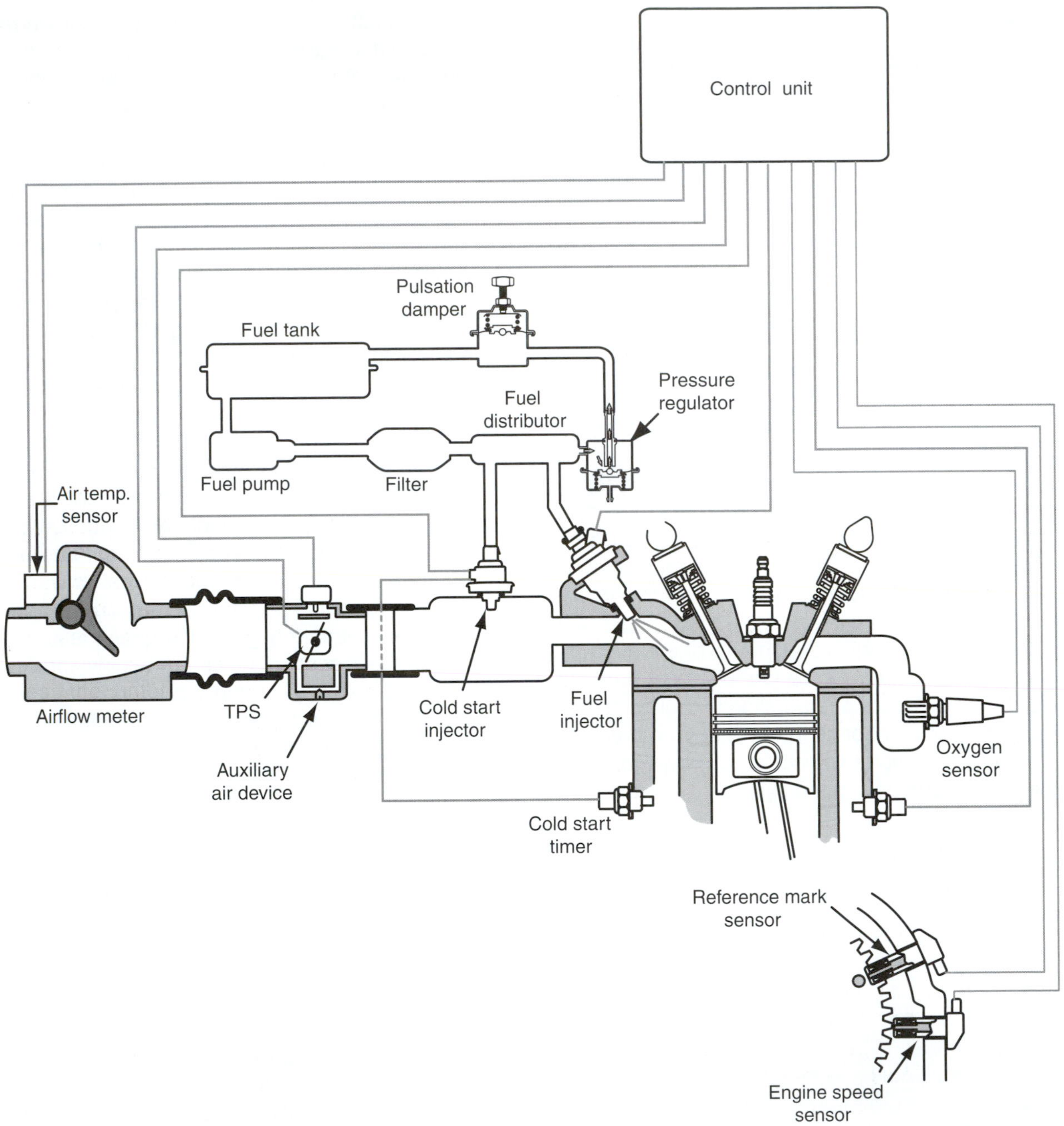

Figure 18–15 The Motronic system with cold-start injector (12).

operating temperature as quickly as possible, some versions of the Bosch Motronic system retard ignition timing during warmup to make the exhaust gas somewhat hotter.

Acceleration

Whenever the driver quickly opens the throttle from an earlier fixed position, the control unit briefly enriches the mixture. This serves the same purpose as the accelerator pump in a carburetor and solves two problems: 1) that air accelerates into the combustion chamber faster than the fuel might, and 2) that it is harder to get the fuel to vaporize in higher pressure intake air. The degree of accelerator enrichment is affected by the engine coolant temperature.

Wide-Open Throttle (WOT)

During wide-open throttle operation, the control unit commands a fixed enriched air/fuel mixture. To avoid fuel quantity miscalculations from airflow sensor fluctuations, at this speed the fuel quantity is calculated based on engine speed. The objectives of the mixture strategy at wide-open throttle is to produce maximum engine torque without allowing detonation and physical damage. Emissions quality is not specifically optimized during wide-open throttle operation because WOT use is ordinarily restricted to driving conditions where safety considerations (passing, perhaps) outweigh the disadvantages of momentary adverse emissions.

Deceleration (Overrun)

When the vehicle is decelerated, the control unit tapers the fuel delivery to zero and retards the ignition timing. Retarding the timing provides better engine braking and reduces the emission of hydrocarbons in the event there is still residual fuel on the intake port walls.

If the engine speed falls below a specified rpm or if the driver opens the throttle, the control unit returns fuel injection and spark advance to the appropriate level. This occurs gradually over a programmed number of engine revolutions rather than suddenly, for driveability and a smooth transition. On some Bosch systems, there is no fuel cutoff during deceleration, and on these a fuel-enrichment program is used to avoid engine bucking and misfire as well as high hydrocarbon emissions from misfires.

Closed Loop

The Bosch Motronic system, like other engine control systems, goes into closed loop once the engine coolant temperature and the oxygen sensor reach normal operating temperature. Bosch and European manufacturers' technical literature frequently refers to the oxygen sensor as a "lambda sensor." While the oxygen sensors work the same way, a different description is used for the optimal air/fuel ratio. Until recently, European engineers identified the ideal ratio by the Greek letter **lambda**. In this country the ideal mixture was always called the *stoichiometric* ratio. Both terms mean the same thing and both oxygen sensors work the same.

INPUTS

Airflow Meter

The Bosch airflow meter used on the Motronic system is the same as described in the pulsed systems section of this chapter. The shaft of the sensor moves the wiper of a potentiometer designed to provide accurate voltage signals to the control unit, Figure 18–16. This unit is designed to provide consistent return signals in spite of aging and rapid changes in temperature.

The control unit can calculate the pulse width for each cylinder by comparing the information from crankshaft position and speed with airflow meter data. The injection then occurs during the appropriate cylinder's intake stroke.

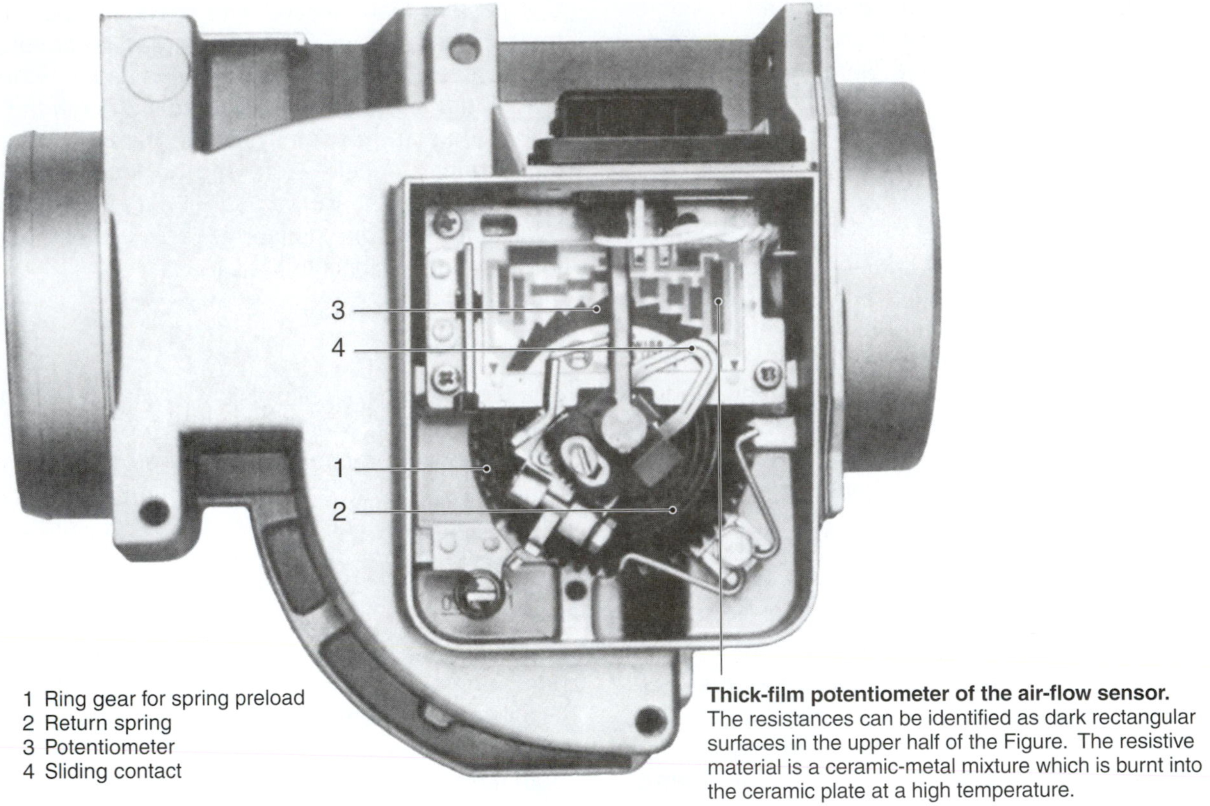

3
4
1
2
1 Ring gear for spring preload
2 Return spring
3 Potentiometer
4 Sliding contact
Thick-film potentiometer of the air-flow sensor.
The resistances can be identified as dark rectangular surfaces in the upper half of the Figure. The resistive material is a ceramic-metal mixture which is burnt into the ceramic plate at a high temperature.

Figure 18–16 Airflow meter. *(Reprinted with permission from Robert Bosch Corporation.)*

Engine Speed Sensor

The control unit senses engine speed using a pulse-generating device that uses the flywheel teeth as the pulse-triggering device, Figure 18–17.

Reference (Crankshaft Position) Sensor

The crankshaft position or reference sensor works like the engine speed sensor except that its trigger is usually a single pin on the flywheel. This sensor provides the control unit with information about crankshaft position.

Throttle Valve Switch

The throttle valve switch contains two sets of electrical contact points, Figure 18–18. Each set

Reference mark sensor
Flywheel
Reference pin
Pickup coil
Engine speed sensor
Permanent magnet

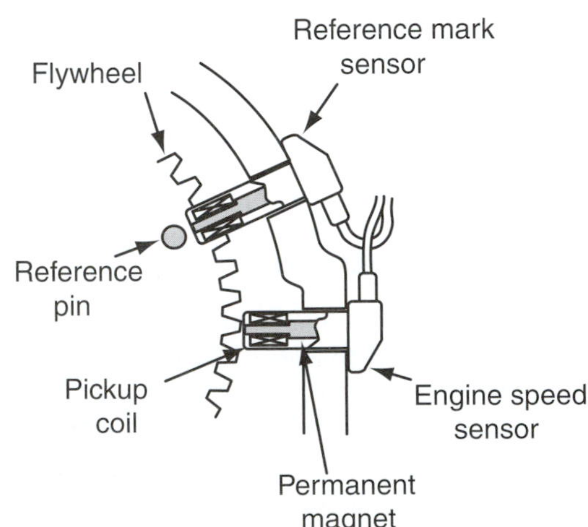

Figure 18–17 Engine speed and reference mark sensors.

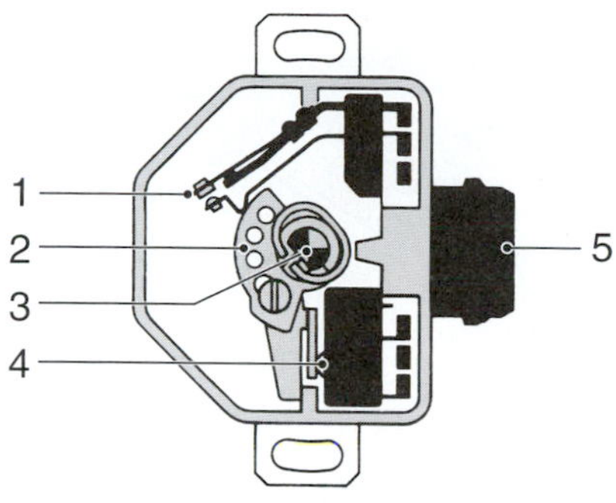

Figure 18–18 Throttle valve switch. *(Reprinted with permission from Robert Bosch Corporation.)*

has a separate circuit and receives a reference voltage from the control unit. This circuit enables the control unit to determine if each of the switches is open or closed. When the throttle is closed, the idle contacts close, completing the circuit for that switch. During part-throttle operation (all normal driving modes), both switches are electrically open. At wide-open throttle the full load contacts close, completing that switch's circuit. The control unit can thus determine three driving conditions from the one unit. The throttle valve switch (containing both switches) is attached to the end of the throttle shaft.

Coolant Temperature Sensor

The coolant temperature sensor is a thermistor bolted into the water jacket near the thermostat housing. Its operation is similar to those on domestic vehicles.

Air Temperature Sensor

The intake air temperature sensor is a thermistor mounted in the intake opening of the airflow sensor, Figure 18–19. Its signal indicates the temperature of the intake air.

Oxygen Sensor (Lambda Sensor)

As previously explained, the earlier Bosch systems use oxygen sensors that work like familiar domestic units, but the literature describes them as *lambda* sensors, using the term commonly used by German engineers to indicate air/fuel stoichiometry.

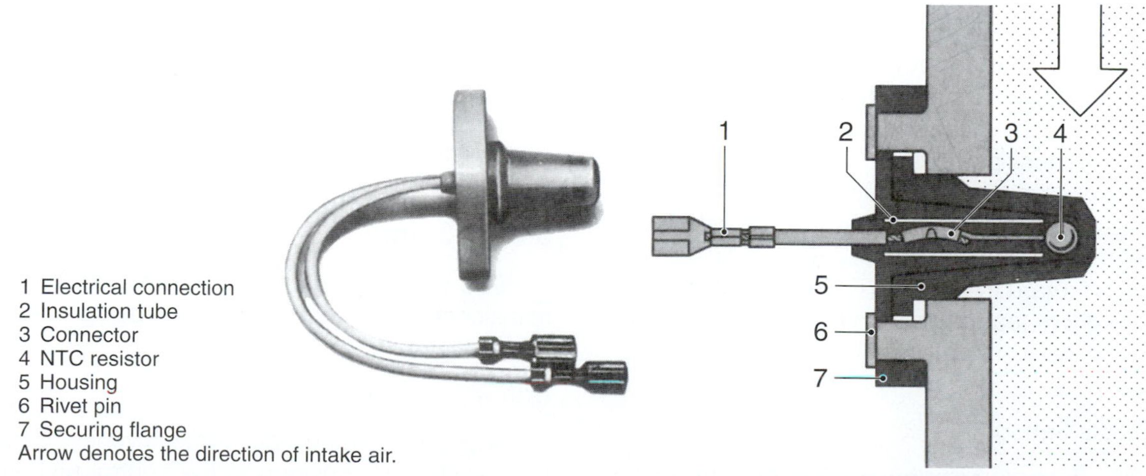

Figure 18–19 Air temperature sensor. *(Reprinted with permission from Robert Bosch Corporation.)*

Lambda and Air/Fuel Ratio

German engineers have long used the Greek letter lambda (L) to represent a ratio reflecting the air/fuel ratio. The relationship lambda represents is:

Lambda = actual inducted air quantity ÷ theoretic air requirement

The theoretical air requirement is 14.7 units (by weight). If the actual air quantity used was 14.7, then lambda = 1. Using this method of expression, an air/fuel ratio richer than 14.7:1 is expressed as a number less than 1. For example, 0.9 is equivalent to 13.23:1 air/fuel ratio (13.23/14.7 = 0.9). An air/fuel ratio leaner than 14.7:1 is expressed as a number greater than 1. For example, 1.03 is equivalent to 15.14:1 air/fuel ratio (15.14/14.7 = 1.03).

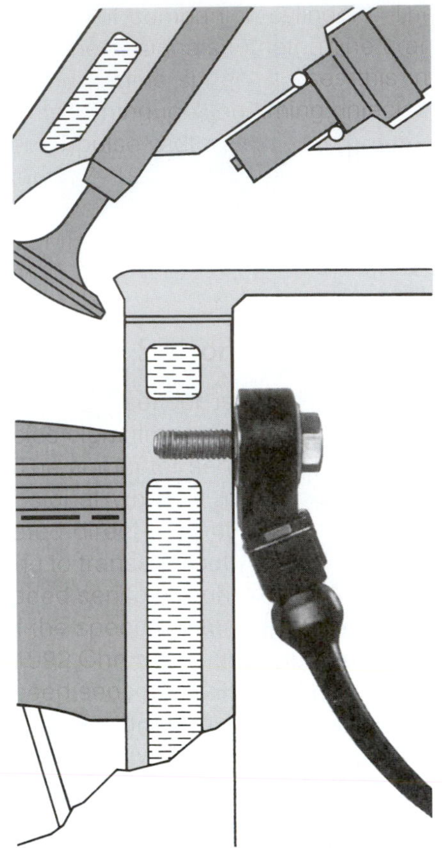

Figure 18–20 Knock sensor. *(Reprinted with permission from Robert Bosch Corporation.)*

Knock Sensor

The Bosch Motronic system uses a piezoelectric knock sensor, Figure 18–20. The sensor generates a voltage signal in response to most engine vibration frequencies, but the signal is strongest at frequencies resonant with detonation knock frequencies. At a vibration between 5 and 10 kiloHertz, the control unit recognizes knock and begins retarding the spark. The knock sensor is located on the block between two cylinders. On some engines, particularly V-type engines, two knock sensors are used. Combined with the information the computer has about crankshaft position, this can identify the individual cylinder with knock, to retard the ignition timing to that cylinder selectively (a feature only on later versions of the Motronic system).

Altitude Sensor

The altitude-sensing device or barometric pressure sensor is an aneroid unit attached to the wiper of a potentiometer similar to the one described in the Inputs section of the introductory Ford chapter, Chapter 13. On most Bosch-equipped vehicles, it is located in the air cleaner.

Starter Signal

The cranking control circuit provides a signal to the control unit when the starter is being cranked.

Battery Voltage

A system voltage input from the vehicle's electrical system is provided to the control unit so it can monitor battery and charging system voltage.

OUTPUTS

Generally Motronic systems control the operation of the fuel injectors, the engine speed limiter, the fuel pump relay, ignition timing and dwell. Specific vehicle manufacturers may use other features designed for the particular car.

Diagnostic & Service Tip

A common source of vacuum leaks, or "false air" as the Bosch literature sometimes calls it (because it does not go through the airflow meter), is around the injectors. The seal is usually a single O-ring. With heat and age, these O-rings can crack and cause a vacuum leak.

It is fairly common for an injector O-ring to leak after the injector is removed and reinstalled, even if it was not leaking air before, because the removal and replacement disturbed the seal. While it is not always necessary to replace O-ring seals when reinstalling an injector, they should be inspected carefully and replaced if necessary. Engine oil is generally regarded as the best O-ring installation lubricant, though at least one manufacturer favors 80-weight gear lubricant. Failure to inspect, and if necessary replace, the injector O-ring can allow a technician to build another new problem into a problem car.

Also, air-shrouded injectors are used by Volkswagen and several other carmakers on some of their models. Idle air does not pass through the regular intake channels but through a special idle air passage that leads to collars around individual injectors. This specially routed idle air keeps the airflow high where air meets fuel and improves vaporization of the fuel. This idle air shroud plumbing, however, constitutes another source of potential vacuum leaks. On early Volkswagen models, a plastic elbow near the idle air controller often developed cracks or splits that were very hard to spot.

Injectors

The injectors used on Bosch systems are like those used in domestic built multipoint fuel injection systems. In fact, domestic manufacturers have historically used Bosch injectors for their multipoint injection systems.

The Bosch injectors are solenoid-operated, pintle-type valves. In open-loop operation, the control unit calculates pulse width based on engine load and engine speed; engine load is calculated from airflow and engine speed information. Figure 18–21 shows a three-dimensional graph or "map" correlating load, speed and air/fuel ratio for a given engine application. Air/fuel ratio is expressed as lambda, explained earlier, but the translation into more familiar domestic stoichiometric terms is straightforward.

This map is stored digitally in the control unit's read-only memory. It functions as the look-up table to determine pulse width. The arrows indicate increased load, speed and richness. Each line intersection represents a relative value of the desired air/fuel ratio.

The control unit can, however, modify the pulse width value represented by each line intersection in response to the following information:

- engine temperature.
- throttle position as indicated by the throttle switch. The control unit also employs a special mixture richening pulse width command at warm WOT to prevent combustion detonation.
- air density. The airflow sensor input must be corrected at higher elevations to allow for the thinner air. This correction is made with input from the air temperature sensor and, on the vehicles so equipped, the altitude sensor.
- battery voltage. Just as on domestic fuel injection systems, the amount of fuel that flows through the injector is affected by the injector's response time, which is directly reflective of charging system voltage. The control unit can detect this voltage and modify the pulse width within a certain range to correct the fuel delivery.

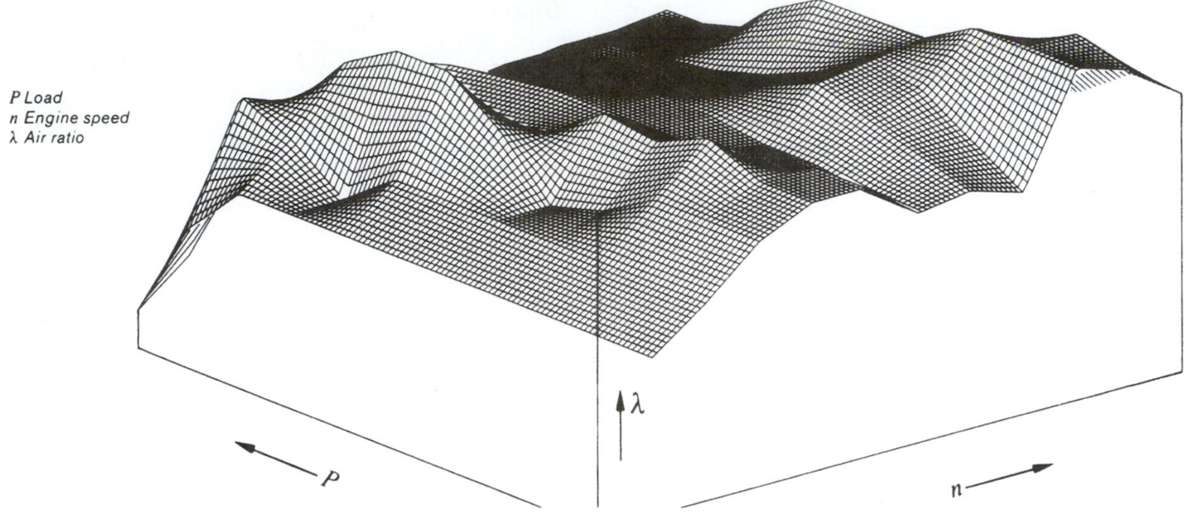

Figure 18–21 Pulse width control map. *(Reprinted with permission from Robert Bosch Corporation.)*

Engine Governor. For each engine with a Bosch control system, there is a specific maximum allowable engine speed. If the engine speed exceeds that limit by more than 80 rpm, the control unit will shut off fuel injection until the engine falls 80 rpm below the limit, Figure 18–22. This "redline" limit is to prevent mechanical damage to the engine from overrevving. The system uses fuel cutoff rather than spark override to keep from dumping excessive hydrocarbons into the atmosphere.

Limiting maximum engine speed $\mathbf{n}_0$ *by suppression of fuel-injection pulses.*

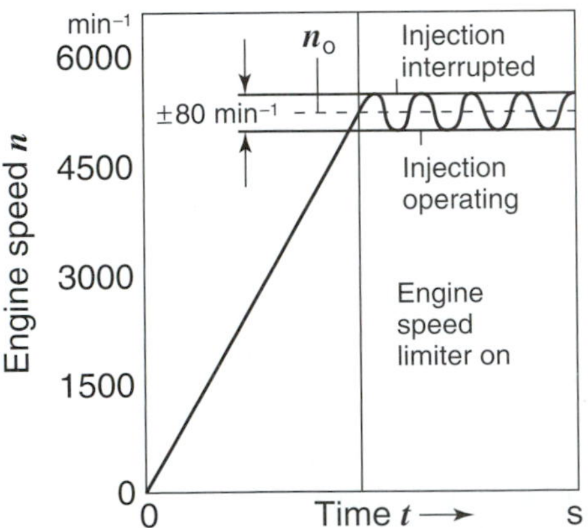

Figure 18–22 Engine speed governor. *(Reprinted with permission from Robert Bosch Corporation.)*

Please note, however, that the engine governor function can only protect the engine from mechanical damage caused by powered overrevving. If a driver downshifts into a lower gear at high speed, the governor can shut off the fuel, but driven by vehicle inertia, the engine could still spin fast enough to break something, probably a valve or connecting rod.

Fuel Pump Relay

As on most fuel injection systems, the fuel pump is indirectly controlled on Motronic systems through a fuel pump relay. Power is available at the relay whenever the ignition switch is turned on. The control unit then grounds or ungrounds the relay's coil to activate or shut off the fuel pump. If the engine does not turn at a minimum specified speed, the control unit will shut off the relay to reduce the risk of fire. As on many other fuel injection systems, the high-pressure fuel pump is located inside the fuel tank, submerged in fuel. Earlier systems employed a fuel pump just outside the tank, on the line to the engine compartment and just ahead of the fuel filter (some have the filter in the engine compartment).

Fuel Pressure Regulator. The Bosch fuel pressure regulator, like most others, uses intake manifold absolute pressure to maintain a constant pressure differential between the fuel in the fuel rail and the air in the intake manifold, Figure 18–23.

Fuel Pressure Pulsation Damper. The pulsation damper, also in Figure 18–23, works to eliminate the noise of fuel pulses from the pump. While it looks much like the pressure regulator, it is not connected to manifold vacuum and it functions as a surge accumulator to absorb the pressure pulses from the opening and closing of the injectors. The pulsation damper is in the return line downstream from the pressure regulator. The higher pressure Bosch systems, as well as those on four-cylinder engines, are more likely than the low pressure systems or those on five-plus cylinder engines to use pressure dampers because the effects of the pulses are greater in them. Higher pressures are often preferred by carmakers because the fuel is more finely atomized under higher pressure, other things being equal.

Ignition Timing

The control unit operates the ignition coil by ground switching the primary windings, Figure 18–24. Like fuel injection pulse width, ignition timing is based on engine load and speed. Figure 18–25

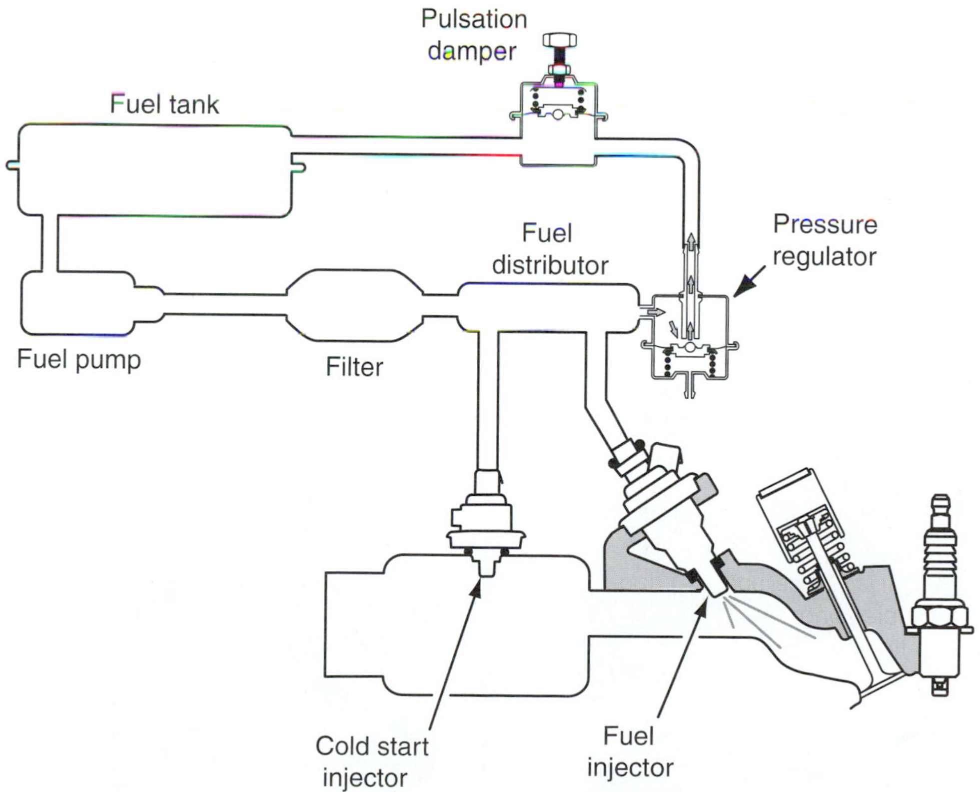

Figure 18–23 Fuel pressure regulator (right) and pulsation damper (left).

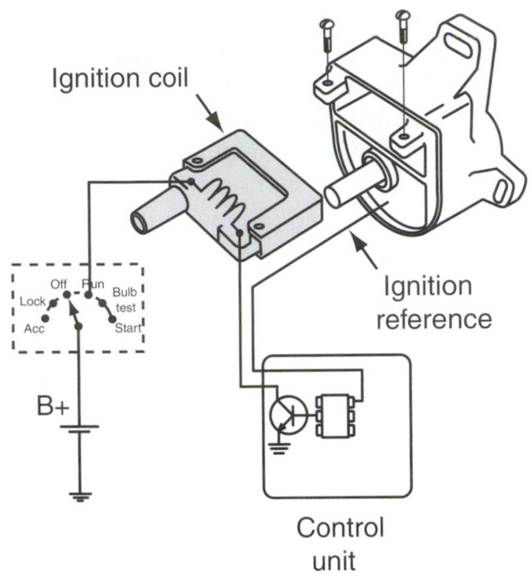

Figure 18–24 Ignition circuit.

shows a three-dimensional map of spark advance for one engine. Arrows point in the directions of increased load, speed, and spark timing advance. The line intersections represent the spark advance for all combinations of engine speed and load. These values are stored digitally as a look-up table in the control unit's read-only memory. The control unit reviews sensor inputs and recalculates spark timing anew between each cylinder firing. The control unit can modify the values in the read-only map in response to inputs from:

- coolant temperature sensor.
- intake air temperature sensor.
- throttle position switch (during WOT operation, timing is set for maximum torque with retard applied only to prevent detonation).
- altitude sensor (if present).

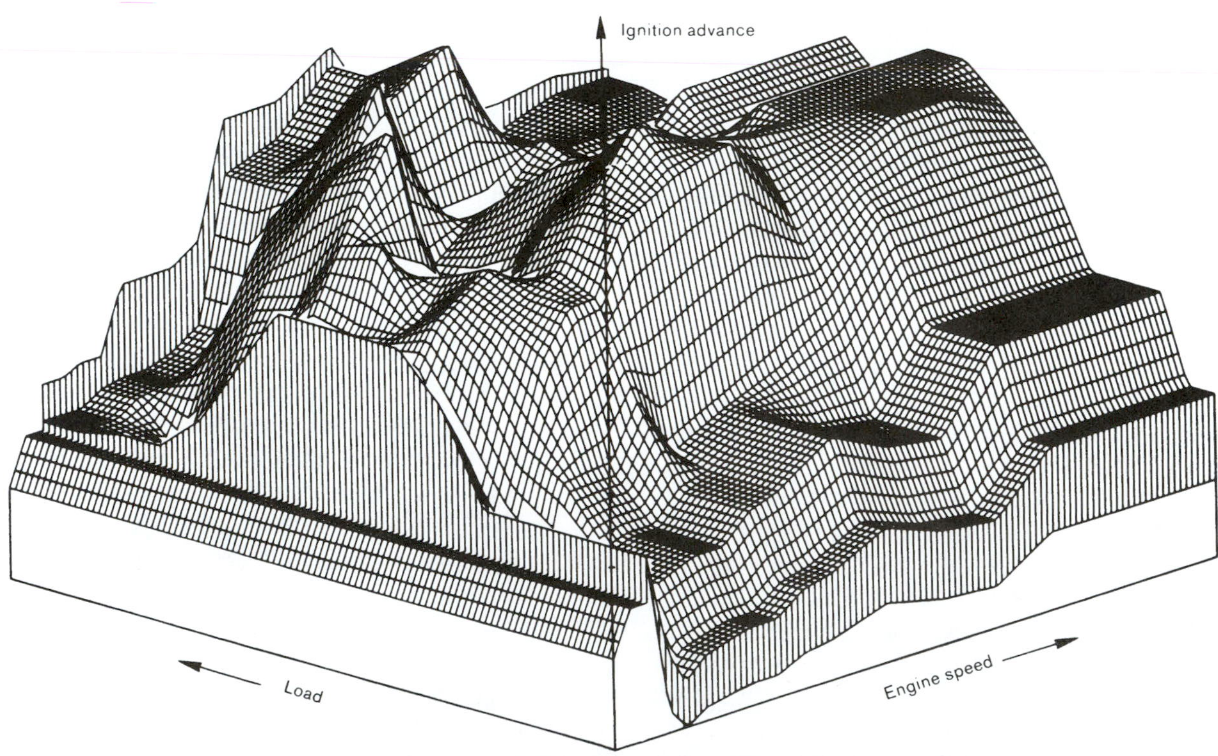

Figure 18–25 Ignition timing map. *(Reprinted with permission from Robert Bosch Corporation.)*

The control unit modifies the spark advance according to the information from either the coolant temperature sensor or the intake air temperature sensor. This modification, of course, depends on the driving conditions. At starting crank, timing depends entirely on engine speed and coolant temperature. At low cranking speeds and low temperatures, the timing is close to TDC. At higher cranking speeds and with warmer engines, the timing is only slightly advanced. With high cranking speed even at low temperature, timing is more advanced. Each of these combinations improves starting and poststartup driveability.

Once the engine starts cold, timing is advanced significantly for a brief period to reduce the need for fuel enrichment. Once the engine is warm enough to reduce the likelihood of stalling, the ignition timing is retarded on some vehicles to speed up engine warmup, even at the cost of lower delivered engine torque. Increased exhaust temperature helps warm the oxygen sensor and catalytic converter more quickly.

On earlier systems, during warm idle, ignition timing is used to control idle speed. With no additional load, transmission in neutral and air conditioning off, the timing can be retarded slightly and still maintain the desired idle speed. This keeps combustion chamber surfaces hotter to reduce the production of unburned hydrocarbons. If there is a load such as the transmission going into gear, the idle speed begins to slow. The control unit responds by advancing ignition timing enough to achieve more torque and bring the idle speed back to the desired speed. This saves opening the throttle and reduces fuel consumption. On applications with the rotary idle adjuster, idle air increases if advancing the spark timing alone is not enough to keep the idle speed at the desired setting.

At wide-open throttle, normal or higher engine coolant and/or intake air temperature cause the control unit to reduce spark timing at those load combinations where there is highest probability of spark knock. If at a specific load there is acceleration short of WOT, the control unit retards the existing spark advance and then gradually begins to restore it. This reduces both spark knock and the production of oxides of nitrogen.

Spark Knock Control

On earlier systems the spark knock control was separate from the Motronic unit. Later it was incorporated into the control unit. The later versions, controlled by the computer, allow a more detailed control of spark advance. When the knock sensor indicates a detonation signal, the control unit recognizes the knock and immediately reduces spark advance until the knock disappears. Then it gradually readvances timing toward the original value until the knock returns.

Dwell Control

The Motronic control unit manages ignition coil dwell in the same way as the General Motors HEI system. As engine speed increases, dwell increases to ensure complete magnetic saturation of the coil. The Bosch system also considers charging system voltage when calculating dwell.

Coil Peak Current Cutoff

If the ignition coil remains on with the engine stopped, both the ignition coil and the driver (transistor) in the control unit may overheat or burn out. To avoid this problem, the control unit automatically turns off the ignition coil anytime engine rpm drops below a specified speed, usually about 30 rpm, well below a successful starting crank speed.

Idle Speed Control

Bosch Motronic systems use two different methods to control idle speed.

Auxiliary Air Device. On some applications, warm idle speed is controlled by an idle air bypass device in the idle air bypass passage. After a cold start and during warmup, idle speed is increased by the air routed through an auxiliary air valve. A bimetallic strip inside the valve holds the valve open at low temperature, Figure 18–26. With the passage open, additional air bypasses the throttle blade. This air has, however, passed the airflow sensor so the computer knows the correct amount of fuel to inject. In addition, once the ignition is turned on, voltage is applied to a heating

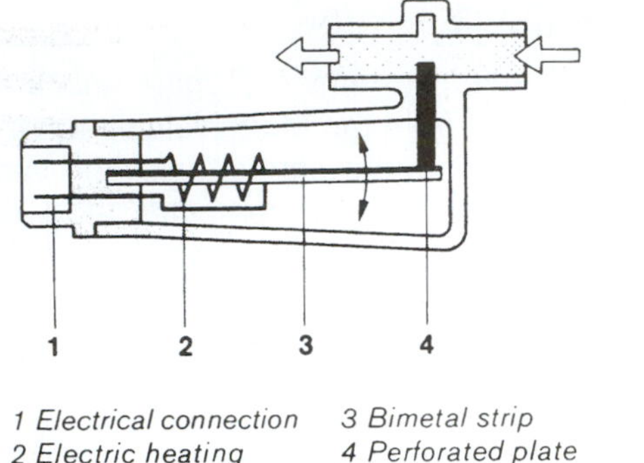

Figure 18–26 Auxiliary air bypass device schematic. *(Reprinted with permission from Robert Bosch Corporation.)*

element wrapped around the bimetallic strip. As the strip warms up, it gradually closes the valve and blocks the flow of auxiliary air, slowing the idle as the engine warms. Once the engine is completely up to operating temperature, its heat is enough to keep the valve closed.

Rotary Idle Control Valve. Other applications use the rotary idle control valve, controlled directly by the computer, Figure 18–27. This valve controls engine idle speed regardless of engine coolant temperature. At the end of an armature shaft, a rotary valve or slider valve can move in the idle air bypass, Figure 18–28. The DC electric motor has a permanent magnetic field and two armature windings. By controlling the voltage applied to the two windings, the control unit can incrementally rotate the armature and thus position the rotary valve to control idle air bypass. The control unit's memory has a programmed idle speed that it wants to maintain for any given engine coolant temperature. It first uses ignition timing to adjust idle speed; it then uses the rotary idle adjuster to bring idle speed to the desired rpm if the timing adjustments have been insufficient.

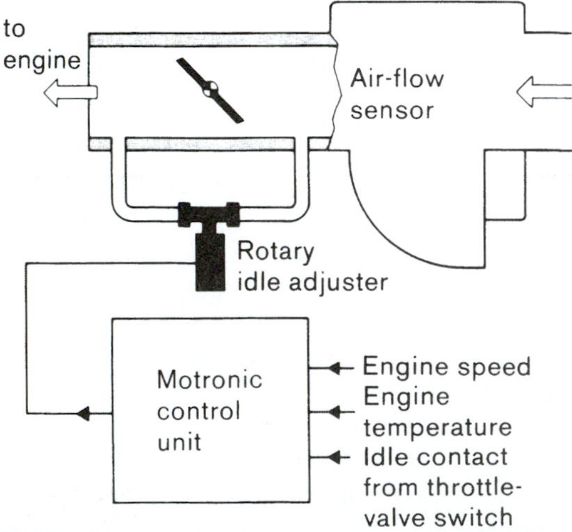

Figure 18–27 Rotary idle adjuster circuit. *(Reprinted with permission from Robert Bosch Corporation.)*

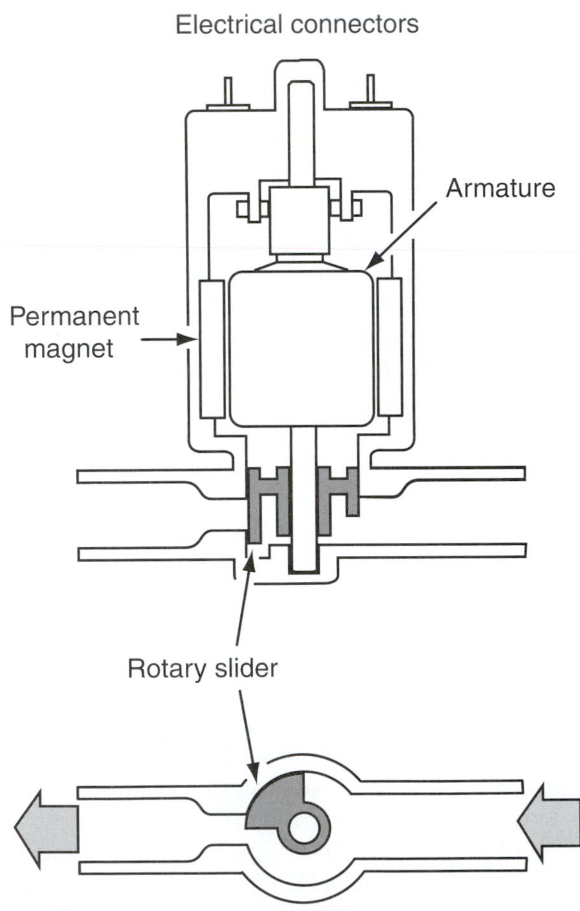

Figure 18–28 Rotary idle adjuster motor.

Turbo Boost Control

Bosch Motronic systems use a combination of spark retard and boost pressure reduction to control spark knock on turbocharged engines. Particularly on smaller engines, Bosch engineers feel reducing boost alone puts unnecessary limits on engine performance. They also find that just using spark retard to control knock is a poor choice because just retarding ignition timing raises exhaust temperature. A temperature rise can be threatening to the already high temperature turbocharger turbine. Using a combination of some of both measures is the strategy they prefer. Spark retard is the fastest measure (provided the knock is caused by spark and not by preignition): it can be applied to the very next power stroke after the detection of knock. It takes several intake strokes even after the wastegate opens before the manifold pressure begins to come down because the turbocharger must slow and the intake air must start to cool. As boost pressure does come down, however, the Motronic system starts to readvance the ignition timing to retain as much engine torque as possible. Later systems include the capacity to retard ignition timing to individual cylinders, as described in earlier chapters on domestic systems.

EGR Valve Control

Early European Bosch-equipped vehicles did not use EGR systems to control the production of oxides of nitrogen (NO_x) at high combustion temperatures. Instead they used the reduction feature of the catalytic converter and the residual exhaust retained by the valve overlap. This method, however, does not allow the precise controls available with an electronically controlled EGR valve, so later versions employ EGR systems functionally similar to those used on most domestic-built systems.

Evaporative Emissions Control

Motronic system vehicles use an evaporative emission system similar to those on domestic vehicles. Fuel vapors from the fuel tank are stored in a charcoal canister, Figure 18–29. Once the

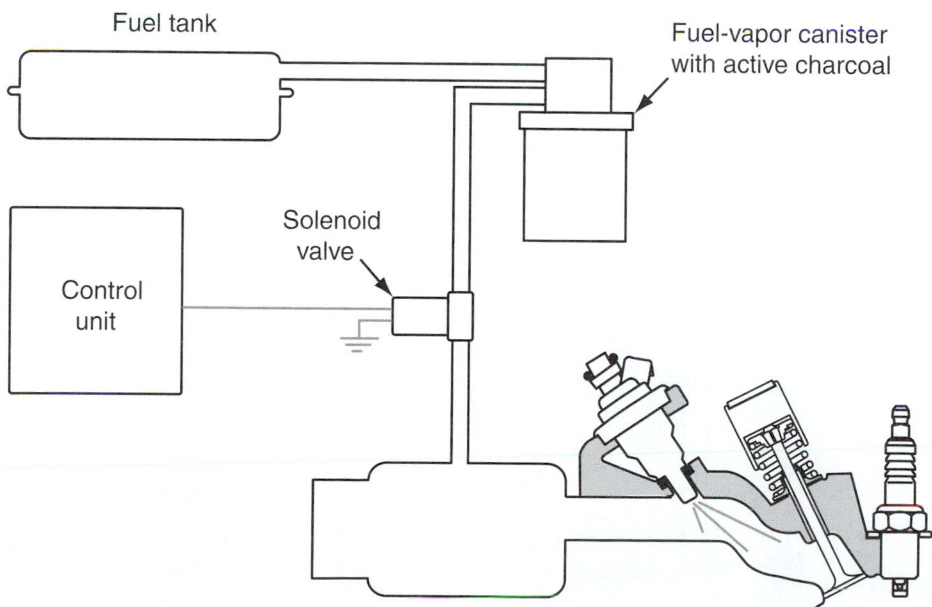

Figure 18–29 Evaporative emissions control system.

system goes into closed loop, the control unit activates a solenoid, opening a valve in the purge line connecting the canister to the intake manifold. As usual, precise mixture control then follows from the oxygen sensor signal.

Electronic Transmission Control

Later versions of the Motronic system include sensors to indicate the transmission state. A vehicle speed sensor, transmission gear selection sensor and kickdown sensor provide the control unit with this information. On some systems, the control unit determines the automatic transmission's hydraulic pressure and shift points (solenoid valves replacing traditional spool-type shift valves), Figure 18–30. Shift point look-up tables (one for economy, one for performance, one for manual shifting) stored in the control unit's read only memory provide optimum transmission shifting according to the programming look-up chart selected by the driver.

Smooth shifts with reduced clutch slip and wear are achieved, especially during WOT operation, by momentarily retarding ignition timing to reduce engine torque output during the shift. Motronic transmission control can thus improve fuel economy, shift quality, transmission torque capacity and expected transmission life.

Start-Stop Control

Another optional feature that can be added to a Motronic system is the start-stop control. This feature is on some European cars, and may become available on domestic cars. This applies

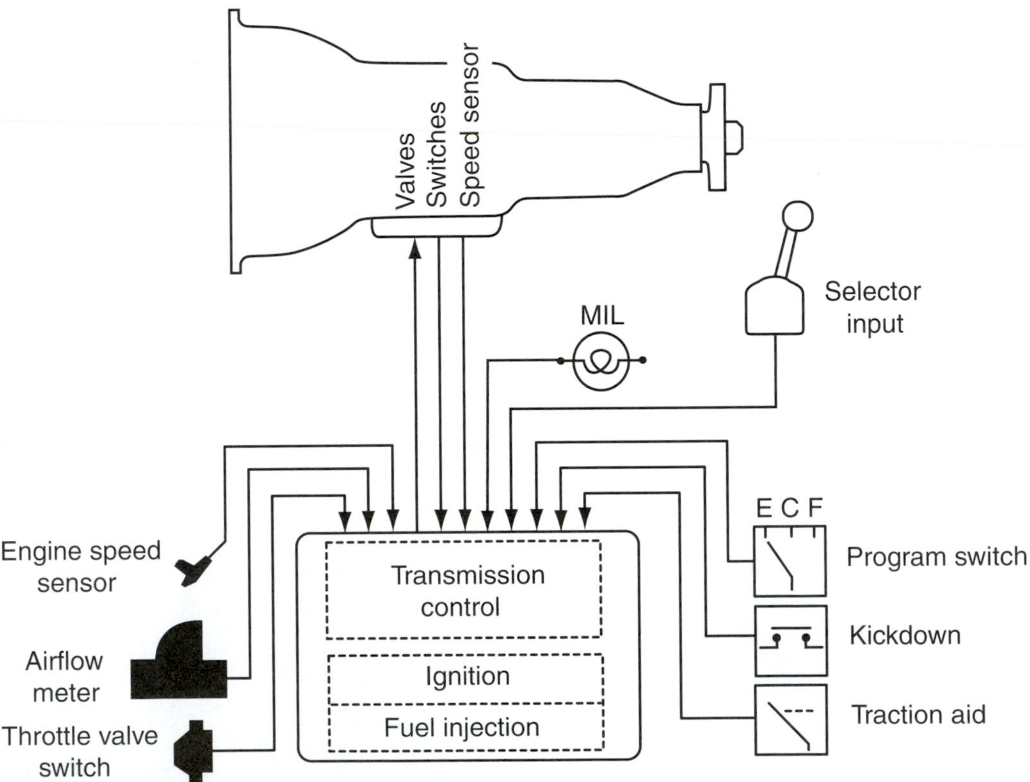

Figure 18–30 Electronic transmission control.

only to vehicles with manual transmissions for safety reasons. This feature helps conserve fuel and reduce emissions.

The start-stop control uses a vehicle speed sensor, a clutch pedal position sensing switch and a separate control module. If vehicle speed falls below about 1 mph with the clutch pedal depressed, the module signals the control unit, which shuts off the fuel. As long as the driver keeps the clutch pedal down, the module will activate the cranking circuit and fuel injection will resume as soon as the driver presses the accelerator. This system does not work until the engine is warm.

Variable Displacement System

Another feature Bosch uses on some vehicles is a variable displacement system not unlike Cadillac's modulated displacement system. The system includes:

- deactivating cylinders by shutting off the injectors to selected cylinders.
- having the intake and exhaust valves of the deactivated cylinders continue functioning.
- providing intake and exhaust manifold valving that allows the exhaust from working cylinders to be passed through the deactivated cylinders to keep them up to normal operating temperature.

✔ SYSTEM DIAGNOSIS AND SERVICE

The vehicle manufacturer establishes the diagnostic procedures for each vehicle, so procedures vary somewhat from one make to another. The diagnostic procedures here are representative of Bosch Motronic systems.

Troubleshooting Guide

The troubleshooting guide consists of a list of driveability complaints such as: Will Not Start Cold, Erratic Idle During Warmup and Backfires. Each complaint is accompanied by a list of possible causes. The possible causes should be checked either in the order of easiest to check, or of most likely to cause the problem, based on the technician's familiarity with the vehicle and the problem. It is important that all components or systems identified in the possible cause list for the complaints that are not part of the Motronic system are working properly before proceeding to the tests for the Motronic system.

Motronic Test Section

The service manual is likely to have a section that contains test procedures for each component of the Motronic system. The tests in this section should not be conducted until the engine is at normal operating temperature. In testing components and their circuits, remember that connections cause more problems than components.

SUMMARY

In this chapter we have learned about the history of Bosch fuel injection systems; we have considered the two mechanically distinct kinds of Bosch engine management systems, pulsed and constant fuel injection. We have seen the differences between the K-Jetronic (constant flow from the injectors), D-Jetronic (with pulsed injectors controlled principally by the intake manifold pressure), L-Jetronic (with fuel delivery determined by an air-door sensor) and finally Motronic (a complete engine management system, controlling mixture as well as ignition timing and other functions).

We have reviewed the difference between today's Motronic systems and the predecessor systems. The different operating conditions and how they work in the different operational modes were covered. We have seen several ways the control unit calculates engine load. We have considered the major input sources and output actuations, as well as the major functions. Finally, we have looked at the diagnostic approaches.

▲ DIAGNOSTIC EXERCISE

A vehicle with a Motronic system comes into the shop for having failed an emission test for high CO readings. The oxygen sensor is verified to be working correctly and is reporting the rich exhaust condition to the engine computer. To check the computer's reaction to the oxygen sensor signal, the pulse width at the fuel injectors is measured. The pulse width on-time is low, about 1 millisecond at idle. What types of faults should be checked for?

REVIEW QUESTIONS

1. The Bosch company has been involved with fuel injection since which of the following?
   A. 1920s
   B. 1950s
   C. 1970s
   D. 1980s
2. *Technician A* says that the fuel injectors in a Bosch pulsed injection system are electronically controlled by a computer.
   Technician B says that the fuel injectors in a Bosch continuous-injection system spray fuel continuously when the fuel distributor provides enough fuel pressure to the injectors.
   Who is correct?
   A. A only
   B. B only
   C. Both A and B
   D. Neither A nor B
3. *Technician A* says that a Motronic system computer may control the fuel injection system, the spark management system and certain other vehicle functions such as the automatic transmission.
   Technician B says that the D-Jetronic, K-Jetronic, KE-Jetronic, L-Jetronic and LH-Jetronic systems are designed to control the fuel injection system only.
   Who is correct?
   A. A only
   B. B only
   C. Both A and B
   D. Neither A nor B
4. Which of the following systems uses manifold pressure as the principle input for calculating engine load and, therefore, for determining how much fuel to inject?
   A. D-Jetronic
   B. L-Jetronic
   C. LH-Jetronic
   D. KE-Jetronic
5. Which of the following systems uses an air-flow meter with a movable door and potentiometer assembly as the principal input for calculating engine load and, therefore, for determining how much fuel to inject?
   A. D-Jetronic
   B. L-Jetronic
   C. LH-Jetronic
   D. KE-Jetronic
6. Which of the following systems uses a hot-wire air mass sensor as the principle input for calculating engine load and, therefore, for determining how much fuel to inject?
   A. D-Jetronic
   B. L-Jetronic
   C. LH-Jetronic
   D. KE-Jetronic
7. *Technician A* says that the main power relay in the Bosch Motronic system contains a diode used to protect the control unit against voltages surges and accidental polarity reversals.
   Technician B says that it is important to make sure that the Bosch Motronic control unit is properly mounted because it is designed to dissipate heat through its mounting brackets.
   Who is correct?
   A. A only
   B. B only
   C. Both A and B
   D. Neither A nor B

8. *Technician A* says that the use of a cold start fuel injector allows the fuel more time to vaporize while attempting to start a cold engine, thus making starting a little easier.
   Technician B says that a cold start injector may form rust internally due to trapped moisture during periods of nonuse such as the summer months, then may be plugged with contamination when cold weather returns, resulting in difficult cold starts.
   Who is correct?
   A. A only
   B. B only
   C. Both A and B
   D. Neither A nor B
9. *Technician A* says that the primary objective of the Bosch Motronic control unit during wide-open throttle operation is to keep engine emissions low.
   Technician B says that the Bosch Motronic control unit primarily calculates wide-open throttle fuel enrichment based on engine speed instead of airflow, due to the fluctuations of the airflow sensor during heavy load conditions.
   Who is correct?
   A. A only
   B. B only
   C. Both A and B
   D. Neither A nor B
10. *Technician A* says that some Bosch Motronic control units are programmed to reduce fuel delivery to zero during vehicle deceleration until engine speed falls below a specified rpm.
    Technician B says that a Bosch Motronic control unit is programmed to increase spark advance during vehicle deceleration until engine speed falls below a specified rpm.
    Who is correct?
    A. A only
    B. B only
    C. Both A and B
    D. Neither A nor B
11. *Technician A* says that a Bosch Motronic system will not enter closed loop until the engine coolant and the oxygen sensor have both reached their normal operating temperature.
    Technician B says that a Bosch Motronic system is not a feedback system and operates in open loop under all driving conditions.
    Who is correct?
    A. A only
    B. B only
    C. Both A and B
    D. Neither A nor B
12. What is a Lambda sensor?
    A. An Engine Coolant Temperature (ECT) sensor
    B. An Intake Air Temperature (IAT) sensor
    C. An oxygen sensor (O_2S)
    D. An airflow meter or an air mass sensor
13. On a Bosch Motronic system, during what engine operating condition will the throttle valve switch have both switch contacts electrically open?
    A. Idle
    B. Part-throttle
    C. Wide-open throttle
    D. Only after engine shutdown
14. On a Bosch Motronic system, the control unit responds to the knock sensor by retarding spark timing when the knock sensor recognizes vibration that is between which of the following amounts?
    A. 500 and 800 Hz
    B. 2 and 4 kHz
    C. 5 and 10 kHz
    D. 12 and 16 kHz
15. *Technician A* says that the fuel injectors used on Bosch Motronic systems are similar to those used on General Motors and Ford vehicles because Bosch generally purchases their fuel injectors from General Motors and Ford.
    Technician B says that in order to prevent engine damage caused by overrevving, the Bosch Motronic control unit will disable the spark from the ignition system if engine rpm exceeds its programmed limit by more than 80 rpm.
    Who is correct?

A. A only
B. B only
C. Both A and B
D. Neither A nor B

16. Why is the fuel pressure regulator on the Bosch Motronic system indexed to manifold absolute pressure?
    A. So that it can keep the fuel rail pressure constant under all driving conditions.
    B. So that it can maintain a constant pressure differential between the fuel in the fuel rail and the air in the intake manifold under all driving conditions.
    C. So that it can vary the flow rate of the fuel injectors as a method of controlling the air/fuel ratio.
    D. So that it can compensate for poor fuel delivery at wide-open throttle in the event that the throttle valve switch were to fail.
17. In order to improve ease of starting, a Bosch Motronic control unit slightly advances the spark timing during engine start during which of the following conditions?
    A. At higher cranking speeds
    B. At lower cranking speeds
    C. With warmer engine temperature
    D. Both A and C
18. On a Bosch fuel-injected engine, the computer may control idle speed by all of the following methods *except* which of the following?
    A. Adjusting ignition timing to control idle rpm
    B. Using a DC motor to vary the physical position of the throttle plate
    C. Using an auxiliary air valve with a bimetallic strip to control throttle bypass air
    D. Using a DC motor-operated rotary control valve to control throttle bypass air
19. *Technician A* says that Bosch systems may retard spark timing in order to control spark knock on a turbocharged application.
    Technician B says that Bosch systems may reduce boost pressure in order to control spark knock on a turbocharged application.
    Who is correct?
    A. A only
    B. B only
    C. Both A and B
    D. Neither A nor B
20. *Technician A* says that on some Bosch Motronic systems, the control unit determines transmission shift points through the use of solenoid valves that replace the traditional spool-type shift valves.
    Technician B says that on some Bosch Motronic systems, the control unit is able to smooth automatic transmission shifts and reduce clutch slip by momentarily retarding ignition timing, particularly during wide-open throttle operation.
    Who is correct?
    A. A only
    B. B only
    C. Both A and B
    D. Neither A nor B

Chapter 19

Asian Computer Control Systems

OBJECTIVES

Upon completion and review of this chapter, you should be able to:

- ❑ Describe the sensors and actuators associated with Nissan's ECCS system.
- ❑ Describe the sensors and actuators associated with Toyota's TCCS system.
- ❑ Describe the sensors and actuators associated with Honda's PGM-FI system.
- ❑ Explain the differences between the Toyota VVT-i system and the Honda VTEC and VTEC-E systems.
- ❑ Understand the basic diagnostic concepts associated with OBD II diagnosis of these systems.

KEY TERMS

Ceramic Zirconia Oxygen Sensor
EGR Temperature Sensor
EVAP System Pressure Sensor
Flow Bench
Front and Rear Oxygen Sensors
Fuel Temperature Sensor
Fuel Trim Compensation Factor
Hot-Film Air Mass Sensor
Two-Trip Malfunction Detection

In this chapter, we concentrate on Toyota, Nissan and Honda computer control systems. Although these are not the only Asian cars imported to North America, they are representative, and common; and their control systems are typical of many of the other Asian manufacturers' products. Because of the widespread use of shared components by Japanese manufacturers, an understanding of one company's systems often provides an understanding of the systems used by others. Further, while there are Asian vehicles manufactured in Korea, these control systems are basically Japanese and the overwhelming number of such imports come unmodified (or simplified) from Japanese cars, so this is where our concentration shall be. In each case, we consider only late model vehicles, not the earlier and simpler systems. Many earlier vehicles used almost off-the-shelf Bosch controls or those built under Bosch license. Early feedback carburetor systems work similarly to domestic computer-controlled carburetor systems.

OBD II standardization of these systems has also minimized the differences between the systems as far as the technician is concerned. That is, how a computer reads its inputs, processes the information, and controls the outputs is essentially the same among different makes of vehicles. How a sensor or actuator works and how it should be pinpoint tested by a technician is also similar among makes. For example, the technician's ability to perform a sweep test on a throttle position sensor to check for glitches applies equally to all makes of vehicles. What is most

Diagnostic & Service Tip

In the last years that feedback carburetors were used on many Japanese cars, the carburetors became much more complicated in order to consistently deliver a stoichiometric air/fuel mixture to the cylinders. Most feedback carburetors were precisely adjusted on specialized equipment called **flow benches** at the factory to meet very narrow tolerances. If you see an adjustment screw marked with paint on one of these carburetors, do not break the paint loose or attempt to adjust the carburetor using this screw. It is very difficult to get the adjustment right with a flow bench and almost impossible without one. In addition, the flow and angle specifications are not published. Never touch a painted adjustment screw on a feedback carburetor, or you will almost certainly not be able to get the unit to work properly again.

often different from one make to another is the diagnostic program that has been programmed into a given PCM by the engineer. Prior to OBD II standards, the technician had to rely on the service manual for specific instructions on how to trigger the diagnostic procedures on a given vehicle. Under OBD II standards, the technician simply connects an OBD II-compliant scan tool to the Diagnostic Link Connector (DLC) and then selects the appropriate function from the scan tool's menu.

NISSAN: ELECTRONIC CONCENTRATED CONTROL SYSTEM (ECCS)

Typical late-model Nissan ECCS applications use a **hot-film air mass sensor** as the principal input to the computer to determine the fuel pulse width and spark advance. The computer, of course, controls the fuel mixture and ignition based on information stored in its memory as well as inputs from all its other sensors. For a general overview of the entire system, see Figure 19–1.

Nissans use both sequential injection and group injection, or simultaneous injection. The sequential mode injects fuel once every complete engine cycle (two crankshaft revolutions) during the corresponding cylinder's intake stroke. The group injection mode sprays all the injectors at once, twice per complete engine cycle. The group injection mode is used when the engine is starting or if the engine is in fail/safe mode or if the camshaft position sensor signal is lost and the engine is running on the crankshaft position sensor signal alone.

The Nissan control system does not set the ignition timing consistently at the edge of knock and use the knock sensor signal as a routine control input. Instead, the spark advance map is scaled to avoid knock under all but the extremes of load, engine temperature and low fuel octane equivalency.

Besides normal conditions, the computer also modifies ignition timing under special circumstances:

- during starting
- during warmup
- during idle
- if the engine is overheated
- during acceleration

The Nissan computer control system employs special fuel enrichment measures under many of the same specific circumstances:

- during warmup
- during cranking and starting
- during acceleration
- when the engine is overheated
- during high load and high vehicle speed conditions

The system employs special lean fuel mixtures (including complete fuel shutoff in some cases) under other circumstances:

- during deceleration
- during low-load, high vehicle speed conditions

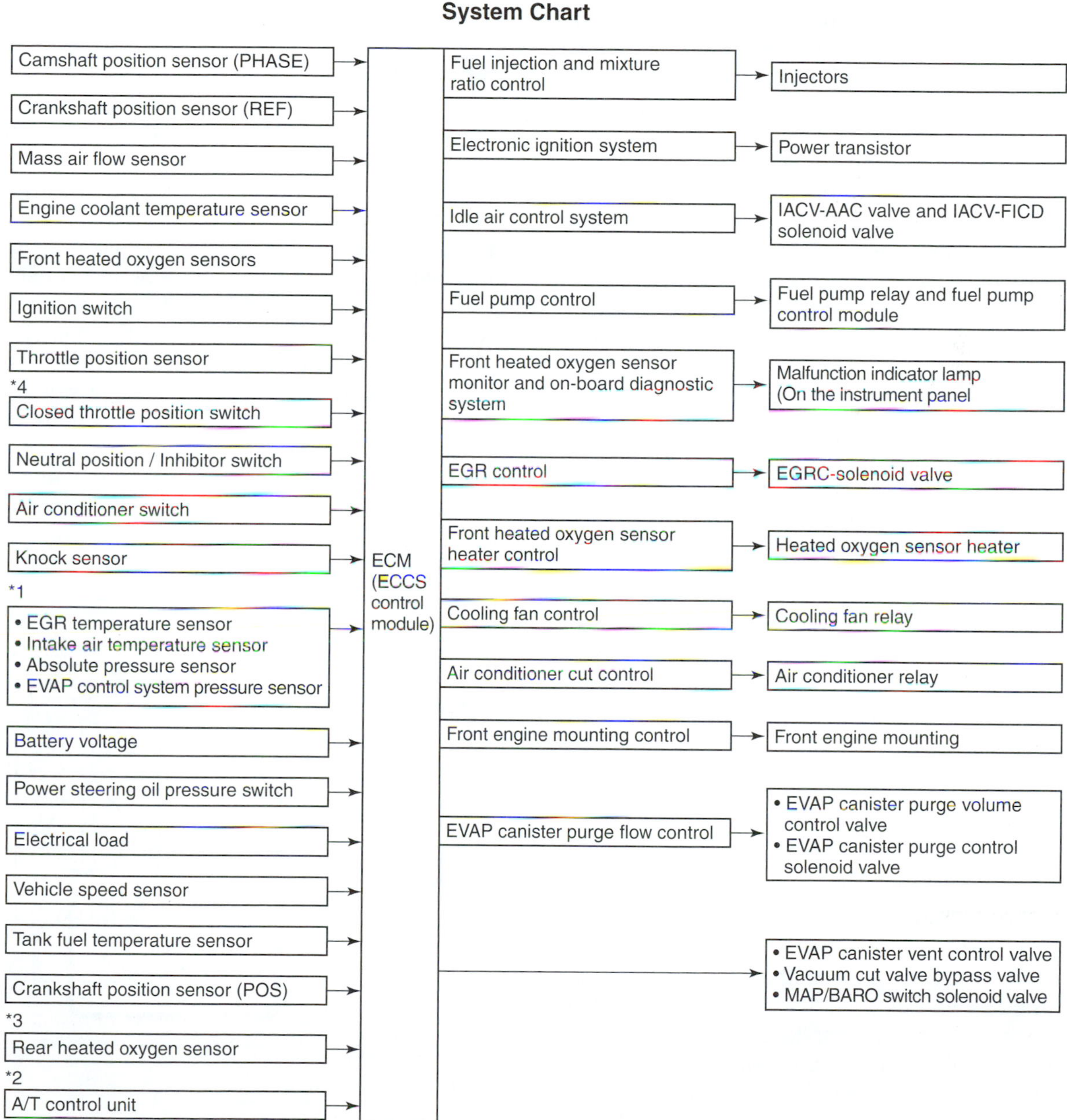

Figure 19–1 Nissan system overview. *(Courtesy of Nissan Motors.)*

Deceleration fuel cutoff will stop as the engine falls below a speed of 2,200 rpm with no load. Fuel cutoff does not begin unless the unloaded engine is above approximately 2,700 rpm.

Mixture Ratio Self-Learning. The feedback control system employs a self-learning ability, similar to systems used on domestic cars. Like them, the Nissan system monitors the success of its pulse-width modification measures to keep the air/fuel mixture stoichiometric by input from the oxygen sensor. The measures used are called *fuel trim*. The computer can modify its working copy of the hardwired mixture-to-pulse-width map depending on the results it sees from the oxygen sensor. It uses both a short-term and a long-term **fuel trim compensation factor**, similar to GM's fuel trim (integrator and block learn). The short- term factor corrects for temporary circumstances requiring fuel mixture adjustment; the long-term factor comes into play for gradual engine component wear and changes in the environment, such as altitude or seasonal climactic change.

Air-Conditioning Cutout. To enhance engine performance and/or to preserve air-conditioner components, the air-conditioning compressor is automatically disengaged regardless of the control settings if:

- the throttle is fully open.
- during startup cranking.
- at high engine speeds (to protect the compressor).

Evaporative Emissions System. The Nissan evaporative emissions system, like most other carmakers, uses a charcoal canister that stores fuel vapors from the tank to prevent their escape into the atmosphere. The system depends on the vapor pressure/vacuum relief valve in the fuel filler cap, and the vapor recovery shutter in the filler tube. The computer activates the EVAP canister purge volume control valve to operate the system under normal driving conditions. The fuel mixture pulse width adjustments are made in response to oxygen sensor information since there is no way for the computer to know how much fuel vapor, if any, is coming through the vapor canister.

INPUTS

Camshaft Position Sensor (CMP). The camshaft position sensor, Figure 19–2, is in the front engine cover facing the camshaft sprocket. This information distinguishes which cylinder is at the firing position. The camshaft position sensor is a coil-and-magnet pulse generator: it consists of a permanent magnet wound in an electrical coil. When the gap between the camshaft sprocket and the magnet changes, the magnetic field changes and generates a voltage signal in the sensor output line. The computer uses the camshaft position signal to sequence the fuel injectors. At the beginning of the startup cranking period, the camshaft position signal has not yet occurred, and all the injectors are simultaneously pulsed for engine startup.

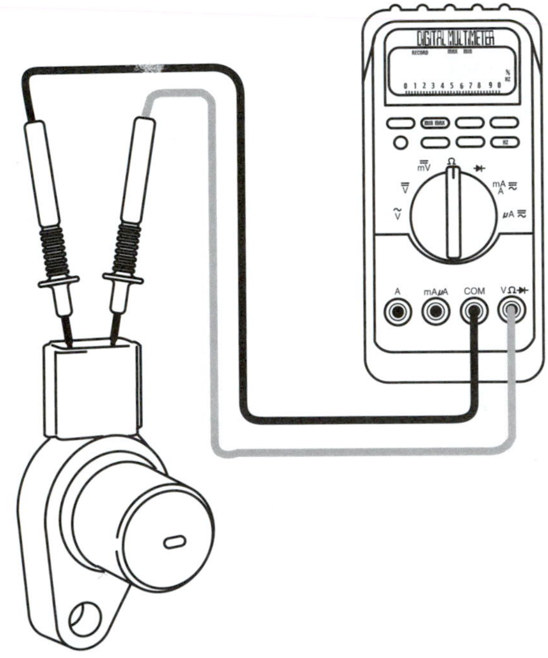

Figure 19–2 Camshaft position sensor (PHASE).

Camshaft position sensors can be checked for shorts and opens with an ohmmeter, as well as for residual magnetism at the tip of the sensor with a strip of ferrous metal or a screwdriver tip.

Crankshaft Position Sensor (CKP). The REF crankshaft position sensor, Figure 19–3, is on the upper oil pan facing the crankshaft accessory belt pulley. It signals the computer when the TDC position for individual cylinders occurs. This sensor is a coil-and-magnet inductive signal generator producing a signal that varies in voltage with engine speed. Resistance through its coil should range between 470 and 570 ohms at room temperature.

Mass Airflow Sensor. The hot film air mass sensor is located in the intake air duct between the air filter and the throttle, Figure 19–4. It should indicate a signal voltage of below 1 volt with the ignition on and the engine off, and should show between 1.0 and 1.7 volts at idle. With the engine running at higher speeds and under load, the signal voltage can rise to as much as 4.0 volts. Because of the sensitivity of the sensor to disturbances in the airflow, a frequent check should be done to make sure there is no dust or foreign material on or near the sensor.

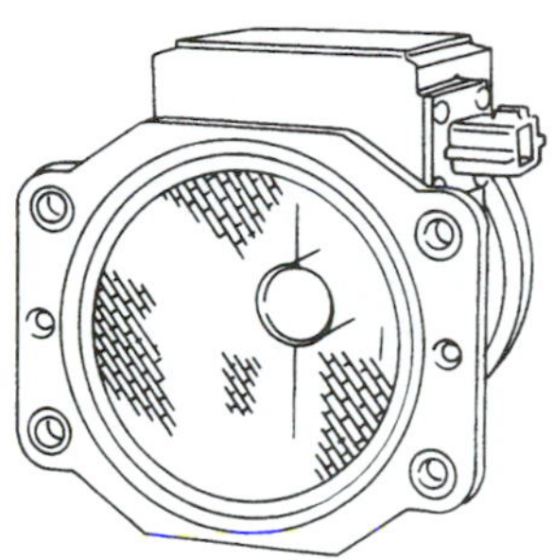

Figure 19–4 Mass airflow sensor. *(Courtesy of Nissan Motors.)*

Engine Coolant Temperature (ECT) Sensor. As with other sensors, the ECT sensor modifies a reference signal from the PCM to correspond to the monitored parameter, Figure 19–5. The resistance of the thermistor in the sensor reduces as the engine temperature increases. Information from the ECT sensor plays a significant role in engine management, because a richer mixture is needed for cooler temperatures to make sure there is enough vaporized fuel to burn. At operating temperatures, the system switches to a closed-loop control system, using the signal from the oxygen sensor. The ECT sensor, however, is critical to the determination of when the system goes into closed loop.

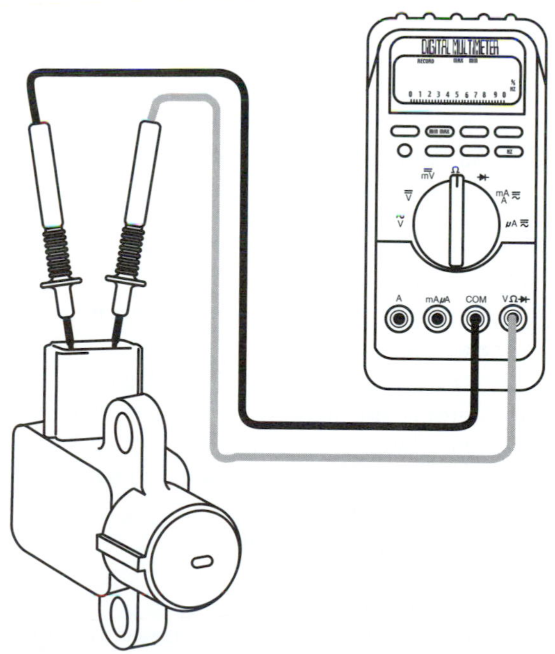

Figure 19–3 Crankshaft position sensor (CKP).

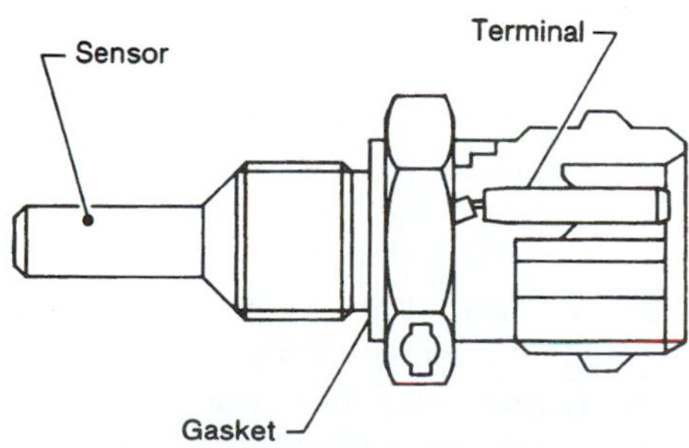

Figure 19–5 Engine coolant temperature sensor. *(Courtesy of Nissan Motors.)*

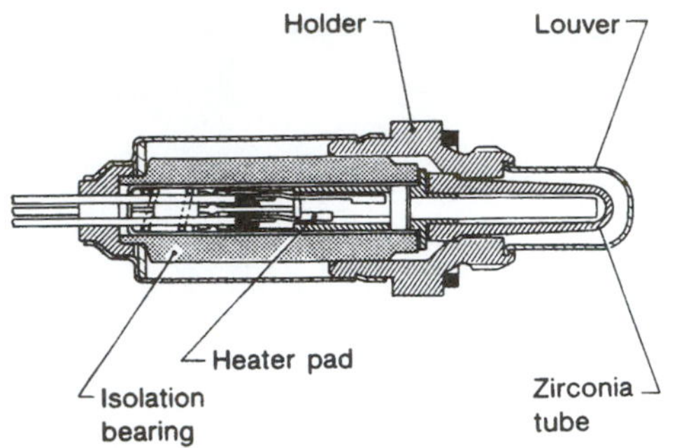

Figure 19–6 Heated oxygen sensor. *(Courtesy of Nissan Motors.)*

Front Heated Oxygen Sensors. Nissans use **ceramic zirconia oxygen sensors**, Figure 19–6. While the operation is similar to other types of sensors, the amplitude of the signal is slightly larger, with the signal switching from about 1 volt to almost zero as the mixture cycles from lean to rich and back across the stoichiometric ratio. The computer gets the sensor's output signal and constantly readjusts the injector pulse width to correct the air/fuel mixture and keep the signal cycling. On V-type engines, two front oxygen sensors are used, one on each bank; and the computer controls the mixture delivered to each bank independently of that delivered to the other. The oxygen sensors are heated to bring the system to closed loop as quickly as possible and to prevent the system from falling out of closed loop should the vehicle idle for a long time.

Diagnostic & Service Tip

Very frequently the problem with an oxygen sensor is not that it has failed completely, but that it has begun to respond too slowly to changes in the residual exhaust oxygen. Once it does so, the computer will not be able to properly cycle the injector pulse width to keep the mixture stoichiometric. There is no specific number of cycles per unit of time that an oxygen sensor should register, but the newer systems should work more quickly than the older. If in doubt, monitor the signal cycling of a "known good" sensor and compare it to one you suspect.

Diagnostic & Service Tip

Oxygen sensors, particularly zirconia-based oxygen sensors, are relatively brittle and delicate components because of the ceramic element. If an oxygen sensor is dropped much more than a foot to a hard surface such as a concrete shop floor, it will probably sustain an invisible internal crack and will have to be replaced. An oxygen sensor is much more delicate than a spark plug as well as more expensive, so they should always be handled carefully.

Ignition Switch. When the ignition switch is turned on, the computer becomes active and sends reference voltage to the sensors that require it. It also begins monitoring the return signals and prepares to energize the appropriate actuators.

Throttle Position Sensor. The computer learns about accelerator pedal movement and position from the throttle position sensor, Figure 19–7. As with many other linear measurement sensors, it receives a reference voltage from the computer, routes it through a variable resistor, and returns a corresponding voltage to the computer. Throttle position signals range from about 0.5 volt to about 4.6 volts from closed throttle to wide-open throttle.

Closed Throttle Position Switch. The closed-throttle position switch, Figure 19–7, tells the computer when the driver has removed his or her foot from the accelerator pedal. The computer then knows to control the idle speed depending on information from its other sensors. The closed-throttle switch is either on or off.

Park/Neutral Switch. The park/neutral switch allows the starter circuit to complete only with the transmission in one of these two gears. It also sends a signal to the computer when the

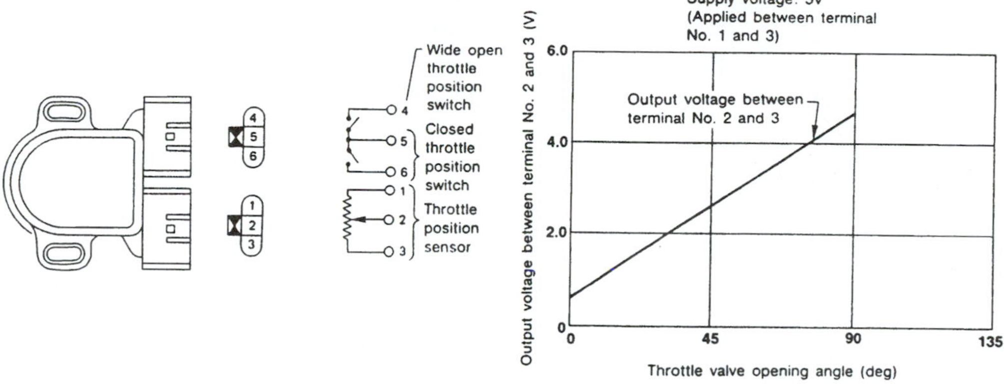

Figure 19–7 Throttle position sensor. *(Courtesy of Nissan Motors.)*

vehicle is in gear. This signal is used for idle control, spark timing and sometimes fuel mixture.

Air-Conditioner Switch. When the air conditioning is switched on, the computer receives a signal that it uses to determine proper idle speed.

Knock Sensor. The Nissan knock sensor is a piezoelectric crystal type, sending a voltage signal to the computer in the event it sustains vibrations at a frequency characteristic of detonation. If the computer receives that signal, it retards ignition timing until the knock disappears. As already mentioned, the Nissan spark advance map is designed to prevent knock except under unusual fuel quality, load or temperature conditions.

The knock sensor can be electrically inspected for shorts or opens. Nissan literature calls for an ohmmeter capable of measuring up to 10 megohms. Like oxygen sensors, knock sensors are brittle and should not be reused if they have been dropped on a hard surface.

EGR Temperature Sensor. The **EGR temperature sensor**, Figure 19–8, reports changes in the EGR passage temperature. While this information does not directly control engine management functions, it is used for emissions diagnosis. If the computer sees no signal or the wrong signal from the EGR temperature sensor, it indicates that insufficient exhaust is recirculating through the EGR system, leading to excessive production of NO_x in the exhaust. The sensor is a

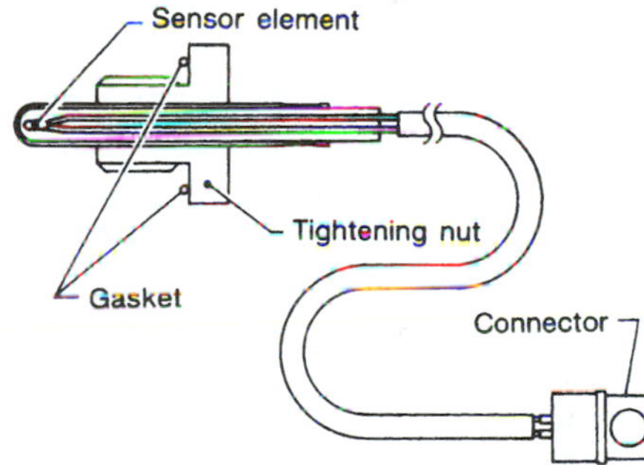

Figure 19–8 EGR temperature sensor. *(Courtesy of Nissan Motors.)*

thermistor, changing its resistance and thus the return signal with temperature changes.

Intake Air Temperature Sensor. The intake air temperature sensor, Figure 19–9, mounts in the air cleaner housing and transmits a voltage signal corresponding to the intake air temperature. The sensor uses a thermistor that is responsive to changes in temperature. As temperature rises, the resistance of the thermistor goes down.

Absolute Pressure Sensor. The Nissan absolute pressure sensor, Figure 19–10, connects to the MAP/BARO switch solenoid through a hose. It detects both ambient pressure and intake manifold pressure. As the pressure rises, the voltage signal rises.

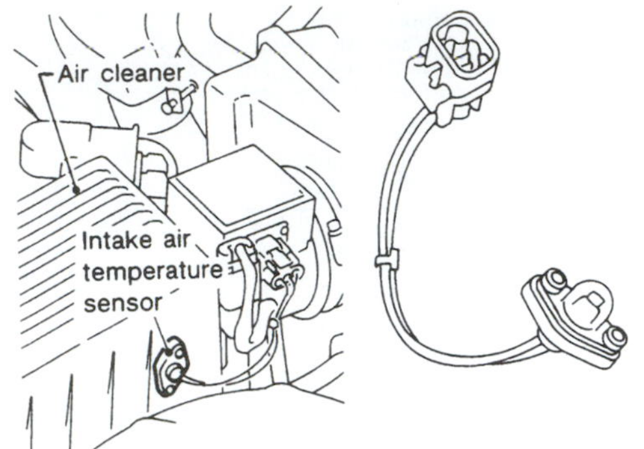

Figure 19–9 Intake air temperature sensor. *(Courtesy of Nissan Motors.)*

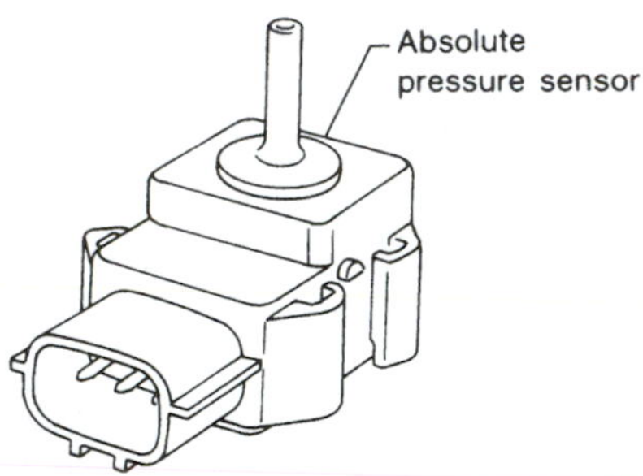

Figure 19–10 Absolute pressure sensor. *(Courtesy of Nissan Motors.)*

EVAP System Pressure Sensor. To monitor the effectiveness of the EVAP control system, the computer receives a signal from the **EVAP system pressure sensor**, Figure 19–11. While this does not directly affect engine control functions, it does play a role in on-board emissions diagnosis. This sensor will indicate if there is a leak in the vapor canister system.

This sensor cannot distinguish a leak in the system from a loose fuel filler cap, which will trigger the same trouble code and light the malfunction indicator lamp (MIL). Check for a properly functioning fuel filler cap first, including a pressure test of the cap, whenever this sensor indicates a problem.

Battery Voltage/Electrical Load. The computer constantly monitors the voltage received from the battery and charging system and will illuminate the check engine light if it falls below proper voltages. This signal is used to modify the injector pulse width and the ignition coil dwell if system voltage falls below the standard charging system voltage.

Power Steering Oil Pressure Switch. The power steering oil pressure switch, Figure 19–12, is on the power steering high pressure tube. When a power steering load occurs, it triggers this switch, signaling to the computer to increase the idle speed setting. This switch conducts battery voltage and can be checked for continuity with an ohmmeter.

Vehicle Speed Sensor. The vehicle speed sensor, Figure 19–13, is in the transaxle on the output shaft. Its pulse generator sends a signal corresponding to vehicle speed to the speedometer, which conveys the signal to the computer.

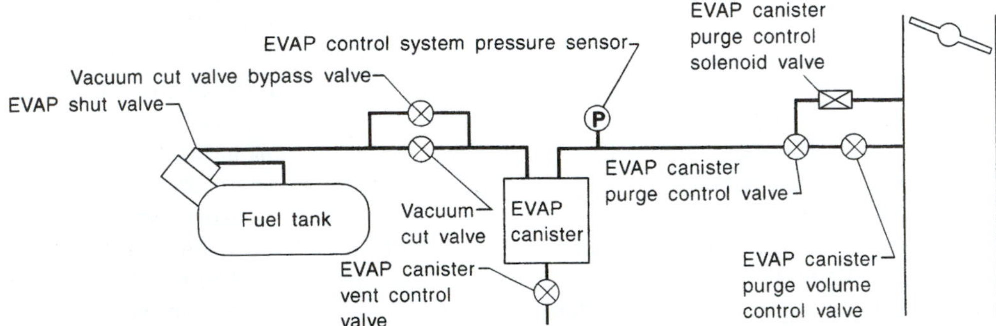

Figure 19–11 Evaporative emission system diagram. *(Courtesy of Nissan Motors.)*

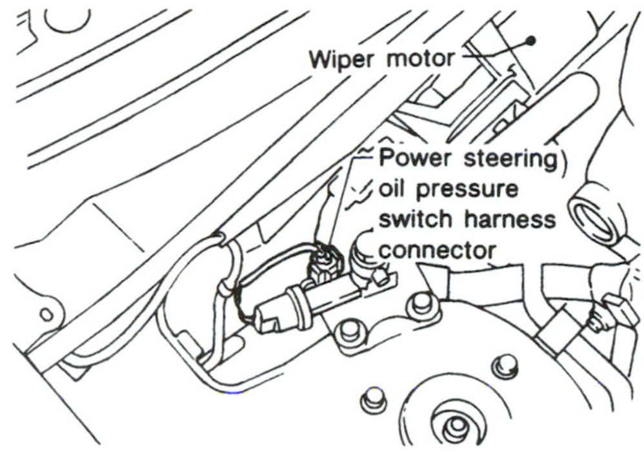

Figure 19–12 Power steering pressure switch. *(Courtesy of Nissan Motors.)*

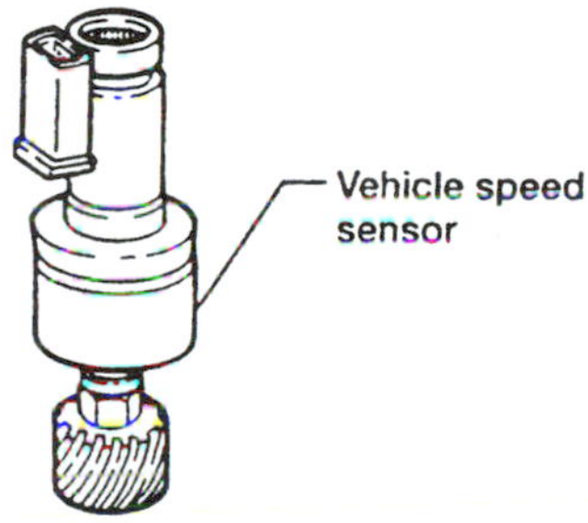

Figure 19–13 Vehicle speed sensor. *(Courtesy of Nissan Motors.)*

This information is used for both fuel pulse width and ignition timing calculations.

Fuel Temperature Sensor. The system uses a thermistor-type **fuel temperature sensor** to monitor the temperature of the fuel in the tank. The sensor is accessible under the rear seat, Figure 19–14. The fuel temperature sensor receives a reference voltage from the computer and returns to it a voltage signal corresponding to the temperature of the fuel, from 3.5 volts when the fuel is at 68°F to 2.2 volts when the fuel reaches 122°. Information from this sensor indicates changes in fuel viscosity and its inclination to vapor lock. Nissan uses this sensor in conjunction with the EVAP pressure sensor. In addition to fuel temperature information, this sensor also helps the PCM determine when to run the fuel system leak detection software.

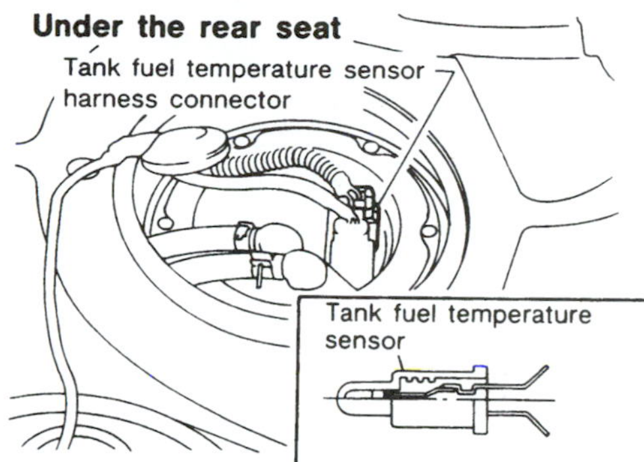

Figure 19–14 Tank fuel temperature sensor. *(Courtesy of Nissan Motors.)*

Crankshaft Position Sensor (CKP). The crankshaft position sensor, Figure 19–15, plays a familiar role in spark advance timing, fuel injection timing and in the newest vehicles emissions monitoring. It is located on the flywheel end of the oil pan and signals one degree changes in the flywheel position to the computer. In these later cars, the crankshaft position sensor is the principal source of information about cylinder misfire. When a cylinder misfires, whether from spark, fuel or mechanical problems, the engine slows slightly at the point when that cylinder's power stroke should occur. At idle, this slowing is perceptible as rough running, but the crankshaft position sensor can report the miss to the computer even at higher speeds.

The emissions-related reason for tracking cylinder miss is that if the cylinder does not fire, it does not consume the fuel that entered with the intake stroke. The first consequence would be a considerable overheating of the catalytic converter from all the increased fuel and oxygen. The second consequence would be a failed catalytic converter and the third would be the dumping of raw hydrocarbons into the air after the catalytic converter could no longer burn them up.

Once the misfiring cylinder is identified by the crankshaft position sensor, the computer can then shut off the fuel injector to that cylinder. This is not, of course, a complete countermeasure, because the oxygen is still pumped through the

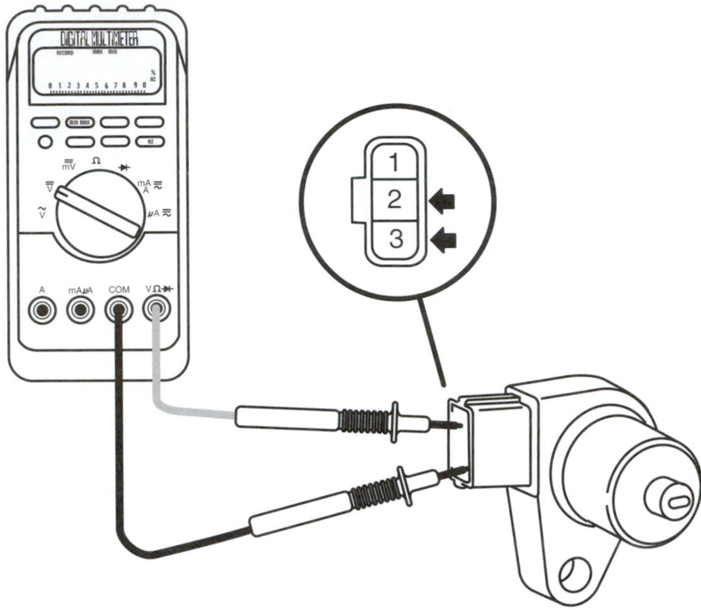

Figure 19–15 Crankshaft position sensor (POS).

cylinder and into the exhaust. This means the oxygen sensor's reports will be inaccurate (overlean) and the computer will drive the system rich. So this is one of the few circumstances when the computer turns on the check engine light the first time the problem occurs, without waiting for it to recur on a subsequent trip.

Rear Heated Oxygen Sensor. Nissans that conform to the OBD II regulations include a rear heated oxygen sensor. This sensor works in the same way as the front sensor or sensors, but is used primarily to monitor the effectiveness of the catalytic converter. As long as the catalytic converter is working properly, the rear heated oxygen sensor's signal should remain almost flat or only slowly cycling over its output range. The computer tracks the signal output and will set a code should the signal mirror that of the front sensor. If the front sensor fails, the computer will make use of the rear sensor to control air/fuel mixture as closely as possible.

Automatic Transmission Diagnostic Communication Line. There is a communications link between the automatic transmission control unit and the engine control computer that conveys information related to transmission malfunctions. Any such information erased from the transmission unit must be separately erased from the engine control computer.

OUTPUTS

Fuel Injectors. The multiport fuel injectors are Bosch-type, as are used on many vehicles. Activated by a constant power source grounded through the computer, each injector opens by lifting its pintle off the seat and allowing fuel to spray through the nozzle for a fixed amount of time, calculated by the computer on the basis of all the system inputs.

IACV-AAC Valve and IACV-FICD Solenoid Valve. The idle air control valve—auxiliary air control (IACV-AAC) valve, controls throttle bypass air that determines the engine's idle speed, Figure 19–16. The computer activates this com-

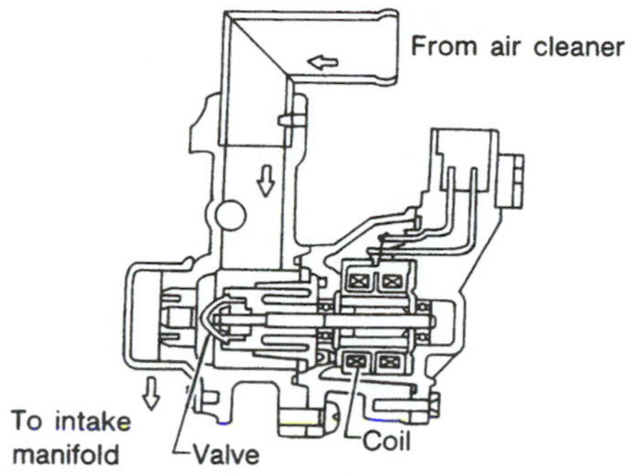

Figure 19–16 Idle air control valve (IACV). *(Courtesy of Nissan Motors.)*

ponent according to inputs from the engine coolant temperature sensor, electric load sensor, air-conditioning switch and power steering pressure switch. Once removed from the engine, the operation of this valve can be checked by cycling the ignition switch on and off to observe whether the valve shaft moves smoothly forward and backward according to the ignition switch position. Removal and reinstallation of the valve should only be done with the ignition switch off.

The IACV-FICD solenoid combines with the IACV-AAC valve to constitute the idle air adjusting unit. The IACV-FICD solenoid is strictly an on-off solenoid moving a plunger in the air passage, Figure 19–17. It can be checked separately from the vehicle for smooth function.

Fuel Pump Relay and Fuel Pump Control Module. The fuel pump relay is activated for one second after the ignition switch is first turned on, to pressurize the fuel system before starting the engine, Figure 19–18. The computer will leave the relay energized as long as it receives input from the crankshaft position (CKP) sensor at the flywheel. The fuel pump control module adjusts voltage supplied to the fuel pump to control the amount of fuel flow, Figure 19–19. It reduces the voltage to approximately 9.5 volts except during:

- engine startup cranking.
- when the engine coolant is below 45°F.
- within 30 seconds after startup with a warm engine.
- under high load and high speed conditions.

Ignition Coil Power Transistor. Late model Nissans use one coil per cylinder in a direct ignition system, Figure 19–20. If a bad coil is suspected,

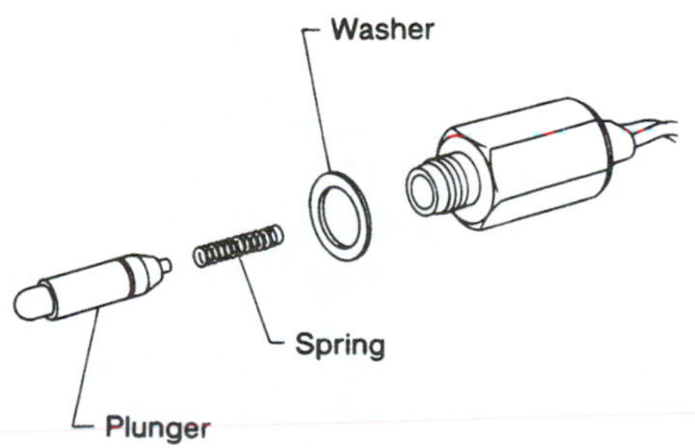

Figure 19–17 IACD-FICD solenoid valve. *(Courtesy of Nissan Motors.)*

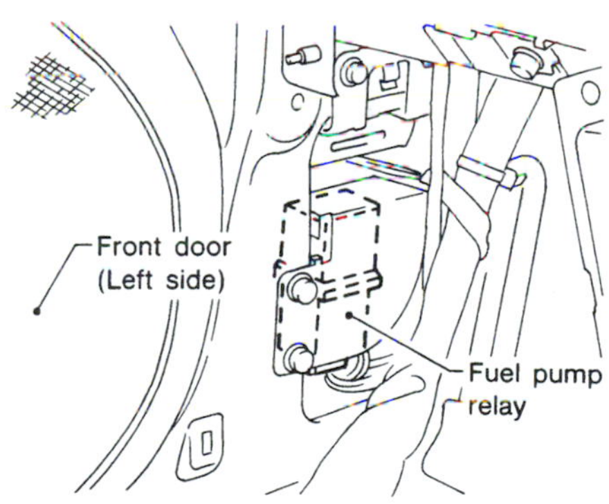

Figure 19–18 Fuel pump relay. *(Courtesy of Nissan Motors.)*

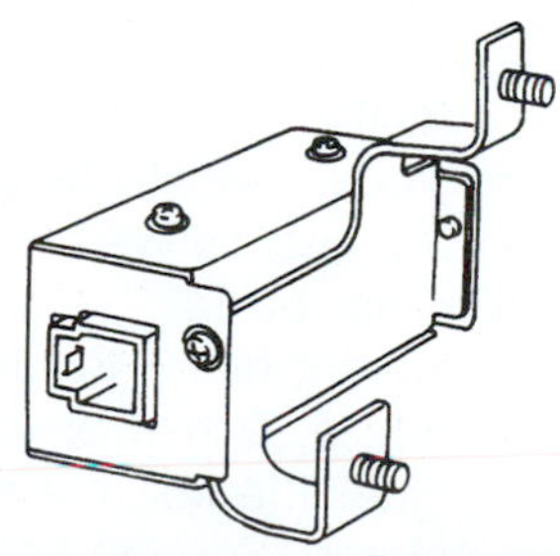

Figure 19–19 Fuel pump control module. *(Courtesy of Nissan Motors.)*

Figure 19–20 Nissan ignition coil.

moving the coil from one cylinder to another to see whether the cylinder miss follows is the accepted technique. No internal repairs are possible; replacement is the only service.

Malfunction Indicator Lamp. The control unit turns on the Malfunction Indicator Lamp (MIL) when it detects a fault in one of the engine's control or actuator circuits. Most malfunctions will have to occur in two successive trips before the computer will turn on the MIL, but certain malfunctions that represent either failure of the engine to run properly or a risk of damage to a component will trigger the MIL the first time the problem is encountered. See the service manual to determine which problems fall into which category for specific models and years.

EGR Solenoid Valve. The EGR solenoid works in response to signals from the PCM. When the PCM grounds the circuit, the coil in the solenoid is energized, and a plunger then moves to cut the vacuum signal from the throttle body to the EGR valve. When the PCM sends an off signal, vacuum passes through the solenoid valve to the EGR valve. The system is thus designed to be fail-safe with the EGR valve responding to manifold vacuum.

Function of the EGR valve is monitored by the EGR temperature sensor, Figure 19–21. The computer can distinguish between EGR flow when there should be none and no EGR flow when there should be some by this temperature sensor, and it sets different codes depending on which condition it detects.

The EGR valve can also be checked manually. If it doesn't move smoothly, it may have mechanical problems.

Fuel Pump Pressure. Fuel pump pressure with zero vacuum at the MAP sensor, should be approximately 43 psi. At idle, the fuel pressure should read approximately 34 psi.

Fast Idle Cam. Instead of a stepper motor, some Nissans use a fast idle cam on the throttle linkage. The fast idle cam works by a thermostatically heated spring.

Oxygen Sensor Heaters. To ensure that the engine control system can go into closed loop as soon as possible, both the **front and rear**

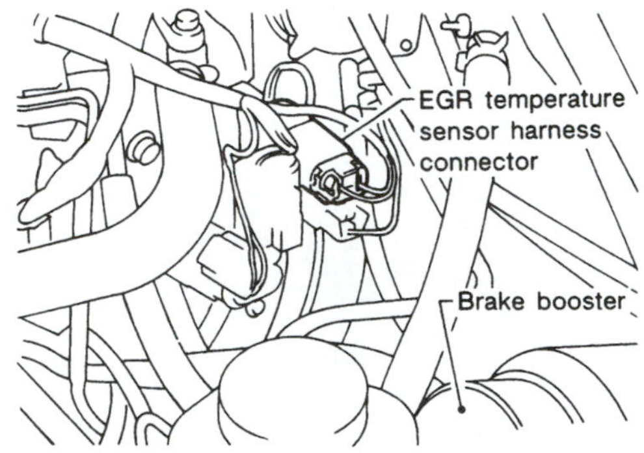

Figure 19–21 EGR temperature sensor. *(Courtesy of Nissan Motors.)*

oxygen sensors include heater circuits to get them to operating temperature as quickly as possible. The computer will store different trouble codes for a failure of the heater circuit or from a failure of the oxygen sensor output circuit.

Cooling Fan Relay. The engine control computer operates the radiator cooling fans through a relay depending on the engine coolant temperature, vehicle speed and whether the air conditioner is turned on. There are two different fan speeds available to the computer. There are also differences in the fan speed maps for cars built for the California market, though the rest of the cooling system is the same.

Air-Conditioner Relay. When the air conditioner is turned on, the a/c relay provides the engine control computer a signal used to modify the control of the idle speed.

Front Engine Mount Control. To keep the engine vibration to a minimum, the engine control system also electrically varies the dampening effect of the front engine mount. The mount is in a soft setting when the engine is idling, and in a firm setting when the engine is doing work moving the vehicle under load.

EVAP System Controls. The late model Nissans use a more complex evaporative control system, storing fuel vapors in a charcoal canister as has been used for many years, but with a more complex set of controls and monitoring measures. Refer to Figure 19–11 under the Inputs section for a complete system diagram. The basic actuator is the EVAP canister purge control valve, Figure 19–22. When the computer sends an off signal to the control solenoid valve, the vacuum between the intake manifold and the EVAP canister purge control valve is shut off. The EVAP canister purge volume control valve controls the passage of vapors from the EVAP canister purge control valve to the intake manifold, Figure 19–23. The system includes, as was described under the Inputs section in this chapter, an EVAP control system pressure sensor, Figure 19–11, intended to detect leaks in the system or a faulty, loose or missing fuel filler cap. While the EVAP control system differs somewhat from one year to the next and from one model to another, fundamentally they all work the same way, storing fuel vapors in the charcoal canister until the engine, running in closed loop, can vent the canister into the intake manifold and burn the stored vapors. Corrections to the fuel mixture required by the variable concentration of fuel vapors entering the engine fall within the ability of the oxygen sensor feedback control system.

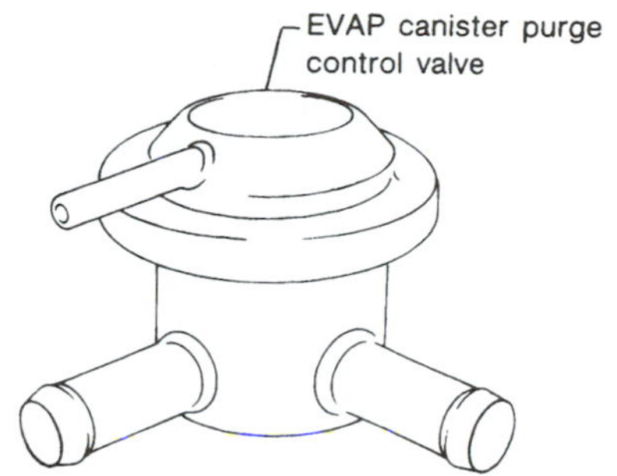

Figure 19–22 EVAP canister purge control valve. *(Courtesy of Nissan Motors.)*

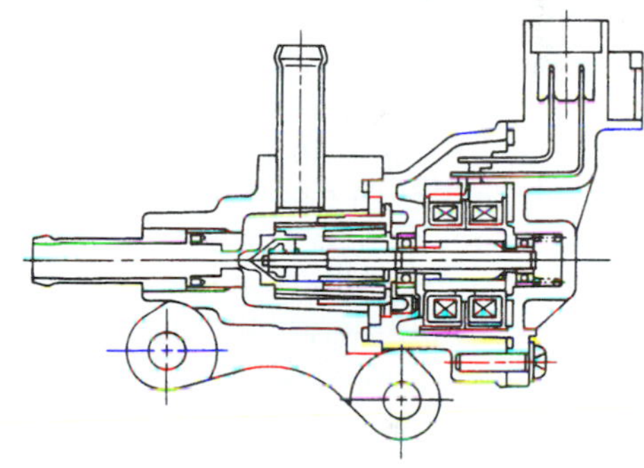

Figure 19–23 EVAP canister purge volume control valve. *(Courtesy of Nissan Motors.)*

Nissan Supercharger

The Nissan Frontier pickup truck 3.3L SOHC V6 engine has a new roots-type three-lobe supercharger for the 2001 model year. Eaton builds this three-lobe rotor supercharger that mounts to the stock 3.3L intake manifold. It is limited to 6 pounds of boost, Figure 19–24.

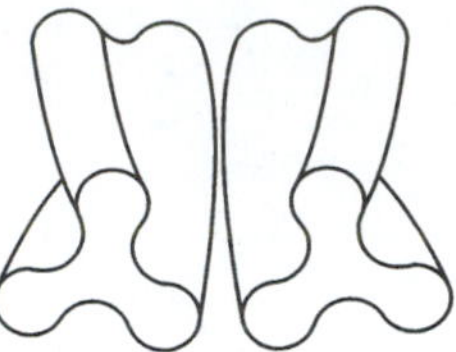

Figure 19–24 Nissan three-lobe rotor supercharger.

Diagnostic & Service Tip

With one coil per cylinder buried in the valve cover on late model vehicles, it is somewhat more difficult to get spark advance and engine speed information without special Nissan tools. Nonetheless, you can remove the coil from the number one cylinder, attach an extension spark plug cable and make the measurements with conventional equipment. Many technicians use a similar technique on the GM Quad-4 engine.

The engine produces 210 horsepower with 245 ft. lb. of torque. A bypass valve diverts the air back into the supercharger during normal driving. The fuel pump flow rate has been increased by 80 percent to meet wide-open throttle demands. The injector flow rate also has been adjusted accordingly. This increased performance has been achieved with only a 10 percent reduction in fuel economy, Figure 19–25.

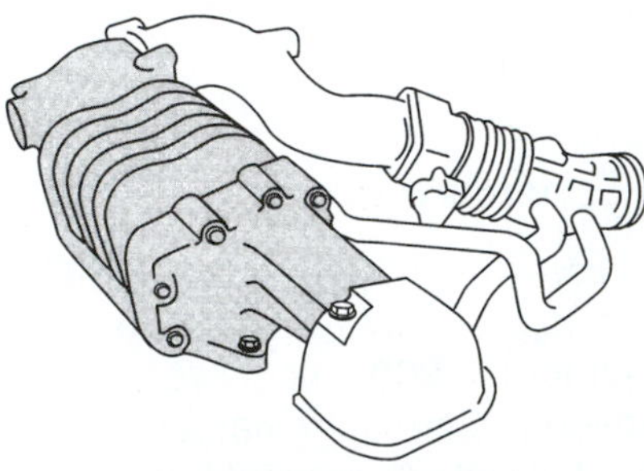

Figure 19–25 Supercharger system.

Nissan Intake Manifold Swirl Control System

The 2001 Nissan Sentra intake manifold has a door in each runner. All doors are mounted on a single shaft. The doors have two positions: rest and fully open, Figure 19–26.

When the door in the runner of the intake manifold is in the rest position, the intake manifold passageway is partially restricted, causing the moving airstream to swirl and increasing air velocity. The increased air velocity improves injected fuel vaporization. The doors are at rest at idle and during low engine speed. At higher engine speed, a single vacuum diaphragm opens the swirl control valves. The PCM turns a swirl control solenoid on or off to apply vacuum to the diaphragm. The restriction in the manifold runner is eliminated when the doors are wide open, improving intake efficiency, Figure 19–27.

Nissan Air-Cleaning Radiator

The California model of the Nissan Sentra is equipped with a new air-cleaning radiator to meet that state's Super Low Emission Vehicle (SULEV)

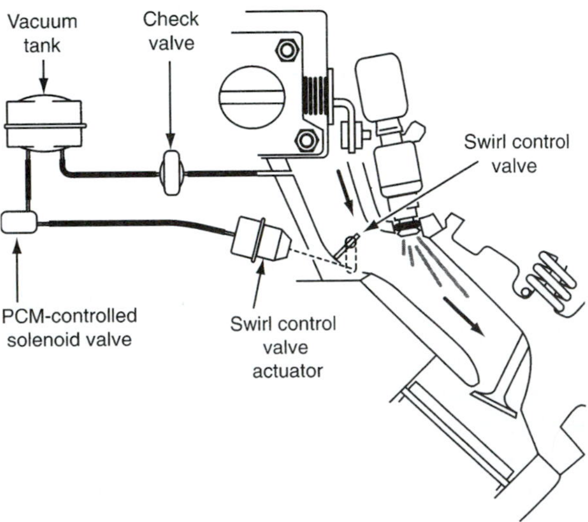

Figure 19–26 Nissan intake manifold swirl control system.

Swirl control operating conditions		
Throttle valve position	Engine rpm	Control solenoid
Closed	Below 3,200 rpm	On
Open	Above 3,200 rpm	Off
Note: Swirl control valves are kept open, at all engine speeds, below 50°F (10°C) and above 131°F (55°C)		

Figure 19–27 Nissan swirl control system operating conditions.

rating. The radiator fins are coated with a special material that converts ozone molecules into oxygen. When the driver drives the vehicle down the street, the cooling fan pulls dirty polluted air into and through the radiator fins. At normal operating temperature, the radiator coating acts as a catalytic converter and cleans the air before it leaves the radiator. This radiator is not available in all parts of the country at this time.

✔ SYSTEM DIAGNOSIS AND SERVICE

Late model vehicles include the OBD II standardized diagnostic system. In addition, they have retained the earlier Nissan self-diagnostics. This older system uses a screwdriver-actuated switch in the computer itself (ordinarily under the front passenger seat, but behind the instrument panel near the glove box on newer vehicles), and displays trouble codes either by flashes of the MIL or, on the earliest systems, by flashing lights on the computer itself. The older systems, of course, are simpler, rely on fewer inputs and outputs and provide less information for diagnosis. Check the shop manual for the specific information corresponding to the model and year of the vehicle you are working on.

Nissans use a **two-trip malfunction detection** logic; that is, for most component circuit failures the computer stores a fault on the first trip in that it is detected and turns on the MIL if it recurs on the next trip (a trip is defined as an engine startup, followed by enough driving to bring the engine to operating feedback control system condition). There are several different ways to erase stored faults in Nissan systems, including changing the computer screwdriver switch from Test Mode II to Mode I or using the scan tool, so read the manual for the specific car before you perform these tests. Disconnecting the battery erases the codes (as well as the self-learning data), but it can take as long as 24 hours of disconnection before the energy stored in the capacitors discharges fully.

TOYOTA COMPUTER-CONTROLLED SYSTEM (TCCS)

As an example of a TCCS, we look at the 1MZ-FE V6 engine in the late model Toyota Camry. Other systems are similar, with appropriate modifications for engine type and vehicle style.

The Toyota system uses a mass airflow sensor to determine how much fuel to inject at a given engine condition. The six fuel injectors are sequentially operated in the firing order of the engine. Spark is delivered through individual coils at each spark plug, similar to the Nissan system described earlier in this chapter.

INPUTS

Mass Airflow Sensor. The mass airflow sensor provides the engine control computer with a fluctuating voltage signal corresponding to the amount of air passing through the air intake system, Figure 19–28.

Heated Oxygen Sensors. The Toyota system uses front and rear oxygen sensors, Figure 19–29, in the same way as the Nissan system described earlier in this chapter. On V6 engines, there are two front oxygen sensors, one for each bank. The front oxygen sensors generate the signals the computer uses for closed-loop fuel mixture control, and the rear sensor is used to

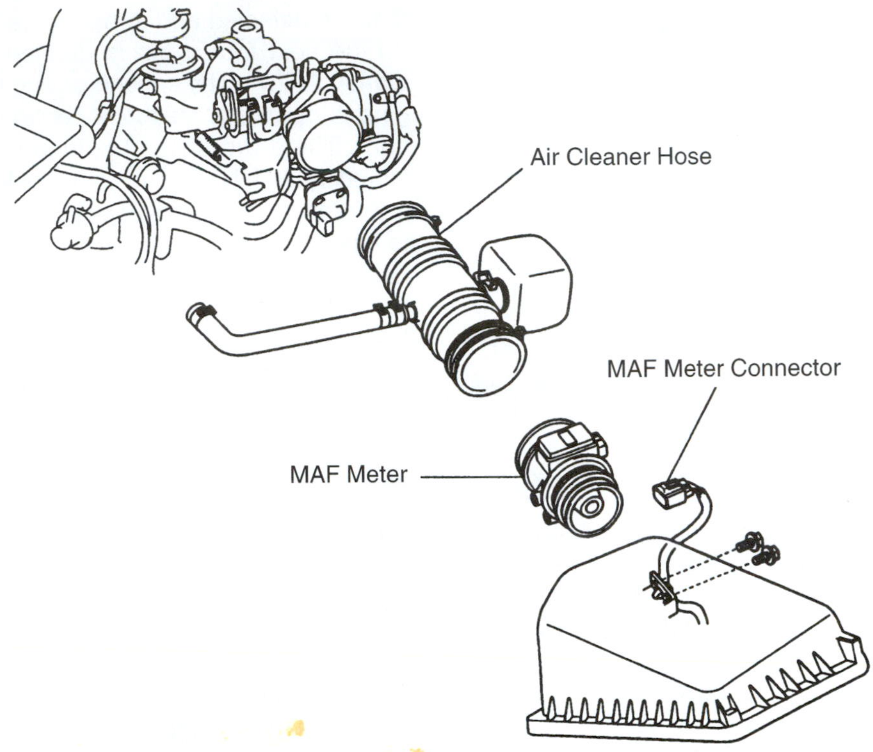

Figure 19–28 Mass airflow sensor (meter).

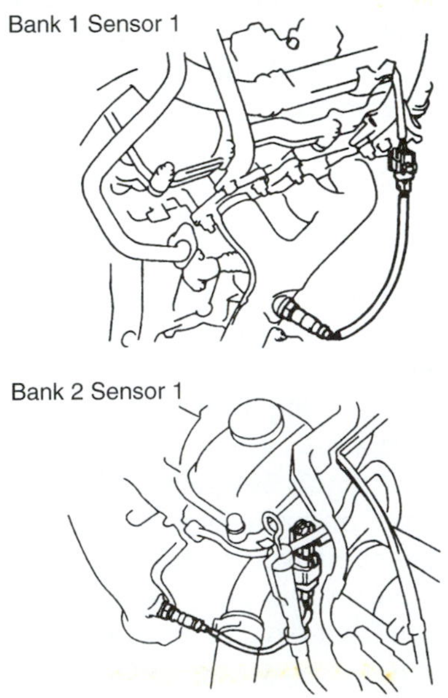

Figure 19–29 Heated oxygen sensor.

determine the effectiveness of the catalytic converter. Should either front sensor fail, the rear sensor signal is used to help maintain fuel trim. The oxygen sensors all include a heater circuit to bring them up to operating temperature as quickly as possible and run the engine in closed loop.

Throttle Position Sensor. The throttle position sensor, Figure 19–30, provides the computer with information about where the driver's foot has positioned the accelerator pedal. On the Toyota system, the throttle position sensor is adjustable using a feeler gauge between the throttle stop screw and stop lever and an ohmmeter on the terminals as shown in the shop manual. A defective throttle position sensor can lead to the wrong idle speed and poor driveability at other speeds.

Engine Coolant Temperature Sensor. The ECT sensor, Figure 19–31, varies its internal resistance inversely with temperature; that is, the warmer it gets, the lower its resistance goes. It threads into the water jacket near the thermostat outlet and can be checked with an ohmmeter. At normal shop temperature, it should indicate approximately 2,000 ohms. The computer uses information from the ECT sensor to calculate fuel mixture, spark advance and ignition timing. Its information also plays a role in determining when to go into closed loop, when to engage the EGR system and when to open the passages to the evaporative emissions charcoal canister.

Knock Sensors. V6 engines use two knock sensors, one on each bank, Figure 19–32; in-line engines use one. The knock sensor, as on other systems, is a piezoelectric design that, when vibrated at a frequency characteristic of knock, sends a signal to the computer, which then retards spark advance until the knock disappears. Spark advance is then readvanced until the knock just reappears. As with most knock sensors, function can be checked by tapping on the engine block adjacent to the knock sensors with a tool while observing the spark advance (warm engine, closed-loop conditions). Knock sensors are relatively delicate and should never be dropped. If dropped to a hard surface, they will frequently sustain damage.

EGR Temperature Sensor. To determine whether its EGR system actuation commands are effective, the computer monitors information from the EGR temperature sensor, Figure 19–33. When the EGR passage opens, hot exhaust gas travels through the channel and the temperature sensor should reflect an increased temperature. This information does not enter into direct control of the engine, but does play an important role in the self-diagnostics of the system's emissions controls.

Camshaft Position (CMP) Sensor. The latest Toyota V6 engines use a direct ignition system with one coil per cylinder rather than a distributor, and hence they require a CMP sensor, Figure 19–34. This sensor is on one of the camshaft sprockets on the accessory drive side of the engine. The purpose of this sensor is to allow the computer to determine which cylinder is

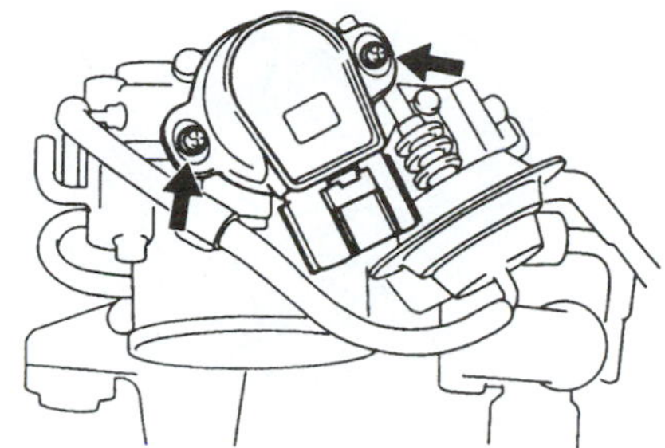

Figure 19–30 Throttle position sensor.

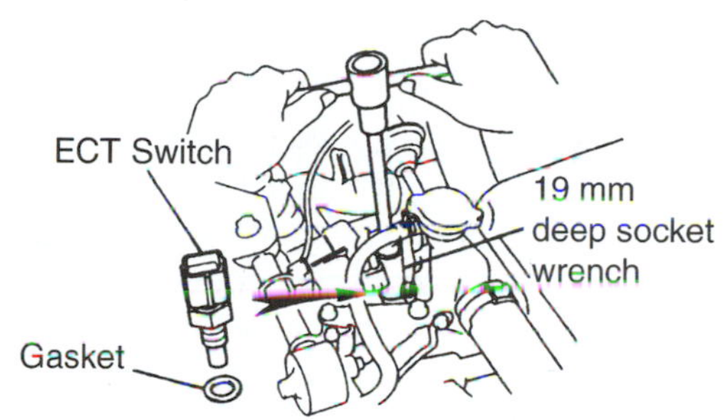

Figure 19–31 Engine coolant temperature sensor.

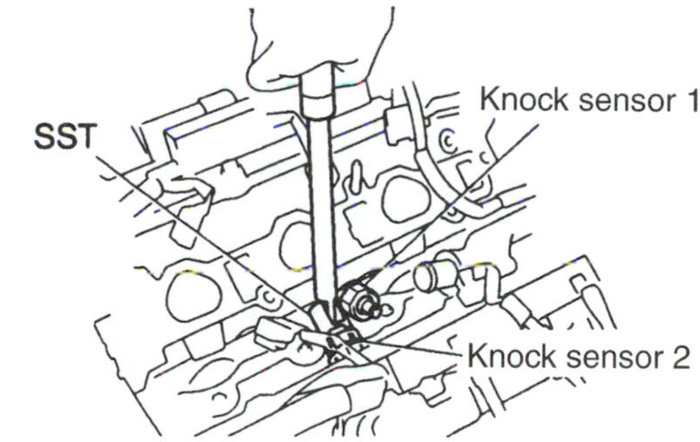

Figure 19–32 Knock sensors.

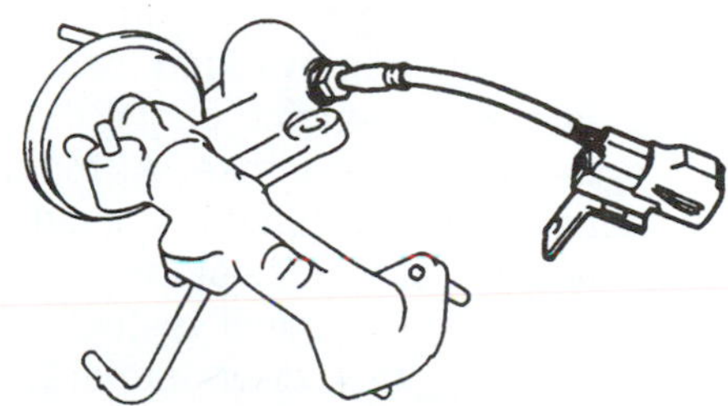

Figure 19–33 EGR gas temperature sensor.

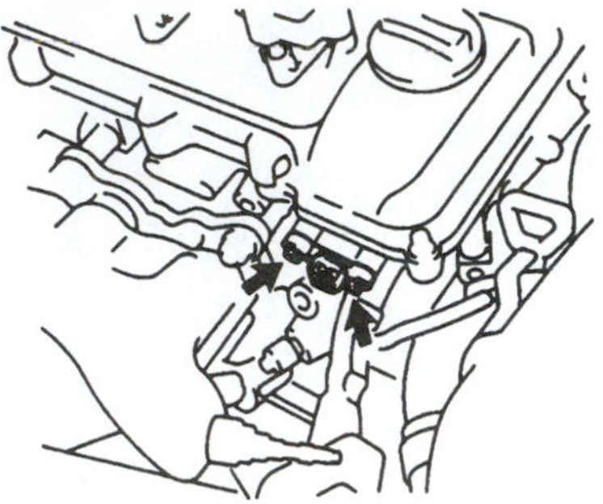

Figure 19–34 Camshaft position sensor.

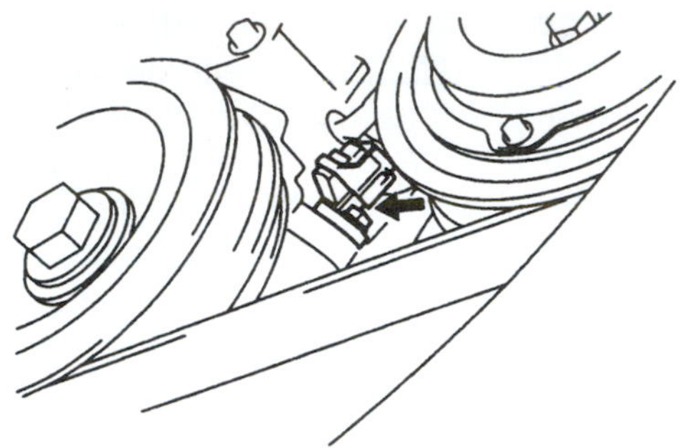

Figure 19–35 Crankshaft position sensor.

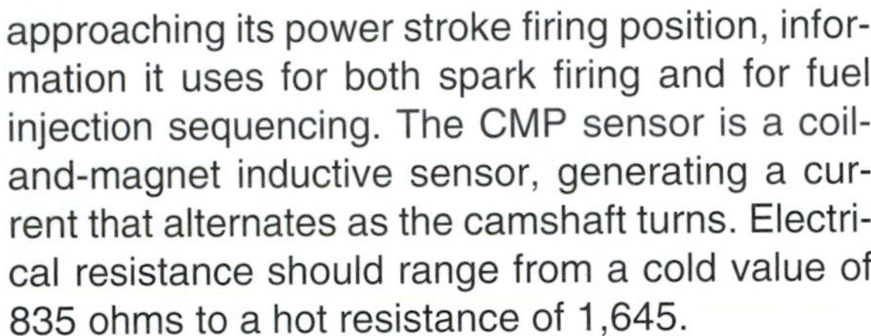

approaching its power stroke firing position, information it uses for both spark firing and for fuel injection sequencing. The CMP sensor is a coil-and-magnet inductive sensor, generating a current that alternates as the camshaft turns. Electrical resistance should range from a cold value of 835 ohms to a hot resistance of 1,645.

Crankshaft Position (CKP) Sensor. The crankshaft position sensor, Figure 19–35, is adjacent to the crankshaft pulley. It provides the computer with information about the position and speed of the crankshaft, information employed for fuel injection, spark timing and idle control as well as functions restricted to on-board diagnostics. Resistance for the magnet-and-coil crankshaft position sensor should range from a cold value of 1,630 ohms to a hot value of 3,225 ohms.

OUTPUTS

Fuel Injectors. The fuel injectors, Figure 19–36, are Bosch-design units, one for each cylinder. The computer sprays them sequentially in the firing order of the engine during the corresponding cylinder's intake stroke. At room temperature, each injector's coil should have approximately 13.8 ohms resistance. Injectors are also subject to mechanical restrictions and blockages, so they should be separately inspected for flow and spray pattern if there is reason to suspect they are not working properly.

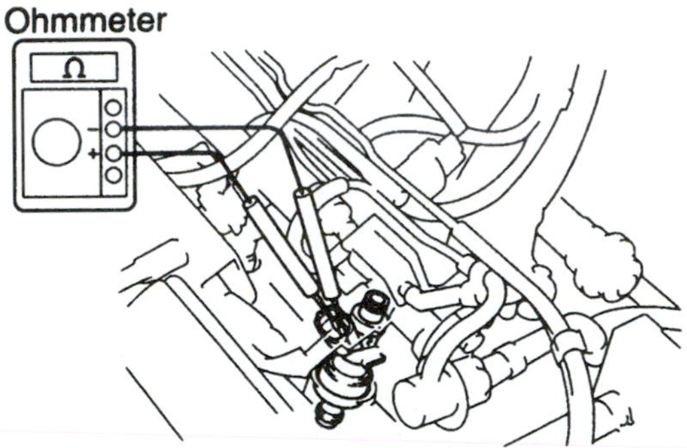

Figure 19–36 Fuel injector resistance check.

Fuel Pump Circuit. The fuel pump on many Toyota systems is under a plate beneath the rear seat, Figure 19–37. The computer energizes a main fuel injection relay, Figure 19–38, which in turn activates the fuel pump. As with most systems, the fuel pump comes on for approximately 1 second after the ignition key is turned on to prime and pressurize the injectors and fuel rail and shuts off if no camshaft position signal is received after the priming second.

The fuel from the pump fills the fuel rail, and the pressure is controlled by the fuel pressure regulator, Figure 19–39. The regulator bolts into the end of the fuel rail and reduces pressure corresponding to intake manifold vacuum, sending unused fuel back to the tank. The fuel pressure regulator, of course, is not directly controlled by the computer, but responds to the pressure from

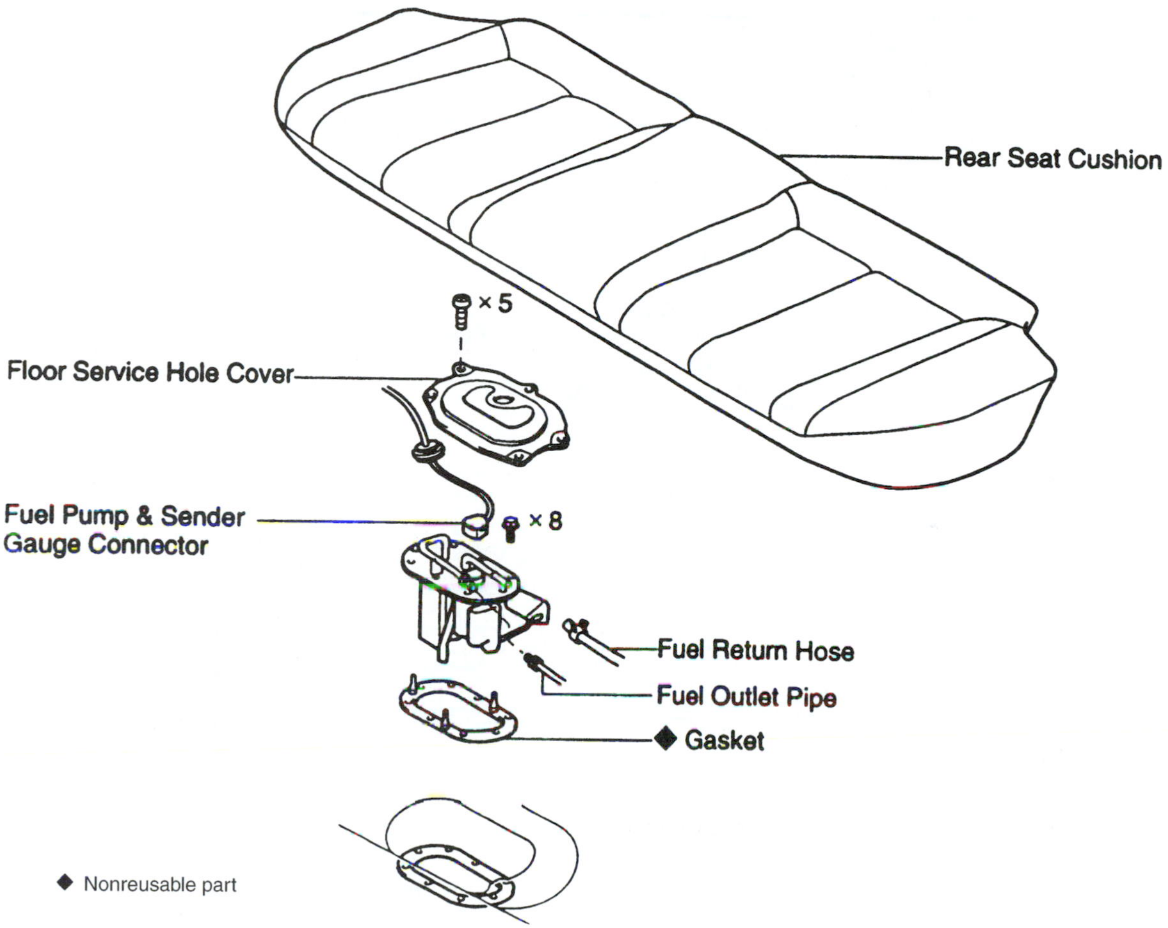

Rear Seat Cushion

× 5

Floor Service Hole Cover

Fuel Pump & Sender Gauge Connector

× 8

Fuel Return Hose

Fuel Outlet Pipe

◆ Gasket

◆ Nonreusable part

Figure 19–37 Fuel pump.

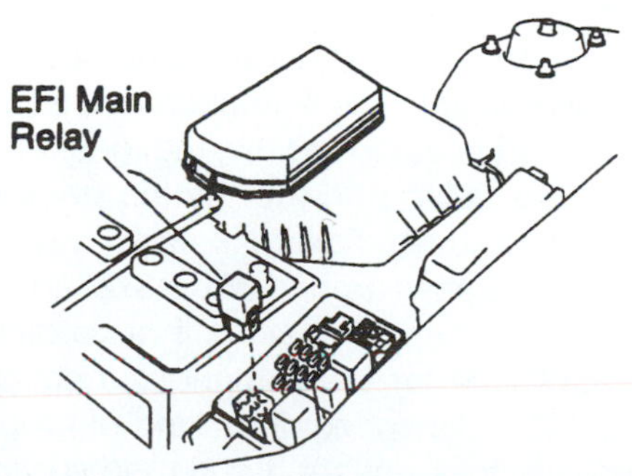

Figure 19–38 EFI (fuel pump) main relay.

the fuel pump and the partial vacuum in the intake manifold.

Toyota has shifted to the returnless fuel system. This change reduces fuel tank emissions caused by the return of fuel from the engine compartment. As in other manufacturers' systems, this returned fuel contains heat, which was absorbed when the fuel was in the engine-mounted fuel injector rail. Fuel in excess of that needed by the engine does not leave the fuel tank. A pressure regulator is located inside the tank above the submersible fuel pump. When the fuel travels outside the tank, it immediately must go to the fuel rail or it must be returned to the fuel tank through the fuel pressure regulator, Figure 19–40.

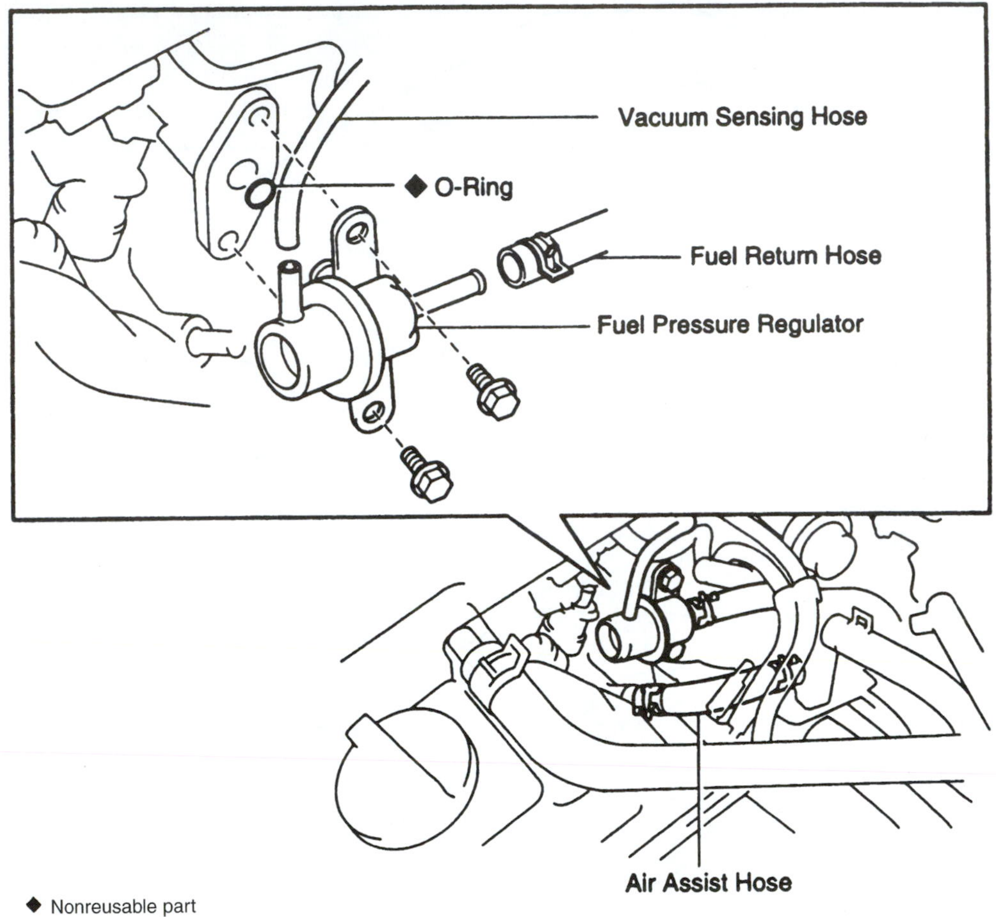

Figure 19–39 Fuel pressure regulator.

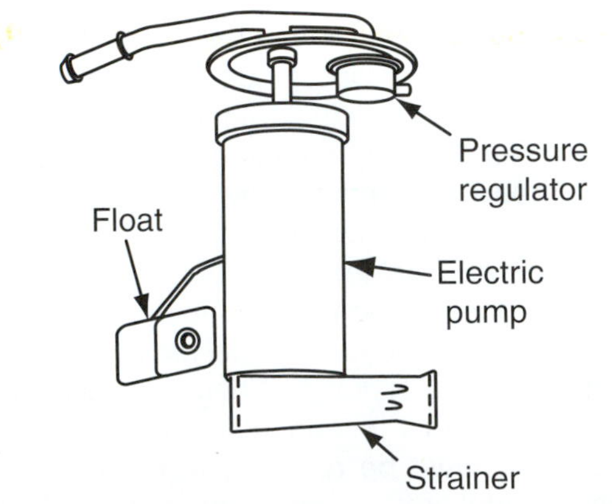

Figure 19–40 Fuel tank module.

Diagnostic & Service Tip

On the Toyota IAC valve and on others of similar design, it is almost always a better service practice to renew the gasket between the valve and the throttle body any time it is removed. Failure to do so can allow "false air" to pass around the gasket and into the intake manifold, making it more difficult or impossible for the computer to correctly control the idle speed. This kind of problem ordinarily does not set a code, so solving it takes more time than the repair is worth.

Idle Air Control System. Figure 19–41 shows the idle air control (IAC) valve. It bolts to the throttle body and provides a bypass channel for air to pass the throttle blade for idle speed control, depending on the engine coolant temperature, electric, and other accessory loads and whether the car is in a drive gear. The IAC valve is easily removed from the throttle body and can be inspected by jumpering the connections to see whether the valve opens and closes properly.

Ignition Coils. The latest V6 engines use a direct ignition system with one coil and driver transistor per pair of cylinders (a waste spark system), Figure 19–42. Ignition coil resistances should be 0.70 to 1.10 ohms through the primary circuit and 10.8 to 17.5 kilohms through the secondary circuit, ranging from cold to hot temperatures. With individual coils of this design, the most time-efficient test for a bad coil is often to merely shift the coil from one cylinder to another to see whether the cylinder miss follows the coil. If it does not, of course, diagnosis should focus on fuel or mechanical aspects of the affected cylinder.

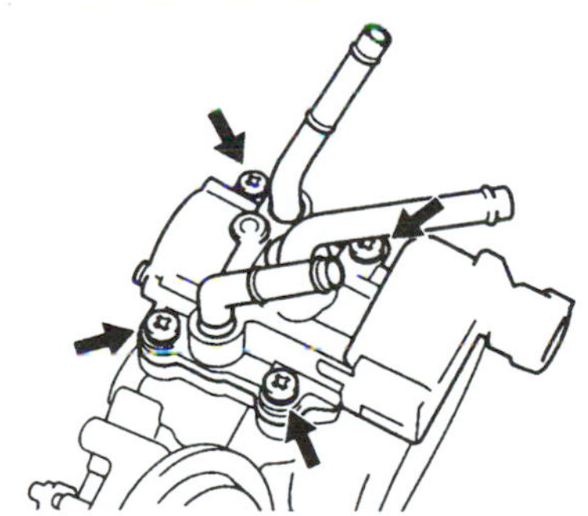

Figure 19–41 Idle air control valve.

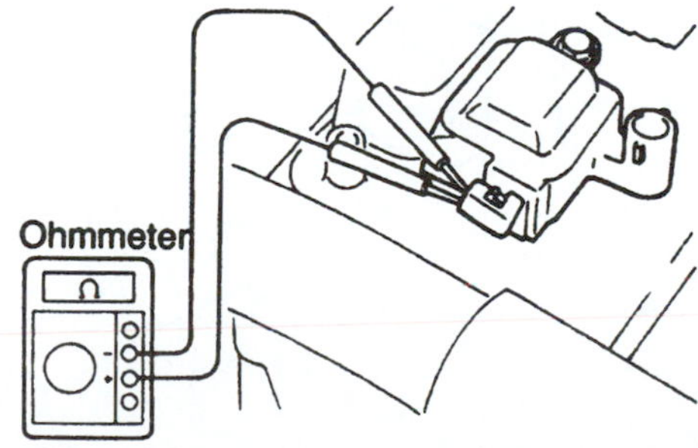

Figure 19–42 Ignition coil resistance check.

Toyota has a waste spark ignition system that looks different from most other waste spark ignition systems. Here is how it works:

This system gets its name from the exhaust cycle. As in the traditional waste spark system, one coil fires two cylinders at the same time. One cylinder is in the compression cycle while the other is in the exhaust cycle. This system uses the three principles of series circuitry. The first principle states that resistance adds together in a series circuit. This means the coil's secondary resistance, the spark plug's internal resistance, the spark plug's gap and the secondary wire resistance all are added together. The second principle states that current is the same in all parts of a series circuit. Therefore, the current flow across both spark plug electrodes in the circuit is the same. The third principle states that voltage drops must equal the power supply. In other words, as electricity travels through each resistance it pays a price. That price is the called voltage drop. The voltage required to power the spark plug during the compression cycle is higher than the voltage required to power the spark plug during the exhaust cycle. However, the coil's electrical potential is so high because of the increased on-time. The ignition reserve is never exceeded.

The waste spark management system has an unusual design. It is an *on-the-coil* system for one side of the engine but not for the other side. The individual coils that control two cylinders are located on one side of the engine. That side uses no wires to deliver the ignition spark. The opposite side of the engine receives ignition spark through traditional high-tension wires, Figure 19–43.

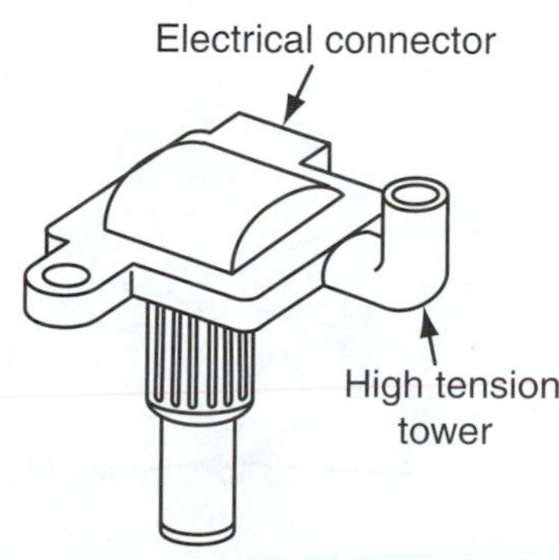

Figure 19–43 Waste spark coil.

Resistances for this coil are:

Primary	.7 to .94 ohms (cold) .85 to 1.10 ohms (hot)	
Secondary	Asian made	10.8 to 14.9 kilohms (cold)
		13.1 to 17.5 kilohms (hot)
	Diamond made	6.8 to 11 kilohms (cold)
		8.6 to 13.7 kilohms (hot)

EGR System. The exhaust gas recirculation system on late model Toyotas uses a vacuum modulator, Figure 19–44, to activate the EGR valve, Figure 19–45. The EGR system should not be in operation at all below an engine coolant temperature of 131°F. The EGR vacuum modulator can be opened and cleaned/inspected for holding vacuum. With the engine warm, check that the modulator passes vacuum along to operate the EGR valve itself and that it opens when actuated. Exact connections to perform these checks can be found in the appropriate shop manual.

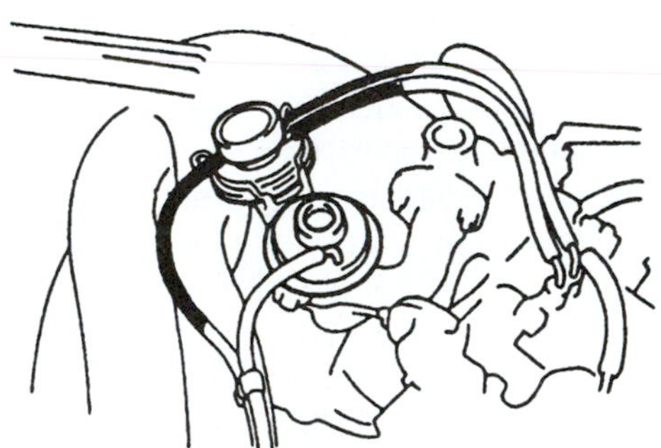

Figure 19–44 EGR vacuum modulator.

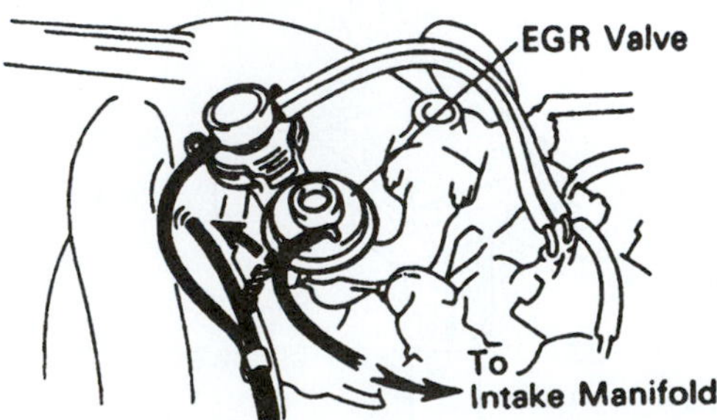

Figure 19–45 EGR valve.

EVAP Control System. As with most manufacturers, Toyota uses a charcoal canister, Figure 19–46, to store fuel vapors and burn them later when the engine is running. With the newer OBD II-compliant vehicles, any detected vapor leaks—including a leaking or missing fuel filler cap—can set a code and turn on the check engine light. While most manufacturers placed the pressure sensor in the top of the fuel tank, Toyota has chosen to locate the pressure sensor at the vapor canister and associated hoses, Figure 19–47.

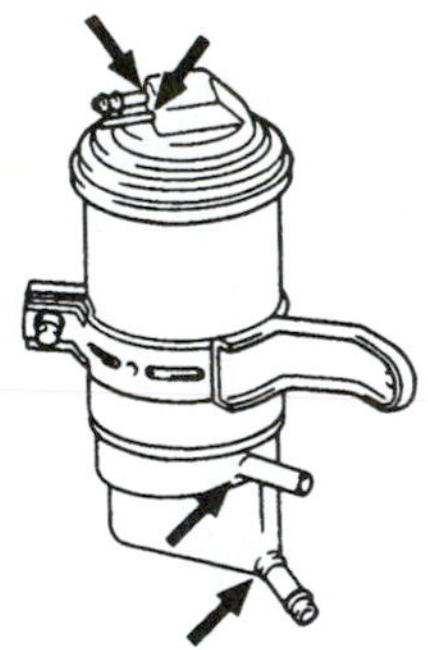

Figure 19–46 EVAP system charcoal canister.

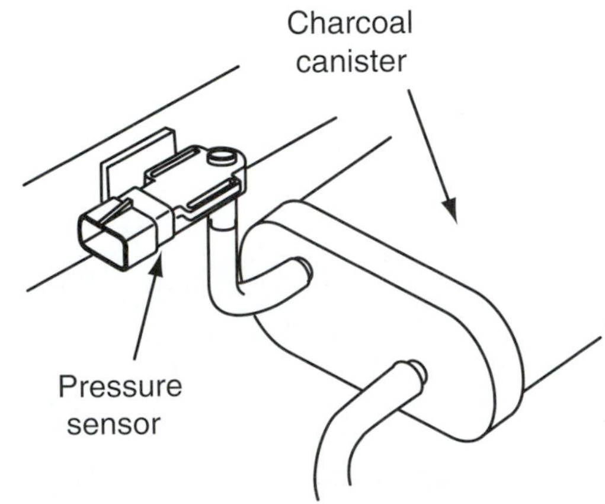

Figure 19–47 EVAP canister.

Variable Valve Timing with intelligence (VVT-i)

Beginning in 1998, a Toyota-badged vehicle was equipped with an engine that has variable valve timing. The recipient of this technology—the Supra and its normally aspirated 3.0-liter, dual overhead camshaft, 24 valve, in-line 6. (The Supra's twin turbocharger-fitted version of this 3.0-liter in-line 6 retains fixed valve timing for 1998, but still manages to develop 320 horsepower and 315 foot-pounds of torque). The Toyota system, named Variable Valve Timing with intelligence or VVT-i, was developed in response to ever more strict emission requirements, but horsepower and torque increase for 1998 too. The 1998 VVT-i equipped engine develops 225 horsepower (5 more than 1997's engine) and 220 foot-pounds of torque at 4,000 rpm (10 foot-pounds more than the previous year, and it peaks 800 rpm sooner).

In the VVT-i system the engine management computer looks at a variety of inputs including coolant temperature, engine speed and engine load. The computer then controls valve overlap (the period of time during which both the intake valve and the exhaust valve are open) by advancing or retarding the intake camshaft. The VVT-i system control devices are an Oil Control Valve (OCV) and a special intake camshaft timing pulley (VVT pulley), Figure 19–48. Changing the intake camshaft "timing" while the engine is operating allows all of the typical factors that must be juggled during camshaft design—low RPM torque, high RPM power, engine smoothness, emissions, and fuel efficiency—to be optimized. A unique feature of the Toyota VVT-i system is that camshaft timing is continuously adjustable; competitive systems allow only two or three distinct camshaft "settings" that are strictly RPM dependent.

The VVT-i sprocket used in the system has a piston with helical splines cut into both its inner and outer surfaces, Figure 19–49. The inner surface meshes with a splined gear attached to the end of the intake camshaft, while the outer surface meshes with a splined cup attached to the camshaft

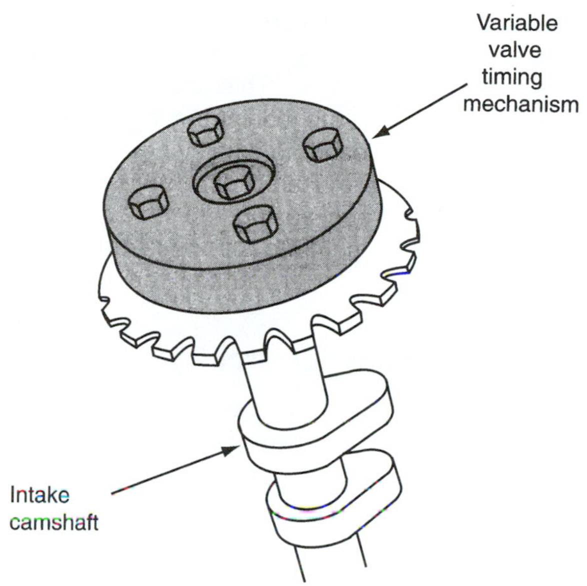

Figure 19–48 Intake camshaft and variable valve timing (VVT) unit.

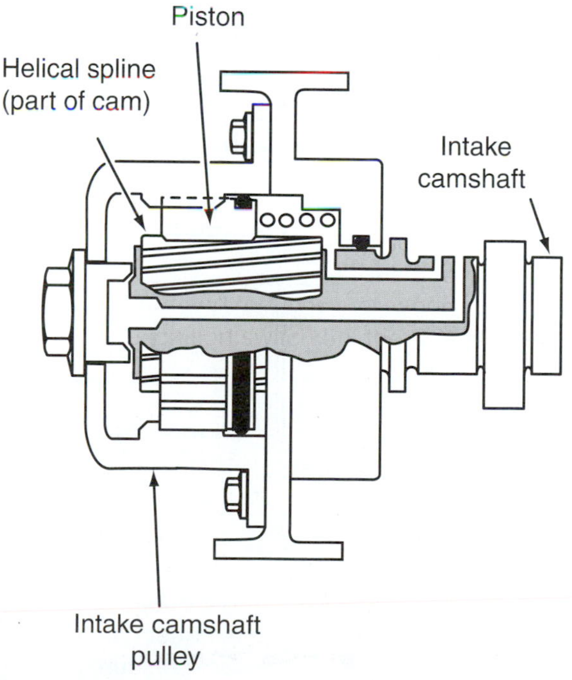

Figure 19–49 VVT-i camshaft timing pulley.

sprocket. Pressurized engine oil is applied to one side of the piston through the OCV, causing the piston to move axially. This axial movement changes the drive sprocket position, relative to the camshaft. The engine management computer modulates the OCV to control camshaft timing, and monitors signals from the camshaft and crankshaft position sensors to determine actual camshaft position.

Two-Way Exhaust Control

Toyota recently introduced a method to control exhaust system backpressure. Exhaust gasses are routed through the entire pathway of the muffler when the engine is at idle. This quiets exhaust noise. When the engine operates at higher speeds, the muffler provides a shorter route through a spring-loaded door for the exhaust gasses to reduce system pressure and improve engine performance. This route, shown in Figure 19–50, is both shorter and noisier.

Supplemental Restraint System (SRS)

The 2000 model year Toyota integrates air bag deployment into the fuel pump drive program. If either the front or side air bag is deployed, the electronic control module stops grounding the fuel pump relay. Figure 19–51 shows that this opens the fuel pump feed circuit and stops the fuel pump.

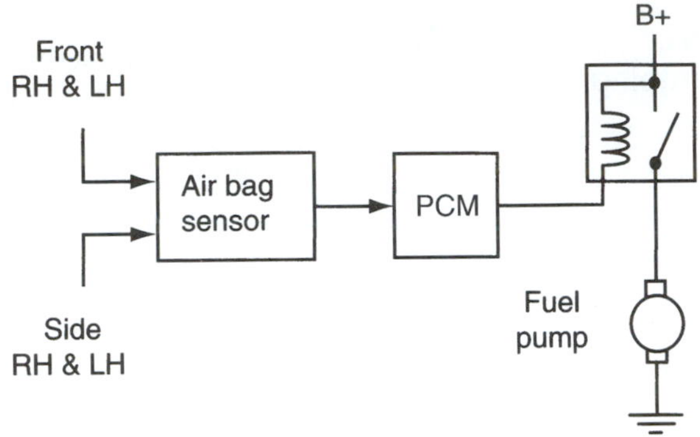

Figure 19–51 In case of an air bag deployment, the electric fuel pump is shut off.

Compressed Natural Gas (CNG)

The 2000 Toyota Camry introduces a natural gas version 5S-FNE-CNG engine. The CNG engine has a new fuel rail design, shown in Figure 19–52, that has an electric fuel shutoff valve at one end, a fuel temperature sender midway down the rail and a fuel pressure sensor and discharge port at the other end.

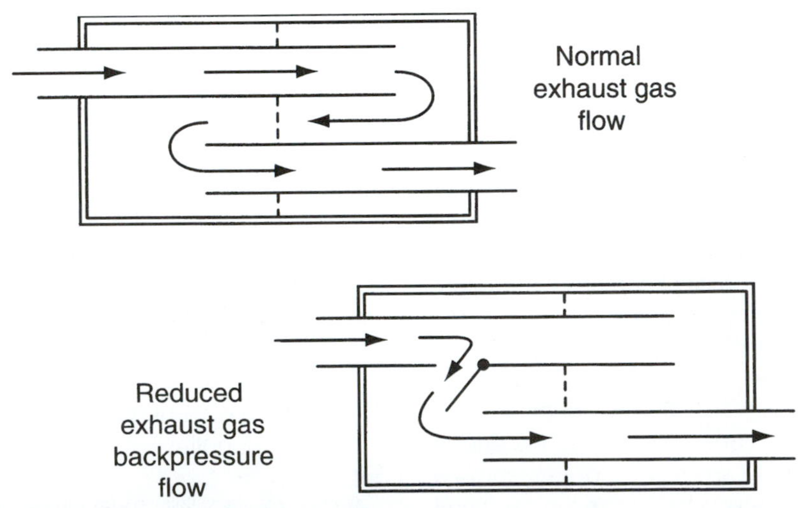

Figure 19–50 Typical two-way exhaust control muffler.

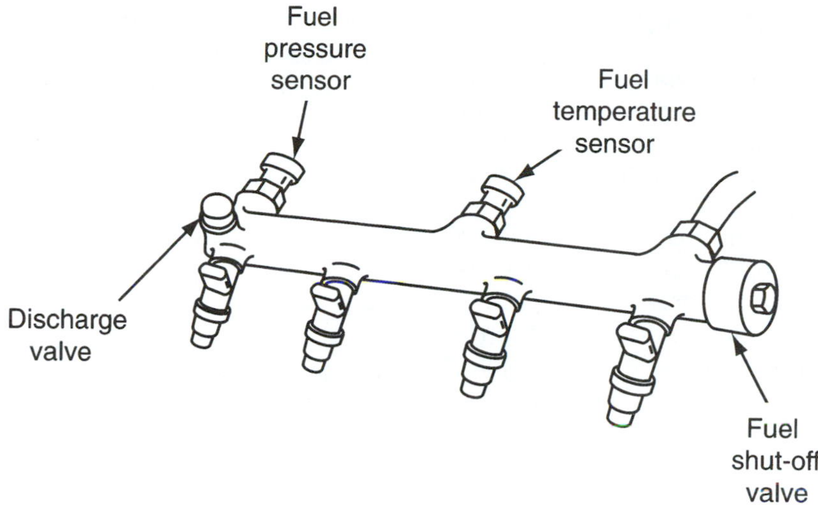

Figure 19–52 The fuel rail used on the compressed natural gas (CNG) vehicle contains a pressure sensor, temperature sensor, shutoff valve and a gauge test point.

The additional hoses, regulators, solenoids and gas shutoff valve are not evident on the Camry until the vehicle is elevated. A large diameter pressurized fuel tank is located at the rear of the vehicle. The fuel tank consumes most of the trunk space.

Active Control Engine Mount (ACEM)

The Avalon V6 engine has a new active oil-filled engine mount. The mount contains a chamber divided by a rubber diaphragm. The chamber, shown in Figure 19–53, has computer-controlled engine vacuum on one side of the diaphragm; the other side of the diaphragm contains damping oil that is used to lessen engine vibration at idle.

At idle or at engine speeds of less than 900 rpm, the computer sends pulses to an electrically operated vacuum control valve. These signals are synchronized to engine rpm. The vacuum travels to the lower side of the rubber diaphragm inside the engine mount. The diaphragm vibrates and transfers that vibration through the damping oil into the rubber mount. The mount cancels out unwanted engine vibration. Figure 19–54 shows the components of the ACEM.

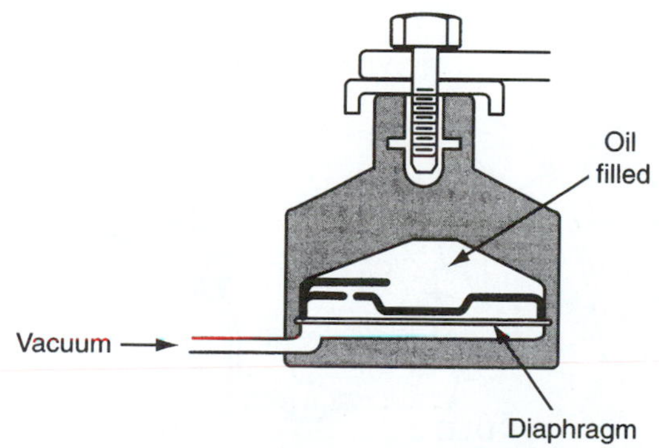

Figure 19–53 A cutaway active engine control mount (AECM).

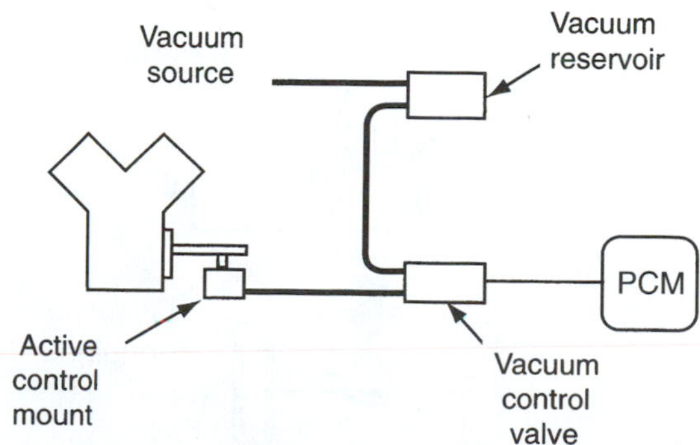

Figure 19–54 Electrical circuit of active engine control mount (AECM).

Toyota Prius—Hybrid Vehicle

The Toyota Prius is a hybrid vehicle. It runs on both gasoline and electrical power. The Prius uses a 1.5L 4-valve DOHC gasoline-powered engine to drive both the vehicle and a high-voltage generator. The generator AC output signal runs through an inverter. The inverter converts AC voltage to DC voltage. The DC voltage charges the 274-volt nickel metal hydride battery pack.

At times, the 274-volt electrical charge can be used to drive a 44-horsepower electric motor. If the motor is powered by electricity, the gasoline engine may or may not run. The inverter converts the DC battery power to the AC energy the electric motor needs. The power flow of the Prius is shown in Figure 19–55.

The Toyota Prius gasoline engine does not have a cranking motor. Instead, the Prius engine uses the electric drive motor to start the gasoline engine. This vehicle has high-voltage circuits that must be deactivated before the technician can work on it. See the manufacturer's instructions for specific deactivation procedures.

✔ SYSTEM DIAGNOSIS AND SERVICE

OBD II Diagnostics. With the latest model Toyotas, the OBD II scanners that can be used on other vehicles can be used on these as well, using the standardized Diagnostic Link Connector in the lower left side of the instrument panel, Figure 19–56. Each vehicle also employs a standard set of diagnostic trouble codes (DTCs) that are common to all manufacturers.

Toyotas use a two-trip fault recording program, as do many other manufacturers. By recording a trouble code in memory the first time it occurs, but not turning on the malfunction indicator light (check engine light) until it recurs in the next successive trip, this reduces the chances of having an excessive number of problems that cause motorists concern. Two-trip fault recording programs also retain the sensitivity to system malfunctions needed to keep the vehicle's emissions performance at its desired state. Almost all component or signal failures that can affect the vehicle's emissions or fuel economy performance will be reflected in the information available on the scan tool. Certain crucial component signals, which would either cause damage to the system or prevent the vehicle from operating, trigger a trouble code on their first occasion.

If the trouble code indicates a problem on a given circuit, remember that it is the entire circuit that has failed, not just the specific component. While finding and checking the component itself is a reasonable first step, Figure 19–57, the problem may actually be in the harness or in the computer itself.

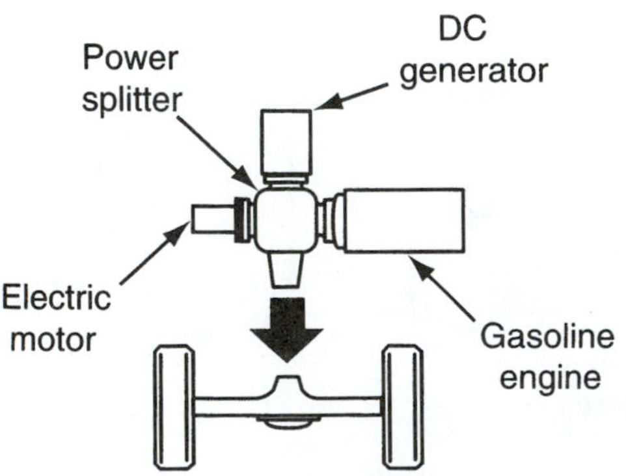

Figure 19–55 Power flow circuit of Prius hybrid vehicle.

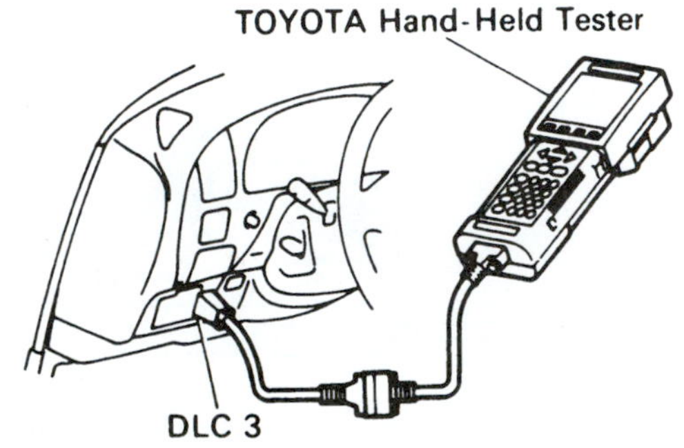

Figure 19–56 OBD II diagnosis with scan tool.

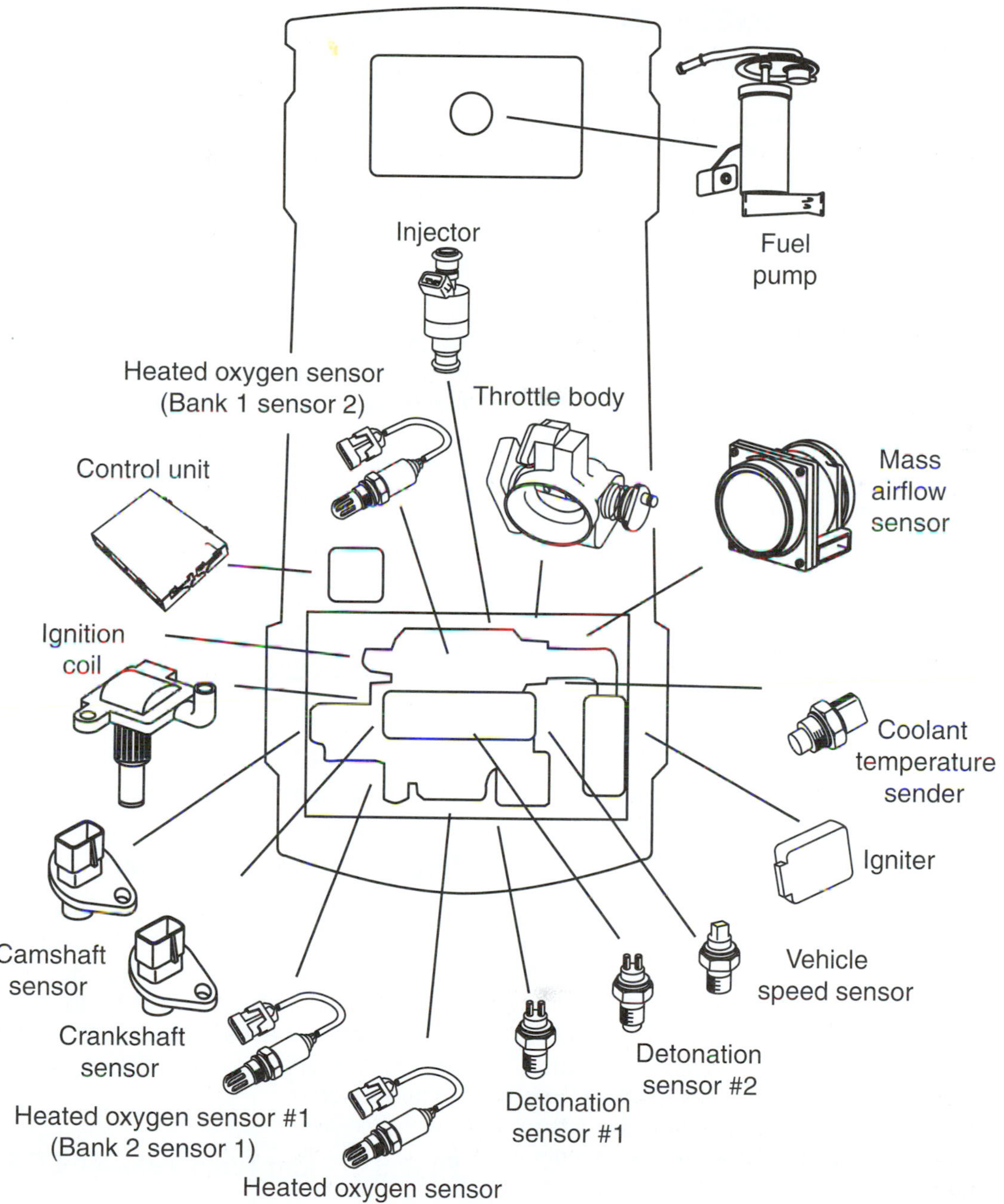

Figure 19–57 Component locations.

HONDA: PROGRAMMED FUEL INJECTION (PGM-FI)

The Honda PGM-FI system is a computerized control system that provides optimum control of the engine's fuel, spark and emission control systems and also provides control of the automatic transaxle. It was introduced in 1985 and has been updated according to OBD I (1988) and OBD II (1996) standards and, since the mid-1990s, allows for bidirectional communication between the PCM and a scan tool.

INPUTS

Manifold Absolute Pressure (MAP) Sensor

On Honda vehicles, an analog MAP sensor is used to measure engine load. On newer models, the MAP sensor is also used to update the barometric pressure values in the PCM's memory whenever the engine is started and during wide-open throttle operation. The MAP sensor receives a reference voltage from the PCM and returns an analog voltage that equates to manifold pressure. As intake manifold pressure is decreased (as vacuum is applied), the MAP sensor voltage is decreased also, similar to the ones found on GM vehicles. The PCM uses the resulting signal, along with other sensor signals and programmed information, to calculate how much air is entering the engine. This calculation is known as the speed density formula. See Chapter 2 for additional information on the speed density formula. Depending on the year and model, Honda MAP sensors may be found in a variety of locations. Newer ones are located attached to the intake manifold to eliminate the need of an attaching vacuum hose.

Barometric Pressure (Baro) Sensor

Honda has traditionally used a full time barometric pressure (Baro) sensor, known prior to OBD II standardization as the pressure absolute (PA) sensor. This sensor is an analog sensor, returning an analog DC voltage to the PCM that is indicative of sensed barometric pressure. This sensor's function has been taken over by the MAP sensor on newer applications, thus eliminating this sensor.

Heated Oxygen Sensors

OBD II applications have a minimum of two heated oxygen sensors. A primary HO_2S is mounted ahead of the catalytic converter to help the PCM control the air/fuel mixture while in closed loop. A secondary HO_2S mounted after the catalytic converter allows the PCM to monitor converter efficiency. Hondas that are certified as low emission vehicles (LEV) or ultralow emission vehicles (ULEV) as well as those that are equipped with a continuously variable transmission (CVT) use a primary oxygen sensor that is able to respond to a wider range of air/fuel ratios than are those on other vehicles. This increases the PCM's accuracy potential when controlling the air/fuel mixture on these vehicles. Therefore, when replacing an oxygen sensor, be careful to use the correct part number.

Throttle Position (TP) Sensor

The throttle position (TP) sensor is a rotary potentiometer that informs the PCM of driver demand. It is found at the throttle body on one end of the throttle shaft and is similar to those found on most other makes. Prior to OBD II standardization, this sensor was referred to as the throttle angle sensor.

Engine Coolant Temperature (ECT) Sensor

The ECT sensor is a negative temperature coefficient (NTC) thermistor and changes its resistance in response to changes in engine temperature. As the engine's coolant warms up, the ECT sensor's resistance decreases, resulting in a decrease of the voltage drop signal that is measured across the sensor by the PCM. Prior to OBD II standardization, this sensor was referred to as the temperature water (TW) sensor.

Intake Air Temperature (IAT) Sensor

The IAT sensor is an NTC thermistor and changes its resistance in response to changes in the temperature of the air entering the intake manifold. As the temperature of the intake air increases, the IAT sensor's resistance decreases, resulting in a decrease of the voltage drop signal that is measured across the sensor by the PCM, similar to the ECT sensor. Prior to OBD II standardization, this sensor was referred to as the temperature air (TA) sensor.

Knock Sensor

The knock sensor, when used, is a piezoelectric crystal used to sense spark knock in the engine, similar to most other makes. The PCM reacts to input from the knock sensor by adjusting ignition timing until the knock just disappears.

Exhaust Gas Recirculation Valve Lift Sensor

The EGR valve lift sensor is a linear potentiometer that is mounted on top of the EGR valve and measures the position of the EGR valve. This sensor equates to the EGR valve position sensor found on other makes such as Ford products with the EEC III and EEC IV systems.

Ignition Pickups—Early Distributor Applications

Crank Angle Sensor. The crank angle sensor consists of two magnetic pickups both contained within the ignition system's distributor housing. Each one monitors a separate reluctor and generates AC voltage pulses. One provides the basic tach reference signal for all cylinders; the other identifies the position of cylinder 1.

Ignition Pickups—Newer Distributor Applications

Crankshaft Position (CKP) Sensor and Top Dead Center (TDC) Sensor. The CKP sensor and the TDC sensor are both permanent magnet sensors that generate AC voltage pulses off of monitored teeth on the crankshaft pulley. The CKP sensor provides the basic tach reference signal and is also used to detect when misfire is present. The TDC sensor is used to identify which piston pair is approaching TDC. The engine will not start if either sensor has failed.

Cylinder Position (CYP) Sensor. The CYP sensor is a permanent magnet pickup that is mounted within the ignition system distributor and equates to the CMP sensor found on other vehicles. It is used for fuel injector sequencing. It produces one AC voltage pulse per distributor/camshaft rotation that identifies the position of cylinder 1. If the CYP sensor were to fail, the PCM would guess at the injector sequence and the engine would continue to run.

Ignition Pickups—Coil-At-Plug Applications

A CKP sensor is used to monitor a 12-tooth reluctor located behind the timing belt drive gear at the crankshaft, Figure 19–58. In addition, two CMP sensors, known as TDC1 and TDC2, are used and are located at the back cover of the cam gear housing, Figure 19–59.

OUTPUTS

Main Relay

The main relay assembly contains two relays. One provides power to the PCM and the fuel injectors whenever the ignition switch is turned to the run position. The other is the fuel pump relay and is under the control of the PCM, Figure 19–60.

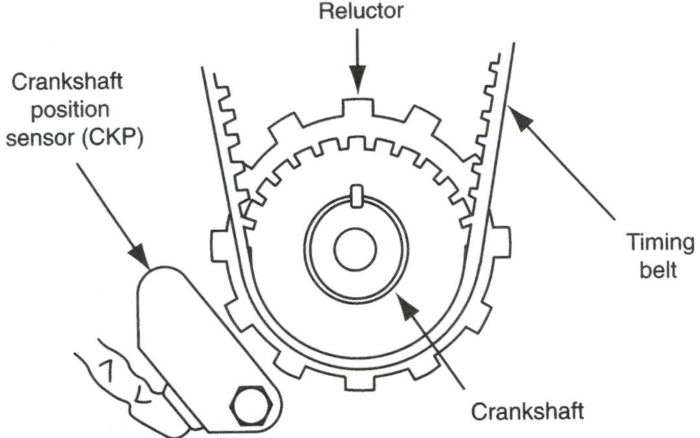

Figure 19–58 Reluctor gear on crankshaft and crankshaft position (CKP) sensor.

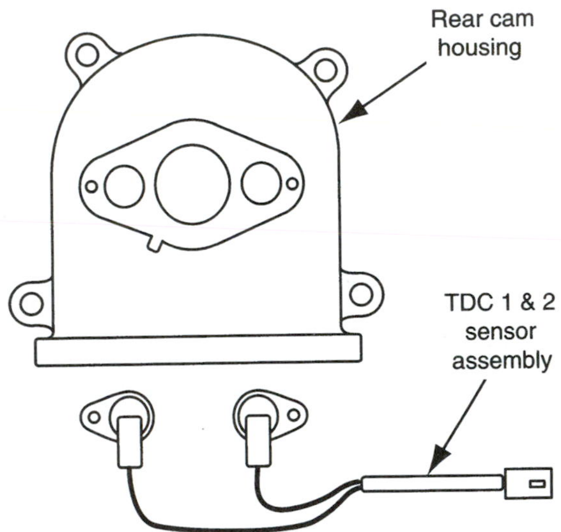

Figure 19–59 Cam housing back cover with TDC1 and TDC2 sensors.

Fuel Injectors

The Honda fuel injectors are low-resistance injectors, typically 1.5 to 3.0 ohms. Whenever the ignition switch is turned to the run position, the main relay provides power to the fuel injectors through external resistors, one per each injector, Figure 19–60. The resistor assembly is typically mounted in front of the left front shock tower.

Fuel Pump Circuit

The PCM energizes the fuel pump relay portion of the main relay assembly in order to run the fuel pump whenever the PCM is receiving a tach reference signal from the ignition system, Figure 19–60. The only exception to this, as is the case with most other makes, is that the PCM energizes the fuel pump relay for 2 seconds when the PCM is initially powered up, whether or not there is a tach reference signal present. This is used to top off the residual fuel rail pressure in order to provide an easy initial start.

Fuel Injection Air (FIA) Control Valve System

A fuel injection air (FIA) control valve system is used on late-model Honda Accord 3.0L and Odyssey applications. The FIA control valve allows filtered air from the air cleaner to be directed into the path of the fuel injectors' spray patterns in order to increase fuel atomization, Figure 19–61. The FIA valve is opened by a ported vacuum signal in order to improve atomization just off idle.

Idle Air Control System

By 1988, Honda PGM-FI applications incorporated an IAC valve. The IAC valve allowed the

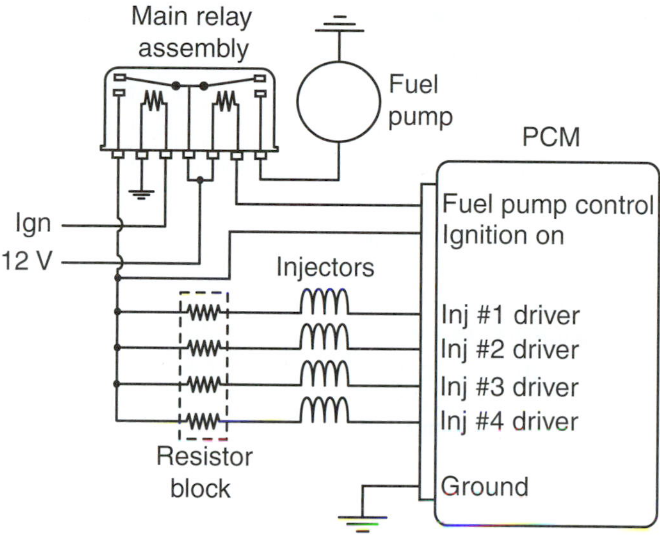

Figure 19–60 Honda main relay circuit.

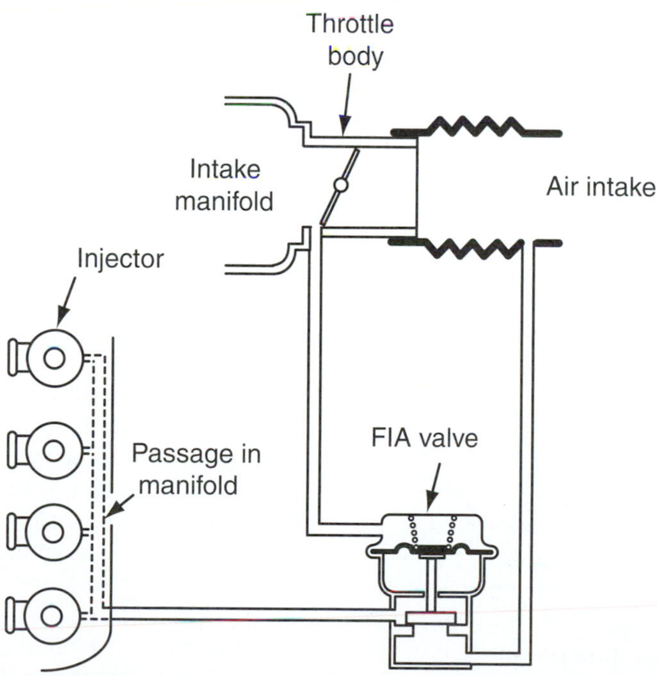

Figure 19–61 Fuel injection air (FIA) control valve system.

PCM to control idle speed by controlling the amount of throttle bypass air.

Ignition Coils

Although Honda applications have traditionally used a single ignition coil and a distributor, several newer Honda applications such as the Honda Accord 6-cylinder engine and the 3.2L Honda Passport engine now use a coil-at-plug design, similar to the coil-on-plug (COP) design found on many other vehicles. In this design, each cylinder has its own ignition coil placed directly over the spark plug so as to eliminate the secondary plug wires, Figure 19–62. The ignition coils in Honda's coil-at-plug system cannot be tested in the traditional method using an ohmmeter because each ignition coil contains an electronic integrated circuit (IC) chip, Figure 19–63. The secondary winding also contains a diode. If the coil has proper power and ground and the PCM's trigger signal is present, but the coil does not produce spark, then the coil is at fault.

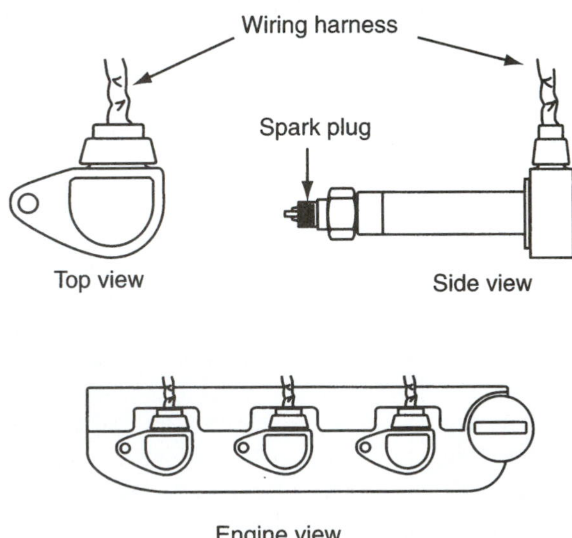

Figure 19–62 Top, side and engine views of ignition coil used in at-plug system.

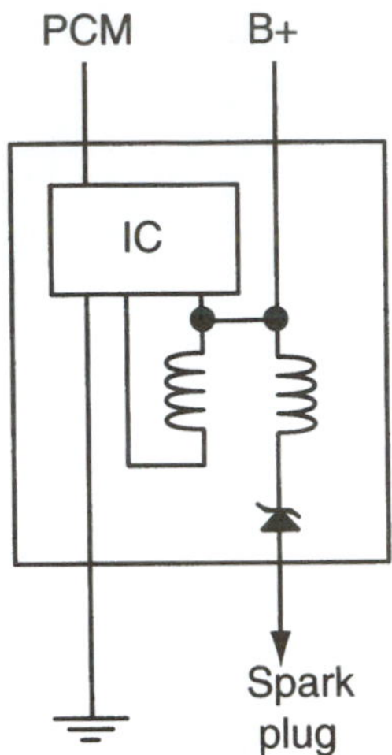

Figure 19–63 Internal schematic of ignition coil used with at-plug system.

EGR System

Honda vehicles that utilize an EGR valve use a closed loop EGR design. A constant vacuum control (CVC) valve receives manifold vacuum and applies between 6 inches of mercury and 10 inches of mercury of vacuum to the EGR valve—more than enough vacuum to hold the EGR wide open. The PCM then controls an EGR control solenoid that is used to vent some of the applied vacuum to atmosphere. Ultimately, the PCM controls EGR flow rates while monitoring the EGR valve lift sensor (discussed in the Inputs portion).

EVAP Control System

Honda's evaporative control system allows the PCM to monitor fuel tank pressure through a pressure sensor, Figure 19–64. When fuel tank pressure exceeds a predetermined value, a two-way valve opens to allow the fuel tank to vent to atmosphere by way of the charcoal canister. The PCM also controls canister purging through a purge solenoid.

Additionally, Honda uses an On-board Refueling Vapor Recovery (ORVR) system. During normal vehicle operation when the fuel cap is

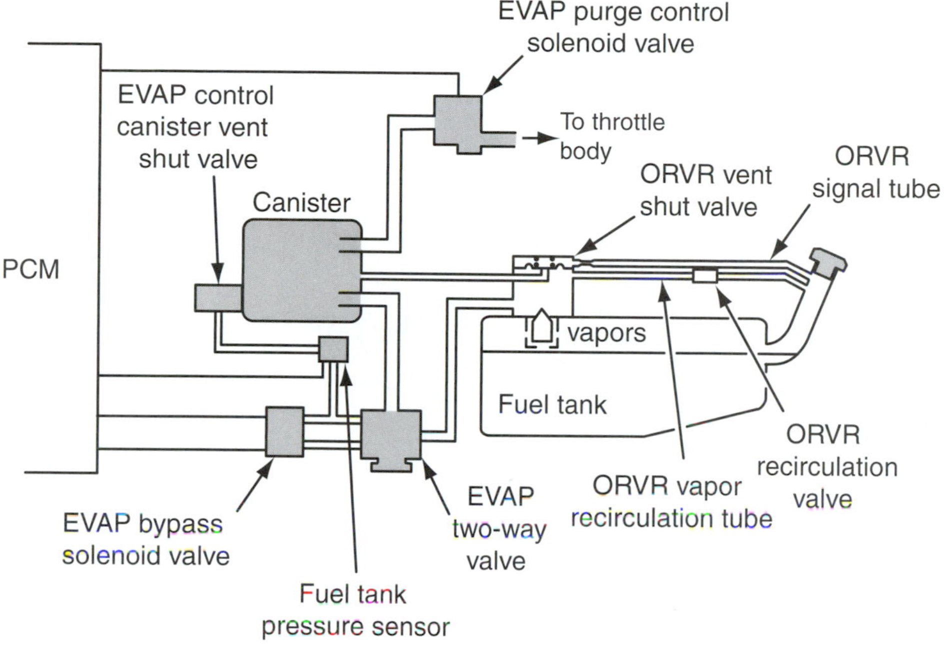

Figure 19–64 On-board Refueling Vapor Recovery (ORVR) and evaporative system

installed, pressure in the filler neck area is equal to the pressure within the main body of the fuel tank. However, once the fuel cap is removed, atmospheric pressure is applied to the area in the neck of the gas tank and is monitored by an ORVR signal tube that controls a normally closed ORVR Vent Shut Valve, Figure 19–64. Meanwhile, pressure within the main body of the fuel tank rises due to the fuel that is entering the tank. The ORVR Vent Shut Valve monitors this pressure as well. The pressure differential created opens the ORVR Vent Shut Valve and allows vapors to be routed to the charcoal canister during refueling.

Variable Valve Timing

Honda applications use two different Variable Valve Timing and Valve Lift Electronic Control systems known as VTEC and VTEC-E. These systems are able to vary intake valve timing, lift and duration.

The VTEC system uses two intake valves per cylinder that are controlled by three rocker arms, Figure 19–65. At low rpms, the rocker arms operate independently. The primary rocker arm is used to open the primary intake valve and is controlled by a low lift, short duration lobe of the intake camshaft. The secondary rocker arm is controlled by an ultralow lift, short duration lobe that opens the secondary intake valve just enough to prevent fuel accumulation in the area of the intake port. A mid (connecting) rocker arm is located between the primary and secondary rocker arms and is controlled by a high lift, long duration lobe. However, at low rpms, the midrocker arm does not mechanically control any valve and simply operates against the tension of a “lost motion” spring assembly. The primary and midrocker arms each contain a synchronizing piston, but at low engine speeds this piston is fully contained within the respective rocker arm due to a return spring and stopper pin assembly located in the secondary piston.

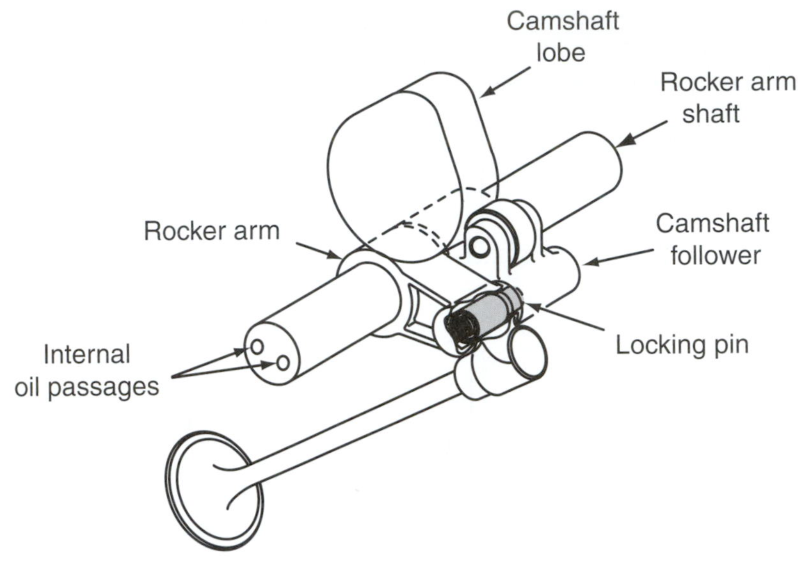

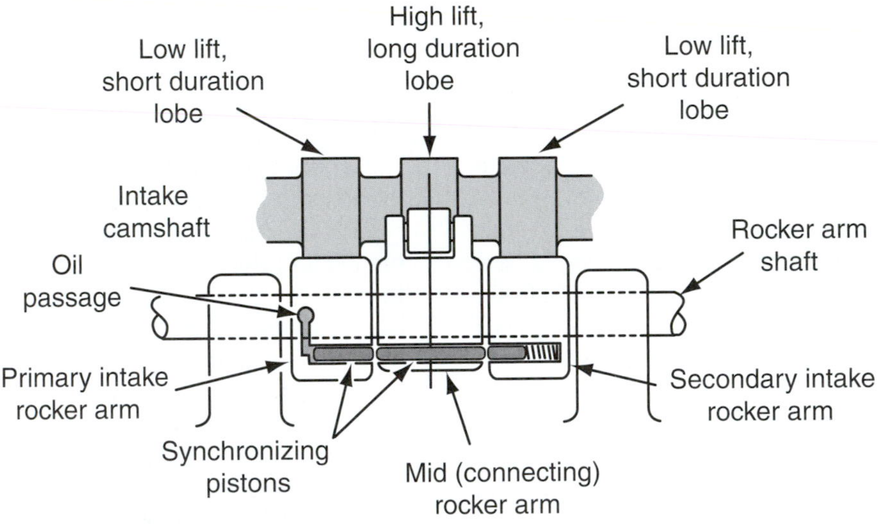

Figure 19–65 VTEC system rocker arms operating independently.

At high engine rpms, the PCM energizes a solenoid that moves a spool valve. The spool valve then directs oil pressure into an oil passage within the primary rocker arm. This oil pressure then moves both synchronizing pistons toward the spring-loaded stopper pin in the secondary rocker arm, thus locking all three rockers together mechanically, Figure 19–66. This causes both the primary and secondary intake valves to be controlled according to the midrocker arm's high lift, long duration cam lobe. Once the engine speed slows down again, the PCM de-energizes the

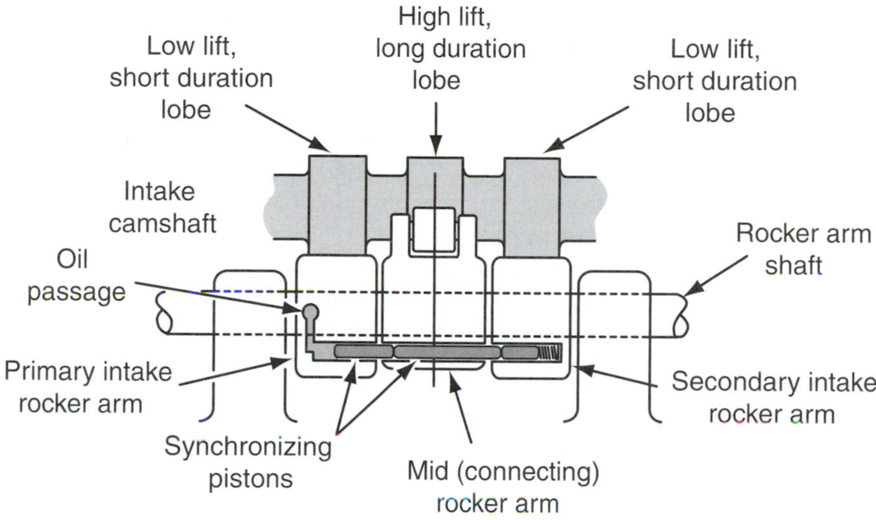

Figure 19–66 VTEC system rocker arms mechanically linked together.

solenoid and allows the applied oil pressure to be vented from the primary rocker arm. This allows the return spring to move both synchronizing pistons back into their respective rocker arms, again allowing independent movement of each rocker arm.

The PCM considers four inputs before energizing the VTEC solenoid. Engine rpm must be between 2,300 and 3,200 rpm on the four-cylinder application, must exceed 3,500 rpm on 3.0L applications, and must exceed 4,900 rpm on certain Prelude applications. Vehicle speed must exceed 19 mph, coolant temperature must exceed 140°F (60°C), and the engine must be under appropriate load.

A VTEC-E is used on some applications and is similar to the VTEC system except that a mid (connecting) rocker arm is not used. A single synchronizing piston within the primary rocker arm causes the primary and secondary rocker arms to be mechanically linked at high engine speeds. Ultimately, the VTEC and VTEC-E systems allow the PCM to provide good low-end performance while still achieving high rpm performance.

Electronic Load Detection (ELD)

Alternator output on Hondas is controlled by the PCM. The PCM reduces field current, which, in turn, reduces alternator output. Output current is based on battery charge and electrical system loads. When the battery is low, or when high electrical demands are made by various systems in the car, the PCM signals the alternator to increase output; when the battery is charged up and there are few electrical system demands, the PCM reduces alternator output. Reduced alternator load results in better fuel economy.

Electronic Throttle Control

The Honda Passport electronic throttle control system is similar to that used by several other manufacturers. This system uses a three-channel accelerator pedal position (APP) sensor to initiate a different resistance signal for each channel. The electronic throttle control system also uses a throttle position feedback sensor (TPS). When the PCM receives the APP signal, it issues a command signal to the electric motor attached to the throttle blade. For more details on how this system works, see Generation III V8s in Chapter 12.

Insight

The Honda Insight demonstrates the possible technology of future drivetrains. In addition to the traditional gasoline-powered engine, the Honda Insight uses electric power to supplement the engine. This combination of gasoline and electric power delivers cleaner emissions and higher mileage.

The Honda Insight has a stator assembly that produces extremely high voltage attached to the rear of the engine. Voltage from the stator winding feeds a battery module in the rear of the vehicle. Figure 19–67 shows a permanent magnet rotor is attached to the rear of the crankshaft. The rotor is powered by the engine when the vehicle is driven down the highway. The rotor and stator work the same way as the alternator does.

Whenever the engine operates at minimal load, such as at steady cruise speed, the battery feeds power back to the stator. This creates strong magnetic fields that attract or repel the permanent magnet rotor. The rotor rotates and drives the powertrain. The electric motor drive does not require driver selection; that is, the driver does not have to choose to have the vehicle powered by electricity. The computer decides when to operate by gasoline power, electric power, or a combination of the two. This combination of gasoline and electric power is fuel efficient; it promises a mileage rate of up to 70 miles per gallon. Further, the Honda Insight has cleaner emissions because when it operates on electricity, it does not burn gasoline and produces no tailpipe emissions.

The technician must be extremely careful when he or she services the Honda Insight because of the high voltage in the electric propulsion system. Honda warns technicians to wear insulated gloves when they service this system. Honda also warns technicians that the rotor assembly contains very strong permanent magnets; people with heart pacemakers should not handle this device. Keep the rotor away from credit and identification cards containing magnetic strips because it will erase all the information encoded in them. Obviously, the rotor must be kept away from the shop computer as well.

To be totally safe while working in the engine compartment, the technician must turn off the high-voltage system, shown in Figure 19–68, and wait for 5 minutes to allow the condensers to discharge. Follow these steps to turn off the high-voltage system:

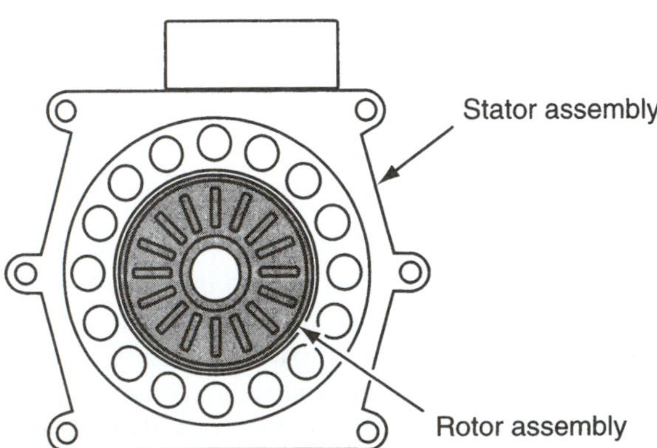

Figure 19–67 Hybrid vehicle rotor and stator assemblies.

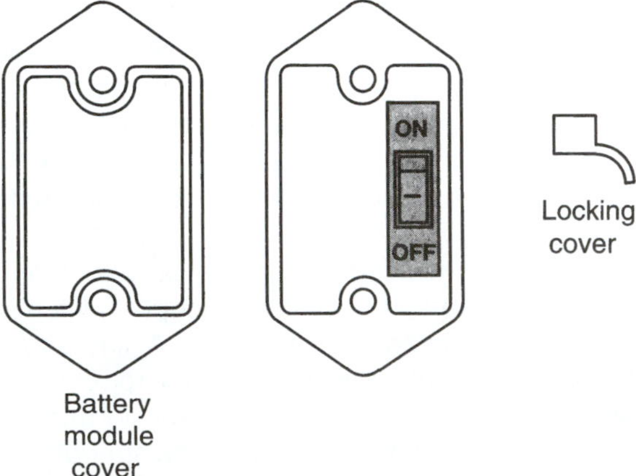

Figure 19–68 The power module safety switch is located under the floor mat in the trunk.

1. Turn the ignition off.
2. Lift the floor mat in the trunk.
3. Remove the battery module switch cover.
4. Remove the locking cover.
5. Turn the battery module off.
6. Reinstall the locking cover in the reverse direction.
7. Wait 5 minutes for the condensers in the system to discharge.

✔ SYSTEM DIAGNOSIS AND SERVICE

Prior to OBD II standardization, Honda PGM-FI systems were able to communicate memory DTCs to the technician through either one or four LEDs that were located on the PCM. With both LED designs, there was no need to *trigger* DTC output as with most other makes. Simply turning on the ignition switch to power up the PCM was all that was required on these early systems in order to read DTC output. When one LED was used, turning on the ignition switch would result in a single flash that did not repeat if codes were not stored in the PCM's memory. If codes were stored in the PCM's memory, this was followed by repetitive flashes of the LED to indicate the stored DTC(s). The total number of flashes represented the DTC. DTCs 1 through 15 were available, with 15 flashes of the LED representing DTC 15. When four LEDs were used, they were numbered from left to right 8, 4, 2 and 1, similar to the base 10 values of binary code columns. The DTC number was interpreted by adding together the values of any illuminated LEDs. For example, if LEDs 8, 2 and 1 were illuminated, the DTC was code 11.

In the early 1990s, Honda began making DTC output available through pulses of the MIL. This version required a jumper connection to be made at a test connector in order to trigger code output. This application allowed the PCM to illuminate the MIL if a fault within the system was sensed.

On OBD II applications, the PCM will illuminate the MIL by providing a ground and set a DTC in an erasable memory if it senses a fault within the system. As with other OBD II vehicles, a scan tool may be connected to the DLC and then used to pull DTCs from the PCM's memory along with freeze frame data. Additionally, a scan tool may also be used to look at data stream values including sensor values and output commands.

SUMMARY

To focus on Asian engine control systems in this chapter, we have studied late model Nissan, Toyota and Honda vehicles.

The chapter considered the different injection patterns (simultaneous and sequential) employed for startup, patterns to avoid the use of a cold-start injector and patterns to optimize emissions during the engine's dirtiest phase: startup cranking. The different fuel enrichment strategies as well as spark advance map alterations were considered. We saw how the self-learning capacities allowed the systems to adjust fuel mixture and spark timing from recorded experience. And we considered the more elaborate fuel canister vapor absorption and storage systems. We learned the use of an EGR temperature sensor to monitor the effectiveness of the bypass valve's circuit. We learned of the interconnections with the transmission.

These systems use a nonwaste spark direct ignition with one coil for each cylinder. And they use the new tandem oxygen sensors to provide information about the effectiveness of the catalytic converter.

We have also seen how each can be diagnosed using standardized OBD II scan tools, as well as noted the more mundane problems that can occur even with high-tech systems such as these.

▲ DIAGNOSTIC EXERCISE

A customer brings in a late-model Nissan with the check engine light on. There are no driveability complaints, either from the customer or noticeable to a technician on a test drive. When checked with the scan tool at the OBD II diagnostic link connector, the computer indicates a problem in the charcoal canister purge system. What is the most likely cause of this problem? How long should the repair take if the most likely cause is the real one?

REVIEW QUESTIONS

1. *Technician A* says that a throttle position sensor on a Nissan ECCS application returns a voltage back to the PCM from about 0.5 volts at idle to as high as 4.6 volts at WOT.
   Technician B says that some Nissan ECCS applications use an absolute pressure sensor to detect both ambient pressure and intake manifold pressure.
   Who is correct?
   A. A only
   B. B only
   C. Both A and B
   D. Neither A nor B
2. How much voltage should Nissan's analog mass airflow (MAF) sensor return to the PCM at idle?
   A. 0.4 to 0.6 volts
   B. 1.0 to 1.7 volts
   C. 2.0 to 2.5 volts
   D. About 4.0 volts
3. What method is used on Nissan ECCS applications to monitor EGR flow?
   A. A pressure sensor is used to monitor the exhaust pressure just before the EGR valve but after a restriction placed in the exhaust tube.
   B. A pressure sensor is used to monitor intake manifold pressure while opening and closing the EGR valve during idle conditions.
   C. A temperature sensor is placed into the EGR passage.
   D. The oxygen sensor is monitored while opening and closing the EGR valve during idle conditions.
4. All *except* which of the following can result from a cylinder misfire?
   A. Production of too much CO_2 in the engine
   B. Overheating of the catalytic converter
   C. Catalytic converter failure
   D. Dumping of raw hydrocarbons into the atmosphere
5. *Technician A* says that some Nissan ECCS applications allow the PCM to electrically soften the dampening effect of the front engine mount so as to minimize engine vibration when the engine is idling.
   Technician B says on Nissan ECCS applications, the PCM may control the operation of the cooling fan.
   Who is correct?
   A. A only
   B. B only
   C. Both A and B
   D. Neither A nor B
6. Fuel pressure on port fuel-injected Nissan applications will typically vary between which of the following?
   A. 9 and 13 psi
   B. 20 and 29 psi
   C. 34 and 43 psi
   D. 46 and 55 psi
7. What component has been added only to California versions of the Nissan Sentra in order to help it meet the state's super low emission vehicle (SULEV) rating?
   A. Supercharger
   B. Electrically controlled front engine mount
   C. Rear heated oxygen sensor
   D. Air-cleaning radiator
8. Which of the following meters should be used to monitor a mass airflow (MAF) sensor signal on a Toyota TCCS application?
   A. Voltmeter
   B. Hertz meter
   C. Duty cycle meter
   D. Pulse width meter
9. What method is used on Toyota TCCS applications to monitor EGR flow?
   A. A pressure sensor is used to monitor the exhaust pressure just before the EGR valve but after a restriction placed in the exhaust tube.

B. A pressure sensor is used to monitor intake manifold pressure while opening and closing the EGR valve during idle conditions.
C. A temperature sensor is placed into the EGR passage.
D. The oxygen sensor is monitored while opening and closing the EGR valve during idle conditions.

10. An ohmmeter is used to test a camshaft sensor on a late-model Toyota TCCS application. With the meter's selector set to the 2K scale, a reading of 1.213 is seen on the display.
Technician A says that this reading indicates that the camshaft sensor needs to be replaced.
Technician B says that this reading indicates that the camshaft sensor's resistance is within specifications for a typical late-model Toyota.
Who is correct?
A. A only
B. B only
C. Both A and B
D. Neither A nor B

11. *Technician A* says that the fuel pump on a Toyota TCCS application is controlled by the ignition switch only and should always run whenever the ignition switch is turned to the run position.
Technician B says that some late-model Toyotas use a returnless fuel injection system with the pressure regulator located within the fuel tank.
Who is correct?
A. A only
B. B only
C. Both A and B
D. Neither A nor B

12. *Technician A* says that the EGR valve on a Toyota TCCS application is designed to open only during cold engine operation.
Technician B says that the EGR valve on a Toyota TCCS application is operated electrically and does not use vacuum to operate.
Who is correct?
A. A only
B. B only
C. Both A and B
D. Neither A nor B

13. *Technician A* says that Toyota's VVT-i system allows the PCM the ability to advance or retard the timing of the intake camshaft.
Technician B says that when working on a 2001 Toyota Prius hybrid vehicle, the high-voltage circuits must be deactivated before performing any work on it.
Who is correct?
A. A only
B. B only
C. Both A and B
D. Neither A nor B

14. A 1998 Toyota comes into the shop with an engine performance complaint.
Technician A says that the PCM on this vehicle has no ability to store DTCs in memory.
Technician B says that the PCM on this vehicle will store DTCs in memory, but that they have to be pulled using manual methods because this system is not designed to allow communication between a scan tool and the PCM.
Who is correct?
A. A only
B. B only
C. Both A and B
D. Neither A nor B

15. Honda's PGM-FI system uses an EGR Valve Lift sensor on top of the EGR valve. What is this sensor?
A. A potentiometer
B. An NTC thermistor
C. A piezoresistive sensor
D. A piezoelectric sensor

16. The "main relay" on a Honda PGM-FI is used to provide power to all *except* which of the following?
A. PCM
B. Cooling fan
C. Fuel injectors
D. Fuel pump

17. What is the primary purpose of Honda's Fuel Injection Air (FIA) control valve system?
A. To allow the PCM to control idle speed
B. To increase WOT engine performance
C. To reduce NO_x emissions
D. To increase fuel atomization

18. *Technician A* says that Honda's VTEC system allows the primary and secondary intake valve rocker arms to be mechanically linked to a central high-lift rocker arm at high engine speeds.
    Technician B says that Honda's VTEC system allows the PCM to control when the primary, secondary, and mid (connecting) rocker arms will be mechanically linked together.
    Who is correct?
    A. A only
    B. B only
    C. Both A and B
    D. Neither A nor B
19. *Technician A* says that the PCM on late-model Honda PGM-FI applications controls alternator field current in order to control charging system output.
    Technician B says that some Honda applications allow the PCM to control throttle position electronically based on input from an Accelerator Pedal Position (APP) sensor.
    Who is correct?
    A. A only
    B. B only
    C. Both A and B
    D. Neither A nor B
20. *Technician A* says that when working on a Honda Insight, it is safe to perform any work within the engine compartment immediately after turning off the high-voltage system.
    Technician B says that when working on a Honda Insight, it is important to turn off the high-voltage system and then to wait at least 5 minutes for the condensers to discharge before performing any work in the engine compartment.
    Who is correct?
    A. A only
    B. B only
    C. Both A and B
    D. Neither A nor B

Chapter 20

Electronically Controlled Diesel Engine Systems

OBJECTIVES

Upon completion and review of this chapter, you should be able to:

- ❏ Understand the basic principles of how diesel engines are similar to and different from gasoline-fueled engines.
- ❏ Describe the fundamentals of the General Motors 6.5-liter diesel engine control system.
- ❏ Describe the fundamentals of the Ford/Navistar/International 7.3-liter diesel engine control system.
- ❏ Describe the fundamentals of the Volkswagen 1.9-liter TDI diesel engine system.

KEY TERMS

Annulus Ring
Compression Ignition (CI)
Direct Injection
Glow Plugs
Injection Pressure
Injection Timing Stepper Motor
Lift Pump
Prechamber
Pumping Losses
Ricardo Chamber
Spark Ignition (SI)
Transfer Pump
Unthrottled
Vacuum Pump

Rudolph Diesel invented the compression-ignition engine named for him almost 100 years ago, and for special applications requiring exceptional durability or exceptional fuel economy or other circumstances, the hardy oil-fueled, fuel-injected engine has been the powerplant of choice.

DIESEL AND GASOLINE ENGINES

Since all the other engines considered in this book are gasoline-fueled, following are the main differences between a gasoline engine and a diesel engine, as well as the features that make a diesel more suitable to specific applications (and less suitable to others).

First, there is no electric coil spark system in a diesel engine. Combustion of the diesel fuel occurs entirely because of the temperature of the air into which it is sprayed. A gasoline engine is called **spark ignition (SI)**; a diesel engine is called **compression ignition (CI)**.

Second, there is no throttle plate in a diesel engine. The diesel engine draws as much air as possible into each cylinder for each engine cycle, regardless of load or the position of the accelerator pedal (more properly called the fuel pedal on a diesel engine). Power output is determined entirely by the amount of fuel injected for each cylinder event. Because of its unthrottled intake, a diesel engine "breathes" much more air than an equivalent gasoline-fueled engine except when the latter is at wide-open throttle. The diesel engine, however, ordinarily does not burn all of its oxygen; it burns only what it has fuel for and just pumps the rest through. Mixture control is absolutely important for a gasoline-fueled engine; it makes little or no difference for a diesel.

Third, in a diesel engine, the fuel oil injection is sprayed directly into the combustion chamber (or into an adjacent portion of that chamber, either a **Ricardo chamber** or a **prechamber**). The fuel begins to burn as soon as injection begins. No fuel is introduced into the intake air stream during its passage through the air intake, filter, tubing or manifold runners. Nothing but air passes the intake valve.

Fourth, compression ratios in diesel engines are much higher than in gasoline-fueled engines, typically between 17 to 1 and 24 to 1 (gas engine compression ratios are usually 8 or 9 to 1).

The higher compression ratio is one of the reasons for a diesel engine's greater efficiency: more air per cylinder event allows more torque per power stroke. In addition, diesel fuel has about 30 percent more energy per gallon than gasoline, and more diesel fuel can be produced from a barrel of crude oil than any other kind of fuel. For certain applications, specifically (but not only) marine, diesel fuel has the considerable advantage of being much less flammable than gasoline and highly resistant to evaporation and the formation of vapors.

Finally, diesel engines have much lower **pumping losses** since they do not develop manifold vacuum. A gasoline engine running at any setting less than wide-open throttle must pull every piston on the intake stroke down against partial vacuum through the intake manifold. These "pumping losses" can mean less horsepower available at idle or part-throttle in gasoline-powered engines.

A diesel engine, therefore, is ordinarily much more economical to operate at partial throttle settings where most driving is done. A diesel can idle using almost no fuel for very long periods of time, though its increased efficiency can allow the engine to cool and carbon to form in the combustion chambers if that is done regularly. This is the reason many bus and truck companies use a high idle when parked with the engine unloaded: there is virtually no decrease in fuel economy, but the combustion chambers are kept hot and clean.

A diesel engine does not generally have problems with either detonation (knock) or overheating. Since there is no fuel in the combustion chamber prior to fuel injection, there is nothing to ignite or detonate. If there is knock, it is either from a mistimed fuel injection pump, fuel of the wrong cetane rating (equivalent to octane for gasoline) or a leaking injector. On diesel engines with very considerable piston ring wear, there is occasionally the phenomenon of the "runaway" diesel, which starts to run on engine lubricating oil leaking around the piston rings. Under such a condition, each combustion event will have knock, and the only way to stop the engine is to choke off the intake air, plug the exhaust or stop the engine with the drivetrain in gear using the vehicle's brakes (not always an option with automatic transmissions).

Diesels are very resistant to overheating because of their thermal efficiency. Except for vehicles designed to move very slowly, such as garbage trucks, most diesel engines can use a much smaller radiator and cooling system than an equivalently powered gasoline-fueled engine needs. A clogged radiator or burst hose, of course, will still allow overheating.

There are two kinds of diesel combustion chambers used on modern engines, prechamber (or Ricardo chamber) and **direct injection**. The prechamber is a small, separate and removable chamber recessed in the cylinder head above the piston. The injector and the glow plug are in the prechamber. The advantages of the prechamber design are easier starting and somewhat smoother and quieter operation. The disadvantage of the prechamber design is reduced fuel economy because of more combustion chamber surface to conduct heat into the water jacket, and the work (and thus friction) needed to force the air into, and the combustion gases out of, the prechamber.

The direct injection diesel is about 15 percent more efficient than prechamber designs, chiefly because of the elimination of the two factors just mentioned. Direct-injection diesels are most com-

Diagnostic & Service Tip

A major incentive for the development of electronically controlled diesel engines has been the elimination of diesel smoke. Due to the excess oxygen present in diesel combustion chambers, these engines actually have less toxic emissions than equivalently powered gasoline engines. However, diesel emissions are visible and thus more objectionable to people.

Even with the electronic controls, the three kinds of diesel smoke will still provide evidence to the technician about what is wrong with the system.

Blue smoke means the same as with gasoline-powered engines: too much engine lubricating oil is getting around either the piston rings or the valves and is burned in the combustion chambers. Major engine repair is the fix.

Black smoke indicates a lack of oxygen getting into the combustion chambers. The most common cause of this is a clogged intake air filter, though anything that restricts the flow of air into the engine would have the same effect. A major objective of the electronic controls is to provide the computer with some way to estimate airflow, so the system can reduce the amount of fuel injected. Thus, on a properly functioning electronic diesel, a clogged air filter will result in a noticeable drop in engine power before it results in black smoke. A leaking injector, however, can cause black smoke with any kind of diesel.

White smoke is fuel, unburned and blown through the exhaust. It is a result of insufficient heat in the combustion chamber to ignite the fuel. The most common cause of white smoke is an engine that is too cold, one or more defective **glow plugs** (electric resistance heaters extending into the combustion chamber) or a thermostat stuck open. An engine that is blowing white smoke is also contaminating its lubricating oil with fuel, and the reduced viscosity of the oil can cause rapid wear of engine parts.

mon on larger commercial vehicles, but the small 1.9-liter Volkswagen diesel described in this chapter is a direct-injection design, as is the medium-size 7.3-liter International/Ford. Ordinarily direct-injection diesels have the combustion chamber recessed into the top of the piston rather than into the cylinder head. The fuel injection is to the center of the piston, adjacent to the glow plug.

Because a diesel engine uses the heat of the rapidly compressed intake air to bring the temperature up to ignition temperature, many diesels require some auxiliary source of heat when started cold. For all diesel engines, there is a temperature below which it is physically impossible for them to start without auxiliary heat, usually around zero degrees Fahrenheit.

The most common of these auxiliary heat sources is the glow plug. A glow plug is nothing more than an electric heater shaped into a rod and extending into the combustion chamber. Most systems use some sort of coolant temperature sensor to determine how long the glow plugs must be powered to reach a suitable fuel ignition temperature, and this period can range up to as much as a minute or more if the temperature is very low. Other engines, notably the Cummins engine used in certain Dodge pickups, use an electric heating grid in the intake manifold to heat the incoming air.

In either case, this explains why the diesel requires a much larger battery to start than an equivalently powerful gasoline engine. Between the considerable power drain of the glow plugs or heat grid and the need to use a much more powerful starter motor to create the much higher compression pressures (for all diesel engines, there is a certain combination of combustion chamber temperature and cranking rpm that must be exceeded if the engine is to start), diesels often have multiple batteries to get started. Sometimes these batteries are wired with a series-parallel switch to ensure maximum cranking speed. Others use compressed air pneumatic turbine starters, which are relatively immune to temperature problems.

Once running, most diesels can keep running even with a complete electrical system failure as long as they do not run out of fuel, lubricating oil pressure or coolant.

At least this was true with purely mechanical diesels prior to those we are considering in this chapter. Complete electrical failure, of course, is extremely rare and disables most engine monitoring equipment, so a driver would do well to stop for repairs even with the older-type engines unless there is a genuine emergency.

Given all the things that older, purely mechanical diesel engines could do, two questions arise: why do gasoline engines still exist, and what is there to control with a computer?

Gasoline engines still exist primarily because diesels do not have the advantages of lightness and moderate manufacturing cost that gasoline engines do. The typical diesel engine is built much more robustly (and thus more expensively) than the typical gasoline engine.

Also, a diesel engine can often weigh almost twice as much and cost three times as much as a gasoline engine for a given power output. It will get better fuel economy and last longer.

The principal reasons for computer controls of diesel engines is to improve emissions and fuel economy. While the most important feature of gasoline engine emissions control is the air/fuel mixture, a diesel engine always runs with more than enough air (unless as we saw, the air filter has become clogged with dirt). What influences the diesel's emissions more than anything else is the exact point of injection timing. In addition, on some older mechanical-injection engines, fuel delivery could go quickly to maximum when the driver's foot pushed the fuel pedal to the floor, but it took slightly longer for the air induction system to reach higher volumetric efficiency levels.

GENERAL MOTORS 6.5-LITER DIESEL: SYSTEM OVERVIEW

Like gasoline-fueled vehicles, the 6.5-liter GM diesel allows technicians to perform DTC-based diagnostics using a scan tool. Of course, diesel systems are fundamentally different, so you will need the proper shop manual to find the appropriate codes, which are, of course, fewer and simpler than the hundreds available for gasoline-fueled vehicles. They are also, of course, different.

The 6.5-liter GM diesel is a V-8 type engine, with pushrods operating the valves through rocker arms. Its compression ratio is 21.3 to 1. Some applications of the engine use a turbocharger, particularly those for heavier vehicles. Turbocharger boost is ordinarily controlled between 2 and 8 psi at peak torque. Most (but not all) call for a diesel catalytic converter on the exhaust system. Sometimes, a diesel catalytic converter has been described as more of a ceramic burn-off filter, but for practical purposes, it means the same thing as it does on gasoline-fueled vehicles. Because of the diesel engine's **unthrottled** free breathing, of course there is no need for an oxygen sensor feedback system. The amount of oxygen in the exhaust corresponds to nothing about the completeness of the fuel burning.

The GM 6.5-liter diesel engine uses a high-swirl precombustion chamber to mix the air and fuel as completely as possible. Like most diesels, its maximum torque output comes at a fairly low engine speed, 1,700 rpm, allowing a long power stroke for the most complete fuel burning. Maximum horsepower occurs at about double the maximum torque speed.

Prechamber Operation. Intake and exhaust strokes on a prechamber diesel are identical to those on gasoline-fueled engines, but we will look briefly at the compression and power strokes for the differences. In Figure 20–1, the compression stroke, you can see how the greatly

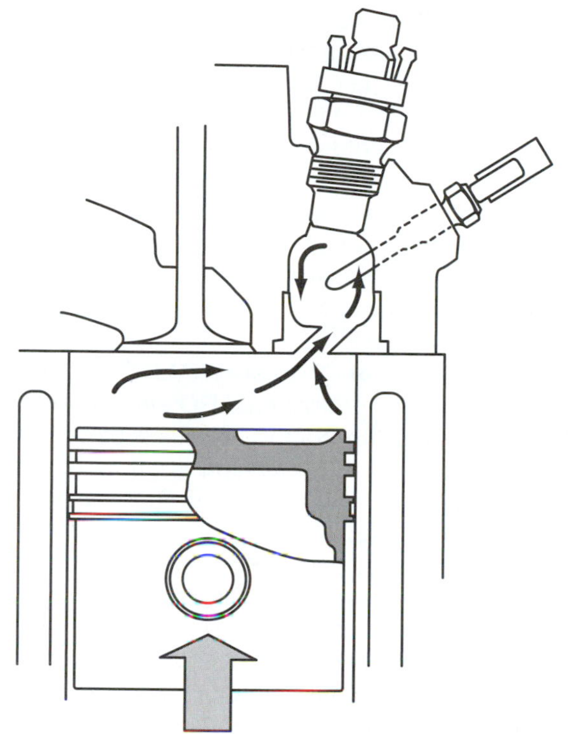

Figure 20–1 Compression stroke with a prechamber diesel.

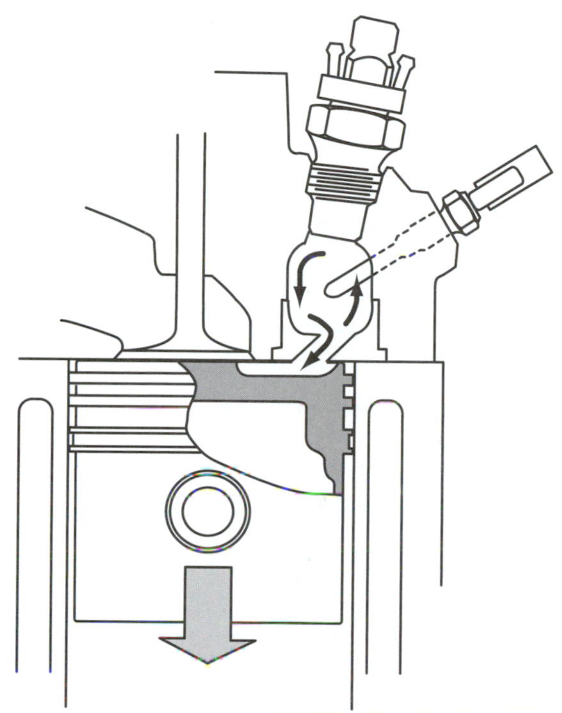

Figure 20–2 Power stroke with a prechamber diesel.

compressed and heated air is forced into the small, precisely machined prechamber. The turbulence of the air in the prechamber promotes atomization of the fuel.

During the power stroke, Figure 20–2, fuel injection has begun and the mixture is burning in the prechamber. The expanding gas blows out of the prechamber and into the main combustion chamber, forcing the piston down and turning the crankshaft. Notice there is almost no room between the piston top and the cylinder head except in the small flame channel recess just below the prechamber. This narrow space above the piston sometimes allows carbon buildup, which can make audible hammering when the engine is started cold, before the block can warm up and expand from the heat. Since fuel injection continues for a brief period of the power stroke after the beginning of combustion, you can see why **injection pressure** must be much higher than for gasoline-fueled injectors. Typically a diesel injector sprays fuel at several thousand psi pressure.

WARNING: Because diesel injection pressures are so high, never allow the injector to spray anywhere close to your skin, either from the pump on the engine or with an injection pressure tester. On this and all other diesel injection systems, the pressure is high enough to easily blow the fuel through the skin and tissue. This kind of accident can result in severe blood poisoning and may even require amputation of the affected finger or hand.

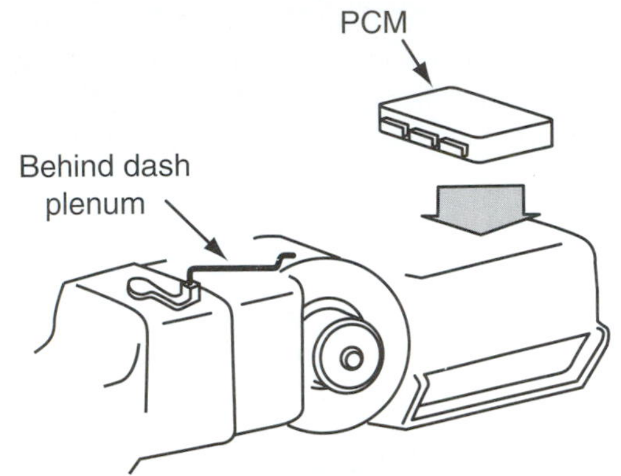

Figure 20–3 Computer location.

The 6.5-liter GM diesel uses a computer, Figure 20–3, to control various functions of engine management. The computer is ordinarily located in the passenger compartment of the vehicle. The computer controls the following system functions:

- fuel quantity
- injection timing
- glow plugs
- turbocharger boost
- EGR valve
- cruise control
- transmission control
- self-diagnostics

The inputs and outputs of the system are not tied one-to-one, but are interrelated as with gasoline-fueled systems. Hence the coolant temperature sensor and the accelerator pedal position sensor both can affect fuel volume. Load and engine speed both affect how the fuel injection timing is determined.

INPUTS

Engine Coolant Temperature Sensor. The engine coolant temperature sensor is the familiar type, with a resistance that goes down as the temperature goes up. This variable resistance modifies the 5-volt reference voltage into a lower signal voltage returned to the computer. The sensor threads into the water jacket next to the thermostat outlet. Some 6.5-diesels use a dual-thermostat system to better control coolant flow, but this does not affect the way the coolant temperature sensor works or mounts.

Intake Air Temperature Sensor. The intake air temperature sensor is similarly a variable resistor, losing resistance as the air temperature goes up. It bolts into the air horn just upstream of the intake manifold.

EGR Control Pressure/BARO Sensor. The EGR control pressure/BARO sensor, Figure 20–4, provides the computer with a barometric reading when the EGR system is off, and information about the pressure in the EGR actuation vacuum lines when it is on.

Turbocharger Boost Sensor. On vehicles equipped with a turbocharger, the turbocharger boost sensor mounts to the top of the air intake horn, Figure 20–5. This provides the computer with information about how much fuel can be injected and how to actuate the wastegate controls.

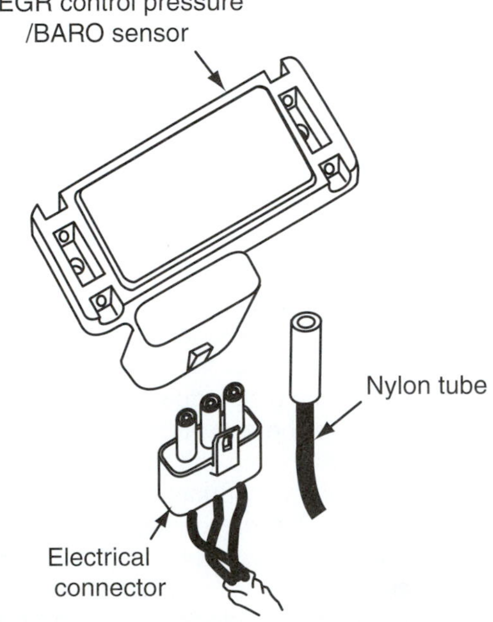

Figure 20–4 EGR control pressure/BARO sensor.

Accelerator Pedal (Fuel Pedal) Position Sensor. The fuel pedal position sensor, Figure 20–6, is similar to the throttle position sensor in gasoline-fueled engines. On the diesel engine, however, it mounts to the fuel pedal lever, inside the passenger compartment. More importantly, it replaces the mechanical linkage previously used to convey the driver's intentions to the injection pump. This system is a true drive-by-wire arrangement. The fuel pedal position sensor is actually three independent variable resistors in one box. Each one gets a 5-volt reference signal from the computer and modifies that according to the fuel pedal position. This sensor means the system is basically a drive-by-wire system, and it has deep fail-safe modes. If there is a problem with the accelerator pedal position sensor, the following results occur:

- if one sensor fails, the computer stores a trouble code, but does not turn the check engine light on, and the vehicle operates normally.
- if two sensors fail, the computer turns on the check engine light and stores a code. The vehicle will operate at noticeably reduced power.
- if all three sensors fail, the check engine light comes on and the trouble codes are stored. In addition, the engine will run only at idle speed. Due to the considerable low-speed torque of a diesel engine, the vehicle may be able to move along the road, even up hills, but slowly.

Optical Sensor. The optical sensor, Figure 20–7, is located in the fuel pump, and is the most important component in the diesel engine system. The optical sensor is responsible not only for the high-pressure injection of the fuel, but also for

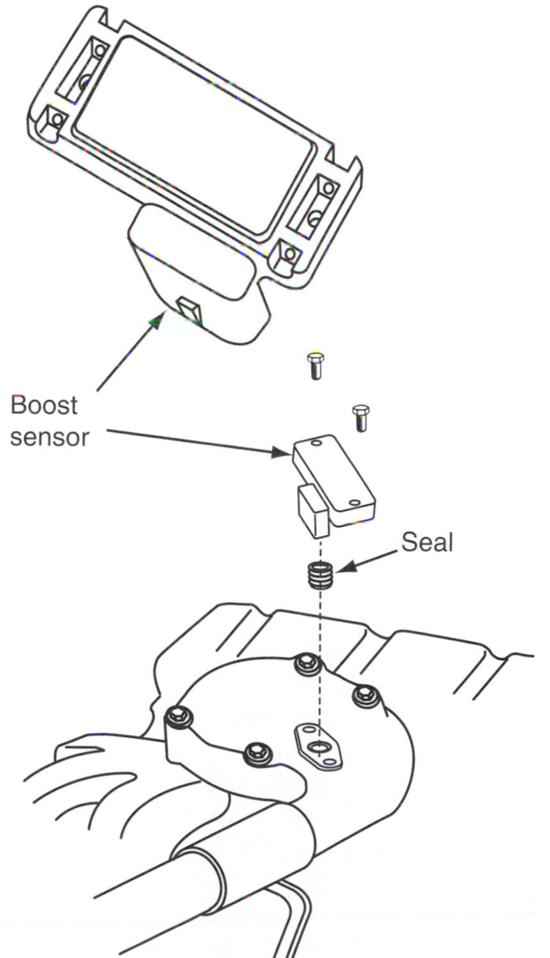

Figure 20–5 Boost sensor.

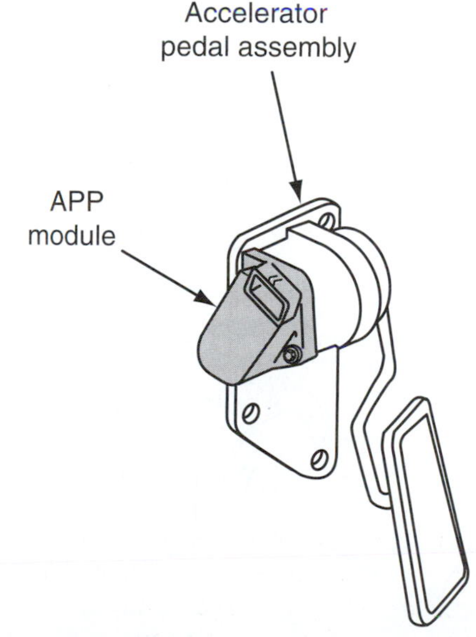

Figure 20–6 Accelerator pedal position sensor.

the timing of the injection—the most significant variable in diesel engine performance, both in fuel economy and in emissions quality.

The sensor consists of two optical receivers separated from their emitters by a thin disk. The disk has two sets of notches, the outside one with 512 notches, and the inside one with 8. The inner notches include one larger notch to identify cylinder position. Since the fuel pump turns at camshaft rather than crankshaft speed, the two signals from this provide information about the position of the crankshaft, pistons and camshaft.

The computer sends the optical sensor a 5-volt reference signal, which is then modified by the movement of the notched disk in the fuel pump, and returned as two information signals—a high resolution signal from the 512-notch ring, and a camshaft position signal. If either signal is out of calibration, the computer sets a trouble code.

Fuel Temperature Sensor. Diesel fuel viscosity is affected more by temperature than is gasoline, so the system includes a fuel temperature sensor, Figure 20–7. This sensor is part of the optical sensor.

Crankshaft Position Sensor. The crankshaft position sensor is in the front engine cover, Figure 20–8. This sensor is a Hall-effect signal generator, receiving a 5-volt reference from the computer. As long as no reluctor vane lines up with the sensor tip, the sensor's magnetic field passes through the Hall-effect switch and turns it off, leaving the reference signal at 5 volts. As the reluctor vane lines up, the magnetic field moves to pass through the vane instead, turning on the Hall-effect switch and dropping the reference voltage to near zero. Erratic signals, shorts and opens in the sensor circuit are all recorded by the computer as trouble codes.

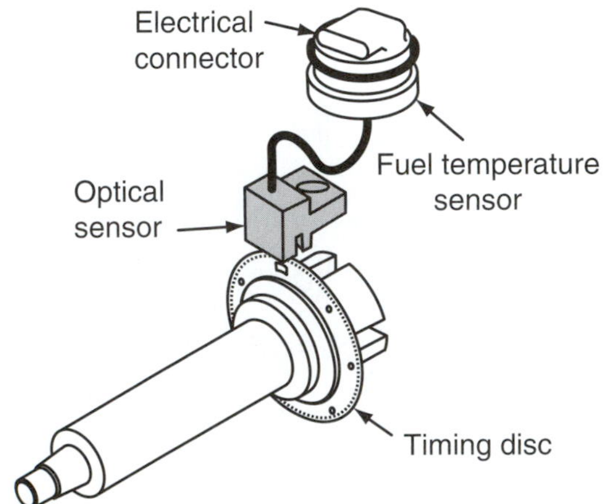

Figure 20–7 Optical sensor/fuel temperature sensor.

Diagnostic & Service Tip

Many shops with extensive GM diesel experience report more problems with the optical sensor than with any other part of the GM diesel system. Checking for proper operation is an important test to perform early in any troubleshooting procedure on these engines. The sensors inside the injector pump may be damaged by any gasoline coming into contact with them.

Vehicle Speed Sensor. The vehicle speed sensor is the same type unit as on gasoline-fueled vehicles. Like those systems, late models use a signal buffer and may include transfer case low range signal switches and separate transmission input and output speed sensors for four-wheel-drive units.

Cruise Control System. Driver controls for the diesel cruise control are identical to those for gasoline engines. Yet since the diesel engine's power comes entirely from the amount of fuel injected, the actuator works differently, moving linkage to increase or reduce fuel delivery instead of opening or closing a throttle. The system includes the usual disabling functions if the clutch or brake pedal is moved. Some systems employ a vacuum pump to work a vacuum servo. The pump is needed because the engine does not inherently develop any useful vacuum.

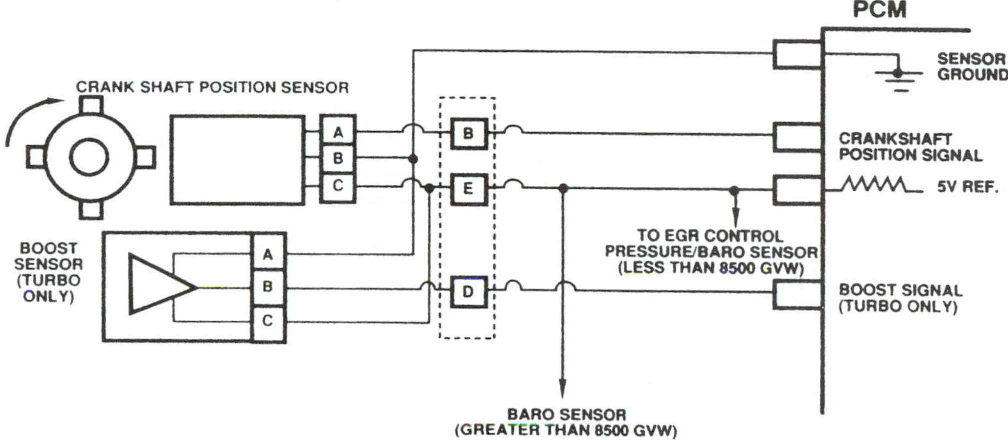

Figure 20–8 Crankshaft position sensor. *(Courtesy of General Motors Corporation, Service Technology Group.)*

A/C Signal. When the air conditioning is turned on, the computer receives a signal that it uses for idle speed control. At full power settings, the computer may also disengage the compressor clutch.

Self-Diagnostic Command. Since the 6.5-liter GM diesel meets OBD II requirements, it can be diagnosed with a scan tool through the standard connector. While codes are different for a diesel engine, the procedure for calling them up is the same. Check a shop manual for an explanation of each of the stored trouble codes.

Glow Plug Feedback Signal. Since this diesel engine uses glow plugs for auxiliary heat when starting a cold engine, the computer receives a feedback signal from the glow plugs when they are turned on. The computer will calculate from several inputs, most importantly the engine coolant temperature and engine speed, how long to turn the glow plugs on and when to turn them off. They usually continue to heat from the time the ignition switch is first turned on until shortly after the engine starts and is running smoothly.

Injection Pulse Width Feedback Signal. The computer controls fuel injection through the fuel solenoid driver (see the following Outputs section). The fuel solenoid driver also returns a pulse-width modulated signal back to the computer indicating when the injector plunger closes. This provides the computer with information about how much fuel was injected in this combustion cycle. The computer employs a calibrated injection pump-mounted resistor to determine fuel delivery rates, the value of which is retained in the computer's memory. If the computer memory has been erased or the computer itself replaced, it will relearn this resistance in the next running cycle.

If the injection pulse width feedback signals are out of range, the computer stores the corresponding trouble code in its memory for recall with a scan tool.

OUTPUTS

Fuel Injection Pump. The heart of any diesel is the high-pressure fuel injection pump, Figure 20–9. The fuel injection pump forces a very fine spray of fuel into the combustion chamber to initiate and continue the power stroke. This pressure may be as high as 3,000 psi, in order to properly atomize the fuel and to overcome the combustion pressure.

The injection pump is actually a series of pumps. Fuel from the frame-mounted electric **lift pump** first arrives at the **transfer pump**, a vane-type pump that varies the pressure it delivers according to engine speed. At idle, transfer pump

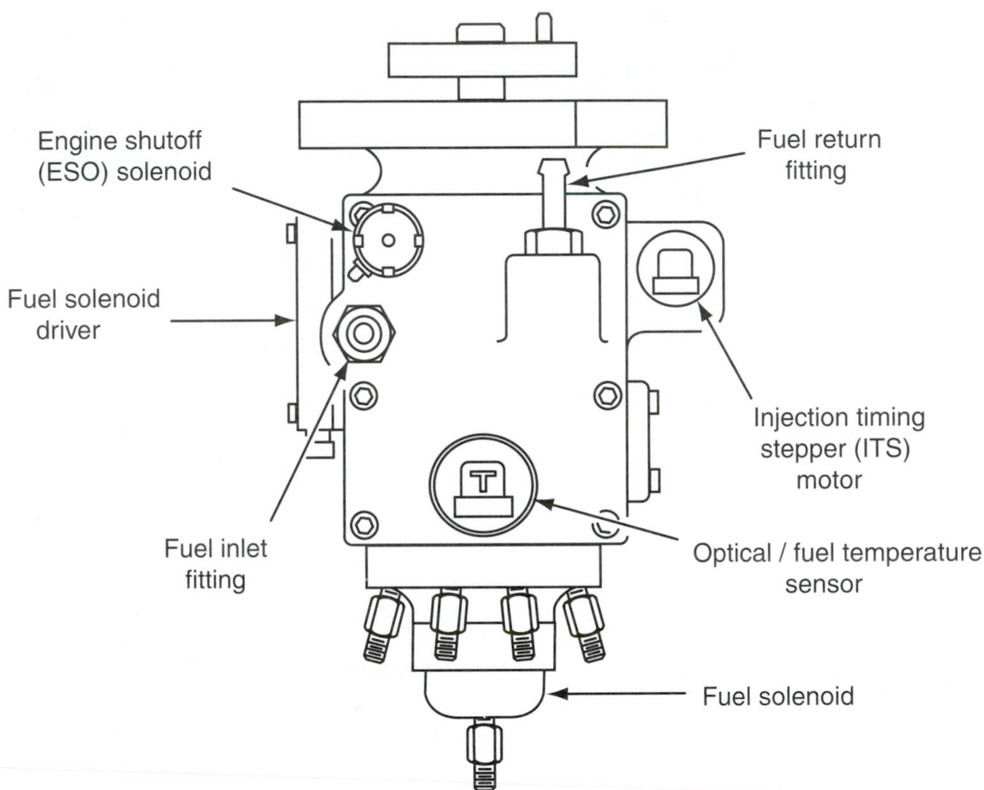

Figure 20–9 Electronic fuel injection pump.

outlet pressure is 20 to 30 psi. At high engine speed, the pressure may be over 100 psi. A relief valve will open if the pressure reaches 125 psi. The pressure regulator also includes a viscosity compensating port. Pressure relief fuel returns to the transfer pump inlet passage.

Injection. From the transfer pump, fuel travels to the charging **annulus ring**. This component has eight slots, one per cylinder. From it, fuel moves to two places: the pump housing and the pumping plungers of the rotor.

The metering of the fuel into the pumping plungers is the most important part of understanding how the 6.5-liter diesel works. Fuel enters from two places: the inlet ports in the annulus rotor and the control valve. As the rotor spins, the annulus slots move into and out of alignment with the inlet slots. When they are aligned, fuel under transfer pump pressure enters the chamber. When the slots are not aligned, fuel flow stops. The second source of fuel is through the control valve. When the fuel solenoid is off, the valve opens, and fuel can run through the center of the pump shaft into the plunger chamber. The fuel that enters the plunger chamber has been metered for injection delivery.

The actual injection is driven by the plungers. They ride on low friction shoe-and-roller assemblies in an eight-lobed cam ring. When the shoe-and-roller assemblies are in the valleys between the lobes, the plungers are fully open. When the shoe-and-roller assemblies ride up the lobe, the plungers move inward and force the fuel through the injector lines to the injectors, Figure 20–10. This fuel metering occurs eight times for each revolution of the pump rotor, for each complete engine cycle of two crankshaft rotations.

The computer controls the amount of fuel injected by determining when to turn the fuel sole-

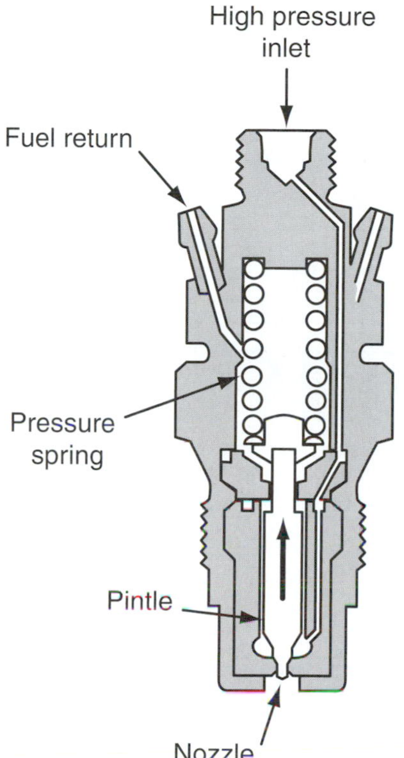

Figure 20–10 Diesel fuel injector.

noid on and off. The longer it is on, the more fuel is injected; the shorter, the less. At warm idle, the solenoid is on for only a brief moment, enough to inject a small droplet of fuel. At high power settings, the solenoid remains on for most of the rotor motion, injecting a maximum amount of fuel. After the injectors have sprayed the amount of fuel the computer has determined is correct for the current conditions, it disengages the solenoid and spills the fuel back into the fill/spill chamber.

These steps can be seen graphically in Figures 20–11 to 20–14.

Idle Quality. One of the important advantages of electronic control of injection volume is the improvement in idle quality. On mechanical diesel injection engines, it is difficult to manufacture all the combustion chambers to have exactly the same compression ratio. In addition, variations in the amount of fuel injected are greatest at idle speeds because of manufacturing tolerances in the injection pump. The slight variations in compression ratio and idle fuel delivery cause a much rougher idle on a mechanically injected diesel than on a gasoline engine. With electronic controls the computer can monitor idle speed variations and inject different amounts of fuel to each cylinder to make the idle speed much smoother.

Injection Timing. Injection volume is just one of the parameters controlled. The other major one is injection timing. Just as with gasoline engines, combustion must start earlier in the compression stroke as the engine turns faster, and there are different optimal injection timing points with different loads.

This system uses an **injection timing stepper motor**, Figure 20–15 on page 592, to achieve this timing control. The stepper motor uses a sliding piston that rotates the injection cam ring with a pin. As the piston moves in its bore, it turns the cam ring to the position determined by the computer. The stepper motor does not actually move the piston; instead it moves a servo valve plunger that uses transfer pump pressure to move the advance piston itself, Figure 20–16 on page 593. When the computer wants more injection advance, the stepper motor retracts in steps. This moves its pivot arm, and the control lever moves away from the servo valve. Spring pressure moves the valve away from the advance passage, and pressurized fuel enters the passage and pushes the piston in the advance direction. When the computer wishes to retard injection timing, it extends the stepper motor arm, pushes the control lever in, and reverses the hydraulic force on the advance piston. The layout is similar to the movement of a spool valve in an automatic transmission or a power steering system.

Engine Shutoff (ESO) Solenoid. The only way to stop a running diesel engine other than stalling it is to shut off either air or fuel. The computer controls the engine shutoff solenoid mounted on the fuel pump. As long as this solenoid is actuated, fuel can flow to the injectors. When it is shut off, the fuel passage is blocked and the engine stops immediately.

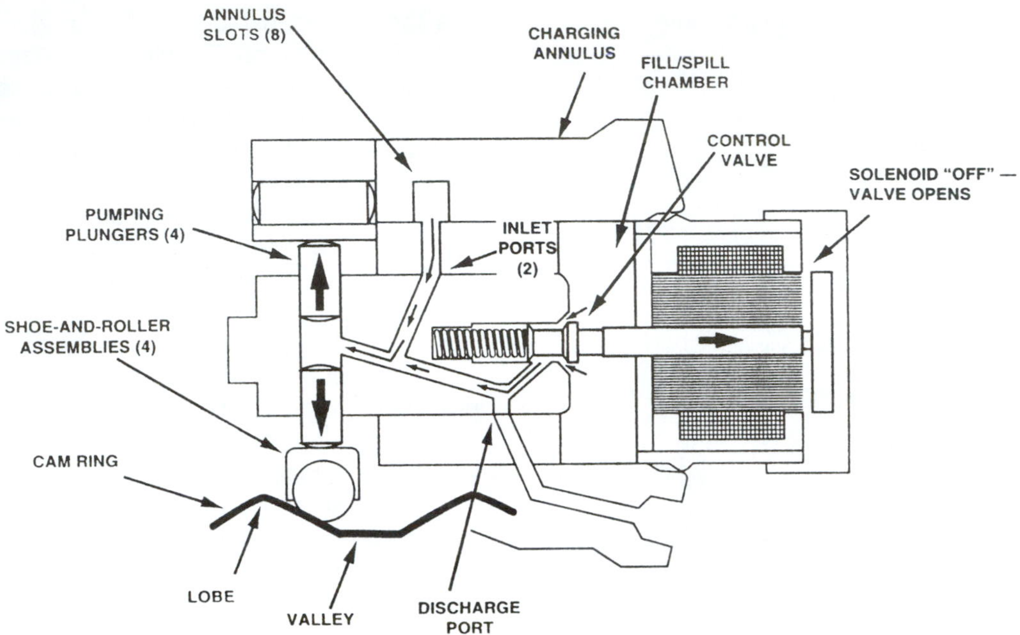

Figure 20–11 Fuel injection metering sequence. *(Courtesy of General Motors Corporation, Service Technology Group.)*

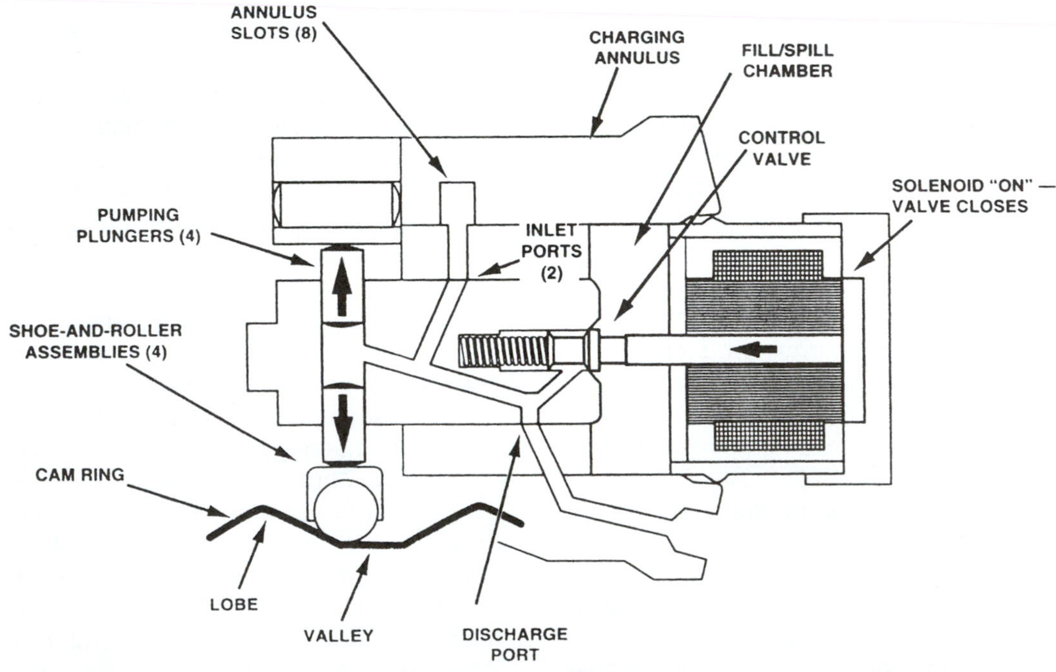

Figure 20–12 Fuel injection metering sequence. *(Courtesy of General Motors Corporation, Service Technology Group.)*

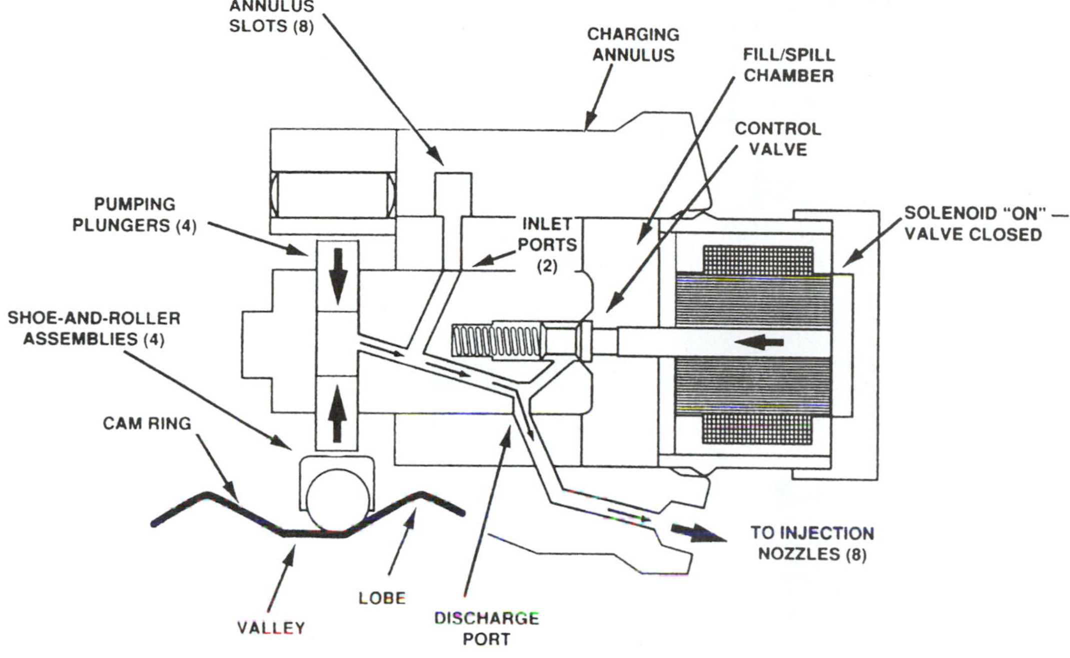

Figure 20–13 Fuel injection metering sequence. *(Courtesy of General Motors Corporation, Service Technology Group.)*

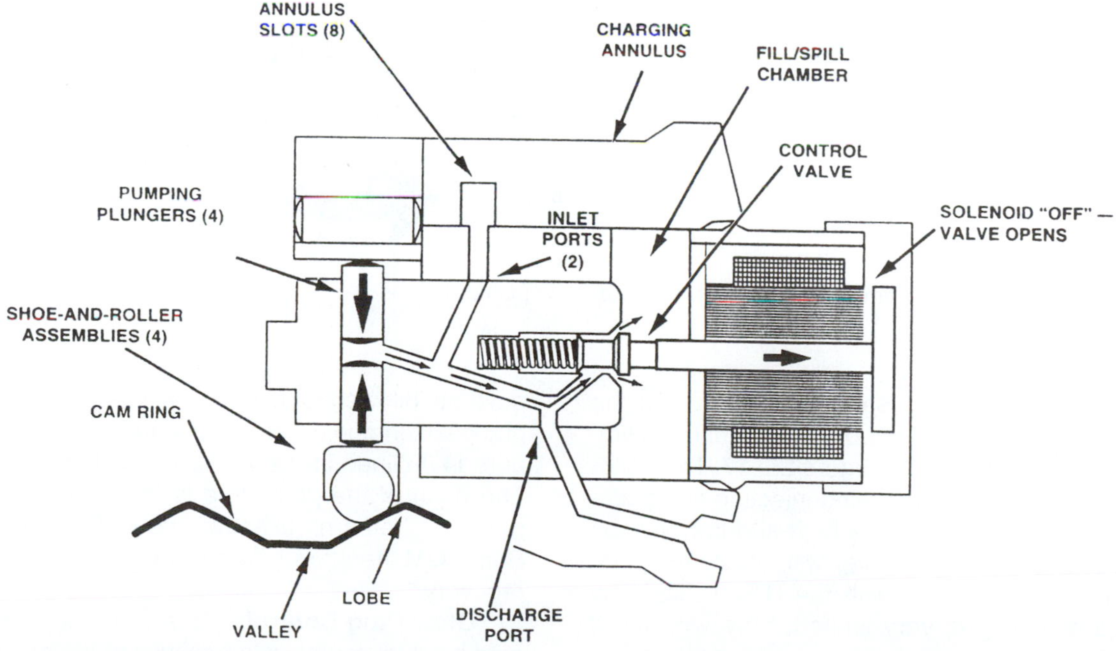

Figure 20–14 Fuel injection metering sequence. *(Courtesy of General Motors Corporation, Service Technology Group.)*

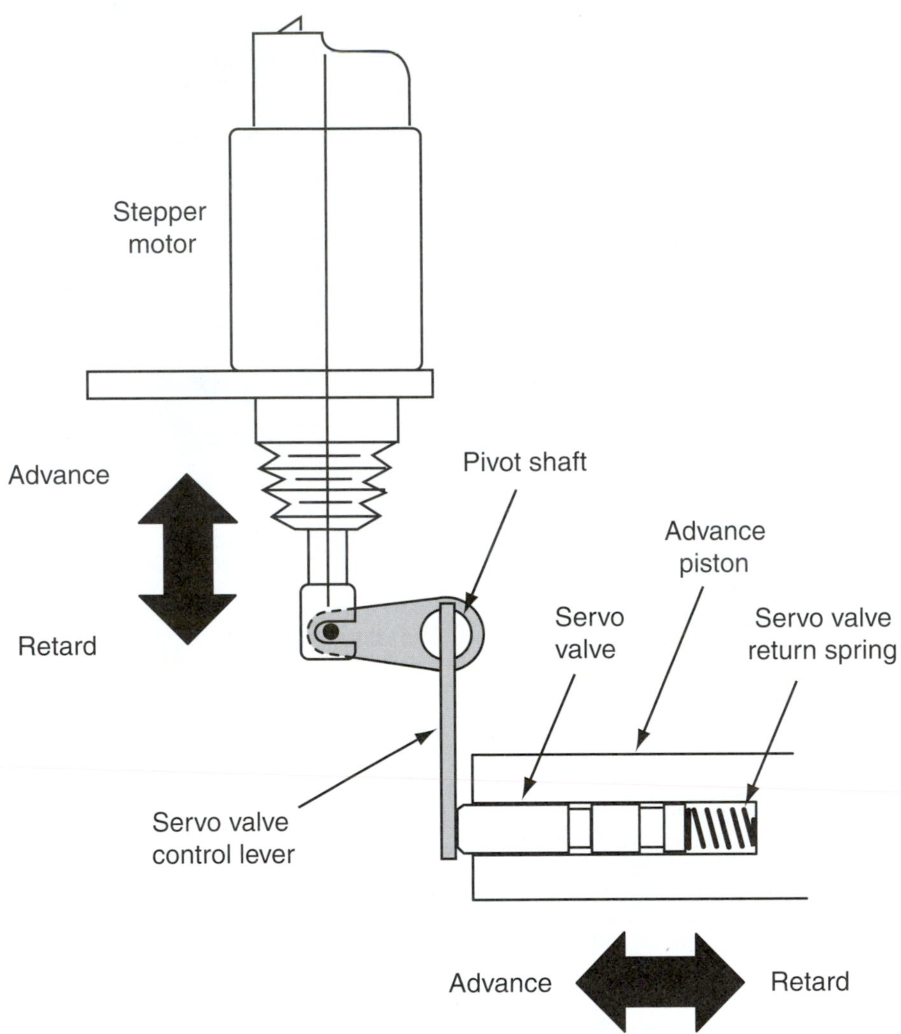

Figure 20–15 Injection timing stepper motor and advance piston.

Injection Timing Stepper Motor. On the right side of the injection pump is the injection timing stepper (ITS) motor, Figure 20–17. The motor contains two coils controlling injection pump timing through four circuits, a high and low position for each coil. The computer will, of course, set a code for any problem with the ITS circuits. The range of timing is very limited, however, and if the computer, the timing gears, the front engine cover, the crankshaft position sensor, or anything else alters injection timing, a special procedure must be followed to mechanically set the injection pump within one degree of accuracy. This procedure is outlined in the appropriate shop manual. The point of the procedure is to allow the computer to "relearn" exactly where TDC occurs. Some GM literature refers to this as "TDC offset recovery."

Glow Plug Control. It is frequently hard to start a diesel engine cold because of the relatively slow crank provided by the starter and the cold combustion chamber walls which may not allow

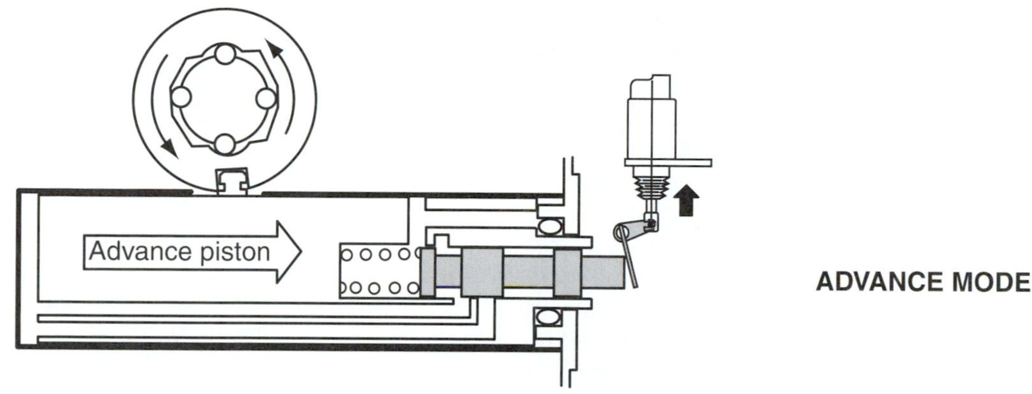

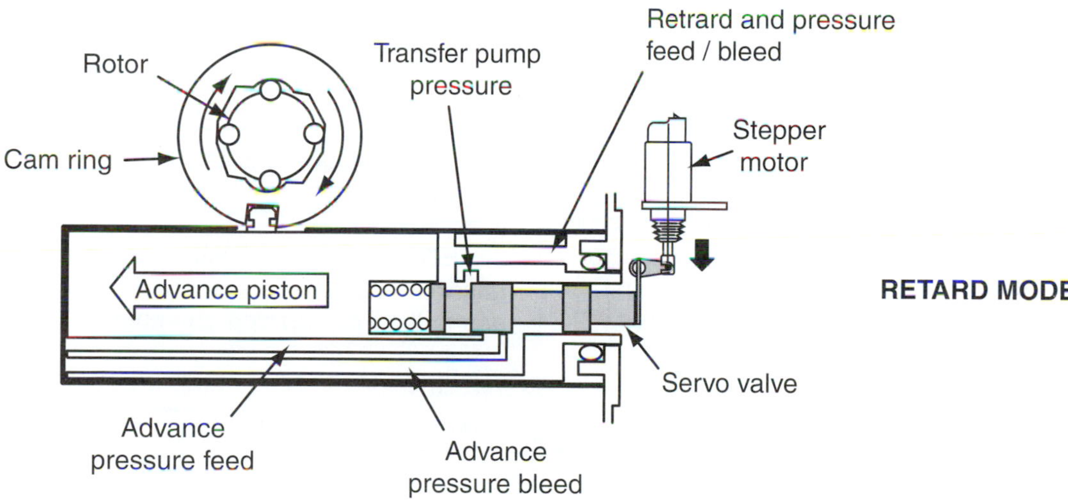

Figure 20–16 Advance piston operation.

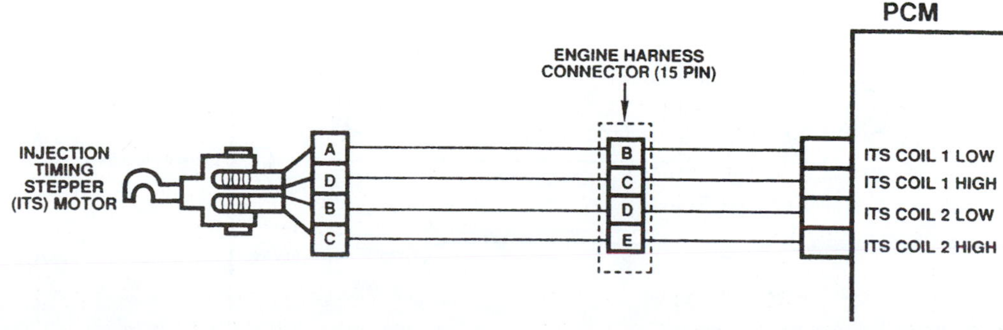

Figure 20–17 Injection timing stepper motor operation. *(Courtesy of General Motors Corporation, Service Technology Group.)*

the compressed air to reach the ignition flash point of the injected fuel. The 6.5-liter GM diesel uses eight glow plugs to assist in providing this initial heat, Figure 20–18.

The computer controls the activation of the glow plugs based on information from the engine coolant temperature sensor, the crankshaft position sensor (for engine rpm at crank and just after startup), and the feedback signal from the glow plugs themselves. It activates the high-current glow plugs through a special high-current relay, the glow plug relay, Figure 20–19. The computer also operates the glow plug lamp on the dashboard that informs the driver when to begin engine cranking. The wait period varies depending on the temperature signal.

CAUTION: According to GM, never install a jumper across the high-current terminals of the glow plug relay, or immediate damage to the glow plugs can result. Glow plugs can be tested for continuity to determine if one has burned out, but an amperage measurement is not practical because most designs use less current as they get warmer, so the results are not relevant. On earlier (1994 and 1995) models, an amperage draw of 14 amps per glow plug, or 55 amps per engine bank, is normal.

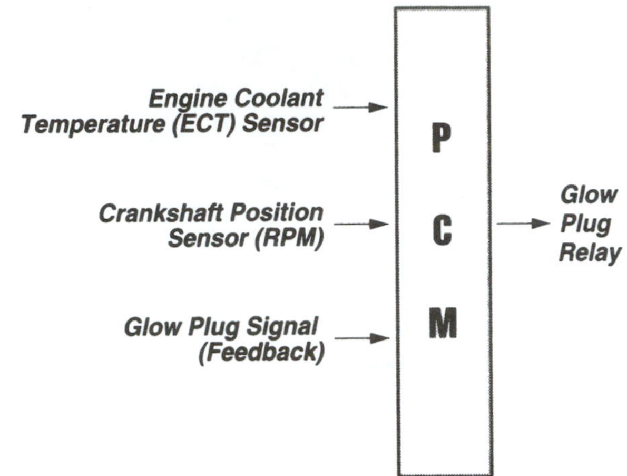

Figure 20–19 Glow plug relay schematic. *(Courtesy of General Motors Corporation, Service Technology Group.)*

EGR System. Diesel engines, just like gasoline engines, can produce undesirable NO_x gas in their combustion chambers if combustion

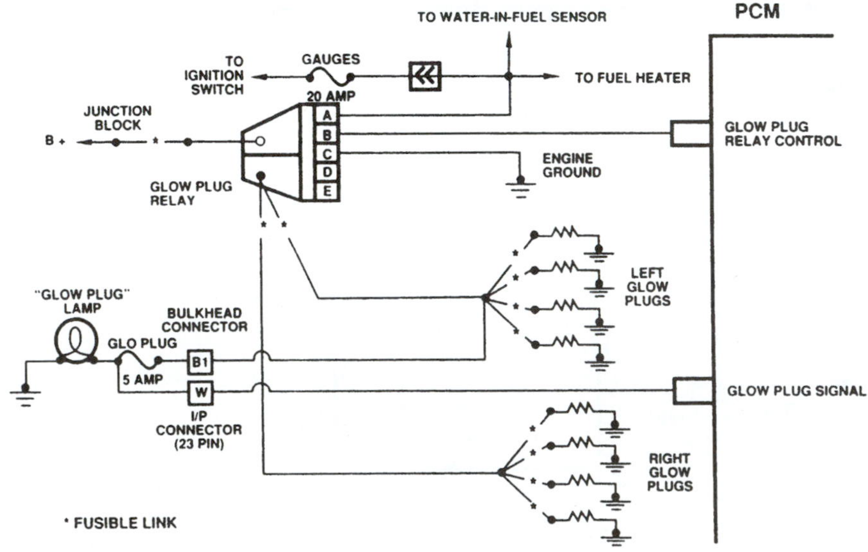

Figure 20–18 Glow plug circuit. *(Courtesy of General Motors Corporation, Service Technology Group.)*

temperatures go much above 2,500°F. As in gasoline engines, exhaust gas is metered back through the intake system to provide an inert gas that will keep the peak temperatures below the NO_x formation levels. Inputs for these calculations come from the accelerator pedal position sensors, the crankshaft position sensor, the EGR control pressure/BARO sensor and the engine coolant temperature sensor.

The EGR control pressure/BARO sensor mounts on the cowl, and monitors the absolute pressure in the EGR vacuum line. The computer uses this input for a barometric pressure reading when the EGR system is off, just after the key is turned on but before the engine starts, for example. If the EGR control pressure sensor reports a vacuum too high or too low, the computer will shut off the EGR system and set the appropriate trouble code.

Actual metering of recirculated exhaust gas is done by the EGR valve, controlled by the EGR solenoid and EGR vent solenoid, Figures 20–20, 20–21, and 20–22. The computer activates these solenoids by providing a ground path for their voltage, applied through the ignition switch. The EGR solenoid cycles to apply a desired amount of vacuum to open the EGR valve. The EGR vent solenoid dumps any vacuum in the line to close the EGR valve during EGR-off conditions. Failure of either solenoid will trigger a corresponding trouble code and turn on the check engine light.

Boost Control System. The system uses a wastegate to control the boost provided by the turbocharger in accord with information the computer gets from the accelerator pedal position sensors, the boost sensor, and engine speed information from the crankshaft position sensor.

Not all applications of the 6.5-liter GM diesel use a turbocharger, and when it is not installed, this information/actuation circuit is not part of the system. The boost sensor is very similar to a MAP/BARO sensor on a gasoline-fueled engine: it is a resistor that changes according to the sensed pressure. As pressure increases, resistance goes down. The computer uses this information to calculate how much fuel can be delivered to the engine.

The engine uses a special **vacuum pump** to actuate the wastegate, as well as the other vacuum-operated devices, the EGR solenoid, and EGR vent solenoid. A special vacuum pump is needed on a diesel engine because the engine does not develop vacuum behind a throttle, as do gasoline-fueled engines. The computer tracks the intake manifold pressure by the pressure sensor

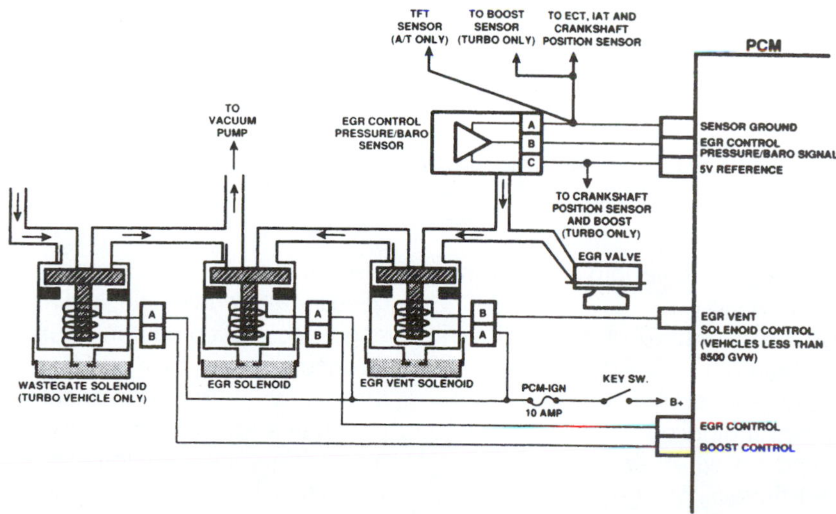

Figure 20–20 EGR valve circuit operation. *(Courtesy of General Motors Corporation, Service Technology Group.)*

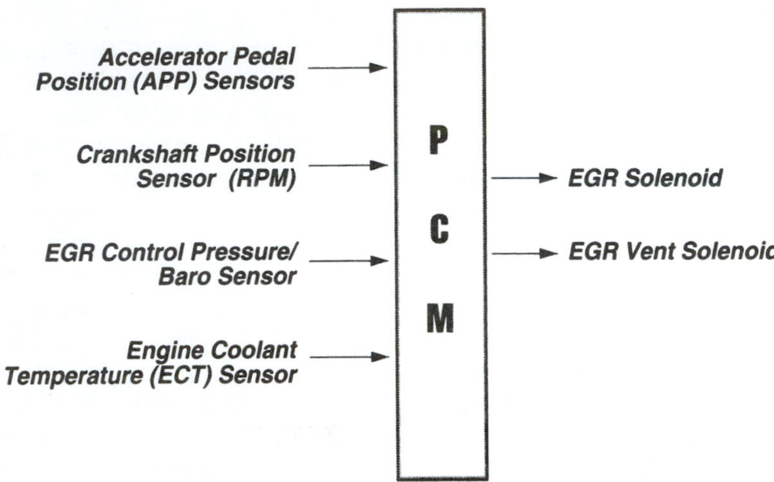

Figure 20–21 EGR control inputs and outputs. *(Courtesy of General Motors Corporation, Service Technology Group.)*

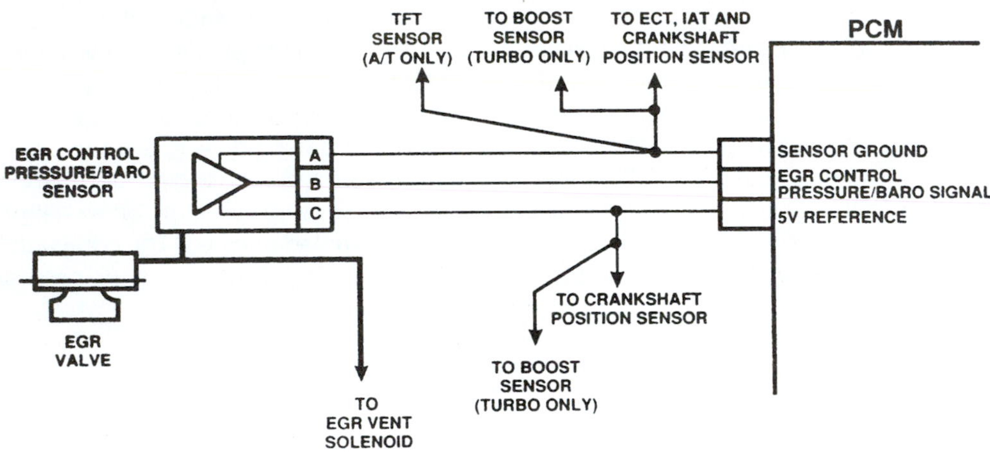

Figure 20–22 EGR control pressure/BARO sensor circuit operation. *(Courtesy of General Motors Corporation, Service Technology Group.)*

and cycles the wastegate solenoid to prevent overboost, Figure 20–23.

Transmission Controls. Automatic transmissions mated to the 6.5-liter GM diesel are also controlled by the engine computer. There are slight differences depending on which transmission is installed and on whether the vehicle is four- or two-wheel-drive. The torque converter lockup works the same way as on gasoline-fueled engines. On many late-model diesel vehicles, the work of the transmission control module, as in a gasoline-fueled vehicle, is directed entirely by the engine control computer. There is a significant difference between two- and four-wheel-drive in that the four-wheel-drives include a transmission output shaft speed sensor to enable the computer to track differences between transfer case settings. In the two-wheel-drive vehicle, only the single vehicle speed sensor is required. As with many other newer GM vehicles, there is a vehicle speed

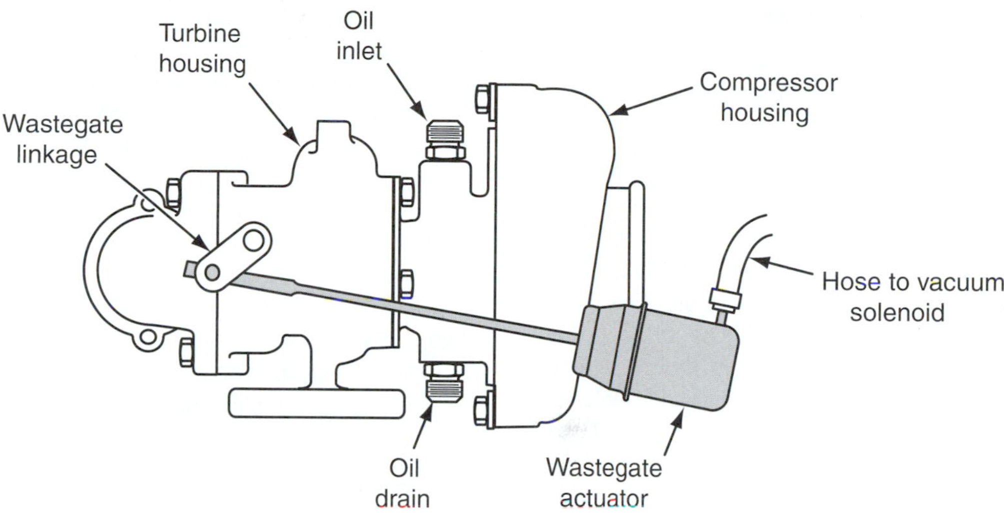

Figure 20–23 Turbocharger wastegate and actuator.

sensor buffer module, essentially a VSS signal interpreter which provides vehicle speed information to the computer, the antilock brake system and the speedometer and odometer.

The fundamental differences between the transmission controls for the diesel engine and those for the gasoline-fueled engine consist in the diesel's greater torque at lower speeds. This changes the shift points to keep the engine at a more favorable rpm. Because of this greater low-speed torque, the diesel can also stay in a given gear longer than an equivalent gasoline-fueled engine.

Lift Pump. While the diesel engine mechanically drives the injection pump, there is another pump to prime the injection pump. The **lift pump**, Figure 20–24, is very similar to fuel pumps on gasoline engines and provides relatively low (15 psi) pressure to purge any air from the injection pump circuits. The lift pump is turned on by the ignition switch and kept on by a switch circuit through the oil pressure sender. Late model vehicles use a computer-controlled fuel pump relay, and the oil pressure sender is redundant.

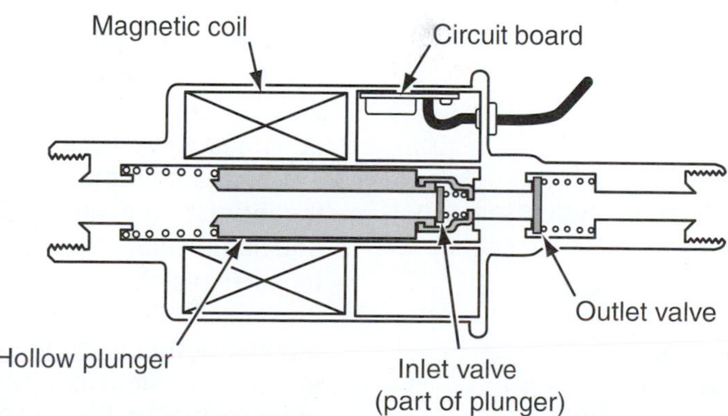

Figure 20–24 Lift pump.

FORD POWERSTROKE/ INTERNATIONAL/NAVISTAR 7.3 DIRECT INJECTION DIESEL: SYSTEM OVERVIEW

Certain Ford trucks are available with the Navistar/International 7.3-liter direct injection diesel engine. This engine is built for Ford by International and is used on a number of other medium-duty trucks as well. Like the GM diesel described earlier, it is a compression-ignition engine. We will not repeat the information about the differences between gasoline and diesel engines here, but will assume the reader understands them.

The Ford/International engine is a direct injection design; that is, there is no prechamber above the cylinder head. Instead, there is a small cavity in the piston top into which the fuel is directly sprayed. The glow plugs also extend into this cavity. As is typical for many direct injection diesels, the compression ratio is somewhat lower than for a prechamber design. This engine has a compression ratio of approximately 18 to 1.

This system is unique among current electronically controlled diesels because it does not use an injection pump. Instead injection is done electronically by the individual injectors. It is also unique in the injection pressure developed; under conditions of high power output, injection can reach as high as 21,000 psi! The precaution mentioned earlier about not allowing injected fuel to come anywhere near people is 10 times more pronounced with these injectors.

INPUTS

The Ford/International system uses eight basic sensors on which to base the computer's calculations. These are:

- fuel pedal position sensor.
- camshaft position sensor.
- injection control pressure sensor.
- boost pressure sensor.
- oil temperature sensor.
- oil pressure sensor.
- coolant temperature sensor.
- ambient air temperature sensor.
- barometric pressure sensor.
- exhaust backpressure sensor.

Most of these sensors are tied to the OBD II diagnostic program in Ford vehicles (not applicable, however to medium-duty International trucks). When the computer detects an absent or out-of-range signal, it will set a code, as described in the appropriate shop manual.

For a system schematic, see Figure 20–25.

OUTPUTS

The engine drives a seven-piston fixed displacement axial pump, pressurizing the engine oil to a pressure between 450 and 2,750 psi during normal operation. This is not an injection pump; it merely provides the individual injectors with highly pressurized engine oil. The computer selects the desired pressure from its various inputs, particularly engine coolant temperature and engine speed. The colder the engine or the greater the load and speed, the higher the selected pressure will be. The desired pressure is obtained by electronic control of the rail pressure control valve (RPCV), Figure 20–26.

The computer then actuates the injectors by commands to the injector drive module (IDM), basically a heavy-duty solid-state relay bank.

Electronic Injectors. The key to the system is in the electronic/hydraulic injectors, Figure 20–27. These work by a combination of electronic control and hydraulic power multiplication.

Fuel enters the injectors at the lower end in the injector tip charging area. The high-pressure engine oil enters the injector at the top in the intensifier piston space. Fundamentally, the way the injection pressure is built is this: a small amount of fuel to be injected waits in the space under the intensifier piston. When the computer actuates the injector, the drive module triggers the electronic solenoid at the top of the injector, raising the poppet valve. This simultane-

International HEUI* Fuel System Operation

*Hydraulically Activated Electronically Controlled Unit Injection

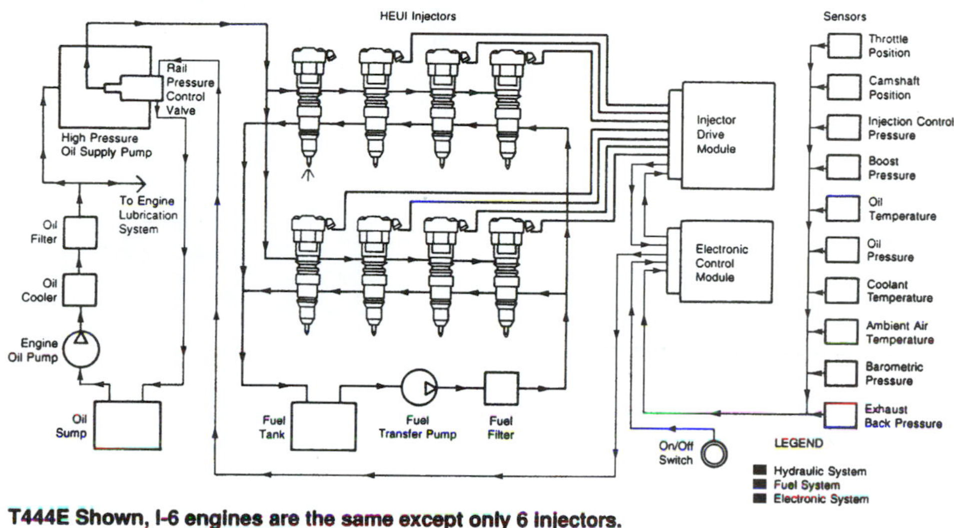

Figure 20–25 System schematic. *(Courtesy of Navistar International, Engine Division.)*

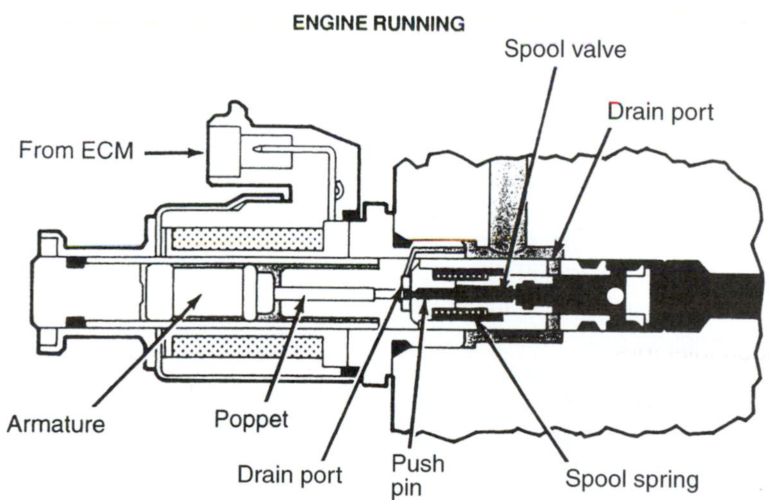

Figure 20–26 Rail pressure control valve. *(Courtesy of Navistar International, Engine Division.)*

ously closes the drain passage and opens the passage to the rail, pressurized as described previously.

Since the diameter of the top of the intensifier piston is seven times larger than the diameter of the lower fuel delivery piston, the pressure is multiplied by the same factor. So if 500 psi is applied to the top of the piston, 3,500 psi is applied to the injector tip. If 3,000 psi is applied to the top, 21,000 psi is applied to the tip.

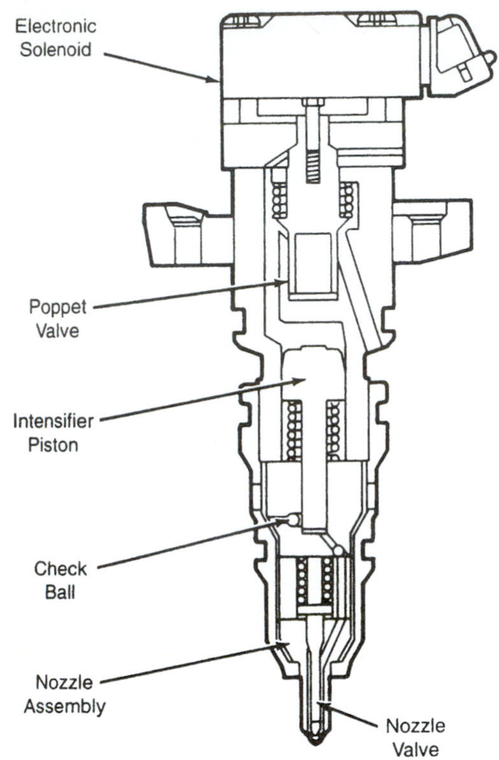

Figure 20–27 Electronic/hydraulic injector. *(Courtesy of Navistar International Transportation Corp.)*

Since the only thing activating these injectors is the computer's command, variations in delivery volume and injection advance timing are simply and entirely a matter of internal computer software. No moving parts have to shift or adjust or rotate except for the computer's sensors.

When the computer deactivates the solenoid, the springs return the intensifier piston and the nozzle needle valve to their rest positions, refilling each chamber with fuel for the next injection.

VOLKSWAGEN 1.9-LITER TDI DIESEL: SYSTEM OVERVIEW

Volkswagen has built small fuel-efficient diesel cars for many years, but the new 1.9-liter turbocharged direct-injection diesel is their most advanced compression ignition product yet, Figure 20–28.

Figure 20–28 VW 1.9-liter TDI. *(Courtesy of Volkswagen of America, Inc.)*

This small four-cylinder engine is a fully-electronic direct-injection design (all their previous diesels used prechambers). Very similar to the GM in basic design (except for its direct injection), the system is a drive-by-wire system with injection quantity and advance determined by the computer from driver commands and sensor inputs. The car includes a diesel catalytic converter to burn off residual hydrocarbons left in the exhaust. The catalytic converter, as is universal on diesels so equipped, does not reduce NO_x, but only the hydrocarbons.

INPUTS

Besides the usual sensors for engine coolant temperature, boost pressure and intake air temperature, the VW TDI system uses a hot-wire air mass sensor, just as used on cars, to measure exactly how much air is entering the engine, Figure 20–29. The other unusual sensor is a needle lift sensor on cylinder 3, which indicates to the computer exactly

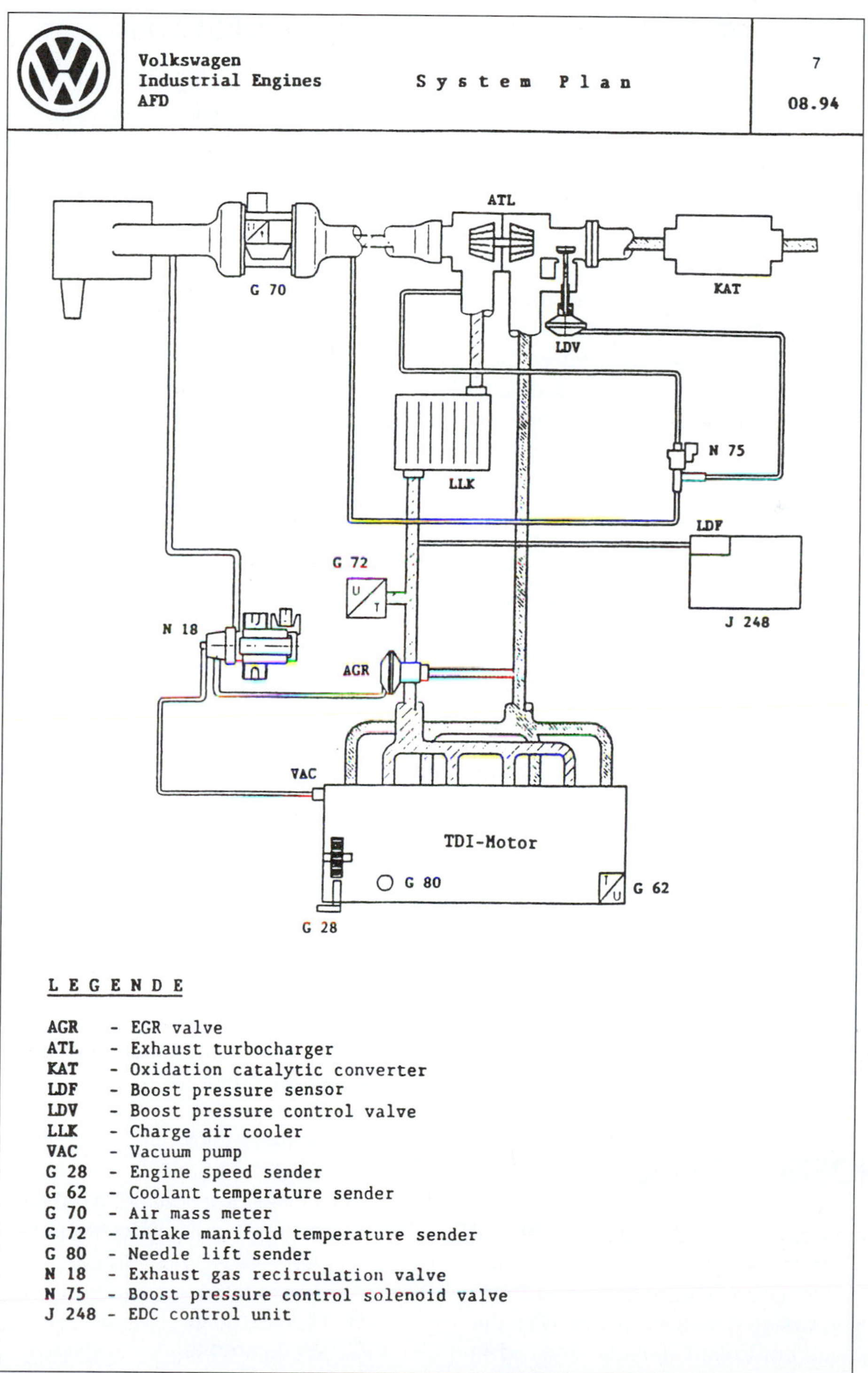

Figure 20–29 System schematic with air mass sensor. *(Courtesy of Volkswagen of America, Inc.*

what the achieved timing advance is and enables it to correct for this as needed.

SUMMARY

In this chapter, we have covered the basics of the General Motors 6.5-liter electronically controlled diesel engine, the Ford/Navistar/International 7.3-liter electronically controlled diesel and the Volkswagen 1.9-liter TDI engine. We have also considered the basic differences between diesel engines and gasoline-fueled engines, as well as diagnostic approaches to the former.

Each of these engines includes a novel drive-by-wire accelerator system, with no mechanical linkage between the accelerator pedal and the engine. Instead power demand information gets to the computer through variable resistance potentiometers, similar to a throttle position sensor on a gasoline engine.

In each engine, the computer controls the injection quantity and injection timing from the injectors. In the GM and VW engines, the computer acts on solenoids and a stepper motor at the injection pump; in the International/Ford system, the injection is entirely electrical with the computer determining pressure, quantity (through pulse width) and timing for each cylinder for each power stroke.

Because each of these systems is installed in vehicles required to be OBD II-compliant, diagnosis is possible using the standardized scan tool, a DVOM and for some sensors, a lab scope. Codes are stored in the same way as for gasoline-fueled engines, except the codes are specific to diesel engines instead of gasoline.

▲ DIAGNOSTIC EXERCISE

Diesel engine problems fall into three general categories: no-starts, rough running and smoke, black or white. Describe what differences there would be in diagnosing these problems with the new electronically controlled diesels covered in this chapter.

REVIEW QUESTIONS

1. Compression ratios on a diesel engine will typically vary between which of the following?
   A. 8:1 and 11:1
   B. 11:1 and 15:1
   C. 17:1 and 24:1
   D. 30:1 and 32:1
2. When comparing the energy content of a gallon of diesel fuel to a gallon of gasoline, diesel fuel is said to have about which of the following?
   A. 20 percent less energy than a gallon of gasoline
   B. 10 percent less energy than a gallon of gasoline
   C. The same energy as a gallon of gasoline
   D. 30 percent more energy than a gallon of gasoline
3. *Technician A* says that the advantages of a diesel engine over a gasoline engine are that a diesel engine of a similar power output gets better fuel economy and lasts longer.
   Technician B says that the advantages of a gasoline engine over a diesel engine are that a gasoline engine of a similar power output weighs less and costs less to manufacture.
   Who is correct?
   A. A only
   B. B only
   C. Both A and B
   D. Neither A nor B
4. On a diesel application, what type of smoke in the exhaust is likely to result from insufficient heat in the combustion chamber?
   A. Blue smoke
   B. Black smoke
   C. White smoke
   D. All of the above
5. On a diesel application, what type of smoke in the exhaust is likely to result from a clogged air filter?
   A. Blue smoke
   B. Black smoke
   C. White smoke
   D. All of the above

6. On a diesel application, what type of smoke in the exhaust is likely to result from engine lubricating oil getting into the combustion chamber?
   A. Blue smoke
   B. Black smoke
   C. White smoke
   D. All of the above
7. *Technician A* says that advantages of incorporating a prechamber design into a diesel engine are easier starting and smoother and quieter engine operation.
   Technician B says that an advantage of incorporating a prechamber design into a diesel engine is increased fuel economy.
   Who is correct?
   A. A only
   B. B only
   C. Both A and B
   D. Neither A nor B
8. What is the primary purpose of a glow plug in a diesel application?
   A. During normal hot engine performance, it provides a backup method of igniting the diesel fuel in the event that the spark ignition system components were to fail.
   B. It provides a source of auxiliary heat within the combustion chamber in order to help start a cold diesel engine.
   C. It is used to preheat the catalytic converter prior to starting a cold diesel engine.
   D. It is used to burn off any excess diesel fuel following engine shutdown.
9. When working with diesel fuel injection systems, it is important to follow proper safety procedures because if a fuel injector is inadvertently allowed to spray fuel directed at, or near, your skin, it may do which of the following?
   A. Spray fuel through your skin and tissue
   B. Result in severe blood poisoning
   C. Require amputation of the affected finger or hand
   D. All of the above
10. A GM 6.5-liter electronically controlled diesel engine has various driveability complaints.
    Technician A says that an OBD II scan tool should be connected to the Diagnostic Link Connector (DLC) under the instrument panel in order to help diagnose the complaints.
    Technician B says that an OBD II scan tool will be of no value on a diesel application because this type of scan tool is designed to communicate with PCMs on gasoline engine applications only.
    Who is correct?
    A. A only
    B. B only
    C. Both A and B
    D. Neither A nor B
11. On a GM 6.5-liter electronically controlled diesel engine, what sensors are located within the fuel pump?
    A. Fuel temperature sensor only
    B. Fuel temperature sensor and engine coolant temperature sensor
    C. Fuel temperature sensor and BARO sensor
    D. Fuel temperature sensor and optical sensor
12. A GM 6.5-liter electronically controlled diesel engine will run, but does not respond to the fuel pedal (accelerator pedal) at all.
    Technician A says that this symptom may result from the failure of any one of the sensors within the accelerator pedal position sensor assembly.
    Technician B says that the computer itself may have failed.
    Who is correct?
    A. A only
    B. B only
    C. Both A and B
    D. Neither A nor B
13. *Technician A* says that on a GM 6.5-liter electronically controlled diesel engine, the coolant temperature and intake air temperature sensors operate differently than the same sensors on a gasoline engine.
    Technician B says that on a GM 6.5-liter electronically controlled diesel engine, the EGR control pressure sensor is also used to provide the PCM with barometric pressure readings when the EGR system is turned off.

Who is correct?
A. A only
B. B only
C. Both A and B
D. Neither A nor B

14. On a GM 6.5-liter electronically controlled diesel engine, how many notches (slits) are monitored by the high-resolution signal from the optical sensor?
A. 180
B. 360
C. 512
D. 720

15. When controlling the glow plugs on a GM 6.5-liter electronically controlled diesel engine, the PCM primarily uses information from all of the following sensors *except* which of the following?
A. Engine coolant temperature sensor
B. Crankshaft position sensor
C. Feedback signal from the glow plugs
D. Accelerator pedal position sensor

16. On a GM 6.5-liter electronically controlled diesel engine, how is the engine shutoff solenoid used to shut down an operating engine?
A. By shutting off the air supply
B. By shutting off the fuel supply
C. By turning off the spark ignition to the spark plugs
D. Both A and B

17. *Technician A* says that GM 6.5-liter electronically controlled diesel engine computers receive input from a single VSS in two-wheel drive applications.
Technician B says that GM 6.5-liter electronically controlled diesel engine computers may receive input from a transmission output shaft speed sensor to allow it to track transfer case settings on four-wheel drive applications.
Who is correct?
A. A only
B. B only
C. Both A and B
D. Neither A nor B

18. *Technician A* says that the Ford/Navistar/International 7.3-liter diesel engine is a direct injection design and uses no prechamber above the cylinder head.
Technician B says that the Ford/Navistar/International 7.3-liter diesel engine has a compression ratio of 24:1.
Who is correct?
A. A only
B. B only
C. Both A and B
D. Neither A nor B

19. *Technician A* says that a Ford/Navistar/International 7.3-liter diesel engine uses an electronic/hydraulic injector, which uses an internal intensifier piston to multiply fuel pressure by seven times.
Technician B says that a Ford/Navistar/International 7.3-liter diesel engine may develop fuel pressures as high as 21,000 psi.
Who is correct?
A. A only
B. B only
C. Both A and B
D. Neither A nor B

20. *Technician A* says that an electronically controlled VW 1.9-liter TDI diesel engine uses a prechamber design.
Technician B says that the primary purpose of the catalytic converter on a VW 1.9-liter TDI diesel engine is to reduce NO_x emissions.
Who is correct?
A. A only
B. B only
C. Both A and B
D. Neither A nor B

Appendix A

Approach to Diagnostics

You might be asking yourself, Now that I have read this book, what do I do? Sometimes it can be difficult to put this type of theory into motion. This appendix is designed to help the reader understand how to convert the information contained in this book into hands-on diagnostics.

COLLECTING THE EVIDENCE

Information Gathering

First, as the vehicle is brought into your shop, gather information from the customer and record it on the repair order. Include as much information as possible about the conditions under which the symptom appears. It is also important to find out if the vehicle has been into a shop for the same symptom prior to this service. And, if you listen carefully, what the customer states about the problem may provide clues as to the type of problem the vehicle has, such as "This problem has been happening *ever since*. . . . That "ever since" might be just the clue you need.

Symptom Verification

The first real step of diagnosis is to attempt to verify the symptom. If the symptom can be verified, you are likely dealing with a hard fault. If the symptom cannot be verified, you are likely dealing with a soft fault.

Visual Inspection

A thorough visual inspection of the system(s) associated with the symptom should be performed. If done properly and conscientiously, problems can often be identified at this time. Additional visual inspections may need to be performed as the area of the fault is further narrowed down through diagnostic processes such as code pulling and gas analysis.

Technical Service Bulletins (TSBs)

After symptom verification and determining whether the fault is hard or soft, you should next check the manufacturer-issued technical service bulletins or TSBs. These are usually available either through electronic information systems or directly from the manufacturer. A TSB is a manufacturer-suggested repair for a problem they are already aware of. It may involve a simple fix or it may involve something as in-depth as replacing a PCM, a PROM chip in the PCM, or even reflashing of the PCM. Make sure that you use the latest TSB issued for the particular symptom that you are trying to repair, as multiple TSBs have sometimes been issued in attempting to correct one problem. Ultimately, access to TSBs is critical to a technician attempting to diagnose engine performance problems and can lead to increased diagnostic effectiveness.

HARD FAULT DIAGNOSIS

Your initial goal in diagnosing a fault of any type should be to narrow down the problem area associated with the symptom to as small an area as possible. You may take several approaches to accomplish this. Failure to reduce the possibilities can result in extreme frustration, as most technicians have experienced at one time or another.

Mental Flowchart

The first approach to narrow down the problem area is verification of the *exact* symptom. When compared to your understanding of how the system is designed to operate, verification can help you narrow down the area in which you need to perform pinpoint tests. In fact, what you are really doing is mentally creating a flowchart approach. For example, if the customer complains that there seems to be a short in the exterior lighting circuits because other circuits seem to energize the running lights (park, tail and side marker lights), the first step would be to gain an understanding of how the system is designed to operate. If it is an unfamiliar system, this can be done through the electrical schematics associated with the system. (Ever notice how most flow charts also include electrical schematics?) The next step would be to attempt to verify the *exact* symptom. This can be accomplished by operating each of the exterior lighting systems while noting the results. For example, if the exterior running lights illuminate when the left directional signal is activated, but not so with the right directional, it should hold true that the short is on the left side of the vehicle. And if the exterior running lights illuminate when the brake pedal is depressed, the short is likely at the rear of the vehicle. In this way, verification of the exact symptom can be used to narrow down the problem area, in this case to a bulb at the left rear of the vehicle that has allowed the filaments to touch each other. This type of approach ultimately results in less frustration in diagnosis of any system than simply tearing harnesses and components apart throughout the entire vehicle.

Help from an On-Board Computer

A second approach to narrowing down the problem area is to let the on-board computers help you. Modern on-board computers are programmed to allow the technician to pull memory codes, perform certain self-tests and perform other functional tests, using either manual methods to trigger these functions or with the help of a scan tool (see Chapters 3 and 4). Additionally, a scan tool allows you to view data stream information concerning input values and output commands. And these features do not just apply to engine control systems, but also to other electronic systems on today's vehicles. For example, a scan tool may also be used to pull memory codes, initiate a self-test, or look at data stream information on many antilock brake systems, body control systems, instrument clusters, and so on. Or, certain nonengine systems may allow the technician to trigger memory code pulling or self-tests manually, such as an Electronic Automatic Temperature Control (EATC) system, by pushing various combinations of buttons on the climate control head. If the computer is programmed to be able to help you, take advantage of it.

Exhaust Gas Analyzer

A third approach to narrowing down the problem area when diagnosing an engine system for poor performance, poor fuel mileage and failed emission tests is to use an exhaust gas analyzer to capture exhaust readings at both idle and 2,500 rpm, and then use these readings to give you an idea of the type of problem that you're attempting to diagnose (see Chapter 5). An exhaust gas analyzer becomes an invaluable tool when attempting to minimize the frustration factor.

Do Not Overlook the Basics

When warranted, the basic engine systems, compression, ignition and air/fuel delivery, must not be forgotten. A vacuum gauge can be used to quickly give the technician ideas of possible fault areas that should be tested further. And a cylinder power balance test (performed either with an ignition scope or within the realm of a PCM-programmed functional test) can quickly identify cylinders that may be lacking compression, spark or fuel delivery. If one or two cylinders are identified as weak, the following areas should be looked at.

Compression. Performing a compression test on each of the cylinders while the starter is cranking the engine and the throttle is held open (known as a *cranking compression test*) tests for cylinder seal. For example, a burned valve could easily cause a particular cylinder to fail this test due to the leak it creates. However, if all cylinders pass this test, and spark and fuel delivery are verified with an exhaust gas analyzer, a running compression test should be performed. To perform this test, all spark plugs must be reinstalled except the one in which the compression gauge is connected. Then the engine is started. Once the engine is running, momentarily release the compression gauge pressure. Now note the compression pressure at idle, 1,500 rpm and on a snap throttle. At idle and 1,500 rpm, the pressure reading will be about one-half of the cranking compression due to the closed throttle and the faster piston speed. Because of the faster piston speed, this is a very poor test to check cylinder seal, but it is an excellent test to check and compare each cylinder's volumetric efficiency (the cylinder's ability to breathe in and out). It is difficult to apply a hard specification to this test because of different camshaft grinds, but the cylinders should be simply compared. One or two cylinders that show lower readings than the rest indicate an inability to breathe in because of an improper opening of the intake valve, and one or two cylinders that show higher readings than the rest indicate an inability to breathe out because of an improper opening of the exhaust valve. This inability of the engine to breathe properly is caused by problems in the valvetrain such as loose rocker arms, worn push rods/lifters or a worn cam lobe. But because these valves still seal when closed, the cranking compression test does not identify the problem. One point to note: have some spare Schrader valves on hand for your compression gauge when performing running compression tests.

Ignition. If the engine cranks, but does not start, both secondary ignition and primary ignition should be checked. If the tach reference signal is not present, either from a distributor or from a crankshaft sensor, neither secondary spark nor fuel injection pulses will be present. If the engine starts and runs but has performance problems or misfire present, the secondary ignition waveform should be checked on an ignition oscilloscope. However, excepting early contact point systems, there is not really much reason today to be concerned with the primary scope pattern if the engine starts and runs, as most failures here result in a no-start situation. The exception is with multiple coil systems (either waste spark or coil-on-plug). If one coil fails to fire its spark plug(s), it could be due to the loss of the primary ignition signal to the affected coil, but the engine could still run using the remaining coils.

Air/Fuel Delivery. This area includes the common maintenance items such as air and fuel filters and the PCV valve and crankcase ventilation filter. Also, if the engine has failed a cylinder power balance test, measure the exhaust emissions with an exhaust gas analyzer at both idle and 1,500 rpm. Then turn off the engine and disable the spark for the failed cylinder. On distributor and waste spark systems, simply use a jumper wire to short one spark plug wire to or from engine ground. On coil-on-plug (COP) systems, simply disconnect the primary connector from the coil for the failed cylinder. Now restart the engine. If unburned hydrocarbons increase dramatically from their previous values, the failed cylinder has both spark and (at least some) fuel present. If this is the case, then cranking and running compression tests should be performed (if they have not already been).

Pinpoint Testing

To isolate any remaining hard faults, visualize the electrical circuit in three parts: the source (battery or alternator), the circuitry and the load component. The source is the first component that should be verified. For example, if the starter cranks slowly, check the battery's open circuit voltage and then load test it. It is a good idea to test the battery's open circuit voltage for symptoms associated with smaller load components, even if the starter cranks okay, due to the affects of low open circuit voltage on computer operation. Also, do not overlook the effects of excessive alternator ripple on computer-controlled systems. To test this, connect a DVOM between the battery positive and negative terminals while scaled to an AC voltage scale. With the engine running, AC voltage should stay well below 500 millivolts. Excessive alternator ripple can, of course, also be seen with a DSO (lab scope).

To test the circuitry, use a DVOM on a DC voltage scale to perform voltage drop tests on both the positive and negative sides of the circuit (see the section about voltage drop testing in the Introduction). This verifies the integrity of the circuitry used to connect the source to the load component. While 100 millivolts per connection is allowed in normal circuits (and even 200 millivolts per connection in starter circuits due to the high-current, low-resistance designs of these circuits), a power or ground circuit that is used to power up a PCM (or other computer) should never exceed 100 millivolts *total* voltage drop in either circuit. This tight specification is due to the high-resistance, low-current flow designs of these circuits. Many good PCMs are replaced simply because of a poor power or ground circuit. Therefore, after verifying each of the input and output circuits, but before replacing the PCM, you should get in the habit of using an electrical schematic to identify each of the PCM's power and ground terminals. Then with the ignition on, perform voltage drop tests between the battery and the PCM.

Ultimately, the load component can be tested in the following methods. If the load component is an electronic control module, simple elimination of possible faults elsewhere in the circuitry through voltage drop testing is used most often to narrow the problem down to the control module itself. Other load components can be tested electrically and functionally. Electrical tests involve the use of an ohmmeter to measure a solenoid's (or relay's) resistance in the windings, or jumper wires to energize the component. Functional testing can be done using other methods, for example, using a vacuum pump and vacuum gauge to test a vacuum solenoid.

Other tools that can make pinpoint testing easier include DSOs, logic probes and breakout boxes (see Chapter 4). DSOs can show voltage waveforms, logic probes can quickly check for the presence of ignition primary and fuel injector signals and breakout boxes are used to improve computer circuit accessibility.

Finally, the use of flowcharts and electrical schematics are invaluable when it comes to pinpoint testing. A flowchart is written according to the design engineer's understanding of system operation. An electrical schematic helps you as a technician to understand system operation.

SOFT FAULT DIAGNOSIS

The question is often asked, How can I pinpoint test a fault when the symptom does not appear to be present? There are several answers to this question. Certainly, past experience on like models with similar problems counts for something. Unfortunately, this approach often results in improper repairs and unnecessary parts replacement. A better approach is suggested in the following paragraphs.

The first step is to determine if the PCM has stored any DTCs in memory that could direct you to a probable problem area. At the very least, a memory DTC can help you determine if the reason for an "intermittent engine shutdown until cooled" is the likely result of the ignition system or the fuel system. However, additional tests can be performed to pinpoint test the exact cause, even when the symptom does not appear to be present.

If the soft fault is in the circuitry, performing voltage drop tests of the circuitry will locate the fault, even when the circuit appears to be functioning normally. Remember, voltage drop testing is used to identify excessive unwanted resistance in the circuitry. This excessive unwanted resistance increases with an increase in temperature to the point of creating a symptom. Likewise, this resistance decreases with a decrease in temperature, to the point that the circuit again becomes operational. But the excessive unwanted resistance does not totally disappear, even when the circuit appears operational and the symptom does not appear to be present. Therefore, if a soft fault is in the circuitry, voltage drop testing can identify the circuit problem even without the symptom present.

If the soft fault is in the load component, using a DSO (lab scope) and a low-current probe to sample the current waveform of the load component can identify faults in the load component that can cause intermittent operation (see Chapter 4). In this way, a load component can be proved to be defective, even though it may still be operational while in your shop. This technology allows you to show your customer a printed waveform that proves that a load component is not being replaced needlessly. It can even allow you to identify a load component that is ready to fail, even before symptoms have been encountered.

VERIFYING PCM CONTROL OF THE AIR/FUEL RATIO

When diagnosing a complaint of poor engine performance, poor fuel economy or an emission test failure, you will need to verify the ability of the PCM to properly control the air/fuel ratio. This can be done through the primary oxygen sensor's voltage waveform, but only if the oxygen sensor itself has first been tested and proved to be in good condition.

To verify the oxygen sensor, use a DSO to monitor the oxygen sensor's waveform while forcing the air/fuel ratio both rich and lean. This can be done on most vehicles by simply throttling the engine slightly. Acceleration results in a rich air/fuel mixture and deceleration results in a lean air/fuel mixture. If needed, a propane enrichment tool can be used to force the mixture rich. Then suddenly turning off the propane will force the mixture lean. The oxygen sensor should be capable of exceeding 800 millivolts while the air/fuel ratio is rich, then descending to less than 175 millivolts when the air/fuel ratio is lean. It should be able to switch between 300 millivolts and 600 millivolts in either direction in less than 100 milliseconds. And it should never show a negative value. Any failure in any of these areas is cause for replacement. If you replace the oxygen sensor, drive the vehicle about 10 miles to break in the new oxygen sensor before you attempt to test it.

Once the oxygen sensor has been verified, sample the oxygen sensor waveform with a DSO while driving the vehicle on a road test. Be sure to have another technician drive the vehicle so that you can safely monitor the DSO. The oxygen sensor should show a consistent cross-counting of its voltage to both sides of the stoichiometric value. Additionally, fuel trim values should be monitored with a scan tool during the road test (see Fuel Trim in Chapter 10). The short-term fuel trim should be reacting to the oxygen sensor's voltage. All of this indicates good computer control of the air/fuel ratio.

If the oxygen sensor's voltage is pegged either high or low, then compare this signal with both the pulse width command to the fuel injectors and the fuel trim values. If the oxygen sensor voltage is pegged lean and the pulse width command and long-term fuel trim both indicate that the PCM is attempting to enrich the air/fuel ratio, the problem is likely mechanical. Likewise, if the oxygen sensor voltage is pegged rich and the pulse width command and long-term fuel trim both indicate that the PCM is attempting to lean out the air/fuel ratio, the problem is also likely mechanical. But if the pulse width command to the fuel injectors and the fuel trim numbers both indicate that the PCM is not reacting properly to the oxygen sensor signal (and is in fact responsible for the pegged rich or lean condition), then the problem is an input signal (faulty input sensor or circuit), high-voltage drop in the PCM's power and ground circuits or the PCM itself. (The assumption is made here that there is no false air upstream of the oxygen sensor from either an exhaust leak or a faulty secondary air injection system.)

SUMMARY

This appendix is designed to help the reader put some of the information contained in various earlier chapters of this book into perspective. No one chapter of this textbook is going to provide a quick repair guide to the problems encountered on today's complex vehicles. Rather, it takes an understanding of the concepts covered in the early chapters in combination with the product-specific information covered in the later chapters to be able to effectively diagnose and repair today's modern systems. Even then, in spite of this author's best efforts, none of us will ever fully retain all of the system product-specific information necessary for effective repair and diagnosis. Therefore, a source of product-specific information is as necessary as any other tool that you might use in your business. These service manuals may be paper or electronic, but they are very necessary.

A good technician today does not know it all. Rather, he or she must strive to understand the operating concepts and diagnostic concepts associated with today's modern systems and then know where to look up the product-specific information when an unfamiliar system is encountered. If this textbook has provided the reader with an understanding of the underlying concepts and with the knowledge required to comprehend the paper/electronic service manual literature, then our goal has been achieved.

Appendix B

Terms and Acronyms

This appendix contains figures that provide terminology and acronyms for common (non-manufacturer-specific) terms, OBD II terms, and terms for both domestic and Asian vehicles as well as many of their OBD II counterparts as applicable. The figures list the manufacturers' terms and acronyms in the same order that these manufacturers are addressed through the chapters of this book.

Term	Acronym	Term	Acronym
Air Conditioning	A/C	Ohms	Ω
Alternating Current	AC	Oxides of Nitrogen	NO_x
Amps	A	Oxygen	O_2
Analog Volt/Ohm Meter	AVOM	Park/Neutral	P/N
Carbon Dioxide	CO_2	Ported Vacuum Switch	PVS
Carbon Monoxide	CO	Positive Crankcase Ventilation	PCV
Digital Volt Meter	DVM	Programmable Read-Only Memory	PROM
Digital Volt-Ohmmeter	DVOM	Pulse Width Modulated	PWM
Direct Current	DC	Random-Access Memory	RAM
Hydrocarbons	HC	Read-Only Memory	ROM
Ignition	IGN	Revolutions per minute	rpm
Instrument Panel	IP or I/P	Tachometer	TACH
Key On, Engine Crank	KOEC	Thermal Vacuum Switch	TVS
Key On, Engine Off	KOEO	Thermostatic Air Cleaner	TAC
Key On, Engine Running	KOER	Variable Pulse Width	VPW
Miles per hour	mph	Vehicle Identification Number	VIN
Nitrogen	N_2	Volts	V
Normally Closed	N/C	Water	H_2O
Normally Open	N/O	Wide-Open Throttle	WOT

Figure B–1 Common terms and acronyms.

OBD II Term	OBD II Acronym
Air-Conditioning Clutch	ACC
Air-Conditioning Cycling Switch	ACCS
Air-Conditioning Demand	ACD
Barometric Pressure sensor	BARO sensor
Canister Purge	CANP
Camshaft Position sensor	CMP sensor
Crankshaft Position sensor	CKP sensor
Data Link Connector	DLC
Diagnostic Link Connector	DLC
Diagnostic Trouble Code	DTC
Distributor Ignition*	DI
Early Fuel Evaporation	EFE
Electronic Ignition**	EI
Engine Coolant Temperature sensor	ECT sensor
Evaporative Emission Control System	EVAP
Flexible Fuel	FF
Generator (Alternator)	GEN
Heated Oxygen Sensor	HO_2S
Idle Air Control	IAC
Idle Speed Control	ISC
Intake Air Temperature sensor	IAT sensor
Knock Sensor	KS
Long Term Fuel Trim	LTFT
Malfunction Indicator Lamp	MIL
Malfunction Indicator Light	MIL
Manifold Absolute Pressure sensor	MAP sensor
Mass Air Flow sensor	MAF sensor
Mixture Control solenoid	MC solenoid
Multiport Fuel Injection	MFI
Oxidation Catalyst	OC
Oxygen Sensor	O_2S
Power Steering Pressure	PSP
Powertrain Control Module	PCM
Pulsed Secondary Air Injection Reaction system	PAIR system
Secondary Air Injection Reaction system	AIR system
Sequential Fuel Injection	SFI
Short Term Fuel Trim	STFT
Speed Density	SD
Three-Way Catalyst	TWC
Throttle Body Injection	TBI
Throttle Position Sensor	TP sensor or TPS
Torque Converter Clutch	TCC
Transmission Control Module	TCM
Vehicle Speed Sensor	VSS

**(mechanical spark distribution using a single coil and a distributor)*
***(electronic spark distribution using multiple coils and no distributor)*

Figure B–2 OBD II standardized terms and acronyms.

GM Term	GM Acronym	OBD II Acronym
Air Injection Reaction	AIR	AIR
Assembly Line Communication Link	ALCL	DLC
Assembly Line Diagnostic Link	ALDL	DLC
Barometric Pressure sensor	BARO sensor	BARO sensor
Backpressure EGR	BPEGR	BPEGR
Block Learn		LTFT
Charcoal Canister Purge	CCP	CANP
Computer Command Control	CCC or C3	
Computer Controlled Coil Ignition	C3I	
Check Engine Light	CEL	MIL
Climate Control Panel	CCP	
Coolant Temperature Sensor	CTS	ECT sensor
Direct Ignition System	DIS	
Digital Fuel Injection	DFI	
Early Fuel Evaporation	EFE	EFE
Electronic Climate Control	ECC	
Electronic Climate Control Panel	ECCP	
Electronic Control Module	ECM	PCM
Electronic Fuel Injection	EFI	
Electronic Spark Control	ESC	
Electronic Spark Timing	EST	
Evaporative Emission Control System	EECS	EVAP
Exhaust Gas Recirculation	EGR	EGR
Fuel Data Panel	FDP	
Heated Oxygen Sensor	HO_2S	HO_2S
High Energy Ignition	HEI	
Idle Air Control	IAC	IAC
Idle Load Compensator	ILC	
Idle Speed Control	ISC	ISC
Integrated Direct Ignition	IDI	
Integrator		STFT
Manifold Absolute Pressure sensor	MAP sensor	MAP sensor
Manifold Air Temperature sensor	MAT sensor	IAT sensor
Mass Air Flow sensor	MAF sensor	MAF sensor
Mixture Control solenoid	M/C solenoid	MC solenoid
Multiport Fuel Injection	MFI	MFI
Oxygen Sensor	O_2S	O_2S
Port Fuel Injection	PFI	MFI
Power Steering Switch	PSSw	PSP
Pulse Air Injection Reaction	PAIR	PAIR
Sequential Fuel Injection	SFI	SFI
Throttle Body Injection	TBI	TBI
Torque Converter Clutch	TCC	TCC
Throttle Position Sensor	TPS	TP sensor or TPS
Tuned Port Injection	TPI	
Vehicle Control Module	VCM	
Vehicle Speed Sensor	VSS	VSS

Figure B–3 General Motors terms and acronyms.

Ford Term	Ford Acronym	OBD II Acronym
Adaptive Strategy		LTFT/STFT
Air Charge Temperature sensor	ACT sensor	IAT sensor
Air-Conditioning Clutch	ACC	
Air-Conditioning Cycling Switch	ACCS	
Air-Conditioning Demand	ACD	
Backpressure EGR	BPEGR	BPEGR
Barometric Pressure sensor	BP sensor	BARO sensor
Brake On/Off switch	BOO switch	
Canister Purge	CANP	CANP
Central Fuel Injection	CFI	TBI
Clutch Engaged Switch	CES	
Computer Controlled Dwell	CCD	
Crankshaft Position sensor	CP sensor	CKP sensor
Cylinder Identification	CID	
Delta Pressure Feedback EGR sensor	DPFE sensor	
Differential Pressure Feedback EGR sensor	DPFE sensor	
Distributorless Ignition System	DIS	
Dual Plug Inhibit	DPI	
EGR Control solenoid	EGRC solenoid	
EGR Valve Position sensor	EVP sensor	
EGR Valve Regulator solenoid	EVR solenoid	
EGR Vent solenoid	EGRV solenoid	
Electronic Control Assembly	ECA	PCM
Electronic Distributorless Ignition System	EDIS	
Electronic Engine Control	EEC I, II, III, IV, V	
Electronic Fuel Injection	EFI	MFI
Engine Coolant Temperature sensor	ECT sensor	ECT sensor
Exhaust Gas Oxygen sensor	EGO sensor	O_2S
Exhaust Gas Recirculation	EGR	EGR
Failure Mode Effects Management	FMEM	
Feedback Carburetor Actuator	FBCA	
Fuel Pump Monitor	FPM	
Heated Exhaust Gas Oxygen sensor	HEGO sensor	HO_2S
Idle Speed Control	ISC	ISC
Idle Tracking Switch	ITS	
Ignition Diagnostic Monitor	IDM	
Inlet Air Solenoid	IAS	
Inferred Mileage Sensor	IMS	
Integrated Relay Control Module	IRCM	
Integrated Vehicle Speed Control	IVSC	
Keep-Alive Memory	KAM	

Figure B–4 Ford Motor Company terms and acronyms.

Ford Term	Ford Acronym	OBD II Acronym
Keep-Alive Power	KAPWR	
Key On, Engine Off	KOEO	
Key On, Engine Running	KOER	
Knock Sensor	KS	KS
Limited Operational Strategy	LOS	
Malfunction Indicator Lamp	MIL	MIL
Manifold Absolute Pressure sensor	MAP sensor	MAP sensor
Mass Air Flow sensor	MAF sensor	MAF sensor
Microprocessor Control Unit	MCU	
Neutral Drive Switch	NDS	
Neutral Gear Switch	NGS	
New Generation Star Tester	NGS	
Octane Adjust	OCT ADJ	
Power Steering Pressure Switch	PSPS	PSP
Pressure Feedback EGR sensor	PFE sensor	
Profile Ignition Pickup	PIP	
Programmable Speedometer/ Odometer Module	PSOM	
Self-Test Automatic Readout tester	STAR tester	
Self-Test Connector		DLC
Self-Test Input	STI	
Self-Test Output	STO	
Sequential Electronic Fuel Injection	SEFI	SFI
Service Bay Diagnostic System	SBDS	
Signal Return	SIG RTN	
Spark Angle Word	SAW	
Spark Output	SPOUT	
Speed Control Command Switches	SCCS	
Temperature Compensated Pump	TCP	
Thermactor Air Bypass	TAB	
Thermactor Air Diverter	TAD	
Thick Film Integrated	TFI	
Throttle Air Bypass	TAB	
Throttle Kicker	TK	
Throttle Position sensor	TP sensor	TP sensor or TPS
Vane Air Flow meter	VAF meter	
Vane Air Temperature sensor	VAT sensor	IAT
Variable Reluctance Sensor	VRS	
Variable Venturi	VV	
Variable Voltage Choke	VVC	
Vehicle Power	VPWR	
Voltage Reference	VREF	

Figure B–4 (*continued*)

DaimlerChrysler Term	DaimlerChrysler Acronym	OBD II Acronym
Actuator Test Mode	ATM	
Adaptive Fuel Control		LTFT/STFT
Automatic Idle Speed	AIS	IAC
Automatic Shut Down relay	ASD relay	
Backpressure EGR	BPEGR	BPEGR
Barometric Pressure Read solenoid	BARO Read	
Charge Temperature sensor		IAT sensor
Coolant Temperature Sensor	CTS	ECT sensor
Diagnostic Readout Box	DRB I, II, III	
Direct Ignition System	DIS	
Distance Sensor	DS	VSS
Electronic Lean Burn	ELB	
Electronic Spark Advance	ESA	
Electronic Spark Control	ESC	
Engine Control Unit	ECU	PCM
Exhaust Gas Recirculation	EGR	EGR
Heated Oxygen sensor	HO_2S	HO_2S
Jeep/Truck Engine Controller	JTEC	
Knock Sensor	KS	KS
Manifold Absolute Pressure sensor	MAP sensor	MAP sensor
Mass Air Flow sensor	MAF sensor	MAF sensor
Next Generation Controller	NGC	
Oxygen sensor	O_2S	O_2S
Power Loss Lamp	PLL	MIL
Sequential Fuel Injection	SFI	SFI
Single Board Engine Controller	SBEC I, II, III, IIIa	
Single Module Engine Controller	SMEC	
Throttle Body Injection	TBI	TBI
Throttle Position Sensor	TPS	TP sensor or TPS

Figure B–5 DaimlerChrysler Corporation terms and acronyms.

Nissan Term	Nissan Acronym	OBD II Acronym
Absolute Pressure Sensor		MAP/BARO
Adaptive Fuel Control	ALPHA	LTFT/STFT
Auxiliary Air Control valve	AAC valve	IAC
Intake Air Temperature sensor	IAT sensor	IAT sensor
Camshaft Position sensor	PHASE	CMP sensor
Crankshaft Position sensor	REF/POS	CKP sensor
Electronic Concentrated Control System	ECCS	
Engine Coolant Temperature sensor	ECT sensor	ECT sensor
Engine Control Unit	ECU	PCM
Exhaust Gas Recirculation	EGR	EGR
Heated Oxygen sensor	HO_2S	HO_2S
Idle Air Control Valve	IACV	IAC
Mass Air Flow sensor	MAF sensor	MAF sensor
Oxygen sensor	O_2S	O_2S
Sequential Fuel Injection	SFI	SFI
Throttle Position Sensor	TPS	TP sensor or TPS

Figure B–6 Nissan terms and acronyms.

Toyota Term	Toyota Acronym	OBD II Acronym
Intake Air Temperature sensor	IAT sensor	IAT sensor
Engine Coolant Temperature sensor	ECT sensor	ECT sensor
Engine Control Unit	ECU	PCM
Exhaust Gas Recirculation	EGR	EGR
Heated Oxygen sensor	HO_2S	HO_2S
Manifold Absolute Pressure sensor	MAP sensor	MAP sensor
Mass Air Flow sensor (airflow meter)	MAF sensor	MAF sensor
Oxygen sensor	O_2S	O_2S
Sequential Fuel Injection	SFI	SFI
Toyota Computer-Controlled System	TCCS	
Throttle Position Sensor	TPS	TP sensor or TPS

Figure B–7 Toyota terms and acronyms.

Honda Term	Honda Acronym	OBD II Acronym
Constant Vacuum Control valve	CVC valve	
Crank Angle sensor	CA sensor	CKP sensor
Electronic Air Control Valve	EACV	IAC
Engine Control Unit	ECU	PCM
Exhaust Gas Recirculation	EGR	EGR
Heated Oxygen sensor	HO_2S	HO_2S
Idle Mixture Adjuster	IMA	MC solenoid
Manifold Absolute Pressure sensor	MAP sensor	MAP sensor
Mass Air Flow sensor	MAF sensor	MAF sensor
Oxygen sensor	O_2S	O_2S
Pressure Absolute sensor	PA sensor	BARO sensor
Programmed Fuel Injection	PGM-FI	
Sequential Fuel Injection	SFI	SFI
Temperature Air sensor	TA sensor	IAT sensor
Temperature Water sensor	TW sensor	ECT sensor
Throttle Angle Sensor	TAS	TP sensor or TPS

Figure B–8 Honda terms and acronyms.

Glossary

A4LD transmission (automatic four-speed light-duty) — An overdrive transmission featuring a PCM-controlled torque converter clutch for rear-wheel drive vehicles.

AC — A parts manufacturing and distribution division of General Motors.

Accumulator — A reservoir located between the evaporator (under dash unit) and the compressor of an air-conditioning system. It allows refrigerant and any remaining liquid to separate. Also used to describe a component in any hydraulic circuit (like an automatic transmission) that moderates the application of some clutch or band.

Active voltage — A voltage that is either actively pulled high or low by a computer circuit.

Actuator — Any mechanism, such as a solenoid, relay or motor, that when activated causes a change in the performance of a given system or circuit.

Adaptive memory — The ability of an engine control computer to assess the success of its actuator and sensor signals, and modify its internal calculations to correct them if the desired result is not achieved.

Adaptive strategy — The internal programs by which the Ford PCM learns from the results of its monitoring of sensors and activation of actuators, and can then modify its programming to better achieve driveability, emissions and fuel economy objectives.

Alternate fuel compatibility — The ability of an engine to run on various fuels. The first step is the use of methanol- and ethanol-compatible seals and other parts in the fuel delivery system. The use of entirely different fuels, such as natural gas, is predicted to be the next step toward fuel economy.

Ampule — A strong glass tube that forms the body of an electrical component such as a vacuum tube.

Analog — A signal that continuously varies within a given range and with time used for the change to occur.

Aneroid — An accordionlike capsule that expands or contracts in response to external pressure changes.

Annulus ring — On the 6.5-liter GM diesel and others with similar injection pumps, a perforated ring in the injection pump that delivers fuel to the injection pump.

Anode — A conductor with positive voltage potential applied.

Arbitration — The process of determining which node can continue to transmit data on the bus when two or more nodes begin transmitting data at the same time.

Aspirator — A device using airflow through a restriction in a tube to create a vacuum in another tube.

Asynchronous — Data sent on a data bus intermittently and as needed, rather than in a steady stream.

Aurora/Northstar — An engine model for Oldsmobile and Cadillac, incorporating virtually all features on earlier computer control systems.

AXOD (automatic transaxle overdrive) — A four-speed automatic transmission featuring a PCM-controlled torque converter clutch used in some front-wheel-drive vehicles.

Background noise — Modern knock sensors are effectively microphones informing the computer of all the engine's internal sounds, which the computer then sorts out for frequencies characteristic of knock. By tracking the increase in background noise with higher engine speed and load, the computer can test the effectiveness of the newer type of knock sensor.

Barometric pressure — The actual ambient pressure of the air at the engine. Because this pressure varies somewhat with the weather and altitude, the information is needed to correctly determine air/fuel mixture.

Barometric pressure sensor — A pressure sensor used to measure atmospheric pressure. Its signal to the PCM may be analog (a variable DC voltage) or digital (a variable digital voltage frequency).

Baud rate — A variable unit of data transmission speed (as one bit per second). In a stream of binary signals, one baud is one bit per second. Baud refers to the speed at which information can be sent and received over a data bus. The limitation comes from the sending and receiving units, not from the wire itself. Many modern automotive computers can transmit and receive at a baud in excess of 10,000. The term *baud* is in honor of the Frenchman J. M. E. Baudot, who is responsible for much of the early development in unit data transmission.

Binary code — This is the form of all the signals transferred on a data bus. It consists of a series of on/off signals corresponding to the digits of the binary numbers, each with a designated meaning. Binary code transfer is very fast.

Bit — A single piece of information in a binary code system; somewhat comparable to a single letter within a word or a single digit within a number.

Bleed — A controlled leak.

Block learn — The adaptive memory systems used on General Motors cars. As the vehicle ages and driving conditions change (like seasonal weather or altitude), the computer can make immediate and semipermanent changes to the fuel mixture and spark maps in its keep-alive memory to optimize driveability, emissions and fuel economy.

Breakout box — A diagnostic tool using special connectors and a terminal box that enables the technician to check voltages on all circuits of a harness while the system is in operation.

Byte — Eight pieces of binary code information strung together to make a complete unit of information. Certain newer on-board computers recognize either 16 or 32 bits of information as representing a byte (known as 16-bit or 32-bit computers).

C3I — A name used by General Motors to describe their three-coil ignition system with a module instead of a distributor.

CALPAK — A removable part of the PCM that plugs in like the PROM and is primarily responsible for storing information regarding fuel injection calibration. Used on the GM 2.0-liter engine.

Capacitor — An electric storage device used in most pressure sensors by Ford EEC IV systems. The unit's capacitance is transformed into a frequency signal internally and sent to the PCM.

Carbon dioxide (CO_2) — A molecule that contains one carbon atom and two oxygen atoms. A nontoxic emission produced by proper combustion taking place in the engine with enough oxygen present. CO_2 levels are an indication of combustion efficiency with high CO_2 levels desired.

Carbon monoxide (CO) — An oxygen-starved molecule that contains one carbon atom and one oxygen atom. A toxic emission produced by combustion taking place in the engine with a shortage of oxygen present. A CO molecule will eventually seek out an oxygen atom and transform itself to CO_2. In the interim, exposure to CO can cause headaches, dizziness, nausea and death. CO has no taste, odor or color and cannot be sensed by human senses.

Catalyst — An agent that causes a chemical reaction between other agents without being changed itself.

Cathode — A conductor with negative voltage potential applied.

Central Multiport Fuel Injection (CMFI) — A fuel injection system used by General Motors on Vortec engines that delivers fuel from a single, centrally located fuel injector through nylon tubes and spring-loaded poppet valves to spray fuel on the back side of each cylinder's intake valve. As a result, only air is moving through the intake manifold, similar to most port-injected engines.

Central Sequential Fuel Injection (CSFI) — A fuel injection system used by General Motors on Vortec engines that is similar to the CMFI system except that multiple injectors are used to deliver fuel to each cylinder by way of nylon tubes and poppet valves. The injector solenoids are centrally located between the upper and lower halves of the intake manifold.

Ceramic zirconia oxygen sensor — One of the two general types of oxygen sensors, it uses ceramic zirconia as a solid electrolyte in what is essentially a low-current, low-voltage cell (as in a battery) to generate a voltage signal corresponding to the amount of oxygen remaining in the exhaust. The oxygen sensor must become hot before it starts to work.

Clamping diode — A diode placed across a coil or winding to trap the voltage spikes produced when the device is turned off.

Closed loop — The triangular relationship between a computer, an actuator and a sensor that allows the computer to monitor the results of its own actuator control through a feedback signal from the sensor (or circuit). This concept may be used with such things as air/fuel ratio control, EGR control, fuel pump and ignition system monitoring, Electronic Automatic Temperature Control (EATC) systems, antilock brake systems and speed control systems.

Closed loop/open loop — Closed loop is the mode in which the computer adjusts the air/fuel mixture in response to signals from the oxygen sensor, which in turn responds to the changed mixture ratio. Open loop refers to control of the mixture based on other information inputs, but not the oxygen sensor signal.

Compression ignition (CI) — As opposed to spark ignition, a compression ignition engine is a diesel engine. It ignites the air/fuel mixture entirely by the heat generated by the compression stroke. Since the typical diesel engine has a compression ratio of 17- to 23-to-1, the temperature generated is usually between 300 and 450°F. Fuel is injected into the compression-ignition combustion chamber at the point flame propagation should begin.

Concentration sampling — A method of gas analysis used by most gas analyzers in automotive shops today. It has a probe that is inserted into the exhaust pipe. The probe allows the analyzer to sample only a small percentage of the total exhaust gas volume. Each of the gasses is measured as a percentage, or concentration, of the total sample. The resulting measurements equate to either percent or parts-per-million (PPM), with PPM simply being a smaller percentage.

Constant volume sampling (CVS) — A method of gas analysis that requires that all of the exhaust is captured and measured along with enough ambient air to create a known volume of gasses. Then the density of this known volume is calculated. Finally, the concentration (percentage) of each gas is calculated in order to figure how many grams of each gas is being expelled from the exhaust system. The analytical computer is also able to use dynamometer readings to determine the distance that the vehicle has traveled in any instant in time or over a longer time period. At this point each gas is calculated as grams per mile (GPM).

Continuous injection — One of the two kinds of fuel injector fuel spray techniques, the other being pulsed injection. A continuous injector (basically limited to the Bosch K-Jetronics systems) sprays fuel whenever the engine is running. The quantity of fuel injected is determined not by the injectors, but by the fuel distributor, a separate component.

Current ramping — The use of a lab scope (Digital Storage Oscilloscope or DSO) to look at current waveforms of load components or actuators, using a low current probe to capture the waveform. The name comes from the fact that current builds slowly when turned on, causing an upslope, or upward ramp, to appear on the waveform.

Data bus network — The data bus network is the information transfer harness between the elements on the multiplexing system. In some vehicles, this consists of a special pair of twisted wires going to each element. The wires are twisted to prevent introduction of spurious signals from electrical fields generated by speed sensors, horn and headlight circuits, radios (particularly high power or CB) and spark plug cables. Many newer systems use a single-wire data bus design. The serial data on a single-wire bus has a waveform that is trapezoidal in shape. This allows the nodes that are communicating on the bus to distinguish between the sharp change of an induced voltage spike and a binary bit of information.

Data stream — The communication between a scan tool and an on-board computer when the scan tool is connected to the Data Link Connector (DLC) and the ignition switch is turned on. The communication is in the form of binary code (serial data) and allows either one-way or two-way communication between the scan tool and the on-board computers.

Demultiplexor — An electronic chip that, when built into a node, gives it the ability to receive and decipher messages on a data bus.

Detonation — Uncontrolled, rapid burning or explosion of the air/fuel charge that results in cylinder knock.

Diagnostic Routines — A section in the service manual that lists possible causes of a driveability complaint. It is the beginning point of the diagnostic procedure once the complaint has been verified.

Diagnostic trouble code (DTC) — A set of alphanumeric sequences the computer stores in its memory when it senses failures in certain sensor or actuator circuits. DTC is the term used in OBD II literature; trouble codes or fault codes were the terms used before OBD II.

Digital — A voltage signal that is either of two voltage values, but does not vary in a continuous fashion between them. Typically these signals will be a voltage that is turned either on or off, but may also vary between two voltage values that are both above ground. On a lab scope, the change of voltage (either high or low) shows a sharp, vertical edge.

Digital storage oscilloscope (DSO) — A lab scope that displays a computer-generated waveform that may be frozen, measured, stored, downloaded to a PC and printed out. The on-screen adjustments include coupling, ground level, volts per division, time per division and trigger characteristics and are similar to an analog lab scope. A DSO is most often used to capture voltage waveforms over the time shown on the screen, but may also be configured with a low-current probe to capture current waveforms (known as *current ramping*). A DSO is designed to perform pinpoint tests of a circuit or component.

Digital volt-ohmmeter (DVOM) — A diagnostic tool that allows the technician to view numerical readings relating to voltage, amperage and resistance measurements. Many DVOMs can also be used to measure frequency, duty cycle, pulse width, rpm and temperature. High-end units also may include a bar graph in order to improve update resolution. Autoscaling DVOMs are designed to rescale automatically according to the values being measured. A DVOM is designed to perform pinpoint tests of a circuit or component.

Direct ignition/waste spark ignition — A direct ignition system eliminates the ignition distributor and fires spark plugs directly from the coil. If the direct ignition system is of the waste spark type, each coil fires two spark plugs simultaneously. One spark occurs at the end of its cylinder's compression stroke just before the power stroke; the other spark occurs at the cylinder 360 degrees apart in firing order from the first, during the end of that cylinder's exhaust stroke. Each spark is of opposite polarity. Nonwaste spark direct ignition systems use one coil per spark plug.

Direct injection — A type of diesel engine in which the fuel is injected directly into the center of the combustion chamber rather than into a prechamber or Ricardo chamber. Direct injection diesels are, because of the reduced area for heat to transfer into the cooling system, more fuel efficient than other types of diesel.

Divert mode — When the PCM directs the air management system to dump air into the air cleaner or to the atmosphere rather than to the exhaust system.

Drive cycle — A series of operating conditions that allows the PCM to complete all of the OBD II emission-related monitors.

Driveability — Those factors, including ease of starting, idle quality and acceleration without hesitation that affect the ease of driving and reliability.

Dualjet — A two-venturi carburetor consisting basically of the primary bores and circuits of a Quadrajet, usually referred to by General Motors as E2ME. The last "E" means it is equipped for CCC systems.

Duraspark — An electronic ignition system developed by Ford. Three generations of the system have evolved: Duraspark I, II and III.

Duty cycle — The portion of time during each cycle when an electrical device is turned on, when the device is cycled at a fixed number of cycles per unit of time.

E-cell — A module containing a cathode and an anode, each of which is gradually sacrificed as current passes through them, eventually causing the circuit to open.

EEPROM — Electronically erasable programmable read-only memory, an improved and updated kind of PROM. Like the PROM, it can retain tabular information for the PCM even if power is withdrawn from the computer. Unlike the PROM, this information can be changed by the computer. Also, unlike the PROM, the EEPROM is a nonreplaceable part of the computer.

EGR temperature sensor — On late model cars, particularly those that are OBD II-compliant, the EGR temperature sensor serves to signal the computer that recirculated exhaust gas flows when it sends the command for it to do so. This sensor makes use of the fact that the exhaust gas is very hot, so the temperature of the passage provides a reliable indicator of exhaust flow.

Electrical schematic — A map of an electrical system's circuitry that can be used by the technician as a diagnostic aid. It is most often used to increase the level of understanding associated with an unfamiliar system.

Engine calibration — Adjustments or settings that produce desired results such as spark timing, air/fuel mixture and EGR control.

Engine calibration assembly — The portion of a Ford PCM that contains the necessary programming to fine-tune the PCM's commands to the specific needs of a particular vehicle, similar to a PROM chip on a General Motor's vehicle. On EEC I, EEC II and EEC III systems, the calibration assembly is attached externally to the PCM, is removable and contains an octane adjustment switch. On EEC IV and EEC V systems, the calibration assembly is a permanent part of the PCM.

Engine metal overtemp mode — A unique feature of the Aurora/Northstar engine, whereby under circumstances of very high engine temperature (if for example the coolant has leaked out), the computer selectively disables four of the fuel injectors to prevent a temperature that would cause permanent mechanical damage to engine components.

EVAP system pressure sensor — Used on many OBD II-compliant vehicles to test the fuel vapor canister system for leaks. On many vehicles, special plumbing allows the MAP sensor to provide this function during the period between ignition on and engine running.

Exhaust gas analyzer — A diagnostic machine that samples and displays measurements of automotive emission readings including HC, CO, CO_2 and O_2 for diagnostic purposes. Additionally, some analyzers also measure and display readings for NO_x.

Fail-safe — A design feature of a system to ensure continued operation at reduced effectiveness or system capacity in the event of certain component or system failures, or to prevent damage to components by disabling or deactivating certain other system components in the event of certain system failures.

Feedback — Refers to a signal from a circuit or sensor that allows a computer to monitor the results of its own control in a closed-loop fashion.

Flow bench — A specialty tool, chiefly used at carmakers' factories, to properly set feedback carburetors to deliver proper air/fuel mixtures. The settings determined on a flow bench are normally marked with paint to indicate that they are not to be changed.

Flow charts — A series of diagnostic charts that are designed to guide the technician through the necessary pinpoint tests in order to identify the exact cause of a fault.

Freeze frame and Snapshot — Another feature of the new systems, these are computer capacities that make fault diagnosis clearer. Whenever a diagnostic trouble code is set, the computer records the information from all the other sensors and actuators that might be relevant to the fault. When this information is called from memory, there is often a useful clue as to what went wrong.

Front and rear oxygen sensors — To ensure that the engine control system can go into closed loop quickly, both the front and rear oxygen sensors include heater circuits to get them to operating temperature quickly as well.

Fuel temperature sensor — On some cars, a sensor mounted in or near the fuel tank to inform the computer of the fuel temperature. This information is used to calculate fuel viscosity and susceptibility to vapor lock.

Fuel trim — The ability of the PCM to modify the base fuel calculation in order to maintain a stoichiometric air/fuel ratio while in closed loop. Fuel trim is divided into both a short-term fuel trim (STFT) and a long-term fuel trim (LTFT).

Functional test — A test (that has been programmed into a computer) designed to aid the technician in diagnosing faults with the controlled system. For example, a functional test may allow the technician to energize certain actuators as desired.

Gallium arsenate crystal — A semiconductor material that changes its conductivity when exposed to a magnetic field.

Glow plugs — Starting aids for diesel engines, these are electrical resistance heaters extending into the combustion chamber. By heating the glow plugs, the high temperature required for compression ignition can be achieved even when the engine is cold.

Grams per mile (gpm) — A measurement of exhaust gasses using a constant volume sampling (CVS) method in conjunction with a dynamometer.

Grid — A fine wire mesh that is placed between the anode and cathode segments of a Vacuum Fluorescent Display (VFD).

Hall effect switch — A magnetic switching device often used to signal crankshaft position and speed to the computer.

Hard fault — A fault that currently exists.

Hard wiring — Wiring designed to carry only one type of signal 100 percent of the time, unlike a data bus that can transmit multiple messages across it. For example, a throttle position sensor (TPS) signal wire is traditionally always used to communicate information about throttle position to the PCM 100 percent of the time.

High-pressure cutout switch — A pressure-operated switch in the high-pressure side of the air-conditioning system that opens when pressure climbs too high. When the switch opens, it disables the A/C compressor clutch circuit.

Hot-film air mass sensor — A type of air mass sensor using changes in the temperature of the heated element to determine the mass of the air passing into the engine.

Hot-wire airflow sensor — An airflow sensor that keeps a special wire at a fixed number of degrees above ambient temperature in the intake airflow. The amount of current required to maintain the heat is converted into a voltage or frequency signal that conveys information about the mass of the intake air.

Hydrocarbons (HC) — A vehicle emission that equates to unburned gasoline or diesel fuel. The engine may produce excessive HCs due to a total misfire or rich air/fuel conditions. HCs may also be released into the atmosphere by reason of fuel evaporation.

Hz (hertz) — Cycles per second; a measure of frequency.

Idle tracking switch — An on/off switch indicating to the computer that it is in charge of the engine for idle speed.

Impedance — The total circuit resistance including resistance and reactance.

Inductance — The property of an electric circuit by which an alternating current through it produces a varying magnetic field that induces voltages in the same circuit or in a nearby circuit. In the reactive sense, the capacity of an electric circuit to produce a counterelectromotive force when a current through it changes.

Inductive pickup/Hall effect pickup — Two types of rotation sensors are used on many vehicles, the inductive pickup and the Hall effect. Inductive pickups consist of a wire coil wound around a permanent magnet. When a metal tooth passes the tip of the magnet, a small, polarity-reversing alternating current is generated (this current can grow much larger with higher speed). The Hall effect pickup receives a reference voltage from the computer and converts it into an on/off signal. Hall effect sensors do not vary the signal voltage output with speed.

Inertia switch — A Ford trademark safety feature and occasional no-start puzzle. To protect against fuel spray after an accident, the inertia switch disables the fuel pump circuit after a vehicle impact.

Initialization mode — The beginning state of an engine control computer when the ignition switch has been turned to on, but the engine not yet started. The computer performs a variety of internal tests and sensor/actuator checks.

Injection pressure — The pressure of the fuel at the injector's delivery end. On gasoline engines, this is controlled ordinarily by a pressure regulator to under 100 psi. On diesel engines the pressure is much higher, since the fuel must flow into the combustion chamber against the power stroke pressures. Diesel injection pressures can reach almost 30,000 psi.

Injection timing stepper motor — On certain electronically controlled diesel systems, the electric motor used by the computer to set the injection timing. *See also* Stepper motor.

Integrator — A term that refers to General Motors' name for the Short-Term Fuel Trim (STFT) prior to OBD II standardization, referring to the PCM's immediate reaction to the last voltage signal communicated from the oxygen sensor.

Interface — Circuitry that converts both input and output information to analog or digital signals as needed, and filters external circuit voltage to protect computer circuits.

KAM — Keep-alive memory, an electronic memory in the computer that must have a constant source of voltage to continue to exist.

Kilobyte — 1,024 bytes of information in binary code.

Lambda — A Greek letter (λ) used to indicate how far the actual air/fuel mixture deviates from the ideal air/fuel mixture. Lambda equals actual inducted air quantity divided by the theoretical air required (14.7). The actual ideal mixture depends on the characteristics of the fuel and of the air drawn in through the intake system.

Lift pump — An electric fuel pump in or near the fuel tank that pressurizes the fuel slightly (about 15 psi) and delivers it to the main fuel pump. Found on both gasoline and diesel engines. Sometimes called a primer pump.

Light-emitting diode (LED) — A semiconductor device that generates a small amount of light when forward biased (when voltage is applied with the polarity such that current flows across the PN junction).

Limp-in mode — A state of a computer engine management system in which one or more major component circuits have failed and a substitute value is used to retain driveability until the vehicle can be repaired.

Logic module and power module — Early to mid-1980s Chrysler products separated the computer functions into two components, the power module, which carried the higher current and worked the actuators, and the logic module, usually in a more protected place, usually more electrically protected. The logic module assesses sensor inputs and instructs the power module when to inject fuel, fire plugs and work actuators.

Logic probe — A computer-safe test light, using LEDs to indicate if a signal is present. Logic probes are typically used to provide quick checks of primary ignition and fuel injector circuits.

Long-term fuel trim (LTFT) — A long-term adjustment made within the PCM so that the performance of the short-term fuel trim (STFT) will cause the oxygen sensor signal to average between 400 and 500 millivolts. LTFT is designed to modify the base fuel calculation to compensate for wear and aging of the engine and other components that affect the air/fuel ratio. Originally known on General Motors' products as "block learn".

Longitudinal — Parallel to the length of the car (opposed to transverse).

Loop configuration — A configuration of a multiplexing network that connects the data bus to all nodes on the bus in a series circuit.

Malfunction indicator light (MIL) — A lamp on the instrument panel used to alert the driver when the PCM identifies a problem with the engine control system. Known commonly as a Check Engine Light prior to OBD II standardization.

Manifold absolute pressure (MAP)/intake manifold vacuum — Manifold absolute pressure (MAP) is the actual pressure in the intake manifold, regardless of the difference between it and ambient air pressure. Intake manifold vacuum, the traditional concept, refers to the relative difference in pressure between the two. Intake manifold vacuum is always a relative measure. MAP measurements are not relative.

Master slave — The computer that controls a data bus network.

Maximum authority — A term describing the maximum idle speed the ISC motor can provide.

Megabyte — 1,048,576 (1,024 squared) bytes of information in binary code.

Memory — A computer's information storage capability.

Memory DTC — A diagnostic trouble code (DTC) that has been set in a computer's memory at some point in the past. This type of DTC may represent either a hard fault or a soft fault.

Micron — A unit of measurement that describes the size of a particle one-millionth of an inch in diameter.

Microprocessor — A processor contained on an integrated circuit; processor (central processing unit) that makes the arithmetic and logic decisions in a microcomputer.

Milliamp — One thousandth of an amp, more correctly milliampere.

Millivolt — One thousandth of a volt.

Minimum authority — A term describing the minimum idle speed achievable by retracting the ISC plunger until the throttle lever rests on the idle stop screw.

Misfire detection — Part of the required OBD II program, misfire detection closely monitors engine rpm for the slight variations when combustion misfire occurs. In most cases, the computer disables the injector for a misfiring cylinder (to protect the catalytic converter from overheating) and sets an appropriate DTC.

Modulate — To find a position between two extremes; to alternate between on and off.

Modulated displacement — A system in which a series of PCM-controlled electromechanical devices can selectively disable the rocker arms of particular cylinders, preventing the valves from opening. The engine then operates on eight, six or four cylinders, depending on power requirements.

Monitor — An on-board diagnostic test that an OBD II PCM performs in order to determine if a component or system is within design specifications.

Motronic — The latest of the Bosch engine management systems, Motronic is their first full engine control system, regulating fuel injection and ignition timing and dwell in one unit.

Multiplexing (MUX) — One of the most significant extensions of computer control technology in vehicles, multiplexing connects many or all the control units in a vehicle together over a single wire network, the data bus. Information from all components is sent over the data bus and received by all.

Multiplexor — An electronic chip that, when built into a node, gives it the ability to send messages on a data bus.

Nitrogen (N_2) — A gas that makes up approximately 78 percent of Earth's atmosphere. Ideally, the nitrogen that enters the engine's combustion chambers will go through unchanged.

Node — A computer that is connected to a data bus and has the ability to send and/or receive messages on the bus.

Noise-vibration-harshness (NVH) — A special acoustic diagnosis tool that can identify the frequencies of various noises and vibrations to compare with known frequencies. Useful to determine whether a noise or vibration is engine-speed-relative, drivetrain-speed-relative, wheel-speed-relative or random.

Nonpowered test light — A tool that can be used to quickly check for a voltage potential (differential). Voltage potential should not be judged based on the brightness of the light. Unfortunately, due to its low internal resistance, a test light can act as a jumper wire and unexpectedly damage a computer or deploy an air bag.

Normally aspirated — An engine that uses ambient atmospheric pressure to force air into the engine (opposed to turbocharged and supercharged).

OBD II — A standardized computer control self-diagnostic program, mandated for most vehicles as of the 1996 model year. Nomenclature, trouble codes and even the size, shape and many circuits of the diagnostic connector are standardized across the industry.

On-Demand DTC — A diagnostic trouble code (DTC) that results from a self-test (performance test) and identifies any hard faults that were found within the computer system.

Open loop — Any operational mode in which a computer controls an output circuit, but does not use a feedback signal from a sensor or circuit to monitor the results of its own control. However, the computer may monitor other input signals/sensors during open-loop control that do not pertain to a feedback signal. In the PCM's open-loop control of the air/fuel ratio, the air/fuel mixture is calculated based on coolant temperature, engine speed and engine load without benefit of the oxygen sensor. *See also* Closed loop.

Oxides of nitrogen (NO_x) — A vehicle emission that is a combination of nitrogen (N_2) and oxygen (O_2) that is combined within the engine's combustion chamber or in the exhaust system due to temperatures that exceeded 2,500°F.

Oxygen (O_2) — A gas that makes up approximately 21 percent of Earth's atmosphere. Most of the oxygen that enters the engine's combustion chambers will be used up during combustion.

Palladium (Pd) — A silver-white, ductile, metallic element; number 46 on the periodic table.

Passive voltage — A voltage that is at rest, not being pulled either high or low by a computer circuit.

Pickup coil — A small coil of wire in which the signal of a pulse magnetic generator is produced, such as in many electronic distributors.

Piezoelectric — A crystal that produces a voltage signal when subjected to physical stress such as pressure or vibration.

Piezoresistive — A semiconductor material whose electrical resistance changes in response to changes in physical stress.

Pinpoint test — A test that checks a computer circuit for the exact fault, whether it be a sensor/actuator, an open or short in the circuitry, a poor or open electrical connection or the computer itself. Pinpoint tests may be performed by either following a flowchart or by following an electrical schematic. Following a pinpoint test series, the technician should know exactly what repair needs to be made in order to restore the system back to its "new" condition.

PIP signal — The profile ignition pickup signal (PIP) comes from a crankshaft- or distributor-mounted sensor indicating cylinder position to the computer.

Platinum (Pt) — A heavy, grayish white, ductile, metallic element; number 78 on the periodic table.

Potentiometer — A three-terminal variable resistor that acts as a voltage divider. Frequently used as a position sensor for throttle angle and so on. The potentiometer can be linear or rotary.

Powertrain Control Module (PCM) — The term now used (since OBD II) for the main computer in charge of engine management.

Prechamber — A prechamber is a small cavity in the cylinder head of a diesel engine into which most of the air is squeezed during the compression stroke and into which the fuel is injected. The glow plug, if present, is also in the chamber. A prechamber often has a small diffuser pin in it to turbulate the air and fuel still more, for better combustion. Opposed to direct injection. *See also* Ricardo chamber.

Pressure cycling switch — A pressure-operated switch in the low-pressure side of the air-conditioning system that opens as freon pressure drops below a specific pressure. When the switch opens, the A/C compressor clutch is disabled. The purpose of this switch is to protect the compressor against operation when the system has leaked empty.

Primary and secondary oxygen sensors — The oxygen sensors are normal types, heated for rapid warmup and closed-loop acquisition. What is new is the use of tandem sensors (also called primary and secondary oxygen sensors), one before and one after the catalytic converter. This combination serves as a test of the converter's effectiveness. So long as the downstream, secondary sensor shows a relatively flat signal, whatever its output voltage, the catalyst is working. When the secondary oxygen sensor starts the characteristic mixture dithering like the primary one (still used for fuel mixture trim), the catalyst has become inefficient.

PROM — Programmable read-only memory.

Propagation — As used in automotive technology, the spread of the flame front across the combustion chamber of an engine.

Protocol — A language in which computers communicate with each other over a data bus. A standardized binary code.

Pulsair — The name General Motors applies to the pulse-type air injection system.

Pulsed injection — One of the two kinds of fuel injector fuel spray techniques, the other being continuous injection. With a pulsed injection system, the injectors are pulsed, in some way, according to the tach reference pulse of the ignition system. Therefore, as engine rpm increases, the frequency of the injector pulse also increases. TBI injectors may be pulsed with every tach reference pulse. PFI systems may pulse their injectors simultaneously (all at the same time), in groups (half at a time), in pairs or sequentially in the engine's firing order just before the intake valve opens.

Pulse width and duty cycle — Both pulse width and duty cycle are a measurement of a solenoid's (or signal's) on-time—pulse width as a real-time measurement and duty cycle as a percentage of the total cycle. If the frequency of the cycle is known (as measured with a Hertz meter), dividing the frequency into 1,000 milliseconds (one second) determines the length of one cycle. If this amount is now multiplied by the percentage of duty cycle on-time, the result is the pulse width as a real-time measurement in milliseconds.

Pulse Width Modulated (PWM) — A signal that is controlled according to a real-time measurement, usually in milliseconds or microseconds. This type of signal may be used to control fuel injectors, cooling fans and blower motors. When a serial data bus uses a fixed pulse width, typical with two-wire busses, the binary code is said to be a PWM signal.

Pumping losses — The load on the engine caused by the work needed to pump the air through the system. On gasoline engines, because of the vacuum developed behind the throttle, pumping losses are greatest at the most closed throttle positions, idle and deceleration. The aerodynamic friction of air through the intake system also contributes to a lesser extent to the pumping losses, though at high speeds and open throttle, that becomes the principal source. Diesel engines, because they are unthrottled, have very low pumping losses compared to gasoline engines.

Purge valve — A vacuum-controlled valve controlling the removal of stored HC (fuel) vapors from the charcoal canister.

Quad driver/output driver — A power transistor in the PCM, capable of working either four or seven different actuators (such as injectors or ignition coils). Each quad driver or output driver works when the PCM uses it to ground a component's circuit, completing it.

Quick test — The whole set of procedures that make up the Ford diagnostic procedure.

RAM — Random-access memory.

REF pulse — The basic timing signal sent to the computer, indicating crankshaft position. This signal is modified by the computer to set the actual timing.

Rhodium (Rh) — A white, hard, ductile, metallic element; a member of the noble metals family; number 45 on the periodic table.

Ricardo chamber — Similar to a prechamber, a Ricardo chamber is a small cavity in the cylinder head of a diesel engine into which most of the air is squeezed during the compression stroke and into which the fuel is injected. The glow plug, if present, is also in the chamber. Opposed to direct injection.

ROM — Read-only memory.

Scan tool or Scanner — A diagnostic tool used to retrieve from a vehicle's computer memory any malfunction codes it has retained. On earlier systems, each manufacturer used a scan tool connector and setup unique to that manufacturer (though there are aftermarket scan tools with adapters that can read many of them). With OBD II-compliant vehicles, the scan tool is a standardized tool that can be used for the same purpose on any vehicle. Many scan tools can also be used to get direct measurements of various sensors and to exercise various actuators to check on their effectiveness, also known as functional tests.

Schmitt trigger — A device that trims an analog signal and converts it to a digital signal.

Self-diagnostic/on-board diagnostics — These terms mean essentially the same: the capacity of the computer to monitor its sensors' and actuators' circuits as well as its own internal functions. In a self-diagnostic system, the computer retains in electronic memory any malfunctions it may encounter, stored as codes the technician can recall with a scan tool.

Self-test — The self-diagnostic portion of the Ford quick-test in which the computer tests itself and its circuits for faults.

Sensor — All the inputs and outputs to the computer are either sensors that provide information or actuators that do work. Most sensors are sent a reference voltage which is reduced in a predictable way by whatever parameter they monitor. Some generate voltages of their own.

Serial data — Digitized code known as *binary code*, which is a combination of high and low voltages known as ones and zeroes, used as a language that is communicated between computers or computer components.

Short-term fuel trim (STFT) — The PCMs immediate response to the oxygen sensor signal. That is, when the oxygen sensor indicates a lean mixture, STFT increases its value so as to increase injector on-time. This causes more fuel to be added to the base fuel calculation and enriches the air/fuel mixture. When the oxygen sensor indicates a rich mixture, STFT decreases its value so as to decrease injector on-time. This causes fuel to be subtracted from the base fuel calculation and leans out the air/fuel mixture. Originally known on General Motors' products as "Integrator."

Signal (or voltage signal) — The voltage values that a computer uses as communication signals.

Single-point/multipoint fuel injection — Single-point injection refers to fuel injection using the equivalent of an electric carburetor. On four-cylinder engines, this usually involves a single injector, but on some V-form engines two injectors are combined in one unit. Also called *throttle-body fuel injection*. Multipoint fuel injection employs one injector for each cylinder.

Slave — A dependent node on a data bus.

Snapshot and Freeze frame — Another feature of the new systems, these are computer capacities that make fault diagnosis clearer. Whenever a diagnostic trouble code is set, the computer records the information from all the other sensors and actuators that might be relevant to the fault. When this information is called from memory, there is often a useful clue what went wrong.

Soft fault — An intermittent fault whose symptom does not appear to currently exist.

Spark ignition (SI) — A spark-ignition engine depends on the heat generated by a spark across the electrodes of a spark plug to ignite the air/fuel mixture. Opposed to compression ignition.

Speed density formula — A term referring to a method of calculating the flow rate and density of the intake air. The result of the calculation indicates the air's ability to evaporate fuel.

SPOUT — The Ford Spark Output (SPOUT) signal comes from the PCM and indicates to the ignition module when to fire the next spark plug. The exact timing is calculated on the basis of inputs from all the computer's sensors.

Standardized 16-pin diagnostic connector — On OBD II-compliant vehicles, the diagnostic socket to which the scan tool connects during diagnosis. This socket is standardized for all cars, though there are a number of circuits that can be used by the manufacturer for special purposes.

Star configuration — A configuration of a multiplexing network that connects the data bus to all nodes on the bus in a parallel circuit.

Star tester — A tester designed to test MCU and EEC IV systems that activates and deactivates the self-test procedure and displays a digital readout of the service codes.

Stepper motor — A type of direct-current electric motor with the unique capacity to move an exact number of turns in response to a computer's signal. Typically such a motor, used to adjust the idle speed bypass opening, for example, has a range of 256 different possible positions it can be set to. This allows the computer to open or close to an exact position. Such motors are also increasingly used in some heater door control systems.

Stoichiometric — In automotive terminology stoichiometric refers to an air/fuel ratio in which all combustible materials are used with no deficiencies or excesses; 14.7 parts air to 1 part fuel, by weight.

Thermac — The term used by General Motors to identify the heated air inlet system. The term is a shortened form of thermostatic air cleaner.

Thermistor — A temperature-sensing device whose resistance changes dramatically as the temperature changes. Most thermistors have a negative temperature coefficient (the resistance goes up as the temperature goes down).

Thermo-time switch — An electrical switch usually applied to a bimetallic strip that uses heat to open or close a set of electrical contacts. The amount of heat is controlled, so the time it takes to open or close the contacts is always controlled.

Thick film integrated (TFI) — A term that identifies the type of chip used in the ignition module attached to the side of the distributor for EEC IV system applications.

Throttle-body injection (TBI) — Throttle-body injection uses a single fuel injector in a throttle body (or a pair of two for V-form engines) that delivers the fuel to the intake manifold.

Torque converter lockup clutch — A computer-controlled, hydraulically actuated clutch in the automatic transmission's torque converter that locks the transmission's input shaft to the engine's crankshaft for fuel economy. The computer determines when to send this command on the basis of sensor inputs.

Torque management — A set of capacities the engine control system has, allowing it to reduce engine torque output under specific circumstances. Its first strategy is to retard spark; in more extreme circumstances it can selectively shut off fuel injectors. The traction control system employs part of the torque management system to prevent front wheel spin under power when the pavement is slippery.

Transducer — A device that receives energy from one system and transfers it to another system, usually in a different energy form.

Transfer pump — On some diesel systems, a pump that receives the fuel, either unpressurized from the tank or from the lift pump, and delivers it to the injection pump.

Trigger — The characteristics that determine the instant in time that a DSO begins to display a waveform. These characteristics include trigger source, trigger level, trigger slope and trigger delay (horizontal waveform position).

Trip — A key cycle consisting of Ignition On, Engine Run, Ignition Off and PCM Power Down, during which the enable criteria for a particular diagnostic test are met and the diagnostic test is run by the PCM. A trip is used by the PCM to perform a diagnostic test to confirm a symptom or its repair.

Two-trip malfunction detection — On some vehicles, the computer waits to record certain faults until they appear on two separate driving trips.

Unthrottled — Diesel engines are said to be unthrottled because they have no throttle and draw in as much air as possible for each cylinder cycle. Hence air/fuel ratio is unimportant for the diesel, except that more fuel means more power up to the limits of the available oxygen.

Vacuum control valve (VCV) — A switch opening or closing its ports to supply vacuum in response to coolant temperature change. It is sometimes referred to as a *ported vacuum switch* or a *thermal vacuum switch*. Thermal vacuum switch, however is more accurately used to refer to a switch located in the air cleaner, opening or closing a port in response to air temperature.

Vacuum fluorescent (VF) — A glass tube containing anodes and a cathode. The anodes are coated with a fluorescent material and are positioned so they form all of the segments of alphanumerical characters. As current passes through the cathode, it gives off electrons. If the anodes have positive voltage applied, the electrons given off by the cathode are attracted to them and strike the fluorescent material, causing it to glow.

Vacuum pump — On automotive and light truck diesels, a mechanically driven pump to develop vacuum. It is used for various purposes like operating heater doors, and is necessary on a diesel engine because there is no effective vacuum developed by the engine itself.

Variable pulse width (VPW) — The term assigned to serial data that consists of bits of information that are of more than one length. In this case, a "one" may be either represented as a short high-voltage pulse or as a long low-voltage pulse. And, conversely, a number "zero" may be represented as a short low-voltage pulse or as a long high-voltage pulse. This allows the voltage to switch between high and low voltage values with every bit of information.

Variable valve timing (VVT) — The ability of a computer to vary valve timing through an actuator designed for that purpose so as to get best performance under high load conditions while still getting good engine response at low rpms.

Vehicle control module (VCM) — A General Motors' PCM that, in addition to controlling the engine, transmission and cruise control systems, now also controls the antilock brake system.

Voltage drop test — A more precise substitute for ground resistance tests, the voltage drop test checks through each connection from power to ground to see whether there is a change in voltage through the connection. This test capitalizes on the fact that voltmeters are very accurate at the bottom of their scale, while ohmmeters are less accurate near zero.

Volumetric efficiency (VE) — A measure of cylinder-filling efficiency. VE is expressed as the percentage of atmospheric pressure to which the cylinder is filled at the completion of the compression stroke. This efficiency varies with engine speed; the point of highest volumetric efficiency (which can rise above 100 percent in some cases) is the point of highest engine output torque.

Wastegate — A valve or door that when opened allows the exhaust gas to bypass the exhaust turbine of a turbocharger. The wastegate is used to limit turbocharger boost pressure.

Waste spark — An ignition system using one coil between every complementary pair of cylinders, firing both plugs simultaneously, one at the end of the compression stroke and the other at the end of the exhaust stroke. A waste spark system has a coil for every pair of cylinders.

Water (H_2O) — One of the byproducts of combustion of a hydrocarbon-based fuel within the engine.

WOT — Wide open throttle.

Zirconium dioxide (ZrO_2) — A white crystalline compound that becomes oxygen-ion conductive at about 315°C (600°F).

Index